Research in computational group theory, an active subfield of computational algebra, has emphasized four areas: finite permutation groups, finite solvable groups, matrix representations of finite groups, and finitely presented groups. This book deals with the last of these areas. It is the first text to present the fundamental algorithmic ideas which have been developed to compute with finitely presented groups that are infinite, or at least not obviously finite. The book describes methods for working with elements, subgroups, and quotient groups of a finitely presented group. The author emphasizes the connection with fundamental algorithms from theoretical computer science, particularly the theory of automata and formal languages, from computational number theory, and from computational commutative algebra. The LLL lattice reduction algorithm and various algorithms for Hermite and Smith normal forms are used to study the abelian quotients of a finitely presented group. The work of Baumslag, Cannonito, and Miller on computing nonabelian polycyclic quotients is described as a generalization of Buchberger's Gröbner basis methods to right ideals in the integral group ring of a polycyclic group. Researchers in computational group theory, mathematicians interested in finitely presented groups, and theoretical computer scientists will find this book useful.

ENCYCLOPEDIA OF MATHEMATICS AND ITS APPLICATIONS

EDITED BY G.-C. ROTA

Volume 48

Computation with finitely presented groups

ENCYCLOPEDIA OF MATHEMATICS AND ITS APPLICATIONS

ENCYCLOPEDIA OF MATHEMATICS AND ITS APPLICATIONS

Computation with finitely presented groups

CHARLES C. SIMS

Rutgers University

Published by the Press Syndicate of the University of Cambridge
The Pitt Building, Trumpington Street, Cambridge CB2 1RP
40 West 20th Street, New York, NY 10011-4211, USA
10 Stamford Road, Oakleigh, Melbourne 3166, Australia

First published 1994

Printed in the United States of America

Library of Congress Cataloging-in-Publication Data
Sims, Charles C.
Computation with finitely presented groups / Charles C. Sims
p.   cm. – (Encyclopedia of mathematics and its applications ;
v. 48)
Includes bibliographical references and index.
ISBN 0-521-43213-8
1. Group theory – Data processing.   2. Finite groups – Data
processing.   3. Combinatorial group theory – Data processing.
I. Title.   II. Series.
QA171.S6173   1993
512′.2 – dc20   92-32383
                                              CIP

A catalog record for this book is available from the British Library.

ISBN 0-521-43213-8 hardback

To Annette

Contents

Preface

In 1970, John Cannon, Joachim Neubüser, and I considered the possibility of jointly producing a single book which would cover all of computational group theory. A draft table of contents was even produced, but the project was not completed. It is a measure of how far the subject has progressed in the past 20 years that it would now take at least four substantial books to cover the field, not including the necessary background material on group theory and the design and analysis of algorithms. In addition to a book like this one on computing with finitely presented groups, there would be books on computing with permutation groups, on computing with finite solvable groups, and on computing characters and modular representations of finite groups.

Computational group theory was originated by individuals trained as group theorists. However, there has been a steadily increasing participation in the subject by computer scientists. There are two reasons for this phenomenon. First, group-theoretic algorithms, particularly ones related to permutation groups, were found to be useful in attacking the graph isomorphism problem, a central problem in theoretical computer science. Once computer scientists began looking at group-theoretic algorithms, it was natural for them to attempt to determine the complexity of these algorithms. Second, the techniques and data structures of computer science have proved valuable in improving existing group-theoretic algorithms and in developing new ones.

This book is intended to be a graduate-level text. I have made a deliberate attempt to make the material accessible to students of both mathematics and computer science. The first chapter contains a quick review of elementary group theory, which would not be necessary if the target audience consisted only of graduate students in mathematics. It also contains a discussion of backtrack searches, which should be familiar to any undergraduate computer science major but which are probably not frequently encountered by mathematics majors. I suspect that initially the computer scientists may have an easier time than the mathematicians, since the first half of the book includes a substantial discussion of the theory of automata and rational languages, topics more familiar to computer scientists than to mathematicians. Moreover, the first half requires only relatively elementary results about groups. However, in the second half, deeper results from group theory are required, and here the computer scientists may find the going somewhat more difficult.

Ideally this book should be accompanied by computer software that would permit the reader to experiment with implementations of the procedure discussed. Unfortunately, an appropriate package does not currently exist, although several of the available systems would be useful in connection with certain topics. Writing my own software would have delayed the publication of the book by several years, at least. The lack of computer support has meant that substantial examples and exercises have been largely omitted, since it is only with the help of the proper software that the reader would be able to explore such material successfully.

I have tried to include as many exercises as I could, but some of the later sections are not as well covered as I would have preferred. The exercises are of two types. Some require only a routine application of a technique discussed in that section. Others are intended to provide new insight. It is not always immediately obvious into which category a particular exercise falls, so the reader is encouraged to look carefully at all of the exercises. The more challenging ones are marked with an asterisk. In a few cases, I don't know the answer to the problems.

A substantial effort was made to include as complete a set of references as possible. To assist instructors in providing reading lists, the Bibliography has been divided into two sections, the first containing books on topics related to the material discussed here, and the second listing articles in journals and conference proceedings. References to books are enclosed in brackets, whereas references to articles are enclosed in parentheses. Michael Vaughan-Lee has written a book on the restricted Burnside problem and an article on the efficient computation of products in large p-groups. Both works appeared in 1990. Information about the book [Vaughan-Lee 1990] is in the book section of the Bibliography, but to find publication data for (Vaughan-Lee 1990) one must look in the articles section.

It is not usual for an author of a mathematics text to make evaluative judgments concerning the works in the Bibliography, and I have agonized over my decision to break with this practice. However, I feel an obligation to the reader to state my opinion that the quality of the papers dealing with the computation of Hermite and Smith normal forms is on average noticeably below the level in the other works cited. In a significant number of these papers there are deficiencies in the exposition and even in the validity of the arguments. I shall mention only one example of the problems I found. Given a set of homogeneous linear equations with integer coefficients, there are efforts to describe a basis of integer solutions. The authors frequently fail to make clear whether they are referring to a vector space basis for the set of all rational solutions of the system such that each basis vector has integer components, or to a $\mathbb{Z}$-basis for the subgroup of all integer vectors which are solutions of the system. Having reached my conclusion, I was faced with several options. I could remain silent, I could provide critiques of individual papers, or I could omit the papers I found questionable from the Bibliography. None of these possibilities seemed appropriate. The problems are significant enough that I could not remain silent. There are simply too many papers to permit me to make detailed comments concerning each one. I do not have the time, and the book is already long. I have read through, at least quickly, all of these papers, so I feel obligated to include them in the Bibliography. Moreover, each of them has some merit. I realize that by making these comments and not identifying the papers I find deficient, I am raising doubts about the quality of all of them, including papers with substantial contributions. To the authors of these papers I apologize.

It is traditional for the preface to explain the numbering system used for sections, propositions, examples, exercises, figures, and tables. Section 3.4 is the fourth section of Chapter 3. Sections and chapters carrying an asterisk may be skipped without affecting the continuity of the material. The second proposition in Section 3.4 is referred to as Proposition 4.2 or as Proposition 4.2 of Chapter 3. The end of a proof is signaled by the symbol "□". Within a given section, the propositions, theorems, lemmas, and corollaries are numbers in a single series. Examples and exercises are numbered separately. The numbering for figures and tables always includes the number of the chapter. Thus, Table 1.8.1 is the first table in Section 8 of Chapter 1.

A great many people have assisted in the preparation of this book. John Cannon, George Havas, Joachim Neubüser, and Michael Newman all made suggestions which improved the exposition. Special thanks go to Steve Schibell, who read the entire manuscript and provided detailed comments which were extremely useful. Gretchen Ostheimer and Eddie Lo helped with the proofreading. The staff of Cambridge University Press have been very supportive. David Tranah has quietly but persistently been after me to write a book for Cambridge for roughly a decade. Lauren Cowles has greatly facilitated my dealings with the New York office. Finally, Edith Feinstein and Jim Mobley have shown remarkable tolerance for my somewhat idiosyncratic style.

Considerable effort has been expended in trying to get the algorithm descriptions in this book right. However, errors almost certainly remain, and I would welcome information about any problems which are discovered. Other comments are also encouraged. My current address for electronic mail is sims@math.rutgers.edu.

Charles Sims

Introduction

This book describes computational methods for studying subgroups and quotient groups of finitely presented groups. The procedures discussed belong to one of the oldest and most highly developed areas of computational group theory. In order to better understand the context for this material, it is useful to know something about computational group theory in general and its place within the area of symbolic computation.

The mathematical uses of computers can be divided roughly into numeric and nonnumeric applications. Numeric computation involves primarily calculations in which real numbers are approximated by elements from a fixed set of rational numbers, called floating-point numbers. Such computation is usually associated with the mathematical discipline numerical analysis. Examples of numerical techniques are Simpson's rule for approximating definite integrals and Newton's method for approximating zeros of functions.

One nonnumeric application of computers to mathematics is symbolic computation. Although it is impossible to give a precise definition, symbolic computation normally involves representing mathematical objects exactly and performing exact calculations with these representations. It includes efforts to automate many of the techniques taught to high school students and college undergraduates, such as the manipulation of polynomials and rational functions, differentiation and integration in closed form, and expansion in Taylor series.

The term "computer algebra" is frequently used as a synonym for "symbolic computation". The books [Akritas 1989], [Buchberger, Collins, & Loos 1983], [Davenport, Siret, & Tournier 1988], [Della Dora & Fitch 1989], and [Geddes, Czapor, & Labahn 1992] all have this phrase in their titles. Although the term "computer algebra" is well established, it conflicts somewhat with current usage within mathematics, where "algebra" usually is used in the narrower sense of "abstract algebra", the study of algebraic structures such as groups, rings, fields, and modules. The word "computer" in the phrase "computer algebra" is also not quite accurate. It is true that much of what is done is motivated by the existence of computers. Nevertheless, the algebraic algorithms which have been developed represent substantial mathematical achievements, whose importance is not dependent entirely on their being incorporated into computer programs. Many of the algorithms, including a number presented in this book, can be useful in calculations carried out by hand.

Within symbolic computation there is a rapidly expanding area of computational (abstract) algebra, which is the study of procedures for manipulating objects from abstract algebra with particular concern for practicality. Computational group theory is the part of computational algebra which considers problems related to groups. Other flourishing subfields of computational algebra are computational number theory, which is described in such books as [Pohst & Zassenhaus 1989] and [Bressoud 1989], and computational algebraic geometry, where the software package Macaulay of David Bayer and Michael Stillman has had a major impact.

Symbolic computation is at the border between mathematics and computer science. The objects being manipulated are mathematical. However, the algorithmic ideas often have come from computer science, and individuals who identify themselves as computer scientists have made important contributions to the subject. This book emphasizes the close connection between techniques for investigating finitely presented groups and such topics as formal language theory and critical-pair/completion procedures, which are now considered to be important parts of computer science.

A point of continuing debate is the role of complexity theory in symbolic computation. The traditional complexity measure in theoretical computer science is asymptotic worst-case complexity. For users of symbolic software, worst-case analyses are often too pessimistic. Of much more relevance is average-case complexity. However, average-case analyses are lacking for many of the most important algebraic algorithms. Moreover, there are cases in which no agreement has been reached on what the average case is. Symbolic computations often require a great deal of computer memory. Frequently one has to tailor a program to fit a specific problem in order to get a solution on a particular computer. In this situation, it does not make sense to talk about asymptotic behavior.

In computation with finitely presented groups there is an even more fundamental difficulty with complexity. Many problems connected with such groups have been shown not to have general algorithmic solutions. There are procedures for trying to solve some of these problems, but the procedures terminate only if the solution has a particular form. In the remaining cases, the procedures continue indefinitely. There is no version of complexity analysis which adequately handles this situation. Ideally, one would want a program to give up quickly if it is not going to get an answer and to work hard when a solution can be found. However, because usually there is no quick way to determine from the input data which outcome is likely, deciding when to throw in the towel is difficult.

Although we do not have an adequate theoretical framework on which to base comparisons of group-theoretic procedures and their computer implementations, we still need to make decisions about which techniques to use on a particular problem. Frequently, all that we can do is apply competing methods to a selection of test problems and compare the results. Experimental evidence is better than nothing, but one must be very careful about drawing conclusions from such evidence. It is hard to be sure that the sample problems are truly representative of the class of problems under consideration.

A computational problem in group theory typically begins "Given a group G, determine ...". Our ability to solve the problem depends heavily on the way G is given. There are three methods commonly used to specify a group G:

(a) One may describe G as the subgroup generated by an explicit finite list of elements in some other group H which is considered to be well-known.

(b) One may define G to be the group of automorphisms or symmetries of some combinatorial or algebraic object.

(c) One may give a finite presentation for G by generators and relations.

In (a), the group H could be the symmetric group on a finite set, the group of invertible n-by-n matrices over a commutative ring, or a free group of finite rank. Examples of the types of objects which might arise in (b) are graphs, block designs, and, in the case of Galois groups, finite extensions of one field by another. Problems related to groups given as in (c) are the main subject of this book, although finitely generated subgroups of free groups are also discussed at some length. The greatest successes in computational group theory so far have come in connection with permutation groups on finite sets, finite solvable groups, and finitely presented groups.

The study of groups given by presentations is called combinatorial group theory. The primary references on the subject are the books [Magnus, Karrass, & Solitar 1976] and [Lyndon & Schupp 1977]. The history of the subject given in [Chandler & Magnus 1982] also provides useful insights. Many examples of finite presentations of groups can be found in [Coxeter & Moser 1980].

As noted earlier, the computer science literature is also relevant. The three volumes by Donald Knuth contain a great deal of material related to our topic. Volume 2 is a particularly valuable reference. Other books describing the fundamental algorithms of computer science with applications to computational group theory are [Aho, Hopcroft, & Ullman 1974] and [Sedgewick 1990].

Because this is one of the first texts to cover a major area within computational group theory, some remarks on the history of the field have been included. Most chapters conclude with a brief section of historical notes. More information, particularly about the history of general group theory, combinatorial group theory, and abstract algebra, can be found in [Dieudonné 1978], [Chandler & Magnus 1982], [Waerden 1985], and [Wussing 1984]. Computational problems related to groups have been studied for at least 150 years. Algorithmic questions have been a part of combinatorial group theory from its beginning. In fact, the algorithms discussed in Chapter 8 are based directly on techniques developed in the middle of the nineteenth century, before presentations of groups were defined and before the concept of an abstract group was clarified.

Perhaps the first references to the potential of computers for group-theoretic calculations came from Alan Turing and Maxwell Newman. At the end of World War II, Turing began work on the design of an "automatic computing engine". As quoted in [Hodges 1983], Turing suggested in a document circulated at the end of 1945 that one of the tasks which might be assigned to this engine was the enumeration of the groups of order 720. In connection with the inauguration in 1951 of a computer at Manchester University, Newman described a probabilistic method by which computers might be able to obtain a crude estimate of the number of groups of order 2^8 (M. H. A. Newman 1951).

During the 1950s, group-theoretic calculations using machines were carried out by several researchers. However, it is reasonable to say that the field of computational group theory came into existence with the work of Joachim Neubüser, who was the first individual to make the application of computers to group theory his primary professional activity. Neubüser's first paper appeared in 1960.

By the late 1960s, the use of computers in abstract algebra was sufficiently common for a conference on the subject to be organized in Oxford, August 29 to September 2, 1967. The proceedings [Leech 1970] of that conference contained 35 papers, at least 20 of which dealt with groups. In his survey (Neubüser 1970) in those proceedings, Neubüser listed roughly 130 publications related to the use of computers in group theory. Not all the contributions to computational group theory which grew out of the Oxford conference were reflected in the proceedings; see the notes at the end of Chapters 5 and 11.

During the 1970s, computational group theorists participated frequently in conferences devoted primarily to traditional computer algebra. Thus the proceedings of the Second Symposium on Symbolic and Algebraic Manipulation sponsored by the Association for Computing Machinery (ACM) in 1971, the 1976 ACM Symposium on Symbolic and Algebraic Computation, and EUROSAM '79, an international symposium held in France, all contain important papers on computational group theory. More recently, computational group theory has been treated for the purpose of organizing conferences as a field of its own or as a subfield within group theory. The first conference devoted exclusively to computational group theory was organized in Durham in 1982 by the London Mathematical Society. The proceedings [Atkinson 1984] of the Durham meeting give a good indication of the breadth of the field.

Let us conclude this introduction with a brief survey of the software currently available for symbolic computation. This is an area of rapid change, so the information given here can be expected to become outdated fairly quickly. The classical symbolic systems provide facilities for working with individual polynomials, rational functions, matrices, Taylor series, and similar objects, primarily with coefficients taken from the real numbers. Among the classical systems currently available are REDUCE and MACSYMA, which have been around for some time, and Maple and Mathematica, which are considerably newer. Axiom (formerly SCRATCHPAD), developed at IBM's Yorktown Heights laboratory, is a system which allows the user some flexibility to work with elements of more general algebraic systems. A new version, written in C, of George Collins's system SAC2 has recently been introduced. Henri Cohen and collaborators have created a package called Pari, which is aimed primarily at number theory but contains an implementation of the LLL algorithm discussed in Chapter 8. The system Macaulay referred to earlier supports computation in algebraic geometry and commutative algebra.

Although quite a bit of group-theoretic software is available, no single package contains implementations of all the procedures discussed in this book. The most mature system for group-theoretic computation is Cayley, developed at Sydney University by John Cannon. A recent addition is the system GAP designed under the direction of Joachim Neubüser at the Technische Hochschule in Aachen. Various programs of a more specialized nature have been written at the Australian National University in Canberra under the leadership of Michael Newman. George Havas of the University of Queensland in Brisbane has produced free-standing implementations of several of the procedures presented here. David Epstein and Derek Holt at the University of Warwick have produced implementations of the Knuth-Bendix procedure for strings described in Chapter 2, as well as a number of other programs for studying "automatic groups" and for exploring the possibility that two given finitely presented groups are isomorphic.

1

Basic concepts

It was not possible to make the exposition in this book self-contained and keep the book to a reasonable length. In a number of places, results from various parts of mathematics and theoretical computer science are stated without proof. This chapter reviews several of the most fundamental concepts with which the reader is assumed to be somewhat familiar. More material on sets, monoids, and groups can be found in algebra texts like [Sims 1984] and books on group theory such as [Hall 1959]. Two standard sources for topics in computer science are [Knuth 1973] and [Aho, Hopcroft, & Ullman 1974]. After a brief discussion of some concepts in set theory, we define monoids, summarize some important facts about groups, consider presentations of monoids and groups by generators and relations, contemplate some sobering facts about our ability to compute using presentations, agree upon a method for describing computational procedures, look at some basic algorithms for computing with integers, and study an important search technique. Readers to whom most of these topics are familiar may wish to proceed directly to Chapter 2 and refer back to this chapter as needed.

1.1 Set-theoretic preliminaries

The concept of a set is central to the formal exposition of mathematics. Despite the importance of set theory, most mathematics texts do not make explicit the axioms for set theory being used. Instead, they adopt a "naive" set theory based largely on intuition.

This book will be no exception. We shall assume that the reader has a basic understanding of *sets* and *set membership*, as well as the operations of *union*, *intersection*, and *difference* of sets. The *empty set* will be denoted $\emptyset$. If A is a *subset* of B, then we shall write $A \subseteq B$ or $B \supseteq A$. To indicate that A is a *proper subset* of B, we shall write $A \subset B$ or $B \supset A$. The *cartesian product* of sets X and Y is the set $X \times Y$ of all ordered pairs (x, y) with

x in X and y in Y. The *diagonal* of $X \times X$ is $D = \{(x,x) \mid x \in X\}$. For a finite set A, the *cardinality* of A will be denoted $|A|$.

A *relation* from X to Y is a subset R of $X \times Y$. If $X = Y$, then we say that R is a relation on X. If (x,y) is in R, then we sometimes write xRy. Suppose A is a subset of X. Then AR is defined to be

$$\{y \in Y \mid aRy \text{ for some } a \text{ in } A\}.$$

The *inverse* of R is the relation $R^{-1} = \{(y,x) \mid (x,y) \in R\}$.

Let R be a relation from X to Y and let S be a relation from Y to Z. The *composition* of R and S is the relation T from X to Z which consists of the pairs (x,z) such that there is an element y in Y with the property that xRy and ySz. We write $T = R \circ S$.

An *equivalence relation* on a set X is a relation E on X such that the following conditions hold:

(i) xEx for all x in X.
(ii) If xEy, then yEx.
(iii) If xEy and yEz, then xEz.

These conditions are called *reflexivity*, *symmetry*, and *transitivity*, respectively. Symmetry can be stated as $E^{-1} = E$, and transitivity means that $E \circ E$ is contained in E. Suppose E is an equivalence relation on X. Then for x and y in X, the sets $\{x\}E$ and $\{y\}E$ are either equal or disjoint. Thus $\Pi = \{\{x\}E \mid x \in X\}$ is a *partition* of X. The elements of Π are called the *equivalence classes* of E.

A *function* from X to Y is a relation f from X to Y such that for all x in X the set $\{x\}f$ has exactly one element. We write $f \colon X \to Y$ and refer to X as the *domain* of f. If y is the unique element of $\{x\}f$, then we write $y = xf$, $y = f(x)$, $y = x^f$, or $x \xmapsto{f} y$, depending on the context. The composition of two functions is a function. The term "map" is a synonym for "function". Let E be an equivalence relation on X and let Π be the set of equivalence classes of E. The *natural map* from X to Π is the function π such that $\pi(x) = \{x\}E$ for all x in X.

Let f be a function from X to Y. We say that f is *surjective* or that f maps X *onto* Y if $Xf = Y$. We say that f is *injective* if for all x_1 and x_2 in X the equality $f(x_1) = f(x_2)$ implies that $x_1 = x_2$. A function which is both injective and surjective is said to be *bijective*, or to be a *one-to-one correspondence*. Functions which are surjective, injective, or bijective are called *surjections*, *injections*, or *bijections*, respectively. A *permutation* of a set X is a bijection from X to itself. The *identity function* on a set X is the function e such that $e(x) = x$ for all x in X. Clearly e is a permutation of X.

A convenient way to describe a function on a small set is by a matrix with two rows. The first row lists the domain and the second row gives the images. For example,

$$\begin{pmatrix} 0 & 1 & 2 & 3 & 4 \\ 2 & 4 & 0 & 1 & 3 \end{pmatrix}$$

defines the permutation of $\{0, 1, 2, 3, 4\}$ which takes 0 to 2, 1 to 4, and so on.

Exercises

1.1. Show that the intersection of any nonempty collection of equivalence relations on a set is again an equivalence relation.

1.2. Suppose that $E_1, E_2, \ldots$ is an infinite sequence of equivalence relations on X and E_i is contained in E_{i+1} for $i \geq 1$. Prove that the union of the E_i is an equivalence relation on X.

1.3. Show that for any set X there is a bijection from the set of equivalence relations on X to the set of partitions of X.

1.4. Suppose $f: X \rightarrow Y$, $g: Y \rightarrow Z$, and $h: Y \rightarrow Z$ are functions. Assume that f is surjective and $f \circ g = f \circ h$. Show that $g = h$.

1.2 Monoids

A *semigroup* is a pair $(S, \bullet)$ consisting of a set S and an *associative binary operation* $\bullet$ *on* S. That is, $\bullet$ is a function from $S \times S$ to S, and if we write the image under $\bullet$ of a pair (s, t) as $s \bullet t$, then

$$s \bullet (t \bullet u) = (s \bullet t) \bullet u$$

for all s, t, and u in S. If $s \bullet t = t \bullet s$ for all s and t in S, then $\bullet$ is said to be *commutative*. In this book, multiplicative notation will normally be used for binary operations. Thus $s \bullet t$ will be written st and referred to as the *product* of s and t. When the binary operation is clear from the context, we refer to S as the semigroup. If S is finite, then the *order* of S is $|S|$.

An *identity element* in a semigroup S is an element e of S such that $es = se = s$ for all s in S. There is at most one identity element in S, and if S has an identity element, then S is called a *monoid*. Note that a semigroup may be empty, but a monoid always contains at least one element, its identity element. When multiplicative notation is used, the identity element is normally denoted by 1.

Let M be a monoid with identity element 1. An element u of M is called a *unit* if there is an element v of M such that $uv = vu = 1$. The element v is unique. We call v the *inverse* of u and write $v = u^{-1}$. If u is a unit with inverse v, then v is a unit with inverse u. The inverse of 1 is 1.

Suppose that x and y are units in M. Then

$$(xy)(y^{-1}x^{-1}) = x(yy^{-1})x^{-1} = x1x^{-1} = xx^{-1} = 1.$$

Similarly $(y^{-1}x^{-1})(xy) = 1$, so xy is a unit with inverse $y^{-1}x^{-1}$. Therefore the set of units is closed under multiplication.

If x is an element of a semigroup S and n is a positive integer, then x^n is defined to be $xx\ldots x$, where the product has n factors. If S is a monoid with identity element 1, then we set $x^0 = 1$. If x is a unit and $n < 0$, then x^n is defined to be $(x^{-1})^{-n}$. The usual laws of exponents hold.

A *group* is a monoid in which every element is a unit. If M is a monoid, then the set of units of M is a group. Commutative groups are said to be *abelian*.

Here are some examples of monoids and groups.

Example 2.1. Let Ω be a set. The set $\mathrm{Rel}(\Omega)$ of all relations on Ω is a monoid with composition as the binary operation. The identity element of $\mathrm{Rel}(\Omega)$ is the identity function on Ω. The set $\mathrm{Fun}(\Omega)$ of functions on Ω is also a monoid. The group of units of $\mathrm{Rel}(\Omega)$ is the set $\mathrm{Sym}(\Omega)$ of permutations of Ω, which is called the *symmetric group* on Ω. If Ω is finite and $|\Omega| = n$, then

$$|\mathrm{Rel}(\Omega)| = 2^{n^2}, |\mathrm{Fun}(\Omega)| = n^n, |\mathrm{Sym}(\Omega)| = n!.$$

If $\Omega = \{1, 2, \ldots, n\}$, then $\mathrm{Sym}(\Omega)$ is also denoted $\mathrm{Sym}(n)$.

Example 2.2. Let R be a commutative ring with identity and let n be a positive integer. The set $M_n(R)$ of n-by-n matrices over R is a monoid under matrix multiplication. The identity element of $M_n(R)$ is the n-by-n identity matrix. The group of units of $M_n(R)$ is the *general linear group* $\mathrm{GL}(n, R)$ of matrices whose determinants are units in R.

Example 2.3. Let M be a monoid. For subsets A and B of M let $AB = \{ab \mid a \in A, b \in B\}$. With this operation the set of all subsets of M becomes a monoid.

Example 2.4. The set $\mathbb{Z}$ of integers with addition as the binary operation is a group. The identity element is 0 and the inverse of n is $-n$. (Here we use *additive*, not multiplicative, notation.)

Example 2.5. The set $\mathbb{Z}$ with multiplication as the binary operation is a monoid with identity element 1. Only 1 and -1 are units.

Example 2.6. The set $\mathbb{N}$ of nonnegative integers with addition as the binary operation is a monoid. Only 0 is a unit.

Example 2.7. Let X be a set. A *word* over X is a finite sequence $U = u_1, u_2, \ldots, u_m$ of elements of X. The empty sequence, with $m = 0$, will be denoted ε. The set of all words over X is denoted X^*. If $V = v_1, \ldots, v_n$ is also a word over X, then UV is defined to be the word $u_1, \ldots, u_m, v_1, \ldots, v_n$. With this multiplication, X^* is a monoid with identity element ε, which is the only unit. We identify an element x of X with the corresponding word of length 1, and we write U as $u_1 \ldots u_m$, without commas. The length m of U is denoted $|U|$. Words over some finite set X will be the objects most commonly manipulated by the algorithms discussed in this book.

If A, B, and C are in X^* and $U = ABC$, then A is a *prefix* of U, C is a *suffix* of U, and B is a *subword* of U. If $U = u_1 \ldots u_m$, then any of the words $u_i u_{i+1} \ldots u_m u_1 \ldots u_{i-1}$, $1 \le i \le m$, is called a *cyclic permutation* of U, and $U^\dagger = u_m u_{m-1} \ldots u_1$ is called the *reversal* of U. The term "string" is sometimes used as a synonym for "word". In particular, if $X = \{0, 1\}$, then elements of X^* are frequently referred to as *bit strings*.

Let M be a monoid with identity element 1. A *submonoid* of M is a subset N of M such that the following conditions hold:

(i) 1 is in N.
(ii) If x and y are in N, then xy is in N.

If in addition all elements of N are units and N contains the inverse of each of its elements, then N is a *subgroup* of M. A submonoid N is a monoid under the restriction to N of the binary operation on M. Similarly, subgroups are groups in their own right.

Example 2.8. Let Ω be a set. The set $\text{Fun}(\Omega)$ is a submonoid of $\text{Rel}(\Omega)$, and $\text{Sym}(\Omega)$ is a subgroup of $\text{Rel}(\Omega)$. Let α be an element of Ω. The set of functions f on Ω such that $f(\alpha) = \alpha$ is a submonoid of $\text{Fun}(\Omega)$, and the set of permutations of Ω which fix α is a subgroup of $\text{Sym}(\Omega)$.

Example 2.9. The set $\mathbb{N}$ is a submonoid of the monoids $(\mathbb{Z}, +)$ and $(\mathbb{Z}, \times)$. For any integer n, the set $n\mathbb{Z}$ of multiples of n is a subgroup of $(\mathbb{Z}, +)$.

Proposition 2.1. *Let M be a monoid. The intersection of any nonempty collection of submonoids of M is a submonoid of M. The intersection of any nonempty collection of subgroups is a subgroup.*

Proof. Exercise. □

Let Y be a subset of a monoid M. The set $\mathcal{N}$ of submonoids of M which contain Y is nonempty, since M is in $\mathcal{N}$. By Proposition 2.1, the intersection K of the elements of $\mathcal{N}$ is a submonoid. Clearly K contains Y

and is the smallest submonoid containing Y. We write $K = \text{Mon}\langle Y \rangle$ and call K the submonoid *generated* by Y. It is easy to see that K consists of all elements of M which can be expressed as a product $y_1 \ldots y_t$ with each y_i in Y. By convention, such a product with $t = 0$ is defined to be 1.

Some authors write Y^* for the submonoid generated by Y. Strictly speaking, Y^* is the set of sequences $y_1, \ldots, y_t$ of elements of Y, and $\text{Mon}\langle Y \rangle$ is the set of products of these sequences. We shall normally distinguish between Y^* and $\text{Mon}\langle Y \rangle$. However, in Chapter 3 we shall bow to tradition and write Y^* for $\text{Mon}\langle Y \rangle$.

Example 2.10. Let $\Omega = \{0, 1, 2, 3, 4\}$ and let f be the function on Ω defined by

$$\begin{pmatrix} 0 & 1 & 2 & 3 & 4 \\ 1 & 2 & 3 & 4 & 1 \end{pmatrix}.$$

To determine the submonoid $\text{Mon}\langle f \rangle$ of $\text{Fun}(\Omega)$, we must compute the powers of f. Now $f^0 = 1$ and $f^1 = f$. Also

$$f^2 = f \circ f = \begin{pmatrix} 0 & 1 & 2 & 3 & 4 \\ 2 & 3 & 4 & 1 & 2 \end{pmatrix},$$

$$f^3 = \begin{pmatrix} 0 & 1 & 2 & 3 & 4 \\ 3 & 4 & 1 & 2 & 3 \end{pmatrix}, \quad f^4 = \begin{pmatrix} 0 & 1 & 2 & 3 & 4 \\ 4 & 1 & 2 & 3 & 4 \end{pmatrix},$$

$$f^5 = \begin{pmatrix} 0 & 1 & 2 & 3 & 4 \\ 1 & 2 & 3 & 4 & 1 \end{pmatrix} = f.$$

Thus $f^6 = f^2$, $f^7 = f^3$, and so on. Therefore $\text{Mon}\langle f \rangle = \{f^0, f^1, f^2, f^3, f^4\}$ and $\text{Mon}\langle f \rangle$ has order 5.

Now suppose that Y is a subset of the group of units of a monoid M. The set of subgroups of M which contain Y is nonempty, so the intersection H of these subgroups is the smallest subgroup of M containing Y. It is not hard to see that H is the submonoid of M generated by $Y \cup Y^{-1}$, where $Y^{-1} = \{y^{-1} \mid y \in Y\}$. We shall denote H by $\text{Grp}\langle Y \rangle$.

A group or a monoid is *finitely generated* if there is a finite set which generates it. If G is a group, then G is finitely generated as a group if and only if G is finitely generated as a monoid. For if $G = \text{Mon}\langle X \rangle$, then $G = \text{Grp}\langle X \rangle$, and if $G = \text{Grp}\langle Y \rangle$, then $G = \text{Mon}\langle Y \cup Y^{-1} \rangle$ and $Y \cup Y^{-1}$ is finite if Y is finite. Actually, when Y is finite, we need to add only one element to Y to be sure of getting a monoid generating set for $\text{Grp}\langle Y \rangle$.

Proposition 2.2. *If a group G is generated as a group by n elements, then G is generated as a monoid by $n + 1$ elements.*

Proof. Suppose $x_1, \ldots, x_n$ generate G as a group. Then $x_1, \ldots, x_n, x_1^{-1}$, $\ldots, x_n^{-1}$ generate G as a monoid. Let $y = x_1^{-1} \ldots x_n^{-1}$. Then

$$x_i^{-1} = x_{i-1} x_{i-2} \cdots x_1 y x_n x_{n-1} \cdots x_{i+1},$$

$1 \leq i \leq n$. Therefore G is generated as a monoid by $x_1, \ldots, x_n$ and y. □

Proposition 2.2 was first proved in (Dyck 1882), so the result dates from the very beginning of combinatorial group theory.

A monoid is *cyclic* if it is generated as a monoid by a single element. A group is *cyclic* if it is generated as a group by a single element. If x is an element of a group, then the *order* of x is the order of $\mathrm{Grp}\langle x \rangle$, provided that group is finite. If $\mathrm{Grp}\langle x \rangle$ is infinite, then x is said to have *infinite order*.

Proposition 2.3. *Let M be a finitely generated monoid. Every generating set for M contains a finite generating set.*

Proof. Let X be a finite generating set for M and let Y be any generating set. Each element of X can be expressed as a product of elements of Y. For each x in X choose one such product and let Z be the set of elements of Y which occur in at least one of the chosen products. Then Z is a finite set, and the submonoid generated by Z contains X. Therefore Z generates M.
□

Proposition 2.3 remains true if M is a group and the generating sets are group generating sets.

Let M and N be monoids with identity elements 1_M and 1_N, respectively. A *homomorphism* from M to N is a function $h \colon M \to N$ such that

(i) $h(1_M) = 1_N$.
(ii) $h(xy) = h(x)h(y)$ for all x and y in M.

If N is a group, then condition (i) follows from condition (ii). If x is a unit in M, then $xx^{-1} = 1_M = x^{-1}x$. Therefore

$$h(x)h(x^{-1}) = h(1_M) = 1_N = h(x^{-1})h(x).$$

Thus $h(x)$ is a unit and $h(x)^{-1} = h(x^{-1})$. In particular, if M is a group, then the image of M under h is a subgroup of N. A homomorphism from M to M is said to be an *endomorphism* of M. If a homomorphism $h \colon M \to N$ is a bijection, then h is called an *isomorphism*. In this case, the inverse map h^{-1} is also an isomorphism, and M and N are said to be *isomorphic*. A *semigroup homomorphism* is a map between semigroups satisfying condition (ii).

For any set X, the monoid X^* is called the *free monoid* generated by X. The set $\mathcal{S}$ of nonempty words in X^* is a subsemigroup of X^*, the *free semigroup* generated by X. The following proposition justifies the use of the adjective "free".

Proposition 2.4. *Let X be a set and let M be a monoid. For each function $f\colon X \to M$ there is a unique extension of f to a homomorphism of X^* into M.*

Proof. Let $U = u_1 \dots u_m$ be in X^* with each u_i in X. The only way to define $f(U)$ so that f becomes a homomorphism from X^* to M is to set $f(U)$ equal to $f(u_1) \dots f(u_m)$. It is easy to check that the map so defined is a homomorphism. $\square$

Here are some basic facts relating submonoids and homomorphisms:

Proposition 2.5. *Let $f\colon M \to N$ be a homomorphism of monoids. If H is a submonoid of M and K is a submonoid of N, then $f(H)$ is a submonoid of N and $f^{-1}(K)$ is a submonoid of M. If M and K are groups, then $f^{-1}(K)$ is a subgroup of M.*

Proof. Exercise. $\square$

The notion of an ideal is usually encountered first in the study of rings. However, the concept is also useful in the theory of monoids. Let M be a monoid. A subset I of M is an *ideal* if for all x in I and all y in M the products xy and yx are in I.

Example 2.11. Let Ω be a set. For any integer $k \geq 1$, the set J_k of functions $f\colon \Omega \to \Omega$ such that $|f(\Omega)| \leq k$ is an ideal of $\mathrm{Fun}(\Omega)$.

Example 2.12. Let X be a set. For any integer $k \geq 0$ the set I_k of words U in X^* such that $|U| \geq k$ is an ideal of X^*. More generally, if $\mathcal{U}$ is any subset of X^*, then the set of words having some element of $\mathcal{U}$ as a subword is an ideal of X^*.

If we modify the definition of an ideal to require only that xy be in I when x is in I and y is in M, then the resulting object is called a *right ideal*. Similarly, a *left ideal* is a subset closed under multiplication on the left by elements of M.

Example 2.13. Let Γ be a subset of a set Ω. The set of functions which map Ω into Γ is a left ideal of $\mathrm{Fun}(\Omega)$. The set of functions on Ω which are not injective is a right ideal of $\mathrm{Fun}(\Omega)$.

Example 2.14. Let $\mathcal{U}$ be a subset of X^*. The set of words having some element of $\mathcal{U}$ as a prefix is a right ideal of X^*.

Let I be an ideal of a monoid M. A *generating set* for I is a subset Y of M such that

$$I = MYM = \{xyz \mid x, z \in M, y \in Y\}.$$

Since we may take $x = z = 1$, it follows that Y is a subset of I. A generating set for a right ideal J of M is a subset T of J such that $J = TM$. Generating sets for left ideals are defined in an analogous manner.

A *minimal generating set* for an ideal I is a generating set Y which does not properly contain any other generating set. An ideal may have no minimal generating sets, and it may have many. However, in X^* minimal generating sets exist and are unique.

Proposition 2.6. *Let I be an ideal of X^* and let $\mathcal{U}$ be the set of all elements of I which do not contain elements of I as proper subwords. Then $\mathcal{U}$ generates I, and $\mathcal{U}$ is a subset of every generating set for I. In particular, $\mathcal{U}$ is the unique minimal generating set for I.*

Proof. Let U be an element of I. Choose a subword V of U of minimal length such that V is in I. Then V is in $\mathcal{U}$, and U is in the ideal generated by $\mathcal{U}$. Therefore $\mathcal{U}$ is a generating set for I. Now let $\mathcal{V}$ be any generating set for I and let U be in $\mathcal{U}$. There exist words A and B in X^* and V in $\mathcal{V}$ such that $U = AVB$. By the definition of $\mathcal{U}$ we must have $A = B = \varepsilon$. Therefore U is in $\mathcal{V}$. $\square$

Proposition 2.7. *Let $I_1 \subseteq I_2 \subseteq \cdots$ be an infinite sequence of ideals in X^*. Assume that there is an integer m such that the minimal generating set of each of the I_j has at most m elements. Then there is an integer n such that $I_j = I_n$ for $j \geq n$.*

Proof. Let

$$I = \bigcup_j I_j.$$

It is easy to see that I is an ideal. Let $\mathcal{U}$ be the minimal generating set for I, and suppose that it is possible to find distinct elements $U_1, \ldots, U_{m+1}$ in $\mathcal{U}$. There is an index j such that I_j contains all of the U_i. Let $\mathcal{V}$ be the minimal generating set for I_j. By assumption, not all of the U_i are in $\mathcal{V}$, and so some U_i contains an element V of $\mathcal{V}$ as a proper subword. But V is in I, so U_i cannot be in $\mathcal{U}$. Therefore $\mathcal{U}$ is finite and $|\mathcal{U}| \leq m$. Thus there is an index n such that I_n contains $\mathcal{U}$. But then $I_n = I$ and $I_j = I_n$ for $j \geq n$.

$\square$

Proposition 2.6 has analogues for right and left ideals of X^*. The minimal generating set for a right ideal J is the set of elements of J which do not contain other elements of J as prefixes. For left ideals, substitute "suffixes" for "prefixes".

Exercises

2.1. Suppose an element u of a monoid has a right inverse v and a left inverse w. Thus $uv = 1 = wu$. Show that u is a unit.

2.2. Suppose G is a group. Prove that the only element u of G with $u^2 = u$ is the identity element.

2.3. Let F be a finite field with q elements. Show that $|\operatorname{GL}(n,F)|$ is

$$(q^n - 1)(q^n - q) \ldots (q^n - q^{n-1}).$$

2.4. Let M and N be monoids. Define a binary operation on $M \times N$ by $(a,x)(b,y) = (ab, xy)$. Show that $M \times N$ is a monoid and the functions $a \mapsto (a,1)$, $x \mapsto (1,x)$, $(a,x) \mapsto a$, and $(a,x) \mapsto x$ are all homomorphisms. Prove that $M \times N$ is a group if both M and N are groups. The monoid $M \times N$ is the *direct product* of M and N.

2.5. Prove that up to isomorphism there is one infinite cyclic monoid and one infinite cyclic group.

2.6. Show that any finite cyclic group is a cyclic monoid.

2.7. Let n be a positive integer. Prove that up to isomorphism there are n cyclic monoids of order n, exactly one of which is a group.

2.8. Show that a submonoid of a finite group is a subgroup.

2.9. Suppose that X is a set with at least two elements. Prove that X^* has an infinite, strictly increasing sequence of ideals.

2.10. Explain why the term "generating set" is preferable to "set of generators".

1.3 Groups

In this section we summarize some basic facts about groups. Because this is intended to be a review of familiar material, most proofs are omitted. They can be found in introductory algebra texts and books on group theory.

Suppose that x is an element of a group G and x has finite order n. Then $x^n = 1$ and $x^m = 1$ if and only if n divides m. If all elements of G have finite order and these orders are bounded above, then the least common multiple e of these orders is called the *exponent* of G. An alternative definition of the exponent is the smallest positive integer e such that $g^e = 1$ for all g in G.

Let G be a group and let H be a subgroup of G. The set $\mathcal{R} = \{Hx \mid x \in G\}$ is a partition of G. The elements of $\mathcal{R}$ are the *right cosets* of H. When $\mathcal{R}$ is finite, its cardinality is denoted $|G : H|$ and is called the *index* of H in G. If G is finite, then $|G| = |G : H||H|$. The set $\mathcal{L} = \{xH \mid x \in G\}$ is also a partition of G. Its elements are the *left cosets* of H in G. There is a one-to-one correspondence between $\mathcal{R}$ and $\mathcal{L}$. If $\mathcal{R} = \mathcal{L}$, then H is said to be a *normal* subgroup of G and we write $H \triangleleft G$. If H is normal, then $\mathcal{R}$ can be made into a group by defining $(Hx)(Hy)$ to be Hxy. This group is denoted G/H and is called the *quotient group* of G by H.

Proposition 3.1. *Let H be a subgroup of a group G. The following are equivalent:*

(1) $H \triangleleft G$.
(2) $Hx = xH$ *for all x in G.*
(3) $x^{-1}Hx \subseteq H$ *for all x in G.*

Let $f \colon G \to H$ be a homomorphism of groups. The *kernel* of f is $f^{-1}(1) = \{x \in G \mid f(x) = 1\}$, which is a normal subgroup of G. Suppose that N is a normal subgroup of G which is generated by a subset X and Y is a subset of G whose image under the natural homomorphism from G to G/N generates G/N. Then $G = \mathrm{Grp}\langle X \cup Y \rangle$.

Two elements x and y of a group G are *conjugate* if there is an element z such that $y = z^{-1}xz$. Conjugacy is an equivalence relation on G. Its equivalence classes are called *conjugacy classes*. Let X be a subset of G. The subgroup N generated by all of the conjugates of the elements of X is normal in G. It is the smallest normal subgroup of G containing X. We call N the *normal closure* of X in G and denote it $\mathrm{Grp}\langle X^G \rangle$.

The notion of conjugacy is extended to subgroups. Two subgroups H and K of G are conjugate if there is an element z of G such that $K = z^{-1}Hz$. A subgroup is normal if and only if it is conjugate only to itself. Conjugate subgroups are isomorphic.

An isomorphism of a group G with itself is called an *automorphism* of G. The set $\mathrm{Aut}(G)$ of all automorphisms of G is a subgroup of $\mathrm{Sym}(G)$ and is called the *automorphism group* of G. Given an element g of G, the map $\varphi_g \colon G \to G$ taking u to $g^{-1}ug$ is an automorphism, the *inner automorphism* induced by g. The map $\varphi \colon g \mapsto \varphi_g$ is a homomorphism of G into $\mathrm{Aut}(G)$. The image of φ is the group $\mathrm{Inn}(G)$ of inner automorphisms of G. The kernel of φ is the *center* $Z(G)$, the set of elements in G which commute with every element of G.

Of particular importance are the three isomorphism theorems:

Proposition 3.2. *Let $f \colon G \to H$ be a surjective group homomorphism with kernel N. Then H is isomorphic to G/N. There is a one-to-one correspondence between subgroups of H and subgroups of G containing N.*

Proposition 3.3. *Let K be a subgroup of a group G and let $N \triangleleft G$. Then*

(1) $NK = KN$ *is a subgroup of G.*

(2) $K \cap N \lhd K$.

(3) $(KN)/N$ *is isomorphic to* $K/(K \cap N)$.

Proposition 3.4. *Let* H *and* K *be normal subgroups of a group* G *and assume that* $K \subseteq H$. *Then* H/K *is a normal subgroup of* G/K, *and* $(G/K)/(H/K)$ *is isomorphic to* G/H.

Suppose that Ω is a set. A subgroup G of $\mathrm{Sym}(\Omega)$ is called a *permutation group on* Ω. If g is in G and α is in Ω, then the image of α under g will be written α^g. Since permutations act on the right, the product gh is defined by $\alpha^{(gh)} = (\alpha^g)^h$. The set $\alpha^G = \{\alpha^g \mid g \in G\}$ is the *orbit* of α under G. The orbits of G partition Ω. The *stabilizer* of α in G is the subgroup $G_\alpha = \{g \in G \mid \alpha^g = \alpha\}$. If α^G is finite, then $|\alpha^G| = |G : G_\alpha|$. Two elements of G map α to the same point if and only if they lie in the same right coset of G_α. If $\alpha_1, \ldots, \alpha_r$ are distinct elements of Ω, then the *cycle* $(\alpha_1, \ldots, \alpha_r)$ is the permutation of Ω which maps α_i to α_{i+1}, $1 \leq i < r$, maps α_r to α_1, and fixes all other points of Ω. If Ω is finite, then an element g of $\mathrm{Sym}(\Omega)$ can be written as a product of *disjoint* cycles, no two of which move the same point. This product is essentially unique. If the number of cycles of even length in the cycle decomposition of g is even, then g is called an *even permutation*. The set $\mathrm{Alt}(\Omega)$ of all even permutations of Ω is a subgroup of index 2 in $\mathrm{Sym}(\Omega)$ and is called the *alternating group* on Ω. If $\Omega = \{1, \ldots, n\}$, then $\mathrm{Alt}(\Omega)$ is also denoted $\mathrm{Alt}(n)$.

Let G be a permutation group on Ω and let H be a permutation group on Δ. We say G and H are *isomorphic as permutation groups* or *permutation isomorphic* if there is an isomorphism σ from G to H and a bijection τ from Ω to Δ such that $\tau(\alpha^g) = \tau(\alpha)^{\sigma(g)}$ for all α in Ω and all g in G. The subgroups $\mathrm{Grp}\langle(1,2,3,4,5,6)\rangle$ and $\mathrm{Grp}\langle(1,2,3)(4,5)(6)\rangle$ of $\mathrm{Sym}(6)$ are both cyclic of order 6 and hence isomorphic. However, the first group has only one orbit, whereas the second has three orbits. This implies that these groups are not permutation isomorphic.

An important class of groups is the one consisting of the groups which satisfy the *ascending chain condition on subgroups*. This means that there is no strictly increasing sequence $H_1 \subset H_2 \subset \cdots$ of subgroups.

Proposition 3.5. *A group* G *satisfies the ascending chain condition on subgroups if and only if all subgroups of* G *are finitely generated.*

Proof. Suppose first that G has a subgroup H which is not finitely generated. Choose x_1 in H. Then $H_1 = \mathrm{Grp}\langle x_1 \rangle$ is not H, so we may choose x_2 in $H - H_1$. Now $H_2 = \mathrm{Grp}\langle x_1, x_2 \rangle$ properly contains H_1 and is not H. Therefore we may choose x_3 in $H - H_2$. Continuing in this way, we construct an infinite sequence $x_1, x_2, \ldots$ of elements such that the groups $H_i = \mathrm{Grp}\langle x_1, \ldots, x_i \rangle$ form a strictly increasing sequence.

Now suppose that all subgroups of G are finitely generated, and let $H_1 \subset H_2 \subset \cdots$ be an infinite, strictly increasing sequence of subgroups in G. The union H of the H_i is a subgroup, and H is generated by a finite set X. Each element of X is in some H_i. Thus we may find an index n such that $X \subseteq H_n$. But then $H_n = H$, so $H_i = H_n$ for all $i \geq n$. $\square$

The final result of this section describes the subgroups of $(\mathbb{Z}, +)$.

Proposition 3.6. *Every subgroup of $(\mathbb{Z}, +)$ is cyclic and consists of the multiples of a unique nonnegative integer n.*

The quotient of $\mathbb{Z}$ by $n\mathbb{Z}$ is denoted $\mathbb{Z}_n$.

Exercises

3.1. Let H be a subgroup of a group G. Show that a subset U of G is a right coset of H if and only if $\{u^{-1} \mid u \in U\}$ is a left coset of H.

3.2. What is the subgroup of $(\mathbb{Z}, +)$ generated by 4 and 6?

3.3. Let $g = (1,3,5,7)(2,4,6,8)$ and $h = (1,4,7)(2,3,6)$. Express the permutation gh as a product of disjoint cycles.

3.4. Let G be a finite cyclic group of order n. Show that G has exactly one subgroup of each order dividing n.

3.5. Suppose that G is an abelian group. Prove that the set of elements of G with finite order is a subgroup of G.

3.6. Let φ be an automorphism of a group G and let N be a subgroup of G. Show that $\varphi(N)$ is isomorphic to N and that, if N is normal in G, then $\varphi(N)$ is normal in G and G/N is isomorphic to $G/\varphi(N)$.

3.7. Prove that $\mathrm{Inn}(G)$ is a normal subgroup of $\mathrm{Aut}(G)$ for any group G.

3.8. Suppose that G is a group generated as a monoid by a set X and H is a subgroup of G. Show that H is normal in G if and only if $x^{-1}Hx \subseteq H$ for every x in X.

3.9. Suppose N is a normal subgroup of a group G. Prove that $\{(ug, g) \mid u \in N, g \in G\}$ is a subgroup of $G \times G$ and that this construction establishes a one-to-one correspondence between the set of normal subgroups of G and the subgroups of $G \times G$ which contain the diagonal subgroup.

1.4 Presentations

Let M be a monoid. A *congruence* on M is an equivalence relation $\sim$ on M which is compatible with the multiplication in M in the sense that whenever x, y, and z are elements of M and $x \sim y$, then $xz \sim yz$ and $zx \sim zy$.

Proposition 4.1. *Let $f \colon M \to N$ be a homomorphism of monoids. For x and y in M define $x \sim y$ to mean $f(x) = f(y)$. Then $\sim$ is a congruence on M.*

Proof. Exercise. $\square$

It turns out that every congruence $\sim$ on a monoid M can be constructed from a homomorphism as in Proposition 4.1. Let Q be the set of equivalence classes of $\sim$, and for x in M let $[x]$ be the element of Q containing x. The rule $[x][y] = [xy]$ defines an associative binary operation on Q with $[1]$ as an identity element. Thus Q is a monoid, the *quotient monoid* of M modulo $\sim$. The natural map $x \mapsto [x]$ is a homomorphism of M onto Q.

Example 4.1. Let f be the function on $\{0, 1, 2, 3, 4\}$ given by

$$\begin{pmatrix} 0 & 1 & 2 & 3 & 4 \\ 1 & 2 & 3 & 4 & 1 \end{pmatrix}.$$

In Example 2.10 we saw that $M = \mathrm{Mon}\langle f \rangle$ has order 5 and $f^5 = f$. The sets $\{1\}$, $\{f, f^3\}$, and $\{f^2, f^4\}$ are the equivalence classes of a congruence $\sim$ on M. This can be seen by noting that for each of these sets C the product $Cf = fC$ is contained in another one of these sets. The quotient monoid Q of M modulo $\sim$ has order 3 and is generated by $u = [f]$, which satisfies $u^3 = u$.

In the case of a congruence on a group G, the class $[1]$ is a normal subgroup of G, and the equivalence classes of $\sim$ are the cosets of $[1]$. The quotient Q is a group. As Example 4.1 shows, congruences on monoids are not necessarily determined by the equivalence class containing the identity element.

Proposition 4.2. *Let M be a monoid and let S be a subset of $M \times M$. The intersection $\sim$ of all congruences on M containing S is a congruence.*

Proof. There is at least one congruence on M containing S, namely $M \times M$, in which any two elements of M are congruent. By Exercise 1.1, $\sim$ is an equivalence relation on M. Clearly $s \sim t$ for all (s, t) in S. Suppose $x \sim y$. Then for each z in M and each congruence $\equiv$ containing S, we know that $x \equiv y$, so $xz \equiv yz$ and $zx \equiv zy$. Thus $xz \sim yz$ and $zx \sim zy$. Therefore $\sim$ is a congruence. $\square$

The congruence $\sim$ of Proposition 4.2 is called the congruence *generated* by S.

A *right congruence* on a monoid M is an equivalence relation $\sim$ on M such that $x \sim y$ implies that $xz \sim yz$ for all z in M. A *left congruence* is defined analogously. Only minor modifications of the preceding discussion are needed to define the concepts of the right congruence generated by a subset S of $M \times M$ and the left congruence generated by S.

Proposition 4.3. *Let M be a monoid and let Q be the quotient of M modulo the congruence $\sim$ generated by a subset S of $M \times M$. Let $f: M \to N$ be a monoid homomorphism such that $f(s) = f(t)$ for all (s, t) in S. Then there* *is a unique homomorphism* $g: Q \to N$ *such that* $f = \pi \circ g$, *where* π *is the natural map from* M *to* Q.

Proof. For x and y in M, define $x \equiv y$ to mean that $f(x) = f(y)$. By Proposition 4.1, $\equiv$ is a congruence on M, and by assumption, $\equiv$ contains $\mathcal{S}$. Therefore $\equiv$ contains $\sim$. Thus if $x \sim y$, then $f(x) = f(y)$. Therefore there is a well-defined map $g: Q \to N$ taking $\pi(x)$ to $f(x)$ for all x in M. Clearly $f = \pi \circ g$. To show that g is a homomorphism, we note that g takes $\pi(1)$, the identity of Q, to $f(1)$, the identity of N. Moreover,

$$g(\pi(x)\pi(y)) = g(\pi(xy)) = f(xy) = f(x)f(y) = g(\pi(x))g(\pi(y)). \quad \square$$

Let X be a set and let $\mathcal{R}$ be a subset of $X^* \times X^*$. The monoid $\mathrm{Mon}\langle X \mid \mathcal{R}\rangle$ is defined to be the quotient monoid Q of X^* modulo the congruence generated by $\mathcal{R}$. The pair $(X, \mathcal{R})$ is said to be a *monoid presentation* for Q and for any monoid isomorphic to Q. The presentation is *finite* if both X and $\mathcal{R}$ are finite. A monoid M is *finitely presented* if M has a finite presentation. If $X = \{x_1, \ldots, x_s\}$ and $\mathcal{R}$ consists of the pairs (U_i, V_i), $1 \le i \le t$, then Q is sometimes written as

$$\mathrm{Mon}\langle x_1, \ldots, x_s \mid U_1 = V_1, U_2 = V_2, \ldots, U_t = V_t\rangle.$$

Here the equations $U_i = V_i$ are called *defining relations* for Q. Under the natural map from X^* to Q, the words U_i and V_i map to the same element. In general, if $f: X^* \to M$ is a monoid homomorphism and U and V are words in X^* such that $f(U) = f(V)$, then we say that the *relation* $U = V$ holds in M (relative to f). If $f(U)$ is the identity element of M, then we normally speak of the relation $U = 1$, rather than the more consistent $U = \varepsilon$. Elements of $\mathrm{Mon}\langle X \mid \mathcal{R}\rangle$ are equivalence classes of words. The equivalence class containing U will usually be denoted $[U]$. There is a tendency to blur the distinction between U and $[U]$. We shall normally be fairly careful to distinguish between a word and the equivalence class containing that word. However, sometimes the effort required actually gets in the way of a clear exposition. In these cases we shall lapse into a controlled ambiguity between words and equivalence classes.

Example 4.2. Let $X = \{x\}$ and $\mathcal{R} = \{(x^6, x^3)\}$, and let $\sim$ be the congruence on X^* generated by $\mathcal{R}$. Since $x^6 \sim x^3$, we have $x^7 \sim x^4$, $x^8 \sim x^5$, and in general $x^i \sim x^{i-3}$ for $i \ge 6$. Therefore each equivalence class of $\sim$ contains an element x^i with $0 \le i \le 5$. Thus the order of $Q = \mathrm{Mon}\langle X \mid \mathcal{R}\rangle$ is at most 6. Let f be the function on $\{0, 1, 2, 3, 4, 5\}$ defined by

$$\begin{pmatrix} 0 & 1 & 2 & 3 & 4 & 5 \\ 1 & 2 & 3 & 4 & 5 & 3 \end{pmatrix}.$$

It is easy to check that $M = \mathrm{Mon}\langle f \rangle$ has order 6 and $f^6 = f^3$. By Proposition 4.3, there is a homomorphism h from Q to M taking the equivalence class $[x]$ to f. Since f generates M, the map h is surjective and $|Q| \geq |M| = 6$. Therefore $|Q| = 6$ and Q is isomorphic to M.

Example 4.3. Let $X = \{a, b\}$ and $\mathcal{R} = \{(ba, ab), (a^4, a^2), (b^3, a^3)\}$. Since $ba \sim ab$, every equivalence class of $\sim$ contains a word of the form $a^i b^j$. Since $b^3 \sim a^3$, we may assume that $0 \leq j \leq 2$, and since $a^4 \sim a^2$, we may assume that $0 \leq i \leq 3$. Therefore $Q = \mathrm{Mon}\langle X \mid \mathcal{R} \rangle$ has order at most 12. Let f and g be the following functions on $\{0, 1, 2, 3, 4, 5, 6, 7, 8, 9, 10, 11\}$:

$$f = \begin{pmatrix} 0 & 1 & 2 & 3 & 4 & 5 & 6 & 7 & 8 & 9 & 10 & 11 \\ 1 & 3 & 4 & 6 & 7 & 8 & 3 & 9 & 10 & 7 & 11 & 10 \end{pmatrix},$$

$$g = \begin{pmatrix} 0 & 1 & 2 & 3 & 4 & 5 & 6 & 7 & 8 & 9 & 10 & 11 \\ 2 & 4 & 5 & 7 & 8 & 6 & 9 & 10 & 3 & 11 & 6 & 3 \end{pmatrix}.$$

Then $f \circ g = g \circ f$, $f^4 = f^2$, and $f^3 = g^3$. Therefore there is a homomorphism of Q onto $M = \mathrm{Mon}\langle f, g \rangle$. The 12 elements $f^i \circ g^j$ with $0 \leq i \leq 3$ and $0 \leq j \leq 2$ map 0 to different points and so are distinct elements of M. Therefore Q is isomorphic to M, and both have order 12.

Proposition 4.4. *Suppose $\mathcal{R}$ and $\mathcal{S}$ are subsets of $X^* \times X^*$ with $\mathcal{R} \subseteq \mathcal{S}$. There is a homomorphism of $\mathrm{Mon}\langle X \mid \mathcal{R} \rangle$ onto $\mathrm{Mon}\langle X \mid \mathcal{S} \rangle$.*

Proof. Let $\sim$ and $\equiv$ be the congruences on X^* generated by $\mathcal{R}$ and $\mathcal{S}$ respectively. Then $\mathcal{R}$ is contained in $\equiv$, so $\sim$ is contained in $\equiv$. Therefore each $\sim$-class is contained in a unique $\equiv$-class. The corresponding map of $\mathrm{Mon}\langle X \mid \mathcal{R} \rangle$ to $\mathrm{Mon}\langle X \mid \mathcal{S} \rangle$ is easily seen to be a homomorphism. □

We shall occasionally use semigroup presentations. Recall that the free semigroup S on X is $X^* - \{\varepsilon\}$. We defined congruences on monoids, but the same definition works for semigroups as well. Let $\mathcal{R}$ be a subset of $S \times S$ and let $\sim$ be the congruence on S generated by $\mathcal{R}$. The semigroup defined by the pair $(X, \mathcal{R})$ is the quotient $S = \mathrm{Sem}\langle X \mid \mathcal{R} \rangle$ of S by $\sim$. Let $\equiv$ be the congruence on X^* generated by $\mathcal{R}$. It is not hard to show that $\equiv$ is the union of $\sim$ and $\{(\varepsilon, \varepsilon)\}$. Therefore $M = \mathrm{Mon}\langle X \mid \mathcal{R} \rangle$ is obtained from S by adding an extra element 1 to S and defining $1u = u1 = u$ for all u in M.

Semigroup presentations are convenient when the generating set contains the identity element.

Example 4.4. The semigroup $S = \mathrm{Sem}\langle a, e \mid a^3 = e, ae = ea = a \rangle$ is a group of order 3, while $M = \mathrm{Mon}\langle a, e \mid a^3 = e, ae = ea = a \rangle$ has order 4 and is not a group. To get a monoid presentation for S on the generators a and e we must add a relation like $e = 1$.

We have defined monoid and semigroup presentations. It is now time to define group presentations. Let X be a set and let $X^\pm = X \times \{1, -1\}$. We shall denote (x, α) in $X^\pm$ by x^α and identify x with x^1. The monoid $(X^\pm)^*$ will be denoted $X^{\pm*}$. Let $\mathcal{R}$ be the set of pairs of the form $(x^\alpha x^{-\alpha}, \varepsilon)$ with x in X and α in $\{1, -1\}$, and set $F = \text{Mon}\langle X^\pm \mid \mathcal{R} \rangle$. The element of F containing a given word U will be denoted $[U]$. We call F the *free group* generated by X. This term is justified by the following proposition.

Proposition 4.5. *The monoid F is a group. If $f\colon X \to G$ is any map of X into a group G, then there is a unique homomorphism of F into G taking $[x]$ to $f(x)$ for each x in X.*

Proof. To show that F is a group, it suffices to show that all the elements in some monoid generating set for F are units. Clearly F is generated as a monoid by $\{[x^\alpha] \mid x \in X, \alpha \in \{1, -1\}\}$. Since

$$[x^\alpha][x^{-\alpha}] = [x^\alpha x^{-\alpha}] = [\varepsilon] = 1$$

and

$$[x^{-\alpha}][x^\alpha] = [x^{-\alpha}x^\alpha] = [\varepsilon] = 1,$$

the element $[x^{-\alpha}]$ is an inverse for $[x^\alpha]$. Therefore F is a group. Now let $f\colon X \to G$ be a function. Extend f to a map of $X^\pm$ into G by defining $f(x^{-1})$ to be $f(x)^{-1}$, the inverse in G of $f(x)$. Then extend f further to a homomorphism of $X^{\pm*}$ into G. Since

$$f(x)f(x)^{-1} = f(x)^{-1}f(x) = 1$$

in G, the defining relations of F hold in G relative to f. Therefore there is a unique homomorphism $g\colon F \to G$ such that g takes $[U]$ to $f(U)$ for each word U in $X^{\pm*}$. It is not hard to see that g is the only homomorphism of F into G mapping $[x]$ to $f(x)$ for each x in X. $\square$

We shall encounter free groups frequently, and it will be useful to have a name for $\mathcal{R}$. Since $\mathcal{R}$ is the set of monoid defining relations for the free group F generated by X, we shall write $\mathcal{R} = \text{FGRel}(X)$. The congruence on $X^{\pm*}$ generated by $\mathcal{R}$ is called *free equivalence*. A word U in $X^{\pm*}$ is *freely reduced* if U contains no subword of the form $x^\alpha x^{-\alpha}$. If X is finite, then $|X|$ is called the *rank* of F. It is conceivable that this definition of rank is ambiguous in the sense that F might be isomorphic to the free group generated by a set Y with $|Y| \neq |X|$. However, this cannot happen. See Exercise 4.3.

Let U be a word in $X^{\pm*}$. Thus $U = x_1^{\alpha_1} \ldots x_m^{\alpha_m}$, where the x_i are in X and the exponents α_i are ± 1. Define U^{-1} to be $x_m^{-\alpha_m} \ldots x_1^{-\alpha_1}$. Then the inverse of $[U]$ in F is $[U^{-1}]$. Of course, U^{-1} is not an inverse for U in $X^{\pm*}$ unless $m = 0$. Clearly $(U^{-1})^{-1} = U$.

Suppose $\mathcal{S}$ is a subset of $X^{\pm*} \times X^{\pm*}$. By $\mathrm{Grp}\langle X \mid \mathcal{S} \rangle$ we shall mean $G = \mathrm{Mon}\langle X^{\pm} \mid \mathcal{R} \cup \mathcal{S} \rangle$, where $\mathcal{R} = \mathrm{FGRel}(X)$. By Proposition 4.4, there is a homomorphism of F onto G, and as remarked in Section 1.2, G is a group. We call $(X, \mathcal{S})$ a *group presentation* for G. The notation $\mathrm{Grp}\langle x_1, \ldots, x_s \mid U_1 = V_1, \ldots, U_t = V_t \rangle$ is often used for $\mathrm{Grp}\langle X \mid \mathcal{S} \rangle$ when $X = \{x_1, \ldots, x_s\}$ and $\mathcal{S} = \{(U_i, V_i) \mid 1 \leq i \leq t\}$.

Proposition 4.6. *Let $\pi : X^{\pm*} \to F$ be the natural map. Then $\mathrm{Grp}\langle X \mid \mathcal{S} \rangle$ is isomorphic to F/N, where N is the normal closure in F of the elements $\pi(V)^{-1}\pi(U)$ with (U, V) in $\mathcal{S}$.*

Proof. The proof consists of "diagram chasing" in commutative diagrams based on Proposition 4.3. The details are left as an exercise. $\square$

We have identified X with the subset $X \times \{1\}$ of $X^{\pm}$. Thus X^* is considered to be a subset of $X^{\pm*}$. In this context, an element of $X^* - \{\varepsilon\}$ is called a *positive word* in $X^{\pm*}$.

Let $\mathcal{S}$ be a subset of $X^* \times X^*$. In general, $\mathrm{Mon}\langle X \mid \mathcal{S} \rangle$ and $\mathrm{Grp}\langle X \mid \mathcal{S} \rangle$ are quite different objects. However, when $\mathrm{Mon}\langle X \mid \mathcal{S} \rangle$ is a group, they are the same.

Proposition 4.7. *Suppose $\mathcal{S}$ is a subset of $X^* \times X^*$. If $G = \mathrm{Mon}\langle X \mid \mathcal{S} \rangle$ is a group, then $\mathrm{Grp}\langle X \mid \mathcal{S} \rangle$ is isomorphic to G.*

Proof. The proof consists of more diagram chasing. $\square$

Example 4.5. Let $X = \{x\}$ and $\mathcal{S} = \{(x^6, x^3)\}$. In Example 4.2 we saw that $M = \mathrm{Mon}\langle X \mid \mathcal{S} \rangle$ has order 6. Let us determine the order of $G = \mathrm{Grp}\langle X \mid \mathcal{S} \rangle$. Since M is not a group and G is a group, it is clear that M and G must be nonisomorphic. By definition, $G = \mathrm{Mon}\langle X^{\pm} \mid \mathcal{T} \rangle$, where $X^{\pm} = \{x, x^{-1}\}$ and $\mathcal{T} = \{(xx^{-1}, \varepsilon), (x^{-1}x, \varepsilon), (x^6, x^3)\}$. Let $\sim$ be the congruence on $X^{\pm*}$ generated by $\mathcal{T}$. Then $x^6 \sim x^3$, so $x^6 x^{-1} \sim x^3 x^{-1}$. But $xx^{-1} \sim \varepsilon$, and hence $x^6 x^{-1} = x^5 xx^{-1} \sim x^5$. Similarly $x^3 x^{-1} \sim x^2$. Thus $x^5 \sim x^2$. Applying the same kind of arguments, we find that $x^4 \sim x$, $x^3 \sim \varepsilon$, and $x^2 \sim x^{-1}$. Since $x^{-1} \sim x^2$, we can find in each $\sim$-class an element x^i with $i \geq 0$. Since $x^3 \sim \varepsilon$, we may assume that $0 \leq i < 3$. Therefore the order of $\mathrm{Grp}\langle X \mid \mathcal{S} \rangle$

is at most 3. If f is the map

$$\begin{pmatrix} 0 & 1 & 2 \\ 1 & 2 & 0 \end{pmatrix}$$

on $\{0, 1, 2\}$, then f is a permutation and $\mathrm{Mon}\langle f \rangle = \mathrm{Grp}\langle f \rangle$ is a group. Since $f^3 = 1$ and so $f^6 = f^3$, we find that $\mathrm{Grp}\langle X \mid \mathcal{S} \rangle$ is isomorphic to $\mathrm{Grp}\langle f \rangle$ and hence has order 3.

Example 4.6. Let $X = \{a, b\}$ and $\mathcal{S} = \{(ba, ab), (a^4, a^2), (b^3, a^3)\}$. Then $X^{\pm} = \{a, a^{-1}, b, b^{-1}\}$ and $\mathrm{Grp}\langle X \mid \mathcal{S} \rangle$ is $\mathrm{Mon}\langle X^{\pm} \mid \mathcal{T} \rangle$, where $\mathcal{T}$ consists of the following pairs:

$$(aa^{-1}, \varepsilon), \quad (a^{-1}a, \varepsilon), \quad (bb^{-1}, \varepsilon), \quad (b^{-1}b, \varepsilon), \quad (ba, ab), \quad (a^4, a^2), \quad (b^3, a^3).$$

Let $\sim$ be the congruence on $X^{\pm *}$ generated by $\mathcal{T}$. Then $a^4 \sim a^2$, so $a^3 \sim a^4 a^{-1} \sim a^2 a^{-1} \sim a$. Similarly $a^2 \sim \varepsilon$ and $a \sim a^{-1}$. Also $a \sim a^3 \sim b^3$, so every $\sim$-class contains an element b^i. Since $b^6 \sim b^3 b^3 \sim a^3 a^3 = a^2 a^2 a^2 \sim \varepsilon$ and $b^{-1} \sim b^6 b^{-1} \sim b^5$, we may assume that $0 \leq i < 6$. Therefore $G = \mathrm{Grp}\langle X \mid \mathcal{S} \rangle$ has order at most 6. We leave it as an exercise to show that $|G| = 6$.

Example 4.7. Let $G = \mathrm{Mon}\langle x, y \mid x^2 = 1, y^3 = 1, (xy)^4 = 1 \rangle$. In G, the images of x and y have inverses, so G is a group. Under the homomorphism of $\{x, y\}^*$ into $\mathrm{Sym}(4)$ taking x to $(1,2)$ and y to $(2,3,4)$, the defining relations of G are satisfied. Thus there is a corresponding homomorphism $f: G \to \mathrm{Sym}(4)$. Let H be the subgroup $\{h \in G \mid 1^{f(h)} = 1\}$. Since G acts transitively on $\{1, 2, 3, 4\}$, the orbit of 1 has four elements and hence $|G : H| = 4$. Since the images under f of 1, x, xy, and xy^2 map 1 to 1, 2, 3, and 4, respectively, these elements are right coset representatives for H in G.

Let $(X, \mathcal{R})$ be a monoid presentation for a monoid M. It is frequently convenient to be able to find other presentations for M. We shall describe four constructions of such presentations. Let $\sim$ be the congruence on X^* generated by $\mathcal{R}$. If U and V are elements of X^* such that $U \sim V$, then we say that (U, V) is a *consequence* of $\mathcal{R}$.

Proposition 4.8. *Suppose (U, V) is a consequence of $\mathcal{R}$. Put $\mathcal{S} = \mathcal{R} \cup \{(U, V)\}$. Then $(X, \mathcal{S})$ is a presentation for M.*

Proof. Clearly $\mathcal{R}$ and $\mathcal{S}$ each generate $\sim$, so $\mathrm{Mon}\langle X \mid \mathcal{R} \rangle$ and $\mathrm{Mon}\langle X \mid \mathcal{S} \rangle$ are identical. $\square$

Proposition 4.9. *Suppose (U, V) is an element of $\mathcal{R}$. If (U, V) is a consequence of $\mathcal{S} = \mathcal{R} - \{(U, V)\}$, then $(X, \mathcal{S})$ is a presentation for M.*

Proof. Again $\mathcal{R}$ and $\mathcal{S}$ generate the same congruence on X^*. $\square$

Proposition 4.10. *Suppose U is in X^* and y is an object not in X. Set $Y = X \cup \{y\}$ and put $\mathcal{S} = \mathcal{R} \cup \{(y, U)\}$. Then $(Y, \mathcal{S})$ is a presentation for M.*

Proof. This proof again involves some diagram chasing, which is left as an exercise. $\square$

Proposition 4.11. *Suppose $\mathcal{R}$ contains an element (y, U), where y is in X and y does not occur in U. Let Y be $X - \{y\}$ and let f be the homomorphism of X^* into Y^* which is the identity on Y and maps y to U. Let $\mathcal{S}$ be the set of pairs $(f(A), f(B))$, where (A, B) ranges over of the elements of $\mathcal{R} - \{(y, U)\}$. Then $(Y, \mathcal{S})$ is a presentation for M.*

Proof. Exercise. $\square$

The constructions of new presentations described in Propositions 4.8 to 4.11 are called *Tietze transformations*. They have been defined for monoid presentations. Similar constructions can be made for group presentations.

Example 4.8. Let us go back over Example 4.5 using the language of relations and Tietze transformations. We start with generators x and x^{-1} and relations $xx^{-1} = 1$, $x^{-1}x = 1$, and $x^6 = x^3$. We saw that $x^3 = 1$ is a consequence of these relations, so we can add $x^3 = 1$ to the set of relations by Proposition 4.9. Now the relation $x^6 = x^3$ is a consequence of the other relations, so we may delete $x^6 = x^3$ by Proposition 4.10. Next $x^{-1} = x^2$ is a consequence of our relations, so by Proposition 4.11 we may use this relation to eliminate x^{-1} from the presentation. This leaves the single generator x and the single defining relation $x^3 = 1$.

Exercises

4.1. Let G be a group and let $\sim$ be a right congruence on G. Show that $H = \{g \in G \mid g \sim 1\}$ is a subgroup of G and that the congruence classes of $\sim$ are the right cosets of H in G.

4.2. Suppose that $\sim$ in Exercise 4.1 is generated by a subset $\mathcal{S}$ of $G \times G$. Prove that $H = \mathrm{Grp}\langle gh^{-1} \mid (g, h) \in \mathcal{S} \rangle$.

4.3. Let X be a finite set and let F be the free group generated by X. Show that the number of homomorphisms of F into a given cyclic group of order 2 is $2^{|X|}$. Thus F determines its rank $|X|$ uniquely.

4.4. Prove that $\mathrm{Grp}\langle x \mid x^n = 1 \rangle$ is isomorphic to $\mathbb{Z}_n$.

4.5. Show that $\text{Mon}\langle x, y \mid xy = yx, \, x^5 = x, \, y^4 = y^2, \, x^2y^2 = x^3y \rangle$ is finite and determine its order.

4.6. Find a sequence of Tietze transformations which takes the monoid presentation $a^2 = b^2 = (ab)^3 = 1$ on generators a and b to the presentation $a^2 = c^3 = (ac)^2 = 1$ on generators a and c.

4.7. Let G be a group. Show that G is finitely presented as a group if and only if G is finitely presented as a monoid.

4.8. Let $M = \text{Mon}\langle a, b \mid ab = 1 \rangle$. Prove that every element of M can be expressed uniquely as $[b^i a^j]$. Does M have any nontrivial units?

4.9. Show that $\text{Mon}\langle a, b \mid abba = 1 \rangle$ is a group.

4.10. Suppose that A and B are in $X^{\pm*}$. Show that A is freely reduced if and only if A^{-1} is freely reduced and that $(AB)^{-1} = B^{-1}A^{-1}$.

4.11. Let X be a set, let $\mathcal{R}$ be a subset of $X^* \times X^*$, and let $\sim$ be the congruence on X^* generated by $\mathcal{R}$. Show that two words U and V are in the same $\sim$-class if and only if there is a sequence of words $U = U_0, U_1, \ldots, U_r = V$ such that for $0 \le i < r$ there are words A, B, P, and Q such that $U_i = APB$, $U_{i+1} = AQB$, and either (P, Q) or (Q, P) is in $\mathcal{R}$.

4.12. Suppose in Exercise 4.11 that I is an ideal of X^* containing the left and right components of all the elements of $\mathcal{R}$. Show that I is a union of $\sim$-classes.

1.5 Computability

The main theme of this book is devising techniques for studying the structure of groups for which finite presentations are known. Unfortunately, we must recognize at the outset that many questions we would like to ask about finitely presented groups have been shown not to have general, algorithmic solutions. This means more than the assertion that nobody has been able to find an algorithm valid for all finitely presented groups. It means that no such algorithms will ever be found.

One of the earliest and most basic of these unsolvability results concerns the word problem. Let $M = \text{Mon}\langle X \mid \mathcal{R} \rangle$ be a finitely presented monoid. The *word problem* for M is to decide, given two words U and V in X^*, whether U and V define the same element of M. This is the same as deciding whether $U \sim V$, where $\sim$ is the congruence on X^* generated by $\mathcal{R}$. It has been shown that there is no algorithm which takes as input X, $\mathcal{R}$, U, and V and returns "yes" or "no" according as U and V do or do not define the same element of M.

It might appear plausible that we are asking too much to find one algorithm which will solve the word problem for all finite presentations. It is conceivable that no universal algorithm exists, but for any given presentation an algorithm can be devised. However, it has been shown that there are specific finitely presented monoids with unsolvable word problems.

Faced with this news, we might conclude that monoids are very "wild" objects, so the unsolvability of the word problem should have been guessed. However, if we restrict ourselves to groups, then surely everything will be nice. Again we are too optimistic. There are finitely presented groups with unsolvable word problems.

We cannot *decide* whether two words U and V define the same element of $M = \mathrm{Mon}\langle X \mid \mathcal{R}\rangle$. However, if U and V do define the same element, then we can *verify* this fact. The definition of the congruence $\sim$ generated by $\mathcal{R}$ is sufficiently concrete to allow us to list the words in the equivalence class containing U.

Proposition 5.1. *Let A and B be words in X^*. Then $A \sim B$ if and only if there is a sequence of words*

$$A = A_0, A_1, \ldots, A_t = B$$

such that for $0 \leq i < t$ the words A_i and A_{i+1} have the form CPD and CQD, respectively, where (P,Q) or (Q,P) is in $\mathcal{R}$.

Proof. Let us write $A \equiv B$ if there is a sequence of the specified type. Then $\equiv$ is easily seen to be an equivalence relation. If (P,Q) is in $\mathcal{R}$, then $P \sim Q$, so $CPD \sim CQD$ for all C and D in X^*. Therefore, if $A \equiv B$, then $A \sim B$. It follows that $\equiv$ is contained in $\sim$. Suppose $A \equiv B$, so there is a sequence $A_0, \ldots, A_t$ as in the proposition. If U is any word, then the sequence

$$AU = A_0U, A_1U, \ldots, A_tU = BU$$

shows that $AU \equiv BU$. Similarly, $UA \equiv UB$. Therefore $\equiv$ is a congruence. If (P,Q) is in $\mathcal{R}$, then the sequence P, Q has the required form, so $P \equiv Q$. Thus by the definition of $\sim$, we know that $\equiv$ contains $\sim$. Therefore $\equiv$ and $\sim$ are the same relation. $\square$

To list all words W such that $U \sim W$, we start with U, then list all words which can be reached by a sequence of length 1, then those which can be reached by a sequence of length 2, and so on. Eventually every word in $[U]$ will appear on the list. Thus if $U \sim V$, then V will be listed and we can stop the procedure. If $[V] \neq [U]$, then the procedure will never terminate. The unsolvability of the word problem means that although we can list the elements of $[U]$, we cannot always list the elements of the complement $X^* - [U]$, for if it were possible to list the complement, then we could start both listing procedures, and eventually V would appear on one list or the other. We would then know whether or not $U \sim V$.

A proof of the unsolvability of the word problem is beyond the scope of this book. See Chapter 12 of [Rotman 1973] for an accessible proof. However, assuming this result, we can easily deduce that a number of other questions cannot be answered algorithmically. For example, suppose $G = \mathrm{Grp}\langle X \mid \mathcal{R}\rangle$ is a finitely presented group. Given words U and V in $X^{\pm *}$, it is natural to ask whether the elements of G defined by U and V are conjugate. If we could solve this *conjugacy problem*, then taking $V = \varepsilon$, we could decide whether U defines the identity element of G. But if g and h are elements of a group, then $g = h$ if and only if $gh^{-1} = 1$. Thus if we can decide equality with 1, then we can decide equality in general, and so solve the word problem.

There is another problem which should be mentioned along with the word and conjugacy problems. This is the *generalized word problem* or the *subgroup membership problem*. Suppose we have a finite presentation of a group G and $U_1, \ldots, U_m, V$ are words defining elements $u_1, \ldots, u_m, v$ of G. Let $H = \mathrm{Grp}\langle u_1, \ldots, u_m \rangle$. The subgroup membership problem is to decide whether v is in H. An extension of the problem would be to express v explicitly as a product of the elements $u_1, u_1^{-1}, \ldots, u_m, u_m^{-1}$, when v is in H. This problem is also not solvable algorithmically for all finite presentations of groups, for if it were, then we could take $m = 0$, or $m = 1$ and $U_1 = \varepsilon$. Then H would be trivial, and we would be able to decide whether V defined the identity element.

There are many other problems concerning finitely presented groups which have been shown to be unsolvable. For example, it is not in general possible to decide whether a finitely presented group is finite, or infinite, or trivial.

One final remark needs to be made. Although we shall often be using group presentations, we shall sometimes be using monoid presentations of groups which are not group presentations. If $(X, \mathcal{R})$ is a finite presentation for a monoid M, then there is no algorithm for deciding whether M is a group. This result was first proved in (Markov 1951). See also (Narendran, Ó'Dúnlaing & Otto 1991). If M is a group, then there is an algorithm for computing, given a word U, a word V such that $[V] = [U]^{-1}$, but the algorithm is very unsatisfactory. Let $W_1, W_2, \ldots$ be the words in X^* listed in some order. As noted above, we can construct computers $C_1, C_2, \ldots$ such that C_i lists the elements of $[UW_i]$. If M is a group then some C_i will eventually print out ε and we can take $V = W_i$.

We do not actually need infinitely many computers to carry out this computation of $[U]^{-1}$. We assemble C_1 and set it running on UW_1. Then we assemble C_2, and when we are finished, we look to see if C_1 has printed out ε. If not, we start C_2 running with UW_2 and assemble C_3. Now we look to see if either C_1 or C_2 has printed out ε. If not, we start C_3 running with UW_3 and begin work on C_4. If we continue in this manner, then eventually we will see ε printed out, and we can stop building computers.

In order to avoid the problem just described, we shall always assume that our groups are "obviously" groups, even when they are given by monoid presentations. Usually the criterion in Exercise 5.2 will be adequate to recognize a group.

Example 5.1. Let $X = \{a, b\}$ and $\mathcal{R} = \{(abab, \varepsilon), (baba, \varepsilon)\}$. Then $M = \mathrm{Mon}\langle X \mid \mathcal{R}\rangle$ is a group since it is clear that $[a]^{-1} = [bab]$ and $[b]^{-1} = [aba]$.

Exercises

5.1. Show that it is possible to verify that a finitely presented monoid is a group.

5.2. Let $M = \mathrm{Mon}\langle X \mid \mathcal{R}\rangle$, where $X = \{x_1, \ldots, x_r\}$, and assume that for $1 \leq i \leq r$ there are words P_i and Q_i in X^* and words U_i and V_i in $\{x_1, \ldots, x_{i-1}\}^*$ such that $(x_i P_i, U_i)$ and $(Q_i x_i, V_i)$ are in $\mathcal{R}$. Prove that M is a group.

1.6 Procedure descriptions

In this book, procedures will be defined using an informal description language which looks a lot like Pascal but contains some features borrowed from C, as well as a few original constructs. The level of informality varies with the need for precision. In some cases, procedure definitions are very close to Pascal or C programs. In other cases, they are only outlines of programs. It is not possible to give a formal description of an informal language. However, the brief summary provided by this section should suffice to allow the reader to understand the procedures presented.

Variables are of just two types. Traditional mathematical objects are represented by single letters, perhaps with subscripts and other decorations. Case and font are significant. Thus s, S, $\mathcal{S}$, σ, and Σ may be different objects. Boolean variables, variables which take only the values "true" and "false", are represented by English words, usually past participles, written in lowercase italic characters. Thus *done*, *found*, and *ok* are typical boolean variables. Case is not significant in the reserved words of Pascal, such as "begin", "for", and "if". Reserved words coming at the beginning of a statement will normally be capitalized.

The assignment operator is $:=$. Thus the statement $x := 3$ means "assign the value 3 to x". The ordinary equals sign is a relational operator. For example, $x = 3$ is a boolean expression which is true if x happens to have the value 3 already, but is false otherwise. The other relational operators are $<$, $\leq$, $\geq$, $>$, and $\neq$.

The operations normally part of the Pascal and C languages are extended by other notations defined in the text. Because variables consist of a single letter, the symbol $*$ used for multiplication in both Pascal and C can be omitted. Set union and intersection are denoted by $\cup$ and $\cap$, respectively. Most statements are either assignment statements or control statements. Occasionally statements indicate an action not easily described by an assignment statement. Such statements are usually introduced by "Let". If a is an even integer, then "Let $a = 2b$" is a statement defining the integer b. The value of a is unchanged. Of course this statement is equivalent to $b := a/2$. Statements are separated by semicolons and grouped into compound statements using "Begin" and "End".

The three primary control statements are If-then-else, the For-loop, and the While-loop. The construction

> If condition then statement$_1$
>
> Else statement$_2$

indicates that the boolean condition is to be evaluated. If the condition is true, then statement$_1$ is executed. If the condition is false, then statement$_2$ is executed. The Else-clause may be omitted. The construction

> For $i := 1$ to n do statement

executes statement with $i = 1$, then with $i = 2$, and so on through $i = n$. The construction

> While condition do statement

means that statement is to be repeated as long as the condition remains true. A typical use of the While-loop has the form

> $done :=$ false;
>
> While not $done$ do
>
>   Begin
>
> $\vdots$
>
>   End

The statements enclosed by "Begin" and "End" are repeated as a block until $done$ becomes true.

Sometimes a For-loop will indicate a loop over a finite set, as in

> For x in X do statement

When the order in which the elements of X are taken is important, that order will be described in the accompanying text, or in a comment in the procedure. Comments have the form $(* \ldots *)$.

The flow of computation defined by If-then-else-statements, For-loops, and While-loops can be interrupted by using Goto-statements, Break-statements, and Continue-statements. The statement

> Goto 99

means that the statement with label 99 is to be executed next. A label is prefixed to the beginning of the statement and separated from it by a colon. The use of Goto-statements is considered poor programming practice, but now and then they provide the cleanest method of accomplishing a task within the control structures available.

The statement

Break

indicates that processing of the innermost loop containing the statement is to be terminated, and the statement immediately following the loop is to be executed next. The statement

Continue

means that the remainder of the body of the innermost loop containing the statement is to be skipped. If the loop is a While-loop, then the boolean condition is tested to determine whether another iteration of the body should be begun. In the case of a For-loop, the loop variable is incremented and the test for termination is carried out.

Procedure definitions begin with a Procedure-statement, which gives the name of the procedure and its arguments, if any. The treatment of arguments here is different from the treatment in either Pascal or C. Arguments are of one of two kinds. Input arguments are supplied by the calling procedure and are not changed by the called procedure. The results of the computation are returned in output arguments. When both input and output arguments are present, they are separated by a semicolon.

Pascal and C require that all variables be declared. That is, the type of values which may be assigned to the variables must be stated. Except for the description of arguments, declarations will be omitted. Global variables, variables shared by two or more procedures, are described in the accompanying text. All other variables are local to the procedure in which they are used.

The indexing of arrays, such as vectors and matrices, will be denoted either as in Pascal, $v[i]$ and $A[i,j]$, or using subscripts, v_i and A_{ij}. Indices are usually integers, but may occasionally be elements of other sets.

Here is an example of a procedure which multiplies two square matrices:

Procedure MPROD$(n, A, B; C)$;
Input: n : a positive integer;
               A, B : n-by-n matrices;
Output: C : an n-by-n matrix, the product of A and B;
Begin
    For $i := 1$ to n do

For $j := 1$ to n do begin
   $s := 0$;
   For $k := 1$ to n do $s := s + A[i,k]B[k,j]$;
   $C[i,j] := s$
End
End.

Subprocedures in Pascal are normally defined within the main procedure using them. To facilitate the exposition, subprocedures will be listed and discussed individually. The Pascal rule that subprocedures must be defined before they are used will be followed.

A procedure which computes a single output may be written as a function. The value returned is the last value assigned to the name of the function. The type of the result is indicated as part of the Function-statement. Here is an example of a function which computes the nonnegative greatest common divisor $\gcd(x,y)$ of two integers x and y using the Euclidean algorithm.

Function $\mathrm{GCD}(x,y)$: integer;
Input: x,y : integers;
($*$ The nonnegative gcd of x and y is returned. $*$)
Begin
   $a := |x|$; $b := |y|$;
   While $a \neq 0$ do begin $c := b \bmod a$; $b := a$; $a := c$ end;
   $\mathrm{GCD} := b$
End.

Recall that in Pascal, $b \bmod a$ is the remainder when b is divided by a. The function GCD is invoked as follows:

$$d := \mathrm{GCD}(18, 33)$$

We shall modify the official Pascal definitions of mod and div so that for any integers a and b with $b \neq 0$ the values of $a \bmod b$ and $a \operatorname{div} b$ are integers satisfying

(i) $0 \leq a \bmod b < |b|$.
(ii) $(a \bmod b) + b(a \operatorname{div} b) = a$.

In addition to arrays, some procedures use another kind of data structure called a *stack*. A stack is a method of storing objects using the philosophy "last in, first out". Only the most recently stored object is accessible for retrieval. Once that object has been removed, the next most recently stored object is available. An object is "pushed" onto the stack and "popped" off of the stack. The details of implementing a stack will normally be ignored.

Suppose that we are using a stack to store integers. The result of the statements

Clear the stack;
Push 7 onto the stack;
Push −3 onto the stack;
Push 11 onto the stack;
Pop x off of the stack;
Push 0 onto the stack;
Push −6 onto the stack;

While the stack is not empty do begin
    Pop y off of the stack; Print y
End

is to assign 11 to x and to print out the numbers −6, 0, −3, and 7, in that order.

A *queue* is a storage method based on the philosophy "first in, first out". It is analogous to the way customers wishing to purchase tickets at a box office are served. Queues will be mentioned as alternatives to stacks in some procedures.

1.7 The integers

In Chapters 2 through 7 we shall primarily be manipulating words. However, beginning with Chapter 8, computations with integers will play a very important role, because almost every question about a finitely generated abelian group can be reduced to the calculation of greatest common divisors of integers.

The function GCD was introduced in the previous section. If $d = \text{GCD}(x, y)$, then d is the nonnegative generator for the additive subgroup of $\mathbb{Z}$ generated by x and y. Thus there exist integers r and s such that $d = rx + sy$. The *extended* Euclidean algorithm computes one such pair of integers.

Procedure GCDX$(x, y; d, r, s)$;
Input: x, y : integers;
Output: d : the nonnegative gcd of x and y;
          r, s : integers such that $d = rx + sy$;
Begin
    $u_1 := \text{signum}(x)$; $u_2 := 0$; $a := |x|$;
    $v_2 := \text{signum}(y)$; $v_1 := 0$; $b := |y|$;
    While $b \neq 0$ do begin
        $q := a \text{ div } b$;
        $a := a - qb$; $u_1 := u_1 - qv_1$; $u_2 := u_2 - qv_2$;

Interchange the pairs a and b, u_1 and v_1, u_2 and v_2
End;

$d := a;\ r := u_1;\ s := u_2$
End.

The value of signum(x) is 1 if $x > 0$, 0 if $x = 0$, and -1 if $x < 0$. The call GCDX$(12, -34; d, r, s)$ sets $d = 2$, $r = 3$, and $s = 1$.

The *least common multiple* lcm(x, y) of x and y is the nonnegative generator of $(x\mathbb{Z}) \cap (y\mathbb{Z})$. It is easy to show that

$$\gcd(x, y)\,\mathrm{lcm}(x, y) = |xy|.$$

If $xy \neq 0$, then lcm$(x, y) = |xy|/\gcd(x, y)$. Clearly lcm$(x, y) = 0$ if $xy = 0$.

Two integers a and b are said to be *congruent modulo* a third integer m if m divides $a - b$. In this case we write $a \equiv b \pmod{m}$. Congruence modulo m is an equivalence relation. The set of *congruence classes* modulo m is $\mathbb{Z}_m$, the set of cosets of $m\mathbb{Z}$ in $\mathbb{Z}$. If $m \neq 0$, then $|\mathbb{Z}_m| = |m|$. For a in $\mathbb{Z}$ let $[a] = [a]_m$ denote the congruence class modulo m containing a. Defining $[a] + [b]$ to be $[a + b]$ and $[a][b]$ to be $[ab]$ makes $\mathbb{Z}_m$ a ring. The group U_m of units in $\mathbb{Z}_m$ considered as a monoid under multiplication consists of the classes $[a]$, with a *relatively prime* to m, that is, with $\gcd(a, m) = 1$. If $[a]$ is in U_m and we invoke GCDX$(a, m; d, r, s)$, then $[r]$ is the inverse of $[a]$ in U_m.

Suppose that $m_1, \ldots, m_r$ are pairwise relatively prime positive integers and that $a_1, \ldots, a_r$ are any integers. The Chinese remainder theorem states that the system of simultaneous congruences

$$x \equiv a_1 \pmod{m_1}, \ldots, x \equiv a_r \pmod{m_r}$$

has a solution which is unique modulo $m_1 \ldots m_r$. The case $r = 2$ is solved as follows: Let $1 = \gcd(m_1, m_2) = u_1 m_1 + u_2 m_2$. Then $x = a_2 u_1 m_1 + a_1 u_2 m_2$ satisfies $x \equiv a_i \pmod{m_i}$, $i = 1, 2$. Of course, x should be reduced modulo $m_1 m_2$.

Proposition 7.1. *Suppose that m and n are positive integers and that m divides n. The map f taking $[a]_n$ to $[a]_m$ is a ring homomorphism of $\mathbb{Z}_n$ onto $\mathbb{Z}_m$, and f maps U_n onto U_m.*

Proof. If $a \equiv b \pmod{n}$, then $a \equiv b \pmod{m}$, so f is well defined. It is easy to see that f is a ring homomorphism. If $\gcd(a, n) = 1$, then $\gcd(a, m) = 1$ so f maps U_n into U_m. To show that f maps U_n onto U_m, we use induction on n/m. If there is an integer k with $m < k < n$ such that m divides k and k divides n, then f is that composition of the homomorphisms from $\mathbb{Z}_n$ to $\mathbb{Z}_k$ and from $\mathbb{Z}_k$ to $\mathbb{Z}_m$. By induction, U_n is mapped onto U_k and U_k is mapped onto U_m. Therefore f maps U_n onto U_m.

We are left with the case in which $p = n/m$ is a prime. Suppose that $\gcd(b, m) = 1$. If p divides m, then $\gcd(b, n) = 1$ and $[b]_n$ is in U_n. If p does not divide m, then by the Chinese remainder theorem we can solve the following pair of simultaneous congruences:

$$a \equiv b \pmod{m},$$

$$a \equiv 1 \pmod{p}.$$

Then $[a]_n$ is in U_n and f maps $[a]_n$ to $[a]_m = [b]_m$. In either case $[b]_m$ is in the image of U_n under f. $\square$

The reduction of integer matrices discussed in Chapter 8 and the computation of Gröbner bases described in Chapter 10 can easily lead to very large integers, with hundreds or even thousands of decimal digits. With integers of this size, the traditional algorithms for multiplication and division are not the most efficient. There is no time to delve into this interesting topic here, but (Collins, Mignotte, & Winkler 1983) provides a good introduction.

Exercises

7.1. Show that at all times in GCDX the matrix

$$\begin{bmatrix} u_1 & u_2 \\ v_1 & v_2 \end{bmatrix}$$

has determinant ± 1.

7.2. Compute the multiplicative inverse of 5 modulo 13.

7.3. Solve the simultaneous congruences $x \equiv 2 \pmod 5$, $x \equiv 3 \pmod 7$.

7.4. Show that after the call $GCDX(x, y; d, r, s)$ the inequalities $|r| \leq \max(1, |y|)$ and $|s| \leq \max(1, |x|)$ hold.

1.8 Backtrack searches

One of the principal algorithms discussed in this book, the low-index subgroup algorithm of Section 5.6, uses a technique called backtrack search. Other backtrack searches are described in Sections 3.6 and 4.9. This section provides a brief introduction to backtracking.

In its simplest form, *backtrack searching* is a method for listing all sequences $a_1, \ldots, a_m$ of integers which satisfy some condition $P(a_1, \ldots, a_m)$. Here m may vary from sequence to sequence. The approach is to consider the possibilities for a_1, then for a given a_1 to consider the possibilities for a_2, and so on. In order for backtracking to be effective, it must be possible to formulate auxiliary conditions $P_j(a_1, \ldots, a_j)$ for $j = 1, 2, \ldots$ such that

(i) If $P(a_1, \ldots, a_m)$ is true, then $P_j(a_1, \ldots, a_j)$ is true for $1 \leq j \leq m$.

(ii) Given $a_1, \ldots, a_{j-1}$, there are at most a finite number of values for a_j such that $P_j(a_1, \ldots, a_j)$ is true and there are explicitly computable upper and lower bounds for these values.

(iii) For sufficiently large j the condition $P_j(a_1, \ldots, a_j)$ is never satisfied.

By (ii), given $a_1, \ldots, a_{j-1}$, it is possible to compute the set $S_j(a_1, \ldots, a_{j-1})$ of integers a_j such that $P_j(a_1, \ldots, a_j)$ is true. For simplicity of exposition, we shall assume that if $a_1, \ldots, a_m$ satisfies P, then no proper initial segment $a_1, \ldots, a_j$, $1 \le j < m$, satisfies P.

The following is a generic backtrack search procedure.

```
Procedure SEARCH;
Begin
    m := 1;  S₁ := S₁( );
    While m > 0 do
        If Sₘ ≠ ∅ then begin
            Choose b in Sₘ;  Sₘ := Sₘ - {b};  aₘ := b;
            If P(a₁, ..., aₘ) then write out a₁, ..., aₘ
            Else begin m := m + 1;  Sₘ := Sₘ(a₁, ..., aₘ₋₁) end
        End
        Else m := m - 1   (* Backtrack! *)
End.
```

Example 8.1. Let us use backtrack search to find the partitions of a positive integer n. A *partition* of n is a sequence $a_1, \ldots, a_m$ of positive integers such that $a_1 \ge a_2 \ge \cdots \ge a_m$ and $a_1 + a_2 + \cdots + a_m = n$. A natural choice for $P_j(a_1, \ldots, a_j)$ is

(1) $a_1 \ge a_2 \ge \cdots \ge a_j \ge 1$, and
(2) $a_1 + a_2 + \cdots + a_j \le n$.

With this definition, $S_1( ) = \{1, 2, \ldots, n\}$. If $j > 1$ and $P_{j-1}(a_1, \ldots, a_{j-1})$ is true, then $S_j(a_1, \ldots, a_{j-1})$ is the set of positive integers not exceeding either a_{j-1} or $n - (a_1 + \cdots + a_{j-1})$.

The sequences $a_1, \ldots, a_j$ examined in SEARCH can be displayed conveniently as a tree. The root of the tree is the empty sequence ( ). The nodes immediately below the root correspond to the elements of $S_1( )$. Figure 1.8.1 shows the search tree traversed in obtaining the partitions of 5.

In Example 8.1, every sequence $a_1, \ldots, a_j$ that satisfies $P_j(a_1, \ldots, a_j)$ is extendable to a sequence $a_1, \ldots, a_m$ satisfying $P(a_1, \ldots, a_m)$. This is not always the case.

Example 8.2. Let us define a *special decomposition* of a positive integer n to be a sequence $a_1, \ldots, a_m$ of positive integers such that $a_1 + \cdots + a_m = n$,

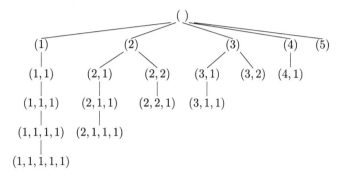

Figure 1.8.1

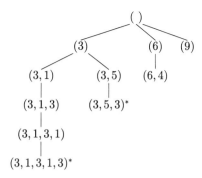

Figure 1.8.2

a_1 is divisible by 3, and $|a_i - a_{i+1}| = 2$ for $1 \le i < m$. Thus $3, 1, 3, 5$ is a special decomposition of 12. A plausible choice for $P_j(a_1, \ldots, a_j)$ is

(1) $a_i > 0$, $1 \le i \le j$, and
(2) a_1 is divisible by 3, and
(3) $|a_i - a_{i+1}| = 2$, $1 \le i < j$, and
(4) $a_1 + \cdots + a_j \le n$.

Figure 1.8.2 shows the search tree examined in determining the special decompositions of $n = 11$. The special decompositions are flagged with an asterisk. With $n = 14$, the tree in Figure 1.8.3 is obtained. The tree has 17 nodes, but only one corresponds to a special decomposition, namely (6,8).

It is sometimes possible to "prune" the search tree by strengthening the conditions $P_j(a_1, \ldots, a_j)$. However, if the new conditions take a long time to check, then the running time may actually increase, even though the number of nodes has decreased.

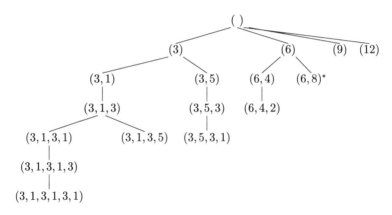

Figure 1.8.3

Example 8.3. Let us look again at the searches of Example 8.2. Suppose that $a_1, \ldots, a_m$ is a special decomposition of n. A simple analysis shows that if a_1 is even, then all the a_i are even and hence n is even. Similarly, if a_1 is congruent to 1 modulo 4, then the sequence modulo 4 is $1, 3, 1, 3, \ldots$ and n is congruent to 1 or 0 modulo 4. If a_1 is congruent to 3 modulo 4, then n is congruent to 0 or 3 modulo 4. These results can be summarized by saying that n is congruent modulo 4 to 0, a_1, or $2a_1 + 2$. Thus condition (2) in Example 8.2 can be replaced by

2'. a_1 is divisible by 3 and n is congruent modulo 4 to 0, a_1, or $2a_1 + 2$.

In addition, condition (4) may be replaced by

4'. If $p = n - (a_1 + a_2 + \cdots + a_j)$, then either $p = 0$ or $p \geq a_j - 2$. If $p > 0$ and $a_j \leq 2$, then $p \geq a_j + 2$.

With these changes, the size of the search tree is substantially reduced. For $n = 11$, we have Figure 1.8.4. With $n = 14$, the tree is as shown in Figure 1.8.5.

Frequently, the size of the search tree is difficult to determine in advance. A technique for getting a quick estimate of the size of the tree is described in (Knuth 1975). The estimate can occasionally be misleading, but in practice it is quite useful.

To guess the size of the tree, we make a number of random "probes" into the tree. A probe is carried out by executing the following procedure:

Procedure PROBE;
Begin
    $n_0 := 1$; $m := 1$; $\mathcal{S}_1 := S_1(\ )$; $n_1 := |\mathcal{S}_1|$;

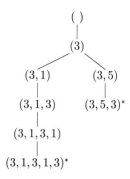

Figure 1.8.4

Figure 1.8.5

While $\mathcal{S}_m \neq \emptyset$ do begin
    Choose b at random from $\mathcal{S}_m$; $a_m := b$;

    If $P(a_1, \ldots, a_m)$ then break
    Else begin
        $m := m + 1$; $\mathcal{S}_m := S_m(a_1, \ldots, a_{m-1})$; $n_m := n_{m-1}|\mathcal{S}_m|$
    End
  End;

  Write out $n_0, \ldots, n_m$
End.

In (Knuth 1975) it is shown that the expected value of n_i is the number of nodes in the search tree which have distance i from the root. The observed n_i may differ substantially from the expected values, but averaging over a number of probes often gives a good picture of the size of the tree.

Example 8.4. Let us use the Knuth technique to explore two special decomposition search trees using the revised auxiliary conditions of Example 8.3. First, suppose that $n = 35$. Table 1.8.1 gives the average values of the n_i obtained on three different trials of 10 probes each. All results are rounded to the nearest integer. The fifth column averages the three trials, and the sixth column gives the actual number of nodes at each distance from the root.

Table 1.8.1

i	Trial 1	Trial 2	Trial 3	Average	Actual
0	1	1	1	1	1
1	2	2	2	2	2
2	2	3	2	2	2
3	4	4	3	4	3
4	7	7	6	7	6
5	14	10	7	10	9
6	26	14	14	18	14
7	42	16	18	25	19
8	51	10	29	30	25
9	51	13	32	32	30
10	19	19	32	23	30
11	19	19	32	23	34
12		26	26	17	22
13		38	26	21	23
14		26		9	8
15		26		9	8
16					1
17					1

Now let us look at $n = 60$. Table 1.8.2 gives the results of trials of 20, 50, and 200 probes. The true values are also given.

Exercises

8.1. Find all partitions of 7 using SEARCH, and construct the search tree examined.

8.2. Find the special decompositions of 15 using SEARCH. First use the auxiliary conditions of Example 8.2. Then use the modified conditions of Example 8.3. In each case construct the search tree.

8.3. Devise auxiliary conditions for conducting a backtrack search for all permutations of $\Omega = \{1, 2, \ldots, n\}$ which do not fix any element of Ω. A permutation f is identified with the sequence $a_1, \ldots, a_n$, where a_i is the image of i under f. Carry out the search for $n = 4$ and describe the search tree.

1.9 Historical notes

The concept of a group developed slowly over a long period of time. The work of Evariste Galois is often cited as the beginning of group theory as a separate area of mathematics, but group-theoretic ideas and examples of groups occurred well before Galois. A number of results in group theory were obtained before the definition of an (abstract) group reached its final form. Arthur Cayley came close to the definition in two attempts (Cayley 1854, 1878). Finitely generated groups were defined in (Dyck 1882), which also contained the definition of a presentation by generators and relations. Abstract finite groups were defined in (Weber 1882). The definition of an arbitrary abstract group did not appear in essentially its modern form

Table 1.8.2

i	Trial 1	Trial 2	Trial 3	Actual
0	1	1	1	1
1	11	11	11	11
2	17	13	14	13
3	25	20	22	20
4	36	26	32	32
5	51	32	45	42
6	79	41	70	63
7	123	63	110	96
8	194	106	170	144
9	282	145	238	187
10	458	239	319	274
11	563	310	417	366
12	528	380	602	503
13	282	169	542	466
14	422	282	859	730
15	563	338	732	674
16	282	338	1014	994
17		451	591	707
18		901	957	1110
19		901	732	669
20		1802	1464	996
21		1802	1577	478
22			1802	664
23			451	243
24			901	310
25				79
26				92
27				14
28				15
29				1
30				1

until (Weber 1893). Note that Sylow's theorems date from 1872 and were originally results about finite permutations groups. The word problem, the conjugacy problem, and the isomorphism problem were formulated in (Dehn 1911), but the idea of trying to determine properties of a group given by a finite presentation was already familiar by that time.

The existence of a finitely presented semigroup S with unsolvable word problem was first proved by Emil Post in (Post 1947). The proof is based on the fact that there is a Turing machine T with unsolvable halting problem. That is, given a finite sequence I of input symbols, there is no algorithm for deciding whether T will stop with input I. The presentation for S encodes a description of T.

Credit for showing that the word problem for groups is unsolvable is usually shared between Petr Novikov and William Boone. The first published proof was (Novikov 1955). During the years 1954–57 Boone developed an alternative proof in a sequence of papers which concluded with (Boone 1957). Other algorithmic questions about groups were shown to be unsolvable in (Adjan 1957) and (Rabin 1958).

2

Rewriting systems

In this book we shall study a number of computational procedures. One of the most basic of these is the Knuth-Bendix procedure for strings presented in Sections 2.5 to 2.7. The Knuth-Bendix procedure manipulates rewriting systems, which are sets of pairs of words. These systems are used to simplify words systematically in an attempt to solve the word problem for particular finitely presented monoids.

2.1 Orderings of free monoids

Let $M = \text{Mon}\langle X \mid \mathcal{R}\rangle$ be a finitely presented monoid. Elements of M are equivalence classes of words. As we study M, we shall frequently have several words belonging to the same equivalence class and so defining the same element of M. It will be useful to be able to select one of these words as "simpler" or "better" in some sense than the others. The notion of simplicity may depend on the presentation. If for any two distinct words in X^* we have a preference for one over the other, we have an ordering on words.

Let S be a set. A *linear ordering* of S is a transitive relation $\prec$ on S such that for any two elements s and t of S exactly one of the following holds: $s \prec t$, $s = t$, $t \prec s$. Let $\prec$ be a linear ordering of S and let s and t be in S. We write $s \preceq t$ if $s \prec t$ or $s = t$, and we write $t \succ s$ and $t \succeq s$ if $s \prec t$ and $s \preceq t$, respectively. The ordering $\prec$ is a *well-ordering* if there are no infinite sequences $s_1, s_2, \ldots$ of elements of S such that $s_i \succ s_{i+1}$ for all $i \geq 1$. If we have a notion of one word being simpler than another and "is simpler than" is a well-ordering of X^*, then we cannot simplify a word, that is, replace it by a simpler word, indefinitely.

Proposition 1.1. *If $\prec$ is a well-ordering of S, then every nonempty subset of S has a least element.*

Proof. The proof is a routine application of the axiom of choice. Let T be a nonempty subset of S and suppose T does not have a smallest element. Choose s_1 in T. Assume that $s_1, \ldots, s_i$ have been chosen in T with $s_1 \succ s_2 \succ \cdots \succ s_i$. Then s_i is not the smallest element of T, so we may choose s_{i+1} in T with $s_i \succ s_{i+1}$. This process generates an infinite, strictly decreasing sequence in S, which contradicts the assumption that $\prec$ is a well-ordering of S. $\square$

It is easy to define a linear ordering of a finite set. We just list the elements of the set from first to last. A set with n elements has $n!$ linear orderings, all of which are well-orderings. We shall need some techniques for constructing orderings of infinite sets.

Suppose $\prec$ is a linear ordering of a set S and n is a positive integer. We can define a linear ordering on the set S^n of n-tuples of elements of S by saying that $(s_1, \ldots, s_n) \prec (t_1, \ldots, t_n)$ if and only if there is an integer i with $1 \leq i \leq n$ such that $s_j = t_j$ for $1 \leq j < i$ and $s_i \prec t_i$. This ordering is called the *left-to-right lexicographic ordering* of S^n. Note that the same symbol is used for the ordering of S and the ordering of S^n. We shall have too many orderings to have a separate symbol for each one. The symbols $\prec, \preceq, \succeq,$ and $\succ$ will normally be used for any linear orderings which arise.

If we use the ordinary ordering of the set $\mathbb{Z}$ of integers to define the left-to-right lexicographic ordering of $\mathbb{Z}^3$, then

$$(-1, 3, 6) \prec (1, -2, -1) \prec (1, -2, 5) \prec (1, 2, -1).$$

There is also a *right-to-left lexicographic ordering* on S^n. For this ordering we compare s_n and t_n, then s_{n-1} and t_{n-1}, and so on. Thus in the right-to-left lexicographic ordering of $\mathbb{Z}^3$,

$$(1, -2, -1) \prec (1, 2, -1) \prec (1, -2, 5) \prec (-1, 3, 6).$$

Proposition 1.2. *Suppose $\prec$ is a linear ordering of S. The corresponding lexicographic orderings of S^n are linear orderings, which are well-orderings if $\prec$ is a well-ordering on S.*

Proof. Exercise. $\square$

Given a linear ordering $\prec$ of a set X, we can define a number of linear orderings of X^*. Let $U = u_1 \ldots u_m$ and $V = v_1 \ldots v_n$ be in X^*, with each u_i and v_i in X. In the *left-to-right lexicographic ordering* of X^*, we say that $U \prec V$ provided one of the following holds:

(i) $m < n$ and $u_i = v_i$, $1 \leq i \leq m$.
(ii) There is an i with $1 \leq i \leq \min(m, n)$ such that $u_j = v_j$ for $1 \leq j < i$ and $u_i \prec v_i$.

This definition may be restated as follows: $U \prec V$ provided either that U is a proper prefix of V or that some prefix of U is less than the prefix of V of the same length i in the left-to-right lexicographic ordering of X^i.

Proposition 1.3. *If $\prec$ is a linear ordering of X, then the left-to-right lexicographic ordering is a linear ordering of X^*.*

Proof. Exercise. □

If $|X| > 1$, then the left-to-right lexicographic ordering of X^* is not a well-ordering, even if X is well-ordered. If a and b are elements of X and $a \prec b$, then $ab \succ a^2 b \succ a^3 b \succ \cdots$ is a strictly decreasing sequence in X^*.

There is also the *right-to-left lexicographic ordering* of X^*. In this ordering, $U \prec V$ provided U is a proper suffix of V or some suffix of U is less than the suffix of V of the same length i in the right-to-left lexicographic ordering of X^i.

From now on, "lexicographic" without further qualification will mean "left-to-right lexicographic". When right-to-left is meant, it will be explicitly stated.

Another ordering of X^* is the *length-plus-lexicographic ordering*. In this ordering, $u_1 \ldots u_m \prec v_1 \ldots v_n$ provided either that $m < n$ or that $m = n$ and $u_1 \ldots u_m$ comes before $v_1 \ldots v_m$ lexicographically.

Proposition 1.4. *The length-plus-lexicographic ordering is a linear ordering of X^*. If $\prec$ is a well-ordering of X, then X^* is also well ordered.*

Proof. We shall prove the statement about well-ordering, leaving the rest as an exercise. Suppose $\prec$ is a well-ordering on X. Let $U_1 \succ U_2 \succ U_3 \succ \cdots$ be a strictly decreasing sequence of words in X^*. Since $|U_i| \geq |U_{i+1}|$ for all i, from some point on all of the U_i have the same length m. By Proposition 1.2, the lexicographic ordering of X^m is a well-ordering. Therefore the sequence must terminate. □

There is also a right-to-left version of the length-plus-lexicographic ordering.

An ordering $\prec$ of X^* is *translation invariant* if $U \prec V$ implies that $AUB \prec AVB$ for all A and B in X^*. Lexicographic orderings are not translation invariant. For example, if a and b are in X and $a \prec b$, then $a \prec a^2$ lexicographically, but $ab \succ a^2 b$. We say that $\prec$ is *consistent with length* if $U \prec V$ implies that $|U| \leq |V|$. The length-plus-lexicographic orderings are clearly consistent with length.

Proposition 1.5. *The length-plus-lexicographic ordering of X^* is translation invariant.*

Proof. Let $U = u_1 \ldots u_m$ and $V = v_1 \ldots v_n$, and suppose that $U \prec V$. To prove translation invariance, it suffices to prove that $Ux \prec Vx$ and $xU \prec xV$ for all x in X. If $m < n$, then $|Ux| = |xU| < |Vx| = |xV|$. Thus we may assume that $m = n$. Let U and V differ first in the i-th term. Then Ux and Vx also differ first in the i-th terms, which are the same as the i-th terms of U and V. Therefore $Ux \prec Vx$. A similar argument shows that $xU \prec xV$.

$\square$

Proposition 1.6. *In a translation invariant well-ordering of X^* the empty word comes first.*

Proof. Suppose $\varepsilon \succ U$ for some word U. Then by translation invariance $U = \varepsilon U \succ U^2$. Similarly, $U^2 \succ U^3$. Thus

$$\varepsilon \succ U \succ U^2 \succ U^3 \succ \cdots$$

is an infinite, strictly decreasing sequence of words, contradicting our assumption that $\prec$ is a well-ordering.

$\square$

The phrase "translation invariant well-ordering" is somewhat awkward. The term "*reduction ordering*" will be used instead.

We shall now define a family of orderings which will play an important role in Chapters 9 and 11. Let X and Y be disjoint sets and assume that $\prec_X$ and $\prec_Y$ are reduction orderings on X^* and Y^*, respectively. Using these orderings, we shall construct an ordering of $(X \cup Y)^*$. Suppose U is a word in $(X \cup Y)^*$. Then U can be written uniquely as

$$A_0 b_1 A_1 b_2 \ldots A_{r-1} b_r A_r,$$

where the b_i are in Y and the A_i are in X^*. Let V be another word in $(X \cup Y)^*$ and let

$$C_0 d_1 C_1 d_2 \ldots C_{s-1} d_s C_s$$

be the corresponding decomposition of V. Define $U \prec V$ if one of the following holds:

(i) $b_1 \ldots b_r \prec_Y d_1 \ldots d_s$.
(ii) $b_1 \ldots b_r = d_1 \ldots d_s$ and $(A_0, \ldots, A_r)$ comes before $(C_0, \ldots, C_r)$ in the lexicographic ordering of $(X^*)^{r+1}$ defined by $\prec_X$.

The ordering $\prec$ on $(X \cup Y)^*$ will be called the *wreath product of $\prec_X$ and $\prec_Y$* and will be denoted $\prec_X \wr \prec_Y$.

Example 1.1. Let $X = \{a\}$ and $Y = \{b\}$, and let $\prec_X$ and $\prec_Y$ be the orderings by length of X^* and Y^*, respectively. If $\prec$ is $\prec_X \wr \prec_Y$, then

$$a^{100} \prec aba^2 \prec a^2ba \prec b^2a \prec bab \prec ab^2.$$

Example 1.2. Let $X = \{a, a^{-1}\}$ and $Y = \{b, b^{-1}\}$, and let $\prec_X$ and $\prec_Y$ be the length-plus-lexicographic orderings with $a \prec_X a^{-1}$ and $b \prec_Y b^{-1}$. If $\prec$ is $\prec_X \wr \prec_Y$, then

$$aba^{-1}b^{-1}a^2 \prec a^{-1}ba^2b^{-1}a \prec a^2b^{-1}aba^{-1} \prec a^2b^{-1}a^{-1}ba.$$

Proposition 1.7. *The wreath product* $\prec_X \wr \prec_Y$ *is a reduction ordering of* $(X \cup Y)^*$.

Proof. Let $\prec$ denote $\prec_X \wr \prec_Y$. Suppose U, V, and W are words in $(X \cup Y)^*$, and let

$$U = A_0 b_1 A_1 \ldots b_r A_r,$$
$$V = C_0 d_1 C_1 \ldots d_s C_s,$$
$$W = E_0 f_1 E_1 \ldots f_t E_t,$$

where the b_i, d_i, and f_i are in Y and the A_i, D_i, and E_i are in X^*. It is easy to see that exactly one of $U \prec V$, $U = V$, and $V \prec U$ holds. Let us show that $\prec$ is transitive. Suppose $U \prec V$ and $V \prec W$. Then $b_1 \ldots b_r \preceq_Y d_1 \ldots d_s \preceq_Y f_1 \ldots f_t$. If $b_1 \ldots b_r \prec_Y f_1 \ldots f_t$, then $U \prec W$. Suppose $b_1 \ldots b_r = d_1 \ldots d_s = f_1 \ldots f_t$, so $r = s = t$. Then $(A_0, \ldots, A_r)$ comes before $(C_0, \ldots, C_r)$, which comes before $(E_0, \ldots, E_r)$. Therefore $U \prec W$ in this case too, and $\prec$ is a linear ordering.

Let $h : (X \cup Y)^* \to Y^*$ be the homomorphism which is the identity on Y and maps each element of X to ε. Then $b_1 \ldots b_r$ can be written $h(U)$. Suppose $U_1 \succ U_2 \succ \cdots$ is an infinite, strictly decreasing sequence of words in $(X \cup Y)^*$. Then $h(U_1) \succeq_Y h(U_2) \succeq_Y \cdots$, and so, from some point on, all the $h(U_i)$ are the same. Without loss of generality, we may assume that $h(U_i) = b_1 \ldots b_r$ for all i. Let

$$U_i = A_0^{(i)} b_1 A_1^{(i)} \ldots b_r A_r^{(i)}.$$

Then for $i \geq 1$ the $(r+1)$-tuple $(A_0^{(i+1)}, \ldots, A_r^{(i+1)})$ comes before $(A_0^{(i)}, \ldots, A_r^{(i)})$ in the lexicographic ordering of $(X^*)^{r+1}$. But this contradicts Proposition 1.2. Therefore $\prec$ is a well-ordering.

Finally, we must prove translation invariance. Let U and V be as before, and assume that $U \prec V$. First suppose y is in Y. If $h(U) = b_1 \ldots b_r \prec_Y$

$d_1 \ldots d_s = h(V)$, then

$$h(Uy) = b_1 \ldots b_r y \prec_Y d_1 \ldots d_s y = h(Vy),$$

since $\prec_Y$ is translation invariant, so $Uy \prec Vy$. Suppose $h(U) = h(V)$. Then $h(Uy) = h(Vy)$, and $(A_0, \ldots, A_r)$ comes before $(C_0, \ldots, C_r)$. By the definition of the (left-to-right) lexicographic ordering on $(X^*)^{r+2}$, this means that $(A_0, \ldots, A_r, \varepsilon)$ comes before $(C_0, \ldots, C_r, \varepsilon)$. Therefore $Uy \prec Vy$ in all cases. A similar argument shows that $yU \prec yV$.

Now suppose x is in X. Then $h(Ux) = h(U)$ and $h(Vx) = h(V)$. If $h(U) \prec_Y h(V)$, then $Ux \prec Vx$. Suppose that $h(U) = h(V)$. Then $Ux \prec Vx$ if and only if $(A_0, \ldots, A_r x)$ comes before $(C_0, \ldots, C_r x)$. Let i be the first index such that $A_i \prec_X C_i$. If $i = r$, then $A_r x \prec C_r x$, since $\prec_X$ is translation invariant. In any case $(A_0, \ldots, A_r x)$ does come before $(C_0, \ldots, C_r x)$, so $Ux \prec Vx$. Similarly, $xU \prec xV$. Therefore $\prec$ is translation invariant. $\square$

Suppose X, Y, and Z are pairwise disjoint sets and $\prec_X$, $\prec_Y$, and $\prec_Z$ are reduction orderings on X^*, Y^*, and Z^*, respectively. Then $(\prec_X \wr \prec_Y) \wr \prec_Z$ and $\prec_X \wr (\prec_Y \wr \prec_Z)$ are both orderings of $(X \cup Y \cup Z)^*$. It is a straightforward exercise to show that these are the same ordering. Hence we can write $\prec_X \wr \prec_Y \wr \prec_Z$ without ambiguity. Note that $\prec_X \wr \prec_Y$ and $\prec_Y \wr \prec_X$ are both orderings of $(X \cup Y)^*$, but they are not the same. Thus $\wr$ is associative but not commutative.

Let $X = \{x_1, \ldots, x_n\}$ and assume that X is ordered by $\prec$ so that $x_1 \prec x_2 \prec \cdots \prec x_n$. On $\{x_i\}^*$ there is a unique reduction ordering $\prec_i$, namely the ordering by length. The ordering $\prec_1 \wr \prec_2 \wr \cdots \wr \prec_n$ will be called the *basic wreath-product ordering* of X^* determined by $\prec$. This ordering is closely related to the recursive path ordering of Dershowitz. See (Dershowitz 1987).

Questions of decidability arise in connection with orderings of free monoids. Suppose X is a finite set and $\mathcal{R}$ is a finite subset of $X^* \times X^*$. There is no algorithm which can decide if there is a reduction ordering of X^* such that $P \succ Q$ for all (P, Q) in $\mathcal{R}$.

Let us turn now to the question of computing with wreath-product orderings. Let X be the disjoint union of nonempty subsets $X_1, \ldots, X_r$, and for $1 \le i \le r$ let $\prec_i$ be a reduction ordering of X_i. Set $\prec$ equal to $\prec_1 \wr \prec_2 \wr \cdots \wr \prec_r$. Suppose we are given two words U and V in X^*. What is an efficient way to decide whether $U \prec V$, $U = V$, or $V \prec U$? If U and V have a common nonempty prefix so that $U = AB$ and $V = AC$ for words A, B, and C, then we may replace U and V by B and C without changing the answer to our question. Thus we may assume that U and V do not start with the same generator. If either U or V is empty, then the answer is obvious, so we may also assume that both U and V are nonempty.

Let m be the largest index such that U contains a generator in X_m. Let

$$U = A_0 x_1 A_1 \ldots A_{t-1} x_t A_t,$$

where each x_j is in X_m and the A_j are in $(X_1 \cup \cdots \cup X_{m-1})^*$. We shall call m the *degree* of U, $Y = x_1 \ldots x_t$ the *head* of U, and A_0 the *left end* of U. Let n be the degree of V, let Z be the head of V, and let B_0 be the left end of V. Since U and V do not start with the same generator, $U \prec V$ if and only if

(a) $m < n$, or
(b) $m = n$ and $Y \prec_m Z$, or
(c) $m = n$, $Y = Z$, and $A_0 \prec B_0$.

Note that if $m = n$ and $Y = Z$, then A_0 and B_0 do not start with the same generator.

Given U, let the integers $d_1, d_2, \ldots$ and the words $P_1, P_2, \ldots$ and $Q_1, Q_2, \ldots$ be defined as follows: Set $Q_1 = U$, and for $i \geq 1$ let P_i be the head of Q_i, let d_i be the degree of Q_i, and let Q_{i+1} be the left end of Q_i, provided that left end is nonempty. Let d_s, P_s, and Q_s be the last terms in these three sequences. We shall call $d_1, \ldots, d_s$ the sequence of degrees of U and $P_1, \ldots, P_s$ the sequence of heads of U.

Proposition 1.8. *Let U and V be nonempty words in X^* which do not start with the same generator. Let $d_1, \ldots, d_s$ and $P_1, \ldots, P_s$ be the sequences of degrees of U and let $e_1, \ldots, e_t$ and $R_1, \ldots, R_t$ be the sequences of degrees and heads of V. Then $U \prec V$ if and only if one of the following holds:*

(a) $s < t$, and for $1 \leq i \leq s$ we have $d_i = e_i$ and $P_i = R_i$.
(b) There is an integer $i \leq \min(s, t)$ such that $d_j = e_j$ and $P_j = R_j$ for $1 \leq j < i$ and either $d_i < e_i$ or $d_i = e_i$ and $P_i \prec_{d_i} R_i$.

Proof. This is an easy induction on the maximum of d_1 and e_1. □

We can construct the sequences $d_1, \ldots, d_s$ and $P_1, \ldots, P_s$ for U in time linear in $|U|$.

Procedure HEAD$(U; s, d_1, \ldots, d_s, P_1, \ldots, P_s)$;
Input: U : a nonempty word;
Output: s : a positive integer;
             $d_1, \ldots, d_s$: the sequence of degrees of U;
             $P_1, \ldots, P_s$: the sequence of heads of U;

Begin
  $s := 0;\ m := 0;$ Let $U = u_1 \ldots u_k$;
  For $i := 1$ to k do begin
    Let j be the degree of u_i;
    If $j > m$ then begin $s := s + 1;\ m := j;\ c_s := m;\ W_s := u_i$ end
    Else if $j = m$ then $W_s := W_s u_i$
  End;
  For $i := 1$ to s do begin $d_i := c_{s+1-i};\ P_i := W_{s+1-i}$ end
End.

Example 1.3. Let $X_1 = \{u, v\}$, $X_2 = \{w, x\}$, and $X_3 = \{y, z\}$. Suppose

$$U = \underline{vu}x u\underline{wv}\underline{wu}vx y ux vz\underline{wu}\underline{wy}uwxv.$$

Here we have underlined the terms in U whose degrees are at least as large as the degree of any earlier term. The procedure HEAD simply groups these by degree, giving

$$W_1 = vu,$$
$$W_2 = xwwx,$$
$$W_3 = yzy.$$

Thus the sequences of degrees and heads for U are 3, 2, and 1 and yzy, $xwwx$, and vu, respectively.

There are many more ways to define linear orderings of free monoids. One approach is to assign to each generator a nonnegative real number called its *weight*. The weight of a word is then defined to be the sum of the weights of its terms. Words are ordered first by weight, with some other ordering used to break ties.

Exercises

1.1. Show that the operation $\wr$ is associative.
1.2. Suppose that $\prec$ is a reduction ordering on X^*. For U and V in X^*, define $U \prec^\dagger V$ to mean $U^\dagger \prec V^\dagger$, where $U^\dagger$ denotes the reversal of U. Prove that $\prec^\dagger$ is a reduction ordering.
1.3. Let $A = \{a, b\}$ and $B = \{c, d\}$ and define $\prec_A$ and $\prec_B$ to be the length-plus-lexicographic orderings of A^* and B^*, respectively, in which $a \prec_A b$ and $c \prec_B d$. Arrange the following words in increasing order using the ordering $\prec_A \wr \prec_B$ on $(A \cup B)^*$: *cadbc, abcbdabd, acabdba, abcbadbd, acdcb, acbadab.*
1.4. Arrange the words in Exercise 1.3 in increasing order using the basic wreath product ordering $\prec$ of $\{a, b, c, d\}$ in which $a \prec b \prec c \prec d$.

1.5. Let $\prec$ be a reduction ordering on X^*. Prove that $U \prec AUB$ for any words A, B, and U in X^* such that $AB \neq \varepsilon$.

*1.6. Let $X = \{a, b\}$ and let $\mathcal{U}$ be the set of all words in X^* of length at most 3. How many of the 15! linear orderings of $\mathcal{U}$ can be extended to reduction orderings of X^*?

2.2 Canonical forms

Suppose that we are working with a finitely presented monoid M. Although the word problem is in general unsolvable, it is natural to be optimistic and look for a solution in M anyway. Assume that M is given as the quotient monoid of the free monoid X^* modulo the congruence $\sim$. One approach to the word problem is to attempt to choose for each element u of M one word U from among all the words in X^* which define u. Such a choice is called a *canonical form* for u.

Usually we would like the canonical form for u to be the "simplest" word defining u. As we saw in the previous section, reduction orderings provide convenient notions of simplicity. Let us fix a reduction ordering $\prec$ on X^*. For each u in M the set of words defining u is nonempty, so this set has a smallest element U. We define U to be the *canonical form* for u relative to $\prec$. For any word V, let $\overline{V}$ be the canonical form for the $\sim$-class containing V.

It must be emphasized that this definition of canonical form is nonconstructive. Given X, a generating set $\mathcal{R}$ for $\sim$, and the definition of $\prec$, we have at this point no way of computing canonical forms. That is, given a word V, we have no algorithm for computing $\overline{V}$. If we can compute canonical forms, then we can solve the word problem, for $U \sim V$ if and only if $\overline{U} = \overline{V}$. The unsolvability of the word problem tells us that there are presentations for which no algorithm for computing $\overline{V}$ exists.

Proposition 2.1. *If U is the canonical form for an element of M, then subwords of U are canonical forms for elements of M.*

Proof. Let V be a subword of U, so $U = AVB$. If V is not a canonical form, then there is a word W such that $V \succ W$ and $V \sim W$. But then $AVB \sim AWB$ and, since $\prec$ is translation invariant, $AVB \succ AWB$. Thus U is not the first word in its $\sim$-class, contradicting the assumption that U is a canonical form. $\square$

Let (P, Q) be an element of a generating set $\mathcal{R}$ for $\sim$. Replacing (P, Q) by (Q, P) does not change $\sim$, so we may assume that $P \succ Q$. In this case, (P, Q) is called a *rewriting rule* with respect to $\prec$. If every element of $\mathcal{R}$ is a rewriting rule, then $\mathcal{R}$ is called a *rewriting system* with respect to $\prec$.

Suppose that $\mathcal{R}$ is a rewriting system. Let $\mathcal{P}$ be the set of left sides of elements in $\mathcal{R}$, let $\mathcal{N}$ be the ideal of X^* generated by $\mathcal{P}$, and let $\mathcal{C}$ be the complement $X^* - \mathcal{N}$. If U is in $\mathcal{N}$, then there are words A, B, P, and Q such that $U = APB$ and (P, Q) is in $\mathcal{R}$. Let $V = AQB$. Then $P \sim Q$, so $U \sim V$. Also $P \succ Q$, so $U \succ V$. If V is in $\mathcal{N}$, then we can repeat this process and get another word W such that $U \sim V \sim W$ and $U \succ V \succ W$. Since $\prec$ is a well-ordering, this process cannot go on indefinitely, and we eventually produce a word C in $\mathcal{C}$ such that $U \sim C$ and $U \succeq C$. This procedure of replacing subwords which are left sides of rewriting rules with the corresponding right sides is called *rewriting*. Elements of $\mathcal{C}$ are said to be *irreducible* or *reduced* with respect to $\mathcal{R}$, since no rewriting can be performed on them. The canonical form of every $\sim$-class is in $\mathcal{C}$, but a $\sim$-class may contain other elements of $\mathcal{C}$ as well. Suppose (P, Q) is in $\mathcal{R}$. To emphasize that rewriting replaces P by Q, we shall frequently write $P \rightarrow Q$ for (P, Q). More generally, we shall write $U \rightarrow V$ or $U \xrightarrow{\mathcal{R}} V$ if V is *derivable from U in one step using* $\mathcal{R}$ in the sense that there are words A, B, P, and Q such that $P \rightarrow Q$ is in $\mathcal{R}$, $U = APB$, and $V = AQB$.

Let us formalize the rewriting process.

Procedure REWRITE$(X, \mathcal{R}, U; V)$;
Input: X : a finite set;
            $\mathcal{R}$: a finite rewriting system on X^* with respect to a reduction ordering;
            U : a word in X^*;
Output: V : a word in X^* irreducible with respect to $\mathcal{R}$ and defining the same element of Mon $\langle X \mid \mathcal{R} \rangle$ as U;
Begin
  Let $\mathcal{P}$ be the set of left sides in $\mathcal{R}$; $V := U$;
  While V contains a subword in $\mathcal{P}$ do begin
    Let $V = APB$ with $P \rightarrow Q$ in $\mathcal{R}$; $V := AQB$
  End
End.

The reduction ordering mentioned in REWRITE need not be explicitly given. However, its existence is necessary to insure termination.

The description of REWRITE is really only an outline. In carrying out the procedure, a number of choices must be made. First, there may be many occurrences of words in $\mathcal{P}$ as subwords in V. Second, an element of $\mathcal{P}$ may be the left side of more than one rule in $\mathcal{R}$, so there may be many choices for Q. The value of V returned by REWRITE may depend on these choices.

Example 2.1. Let $X = \{x, y\}$ and let $\mathcal{R} = \text{FGRel}(X)$, so $\mathcal{R}$ consists of the rules

$$xx^{-1} \rightarrow \varepsilon, \quad x^{-1}x \rightarrow \varepsilon, \quad yy^{-1} \rightarrow \varepsilon, \quad y^{-1}y \rightarrow \varepsilon,$$

and $F = \text{Mon}\,\langle X^{\pm} \mid \mathcal{R}\rangle$ is a free group of rank 2. Let

$$U = xyx^{-1}xy^{-1}x^{-1}y^{-1}yxy.$$

We can rewrite U using $\mathcal{R}$ in several ways. Here are two possibilities:

$$xy\underline{x^{-1}x}y^{-1}x^{-1}y^{-1}yxy \to xy\underline{y^{-1}}x^{-1}y^{-1}yxy$$
$$\to \underline{xx^{-1}}y^{-1}yxy \to \underline{y^{-1}y}xy \to xy$$

and

$$xyx^{-1}xy^{-1}x^{-1}\underline{y^{-1}y}xy \to xyx^{-1}xy^{-1}\underline{x^{-1}x}y$$
$$\to xyx^{-1}\underline{xy^{-1}}y \to xy\underline{x^{-1}x} \to xy.$$

The underscores highlight the subwords which are replaced (in this case deleted) in passing to the next word. Both of these rewritings produced the same result. Rewriting in $X^{\pm*}$ using the system FGRel(X) is called *free reduction*. We shall develop a simple test in Section 2.3 which will confirm that free reduction always produces canonical forms.

Example 2.2. Let $X = \{a, b\}$ and let $\mathcal{R}$ consist of the rules $a^2 \to \varepsilon$, $b^{10} \to \varepsilon$, and $ba \to ab^4$. This is a rewriting system on X^* with respect to the basic wreath-product ordering with $a \succ b$. Let us rewrite the word baa in two ways. The first way is very easy:

$$b\underline{aa} \to b.$$

The second way takes a little longer:

$$\underline{ba}a \to abbbb\underline{ba} \to abbb\underline{ba}bbbb \to ab\underline{ba}bbbbbbbb$$
$$\to ababb\underline{bbbbbbbbbb} \to a\underline{ba}bb \to \underline{aa}bbbbbb \to bbbbbb.$$

The result is b in the first case and b^6 in the second.

There is an extensive literature on rewriting systems. The books [Le Chenadec 1986], [Benninghofen, Kemmerich, & Richter 1987], and [Jantzen 1988] contain many references. We do not have time to go very deeply into the subject. Our first goal will be to develop a condition on a rewriting system $\mathcal{R}$ which guarantees that the result V returned by REWRITE$(X, \mathcal{R}, U; V)$ depends only on the input word U, not on choices made in the procedure. Then we shall present a procedure which attempts to add rules to $\mathcal{R}$, if necessary, so that the condition is satisfied.

Let $\mathcal{R}$ be a rewriting system on X^* with respect to a reduction ordering $\prec$ and let $\sim$ be the congruence generated by $\mathcal{R}$. Suppose U and V are in X^*. We shall say that V is *derivable from* U *using* $\mathcal{R}$ if there is a sequence of words

$$U = U_0, U_1, \ldots, U_t = V$$

with $t \geq 0$ such that U_{i+1} is derivable from U_i in one step, $0 \leq i < t$. In this case we shall write $U \xrightarrow{*}{\mathcal{R}} V$. Clearly $\xrightarrow{*}{\mathcal{R}}$ is a reflexive and transitive relation on X^*. Also, if $U \xrightarrow{*}{\mathcal{R}} V$, then $U \sim V$ and $U \succeq V$.

Proposition 2.2. *If* $U \xrightarrow{*}{\mathcal{R}} V$, *then* $UW \xrightarrow{*}{\mathcal{R}} VW$ *and* $WU \xrightarrow{*}{\mathcal{R}} WV$ *for all* W *in* X^*.

Proof. Exercise. $\square$

Proposition 2.3. *If* U *and* V *are words in* X^*, *then* $U \sim V$ *if and only if there is a sequence of words* $U = U_0, U_1, \ldots, U_t = V$ *such that for* $0 \leq i < t$ *either* $U_i \xrightarrow{*}{\mathcal{R}} U_{i+1}$ *or* $U_{i+1} \xrightarrow{*}{\mathcal{R}} U_i$.

Proof. Let us write $U \equiv V$ if such a sequence $U_0, \ldots, U_t$ exists. Since $U_i \sim U_{i+1}$, we know that $U \equiv V$ implies that $U \sim V$. It is easy to see that $\equiv$ is an equivalence relation. By Proposition 2.2, $\equiv$ is a congruence. If (P, Q) is in $\mathcal{R}$, then $P \equiv Q$. Therefore by the definition of $\sim$, it follows that $\sim$ is contained in $\equiv$. Thus $\sim$ and $\equiv$ are the same relation. $\square$

There are a number of useful properties which the relations $\xrightarrow{}{\mathcal{R}}$ and $\xrightarrow{*}{\mathcal{R}}$ may possess:

- The Church-Rosser property: If $U \sim V$, then there is a word Q such that $U \xrightarrow{*}{\mathcal{R}} Q$ and $V \xrightarrow{*}{\mathcal{R}} Q$.
- Confluence: If $W \xrightarrow{*}{\mathcal{R}} U$ and $W \xrightarrow{*}{\mathcal{R}} V$, then there is a word Q such that $U \xrightarrow{*}{\mathcal{R}} Q$ and $V \xrightarrow{*}{\mathcal{R}} Q$.
- Local confluence: If $W \xrightarrow{}{\mathcal{R}} U$ and $W \xrightarrow{}{\mathcal{R}} V$, then there is a word Q such that $U \xrightarrow{*}{\mathcal{R}} Q$ and $V \xrightarrow{*}{\mathcal{R}} Q$.

Proposition 2.4. *If the Church-Rosser property holds, then every* $\sim$-*class contains a unique element of* $\mathcal{C}$, *the canonical form for that class.*

Proof. We have already noted that the canonical form for each $\sim$-class is in $\mathcal{C}$. Suppose U and V are elements of $\mathcal{C}$ in the same $\sim$-class. Then by the Church-Rosser property there is a word Q such that $U \xrightarrow{*}{\mathcal{R}} Q$ and $V \xrightarrow{*}{\mathcal{R}} Q$. But since U and V are in $\mathcal{C}$, we must have $U = Q = V$. $\square$

Proposition 2.5. *For any rewriting system $\mathcal{R}$ relative to a reduction ordering $\prec$, the Church-Rosser property, confluence, and local confluence are equivalent.*

Proof. (a) The Church-Rosser property implies confluence. Assume that $W \xrightarrow[\mathcal{R}]{*} U$ and $W \xrightarrow[\mathcal{R}]{*} V$. Then $W \sim U$ and $W \sim V$. Therefore $U \sim V$, and by the Church-Rosser property there is a word Q such that $U \xrightarrow[\mathcal{R}]{*} Q$ and $V \xrightarrow[\mathcal{R}]{*} Q$.

(b) Confluence implies local confluence. Since local confluence is a special case of confluence, this is obvious.

(c) Local confluence implies confluence. Let us say that confluence fails at a word W if there are words U and V such that $W \xrightarrow[\mathcal{R}]{*} U$ and $W \xrightarrow[\mathcal{R}]{*} V$ but no word Q such that $U \xrightarrow[\mathcal{R}]{*} Q$ and $V \xrightarrow[\mathcal{R}]{*} Q$. Let $\mathcal{W}$ be the set of words at which confluence fails, and assume $\mathcal{W}$ is nonempty. Since $\prec$ is a well-ordering, $\mathcal{W}$ has a smallest element W. Suppose that $W \xrightarrow[\mathcal{R}]{*} U$ and $W \xrightarrow[\mathcal{R}]{*} V$. We want to show that there is a word Q such that $U \xrightarrow[\mathcal{R}]{*} Q$ and $V \xrightarrow[\mathcal{R}]{*} Q$. If $U = W$, then we may take $Q = V$. If $V = W$, then we may take $Q = U$. Therefore we may assume that $U \neq W$ and $V \neq W$. There exist words A and B derivable from W in one step such that $A \xrightarrow[\mathcal{R}]{*} U$ and $B \xrightarrow[\mathcal{R}]{*} V$. By local confluence, there is a word C such that $A \xrightarrow[\mathcal{R}]{*} C$ and $B \xrightarrow[\mathcal{R}]{*} C$. Since $A \prec W$, it cannot happen that A is in $\mathcal{W}$. Thus there is a word D such that $U \xrightarrow[\mathcal{R}]{*} D$ and $C \xrightarrow[\mathcal{R}]{*} D$. Therefore $B \xrightarrow[\mathcal{R}]{*} D$. Since $B \prec W$, there is a word Q such that $D \xrightarrow[\mathcal{R}]{*} Q$ and $V \xrightarrow[\mathcal{R}]{*} Q$. But then $U \xrightarrow[\mathcal{R}]{*} Q$ and confluence does not fail at W. Therefore $\mathcal{W} = \emptyset$ and confluence holds.

(d) Confluence implies the Church-Rosser property. Suppose $U \sim V$. We want to show that there is a word Q such that $U \xrightarrow[\mathcal{R}]{*} Q$ and $V \xrightarrow[\mathcal{R}]{*} Q$. By Proposition 2.3, there is a sequence $U = U_0, U_1, \ldots, U_t = V$ such that for $0 \leq i < t$ either $U_i \xrightarrow[\mathcal{R}]{*} U_{i+1}$ or $U_{i+1} \xrightarrow[\mathcal{R}]{*} U_i$. We proceed by induction on t. If $t = 0$, then we may take $Q = U = V$. If $t = 1$, then we may take Q to be the smaller of U and V. Suppose $t \geq 2$. Then $U_1 \sim V$ and by induction there is a word A such that $U_1 \xrightarrow[\mathcal{R}]{*} A$ and $V \xrightarrow[\mathcal{R}]{*} A$. If $U_0 \xrightarrow[\mathcal{R}]{*} U_1$, then we may take Q to be A. Suppose $U_1 \xrightarrow[\mathcal{R}]{*} U_0$. By confluence, there is a word Q such that $U_0 \xrightarrow[\mathcal{R}]{*} Q$ and $A \xrightarrow[\mathcal{R}]{*} Q$. But then $V \xrightarrow[\mathcal{R}]{*} Q$ and we are done. $\square$

Note that our proof of Proposition 2.5 relied heavily on the fact that $\prec$ is a well-ordering. In more general situations, local confluence need not imply confluence.

Proposition 2.6. *If $\mathcal{R}$ is a confluent rewriting system on X^*, then after* REWRITE$(X, \mathcal{R}, U; V)$ *the value of V depends only on $\mathcal{R}$ and U, not on the choices made during the rewriting.*

Proof. By Propositions 2.5 and 2.4, the value of V is the canonical form for the $\sim$-class containing U. $\square$

A rewriting system $\mathcal{R}$ is *reduced* if each right side is irreducible with respect to $\mathcal{R}$, no word is the left side of two different rules, and no left side contains another left side as a proper subword. Equivalently, $\mathcal{R}$ is reduced if for each (P, Q) in $\mathcal{R}$ both P and Q are irreducible with respect to $\mathcal{R} - \{(P, Q)\}$.

Proposition 2.7. *Let $\prec$ be a reduction ordering on X^*. Every congruence on X^* is generated by a unique reduced, confluent rewriting system with respect to $\prec$.*

Proof. Let $\sim$ be a congruence on X^* and let M be the quotient of X^* modulo $\sim$. Let $\mathcal{C} = \{\overline{U} \mid U \in X^*\}$ be the set of canonical forms for the elements of M with respect to $\prec$ and let $\mathcal{N}$ be the ideal $X^* - \mathcal{C}$. Let $\mathcal{P}$ be the unique minimal generating set for $\mathcal{N}$ and set $\mathcal{S} = \{(P, \overline{P}) \mid P \in \mathcal{P}\}$. We shall show that $\mathcal{S}$ is a reduced, confluent rewriting system with respect to $\prec$, and that $\mathcal{S}$ is unique with these properties.

For any word U, we have $U \succeq \overline{U}$. If P is in $\mathcal{N}$, then $P \neq \overline{P}$, so $P \succ \overline{P}$. Thus $\mathcal{S}$ is a rewriting system with respect to $\prec$. Let $\equiv$ be the congruence generated by $\mathcal{S}$. Since $P \sim \overline{P}$ for all P in $\mathcal{P}$, the relation $\equiv$ is contained in $\sim$. But any word which is irreducible with respect to $\mathcal{S}$ is in $\mathcal{C}$. Thus REWRITE$(X, \mathcal{S}, U; V)$ returns V as $\overline{U}$. Therefore $U \equiv \overline{U}$ for all words U, and this implies that $\equiv$ and $\sim$ are the same. By its definition, $\mathcal{S}$ is reduced. If $U \sim V$, then $U \xrightarrow{*}{}_{\mathcal{S}} \overline{U}$ and $V \xrightarrow{*}{}_{\mathcal{S}} \overline{V}$. But $\overline{U} = \overline{V}$, so the Church-Rosser property holds and $\mathcal{S}$ is confluent.

Now suppose that $\mathcal{T}$ is any reduced, confluent rewriting system with respect to $\prec$ which generates $\sim$. Let P be in $\mathcal{P}$. Rewriting P using $\mathcal{T}$ must give $\overline{P}$. Therefore P contains a left side of $\mathcal{T}$ as a subword. But all proper subwords of P are in $\mathcal{C}$. Thus $\mathcal{T}$ contains a rule (P, Q). Since $\mathcal{T}$ is reduced, Q is in $\mathcal{C}$, so $Q = \overline{P}$. Therefore $\mathcal{S} \subseteq \mathcal{T}$. If (U, V) is in $\mathcal{T} - \mathcal{S}$, then U is not in $\mathcal{C}$, and hence U contains a subword P in $\mathcal{P}$. But P is a left side in $\mathcal{S}$, and this contradicts the assumption that $\mathcal{T}$ is reduced. $\square$

Suppose that $\prec$ is a reduction ordering on X^*, $\mathcal{R}$ is a subset of $X^* \times X^*$, and $\sim$ is the congruence generated by $\mathcal{R}$. The reduced, confluent rewriting system with respect to $\prec$ which generates $\sim$ will be denoted $\mathrm{RC}(X, \prec, \mathcal{R})$ or $\mathrm{RC}(X, \prec, \sim)$. If $\mathcal{R}$ is already confluent, it is easy to determine $\mathrm{RC}(X, \prec, \mathcal{R})$.

Proposition 2.8. *Let $\mathcal{R}$ be a confluent rewriting system with respect to $\prec$ on X^*. Let $\mathcal{P}$ be the set of left sides in $\mathcal{R}$ which do not contain any left sides as proper subwords. For P in $\mathcal{P}$ let $\overline{P}$ be the result of rewriting P using $\mathcal{R}$. Then $\mathrm{RC}(X, \prec, \mathcal{R}) = \{(P, \overline{P}) \mid P \in \mathcal{P}\}$.*

Proof. Let $\sim$ be the congruence generated by $\mathcal{R}$. Since $\mathcal{R}$ is confluent, any word which is not first in its $\sim$-class contains a subword which is a left side in $\mathcal{R}$. Thus $\mathcal{P}$ is the minimal generating set for the ideal of "noncanonical forms", and $\mathrm{RC}(X, \prec, \mathcal{R})$ is $\{(P, \overline{P}) \mid P \in \mathcal{P}\}$. $\square$

Proposition 2.9. *Let X be a finite set and let $\sim$ be a congruence on X^* such that the set of $\sim$-classes is finite. For every reduction ordering $\prec$ of X^* the system $\mathrm{RC}(X, \prec, \sim)$ is finite.*

Proof. Let (P, Q) be a rule in $\mathcal{S} = \mathrm{RC}(X, \prec, \sim)$. For $0 \leq i < |P|$ let P_i be the prefix of P of length i. Each P_i is a canonical form, so P_i and P_j are in different $\sim$-classes if $0 \leq i < j < |P|$. Therefore $|P|$ cannot exceed the number of $\sim$-classes. Since no two distinct rules in $\mathcal{S}$ have the same left side, this means that $\mathcal{S}$ is finite. $\square$

If $\mathcal{R}$ is a confluent rewriting system on X^*, then we shall say that $(X, \mathcal{R})$ is a *confluent presentation* for $\mathrm{Mon}\langle X \mid \mathcal{R} \rangle$. In Section 1.5 we remarked that it is in general impossible to decide whether a finitely presented monoid is a group. However, if we have a finite, confluent presentation for the monoid, the situation is different.

Proposition 2.10. *Let $(X, \mathcal{R})$ be a finite, confluent presentation for a monoid M. It is possible to decide whether M is a group.*

Proof. See Exercises 2.2 to 2.4. $\square$

Exercises

2.1. Suppose that $\mathcal{R}$ is a confluent rewriting system with respect to the reduction ordering $\prec$. Show that $\{(P^\dagger, Q^\dagger) \mid (P, Q) \in \mathcal{R}\}$ is confluent with respect to the ordering $\prec^\dagger$ defined in Exercise 1.2.

2.2. Let $(X, \mathcal{R})$ be a finite monoid presentation for a group G. For U in X^* let $[U]$ denote the element of G defined by U. Suppose that (AxB, Q) is in $\mathcal{R}$, where A and B are in X^* and x is in X. Show that $[x]^{-1} = [B][Q]^{-1}[A]$.

2.3. Let X, $\mathcal{R}$, and G be as in Exercise 2.2 and assume that $\mathcal{R}$ is confluent with respect to some reduction ordering on X^*. Set $X_0 = \emptyset$, and for $i \geq 0$ define X_{i+1} to be the union of X_i and the set of x in X for which there is an element (P, Q) of $\mathcal{R}$ such that x occurs in P and Q is in $(X_i)^*$. Prove that $X_m = X$ for some m.

2.4. Use the ideas of Exercises 2.2 and 2.3 to construct a proof of Proposition 2.10.

2.5. Let X be a finite set, let $\prec$ be a length-plus-lexicographic ordering on $X^{\pm*}$, and let $\sim$ be a congruence on $X^{\pm*}$ such that $x^\alpha x^{-\alpha} \sim \varepsilon$ for x in X and α in $\{1, -1\}$. Suppose that (P, Q) is in $\mathrm{RC}(X^\pm, \prec, \sim)$. Show that $|P| \leq |Q| + 2$.

2.3 A test for confluence

Let $\prec$ be a reduction ordering on X^* and let $\mathcal{R}$ be a finite rewriting system on X^* with respect to $\prec$. If $\mathcal{R}$ is confluent, we can solve the word problem in $M = \mathrm{Mon}\langle X \mid \mathcal{R} \rangle$ by using REWRITE to compute canonical forms. In this section we shall develop a test for confluence. This test is valid even when $\mathcal{R}$ is infinite, but in that case there is no guarantee that we can actually carry out the test.

Suppose $\mathcal{R}$ is not confluent and hence not locally confluent. We shall say that local confluence fails at a word W if there are words U and V in X^* such that $W \xrightarrow{\mathcal{R}} U$ and $W \xrightarrow{\mathcal{R}} V$, but there is no word Q for which $U \xrightarrow{*}{\mathcal{R}} Q$ and $V \xrightarrow{*}{\mathcal{R}} Q$.

Proposition 3.1. *Suppose local confluence fails at a word W but does not fail at any proper subword of W. Then one of the following conditions holds:*

(1) W *is the left side of two different rules in* $\mathcal{R}$.
(2) W *is the left side of a rule in* $\mathcal{R}$, *and* W *contains another left side as a proper subword.*
(3) W *can be written* ABC, *where* A, B, *and* C *are nonempty words and* AB *and* BC *are left sides in* $\mathcal{R}$.

Proof. By assumption, there are words A_1, B_1, P_1, Q_1, A_2, B_2, P_2, and Q_2 such that the following hold:

(i) $W = A_1 P_1 B_1 = A_2 P_2 B_2$.
(ii) (P_1, Q_1) and (P_2, Q_2) are in $\mathcal{R}$.
(iii) There is no word derivable from both $U_1 = A_1 Q_1 B_1$ and $U_2 = A_2 Q_2 B_2$.

Suppose first that the occurrences of P_1 and P_2 in W do not overlap. Then we may assume that $W = A_1 P_1 C P_2 B_2$, where $B_1 = C P_2 B_2$ and $A_2 = A_1 P_1 C$. But then $U_1 = A_1 Q_1 C P_2 B_2$ and $U_2 = A_1 P_1 C Q_2 B_2$. Therefore $A_1 Q_1 C Q_2 B_2$ is derivable from both U_1 and U_2, which contradicts our assumption.

Assume now that the occurrences of P_1 and P_2 do overlap. We may assume that $W = A_1 ABC B_2$, where $B \neq \varepsilon$ and one of the following holds:

(a) $P_1 = ABC$ and $P_2 = B$.
(b) $P_1 = AB$, $P_2 = BC$, and both A and C are nonempty.

Suppose $A_1 B_2 \neq \varepsilon$. Then ABC is a proper subword of W and by assumption local confluence does not fail at ABC. Assume (a) holds. There is a word Q derivable from both Q_1 and $A Q_2 C$. But then $A_1 Q B_2$ is derivable from both $U_1 = A_1 Q_1 B_2$ and $U_2 = A_1 A Q_2 C B_2$. Similarly, we get a contradiction if (b) holds.

Thus we have $A_1 = B_2 = \varepsilon$. Suppose (a) holds. If $AC \neq \varepsilon$, then condition (2) of the proposition holds. If $AC = \varepsilon$, then $P_1 = P_2$ and we must have $Q_1 \neq Q_2$, since otherwise $U_1 = U_2$. Therefore condition (1) holds if $AC = \varepsilon$. Finally, (b) is precisely condition (3). $\square$

If W satisfies the conclusion of Proposition 3.1, then we shall say that W is an *overlap of left sides* in $\mathcal{R}$. If condition (3) holds, then W is a *proper overlap*. With a reduced rewriting system, only proper overlaps are possible. Note that conditions (1) and (2) can be combined into a single condition which states that W is a left side of a rule in $\mathcal{R}$ and W contains the left side of some other rule as a subword.

If $\mathcal{R}$ is finite, then the set $\mathcal{W}$ of words which are overlaps of left sides in $\mathcal{R}$ is finite. For each word W in $\mathcal{W}$ we can list the finite set $\mathcal{U}$ of words derivable in one step from W. For each U in $\mathcal{U}$ we can invoke REWRITE$(X, \mathcal{R}, U; V)$. If more than one value of V is obtained, then $\mathcal{R}$ is not confluent, for we have found two words which are irreducible with respect to $\mathcal{R}$ and define the same element of M. If for all U we obtain the same value of V, then local confluence does not fail at W. By performing this test for all W in $\mathcal{W}$, we can decide whether $\mathcal{R}$ is confluent.

Example 3.1. Let X be any set. In Section 1.4 we defined $X^{\pm} = X \times \{1, -1\}$ and wrote x^{α} for (x, α). The free group generated by X is defined to be $F = \text{Mon}\,\langle X^{\pm} \mid \mathcal{R} \rangle$, where $\mathcal{R} = \text{FGRel}(X)$ consists of the rules $x^{\alpha} x^{-\alpha} \to \varepsilon$, where x is in X and α is ± 1. By Proposition 3.1, if local confluence fails, then it fails at one of the words $x^{\alpha} x^{-\alpha} x^{\alpha}$. But there are only two possible ways to rewrite such a word:

$$x^{\alpha} \underline{x^{-\alpha} x^{\alpha}} \to x^{\alpha}, \quad \underline{x^{\alpha} x^{-\alpha}} x^{\alpha} \to x^{\alpha},$$

and the results agree. Therefore $\mathcal{R}$ is confluent, and free reduction, the process of rewriting with respect to $\mathcal{R}$, always produces canonical forms. A word in $X^{\pm *}$ is freely reduced, or irreducible with respect to $\mathcal{R}$, if it contains no subwords of the form $x^{\alpha} x^{-\alpha}$. The freely reduced words are the canonical forms for the elements of F with respect to every reduction ordering of $X^{\pm *}$.

Example 3.2. Let $Y = \{x, y, z\}$ and let $\mathcal{T}$ consist of the rules

$$xyz \to \varepsilon, \quad yzx \to \varepsilon, \quad zxy \to \varepsilon.$$

To check that $\mathcal{T}$ is confluent, we need only test local confluence at the words $xyzx$, $yzxy$, $zxyz$, $xyzxy$, $yzxyz$, and $zxyzx$. At $xyzx$ we have

$$xy\underline{zx} \to x, \quad \underline{xyz}x \to x.$$

Local confluence holds at the other five words as well, so $\mathcal{T}$ is confluent.

It turns out that $G = \text{Mon}\,\langle Y \mid \mathcal{T} \rangle$ is a free group of rank 2. Let $X = \{a, b\}$ and define homomorphisms $f : X^{\pm *} \to Y^*$ and $g : Y^* \to X^{\pm *}$ by setting

$$f(a) = x, \quad f(b) = y, \quad f(a^{-1}) = yz, \quad f(b^{-1}) = zx,$$

$$g(x) = a, \quad g(y) = b, \quad g(z) = b^{-1}a^{-1}.$$

Let $\sim$ be free equivalence on $X^{\pm *}$ and let $\approx$ be the congruence on Y^* generated by T. If (U, ε) is in $\mathrm{FGRel}(X)$, then $f(U) \approx \varepsilon$. Therefore f defines a homomorphism $\overline{f}$ from the free group F generated by X to G. Also, if (V, ε) is in T, then $g(V) \sim \varepsilon$. Thus g defines a homomorphism $\overline{g} : G \to F$. Now $\overline{f}$ maps $[a]$, $[b]$, $[a^{-1}]$, and $[b^{-1}]$ to $[x]$, $[y]$, $[yz]$, and $[zx]$, respectively, and $\overline{g}$ maps these elements to $[a]$, $[b]$, $[bb^{-1}a^{-1}] = [a^{-1}]$, and $[b^{-1}a^{-1}a] = [b^{-1}]$. Therefore $\overline{f} \circ \overline{g}$ is the identity on F. A similar argument shows that $\overline{g} \circ \overline{f}$ is the identity on G, and so $\overline{f}$ and $\overline{g}$ are isomorphisms.

Example 3.3. Let $X = \{x, y, z\}$ and let S consist of the rules

$$x^2 \to \varepsilon, \quad yz \to \varepsilon, \quad zy \to \varepsilon.$$

Local confluence has to be checked at x^3, yzy, and zyz. This is easily done, so S is confluent.

Example 3.4. Let $Y = \{a, b\}$ and let T consist of the rules

$$abab \to \varepsilon, \quad baba \to \varepsilon.$$

To show that T is confluent, we must check local confluence at the following six words:

$$\begin{array}{cc} ababa, & babab, \\ ababab, & bababa, \\ abababa, & bababab. \end{array}$$

For example, we have

$$a\underline{baba} \to a, \quad \underline{abab}a \to a,$$

and

$$ba\underline{baba} \to ba, \quad \underline{baba}ba \to ba.$$

Local confluence holds at the other four words as well, and thus T is confluent.

Now $H = \mathrm{Mon}\,\langle Y \mid T \rangle$ is isomorphic to $K = \mathrm{Mon}\,\langle X \mid S \rangle$ of Example 3.3. Define $f : X^* \to Y^*$ and $g : Y^* \to X^*$ by

$$f(x) = ab, \quad f(y) = a, \quad f(z) = bab,$$
$$g(a) = y, \quad g(b) = zx.$$

Let $\sim$ be the congruence on X^* generated by S and let $\approx$ be the congruence on Y^* generated by T. For (U,ε) in S and (V,ε) in T we have $f(U) \approx \varepsilon$ and $g(V) \sim \varepsilon$. Moreover, $f \circ g$ maps x, y, and z to elements of $[x]$, $[y]$, and $[z]$, respectively. For example,

$$(f \circ g)(z) = g(bab) = zxyzx \sim z.$$

Similarly, $g \circ f$ maps a and b to elements of $[a]$ and $[b]$, respectively. Therefore f and g induce isomorphisms between H and K.

Example 3.5. Let $X = \{a,b\}$ and let S consist of the following rules:

$$
\begin{aligned}
a^2 &\to \varepsilon, & ab^2a &\to bab, \\
b^3 &\to \varepsilon, & baba &\to ab^2, \\
abab &\to b^2a, & b^2ab^2 &\to aba.
\end{aligned}
$$

Since the length of each left side is greater than the length of the corresponding right side, this is a rewriting system with respect to any of the length-plus-lexicographic orderings of X^*. To show that S is confluent, we must test local confluence at 30 words. For example,

$$ab\underline{abba} \to a\underline{bbab} \to babb,$$
$$a\underline{babb}a \to bb\underline{aba}a \to babb.$$

Local confluence holds at these words, so S is confluent. The monoid $G = \mathrm{Mon}\langle X \mid S \rangle$ is a group since $[a]^{-1} = [a]$ and $[b]^{-1} = [b^2]$. In fact G is a finite group. See Exercise 3.3.

Example 3.6. Let $X = \{a,b,c\}$ and let S consist of $\mathrm{FGRel}(X)$ together with the following rules:

$$
\begin{aligned}
ca &\to ac, & ca^{-1} &\to a^{-1}c, & c^{-1}a &\to ac^{-1}, & c^{-1}a^{-1} &\to a^{-1}c^{-1}, \\
cb &\to bc, & cb^{-1} &\to b^{-1}c, & c^{-1}b &\to bc^{-1}, & c^{-1}b^{-1} &\to b^{-1}c^{-1}, \\
ba &\to abc, & ba^{-1} &\to a^{-1}bc^{-1}, & b^{-1}a &\to ab^{-1}c^{-1}, & b^{-1}a^{-1} &\to a^{-1}b^{-1}c.
\end{aligned}
$$

Then S is a rewriting system with respect to the basic wreath-product ordering of $X^{\pm*}$ with $c \prec c^{-1} \prec b \prec b^{-1} \prec a \prec a^{-1}$. To verify the confluence of S, we must test local confluence at 37 words. For example,

$$\underline{cba} \to \underline{ca}bc \to a\underline{cb}c \to abcc,$$
$$\underline{cba} \to b\underline{ca} \to \underline{ba}c \to abcc.$$

The set of canonical forms is the set of words $a^\alpha b^\beta c^\gamma$, where α, β, and γ are arbitrary integers.

Example 3.7. Let $X = \{a, b\}$ and let $\mathcal{S}$ consist of FGRel(X) along with the following rules:

$$ba \to ab, \quad ba^{-1} \to a^{-1}b, \quad b^{-1}a \to ab^{-1}, \quad b^{-1}a^{-1} \to a^{-1}b^{-1}.$$

It is easy to check that $\mathcal{S}$ is a confluent, reduced rewriting system with respect to the length-plus-lexicographic ordering of $X^{\pm *}$ with $a \prec a^{-1} \prec b \prec b^{-1}$. Now let $\mathcal{T}$ consist of FGRel(X) plus the following set of rules:

$$ba \to ab, \quad b^{-1}a \to ab^{-1}, \quad a^{-1}b \to ba^{-1}, \quad a^{-1}b^{-1} \to b^{-1}a^{-1},$$
$$ab^i a^{-1} \to b^i, \quad i = \pm 1, \pm 2, \ldots.$$

It is not hard to see that $\mathcal{T}$ is an infinite, confluent, reduced rewriting system with respect to the length-plus-lexicographic ordering of X^* with $a \prec b \prec b^{-1} \prec a^{-1}$. The systems $\mathcal{S}$ and $\mathcal{T}$ generate the same congruence $\sim$ on X^*. Therefore whether or not the reduced, confluent rewriting system RC$(X, \prec, \sim)$ is finite depends on the ordering $\prec$.

Let us formalize the confluence test based on Proposition 3.1. The function CONFLUENT described next implements one version of the test.

```
Function CONFLUENT(X, R): boolean;
Input: X     : a finite set;
       R     : a finite rewriting system on X* with respect to some
               reduction ordering;
(* True is returned if R is confluent. *)
Begin
   For all (P, Q) in R do
     For all (R, S) in R do
       For all nonempty suffixes B of P do begin
         Let U be the longest common prefix of B and R;
         Let B = UD and R = UE;
         If D or E is empty then begin
           Let P = AB;
           REWRITE(X, R, ASD; V); REWRITE(X, R, QE; W);
           If V ≠ W then begin CONFLUENT := false; Goto 99 end
         End
       End;
     End;
   CONFLUENT := true;
99:End.
```

In CONFLUENT, the conditions of Proposition 3.1 have been revised slightly. See Exercise 3.5. If an appropriate index structure is available, then the operation of CONFLUENT can be speeded up substantially. See Exercise 3.6.

The following example describes some confluent rewriting systems of theoretical interest. These systems may be infinite.

Example 3.8. Let S be a semigroup. In the following discussion it will be necessary to distinguish carefully between elements of S and elements of S^*. If x and y are in S, then "xy" could mean either the sequence x, y, which is a word of length 2 in S^*, or the single element of S which is the product of x and y. To resolve this ambiguity, we shall introduce an explicit symbol $\bullet$ for multiplication in S. Thus $x \bullet y$ is the product of x and y, and xy is the word of length 2.

Let $\mathcal{S}$ consist of the rules $xy \to x \bullet y$ for all x and y in S. The set $\mathcal{S}$ is a rewriting system with respect to any length-plus-lexicographic ordering of S^*. To prove the confluence of $\mathcal{S}$, we must test local confluence at the words xyz. We have the following reductions:

$$xyz \to \underline{xy} \bullet z \to x \bullet (y \bullet z),$$
$$\underline{xyz} \to (x \bullet y)z \to (x \bullet y) \bullet z.$$

Thus local confluence is just the associative law in S. The pair $(S, \mathcal{S})$ is a semigroup presentation for S. If S is a monoid with identity e, then $(S, \mathcal{S})$ is not a monoid presentation for S. To get a monoid presentation, we must add the rule $e \to \varepsilon$. Let $\mathcal{T} = \mathcal{S} \cup \{(e, \varepsilon)\}$. The rewriting system $\mathcal{T}$ is confluent, and the set of canonical forms is $(S - \{e\}) \cup \{\varepsilon\}$. We call $(S, \mathcal{S})$ and $(S, \mathcal{T})$ the *multiplication-table presentations* of S as a semigroup and as a monoid, respectively.

If S is a monoid, then the presentation $(S, \mathcal{T})$ is confluent but not reduced. For example, if $x \bullet y = e$ in S, then the right side of the rule $xy \to e$ is not irreducible. Also, the left side e is a proper subword of other left sides. Let $\sim$ be the congruence generated by $\mathcal{T}$. For any length-plus-lexicographic ordering $\prec$ of S^*, the system $\mathrm{RC}(S, \prec, \sim)$ consists of the following rules:

$$e \to \varepsilon,$$
$$xy \to x \bullet y, \qquad x \neq e, \quad y \neq e, \quad x \bullet y \neq e,$$
$$xy \to \varepsilon, \qquad x \neq e, \quad y \neq e, \quad x \bullet y = e.$$

In Exercise 2.4 of Chapter 1 we constructed the direct product of two monoids. We shall now describe another type of product called the free product. We begin with a proposition about combining two rewriting systems with disjoint sets of generators.

Proposition 3.2. *Suppose X and Y are disjoint sets and $\mathcal{R}$ and $\mathcal{S}$ are, respectively, confluent rewriting systems on X^* and Y^* with respect to reduction orderings on those sets. Then $\mathcal{R} \cup \mathcal{S}$ is a confluent rewriting system with respect to any reduction ordering of $(X \cup Y)^*$ which extends the given orderings.*

Proof. The words described in Proposition 3.1 at which the local confluence of $\mathcal{R} \cup \mathcal{S}$ must be checked are either in X^* or in Y^*.

□

Let M and N be monoids. By taking an isomorphic copy if necessary, we may assume that $M \cap N = \emptyset$. Let $\mathcal{T}_M$ and $\mathcal{T}_N$ be the monoid multiplication-table rewriting systems for M and N, respectively. By Proposition 3.2, $\mathcal{T}_M \cup \mathcal{T}_N$ is a confluent rewriting system. The monoid Mon $\langle M \cup N \mid \mathcal{T}_M \cup \mathcal{T}_N \rangle$ is called the *free product* of M and N and is denoted $M * N$. Canonical forms for $M * N$ are ε and $u_1 u_2 \ldots u_m$, where each u_i is a nonidentity element of M or of N, and for $1 \leq i < m$ either u_i is in M and u_{i+1} is in N or u_i is in N and u_{i+1} is in M. It is clear from the definition that $M * N = N * M$.

Proposition 3.3. *Suppose $(X, \mathcal{R})$ and $(Y, \mathcal{S})$ are monoid presentations for M and N, respectively. If $X \cap Y = \emptyset$, then $(X \cup Y, \mathcal{R} \cup \mathcal{S})$ is a presentation of $M * N$.*

Proof. The proof is primarily diagram chasing and is omitted. □

The operation of forming free products is associative.

Proposition 3.4. *If L, M, and N are monoids, then $(L * M) * N$ and $L * (M * N)$ are naturally isomorphic.*

Proof. Exercise. □

Proposition 3.5. *If M and N are groups, then $M * N$ is a group.*

Proof. Let $f : M \to M * N$ be the obvious homomorphism. The image of M under f is a group. Similarly, the image of N in $M * N$ is a group. Since these images generate $M * N$ as a monoid, it follows that $M * N$ is a group. □

Example 3.9. The free group F generated by a finite set X is defined by monoid relations $x^\alpha x^{-\alpha} = 1$ on generators x^α with x in X and $\alpha = \pm 1$. For x in X, let F_x be the free group generated by $\{x\}$. Then F_x is isomorphic to $\mathbb{Z}$, and F is (isomorphic to) the free product of the groups F_x.

Example 3.10. The group of Example 3.3 is the free product of Mon $\langle x \rangle$, which is isomorphic to $\mathbb{Z}_2$, and Mon $\langle y, z \rangle$, which is isomorphic to $\mathbb{Z}$.

Proposition 3.6. *Let* $X = \{x_1, \ldots, x_s\}$, *let* m *be a positive integer, and let* $\mathcal{R}$ *be the set of all pairs* (P, ε), *where* P *is a cyclic permutation of the word* $(x_1 x_2 \ldots x_s)^m$. *Then* $(X, \mathcal{R})$ *is a reduced, confluent presentation, and* Mon $\langle X \mid \mathcal{R} \rangle$ *is the free product of* $\mathbb{Z}_m$ *and* $s - 1$ *copies of* $\mathbb{Z}$.

Proof. Examples 3.3 and 3.4 do the case $s = m = 2$. The general case is left as an exercise. $\square$

Sometimes the most convenient description of a group G by a confluent rewriting system involves restricting the words used to describe elements of G. Suppose $\mathcal{R}$ is a confluent rewriting system on X^*. It may happen that $M = $ Mon $\langle X \mid \mathcal{R} \rangle$ is not isomorphic to G but there is a subsemigroup S of X^* whose image in M is isomorphic to G. If S is closed under rewriting with respect to $\mathcal{R}$, then we can compute in G by restricting the words used to elements of S. In this situation we shall say that the triple $(X, \mathcal{R}, S)$ is a *restricted presentation* for G. Perhaps the simplest restricted presentation for G is $(G, \mathcal{S}, S)$, where $\mathcal{S} = \{(xy, x \bullet y) \mid x, y \in G\}$ is the set of semigroup relations obtained from the multiplication table of G, and S is the set of nonempty words in G^*.

Exercises

3.1. Show that the following rewriting system is confluent:

$$a^5 \to \varepsilon, \quad b^5 \to \varepsilon, \quad b^4 a^4 \to (ab)^4, \quad (ba)^4 \to a^4 b^4.$$

3.2. Prove confluence in Example 3.7.

3.3. Find the order of G in Example 3.5.

3.4. Let X be a subset of Y and let G be the free group generated by Y. Show that the subgroup of G generated by the image of X is isomorphic to the free group generated by X.

3.5. Let $\mathcal{R}$ be a rewriting system on X^* and let W be a word in X^* which is an overlap of left sides in $\mathcal{R}$. Show that there are words A, D, E, and U such that AUD and UE are left sides in $\mathcal{R}$, $W = AUDE$, $U \neq \varepsilon$, and either D or E is empty.

3.6. Suppose that $\mathcal{R}$ is a finite rewriting system on X^* and for any word B in X^* there is an efficient mechanism for finding all rules (R, S) in $\mathcal{R}$ such that either R is a prefix of B or B is a prefix of R. Modify CONFLUENT so that its main loop has the form

For all (P, Q) in $\mathcal{R}$ do
　For all nonempty suffixes B of P do begin
　　$\vdots$
End

3.7. Show that free groups have no nonidentity elements of finite order. (Hint: Show that every nonidentity element is conjugate to an element $[A]$, where A^2 is a freely reduced word.)

2.4 Rewriting strategies

Let $X = \{a, b, c\}$. In Example 3.6 we established the confluence of the rewriting system $\mathcal{S}$ on $X^{\pm *}$ consisting of $\mathrm{FGRel}(X)$ and the following rules:

$$ca \to ac, \quad ca^{-1} \to a^{-1}c, \quad c^{-1}a \to ac^{-1}, \quad c^{-1}a^{-1} \to a^{-1}c^{-1},$$
$$cb \to bc, \quad cb^{-1} \to b^{-1}c, \quad c^{-1}b \to bc^{-1}, \quad c^{-1}b^{-1} \to b^{-1}c^{-1},$$
$$ba \to abc, \quad ba^{-1} \to a^{-1}bc^{-1}, \quad b^{-1}a \to ab^{-1}c^{-1}, \quad b^{-1}a^{-1} \to a^{-1}b^{-1}c.$$

Rewriting using $\mathcal{S}$ and generalizations of $\mathcal{S}$ described in Section 9.10 was first discussed by P. Hall, who called the process *collection*. Given a word U in $X^{\pm *}$, there usually are many ways of collecting or rewriting U to obtain an irreducible word V. Which choices are made can greatly affect the number of steps and the length of the intermediate words.

One strategy for rewriting with $\mathcal{S}$ which has useful theoretical applications is called *collection to the left*. Chapters 11 and 12 of [Hall 1959] make extensive use of this strategy. In the first step of collection to the left, all occurrences of a and a^{-1} are moved to the left and then canceled to the extent possible. Then occurrences of b and b^{-1} are moved left past all occurrences of c and c^{-1}. Applying this strategy to $U = baba^{-1}b^{-1}a$, we get the following reduction:

$$\underline{ba}ba^{-1}b^{-1}a \to abc\underline{ba}^{-1}b^{-1}a \to ab\underline{ca}^{-1}bc^{-1}b^{-1}a$$
$$\to a\underline{ba}^{-1}cbc^{-1}b^{-1}a \to \underline{aa}^{-1}bc^{-1}cbc^{-1}b^{-1}a \to bc^{-1}cbc^{-1}\underline{b^{-1}a}$$
$$\to bc^{-1}cb\underline{c^{-1}a}b^{-1}c^{-1} \to bc^{-1}\underline{cba}c^{-1}b^{-1}c^{-1} \to bc^{-1}\underline{ca}bcc^{-1}b^{-1}c^{-1}$$
$$\to b\underline{c^{-1}a}cbcc^{-1}b^{-1}c^{-1} \to \underline{ba}c^{-1}cbcc^{-1}b^{-1}c^{-1} \to abc\underline{c^{-1}c}bcc^{-1}b^{-1}c^{-1}$$
$$\to ab\underline{cc^{-1}}bccc^{-1}b^{-1}c^{-1} \to ab\underline{cb}c^{-1}ccc^{-1}b^{-1}c^{-1} \to abbcc^{-1}\underline{ccc^{-1}b}^{-1}c^{-1}$$
$$\to abbcc^{-1}\underline{ccb}^{-1}c^{-1}c^{-1} \to abbcc^{-1}\underline{cb}^{-1}cc^{-1}c^{-1} \to abbc\underline{cc^{-1}b}^{-1}ccc^{-1}c^{-1}$$
$$\to abb\underline{cb}^{-1}c^{-1}ccc^{-1}c^{-1} \to ab\underline{bb}^{-1}cc^{-1}ccc^{-1}c^{-1} \to ab\underline{cc}^{-1}ccc^{-1}c^{-1}$$
$$\to abc\underline{cc}^{-1}c^{-1} \to ab\underline{cc}^{-1} \to ab.$$

However, if we choose a different strategy, we can carry out the reduction with only five applications of the rewriting rules:

$$baba^{-1}\underline{b^{-1}a} \rightsquigarrow baba^{-1}\underline{ab}^{-1}c^{-1} \to ba\underline{bb}^{-1}c^{-1} \to \underline{ba}c^{-1} \to ab\underline{cc}^{-1} \to ab.$$

With this small rewriting system, it is relatively easy in hand computation to avoid long, unnecessary sequences of rewriting steps. However, with large systems having perhaps millions of rules, it is very difficult to decide on a good, let alone an optimum, strategy.

Complicating the situation is the fact that our primary goal is to optimize the running time of the rewriting process, and this is not necessarily the same as minimizing the number of applications of the rules. It might be that we could always find a short reduction, but doing so would take a long time. It is possible that a strategy which employs more steps but is able to make the choice of those steps very rapidly will yield a faster algorithm.

Suppose we are rewriting a word U. Thus we want to write U as APB, where (P,Q) is a rule, and to replace U by AQB. How should we choose the occurrence of the left side P in U? We could select it so that $|P| - |Q|$ is maximal, which would minimize the length of the word AQB. We could also select the occurrence so that AQB is as early as possible in the ordering $\prec$ we are using. Collection to the left makes sense only for rewriting systems with respect to wreath product orderings in which there are left sides $y^\beta x^\alpha$ for all generators x and y with $x \succ y$ and all α, β in $\{1, -1\}$. However, a general strategy which is in the spirit of collection to the left chooses the left side P in U which is as large as possible with respect to $\prec$ and, subject to this, as close to the beginning of U as possible. Unfortunately, each of these strategies requires finding all occurrences of left sides in U, and frequently this is not practical.

Usually the first occurrence of a left side found is the one that is used. In this case, the way one searches for left sides becomes important. One can take each left side in turn and look for occurrences of it anywhere in U. Or one can take a fixed term in U and look for left sides beginning at that term. The terms can be considered from left to right or from right to left. Here is an implementation of a strategy which will be called *rewriting from the left*. The set X of generators and the rules (P_i, Q_i), $1 \le i \le n$, are assumed to be available as global variables.

```
Procedure REWRITE_FROM_LEFT(U; V);
Input:   U     : a word;
Output:  V     : a word, the rewritten form of U;
Begin
   V := ε;  W := U;
   While W ≠ ε do begin
      Let W = xW₁ with x in X;  W := W₁;  V := Vx;
      For i := 1 to n do
         If Pᵢ is a suffix of V then begin
            Let V = RPᵢ;  W := QᵢW;  V := R;
            Break
         End
   End
End.
```

At all times in REWRITE_FROM_LEFT, the word VW is derivable from U. Moreover, V is the longest prefix of VW known to be irreducible with respect to the set of rules. In Section 3.5 we shall examine more efficient ways to test whether a word V has some P_i as a suffix. In REWRITE_FROM_LEFT, we simply test each P_i in turn.

Suppose we take the rewriting system of Example 3.6:

$$a^2 \to \varepsilon, \qquad ab^2a \to bab,$$
$$b^3 \to \varepsilon, \qquad baba \to ab^2,$$
$$abab \to b^2a, \qquad b^2ab^2 \to aba.$$

Here is a sample of the way REWRITE_FROM_LEFT rewrites a word:

$ab\underline{aa}babbbaaabbabbbaba \to a\underline{bb}abbbaaabbabbbbaba \to ba\underline{bbb}aaabbabbbbaba$
$\to b\underline{aa}aabbabbbbaba \to b\underline{aa}bbabbbbaba \to \underline{bbb}abbbbaba$
$\to a\underline{bbb}aba \to \underline{aa}ba \to ba.$

There is an analogous procedure for rewriting from the right. The second rewriting of $baba^{-1}b^{-1}a$ given above illustrates the strategy of rewriting from the right.

Exercises

4.1. Describe a procedure REWRITE_FROM_RIGHT which keeps track of the longest suffix of the word which is known to be irreducible.

4.2. Show that rewriting a word U from the right using the rules (P_i, Q_i), $1 \le i \le n$, can be accomplished by rewriting $U^\dagger$ from the left using the rules $(P_i^\dagger, Q_i^\dagger)$ and then reversing the result.

4.3. Let $\mathcal{R}$ be the rewriting system of Example 3.6. Using $\mathcal{R}$, rewrite each of the following words with collection to the left, rewriting from the left, and rewriting from the right: cba, $(abc)^3$, $ca^{-1}b^{-1}abc^{-1}a$.

4.4. Let $\mathcal{S}$ be the rewriting system of Example 3.5. Using $\mathcal{S}$, rewrite each of the following words with rewriting from the left and rewriting from the right: $(abab^2)^3$, $bbababaabbababb$.

2.5 The Knuth-Bendix procedure

In this section we shall encounter the first of the fundamental procedures which are the subject of this book. The procedure is called the *Knuth-Bendix procedure for strings*. It is based on a much more general procedure described in (Knuth & Bendix 1970).

Let $(X, \mathcal{R})$ be a finite monoid presentation, let $\prec$ be a reduction ordering on X^*, and let $\mathcal{T}$ denote the reduced, confluent rewriting system $\mathrm{RC}(X, \prec, \mathcal{R})$. The main result of this section states that if $\mathcal{T}$ is finite, then $\mathcal{T}$ can be computed from $X, \mathcal{R}$, and an effective definition of $\prec$. By an effective definition of $\prec$ is meant a definition which provides an algorithm for deciding which of two given words comes first. It must be emphasized that there is no procedure for deciding whether $\mathcal{T}$ is finite.

We shall first describe a basic version of the Knuth-Bendix procedure for strings called KBS_1. It has input arguments X, $\mathcal{R}$, and $\prec$, and it returns $\mathcal{T}$, provided $\mathcal{T}$ is finite. Because KBS_1 is not practical for serious computation, a second, more efficient version of the Knuth-Bendix procedure will be given in Section 2.6. Additional improvements will be sketched in Section 2.7. Within KBS_1, a sequence of rules (P_i, Q_i), $1 \le i \le n$, is generated. The procedure REWRITE_FROM_LEFT is used to rewrite a word with respect to $\mathcal{S} = \{(P_i, Q_i) \mid 1 \le i \le n\}$.

The idea behind KBS_1 is quite simple. Suppose we apply CONFLU-ENT to the current set of rules. If the rules are confluent, we stop. Otherwise we find two irreducible words A and B which are equivalent under the congruence $\sim$ generated by $\mathcal{R}$. Assuming the $A \succ B$, we add (A, B) to the set of rules. The only point which needs some care is making certain that for every pair of indices i and j eventually all overlaps of the left sides P_i and P_j are considered.

Two subroutines are used in KBS_1. The subroutine TEST_1 adds a new rule, if necessary, in order to insure that there is a word derivable from two given words using the rules in $\mathcal{S}$.

```
Procedure TEST_1(U, V);
Input: U, V    : words;
Begin
   REWRITE_FROM_LEFT(U; A); REWRITE_FROM_LEFT(V; B);
   If A ≠ B then begin
     If A ≺ B then interchange A and B;
     n := n + 1; P_n := A; Q_n := B
   End
End.
```

The second subroutine, OVERLAP_1, is based on the inner loop of CON-FLUENT in Section 2.3. It checks the overlaps of P_i and P_j in which P_i occurs at the beginning of the word. When failures of local confluence are found, new rules are added. Note the way OVERLAP_1 combines the case in which P_j is a subword of P_i with the case in which P_i has a nonempty suffix B which is also a prefix of P_j.

```
Procedure OVERLAP_1(i, j);
Input: i, j    : positive integers not exceeding n;
Begin
   For k := 1 to |P_i| do begin
     Let P_i = AB with |B| = k;
```

Let U be the longest common prefix of B and P_j;
Let $B = UD$ and $P_j = UE$;
If D or E is empty then TEST_1(AQ_jD, Q_iE)
  End
End.

Here then is the definition of KBS_1.

Procedure KBS_1$(X, \prec, \mathcal{R}; \mathcal{T})$;
Input: X : a finite set;
               $\prec$: a reduction ordering on X^*;
               $\mathcal{R}$: a finite subset of $X^* \times X^*$;
Output: $\mathcal{T}$: RC$(X, \prec, \mathcal{R})$, if it is finite;
($*$ WARNING – TERMINATION MAY NOT OCCUR. $*$)
Begin
  $n := 0$; $i := 1$;
  For (U, V) in $\mathcal{R}$ do TEST_1(U, V);

  While $i \leq n$ do begin
    For $j := 1$ to i do begin
      OVERLAP_1(i, j);
      If $j < i$ then OVERLAP_1(j, i)
    End;

    $i := i + 1$
  End;

  Let $\mathcal{P}$ be the set of P_i such that every proper subword of P_i is
    irreducible with respect to $\mathcal{S}$;
  $\mathcal{T} := \emptyset$;

  For P in $\mathcal{P}$ do begin
    REWRITE_FROM_LEFT$(P; Q)$; Add (P, Q) to $\mathcal{T}$
  End
End.

Proposition 5.1. *If* RC$(X, \prec, \mathcal{R})$ *is finite, then* KBS_1$(X, \prec, \mathcal{R}; \mathcal{T})$ *terminates with* $\mathcal{T} = $ RC$(X, \prec, \mathcal{R})$.

Proof. Suppose a call TEST_1(U, V) is made and the two calls to REWRITE_FROM_LEFT have been completed. Then $\mathcal{S} \cup \{(U, V)\}$ and $\mathcal{S} \cup \{(A, B)\}$ generate the same congruence on X^*. Therefore, after the first For-loop in KBS_1, the sets $\mathcal{R}$ and $\mathcal{S}$ generate the same congruence $\sim$ on X^*. With any later calls TEST_1(U, V), we have $U \sim V$, so at any time during the While-loop in KBS_1 the set $\mathcal{S}$ generates $\sim$. Note also that P_i is irreducible with respect to $\{(P_j, Q_j) \mid 1 \leq j < i\}$.

Suppose that KBS_1 does not terminate. Then the While-loop does not terminate and produces an infinite sequence $(P_i, Q_i), i = 1, 2, \ldots$, of rewriting rules. Let $\mathcal{U}$ be the set of these rules.

Lemma 5.2. $\mathcal{U}$ *is confluent and generates* $\sim$.

Proof. From the preceding remarks, it is clear that $\mathcal{U}$ generates $\sim$. Suppose $\mathcal{U}$ is not confluent, and let W be the first word (with respect to $\prec$) at which local confluence fails. By Proposition 3.1, $W = ABC$, where $B \neq \varepsilon$ and one of the following holds:

(i) $W = P_j$ and $B = P_i$, with $j < i$, and there is no word derivable using $\mathcal{U}$ from both AQ_iC and Q_j.
(ii) $AB = P_i$, $BC = P_j$, and there is no word derivable from both AQ_j and Q_iC.

Suppose (i) holds. At some point in the running of KBS_1 the call OVERLAP_1(j, i) is made. After this call is completed, there is a word R derivable from both AQ_iC and Q_j using the current set $\mathcal{S}$. Since $\mathcal{S}$ is a subset of $\mathcal{U}$, the word R is derivable from both AQ_iC and Q_j using $\mathcal{U}$. Therefore (i) cannot hold.

Similarly (ii) does not hold, for the call OVERLAP_1(i, j) would force the existence of a word derivable from both AQ_j and Q_iC using $\mathcal{U}$. Therefore $\mathcal{U}$ is confluent. $\square$

We now resume the proof of Proposition 5.1. Let $\mathcal{C}$ be the set of canonical forms for $\sim$ and let $\mathcal{P}$ be the set of words P in $X^* - \mathcal{C}$ such that every proper subword of P is in $\mathcal{C}$. Since rewriting an element P of $\mathcal{P}$ using $\mathcal{U}$ must produce the element $\overline{P}$ of $\mathcal{C}$ such that $P \sim \overline{P}$, there is a rule in $\mathcal{U}$ with left side P, and this rule is unique. Let $\mathcal{V}$ be the set of rules (P, Q) in $\mathcal{U}$ such that P is in $\mathcal{P}$. By assumption, $\mathrm{RC}(X, \prec, \mathcal{R})$ is finite. Therefore $\mathcal{P}$ and $\mathcal{V}$ are finite. Let n be the largest integer such that (P_n, Q_n) is in $\mathcal{V}$. If $i > n$, then P_i is not in $\mathcal{C}$ and is irreducible with respect to $\{(P_j, Q_j) \mid 1 \leq j < i\}$. But this is impossible. Therefore $\mathcal{U}$ is finite and the While-loop in KBS_1 terminates.

By Proposition 2.8, the last For-loop defines $\mathcal{T}$ to be $\mathrm{RC}(X, \prec, \mathcal{R})$. $\square$

Let us illustrate the use of KBS_1 with some examples. To assist the reader in following the examples, a description of the source of each rule will be given. New rules are added to $\mathcal{S}$ only in TEST_1, which is called in two places: in the first For-loop of KBS_1 and in OVERLAP_1. To indicate the source of a rule, we shall write one of the following:

Relation i,

Overlap i, j, k.

Table 2.5.1

i	P_i	Q_i	Source
1	aa^{-1}	ε	Relation 1
2	$a^{-1}a$	ε	Relation 2
3	bb^{-1}	ε	Relation 3
4	$b^{-1}b$	ε	Relation 4
5	ba	ab	Relation 5
6	aba^{-1}	b	Overlap 5 1 1
7	$b^{-1}ab$	a	Overlap 4 5 1
8	ba^{-1}	$a^{-1}b$	Overlap 2 6 1
9	$b^{-1}a$	ab^{-1}	Overlap 7 3 1
10	$b^{-1}a^{-1}b$	a^{-1}	Overlap 4 8 1
11	$ab^{-1}a^{-1}$	b^{-1}	Overlap 9 1 1
12	$b^{-1}a^{-1}$	$a^{-1}b^{-1}$	Overlap 10 3 1

The first says that the call to TEST_1 which added the rule occurred on the i-th iteration of the first For-loop in KBS_1. The second indicates that the call to TEST_1 came from OVERLAP_1 when its arguments were i and j, and the variable k within OVERLAP_1 had the specified value.

Example 5.1. Let $X = \{a, b\}$, let $\prec$ be the length-plus-lexicographic ordering of $X^{\pm *}$ with $a \prec a^{-1} \prec b \prec b^{-1}$, and let $\mathcal{R}$ consist of the following rules:

$$aa^{-1} \to \varepsilon, \quad a^{-1}a \to \varepsilon, \quad bb^{-1} \to \varepsilon, \quad b^{-1}b \to \varepsilon, \quad ba \to ab.$$

If we process the elements of $\mathcal{R}$ in the order listed, then the sequence of rules generated by KBS_1 is as given in Table 2.5.1.

The final reduced, confluent rewriting system consists of rules 1 to 5, 8, 9, and 12.

Example 5.2. Let X and $\mathcal{R}$ be the same as in Example 5.1 and let $\prec$ be the length-plus-lexicographic ordering of $X^{\pm *}$ with $a \prec b \prec b^{-1} \prec a^{-1}$. The first 20 rules produced by KBS_1 are listed in Table 2.5.2.

It is relatively easy to show that $P_{2i} = ab^{6-i}a^{-1}$, $Q_{2i} = b^{6-i}$, $P_{2i+1} = ab^{i-3}a^{-1}$, and $Q_{2i+1} = b^{i-3}$ for $i \geq 8$. The procedure KBS_1 never terminates because $\mathrm{RC}(X, \prec, \mathcal{R})$ is infinite, as shown in Example 3.7.

Example 5.3. Let $X = \{a, b\}$, let $\prec$ be the length-plus-lexicographic ordering of X^* with $a \prec b$, and let $\mathcal{R}$ consist of the following pairs:

$$a^2 \to \varepsilon, \quad b^3 \to \varepsilon, \quad (ab)^3 \to \varepsilon.$$

The sequence of rules produced by KBS_1 is given in Table 2.5.3. The final reduced, confluent system consists of rules 1, 2, and 6 to 9. This is the presentation of Example 3.5.

Table 2.5.2

i	P_i	Q_i	Source
1	aa^{-1}	ε	Relation 1
2	$a^{-1}a$	ε	Relation 2
3	bb^{-1}	ε	Relation 3
4	$b^{-1}b$	ε	Relation 4
5	ba	ab	Relation 5
6	aba^{-1}	b	Overlap 5 1 1
7	$b^{-1}ab$	a	Overlap 4 5 1
8	$a^{-1}b$	ba^{-1}	Overlap 2 6 1
9	$abba^{-1}$	bb	Overlap 5 6 1
10	$b^{-1}a$	ab^{-1}	Overlap 7 3 1
11	$ba^{-1}b^{-1}$	a^{-1}	Overlap 8 3 1
12	$abbba^{-1}$	bbb	Overlap 5 9 1
13	$ab^{-1}a^{-1}$	b^{-1}	Overlap 10 1 1
14	$a^{-1}b^{-1}$	$b^{-1}a^{-1}$	Overlap 4 11 1
15	ab^4a^{-1}	b^4	Overlap 5 12 1
16	$ab^{-2}a^{-1}$	b^{-2}	Overlap 10 13 1
17	ab^5a^{-1}	b^5	Overlap 5 15 1
18	$ab^{-3}a^{-1}$	b^{-3}	Overlap 10 16 1
19	ab^6a^{-1}	b^6	Overlap 5 17 1
20	$ab^{-4}a^{-1}$	b^{-4}	Overlap 10 18 1

Table 2.5.3

i	P_i	Q_i	Source
1	a^2	ε	Relation 1
2	b^3	ε	Relation 2
3	$(ab)^3$	ε	Relation 3
4	$babab$	a	Overlap 1 3 1
5	$ababa$	b^2	Overlap 3 2 1
6	$(ba)^2$	ab^2	Overlap 4 2 1
7	$(ab)^2$	b^2a	Overlap 2 4 1
8	ab^2a	bab	Overlap 4 3 2
9	b^2ab^2	aba	Overlap 5 5 1

Example 5.4. Let $X = \{a, b, b^{-1}\}$, let $\prec$ be the length-plus-lexicographic ordering of X^* with $a \prec b \prec b^{-1}$, and let $\mathcal{R}$ consist of the following rules:

$$a^2 \to \varepsilon, \quad bb^{-1} \to \varepsilon, \quad b^3 \to \varepsilon, \quad (ab)^3 \to \varepsilon.$$

This is the same group as in Example 5.3. Note that the monoid relation $b^{-1}b \to \varepsilon$ is not needed since the relation $b^3 \to \varepsilon$ implies that b has an inverse. The relation $bb^{-1} \to \varepsilon$ simply names that inverse. Table 2.5.4 lists the sequence of rules produced by KBS_1. The final system consists of rules 1, 2, 5, 8, 9, 12, and 14 to 18.

Table 2.5.4

i	P_i	Q_i	Source
1	a^2	ε	Relation 1
2	bb^{-1}	ε	Relation 2
3	b^3	ε	Relation 3
4	$(ab)^3$	ε	Relation 4
5	b^2	b^{-1}	Overlap 3 2 1
6	$babab$	a	Overlap 1 4 1
7	$ababa$	b^{-1}	Overlap 4 2 1
8	b^{-2}	b	Overlap 5 2 1
9	$b^{-1}b$	ε	Overlap 5 3 2
10	$baba$	ab^{-1}	Overlap 6 2 1
11	$abab$	$b^{-1}a$	Overlap 3 6 1
12	bab	$ab^{-1}a$	Overlap 6 4 2
13	$b^{-1}ab^{-1}a$	ab	Overlap 7 4 3
14	$b^{-1}ab^{-1}$	aba	Overlap 7 7 1
15	$b^{-1}aba$	bab^{-1}	Overlap 5 10 1
16	$abab^{-1}$	$b^{-1}ab$	Overlap 11 5 1
17	$ab^{-1}ab$	bab^{-1}	Overlap 12 5 1
18	$bab^{-1}a$	$b^{-1}ab$	Overlap 5 12 1

Example 5.5. This example is somewhat longer than the first four. Let $X = \{a, b, c\}$, let $\prec$ be the basic wreath-product ordering of $X^{\pm*}$ with $c \prec c^{-1} \prec b \prec b^{-1} \prec a \prec a^{-1}$, and let $\mathcal{R}$ consist of the following rules:

$$aa^{-1} \to \varepsilon, \quad a^{-1}a \to \varepsilon, \quad bb^{-1} \to \varepsilon, \quad b^{-1}b \to \varepsilon, \quad cc^{-1} \to \varepsilon, \quad c^{-1}c \to \varepsilon,$$

$$ca \to ac, \quad cb \to bc, \quad ba \to abc.$$

The sequence produced by KBS_1 is given in Table 2.5.5. The final set consists of rules 1 to 9, 16, 18 to 20, 27, 36, and 39 to 41, a total of 18 rules. These are the rules of Example 3.6.

Suppose that X is a finite set and $\mathcal{R}$ is a finite, confluent rewriting system on X^* with respect to a reduction ordering $\prec$. Let $\sim$ be the congruence generated by $\mathcal{R}$. There is a straightforward but time-consuming way to verify that $\sim$ has only finitely many classes. Choose a large integer n and show that each of the $|X|^n$ words in X^* of length n contains a left side in $\mathcal{R}$ as a subword. Thus if U in X^* is irreducible with respect to $\mathcal{R}$, then $|U| < n$, and hence $\sim$ has only finitely many canonical forms.

This approach cannot verify that $\sim$ has infinitely many classes. However, using the theory of automata, it is possible to decide whether $\sim$ has infinitely many classes. Automata are the subject of Chapter 3. Section 3.10 describes the use of automata to determine whether a monoid defined by a finite, confluent presentation is finite or infinite.

Table 2.5.5

i	P_i	Q_i	Source
1	aa^{-1}	ε	Relation 1
2	$a^{-1}a$	ε	Relation 2
3	bb^{-1}	ε	Relation 3
4	$b^{-1}b$	ε	Relation 4
5	cc^{-1}	ε	Relation 5
6	$c^{-1}c$	ε	Relation 6
7	ca	ac	Relation 7
8	cb	bc	Relation 8
9	ba	abc	Relation 9
10	aca^{-1}	c	Overlap 7 1 1
11	$c^{-1}ac$	a	Overlap 6 7 1
12	bcb^{-1}	c	Overlap 8 3 1
13	$c^{-1}bc$	b	Overlap 6 8 1
14	$abca^{-1}$	b	Overlap 9 1 1
15	$b^{-1}abc$	a	Overlap 4 9 1
16	ca^{-1}	$a^{-1}c$	Overlap 2 10 1
17	$aba^{-1}cc$	bc	Overlap 9 10 1
18	$c^{-1}a$	ac^{-1}	Overlap 11 5 1
19	cb^{-1}	$b^{-1}c$	Overlap 4 12 1
20	$c^{-1}b$	bc^{-1}	Overlap 13 5 1
21	$ba^{-1}c$	$a^{-1}b^{-1}$	Overlap 2 14 1
22	$aba^{-1}bc$	b^2	Overlap 9 14 1
23	$b^{-1}ab$	ac^{-1}	Overlap 15 5 1
24	$b^{-1}a^2bc^2$	a^2	Overlap 15 7 1
25	$b^{-1}ac$	ab^{-1}	Overlap 15 12 2
26	$c^{-1}a^{-1}c$	a^{-1}	Overlap 6 16 1
27	ba^{-1}	$c^{-1}a^{-1}b$	Overlap 13 16 1
28	$ac^{-1}a^{-1}b^2c^2$	b^2c	Overlap 17 8 1
29	$ac^{-1}a^{-2}bc$	$a^{-1}b$	Overlap 17 16 1
30	$ac^{-1}a^{-1}$	c^{-1}	Overlap 18 1 1
31	$c^{-1}b^{-1}c$	b^{-1}	Overlap 6 19 1
32	$bc^{-1}b^{-1}$	c^{-1}	Overlap 20 3 1
33	$b^{-1}a^{-1}b$	$a^{-1}c$	Overlap 4 21 1
34	$c^{-1}a^{-1}b$	$a^{-1}bc^{-1}$	Overlap 21 5 1
35	$c^{-1}a^{-2}b$	$a^{-2}bc^{-1}$	Overlap 21 16 1
36	$b^{-1}a$	$ac^{-1}b^{-1}$	Overlap 23 3 1
37	$ac^{-1}b^{-1}$	$ab^{-1}c^{-1}$	Overlap 25 5 1
38	$ab^{-1}a^{-1}$	$b^{-1}c$	Overlap 25 10 2
39	$c^{-1}a^{-1}$	$a^{-1}c^{-1}$	Overlap 26 5 1
40	$c^{-1}b^{-1}$	$b^{-1}c^{-1}$	Overlap 31 5 1
41	$b^{-1}a^{-1}$	$a^{-1}b^{-1}c$	Overlap 33 3 1

Exercises

5.1. Let $X = \{a, b\}$ and let $\mathcal{R}$ consist of the rules $a^2 \to \varepsilon$, $b^2 \to \varepsilon$, $(ab)^2 \to \varepsilon$. Apply KBS_1 using the length-plus-lexicographic ordering $\prec$ with $a \prec b$. Repeat the computation using the length-plus-lexicographic ordering with $b \prec a$.

5.2. Suppose that $X = \{x\}$ and $\mathcal{R}$ consists of the rules $x^m \to \varepsilon$ and $x^n \to \varepsilon$, where m and n are positive integers. Show that KBS_1 returns the single rule $x^d \to \varepsilon$, where $d = \gcd(m, n)$. Compare the operation of KBS_1 on this input with the operation of the Euclidean algorithm.

5.3. Let $\mathcal{R}$ be a finite, confluent rewriting system on X^* and assume that $\mathrm{Mon}\langle X \mid \mathcal{R}\rangle$ is finite. Design a backtrack search for the words in X^* which are irreducible with respect to $\mathcal{R}$. Given a positive integer n, apply the technique of Knuth described in Section 1.8 to devise a method for estimating the number of irreducible words of length at most n.

2.6 A second version

The procedure KBS_1 of Section 2.5 retains all rules that are constructed and uses them in the processes of rewriting and forcing local confluence. It is usually better to delete rules which become redundant and maintain a reduced set of rules at all times. In the procedure KBS_2, described below, boolean flags $active[i]$ are used to indicate which rules are still active. If $active[i]$ is true, then (P_i, Q_i) is active and is used in rewriting and forming overlaps. If $active[i]$ is false, then (P_i, Q_i) will play no more part in the operation of the procedure. The procedure REWRITE_FROM_LEFT must be modified to look only for the left sides of active rules. In KBS_2 we retain inactive rules to provide a complete history of the computation. In practice, rules are deleted as soon as they become inactive, and the space freed up is made available for new rules.

Assume that the current set of active rules is reduced. When a new rule (A, B) is added, the set of active rules may no longer be reduced. Suppose that (P_i, Q_i) is one of the active rules different from (A, B). If Q_i contains A as a subword, then we must rewrite Q_i to obtain an irreducible word. If P_i contains A as a subword, then we must remove (P_i, Q_i) from the set of active rules. However, we must make sure that even with (P_i, Q_i) no longer active there is still a word derivable from both P_i and Q_i. To manage all of these revisions to the set of rules, a stack of pairs of words is used. If (U, V) is on the stack, then $U \sim V$, so there must be a word derivable from both U and V. The purpose of the subroutine TEST_2 is to clear this stack.

```
Procedure TEST_2;
Begin
   While the stack is not empty do begin
      Pop (U, V) from the stack;
      REWRITE_FROM_LEFT(U; A);
      REWRITE_FROM_LEFT(V; B);
      If A ≠ B then begin
         If A ≺ B then interchange A and B;
         n := n + 1;  P_n := A;  Q_n := B;  active[n] := true;
         For i := 1 to n − 1 do
```

    If $active[i]$ then
        If P_i contains A as a subword then
            Begin $active[i] :=$ false; Push (P_i, Q_i) onto the stack end
        Else if Q_i contains A as a subword then
            Begin REWRITE_FROM_LEFT$(Q_i; C)$; $Q_i := C$ end
    End
  End
End.

The procedure OVERLAP_1 requires only minimal changes, which reflect the fact that because the set of active rules is reduced, only proper overlaps, that is, overlaps of type (3) in Proposition 3.1, can occur.

Procedure OVERLAP_2(i, j);
Input: i, j : indices of active rules;
Begin
  $m := \min(|P_i|, |P_j|) - 1$; $k := 1$;
  While $(k \leq m)$ and $active[i]$ and $active[j]$ do begin
    Let B be the suffix of P_i of length k;
    If B is a prefix of P_j then begin
      Let $P_i = AB$ and $P_j = BC$; Push (AQ_j, Q_iC) onto the stack;
      TEST_2
    End;
    $k := k + 1$
  End
End.

Here is the second version of the Knuth-Bendix procedure for strings:

Procedure KBS_2$(X, \prec, \mathcal{R}; \mathcal{T})$;
Input: X : a finite set;
         $\prec$: a reduction ordering of X^*;
         $\mathcal{R}$: a finite subset of $X^* \times X^*$;
Output: $\mathcal{T}$: RC$(X, \prec, \mathcal{R})$, if it is finite;
($*$ WARNING – TERMINATION MAY NOT OCCUR. $*$)
Begin
  $n := 0$; $i := 1$;
  For (U, V) in $\mathcal{R}$ do begin push (U, V) on the stack; TEST_2 end;
  While $i \leq n$ do begin
    $j := 1$;
    While $j \leq i$ and $active[i]$ do begin
      If $active[j]$ then begin
        OVERLAP_2(i, j);

If $j < i$ and $active[i]$ and $active[j]$ then OVERLAP_2(j,i)
End;

$\quad j := j+1$
End;

$\quad i := i+1$
End;

Let T be the set of currently active rules
End.

Proposition 6.1. *If* RC$(X, \prec, \mathcal{R})$ *is finite, then* KBS_2$(X, \prec, \mathcal{R}; T)$ *terminates with* $T = $ RC$(X, \prec, \mathcal{R})$.

Proof. This proposition is more difficult to prove than Proposition 5.1, since rules may be modified or deleted. The proof given here is based heavily on [Le Chenadec 1986].

All changes to the set of active rules take place in TEST_2. A rule (P_i, Q_i) is made inactive only when a new rule (P_n, Q_n) is found such that P_i contains P_n as a proper subword. Thus the ideal of X^* generated by the left sides of the active rules increases each time a new rule is added and is not affected by deletions or modifications of existing rules.

Lemma 6.2. *Calls to* TEST_2 *terminate.*

Proof. Suppose at the beginning of a call to TEST_2 there are r active rules and s pairs on the stack. Since pairs are added to the stack only after the deletion of an active rule, the number of active rules at any time during this execution of TEST_2 cannot exceed $r+s$. By Proposition 2.7 of Chapter 1, the ideal generated by the left sides of the active rules stabilizes at some point in this execution of TEST_2. But then no new rules are created, so the stack must eventually become empty. $\square$

Let $\sim$ be the congruence generated by $\mathcal{R}$. The operation of TEST_2 does not change the congruence generated by the active rules and the pairs on the stack. Thus after the completion of the For-loop in KBS_2, at any point in the execution of KBS_2 outside TEST_2 the set of active rules generates $\sim$.

Let S be the set of all pairs which occur as active rules at some point during the operation of KBS_2 and let T be the subset of those elements of S which are never deleted or modified. If KBS_2 terminates, then T is the rewriting system returned by KBS_2. Let (R, S) and (T, U) be distinct elements of T. From some point on, both of these rules are active. Since the set of active rules is reduced, neither R nor S contains T as a subword. Therefore T is reduced.

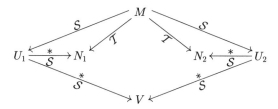

Figure 2.6.1

Lemma 6.3. *Suppose M, N_1, and N_2 are words in X^* such that $M \xrightarrow{T} N_i$, $i = 1, 2$. Then there are words U_1, U_2, and V such that for $i = 1, 2$ we have $M \xrightarrow{S} U_i \xrightarrow{*}{S} N_i$ and $U_i \xrightarrow{*}{S} V$.*

Proof. The statement of the lemma corresponds to Figure 2.6.1. For $i = 1$, 2 there are words A_i, B_i, L_i, and R_i such that $M = A_i L_i B_i$, (L_i, R_i) is in T, and $N_i = A_i R_i B_i$. If the occurrences of L_1 and L_2 in M do not overlap, then we may assume that $A_1 L_1$ is a prefix of A_2. In this case $M = A_1 L_1 C L_2 B_2$ for some word C, and we may take $U_i = N_i$ and $V = A_1 R_1 C R_2 B_2$.

Suppose that the occurrences of L_1 and L_2 do overlap. We may assume that $M = A_1 ABCB_2$, where $L_1 = AB$, $L_2 = BC$, and $B \neq \varepsilon$. At some point in the operation of KBS_2 the overlap ABC was considered. Suppose at that time (L_1, S_1) and (L_2, S_2) were active rules. Because of the call to OVERLAP_2, there is a word T such that $S_1 C \xrightarrow{*}{S} T$ and $A S_2 \xrightarrow{*}{S} T$. Let $U_1 = A_1 S_1 C B_2$, $U_2 = A_1 A S_2 B_2$, and $V = A_1 T B_2$. Then for $i = 1, 2$ we have $M \xrightarrow{*}{S} U_i \xrightarrow{*}{S} V$. Since $S_i \xrightarrow{*}{S} R_i$, we also have $U_i \xrightarrow{*}{S} N_i$. □

Lemma 6.4. *For all words M, N, N_1, and N_2 in X^*, the following hold:*

(a) *If $M \xrightarrow{*}{S} N$, then there is a word V such that $M \xrightarrow{*}{T} V$ and $N \xrightarrow{*}{T} V$.*
(b) *If $M \xrightarrow{*}{T} N_1$ and $M \xrightarrow{*}{T} N_2$, then there is a word V such that $N_1 \xrightarrow{*}{T} V$ and $N_2 \xrightarrow{*}{T} V$.*
(c) *If $M \xrightarrow{*}{S} N_1$ and $M \xrightarrow{*}{S} N_2$, then there is a word V such that $N_1 \xrightarrow{*}{T} V$ and $N_2 \xrightarrow{*}{T} V$.*

Proof. We shall prove (a), (b), and (c) simultaneously by induction. Let M be the first word with respect to $\prec$ at which one of these statements fails.

Suppose (a) fails for M and some word N. Then $N \neq M$, since otherwise we could take $V = M$. Let $M \xrightarrow{S} M_1 \xrightarrow{*}{S} N$ be the derivation of N from M. Then there are words A, B, L, and R such that $M = ALB$, $M_1 = ARB$, and (L, R) is in S. Among the counterexamples to (a), choose one where L is minimal. The sequence of words S such that (L, S) is an active rule is a nondecreasing sequence. Let S be the last such word. Then either (L, S)

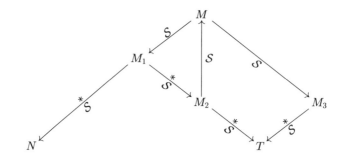

Figure 2.6.2

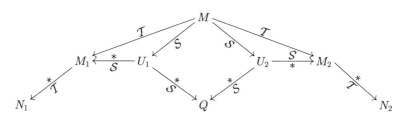

Figure 2.6.3

is in $\mathcal{T}$ or at some point (L, S) is made inactive and placed on the stack. Let $M_2 = ASB$.

Suppose first that (L, S) is in $\mathcal{T}$. Then $M \xrightarrow{*}{\mathcal{T}} M_2$. Since $R \xrightarrow{*}{\mathcal{S}} S$, we have $M_1 \xrightarrow{*}{\mathcal{S}} M_2$. By the minimality of M, (c) holds for the triple M_1, N, M_2. Thus there is a word V such that $N \xrightarrow{*}{\mathcal{T}} V$ and $M_2 \xrightarrow{*}{\mathcal{T}} V$. Since $M \xrightarrow{*}{\mathcal{T}} M_2$, statement (a) holds.

Suppose that (L, S) is not in $\mathcal{T}$. Then at some point (L, S) was placed on the stack. Sometime later, (L, S) was removed from the stack and a rule was added, if necessary, so that there is a word W derivable from both L and S using rules in $\mathcal{S}$ whose left sides are strictly less than L. Let (P, Q) be the first rule used in one derivation of W from L. Then $L = CPD$, where $CD \neq \varepsilon$. Set $M_3 = ACQDB$ and $T = AWB$. Then $M_2 \xrightarrow{*}{\mathcal{S}} T$ and $M_3 \xrightarrow{*}{\mathcal{S}} T$. At this point, the picture is as in Figure 2.6.2. By (c) applied to M_1, N, T, there is a word E such that $N \xrightarrow{*}{\mathcal{T}} E$ and $T \xrightarrow{*}{\mathcal{T}} E$. By our assumption on L, statement (a) holds for M and E, since we can derive E from M by the route $M \xrightarrow{}{\mathcal{S}} M_3 \xrightarrow{*}{\mathcal{S}} T \xrightarrow{*}{\mathcal{T}} E$. The first step uses the rule (P, Q) and $P \prec L$. Therefore there is a word V such that $M \xrightarrow{*}{\mathcal{T}} V$ and $E \xrightarrow{*}{\mathcal{T}} V$. But then $N \xrightarrow{*}{\mathcal{T}} V$, and (a) holds.

Now suppose that (b) fails for M, N_1, and N_2. Clearly we may assume that M, N_1, and N_2 are distinct. For $i = 1, 2$ let $M \xrightarrow{}{\mathcal{T}} M_i$ be the first step in deriving N_i from M using $\mathcal{T}$. By Lemma 6.3, there are words U_1, U_2, and Q such that the derivations in Figure 2.6.3 hold.

By (c) applied to U_1, M_1, and Q, there is a word R such that $M_1 \xrightarrow[\mathcal{T}]{*} R$ and $Q \xrightarrow[\mathcal{T}]{*} R$. By (c) applied to U_2, R, and M_2, there is a word S such that $R \xrightarrow[\mathcal{T}]{*} S$ and $M_2 \xrightarrow[\mathcal{T}]{*} S$. Thus $M_1 \xrightarrow[\mathcal{T}]{*} S$. By (b) applied to M_1, N_1, and S, there is a word T such that $N_1 \xrightarrow[\mathcal{T}]{*} T$ and $S \xrightarrow[\mathcal{T}]{*} T$. Therefore $M_2 \xrightarrow[\mathcal{T}]{*} T$. Finally by (b) applied to M_2, N_2, and T, there is a word V such that $N_2 \xrightarrow[\mathcal{T}]{*} V$ and $T \xrightarrow[\mathcal{T}]{*} V$. Since $N_1 \xrightarrow[\mathcal{T}]{*} V$, we see that (b) holds for M, N_1, and N_2.

Now suppose that (c) fails for M, N_1, and N_2. Clearly $N_1 \neq N_2$. If $M = N_1$ or $M = N_2$, then we can apply (a). Thus we may assume that M, N_1, and N_2 are distinct. By (a), there are words V_1 and V_2 such that $M \xrightarrow[\mathcal{T}]{*} V_i$ and $N_i \xrightarrow[\mathcal{T}]{*} V_i$, $i = 1, 2$. By (b) applied to M, V_1, and V_2, there is a word V such that $V_i \xrightarrow[\mathcal{T}]{*} V$, $i = 1, 2$. Then $N_i \xrightarrow[\mathcal{T}]{*} V$, so (c) holds. $\square$

By part (b) of Lemma 6.4, the system $\mathcal{T}$ is confluent. Since $\mathcal{T} \subseteq \mathcal{S}$, by part (c) of the lemma $\mathcal{S}$ is confluent. A word is irreducible with respect to $\mathcal{S}$ if and only if it is irreducible with respect to $\mathcal{T}$. Therefore the congruences generated by $\mathcal{S}$ and $\mathcal{T}$ have the same canonical forms. Hence $\mathcal{S}$ and $\mathcal{T}$ generate the same congruence, namely $\sim$. Since $\mathcal{T}$ is reduced, $\mathcal{T}$ is $\mathrm{RC}(X, \prec, \mathcal{R})$. Therefore $\mathcal{T}$ is finite. Once all the rules in $\mathcal{T}$ become active, no new rules are added, and the While-loop eventually terminates. Thus KBS_2 terminates and returns $\mathrm{RC}(X, \prec, \mathcal{R})$. $\square$

The difference between the operation of KBS_1 and the operation of KBS_2 is not large for the small examples discussed in Section 2.5.

Example 6.1. With the data of Example 5.1, the procedure KBS_2 constructs a total of 12 rules, all with different left sides. The maximum number active at any one time is 10, and the stack never has more than two pairs on it.

Example 6.2. Using the data of Example 5.4, KBS_2 constructs a total of 17 rules, all with different left sides. This is one less than the number produced by KBS_1. With KBS_2, no more than 11 rules are active at any one time, and the stack never has more than two pairs on it. Thus the storage requirements for a version of KBS_2 which does not retain inactive rules would be less on this example than the storage needs of KBS_1.

Example 6.3. On the data in Example 5.5, KBS_2 generates 34 rules, two of which have the same left side. The maximum number active is 20, and stack never has more than two pairs on it.

Example 6.4. Let $X = \{a, b, b^{-1}\}$, let $\prec$ be the length-plus-lexicographic ordering of X^* with $a \prec b \prec b^{-1}$, and let $\mathcal{R}$ consist of the following pairs:

$$a^2 \to \varepsilon, \quad bb^{-1} \to \varepsilon, \quad b^3 \to \varepsilon, \quad (ab)^7 \to \varepsilon, \quad (abab^{-1})^4 \to \varepsilon.$$

The system $\mathrm{RC}(X, \prec, \mathcal{R})$ has 40 rules. The procedure KBS_1 constructs a total of 81 rules in computing $\mathrm{RC}(X, \prec, \mathcal{R})$. With KBS_2 there are 89 rules generated involving 85 different left sides. However, there are never more than 40 rules active at one time, and the stack never has more than three entries.

Example 6.5. If $\prec$ is a length-plus-lexicographic ordering and $\mathrm{Mon}\,\langle X \mid \mathcal{R} \rangle$ is a group, then the left and right sides of the rules in $\mathrm{RC}(X, \prec, \mathcal{R})$ will have roughly equal lengths. The initial presentation in Example 6.4 does not satisfy this condition. The following presentation is equivalent and more balanced:

$$a^2 \to \varepsilon, \quad bb^{-1} \to \varepsilon, \quad b^2 \to b^{-1}, \quad (b^{-1}a)^3 b^{-1} \to (ab)^3 a,$$
$$(bab^{-1}a)^2 \to (abab^{-1})^2.$$

With this presentation, KBS_1 constructs a total of 56 rules. Using KBS_2, 59 rules with 57 different left sides are constructed, but no more than 40 are active at one time.

Example 6.6. The presentation $\mathcal{R}$ consisting of the pairs

$$a^2 \to \varepsilon, \quad bb^{-1} \to \varepsilon, \quad b^3 \to \varepsilon, \quad (ab)^7 \to \varepsilon, \quad (abab^{-1})^8 \to \varepsilon$$

is frequently used to test programs for studying finitely presented groups. Let $X = \{a, b, b^{-1}\}$ and let $\prec$ be the length-plus-lexicographic ordering of X^* with $a \prec b \prec b^{-1}$. The system $\mathcal{T} = \mathrm{RC}(X, \prec, \mathcal{R})$ has 1026 rules with the longest left side having length 37. The procedure KBS_2 is not really adequate to compute $\mathcal{T}$. To see the problem, consider the following facts: In $\mathcal{T}$ there are 143 left sides which start with a, 457 which start with b, and 426 which start with b^{-1}. There are 444 left sides which end with a, 296 which end with b, and 286 which end with b^{-1}. To prove that $\mathcal{T}$ is confluent, the number of overlaps ABC with $|B| = 1$ which must be considered is

$$444 \times 143 + 296 \times 457 + 286 \times 426 = 320600.$$

The number of overlaps with $|B| = 2$ is 235979. Processing all of these overlaps will take a long time. It turns out that KBS_2 generates many rules which later become redundant, so the running time to determine $\mathcal{T}$ is longer than the time needed to show that $\mathcal{T}$ is confluent. See Section 3.5.

Although KBS_2 is a substantial improvement over KBS_1 and can be practical for small problems, KBS_2 still has weaknesses. One of the biggest weaknesses is the crude manner in which REWRITE_FROM_LEFT searches for a left side which is a suffix of V. This can be improved with an indexing structure for the left sides, such as described in Section 3.5. However, there are other problems as well. One is illustrated in Example 6.6. Another is that KBS_2 does not work hard enough to derive short rules quickly. Suppose that a rule of the form (xP, xQ) has been found and generator x is a unit in $\mathrm{Mon}\,\langle X \mid \mathcal{R} \rangle$. Then $P \sim Q$ and the rule (P, Q) should be immediately deduced. However, it may be a long time before KBS_2 finds this rule. The next section discusses a number of possibilities for further improvements in the Knuth-Bendix procedure.

Exercise

6.1. Apply KBS_2 to the presentations $a^2 \to \varepsilon$, $b^3 \to \varepsilon$, $(ab)^i \to \varepsilon$, $2 \le i \le 5$, and compare the rules produced during the computation with those produced by KBS_1.

2.7 Some useful heuristics

It is impossible to pursue deeply questions about the efficiency of programs implementing the Knuth-Bendix procedure for strings without a discussion of the data structure to be used to store rules. Most of the time used by a computer program executing the Knuth-Bendix procedure is spent in rewriting. The manner in which the active rules are stored and the nature of any other information about the rules which is kept can greatly affect the time needed to rewrite words. It can also affect the time needed to locate overlaps. Because the set of rules is continually changing, it must be possible to update the stored data concerning the rules with as little overhead as possible. Many interesting problems strain both the available cpu and memory resources. A careful analysis is required in order to get a good balance between speed and memory usage.

One way to gain some speed in the rewriting process while not sacrificing very much memory is to sort the rules according to the right-to-left lexicographic order of their left sides. The lexicographic ordering is used here no matter what reduction ordering is being used to order the left and right sides of rules. To see whether a particular word V has a left side as a suffix, one looks V up in this dictionary of left sides using the technique known as binary search. The main drawback of this approach is that when a new rule is found, the entire set of rules must be resorted. Automata, the subject of Chapter 3, provide index structures for the set of rules which permit even faster rewriting and much easier updating. The problem with automata is that they take a substantial amount of space, so their use reduces the size of the rewriting systems which can be stored. Computer scientists have developed a number of other techniques which can be used to design indexes. See, for example, the chapters on searching and string matching in [Sedgewick 1990]. All of these methods have both strengths and weaknesses. Deciding which one is best for a particular problem is often difficult.

No matter what data structure is used for rules, a considerable amount of work is needed to keep the set of active rules reduced when a new rule is found. It is often useful to delay removing redundant rules until several new rules have been found.

In addition to deciding on the data structure to be used to store rules, someone designing a program to carry out the Knuth-Bendix procedure should consider using several heuristics which have been found helpful by others. A word of caution is in order, however. There probably is no such thing as the best Knuth-Bendix program. Given two different programs A and B, it is likely that one can find two sets of input data such that program A does better with the first set and program B does better with the second.

The first two heuristics concern the formation of overlaps. In KBS_1 and KBS_2, the order in which the rules are found completely determines the order in which overlaps are formed. Frequently it is useful to give preference to forming short overlaps. This can be done in several ways. If one is looking for overlaps of left sides AB and BC, one can use such measures as $|ABC|$, $|AB| + |BC|$, and $\max(|AB|, |BC|)$ to define shortness and select short overlaps before long ones. For example, if KBS_2 is run on the data of Example 6.6 and is told to ignore all overlaps ABC with $|ABC| > 44$, it will still construct $\mathrm{RC}(X, \prec, \mathcal{R})$, although it will not have produced a proof of confluence.

One can also use knowledge about the existence of inverses to guide the formation of overlaps, as suggested in [Le Chenadec 1986]. Suppose that a new rule $P \rightarrow Q$ is found, and $P = Ax$, where x is in X, and in the monoid being investigated $[x]$ has a right inverse. Thus there is a rule $xU \rightarrow \varepsilon$, which either is part of the original presentation or has been discovered by the Knuth-Bendix procedure. It seems to be a good idea to give a very high priority to processing the overlap AxU:

$$A\underline{x}U \rightarrow A, \quad \underline{Ax}U \rightarrow QU.$$

Let S be the result of rewriting QU. If $S \prec A$, then the rule $A \rightarrow S$ can be substituted for $P \rightarrow Q$. If $S \succ A$, then we have a new rule $S \rightarrow A$. A special case of this heuristic occurs when Q also ends in x. If $Q = Bx$, then $P \rightarrow Q$ can be replaced immediately by $A \rightarrow B$.

There is a similar heuristic in case a new left side starts with a generator whose image in the monoid has a left inverse. Also, if $P \rightarrow Q$ is a new rule and all the generators in P and Q define units, then the words P^{-1} and Q^{-1} represent the same element of the monoid. Rewriting them can give a useful new rule. (Here P^{-1} is formed by reversing the terms of P and then replacing each term x by a word representing the inverse of $[x]$.)

The procedure BALANCE listed below describes one way to implement heuristics based on the ideas of Le Chenadec. BALANCE is designed to be part of a modified version of KBS_2. The following additional data are assumed available: a subset Y of X of generators known to define units in Mon $\langle X \mid \mathcal{R} \rangle$, for each element y of Y a word U_y defining the inverse of $[y]$, and an integer m which is used to limit the application of the heuristics. Without some type of limit, the use of BALANCE may lead to a failure of termination, even when RC$(X, \prec, \mathcal{R})$ is finite.

The input to BALANCE consists of the left and right sides of a rule which has just been discovered, probably by processing an overlap. The purpose of BALANCE is to try to find a rule which is shorter or more balanced in the sense that its left and right sides have more nearly equal lengths. One rule is returned explicitly. Any other rules produced are pushed onto the stack to be processed later. Although the procedure is valid with any ordering, its use is most natural when the ordering involved is consistent with length.

```
Procedure BALANCE(A, B; C, D);
Input:    A, B      : words, the left and right sides of a new rule;
Output:   C, D      : words, the left and right sides of a possibly more
                       balanced rule;
(* As a side effect, additional pairs of words may be pushed onto the
     stack. *)
Begin
  C := A;  D := B;

  (* Execution of the following "infinite loop" is controlled by the
       Continue-statement and Break-statement. *)

  While true do begin
    Let C = Lx with x in X;

    If x is in Y then
      If D ends in x then
        Begin Let D = Mx;  C := L;  D := M;  Continue end
      Else if |C| + |Uₓ| ≤ m then begin
        REWRITE_FROM_LEFT(DUₓ; P);

        If P ≺ L then
          Begin C := L;  D := P;  Continue end
        Else if P ≻ L then push (P, L) onto the stack
      End;

    Let C = xL with x in X;

    If x is in Y then
      If D starts with x then
        Begin Let D = xM;  C := L;  D := M;  Continue end
      Else if |C| + |Uₓ| ≤ m then begin
        REWRITE_FROM_LEFT(UₓD; P);
```

If $P \prec L$ then
    Begin $C := L$; $D := P$; Continue end
    Else if $P \succ L$ then push (P, L) onto the stack
  End;
   Break
  End
End.

Example 7.1. To illustrate the use of BALANCE, let us find the reduced, confluent rewriting system for the presentation in Example 5.3 using KBS_2 with one small change to TEST_2. In TEST_2 replace the lines

If $A \prec B$ then interchange A and B;
$n := n + 1$; $P_n := A$; $Q_n := B$;

with the following statements:

If $A \prec B$ then interchange A and B;
BALANCE$(A, B; C, D)$;
$n := n + 1$; $P_n := C$; $Q_n := D$;

In this example, $X = \{a, b\}$, and $\mathcal{R}$ consists of the rules

$$a^2 \to \varepsilon, \quad b^3 \to \varepsilon, \quad (ab)^3 \to \varepsilon.$$

We set $Y = X$ and define $U_a = a$ and $U_b = b^2$. The integer m is taken to be large, say $m = 100$. The first step in KBS_2 is to apply TEST_2 to each of the rules in $\mathcal{R}$. With the first two rules, nothing is changed by the call to BALANCE. For example, the call BALANCE$(a^2, \varepsilon; C, D)$ simply looks at the equality $a = a$ twice and returns C as a^2 and D as ε. However, with the third input rule there is a change. When the call BALANCE$((ab)^3, \varepsilon; C, D)$ is made, C is initially set to $(ab)^3$, and D to ε. Then the rightmost b of C is moved to the right end of D as b^2. Since b^2 is irreducible and $ababa \succ b^2$, we set C equal to $ababa$ and D equal to b^2. Now the rightmost a is moved to the right end of D. Again no rewriting is possible, and the order has not been reversed. Thus C is set equal to $abab$, and D to b^2a. Next the rightmost b of C is moved to the right end of D as b^2. This gives us the pair of words aba and b^2ab^2. These are irreducible, but the second is larger than the first. This time we do not change C and D. Instead we push (b^2ab^2, aba) onto the stack. Now the a at the left end of C is moved to the left end of D. This leads to (ab^2a, bab) being pushed onto the stack. No further changes are made on this call to BALANCE. The values $C = (ab)^2$ and $D = b^2a$ are returned to TEST_2, and P_3 and Q_3 are set equal to these words, respectively.

Now TEST_2 pops (ab^2a, bab) off the stack, and the call BALANCE $(ab^2a, bab; C, D)$ is made. Initially $C = ab^2a$ and $D = bab$. When the right a of C is moved to the right end of D, the pair $(baba, ab^2)$ is pushed onto the stack. When the left a of C is moved to the left end of D, the word $abab$ is reduced to b^2a, so we have $P = L$, and nothing happens. The rule (ab^2a, bab) is returned to TEST_2, and this becomes the fourth rule. Now $(baba, ab^2)$ is popped off the stack. On the call to BALANCE with this rule, (b^2ab^2, aba) is again pushed onto the stack. (It would be possible to do some extra bookkeeping and prevent duplicate pairs from being placed on the stack.) The rule $(baba, ab^2)$ is returned unchanged and it becomes the fifth rule. Now (b^2ab^2, aba) is popped off the stack. No changes are made to this rule and no more pairs are added to the stack. The sixth rule becomes (b^2ab^2, aba). Now the last pair (b^2ab^2, aba) is popped off the stack. It is immediately reduced to the equality $aba = aba$, and the call to TEST_2 terminates. Thus at this point we have found all the rules in the final set without storing any redundant ones. The overlaps must still be processed, but no more rules are found.

The next observation is more than a heuristic. It is an improved version of Proposition 3.1, which tells us that certain overlaps can be skipped.

Proposition 7.1. *Let $\mathcal{R}$ be a rewriting system on X^* relative to a reduction ordering $\prec$. Suppose that for all (P, Q) in $\mathcal{R}$ the word P is irreducible with respect to $\mathcal{R} - \{(P, Q)\}$. Assume that $\mathcal{R}$ is not confluent and let W be the first word, with respect to $\prec$, at which confluence fails. Then local confluence fails at W, and $W = ABC$, where $B \neq \varepsilon$, AB and BC are left sides in $\mathcal{R}$, and there are no other occurrences of left sides in W.*

Proof. It is easy to see that local confluence must fail at W. By Proposition 3.1 and our assumption on $\mathcal{R}$, the word W has the form ABC, where $B \neq \varepsilon$, there exist rules $AB \to U$ and $BC \to V$ in $\mathcal{R}$, and there is no word derivable from both AV and UC. Suppose that W contains another occurrence of a left side P. Then P cannot be a subword of AB or BC, nor can these words be subwords of P. Thus $W = DEBFG$, where $A = DE$, $C = FG$, $P = EBF$, and D, E, F, and G are all nonempty.
Let $P \to Q$ be in $\mathcal{R}$. We have the following reductions:

$$\underline{DEBF} \to UF, \quad D\underline{EBF} \to DQ,$$
$$\underline{EBFG} \to QG, \quad E\underline{BFG} \to EV.$$

By the minimality of W, there are words S and T such that $UF \xrightarrow{*}{\mathcal{R}} S$, $DQ \xrightarrow{*}{\mathcal{R}} S$, $QG \xrightarrow{*}{\mathcal{R}} T$, and $EV \xrightarrow{*}{\mathcal{R}} T$. Thus $DQG \xrightarrow{*}{\mathcal{R}} SG$ and $DQG \xrightarrow{*}{\mathcal{R}} DT$. Again by the minimality of W, there is a word L such that $SG \xrightarrow[\mathcal{R}]{*} L$ and $DT \xrightarrow[\mathcal{R}]{*} L$. But then

$$UC = UFG \xrightarrow[\mathcal{R}]{*} SG \xrightarrow[\mathcal{R}]{*} L,$$
$$AV = DEV \xrightarrow[\mathcal{R}]{*} DT \xrightarrow[\mathcal{R}]{*} L,$$

contradicting the assumption that no word is derivable from both UC and AV. $\square$

A good way to use Proposition 7.1 is to find overlaps by a backtrack search. For example, one could start with a left side P and then undertake a backtrack search for words C such that C is irreducible and PC contains exactly two left sides in $\mathcal{R}$ as subwords, one of which is P and the other is a suffix of PC. See Section 3.5. Proposition 7.1 makes it practical to prove the confluence of the rewriting system $\mathcal{T}$ of Example 6.6.

Exercises

7.1. Work through Example 5.1 using the modification to KBS_2 described in Example 7.1. Take $Y = X$, $U_a = a^{-1}$, $U_{a^{-1}} = a$, $U_b = b^{-1}$, $U_{b^{-1}} = b$, and $m = 100$.

7.2. Let X, $\prec$, and $\mathcal{R}$ be as in Proposition 7.1 and let $\mathcal{W}$ be a nonempty subset of X^* which is closed under taking subwords and closed under rewriting with respect to $\mathcal{R}$. Suppose confluence fails at some word in $\mathcal{W}$. Let W be the first word in $\mathcal{W}$ with respect to $\prec$ at which confluence fails. Show that the conclusion of Proposition 7.1 holds for W.

2.8 Right congruences

So far in this chapter we have been studying two-sided congruences. The rewriting-system approach can also be useful with right or left congruences. We shall concentrate on right congruences here, leaving it to the reader to make the obvious modifications for left congruences.

Let $\sim$ be a right congruence on X^* and let $\prec$ be a reduction ordering. Every $\sim$-class has a first element, which we call its *canonical form*. Corresponding to Proposition 2.1, we have:

Proposition 8.1. *If U is a canonical form, then every prefix of U is a canonical form.*

Proof. Exercise. $\square$

If $P \sim Q$ and $P \succ Q$, then we may replace P by Q in a word U and still have an equivalent word only when P occurs as a prefix of U. Thus we have the notion of rewriting in the context of right congruences, but the rewriting takes place only at the left ends of words.

The following proposition allows us to transfer most of our work on two-sided congruences to right congruences.

Proposition 8.2. *Let X be a set and let $\#$ be an object not in X. Set $Y = X \cup \{\#\}$. Suppose $\mathcal{S}$ is a subset of $X^* \times X^*$ and let $\sim$ be the right congruence on X^* generated by $\mathcal{S}$. Let $\mathcal{T} = \{(\#U, \#V) \mid (U, V) \in \mathcal{S}\}$ and let $\equiv$ be the two-sided congruence on Y^* generated by $\mathcal{T}$. Then the $\sim$-classes are in one-to-one correspondence with the $\equiv$-classes in $\#X^*$.*

Proof. The set $\#X^*$ is the set of all words in Y^* which begin with $\#$ and have no other occurrences of $\#$.

Lemma 8.3. *The set $\#X^*$ is a union of $\equiv$-classes.*

Proof. Let U be in X^* and let V be in Y^* with $\#U \equiv V$. Then there is a sequence

$$\#U = U_0, U_1, \ldots, U_m = V$$

of words in Y^* such that for $0 \le i < m$ the word U_{i+1} is obtained from U_i by replacing a subword P with another word Q, where (P, Q) or (Q, P) is in $\mathcal{T}$. Now any such P starts with $\#$ and has no other occurrences of $\#$. Since U_0 has the same form, U_1 must be in $\#X^*$. By induction on m, it follows that V is in $\#X^*$. $\square$

With a slight extension of the argument used in the proof of Lemma 8.3, it can be shown that for all words A and B of X^* we have $A \sim B$ if and only if $\#A \equiv \#B$.

Let us continue the notation of Proposition 8.2.

Proposition 8.4. *Let $\prec$ be a reduction ordering on Y^*. If $\mathcal{S}$ is finite, then $\mathrm{KBS\_2}(Y, \prec, \mathcal{T}; \mathcal{U})$ terminates.*

Proof. Let $\mathcal{V}$ be a rewriting system with respect to $\prec$ contained in $(\#X^*) \times (\#X^*)$. Suppose A and B are left sides in $\mathcal{V}$. Since A and B start with $\#$ and have no other occurrences of $\#$, the only way to have an overlap of these words is for one to be a prefix of the other. Thus by Proposition 3.1, if $\mathcal{V}$ is reduced, then $\mathcal{V}$ is confluent. As noted in the proof of Proposition 6.1, the first For-loop of KBS_2 always terminates with a reduced rewriting system. In the case of the call $\mathrm{KBS\_2}(Y, \prec, \mathcal{T}; \mathcal{U})$, that system will be confluent, so the While-loop will add no new rules and KBS_2 will terminate. $\square$

Let us say that a rewriting system $\mathcal{S}$ on X^* is *prefix confluent* if the corresponding system $\mathcal{T}$ on Y^* is confluent, and *prefix reduced* if $\mathcal{T}$ is reduced. As noted in the proof of Proposition 8.4, if $\mathcal{S}$ is prefix reduced, then $\mathcal{S}$ is prefix confluent.

Corollary 8.5. *Let $\prec$ be a reduction ordering on X^* and let $\sim$ be a finitely generated right congruence on X^*. There is a unique prefix reduced rewriting system $\mathcal{U}$ with respect to $\prec$ such that $\mathcal{U}$ generates $\sim$. Moreover, $\mathcal{U}$ is finite.*

Proof. Let $\mathcal{S}$ be a finite generating set for $\sim$. Form the set $\mathcal{T}$ as in Proposition 8.4 and call KBS_2$(Y, \prec, \mathcal{T}; \mathcal{V})$. Here $\prec$ is any extension to Y^* of the given ordering on X^*. Let $\mathcal{U} = \{(P, Q) \mid (\#P, \#Q) \in \mathcal{V}\}$. Then $\mathcal{U}$ is finite and prefix reduced, and $\mathcal{U}$ generates $\sim$. The uniqueness of $\mathcal{U}$ is established as in the proof of Proposition 2.7. $\square$

The preceding results show that we can handle finitely generated right congruences quite well. However, the right congruences which occur in practice often are not finitely generated, although they are finitely generated modulo some two-sided congruence.

Let $\cong$ be a two-sided congruence on X^* and let M be the quotient monoid of X^* modulo $\cong$. If we have a right congruence $\approx$ on M, then we can pull it back to a right congruence $\sim$ on X^* by defining $U \sim V$ to mean $[U] \approx [V]$, where $[U]$ denotes the $\cong$-class containing U. The right congruence $\approx$ on M is finitely generated if and only if there is a finite subset $\mathcal{S}$ of $X^* \times X^*$ such that $\sim$ is generated by $\mathcal{S}$ and $\cong$. If such a set $\mathcal{S}$ exists and $\cong$ is generated as a two-sided congruence by a finite subset $\mathcal{R}$ of $X^* \times X^*$, then the pair $(\mathcal{R}, \mathcal{S})$ is a finite description of $\sim$, even though $\sim$ may not be finitely generated as a right congruence.

Proposition 8.6. *Let $\mathcal{R}$ and $\mathcal{S}$ be subsets of $X^* \times X^*$, let $\cong$ be the congruence generated by $\mathcal{R}$, and let $\sim$ be the right congruence generated by $\cong$ and $\mathcal{S}$. Set $Y = X \cup \{\#\}$ and $\mathcal{T} = \{(\#P, \#Q) \mid (P, Q) \in \mathcal{S}\}$. The $\sim$-classes are in one-to-one correspondence with the classes in $\#X^*$ of the two-sided congruence $\equiv$ on Y^* generated by $\mathcal{R} \cup \mathcal{T}$.*

Proof. If U is in X^*, then replacing a subword P in $\#U$ with Q, where (P, Q) or (Q, P) is in $\mathcal{R} \cup \mathcal{T}$, yields a word in $\#X^*$. Thus $\#X^*$ is a union of $\equiv$-classes. Similarly, if V is in X^*, then $U \sim V$ if and only if $\#U \equiv \#V$. $\square$

Suppose in the notation of Proposition 8.6 that $\prec$ is a reduction ordering on Y^* such that $\mathcal{U} = \mathrm{RC}(Y, \prec, \mathcal{R} \cup \mathcal{T})$ is finite. Then we may use the Knuth-Bendix procedure to compute $\mathcal{U}$ and compute canonical forms for $\sim$ by rewriting with respect to $\mathcal{U}$.

Example 8.1. Let $X = \{a, b\}$, $\mathcal{R} = \{(ba, ab)\}$, and $\mathcal{S} = \{(ab^4, a^3b^2), (a^7, ab^3)\}$. Set $Y = \{a, b, \#\}$ and let $\prec$ be the length-plus-lexicographic ordering of Y^* with $\# \prec a \prec b$. Then KBS_2$(Y, \prec, \mathcal{R} \cup \mathcal{T}; \mathcal{U})$ terminates with $\mathcal{U}$ as the following rewriting system:

$$ba \rightarrow ab, \qquad \#a^3b^4 \rightarrow \#a^5b^2,$$
$$\#ab^4 \rightarrow \#a^3b^2, \qquad \#a^4b^4 \rightarrow \#a^6b^2,$$
$$\#a^2b^4 \rightarrow \#a^4b^2, \qquad \#a^5b^4 \rightarrow \#a^3b^3,$$
$$\#a^7 \rightarrow \#ab^3, \qquad \#a^6b^4 \rightarrow \#a^4b^3.$$

To compute the canonical form for $U = ba^2b^3a^2b^2a^3b$, we rewrite $\#U$ with respect to $\mathcal{U}$. Using just the rule $ba \rightarrow ab$, we have $\#U \xrightarrow{*}_{\mathcal{R}} \#a^7b^7$. Then we have

$$\#a^7b^7 = \#aaaaaaabbbbbbb \rightarrow \#abbbbbbbbbb \rightarrow \#aaabbbbbbbb \rightarrow$$
$$\#aaaaabbbbbb \rightarrow \#aaabbbbb \rightarrow \#aaaaabbb.$$

Thus the canonical form for U is a^5b^3. Note that $\sim$ has infinitely many classes. The words $\#b^i$ are all irreducible with respect to $\mathcal{U}$, so $\varepsilon, b, b^2, \ldots$ are all canonical forms for $\sim$.

Example 8.2. Let X, $\prec$, and $\mathcal{R}$ be as in Example 8.1, and let $\mathcal{S} = \{(ab^4, a^3b^2)\}$. Since $ab^4 \sim a^3b^2$, for $i \geq 1$ we have

$$a^ib^4 \sim ab^4a^{i-1} \sim a^3b^2a^{i-1} \sim a^{i+2}b^2.$$

It is not hard to show that the set of rules consisting of $ba \rightarrow ab$ and $\#a^ib^4 \rightarrow \#a^{i+2}b^2, i = 1, 2, \ldots$, is reduced and confluent. Thus the canonical forms for $\sim$ with respect to $\prec$ are the words a^ib^j with $j < 4$ or $i = 0$.

We shall now investigate an important special case of the situation described in Proposition 8.6. To do so, we must fix a fair amount of notation:

X, a finite set,
F, the free group generated by X,
$\mathcal{R} = \mathrm{FGRel}(X)$, the standard set of monoid defining relations for F,
$\cong$, free equivalence, the congruence on $X^{\pm*}$ generated by $\mathcal{R}$,
$[U]$, the $\cong$-class containing the word U,
$Y = X^\pm \cup \{\#\}$,
$\prec$, a reduction ordering of Y^*,
$\mathcal{S}$, a finite subset of $X^{\pm*} \times X^{\pm*}$.
$\sim$, the right congruence on $X^{\pm*}$ generated by $\mathcal{S}$ and $\cong$,
H, the subgroup of F generated by the elements $[PQ^{-1}] = [P][Q]^{-1}$ with (P, Q) in $\mathcal{S}$,
$\mathcal{T} = \{(\#P, \#Q) \mid (P, Q) \in \mathcal{S}\}$.

The congruence classes of $\sim$ correspond to the right cosets of H in F.

Proposition 8.7. *If U and V are in $X^{\pm *}$, then $U \sim V$ if and only if $H[U] = H[V]$.*

Proof. As noted above, $\sim$ is the pullback to $X^{\pm *}$ of the right congruence on F generated by the image of $\mathcal{S}$ in $F \times F$. The proposition now follows from Exercise 4.1 in Chapter 1. $\square$

Let $\mathcal{C}$ be the set of canonical forms for $\sim$ with respect to $\prec$. For U in $X^{\pm *}$ let $\overline{U}$ be the element of $\mathcal{C}$ with $U \sim \overline{U}$. Finally, let $\mathcal{P}$ be the set of freely reduced words P such that P is not in $\mathcal{C}$ but every proper prefix of P is in $\mathcal{C}$, and let $\mathcal{S}_1$ be the set of pairs $(P, \overline{P})$ with P in $\mathcal{P}$.

Proposition 8.8. *The sets $\mathcal{P}$ and $\mathcal{S}_1$ are finite, and $\sim$ is generated by $\mathcal{S}_1$ and $\cong$.*

Proof. The right congruence generated by $\mathcal{S}_1$ and $\cong$ is contained in $\sim$. However, we can compute canonical forms for $\sim$ using $\mathcal{S}_1$ and $\mathcal{R}$. Given a word U in $X^{\pm *}$, we can freely reduce U using $\mathcal{R}$. If at this point U is not in $\mathcal{C}$, then there is a prefix P of U which is in $\mathcal{P}$, and $(P, \overline{P})$ is in $\mathcal{S}_1$. We can replace P by $\overline{P}$ and continue rewriting until an element of $\mathcal{C}$ is obtained. Thus $\mathcal{S}_1$ and $\cong$ generate $\sim$. Moreover, if $\mathcal{T}_1 = \{(\#P, \#Q) \mid (P, Q) \in \mathcal{S}_1\}$, then $\mathcal{T}_1 \cup \mathcal{R}$ is a confluent rewriting system on Y^*.

It remains to show that $\mathcal{P}$ and $\mathcal{S}_1$ are finite. If (P, Q) is in $\mathcal{S}_1$, then P and Q are freely reduced and P and Q do not end in the same element of $X^{\pm}$, for if $P = Au$ and $Q = Bu$ with u in $X^{\pm}$, then

$$A \sim Auu^{-1} = Pu^{-1} \sim Qu^{-1} = Buu^{-1} \sim B.$$

Since both A and B are in $\mathcal{C}$, this means $A = B$. Thus $P = Q$, which is not true. Since P and Q are freely reduced and do not end in the same generator, PQ^{-1} is freely reduced.

Lemma 8.9. *Suppose that (P, Q) and (R, S) are in $\mathcal{S}_1$ and $PQ^{-1} = RS^{-1}$. Then $P = R$ and $Q = S$.*

Proof. Either P is a prefix of R or R is a prefix of P. Since proper prefixes of P and R are in $\mathcal{C}$, we must have $P = R$. Then $Q^{-1} = S^{-1}$, so $Q = S$. $\square$

Lemma 8.10. *If u is in $X^{\pm}$ and (Au, Q) is in $\mathcal{S}_1$, then (Qu^{-1}, A) is in $\mathcal{S}_1$.*

Proof. As noted above, Qu^{-1} is freely reduced. Moreover, $Qu^{-1} \sim Auu^{-1} \sim A$ and $Qu^{-1} \neq A$. Since A is in $\mathcal{C}$, it follows that $Qu^{-1} \succ A$. Also, since Q is in $\mathcal{C}$, Qu^{-1} is in $\mathcal{P}$. Thus (Qu^{-1}, A) is in $\mathcal{S}_1$. $\square$

In Lemma 8.10, $(Qu^{-1})A^{-1} = (AuQ^{-1})^{-1}$. Therefore the set $\mathcal{H}$ of words PQ^{-1} with (P,Q) in $\mathcal{S}_1$ is closed under the map $^{-1}$. The image of $\mathcal{H}$ in F generates H. Since H is finitely generated, there is a finite subset $\mathcal{S}_2$ of $\mathcal{S}_1$ such that the set $\mathcal{H}_2$ of words PQ^{-1} with (P,Q) in $\mathcal{S}_2$ is closed under $^{-1}$ and the image of $\mathcal{H}_2$ in F generates H as a monoid. Thus $\sim$ is generated by $\mathcal{S}_2$ and $\cong$. If u is in $X^{\pm}$ and (Au,Q) is in $\mathcal{S}_2$, then $(AuQ^{-1})^{-1} = Qu^{-1}A^{-1}$ is in $\mathcal{H}_2$. By Lemmas 8.9 and 8.10, the pair (Qu^{-1}, A) is in $\mathcal{S}_2$. Let $\mathcal{T}_2 = \{(\#P, \#Q) \mid (P,Q) \in \mathcal{S}_2\}$.

Lemma 8.11. *The rewriting system $\mathcal{T}_2 \cup \mathcal{R}$ is reduced and confluent.*

Proof. Since $\mathcal{T}_2 \cup \mathcal{R}$ is a subset of the reduced system $\mathcal{T}_1 \cup \mathcal{R}$, it follows that $\mathcal{T}_2 \cup \mathcal{R}$ is reduced. Since $\mathcal{R}$ is confluent and $\mathcal{T}_2$ is reduced, we need to consider only overlaps of an element $(\#Au, \#Q)$ in $\mathcal{T}_2$ and an element uu^{-1} in $\mathcal{R}$. Processing this overlap leads to the following reductions:

$$\#A\underline{uu^{-1}} \to \#A, \quad \underline{\#Auu^{-1}} \to \#Qu^{-1} \to \#A.$$

Thus we have confluence. $\square$

Let $\equiv$ be the congruence on Y^* generated by $\mathcal{T} \cup \mathcal{R}$. Since both $\mathcal{T}_1 \cup \mathcal{R}$ and $\mathcal{T}_2 \cup \mathcal{R}$ are reduced, confluent rewriting systems on Y^* which generate $\equiv$, it follows that $\mathcal{T}_1 \cup \mathcal{R} = \mathcal{T}_2 \cup \mathcal{R} = \mathrm{RC}(Y, \prec, \mathcal{T} \cup \mathcal{R})$ is finite, as is $\mathcal{P}$. Therefore the call KBS_2$(Y, \prec, \mathcal{T} \cup \mathcal{R}; \mathcal{U})$ terminates with $\mathcal{U} = \mathcal{T}_1 \cup \mathcal{R}$. This completes the proof of Proposition 8.8. $\square$

Proposition 8.12. *Suppose that $\prec$ is consistent with length. Then $|P| \leq |Q| + 2$ for all (P,Q) in $\mathcal{S}_1$.*

Proof. Let $P = Tu$ with u in $X^{\pm}$. Then $Qu^{-1} \sim T$. Since T is in $\mathcal{C}$ and $T \neq Qu^{-1}$, we have $Qu^{-1} \succ T$. Since $\prec$ is consistent with length, $|Qu^{-1}| \geq |T|$. Thus $|Q| + 1 \geq |P| - 1$ or $|P| \leq |Q| + 2$. $\square$

The computation of $\mathcal{T}_1$ is essentially the process usually referred to as *Nielsen reduction*, at least in the case in which $\prec$ is a length-plus-lexicographic ordering. For any word U in $X^{\pm*}$, let $L(U)$ denote the length of the freely reduced word freely equivalent to U. Let $\mathcal{A}$ be a set of nonempty words in $X^{\pm*}$ which are freely reduced, and set $\mathcal{A}^{-1} = \{A^{-1} \mid A \in \mathcal{A}\}$. The set $\mathcal{A}$ is said to be *Nielsen reduced* if the following additional conditions hold:

(1) $\mathcal{A} \cap \mathcal{A}^{-1} = \emptyset$.
(2) If A and B are in $\mathcal{A} \cup \mathcal{A}^{-1}$ and $L(AB) < |A|$, then $B = A^{-1}$.
(3) If A, B, and C are in $\mathcal{A} \cup \mathcal{A}^{-1}$ and $L(ABC) \leq |A| - |B| + |C|$, then $B = A^{-1}$ or $B = C^{-1}$.

For each pair $\{A, A^{-1}\}$ in $\mathcal{H} = \mathcal{H}_2$ above, pick one representative, and let $\mathcal{A}$ denote the set of words thus selected.

Proposition 8.13. *If $\prec$ is consistent with length, then $\mathcal{A}$ is Nielsen reduced.*

Proof. By construction, elements of $\mathcal{A}$ are nonempty and freely reduced. Moreover, $\mathcal{A} \cap \mathcal{A}^{-1} = \emptyset$, since $A \neq A^{-1}$ for all nonempty, freely reduced words A. Suppose that A and B are in $\mathcal{A} \cup \mathcal{A}^{-1}$ and $L(AB) < |A|$. Let $A = PQ^{-1}$ and $B = RS^{-1}$ with (P, Q) and (R, S) in $\mathcal{S}_1$. Since A and B are freely reduced, a prefix of B of length greater than $|B|/2$ cancels a suffix of A. If R cancels a suffix of Q^{-1}, then R is a prefix of Q, which is not possible since Q is in $\mathcal{C}$. Thus Q^{-1} is canceled, so Q is a proper prefix of R. Let $R = QD$. Then $D \neq \varepsilon$ and D^{-1} is a suffix of P. Suppose $P = ED^{-1}$. Then $ED^{-1} \sim Q$ and $QD \sim S$. But this means that $E \sim ED^{-1}D \sim QD \sim S$. Since both E and S are in $\mathcal{C}$, it follows that $E = S$. Therefore $A = SD^{-1}Q^{-1}$ and $B = QDS^{-1} = A^{-1}$.

Now suppose that A, B, and C are in $\mathcal{A} \cup \mathcal{A}^{-1}$ and $L(ABC) \leq |A| - |B| + |C|$. By the previous argument, at most half of B cancels a suffix of A, and at most half of B cancels a prefix of C. Thus we can write $B = UV^{-1}$, with $|U| = |V|$, and U cancels a suffix of A and V^{-1} cancels a prefix of C. Now $U \sim V$. Suppose $U \succ V$. Then (U, V) is in $\mathcal{S}_1$, and we conclude that $B = A^{-1}$ as above. If $V \succ U$, then (V, U) is in $\mathcal{S}_1$, and a similar argument shows that $B = C^{-1}$. $\square$

The following example shows that the assumption on $\prec$ in Proposition 8.13 is necessary.

Example 8.3. Let $X = \{a, b, c\}$, let $\prec$ be the basic wreath product ordering of Y^* in which $\# \prec a \prec a^{-1} \prec b \prec b^{-1} \prec c \prec c^{-1}$, and let $\mathcal{S}$ consist of the following pairs:

$$b \to a^2, \quad a^2 b^{-1} \to \varepsilon, \quad c \to a^2, \quad a^2 c^{-1} \to \varepsilon.$$

Then $\mathcal{T} \cup \mathcal{R}$ is reduced and confluent. The set $\mathcal{H}$ consists of the words ba^{-2}, $a^2 b^{-1}$, ca^{-2}, and $a^2 c^{-1}$. If $A = ba^{-2}$ and $B = a^2 c^{-1}$, then the free reduction of AB is bc^{-1}, which has length less than A.

Our interest in Nielsen reduced sets is justified by the following proposition.

Proposition 8.14. *If $\mathcal{A}$ is a subset of $X^{\pm *}$ which is Nielsen reduced, then the image of $\mathcal{A}$ in F is a set of free generators for the subgroup it generates.*

Proof. See Theorem 7.3.1 of [Hall 1959] or the discussion in Section 3.2 of [Magnus et al. 1976]. $\square$

Corollary 8.15. *Every finitely generated subgroup of F is free.*

Proof. Let $\mathcal{B}$ be a finite set of reduced words. Take the preceding $\mathcal{S}$ to be the set of pairs (B, ε), with B in $\mathcal{B}$. Then H is the subgroup of F generated by the image of $\mathcal{B}$. Let $\prec$ be a length-plus-lexicographic ordering on Y^*. By Proposition 8.13, there is a Nielsen reduced set $\mathcal{A}$ whose image in F generates H. By Proposition 8.14, H is free. $\square$

More information about the uses of Nielsen reduction can be found in [Hall 1959], [Magnus et al. 1976], and [Lyndon & Schupp 1977].

We close this section with a final comment about Proposition 8.6. It may happen that $\sim$ has only finitely many classes but $\mathrm{RC}(X, \prec, \cong)$ is infinite. In this situation, the Knuth-Bendix procedure will churn along indefinitely, producing longer and longer rules for $\cong$, rules which are unnecessary in computing canonical forms for $\sim$. It is still possible to obtain enough information to compute canonical forms for $\sim$ using the rewriting-system approach. How this is done is described in Section 3.10.

Exercise

8.1. Let $X = \{a, b\}$, let F be the free group generated by X, and let H be the subgroup of F generated by the images of aba^3b^2a, $ab^2a^2b^{-1}$, and $a^{-2}b^{-1}a^{-1}$. Find a Nielsen reduced set of words whose image in F generates H.

3

Automata and rational languages

The theory of formal languages is an important part of theoretical computer science. The theory classifies subsets of free monoids according to the difficulty of deciding whether a given word belongs to the subsets. From this point of view, the simplest subsets are the finite ones. The next simplest are called rational languages. If $\mathcal{L}$ is a rational language, then $\mathcal{L}$ may be infinite, but there is a finite combinatorial object called an automaton with which one can decide whether a word U belongs to $\mathcal{L}$ in time proportional to $|U|$.

One of the first papers to suggest a connection between combinatorial group theory and formal language theory was (Anisimov 1971). Other authors have pursued this topic. See for example (Muller & Schupp 1983, 1985) and (Gilman 1984b, 1987). However, the work which has stirred up the greatest interest in formal language theory among group theorists is [Epstein et al. 1992]. This book has no less than six authors and the intriguing title *Word Processing and Group Theory*. One of its most important contributions is the definition of a class of groups in which the multiplication and comparison of elements can be described using automata. Such groups are said to have an *automatic structure*.

We shall not attempt here an exposition of the theory of groups with an automatic structure. However, the present chapter provides an introduction to a number of applications of automata in combinatorial group theory. Automata can be used as index structures to rewriting systems. Automata also appear in one approach to studying right congruences on finitely presented monoids using the Knuth-Bendix procedure for strings. Automata have been used for over 50 years, usually without the term being explicitly mentioned, in the procedure coset enumeration discussed in Chapters 4 and 5.

To streamline the exposition in this chapter, an unorthodox approach to the subject of rational languages has been adopted. A result usually referred to as the Myhill-Nerode theorem is taken as the definition of a rational language. Traditional expositions of this material may be found in [Aho et al. 1974], [Eilenberg 1974], and [Revesz 1983]. The presentation given here was strongly influenced by [Epstein et al. 1992] and by [Eilenberg 1974].

3.1 Languages

Let X be a set. A *language over* X is simply a subset of X^*, that is, a set of words. In this context, X is called the *alphabet* of the language. Let $\mathcal{L}$ and $\mathcal{M}$ be languages over X. The union $\mathcal{L} \cup \mathcal{M}$, the intersection $\mathcal{L} \cap \mathcal{M}$, and the complement $X^* - \mathcal{L}$ are also languages. The product language $\mathcal{L}\mathcal{M}$ was defined in Section 1.2 to be $\{UV \mid U \in \mathcal{L}, V \in \mathcal{M}\}$. We define $\mathcal{L}^0$ to be $\{\varepsilon\}$, and for a positive integer k we put $\mathcal{L}^k = \mathcal{L}\mathcal{L}\ldots\mathcal{L}$, where the product has k factors.

It is conventional to denote

$$\mathcal{K} = \bigcup_{k \geq 0} \mathcal{L}^k$$

by $\mathcal{L}^*$. As remarked in Section 1.2, $\mathcal{L}^*$ has already been defined as the set of all finite sequences of elements of $\mathcal{L}$, and $\mathcal{K}$ is the set of products of these sequences. That is, $\mathcal{K}$ is the submonoid generated by $\mathcal{L}$. Nevertheless, in this chapter we shall follow tradition and write $\mathcal{L}^*$ for $\mathcal{K}$.

Let $\mathcal{L}$ be a language over X. It is possible to classify elements of X^* according to the ways in which they can be extended to yield elements of $\mathcal{L}$. For W in X^*, set

$$C(\mathcal{L}, W) = \{U \in X^* \mid WU \in \mathcal{L}\} \,.$$

We call $C(\mathcal{L}, W)$ the $\mathcal{L}$-*cone* of W, or simply the *cone* of W when $\mathcal{L}$ is clear from the context.

Example 1.1. If $\mathcal{L} = X^*$, then $C(\mathcal{L}, W) = X^*$ for every W in X^*.

Example 1.2. If $\mathcal{L} = \emptyset$, then $C(\mathcal{L}, W) = \emptyset$ for all W.

Example 1.3. If $\mathcal{L} = \{\varepsilon\}$, then $C(\mathcal{L}, \varepsilon) = \{\varepsilon\}$ and $C(\mathcal{L}, W) = \emptyset$ for every nonempty word W.

Example 1.4. Suppose $X = \{a\}$ and $\mathcal{L} = \{a^{2i} \mid i \geq 0\}$, the set of words in X^* of even length. Then $C(\mathcal{L}, W)$ is $\mathcal{L}$ or $\{a^{2i+1} \mid i \geq 0\}$ according as $|W|$ is even or odd.

For any language $\mathcal{L}$ over X, the cone $C(\mathcal{L}, \varepsilon)$ is $\mathcal{L}$. If W is in X^*, then W is in $\mathcal{L}$ if and only if $C(\mathcal{L}, W)$ contains ε. Also, W is a prefix of one or more elements of $\mathcal{L}$ if and only if $C(\mathcal{L}, W)$ is nonempty.

If $\mathcal{L}$ is a language over X, then $C(\mathcal{L})$ is defined to be $\{C(\mathcal{L}, W) \mid W \in X^*\}$, the set of cones of $\mathcal{L}$. We say that $\mathcal{L}$ is *rational* if X and $C(\mathcal{L})$ are both finite.

Proposition 1.1. *Suppose $\mathcal{L}$ is a finite language over a finite alphabet X. Then $\mathcal{L}$ is rational.*

Proof. Let m be the maximum length of $|U|$ for U in $\mathcal{L}$. If W is in X^* and $|W| > m$, then $C(\mathcal{L}, W) = \emptyset$. Since X is finite, there are only finitely many words W with $|W| \le m$, and hence $C(\mathcal{L})$ is finite. $\square$

The empty set may or may not be a cone of a language $\mathcal{L}$. The empty set is in $C(\mathcal{L})$ if and only if there is a word W which is not a prefix of any element of $\mathcal{L}$. Let $C_t(\mathcal{L}) = C(\mathcal{L}) - \{\emptyset\}$. Elements of $C_t(\mathcal{L})$ will be called *trim cones*. It is easy to see that $C_t(\mathcal{L}) = \emptyset$ if and only if $\mathcal{L} = \emptyset$.

Example 1.4 describes an infinite rational language. As we shall see, all rational languages have nice finite descriptions which make computing with rational languages much simpler than working with arbitrary languages.

Proposition 1.2. *Let $\mathcal{L}$ and $\mathcal{M}$ be rational languages over X. The following languages are rational: $\mathcal{L} \cup \mathcal{M}$, $\mathcal{L} \cap \mathcal{M}$, $X^* - \mathcal{L}$, $\mathcal{L}\mathcal{M}$, $\mathcal{L}^*$.*

Proof. Let W be in X^*. Then

$$C(\mathcal{L} \cup \mathcal{M}, W) = C(\mathcal{L}, W) \cup C(\mathcal{M}, W),$$
$$C(\mathcal{L} \cap \mathcal{M}, W) = C(\mathcal{L}, W) \cap C(\mathcal{M}, W),$$
$$C(X^* - \mathcal{L}, W) = X^* - C(\mathcal{L}, W).$$

Therefore

$$C(\mathcal{L} \cup \mathcal{M}) \subseteq \{\mathcal{U} \cup \mathcal{V} \mid \mathcal{U} \in C(\mathcal{L}), \mathcal{V} \in C(\mathcal{M})\},$$
$$C(\mathcal{L} \cap \mathcal{M}) \subseteq \{\mathcal{U} \cap \mathcal{V} \mid \mathcal{U} \in C(\mathcal{L}), \mathcal{V} \in C(\mathcal{M})\},$$
$$C(X^* - \mathcal{L}) = \{X^* - \mathcal{U} \mid \mathcal{U} \in C(\mathcal{L})\}.$$

In each case the set on the right is finite, so $\mathcal{L} \cup \mathcal{M}, \mathcal{L} \cap \mathcal{M}$, and $X^* - \mathcal{L}$ are rational.

Computing $C(\mathcal{L}\mathcal{M}, W)$ is a little more complicated. Suppose $WU = PQ$, where P is in $\mathcal{L}$ and Q is in $\mathcal{M}$. We have two cases, depending on whether W is a prefix of P or P is prefix of W. Suppose $P = WR$. Then R is in $C(\mathcal{L}, W)$, and Q can be any element of $\mathcal{M}$. If $W = PR$, then U is in $C(\mathcal{M}, R)$. From this it follows that

$$C(\mathcal{L}\mathcal{M}, W) = C(\mathcal{L}, W)\mathcal{M} \cup \bigcup_R C(\mathcal{M}, R),$$

where the union is over those words R with the property that there is a word P in $\mathcal{L}$ such that $W = PR$. Thus the cones of $\mathcal{LM}$ are among the sets which can be constructed as follows: Choose a cone of $\mathcal{L}$ and take the product with $\mathcal{M}$, then form the union of this with the union of a subset of $C(\mathcal{M})$. Only finitely many sets can be constructed this way, so $C(\mathcal{LM})$ is finite.

Finally, suppose U is in $C(\mathcal{L}^*, W)$. Then $WU = P_1 \ldots P_n$, where the P_i are in $\mathcal{L}$. We can write $W = P_1 \ldots P_m Q$, where $Q = \varepsilon$ if $m = n$, and Q is a prefix of P_{m+1} if $m < n$. Then $U = RP_{m+2} \ldots P_n$, where $R = \varepsilon$ if $m = n$, and $P_{m+1} = QR$ if $m < n$. In either case, U is in $C(\mathcal{L}, Q)\mathcal{L}^*$. Therefore

$$ C(\mathcal{L}^*, W) = \left[\bigcup_Q C(\mathcal{L}, Q) \right] \mathcal{L}^*, $$

where the union is over those words Q for which there is a V in $\mathcal{L}^*$ such that $W = VQ$. Since $C(\mathcal{L})$ is finite, there are only finitely many possibilities for $C(\mathcal{L}^*, W)$. $\square$

Propositions 1.1 and 1.2 allow us to construct a great many rational languages. In fact, it can be shown that every rational language can be built up from finite sets by repeated use of Proposition 1.2. We shall not need this fact here. A proof may be found in [Aho et al. 1974].

It will be useful to have an example of a language which is not rational.

Example 1.5. Let $X = \{a, b\}$ and let $\mathcal{L} = \{a^i b^i \mid i \geq 0\}$. Then for $k \geq 1$ we have $C(\mathcal{L}, a^k) = \{a^i b^{i+k} \mid i \geq 0\}$ and $C(\mathcal{L}, a^k b) = \{b^{k-1}\}$. Thus clearly $C(\mathcal{L})$ is infinite and $\mathcal{L}$ is not rational.

Proposition 1.3. *Let $\mathcal{U}$ be a finite subset of X^*. The two-sided ideal $\mathcal{I}$ and the right ideal $\mathcal{J}$ generated by $\mathcal{U}$ are rational languages, as are their complements.*

Proof. By Proposition 1.1, $\mathcal{U}$ is rational. Now $\mathcal{I} = X^* \mathcal{U} X^*$ and $\mathcal{J} = \mathcal{U} X^*$, so both are rational by Proposition 1.2. Also, the complements of $\mathcal{I}$ and $\mathcal{J}$ are rational by Proposition 1.2. $\square$

Corollary 1.4. *If $(X, \mathcal{R})$ is a finite monoid presentation and $\mathcal{R}$ is a confluent rewriting system with respect to a reduction ordering $\prec$, then the set of canonical forms for $\mathrm{Mon}\,\langle X \mid \mathcal{R} \rangle$ relative to $\prec$ is a rational language over X.*

Proof. The set of canonical forms is the complement of the ideal generated by the left sides in $\mathcal{R}$. $\square$

In addition to the closure properties of Proposition 1.2, the set of rational languages over X is closed under taking prefixes.

Proposition 1.5. *Let $\mathcal{L}$ be a rational language over X. The set $\mathcal{P}$ of words which are prefixes of at least one element of $\mathcal{L}$ is a rational language.*

Proof. Let W be in X^*. A word U is in $C(\mathcal{P}, W)$ if and only if WU is in $\mathcal{P}$, which is true if and only if there is a word V such that WUV is in $\mathcal{L}$. But this says that $C(\mathcal{P}, W)$ is the set of prefixes of the elements of $C(\mathcal{L}, W)$. Since $C(\mathcal{L})$ is finite, there are only finitely many possibilities for $C(\mathcal{P}, W)$. □

Exercises

1.1. Which of the following languages are rational over $\{a, b\}$?

(a) $\{a^{3i+1} \mid i \geq 0\}$
(b) $\{b^{i^2} \mid i \geq 1\}$
(c) $\{(ab)^i \mid i \geq 0\}$
(d) $\{a^{2i}b^{3j} \mid i, j \geq 0\}$

1.2. Suppose k is a nonnegative integer and $\mathcal{L}$ is the set of words of length k over a finite alphabet X. For W in X^* describe $C(\mathcal{L}, W)$.

1.3. Let $\mathcal{L}$ be a rational language over X. Show that $C(\mathcal{L}, W)$ is a rational language for each W in X^*.

1.4. Let $\mathcal{L}$ be a language over X. For W in X^* let $B(\mathcal{L}, W) = \{U \in X^* \mid UW \in \mathcal{L}\}$, the *backward $\mathcal{L}$-cone* of W, and let $B(\mathcal{L}) = \{B(\mathcal{L}, W) \mid W \in X^*\}$. For σ in $C(\mathcal{L})$, let $D(\sigma) = \{W \in X^* \mid C(\mathcal{L}, W) = \sigma\}$. Show that

$$B(\mathcal{L}, W) = \bigcup_{W \in \sigma} D(\sigma).$$

Conclude that if $\mathcal{L}$ is rational, then $B(\mathcal{L})$ is finite and

$$|B(\mathcal{L})| \leq 2^{|C(\mathcal{L})|}.$$

1.5. For a language $\mathcal{L}$ over X, put $\mathcal{L}^\dagger = \{U^\dagger \mid U \in \mathcal{L}\}$. (Recall that $U^\dagger$ is the reversal of U.) Prove that $C(\mathcal{L}^\dagger, W) = B(\mathcal{L}, W^\dagger)^\dagger$. Deduce from this that $\mathcal{L}$ is rational if and only if $\mathcal{L}^\dagger$ is rational.

1.6. Suppose that $\mathcal{L}$ is a rational language over X. Show that the set of suffixes of elements of $\mathcal{L}$ and the set of subwords of elements of $\mathcal{L}$ are rational.

3.2 Automata

In this section we shall define combinatorial objects called automata and associate to each automaton a language. In Section 3.3 we shall show that the rational languages are precisely the languages associated with finite automata.

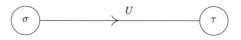

Figure 3.2.1

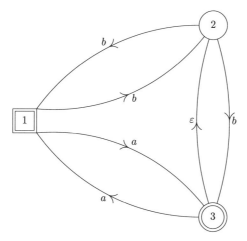

Figure 3.2.2

An *automaton* is a quintuple $(\Sigma, X, E, A, \Omega)$, where Σ and X are sets, A and Ω are subsets of Σ, and E is a set of triples of the form (σ, U, τ), where σ and τ are in Σ and U is a word in X^* of length at most 1. Thus U is either the empty word or an element of X. Elements of Σ are called *states*, A is the set of *initial states*, and Ω is the set of *terminal states*. We may consider (Σ, E) to be a directed graph. If $e = (\sigma, U, \tau)$ is in E, then we think of e as a directed edge going from σ to τ and having label U (Figure 3.2.1). In Chapter 7 we shall consider generalized automata, in which edge labels can be arbitrary words in X^*.

Example 2.1. Let $\Sigma = \{1, 2, 3\}$, $X = \{a, b\}$, $A = \{1\}$, and $\Omega = \{1, 3\}$. Let E consist of the following edges:

$$(1, a, 3), \quad (1, b, 2), \quad (2, b, 1), \quad (2, b, 3), \quad (3, \varepsilon, 2), \quad (3, a, 1).$$

We can picture the automaton $(\Sigma, X, E, A, \Omega)$ as shown in Figure 3.2.2. In this diagram, and other diagrams representing finite automata, the following conventions are used: Initial states are represented by squares, and noninitial states by circles. Terminal states have a double boundary; nonterminal states have a single boundary.

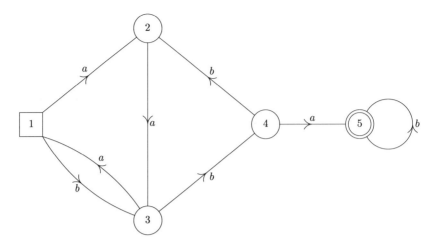

Figure 3.2.3

Example 2.2. Let $f: X \to G$, where X is a set and G is a group. Let H be a subgroup of G, let Σ_S be the set of right cosets of H in G, and let E_S be the set of triples (σ, x, τ), where σ and τ are in Σ_S, x is in X, and $\sigma f(x) = \tau$. The graph (Σ_S, E_S) is called the *Schreier diagram* for G relative to H, X, and f. (Traditionally, this term has been used only when $f(X)$ is a monoid generating set for G.) We shall call $(\Sigma_S, X, E_S, \{H\}, \{H\})$ the *Schreier automaton* for G relative to H, X, and f. When H is the trivial subgroup, so that Σ_S can be identified with G, the terms *Cayley diagram* and *Cayley automaton* are sometimes used.

Example 2.3. We can generalize Example 2.2 to right congruences on monoids. Let $f: X \to M$, where X is a set and M is a monoid. Let $\sim$ be a right congruence on M and let Σ be the set of $\sim$-classes. If σ is in Σ and x is in X, then $\sigma f(x)$ is contained in a unique element τ of Σ. Let E be the set of triples (σ, x, τ), where σ and τ are in Σ, x is in X, and $\sigma f(x) \subseteq \tau$. For any subsets A and Ω of Σ, the quintuple $(\Sigma, X, E, A, \Omega)$ is an automaton. A natural choice is to take $A = \Omega = \{\alpha\}$, where α is the $\sim$-class containing 1. However, other choices might be useful.

Example 2.4. Let $\Sigma = \{1, 2, 3, 4, 5\}$ and let $X = \{a, b\}$. Figure 3.2.3 defines an automaton with eight edges, a single initial state, 1, and a single terminal state, 5.

Our definition of automaton allows Σ and X to be infinite. Computations can normally be performed only with *finite automata*, automata for which Σ and X are finite.

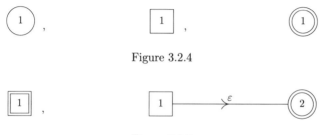

Figure 3.2.4

Figure 3.2.5

Let $\mathcal{A} = (\Sigma, X, E, A, \Omega)$ be an automaton. A *path* in $\mathcal{A}$ is a sequence P of edges in E of the form

$$(\sigma_0, U_1, \sigma_1), \ (\sigma_1, U_2, \sigma_2), \ldots, \ (\sigma_{t-1}, U_t, \sigma_t),$$

where each edge after the first begins at the state at which the previous edge ends. The path P is said to *go from σ_0 to σ_t* and to have *length t*. The state σ_0 is the *starting point*, σ_t is the *endpoint*, and $\sigma_1, \ldots, \sigma_{t-1}$ are the *intermediate states* of P. The *signature* of P is the product $\mathrm{Sg}(P) = U_1 U_2 \ldots U_t$ of the labels on the edges of P. For any state σ, we allow the empty path from σ to σ and define its signature to be ε. In Example 2.1, let P be the following sequence of edges:

$$(1, a, 3), \quad (3, \varepsilon, 2), \quad (2, b, 3), \quad (3, \varepsilon, 2), \quad (2, b, 1).$$

Then P is a path of length 5 from 1 to 1 having signature ab^2.

Let Λ and Φ be subsets of Σ. If P is a path in $\mathcal{A}$ such that the starting point of P is in Λ and the endpoint of P is in Φ, then we shall say that P *goes from Λ to Φ*. If Q is a path in $\mathcal{A}$ such that the starting point of Q is the endpoint of P, then we can *concatenate* P and Q to obtain a path R such that $\mathrm{Sg}(R) = \mathrm{Sg}(P)\,\mathrm{Sg}(Q)$.

The language *recognized* by an automaton $\mathcal{A} = (\Sigma, X, E, A, \Omega)$ is the set $\mathcal{L}$ of signatures of paths in $\mathcal{A}$ from A to Ω. We shall denote $\mathcal{L}$ by $L(\mathcal{A})$. In Example 2.1, the state 1 is both initial and terminal. Thus the path P above shows that ab^2 is in the language $\mathcal{L}$ recognized by the automaton of Example 2.1. It is easy to see that ba is not in $\mathcal{L}$.

Example 2.5. It is not difficult to construct automata which recognize $\emptyset$, $\{\varepsilon\}$, and X^*. Thus $(\emptyset, X, \emptyset, \emptyset, \emptyset)$ recognizes $\emptyset$, as do $(\{1\}, X, \emptyset, \emptyset, \emptyset)$, $(\{1\}, X, \emptyset, \{1\}, \emptyset)$, and $(\{1\}, X, \emptyset, \emptyset, \{1\})$. The last three automata may be pictured as in Figure 3.2.4. The automaton $(\emptyset, X, \emptyset, \emptyset, \emptyset)$ is the *empty automaton* with alphabet X. The automata in Figure 3.2.5 recognize $\{\varepsilon\}$. Let X be a set and let $E = \{(1, x, 1) \mid x \in X\}$. Then $(\{1\}, X, E, \{1\}, \{1\})$

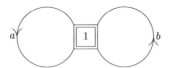

Figure 3.2.6

Figure 3.2.7

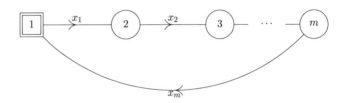

Figure 3.2.8

recognizes X^*. Figure 3.2.6 describes this automaton for the case $X = \{a, b\}$.

Example 2.6. Let $U = x_1 \dots x_m$ be any nonempty word in X^*. Then the automaton in Figure 3.2.7 recognizes $\{U\}$, and the automaton in Figure 3.2.8 recognizes $\{U^i \mid i \geq 0\}$.

Example 2.7. Let $\mathcal{A}$ be the Schreier automaton for a group G relative to a subgroup H and a function $f: X \to G$. We can extend f to be a homomorphism from X^* to G. The language $L(\mathcal{A})$ is the set of words in X^* whose images in G are elements of H.

Two automata $\mathcal{A}_1 = (\Sigma_1, X, E_1, A_1, \Omega_1)$ and $\mathcal{A}_2 = (\Sigma_2, X, E_2, A_2, \Omega_2)$ with the same alphabet X are *isomorphic* if there is a bijection $f : \Sigma_1 \to \Sigma_2$ such that $f(A_1) = A_2$, $f(\Omega_1) = \Omega_2$, and the map $(\sigma, U, \tau) \mapsto (f(\sigma), U, f(\tau))$ of $\Sigma_1 \times X^* \times \Sigma_1$ to $\Sigma_2 \times X^* \times \Sigma_2$ maps E_1 onto E_2.

Proposition 2.1. *Isomorphic automata recognize the same language.*

Proof. Exercise. □

The use of the words "automaton" and "recognize" is intended to convey the idea that automata simulate in some way the operation of certain kinds of computers. It must be emphasized, however, that automata are mathematical, not physical, objects.

We say that an automaton $\mathcal{A} = (\Sigma, X, E, A, \Omega)$ is *deterministic* if the following conditions hold:

(i) $|A| \leq 1$.
(ii) The label on each edge in E is nonempty.
(iii) For each σ in Σ and each x in X there is at most one edge (σ, x, τ) in E.

We can rephrase condition (iii) as follows: If (σ, x, τ) and (σ, x, ρ) are in E, then $\tau = \rho$. The automaton of Example 2.1 is not deterministic, since it contains the edges $(2, b, 1)$ and $(2, b, 3)$ and has an edge with an empty label. The automata of Examples 2.2, 2.4, and 2.5 are deterministic. If $\mathcal{A}$ is deterministic and $A \neq \emptyset$, then A contains a single state. We shall normally denote this unique initial state by α.

A *complete* automaton is one which is deterministic, has exactly one initial state, and satisfies the condition that for each state σ and each x in X there is exactly one edge (σ, x, τ) in E. The Schreier automata of Example 2.2 are complete, but the automaton of Example 2.4 is not, since there are no edges $(2, b, \tau)$ or $(5, a, \tau)$.

Proposition 2.2. *In a deterministic automaton, a path is determined by its starting point and its signature.*

Proof. Let $\mathcal{A} = (\Sigma, X, E, A, \Omega)$ be a deterministic automaton and suppose that P is a path in $\mathcal{A}$ which starts at σ and has signature V. If $V = \varepsilon$, then P must have length 0, since all edge labels have length 1. Now suppose that $V \neq \varepsilon$ and let x be the first term of V. Let (σ, x, τ) be the first edge of P. Then by the definition of deterministic automaton, τ is uniquely determined. Let $V = xW$. The remaining edges of P define a path starting at τ and having signature W. By induction on the length of P, this path is unique. Therefore P is determined by σ and V. $\square$

Suppose that $\mathcal{A} = (\Sigma, X, E, A, \Omega)$ is a deterministic automaton and assume that σ and τ are states in Σ. If U is a word in X^* and U is the signature of a path in $\mathcal{A}$ from σ to τ, then we shall write $\tau = \sigma^U$. The notation does not identify $\mathcal{A}$, which must be clear from the context.

The following procedure can be used to determine whether there is a path which starts at σ and has signature U.

Procedure $\mathrm{TRACE}(\mathcal{A}, \sigma, U; \tau, B, C)$;
Input: $\mathcal{A}$: a finite deterministic automaton $(\Sigma, X, E, A, \Omega)$;
          σ : a state in Σ;

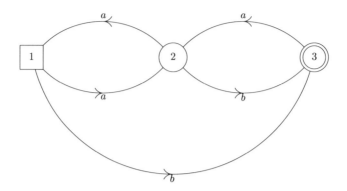

Figure 3.2.9

$$U \qquad : \text{a word in } X^*;$$
Output: $\tau \qquad : \text{a state in } \Sigma;$
$$B, C \qquad : \text{words in } X^* \text{ such that } U = BC, \tau = \sigma^B, \text{ and } |B| \text{ is}$$
$$\text{as large as possible};$$
Begin
 $\tau := \sigma; \ B := \varepsilon; \ C := U;$

 While $C \neq \varepsilon$ do begin
  Let x be the first term of C;
  If τ^x is not defined then break;
  Let (τ, x, ρ) be in E; Let $C = xD$;
  $B := Bx; \ C := D; \ \tau := \rho$
 End
End.

There is a path P in $\mathcal{A}$ starting at σ such that $\text{Sg}(P) = U$ if and only if $\text{TRACE}(\mathcal{A}, \sigma, U; \tau, B, C)$ returns $B = U$, and hence $C = \varepsilon$.

Example 2.8. Suppose that $X = \{a, b\}$ and $\mathcal{A}$ is given by Figure 3.2.9. The result of $\text{TRACE}(\mathcal{A}, 1, aba^2b^2ab; \tau, B, C)$ is to set $\tau = 3$, $B = aba^2b$, and $C = bab$. After $\text{TRACE}(\mathcal{A}, 3, a^3ba^2; \rho, R, S)$, we have $\rho = 1$, $R = a^3b^2a$, and $S = \varepsilon$.

In view of Proposition 2.2, there is a straightforward way of decid-ing membership in $L(\mathcal{A})$ for any finite, deterministic automaton $\mathcal{A} = (\Sigma, X, E, A, \Omega)$. Suppose W is in X^*. If $A = \emptyset$, then $L(\mathcal{A})$ is empty and W is not in $L(\mathcal{A})$. If $A = \{\alpha\}$, then W is in $L(\mathcal{A})$ if and only if the call $\text{TRACE}(\mathcal{A}, \alpha, W; \tau, B, C)$ returns with τ in Ω and $B = W$. We shall show later that membership in $L(\mathcal{A})$ can be decided for any finite automaton $\mathcal{A}$.

The set of edges in a complete, finite automaton is conveniently described by its *transition table*. The rows of the table are indexed by states, and the columns by elements of X. The entry in the σ-th row and x-th column

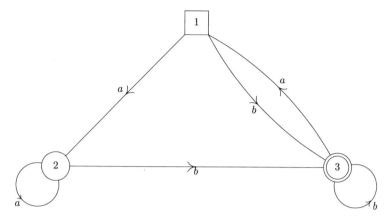

Figure 3.2.10

is the state $\tau = \sigma^x$, that is, the state τ such that (σ, x, τ) is an edge. For example, the transition table

	a	b
1	2	3
2	2	3
3	1	3

describes the edges in the complete automaton shown in Figure 3.2.10.

Suppose that $\mathcal{A}$ is a finite, deterministic automaton which is not complete. We can still use a transition table to describe $\mathcal{A}$, but certain entries will not be defined. On the printed page, undefined entries can be left blank. In a computer, every memory location always has a value, so we must use some value which cannot be interpreted as a state to signal the missing entries in the transition table. Typically, positive integers are used to represent states, and 0 is used to flag undefined entries.

Exercises

2.1. Construct deterministic automata with $X = \{a\}$ which recognize the languages $\{a^2, a^4, a^6\}$ and $\{a^{2i} \mid i > 0\}$. In each case, what is the minimum number of states required?

2.2. Construct the Cayley automaton for the symmetric group Sym(3) where X is $\{(1,2),(2,3)\}$ and f is the identity map.

2.3. Let $\mathcal{A}$ be the automaton of Example 2.4. What is the result of the call TRACE($\mathcal{A}, 1$, $a^2ba^2b; \sigma, A, B$)?

2.4. Let $\mathcal{A}$ be a deterministic automaton and let P and Q be paths in $\mathcal{A}$ with the same starting point. Let R be the longest common initial segment of P and Q. Show that Sg(R) is the longest common prefix of Sg(P) and Sg(Q).

2.5. Suppose $\mathcal{A} = (\Sigma, X, E, A, \Omega)$ is an automaton and $A = \Omega \neq \emptyset$. Show that $L(\mathcal{A})$ is a submonoid of X^*.

2.6. Suppose that k is a nonnegative integer. Let $\Sigma = \{1, 2, \ldots, k+1\}$ and $E = \{(i, x, i+1) \mid 1 \le i \le k, x \in X\}$. Show that $\mathcal{A} = (\Sigma, X, E, \{1\}, \{k+1\})$ recognizes the set of words in X^* of length k. Draw a diagram for $\mathcal{A}$.

2.7. If $|X| = 1$, then we can identify X^* with the set of nonnegative integers under addition. Show that rational sets correspond to sets L obtained as follows: Choose a positive number n and a subset A of $\{0, 1, \ldots, n-1\}$. Choose finite sets B and C of nonnegative integers. Set L equal to the union of C with the set of nonnegative integers i such that i is not in B and $i \bmod n$ is in A.

2.8. Let $\mathcal{A} = (\Sigma, X, E, A, \Omega)$ be an automaton. Define $\mathcal{A}^\dagger$ to be $(\Sigma, X, E^\dagger, \Omega, A)$, where $E^\dagger = \{(\tau, U, \sigma) \mid (\sigma, U, \tau) \in E\}$. Show that $\mathcal{A}^\dagger$ is an automaton and that $L(\mathcal{A}^\dagger) = L(\mathcal{A})^\dagger$, the set of reversals of the elements of $L(\mathcal{A})$.

3.3 Automata, continued

We shall now begin to relate finite automata and rational languages.

Proposition 3.1. *Every language $\mathcal{L}$ over an alphabet X is recognized by a complete automaton whose states are the cones of $\mathcal{L}$.*

Proof. Let Σ be the set $C(\mathcal{L})$ of cones of $\mathcal{L}$. Set $\alpha = C(\mathcal{L}, \varepsilon) = \mathcal{L}$ and put $A = \{\alpha\}$. Let Ω be the set of cones in $C(\mathcal{L})$ which contain ε.

Lemma 3.2. *Let W be a word in X^* and set $\sigma = C(\mathcal{L}, W)$. If x is in X, then $C(\mathcal{L}, Wx) = \{U \in X^* \mid xU \in \sigma\}$.*

Proof. Let U be in X^*. Then U is in $C(\mathcal{L}, Wx)$ if and only if WxU is in $\mathcal{L}$, which is true if and only if xU is in σ. $\square$

Let E be the set of triples (σ, x, τ), where σ is in Σ, x is in X, and $\tau = \{U \in X^* \mid xU \in \sigma\}$. By Lemma 3.2, $\mathcal{A} = (\Sigma, X, E, A, \Omega)$ is an automaton, which is clearly complete. To finish the proof of the proposition, we shall show that $\mathcal{A}$ recognizes $\mathcal{L}$.

Lemma 3.3. *Let $(\sigma_0, x_1, \sigma_1), \ldots, (\sigma_{t-1}, x_t, \sigma_t)$ be a path in $\mathcal{A}$ starting at α. Then $\sigma_t = C(\mathcal{L}, x_1 \ldots x_t)$.*

Proof. If $t = 0$, then $\sigma_t = \alpha = C(\mathcal{L}, \varepsilon)$. Assume $t > 0$. By induction on t, we know that $C(\mathcal{L}, x_1 \ldots x_{t-1}) = \sigma_{t-1}$. By the definition of E and Lemma 3.2, it follows that $\sigma_t = C(\mathcal{L}, x_1 \ldots x_t)$. $\square$

A word U is in $\mathcal{L}$ if and only if $C(\mathcal{L}, U)$ contains ε. Since $\mathcal{A}$ is complete, there is a unique path P in $\mathcal{A}$ such that P starts at α and $\mathrm{Sg}(P) = U$. Let P end at σ. By Lemma 3.3, $\sigma = C(\mathcal{L}, U)$, and σ is in Ω if and only if U is in $\mathcal{L}$. Thus $\mathcal{A}$ recognizes $\mathcal{L}$. $\square$

We shall write $A(\mathcal{L})$ for the automaton $\mathcal{A}$ constructed in the proof of Proposition 3.1.

Corollary 3.4. *Let $\mathcal{L}$ be a rational language over a finite alphabet X. Then $\mathcal{L}$ is recognized by a finite, complete automaton.*

Proof. By Proposition 3.1, $\mathcal{L}$ is recognized by the complete automaton $A(\mathcal{L})$. Since $\mathcal{L}$ is rational, $C(\mathcal{L})$ is finite, and therefore $A(\mathcal{L})$ is finite. □

Proposition 3.5. *If $\mathcal{A}$ is a finite automaton, then $L(\mathcal{A})$ is a rational language.*

Proof. Let $\mathcal{A} = (\Sigma, X, E, A, \Omega)$ and let $\mathcal{L} = L(\mathcal{A})$. Suppose W is in X^* and U is in $C(\mathcal{L}, W)$. There are paths P and Q in $\mathcal{A}$ such that P goes from A to a state σ, Q goes from σ to Ω, $W = \mathrm{Sg}(P)$, and $U = \mathrm{Sg}(Q)$. Conversely, given such paths P and Q, then U is in $C(\mathcal{L}, W)$. Let $\Lambda(W)$ be the set of endpoints of paths in $\mathcal{A}$ which start in A and have W as their signature. Then $C(\mathcal{L}, W)$ is the set of signatures of paths from $\Lambda(W)$ to Ω. Thus $C(\mathcal{L}, W)$ depends only on $\Lambda(W)$. Since Σ is finite, there are only finitely many possibilities for $\Lambda(W)$. Thus $C(\mathcal{L})$ is finite. In fact,

$$|C(\mathcal{L})| \leq 2^{|\Sigma|}. \quad \square$$

Corollary 3.6. *A language $\mathcal{L}$ over a finite alphabet X is rational if and only if it is recognized by a finite automaton.*

Corollary 3.7. *If $\mathcal{A}$ is a finite automaton, then there is a complete, finite automaton $\mathcal{A}_1$ such that $L(\mathcal{A}_1) = L(\mathcal{A})$.*

Proof. By Proposition 3.6, $L(\mathcal{A})$ is a rational language. By Proposition 3.1, $\mathcal{A}_1 = A(L(\mathcal{A}))$ is a complete, finite automaton recognizing $L(\mathcal{A})$.

□

Suppose $\mathcal{A} = (\Sigma, X, E, A, \Omega)$ is a deterministic automaton which is not complete. It is possible to construct a complete automaton $\mathcal{A}_1$ such that $L(\mathcal{A}_1) = L(\mathcal{A})$. Let ∞ be an object not in Σ. Set $\Sigma_1 = \Sigma \cup \{\infty\}$ and let S be the set of pairs (σ, x) with σ in Σ and x in X such that there is no edge (σ, x, τ) in E. Put

$$E_1 = E \cup \{(\sigma, x, \infty) \mid (\sigma, x) \in S\} \cup \{(\infty, x, \infty) \mid x \in X\}.$$

If $A = \emptyset$, then let $A_1 = \{\infty\}$; otherwise, let $A_1 = A$. Finally, set $\mathcal{A}_1 = (\Sigma_1, X, E_1, A_1, \Omega)$.

Proposition 3.8. $L(\mathcal{A}_1) = L(\mathcal{A})$.

Proof. Exercise. □

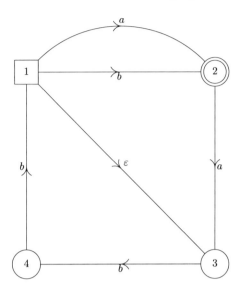

Figure 3.3.1

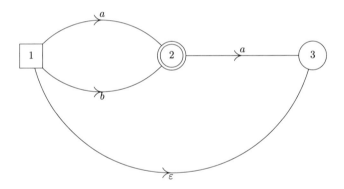

Figure 3.3.2

We call $\mathcal{A}_1$ the *completion* (more correctly *a completion*) of $\mathcal{A}$. If $\mathcal{A}$ is already complete, then $\mathcal{A}$ is said to be its own completion.

Let $\mathcal{A} = (\Sigma, X, E, A, \Omega)$ be any automaton and let Σ_1 be a subset of Σ. Set $A_1 = A \cap \Sigma_1$ and $\Omega_1 = \Omega \cap \Sigma_1$. Define E_1 to be the set of edges (σ, U, τ) in E such that σ and τ are in Σ_1. The automaton $\mathcal{A}_1 = (\Sigma_1, X, E_1, A_1, \Omega_1)$ is called the *restriction* of $\mathcal{A}$ to Σ_1. Every path in $\mathcal{A}_1$ is a path in $\mathcal{A}$, so $L(\mathcal{A}_1) \subseteq L(\mathcal{A})$.

Example 3.1. Suppose $X = \{a, b\}$ and $\mathcal{A}$ is the automaton in Figure 3.3.1. The restriction of $\mathcal{A}$ to $\{1, 2, 3\}$ is shown in Figure 3.3.2.

3.1. Construct a completion of the automaton in Example 2.4.

3.2. Describe the restriction of the automaton in Example 2.4 to $\{1, 3, 4, 5\}$.

3.3. Let X and Y be finite sets and let $f: X^* \to Y^*$ be a homomorphism of monoids. Prove that if $\mathcal{L}$ is a rational language over X, then $f(\mathcal{L})$ is a rational language over Y. (Hint: Let $\mathcal{A}$ be a finite automaton recognizing $\mathcal{L}$. Construct a finite automaton recognizing $f(\mathcal{L})$.)

3.4 The subset construction

Suppose $\mathcal{A}$ is a finite automaton. In Corollary 3.7 we proved the existence of a finite, complete automaton $\mathcal{A}_1$ such that $L(\mathcal{A}_1) = L(\mathcal{A})$. In this section we shall give a construction of one such $\mathcal{A}_1$.

Let $\mathcal{A} = (\Sigma, X, E, A, \Omega)$. For any subset Λ of Σ, let $\overline{\Lambda}$ be the set of all states μ such that there is a path P from Λ to μ with $\mathrm{Sg}(P) = \varepsilon$. Let us call $\overline{\Lambda}$ the *closure* of Λ and define Λ to be *closed* if $\overline{\Lambda} = \Lambda$. It is clear that $\overline{\Lambda}$ is closed.

The automaton $\mathcal{A}_1 = (\Sigma_1, X, E_1, A_1, \Omega_1)$ is constructed as follows: Σ_1 is the set of closed subsets of Σ, $A_1 = \{\overline{A}\}$, and Ω_1 is the set of those elements of Σ_1 which contain a state in Ω. For Λ in Σ_1 and x in X, let Λ^x be the set of states μ such that there is an edge (λ, x, μ) in E with λ in Λ. Put E_1 equal to the set of triples $(\Lambda, x, \overline{\Lambda^x})$ with Λ in Σ_1 and x in X. We shall say that $\mathcal{A}_1$ is obtained from $\mathcal{A}$ by the *subset construction*.

Example 4.1. Before proving that $L(\mathcal{A}_1) = L(\mathcal{A})$, let us apply the subset construction to the automaton in Example 3.1. A set of states is closed provided 3 is in the set whenever 1 is. Thus there are 12 closed sets:

$$
\begin{array}{ll}
\Lambda_1 = \emptyset, & \Lambda_7 = \{2, 4\}, \\
\Lambda_2 = \{2\}, & \Lambda_8 = \{3, 4\}, \\
\Lambda_3 = \{3\}, & \Lambda_9 = \{1, 2, 3\}, \\
\Lambda_4 = \{4\}, & \Lambda_{10} = \{1, 3, 4\}, \\
\Lambda_5 = \{1, 3\}, & \Lambda_{11} = \{2, 3, 4\}, \\
\Lambda_6 = \{2, 3\}, & \Lambda_{12} = \{1, 2, 3, 4\}.
\end{array}
$$

In $\mathcal{A}_1$, we have $A_1 = \{\Lambda_5\}$ and $\Omega_1 = \{\Lambda_2, \Lambda_6, \Lambda_7, \Lambda_9, \Lambda_{11}, \Lambda_{12}\}$.

To illustrate how the edges of $\mathcal{A}_1$ are determined, let us find the value of i such that $(\Lambda_8, b, \Lambda_i)$ is in E_1. The edges of $\mathcal{A}$ starting at states in Λ_8 and having the label b are $(3, b, 4)$ and $(4, b, 1)$. Thus $(\Lambda_8)^b = \{1, 4\}$. The closure of $\{1, 4\}$ is $\{1, 3, 4\} = \Lambda_{10}$, so $(\Lambda_8, b, \Lambda_{10})$ is in E_1. The full set E_1 is given in Table 3.4.1, where i is written for Λ_i.

Proposition 4.1. *If $\mathcal{A}_1$ is obtained from a finite automaton $\mathcal{A}$ by the subset construction, then $L(\mathcal{A}_1) = L(\mathcal{A})$.*

Table 3.4.1

	a	b		a	b
1	1	1	7	3	5
2	3	1	8	1	10
3	1	4	9	6	7
4	1	5	10	2	12
5	2	7	11	3	10
6	3	4	12	6	12

Proof. Suppose first that W is in $L(\mathcal{A})$. There is a path P in $\mathcal{A}$ from A to Ω such that $W = \mathrm{Sg}(P)$. Let the edges of P be $(\sigma_{i-1}, U_i, \sigma_i)$, $1 \le i \le t$, and let $i_1 < i_2 < \cdots < i_s$ be the values of i for which $U_i \ne \varepsilon$. Then U_{i_j} is an element x_j of X, $1 \le j \le s$, and $W = x_1 \ldots x_s$. Since σ_0 is in A, the states $\sigma_0, \sigma_1, \ldots, \sigma_{i_1-1}$ are all in $\Lambda_0 = \overline{A}$, the unique element of A_1. Let $(\Lambda_0, x_1, \Lambda_1)$ be in E_1. Then σ_{i_1} is in Λ_1, and so σ_j is in Λ_1 for $i_1 \le j < i_2$. Let $(\Lambda_1, x_2, \Lambda_2)$ be in E_1. Then σ_j is in Λ_2 for $i_2 \le j < i_3$. Continuing in this manner, we construct a path Q in $\mathcal{A}_1$ consisting of edges $(\Lambda_{j-1}, x_j, \Lambda_j)$, $1 \le j \le s$, such that σ_t is in Λ_s. Now σ_t is in Ω, so Λ_s is in Ω_1. Since $\mathrm{Sg}(Q) = W$, it follows that W is in $L(\mathcal{A}_1)$.

The proof that $L(\mathcal{A}_1) \subseteq L(\mathcal{A})$ is left as an exercise. $\square$

When E has no edges with empty labels, then $|\Sigma_1| = 2^{|\Sigma|}$. To decide membership in $L(\mathcal{A})$ for a single word, it is not necessary to construct all of $\mathcal{A}_1$. See Exercise 4.3.

Exercises

4.1. Apply the subset construction to the automaton of Example 2.1.

4.2. Let $\mathcal{A} = (\Sigma, X, E, A, \Omega)$ be a finite automaton and let Λ be a subset of Σ. Show how to compute $\overline{\Lambda}$ in time at worst linear in $|\Lambda| + |E|$.

4.3. Suppose $\mathcal{A}$ is as in Exercise 4.2 and $U = u_1 \ldots u_s$ is a word in X^*. Show that it is possible to decide whether U is in $L(\mathcal{A})$ in time polynomial in $s + |\Sigma| + |E|$. (Hint: Set $\Lambda_0 = \overline{A}$ and for $1 \le i \le s$ define Λ_i to be $\overline{\Lambda_{i-1}^{u_i}}$. How hard is it to compute Λ_s?)

3.5 Index automata

We come now to our first important application of automata. Let $\mathcal{R}$ be a finite rewriting system on X^* with respect to a reduction ordering $\prec$. An *index automaton* for $\mathcal{R}$ is a complete automaton $\mathcal{A} = (\Sigma, X, E, A, \Omega)$ recognizing the set of words which are reducible with respect to $\mathcal{R}$. Not all index automata for $\mathcal{R}$ are isomorphic. We shall usually impose additional conditions on $\mathcal{A}$ and assume that various supplemental data about $\mathcal{A}$ are available. For example, we want information which will tell us that a specific left side in $\mathcal{R}$ is a subword of a given word W, not just that W is reducible with respect to $\mathcal{R}$.

Let $\mathcal{A}$ be an index automaton for $\mathcal{R}$ and let P be a path in $\mathcal{A}$ from A to some state ω in Ω. Then $\mathrm{Sg}(P)$ is reducible with respect to $\mathcal{R}$ and hence $\mathrm{Sg}(P)$ contains as a subword the left side of some rule in $\mathcal{R}$. It is possible that $\mathrm{Sg}(P)$ contains many left sides. However, for some $\mathcal{A}$ it is possible to identify a rule (L, R) in $\mathcal{R}$ depending only on ω such that $\mathrm{Sg}(P)$ always contains L as a subword. A *rule identifier* for $\mathcal{A}$ is a function $f : \Omega \to \mathcal{R}$ such that for any ω in Ω and any path P from A to ω the left side of $f(\omega)$ is a subword of $\mathrm{Sg}(P)$. As we shall see, it is always possible to find an index automaton for $\mathcal{R}$ which has a rule identifier.

Given an index automaton for $\mathcal{R}$ with a rule identifier, rewriting with respect to $\mathcal{R}$ can be speeded up significantly.

Procedure INDEX_REWRITE($\mathcal{R}, \mathcal{A}, U; V$);
  Input: $\mathcal{R}$: a finite, nonempty rewriting system on X^*;
           $\mathcal{A}$: an index automaton $(\Sigma, X, E, A, \Omega)$ for $\mathcal{R}$ with rule
                 identifier f;
           U : a word in X^*;
  Output: V : a word irreducible with respect to $\mathcal{R}$ and derivable
                 from U using $\mathcal{R}$;
  Begin
    Let α_0 be the unique state in A; $V := \varepsilon$; $W := U$;
    While $W \neq \varepsilon$ do begin
      Let $W = xS$ with x in X; $W := S$; $k := |V|$; $\sigma := (\alpha_k)^x$;
      If σ is not in Ω then begin $\alpha_{k+1} := \sigma$; $V := Vx$ end
      Else begin
        Let $f(\sigma) = (L, R)$;   (* L is a suffix of Vx. *)
        Delete the suffix of V of length $|L| - 1$; $W := RW$
      End
    End
  End.

In order to prove that INDEX_REWRITE works correctly, we must first verify several assertions.

Proposition 5.1. *At all times in* INDEX_REWRITE, *if B is a prefix of V and $i = |B|$, then α_i is the endpoint of the path P in $\mathcal{A}$ which starts in A and has B as its signature.*

Proof. Since $\mathcal{A}$ is complete, P exists and is unique. Initially V is ε and α_0 is the element of A. Thus the conclusion of the proposition holds when V is initialized. If the conclusion holds and a suffix is deleted from V, then the conclusion still holds. The only other place that V or one of the α_i changes is when α_{k+1} is set equal to σ and V is replaced by Vx. Since by induction α_k is the endpoint of the path starting at α_0 and having V as its signature, it follows that the path starting at α_0 and having Vx as its signature ends at σ. $\square$

Proposition 5.2. *At all times V is irreducible with respect to $\mathcal{R}$. If σ is in Ω, then Vx ends with the left side of $f(\sigma)$.*

Proof. Initially V is empty and hence irreducible. If V is irreducible and a suffix of V is deleted, then V is still irreducible. When a generator x is appended to V, the state α_{k+1} is not in Ω, and therefore by Proposition 5.1 Vx is not in the language recognized by $\mathcal{A}$. Thus Vx is irreducible.

Suppose that σ is in Ω. Then by the definition of an index automaton, Vx contains the left side L of $f(\sigma)$ as a subword. Since V is irreducible, L must be a suffix of Vx. $\square$

Proposition 5.3. INDEX_REWRITE *performs as claimed.*

Proof. The sequence of words VW occurring at the beginning of the body of the While-loop is nonincreasing with respect to $\prec$, and each is derivable from the previous one using $\mathcal{R}$. In the body of the While-loop, either VW is replaced by a strictly earlier word or $|W|$ is reduced. Thus the While-loop must terminate. When this occurs, W is empty and V is derivable from U using $\mathcal{R}$. By Proposition 5.1, V is irreducible with respect to $\mathcal{R}$. $\square$

If $\mathcal{R}$ is reduced, then the word V returned by INDEX_REWRITE is the same as that returned by the procedure REWRITE_FROM_LEFT of Section 2.4. Since the inner loop on i in REWRITE_FROM_LEFT has been eliminated, the running time of INDEX_REWRITE is substantially less than the running time of REWRITE_FROM_LEFT.

Now that the usefulness of index automata has been demonstrated, we shall give a construction of an index automaton $\mathcal{A}$ for a given finite, nonempty rewriting system $\mathcal{R}$. The construction of $\mathcal{A}$ is somewhat simpler if $\mathcal{R}$ is reduced, and we assume this to be the case.

Let $\mathcal{L}$ be the set of left sides in $\mathcal{R}$ and let Σ be the set of prefixes of elements of $\mathcal{L}$. The edges of $\mathcal{A}$ are of two kinds. Let E_1 be the set of triples (L, x, L) with L in $\mathcal{L}$ and x in X, and let E_2 be the set of triples (U, x, V), where U is in $\Sigma - \mathcal{L}$, x is in X, and V is the longest suffix of Ux which belongs to Σ. It will be convenient to let E_3 denote the set of edges (U, x, V) in E_2 for which $V = Ux$. Each L in $\mathcal{L}$ is the left side of a unique rule (L, R) in $\mathcal{R}$, and we define $f(L)$ to be that rule. Set

$$I(\mathcal{R}) = (\Sigma, X, E_1 \cup E_2, \{\varepsilon\}, \mathcal{L}).$$

Proposition 5.4. *If $\mathcal{R}$ is a nonempty, finite, reduced rewriting system on X^*, then $I(\mathcal{R})$ is an index automaton for $\mathcal{R}$ and f is a rule identifier for $I(\mathcal{R})$.*

Proof. Suppose that P is a path in $I(\mathcal{R})$ starting at ε and ending at the state V in Σ.

Lemma 5.5. *If V is not in $\mathcal{L}$, then V is the longest suffix of* $\mathrm{Sg}(P)$ *belonging to Σ.*

Proof. Assume that V is not in $\mathcal{L}$. Because of the edges in E_1, once a path reaches an element of $\mathcal{L}$, the path stays at that element. Thus no state involved in P belongs to $\mathcal{L}$. If P is empty, then the conclusion clearly holds. Suppose that P is nonempty, let (U, x, V) be the last edge of P, and let Q be the initial segment of P consisting of all but the last edge. By induction on the length of P, the longest suffix of $W = \mathrm{Sg}(Q)$ belonging to Σ is U. Since (U, x, V) is in E_2, it follows that V is a suffix of Ux and hence V is a suffix of $\mathrm{Sg}(P) = Wx$. Now let T be any suffix of Wx belonging to Σ. Then T has the form Sx, where S is in Σ and S is a suffix of $\mathrm{Sg}(Q)$. Therefore S is a suffix of U and T is a suffix of Ux. Thus T is a suffix of V and V is the longest suffix of $\mathrm{Sg}(P)$ belonging to Σ. $\square$

Now suppose that V is in $\mathcal{L}$ and let Q be the longest initial segment of P which does not involve the state V. The edge of P immediately following Q has the form (U, x, V), where U is not in $\mathcal{L}$. By Lemma 5.5, U is the longest suffix of $\mathrm{Sg}(Q)$ belonging to Σ. Thus V is a suffix of Ux and therefore V is a subword of $\mathrm{Sg}(P)$. From this it follows that every word recognized by $I(\mathcal{R})$ is reducible with respect to $\mathcal{R}$.

Finally, suppose that W is reducible with respect to $\mathcal{R}$. Since $I(\mathcal{R})$ is complete, there is a unique path P in $I(\mathcal{R})$ starting at ε such that $W = \mathrm{Sg}(P)$. It may happen that W contains many elements of $\mathcal{L}$ as subwords. We need to pick out the "first" such subword. Let $W = SUxT$, where S, U, and T are in X^*, x is in X, Ux is in $\mathcal{L}$, and SU has no subwords in $\mathcal{L}$. Define Q to be the initial segment of P with $\mathrm{Sg}(Q) = SU$, and let Q end at the state B. By Lemma 5.5, U is a suffix of B. Let the edge of P following Q be (B, x, V). Since Ux is a suffix of Bx, it follows that Ux is a suffix of V. But Ux is in $\mathcal{L}$ and Ux is not a subword of any other element of $\mathcal{L}$. Therefore $V = Ux$ and $SUx = SV$ is in $L(I(\mathcal{R}))$. Since all further edges of P end at V, we see that W is in $L(I(\mathcal{R}))$. $\square$

Example 5.1. In Example 2.3.5 the rewriting system $\mathcal{S}$ on $\{a, b\}^*$ consisting of the rules

$$a^2 \to \varepsilon, \qquad ab^2a \to bab,$$
$$b^3 \to \varepsilon, \qquad baba \to ab^2,$$
$$abab \to b^2a, \qquad b^2ab^2 \to aba,$$

was shown to be confluent. Let us construct $I(\mathcal{S})$. The elements of Σ are

Table 3.5.1

	a	b		a	b
1	2	3	10	15	9
2	4	5	11	4	16
3	6	7	12	12	12
4	4	4	13	13	13
5	8	9	14	14	14
6	4	10	15	15	15
7	11	12	16	15	17
8	4	13	17	17	17
9	14	12			

$$S_1 = \varepsilon, \qquad S_{10} = bab,$$
$$S_2 = a, \qquad S_{11} = b^2 a,$$
$$S_3 = b, \qquad S_{12} = b^3,$$
$$S_4 = a^2, \qquad S_{13} = abab,$$
$$S_5 = ab, \qquad S_{14} = ab^2 a,$$
$$S_6 = ba, \qquad S_{15} = baba,$$
$$S_7 = b^2, \qquad S_{16} = b^2 ab,$$
$$S_8 = aba, \qquad S_{17} = b^2 ab^2,$$
$$S_9 = ab^2,$$

Since $|\mathcal{L}| = 6$ and $|X| = 2$, there are 12 edges in E_1. Each element of $\Sigma - \{\varepsilon\}$ yields an edge in E_3. For example, $S_8 = S_5 a$, so (S_5, a, S_8) is in E_3. The edges in $E_2 - E_3$ require a little more work to determine. For example, the longest suffix of $S_{11} a = b^2 a^2$ belonging to Σ is $S_4 = a^2$. Thus (S_{11}, a, S_4) is in E_2. The full set of edges in $I(\mathcal{S})$ is described by Table 3.5.1. Here i has been written for S_i. Under this identification of Σ with $\{1, \ldots, 17\}$, the terminal states of $I(\mathcal{S})$ are 4, 12, 13, 14, 15, and 17. The rule identifier f maps 4 to (a^2, ε), 12 to (b^3, ε), and so on.

For our next application of index automata we shall assume that $\mathcal{R}$ is reduced and that $\mathcal{A} = (\Sigma, X, E, A, \Omega)$ is isomorphic to $I(\mathcal{R})$. Let $A = \{\alpha\}$. For each state σ in Σ there is a unique word U of minimal length such that $\sigma = \alpha^U$. Under the isomorphism of $\mathcal{A}$ with $I(\mathcal{R})$ the state σ corresponds to a prefix of some left side in $\mathcal{R}$, and U is that prefix. Set $\ell(\sigma) = |U|$. We shall call ℓ the *length function* for $\mathcal{A}$. Given $\mathcal{A}$, ℓ, and the rule identifier f of $\mathcal{A}$, we can implement an improved version of CONFLUENT using a backtrack search for overlaps satisfying the condition of Proposition 7.1 in Chapter 2.

Function INDEX_CONFLUENT($\mathcal{R}, \mathcal{A}$): boolean;
Input: $\mathcal{R}$: a reduced rewriting system;
     $\mathcal{A}$: an index automaton $(\Sigma, X, E, A, \Omega)$ isomorphic to
            $I(\mathcal{R})$ with length function ℓ and rule identifier f;
(∗ True is returned if $\mathcal{R}$ is confluent and false is returned otherwise. ∗)
Begin
  Let $A = \{\alpha\}$;
  For each rule (U, V) in $\mathcal{R}$ do begin
    Let $U = xS$ with x in X; $\beta_0 := \alpha^S$; $r := 0$; $backtrack :=$ false;
    While $r \geq 0$ and not $backtrack$ do begin
      $backtrack := \ell(\beta_r) \leq r$;
      If $\beta_r \in \Omega$ and not $backtrack$ then begin
        Let $f(\beta_r) = (L, R)$;
        (∗ The suffix of U of length $\ell(\beta_r) - r$ is a prefix of L. ∗)
        Process the overlap of U and L;
        If failure of confluence is detected then begin
          INDEX_CONFLUENT := false; Goto 99
        End;
        $backtrack :=$ true
      End;
      If not $backtrack$ then begin
        Choose x in X; $\beta_{r+1} := (\beta_r)^x$; $Y_{r+1} := X - \{x\}$; $r := r + 1$
      End
      Else
        While $backtrack$ and $r > 0$ do
          If $Y_r \neq \emptyset$ then begin
            Choose x in Y_r; $Y_r := Y_r - \{x\}$; $\beta_r := (\beta_{r-1})^x$;
            $backtrack :=$ false
          End
          Else $r := r - 1$
    End
  End;
  INDEX_CONFLUENT := true
99: End.

Given a left side $U = xS$ in $\mathcal{R}$, we perform a backtrack search for
nonempty words V such that UV has exactly two subwords which are
left sides in $\mathcal{R}$, namely U and L, where L is a suffix of UV and $|L| > |V|$.
Our test condition for a word W to be a proper prefix of such a V is that
SW be irreducible with respect to $\mathcal{R}$ and that SW have a suffix T with
$|T| > |W|$ such that T is a prefix of some rule in $\mathcal{R}$. At any given moment
W is the word $x_1 \ldots x_r$, where x_i is the element of X chosen to construct β_i

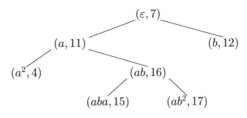

Figure 3.5.1

as $(\beta_{i-1})^{x_i}$. The state β_r is α^{SW} and hence $\ell(\beta_r)$ is the length of the longest suffix T of SW which is a prefix of some left side in $\mathcal{R}$.

Figure 3.5.1 shows the search tree examined by INDEX_CONFLUENT in checking the confluence of the rewriting system S in Example 5.1 when $U = b^3$. Each node of the tree is labeled by the pair (W, β_r), where $r = |W|$. The nodes $(b, 12)$, $(aba, 15)$, and $(ab^2, 17)$ correspond to the overlaps b^4, b^3aba, and b^3ab^2. The node $(a^2, 4)$ does not correspond to an overlap since $\ell(\beta_r) \le r$ in that case.

Using INDEX_CONFLUENT, it is practical to verify the confluence of the rewriting system $\mathcal{T}$ of Example 6.6 in Chapter 2. Only 6129 overlaps need to be processed.

Let us return to the problem of constructing an automaton $\mathcal{A} = (\Sigma, X, E, \{\alpha\}, \Omega)$ isomorphic to $I(\mathcal{R})$. In order to be able to construct the edges in $E_2 - E_3$, it is useful to be able to obtain for any σ in Σ the word $U = U_\sigma$ of length $\ell(\sigma)$ such that $\sigma = \alpha^U$. There are several ways to facilitate this. One could explicitly store U_σ with σ. One could store with σ a pointer to a rule of whose left side U_σ is a prefix. Also, assuming $U_\sigma \ne \varepsilon$, one could write U_σ as Sx with x in X and store the pair (τ, x), where $\tau = \alpha^S$. One could then reconstruct U_σ recursively as $U_\tau x$.

The following procedure constructs an automaton isomorphic to $I(\mathcal{R})$ for a reduced rewriting system $\mathcal{R}$. The method of determining the words U_σ is left unspecified.

Procedure INDEX($\mathcal{R}; \mathcal{A}$);
Input: $\mathcal{R}$: a nonempty, finite, reduced rewriting system on X^*;
Output: $\mathcal{A}$: an automaton isomorphic to $I(\mathcal{R})$ together with its
                  length function ℓ and its rule identifier f;
Begin
    $n := 1$; $\ell(1) := 0$; $\Sigma := \{1\}$; $A := \{1\}$; $\Omega := \emptyset$; $E := \emptyset$;
    (* At all times $\mathcal{A}$ will denote $(\Sigma, X, E, A, \Omega)$. The sets $\Sigma_1, \Sigma_2, \ldots$ are
        all assumed to be empty initially. We first determine the edges
        in E_1 and E_3. *)
    For (L, R) in $\mathcal{R}$ do begin
        Let $L = x_1 \ldots x_s$ with each x_i in X; $\sigma := 1$;
        For $i := 1$ to s do

If σ^{x_i} is defined then $\sigma := \sigma^{x_i}$
Else begin
  $n := n + 1$; $\ell(n) := i$; Add n to Σ and Σ_i;
  Add (σ, x_i, n) to E; $\sigma := n$;
  Record information on how to reconstruct $U_\sigma = x_1 \ldots x_i$
End;
$\Omega := \Omega \cup \{\sigma\}$; $f(\sigma) := (L, R)$;
For x in X do add (σ, x, σ) to E
End;
($*$ Determine the edges in $E_2 - E_3$. $*$)
For x in X do if 1^x is not defined then add $(1, x, 1)$ to E;
$i := 1$;
While $\Sigma_i \neq \emptyset$ do begin
 For σ in Σ_i do begin
  Let $U_\sigma = x_1 \ldots x_i$; $\tau := 1^{x_2 \ldots x_i}$;
  For x in X do if σ^x is not defined then add (σ, x, τ^x) to E
 End;
 $i := i + 1$
End
End.

There are difficulties connected with the use of index automata in the Knuth-Bendix procedure for strings. The problem is that the rewriting system is continually changing as new rules are found and redundant rules are deleted. Updating the index automaton after each change is time-consuming. There are several alternatives which can be used. One can partition $\mathcal{R}$ into a number of smaller sets and construct an index automaton for each set. A change in $\mathcal{R}$ may require that only one or two of the indexes be modified. Another option is to use an index structure containing less information and hence more easily updated. A good choice is an automaton recognizing the set of left sides in $\mathcal{R}$. One can get such an automaton by removing all edges from $I(\mathcal{R})$ except those in E_3. With either of these approaches there is a penalty to be paid in the form of slower rewriting and formation of overlaps. The right balance remains to be determined.

The Rabin-Karp string matching algorithm as described, for example, in [Sedgewick 1990] can be modified to produce an index structure which uses much less space than an index automaton. This Rabin-Karp index cannot be used to locate overlaps, but it permits rewriting to be carried out on average almost as quickly as can be done with an index automaton.

Exercises

5.1. Use the procedure INDEX_REWRITE with the rewriting system $\mathcal{S}$ and the index automaton $I(\mathcal{S})$ constructed in Example 5.1 to rewrite the word bab^2abab^2ab.

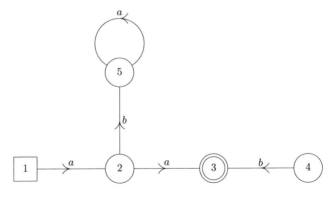

Figure 3.6.1

5.2. Using INDEX_CONFLUENT, find all overlaps of left sides in S satisfying the conclusion of Proposition 7.1 of Chapter 2.

5.3. Construct $I(\mathcal{R})$, where $\mathcal{R}$ is the rewriting system on $\{a,b\}^*$ consisting of the rules $a^5 \to \varepsilon$, $b^5 \to \varepsilon$, $b^4a^4 \to (ab)^4$, and $(ba)^4 \to a^4b^4$.

5.4. Suppose $|\mathcal{R}| = 1$. Show that an automaton isomorphic to $I(\mathcal{R})$ may be constructed much more simply than is done in INDEX.

5.5. Let $\mathcal{R}$ be a reduced rewriting system and let $I_0(\mathcal{R})$ be the automaton obtained from $I(\mathcal{R})$ by deleting all edges except those in E_3. Devise efficient methods of updating $I_0(\mathcal{R})$ when a rule is added to or deleted from $\mathcal{R}$.

3.6 Trim automata

In Section 2.8 we discussed the use of the Knuth-Bendix procedure for strings to investigate right congruences which are finitely generated modulo a finitely generated two-sided congruence. To illustrate the use of automata in such computations, it is necessary to develop a little more theory.

It may happen that some states of an automaton $\mathcal{A}$ are superfluous and may be deleted without affecting the language recognized by $\mathcal{A}$. To see how this can happen, let us consider the automaton $\mathcal{A}$ shown in Figure 3.6.1. The single initial state is 1 and the single terminal state is 3. No path in $\mathcal{A}$ from 1 to 3 passes through either state 4 or state 5. There are paths from 1 to 5, but there is no way to complete any of them to a path to 3. There is no path at all from 1 to 4.

The observations of the previous paragraph suggest the following definitions: Let $\mathcal{A}$ be an automaton and let σ be a state of $\mathcal{A}$. We say that σ is *accessible* if there is a path in $\mathcal{A}$ from some initial state to σ. We say that σ is *coaccessible* if there is a path from σ to some terminal state. Finally, σ is *trim* if it is both accessible and coaccessible. Clearly σ is trim if and only if it occurs on a path from an initial state to a terminal state. In the preceding example, states 1, 2, 3, and 5 are accessible, and states 1, 2, 3, and 4 are coaccessible. The trim states are 1, 2, and 3.

Given a finite automaton $\mathcal{A}$, we can find the set of trim states of $\mathcal{A}$ with the following procedure:

Procedure TRIM($\mathcal{A}; \Lambda$);
Input: $\mathcal{A}$: a finite automaton $(\Sigma, X, E, A, \Omega)$;
Output: Λ : the set of trim states of $\mathcal{A}$;
Begin
  (* Find the set Λ_a of accessible states. *)
  $\Lambda_a := A$; $\Phi := \Lambda_a$;
  While $\Phi \neq \emptyset$ do begin
    Choose φ in Φ; $\Phi := \Phi - \{\varphi\}$;

    For each edge (φ, U, λ) in E do
      If λ is not in Λ_a then begin
        $\Lambda_a := \Lambda_a \cup \{\lambda\}$; $\Phi := \Phi \cup \{\lambda\}$
    End
  End;
  (* Now find the set Λ_c of coaccessible states. *)
  $\Lambda_c := \Omega$; $\Phi := \Lambda_c$;
  While $\Phi \neq \emptyset$ do begin
    Choose φ in Φ; $\Phi := \Phi - \{\varphi\}$;

    For each edge (λ, U, φ) in E do
      If λ is not in Λ_c then begin
        $\Lambda_c := \Lambda_c \cup \{\lambda\}$; $\Phi := \Phi \cup \{\lambda\}$
    End
  End;
  $\Lambda := \Lambda_a \cap \Lambda_c$
End.

Example 6.1. Let us apply TRIM to the automaton constructed in Example 4.1. The accessible states are found to be 1, 3, 4, 5, and 7. All states except 1 are coaccessible. Thus the set of trim states is $\{3, 4, 5, 7\}$. The restriction to the set of trim states is described by Figure 3.6.2.

Although the instructions for the two While-loops in TRIM are very similar, there is an important point to be noticed concerning data structures. In the first loop, we need to know, for a state φ, which edges start at φ. However, in the second loop, we need to know which edges end at φ. The data structure used to represent the edges in the automaton may favor one or the other type of access.

We shall call the restriction of an automaton $\mathcal{A}$ to its set of trim states the *trim part* of $\mathcal{A}$ and denote it $\mathcal{A}_t$. The automaton is *trim* if all states are trim.

Proposition 6.1. *For any automaton $\mathcal{A}$ the languages $L(\mathcal{A})$ and $L(\mathcal{A}_t)$ are the same.*

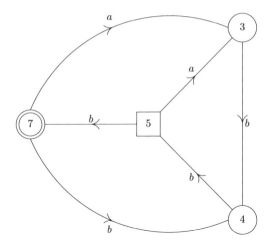

Figure 3.6.2

Proof. Since any path in $\mathcal{A}$ from an initial state to a terminal state passes through only trim states, the result is immediate. □

Occasionally we shall need to consider the *accessible part* $\mathcal{A}_a$ of an automaton, the restriction of $\mathcal{A}$ to its accessible states. If $\mathcal{A}_a = \mathcal{A}$, then $\mathcal{A}$ is *accessible*.

Let $\mathcal{L}$ be a language over X. The states of $A(\mathcal{L})$ are the cones of $\mathcal{L}$. If σ is the cone $C(\mathcal{L}, W)$, then there is a path P in $A(\mathcal{L})$ from the initial state to σ such that $\text{Sg}(P) = W$. Therefore every state of $A(\mathcal{L})$ is accessible. The cone σ is coaccessible if and only if it is nonempty. Thus the set $C_t(\mathcal{L})$ defined in Section 3.1 is the set of trim states of $A(\mathcal{L})$. The trim part of $A(\mathcal{L})$ will be denoted $A_t(\mathcal{L})$.

Proposition 6.2. *Let $\mathcal{A}$ be a trim automaton. Then $L(\mathcal{A}) = \emptyset$ if and only if the set of states of $\mathcal{A}$ is empty. Also, $L(\mathcal{A})$ contains a nonempty word if and only if some edge of $\mathcal{A}$ has a nonempty label.*

Proof. Let $\mathcal{A} = (\Sigma, X, E, A, \Omega)$. We have already remarked that the trim states are precisely the states occurring in paths from A to Ω. Thus if there are trim states, then $L(\mathcal{A}) \neq \emptyset$. Clearly, if $\Sigma = \emptyset$, then $L(\mathcal{A}) = \emptyset$. If (σ, U, τ) is an edge with $U \neq \varepsilon$, then U is a subword of some word in $L(\mathcal{A})$, and therefore $L(\mathcal{A})$ contains a nonempty word. □

Corollary 6.3. *Given a finite automaton $\mathcal{A}$, it is possible to decide whether $L(\mathcal{A}) = \emptyset$ and whether $L(\mathcal{A})$ contains a nonempty word.*

Proof. Using TRIM, we can compute $\mathcal{A}_t$ and apply Proposition 6.2. □

A *circuit* in an automaton is a path which starts and ends at the same point.

Proposition 6.4. *Suppose $\mathcal{A}$ is a finite, trim automaton. Then $L(\mathcal{A})$ is infinite if and only if $\mathcal{A}$ has a circuit with nonempty signature.*

Proof. Let $\mathcal{A} = (\Sigma, X, E, A, \Omega)$. Suppose first that $\mathcal{A}$ has a circuit P which starts and ends at σ and the signature W of P is not ε. Since $\mathcal{A}$ is trim, there is a path Q from A to σ and a path R from σ to Ω. By concatenating Q, any number nonnegative i copies of P, and R, we obtain a path from A to Ω with signature $\mathrm{Sg}(Q)W^i\,\mathrm{Sg}(R)$. All of these words are distinct and are in $L(\mathcal{A})$, so $L(\mathcal{A})$ is infinite.

Now suppose that $L(\mathcal{A})$ is infinite. Let n be the number of edges in $\mathcal{A}$. Since $L(\mathcal{A})$ is infinite, $L(\mathcal{A})$ contains a word W of length greater than n. Let P be a path from A to Ω with $W = \mathrm{Sg}(P)$. Some edge (σ, U, τ) with $U \neq \varepsilon$ must appear more than once in P. Let Q be the subpath of P which starts with the first occurrence of (σ, U, τ) and continues until σ is reached again. Then Q is a circuit, and the signature of Q contains U as a prefix.

$\square$

Corollary 6.5. *Given a finite automaton $\mathcal{A}$, it is possible to decide whether $L(\mathcal{A})$ is infinite.*

Proof. Replace $\mathcal{A}$ by $\mathcal{A}_t$, if necessary, so that $\mathcal{A}$ is trim. Let $\mathcal{A} = (\Sigma, X, E, A, \Omega)$. If σ is in Σ, then the set of signatures of circuits in $\mathcal{A}$ starting at σ is $L(\mathcal{A}_\sigma)$, where $\mathcal{A}_\sigma = (\Sigma, X, E, \{\sigma\}, \{\sigma\})$. By Corollary 6.3, we can decide whether $L(\mathcal{A}_\sigma)$ contains a nonempty word. Applying this test for each σ in Σ, we can decide whether $\mathcal{A}$ has a circuit with nonempty signature. $\square$

When the subset construction of Section 3.4 is applied to an automaton $\mathcal{A}$, it frequently happens that only a few states of the resulting automaton $\mathcal{A}_1$ are accessible, and even fewer states are trim. It is possible to construct directly the accessible part of $\mathcal{A}_1$. The following example illustrates this construction.

Example 6.2. Let $X = \{a, b\}$ and let $\mathcal{U} = \{a^3, b^3, (ab)^2\}$. We shall construct a complete automaton recognizing the ideal $\mathcal{I} = X^*\mathcal{U}X^*$. Figure 3.6.3 gives a nondeterministic automaton $\mathcal{A}$ recognizing $\mathcal{I}$. If we apply the subset construction directly to $\mathcal{A}$, we shall get an automaton $\mathcal{A}_1$ with $2^9 = 512$ states, since all subsets of $\{1, 2, 3, 4, 5, 6, 7, 8, 9\}$ are closed. However, by imitating the first While-loop in TRIM, we can avoid considering most of these states. The initial state is $\Lambda_1 = \{1\}$. The set $(\Lambda_1)^a$ is $\Lambda_2 = \{1, 3, 7\}$, and $(\Lambda_1)^b$ is $\Lambda_3 = \{1, 5\}$. Now we compute $(\Lambda_2)^a$ and $(\Lambda_2)^b$, which turn out

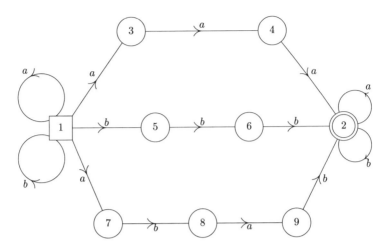

Figure 3.6.3

Table 3.6.1

	a	b			a	b
1	2	3		7	7	10
2	4	5		8	4	10
3	2	6		9	11	9
4	7	5		10	12	9
5	8	6		11	7	10
6	2	9		12	7	10

to be $\Lambda_4 = \{1,3,4,7\}$ and $\Lambda_5 = \{1,5,8\}$, respectively. Continuing in this manner, we find that

$$(\Lambda_3)^a = \Lambda_2, \qquad\qquad (\Lambda_3)^b = \Lambda_6 = \{1,5,6\},$$
$$(\Lambda_4)^a = \Lambda_7 = \{1,2,3,4,7\}, \qquad (\Lambda_4)^b = \Lambda_5,$$
$$(\Lambda_5)^a = \Lambda_8 = \{1,3,7,9\}, \qquad (\Lambda_5)^b = \Lambda_6,$$
$$(\Lambda_6)^a = \Lambda_2, \qquad\qquad (\Lambda_6)^b = \Lambda_9 = \{1,2,5,6\},$$
$$(\Lambda_7)^a = \Lambda_7, \qquad\qquad (\Lambda_7)^b = \Lambda_{10} = \{1,2,5,8\},$$
$$(\Lambda_8)^a = \Lambda_4, \qquad\qquad (\Lambda_8)^b = \Lambda_{10},$$
$$(\Lambda_9)^a = \Lambda_{11} = \{1,2,3,7\}, \qquad (\Lambda_9)^b = \Lambda_9,$$
$$(\Lambda_{10})^a = \Lambda_{12} = \{1,2,3,7,9\}, \qquad (\Lambda_{10})^b = \Lambda_9,$$
$$(\Lambda_{11})^a = \Lambda_7, \qquad\qquad (\Lambda_{11})^b = \Lambda_{10},$$
$$(\Lambda_{12})^a = \Lambda_7, \qquad\qquad (\Lambda_{12})^b = \Lambda_{10}.$$

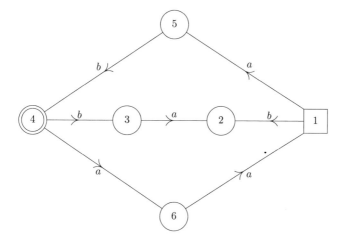

Figure 3.6.4

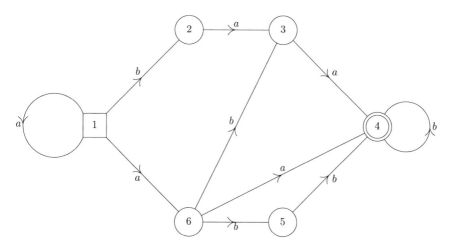

Figure 3.6.5

Thus the edges of the accessible part $\mathcal{A}_2$ of $\mathcal{A}_1$ are given by Table 3.6.1, where i has been written for Λ_i. The set of terminal states of $\mathcal{A}_2$ is $\{7, 9, 10, 11, 12\}$, and it is easily checked that $\mathcal{A}_2$ is trim.

The variation of the subset construction outlined in Example 6.2 will be referred to as the *accessible subset construction*.

Exercises

6.1. Compute the trim part of the automaton in Figure 3.6.4.
6.2. Apply the accessible subset construction to the automaton in Figure 3.6.5.

6.3. Let $\mathcal{A} = (\Sigma, X, E, A, \Omega)$ be a finite, trim automaton. Set $\mathcal{B}_1 = \mathcal{A}$, and for $i \geq 1$ define
$\mathcal{B}_{i+1}$ to be the restriction of $\mathcal{B}_i$ to the set of states in $\mathcal{B}_i$ which have incoming edges in
$\mathcal{B}_i$. Show that there is an integer $k \leq |\Sigma| + 1$ such that $\mathcal{B}_k = \mathcal{B}_{k+1}$. Show that $L(\mathcal{A})$ is
infinite if and only if $\mathcal{B}_k$ has an edge with a nonempty label.

6.4. Let $\mathcal{A}$ be a trim, deterministic automaton such that $L(\mathcal{A})$ is an ideal. Prove that $\mathcal{A}$ is
either empty or complete.

3.7 Minimal automata

A given rational language $\mathcal{L}$ may be recognized by many different finite
automata. In this section we shall show that $A(\mathcal{L})$ and its trim part $A_t(\mathcal{L})$
are the smallest complete automaton and the smallest trim, determinis-
tic automaton, respectively, recognizing $\mathcal{L}$. We shall also describe how to
construct an automaton isomorphic to $A_t(\mathcal{L})$.

To establish the main results of this section, we must consider certain
kinds of maps from the set of states of one deterministic automaton to the
set of states of another deterministic automaton with the same alphabet.
These maps or "morphisms" make various collections of automata into categories. Category theory will not play a major role in our discussions, and
no prior knowledge of categories is assumed. A more complete treatment
of categories of automata can be found in [Eilenberg 1974].

For $i = 1, 2$, let $\mathcal{A}_i = (\Sigma_i, X, E_i, A_i, \Omega_i)$ be a deterministic automaton with
alphabet X. An *expanding morphism* from $\mathcal{A}_1$ to $\mathcal{A}_2$ is a map $f : \Sigma_1 \to \Sigma_2$
such that

   (i) $f(A_1) \subseteq A_2$.
  (ii) $f(\Omega_1) \subseteq \Omega_2$.
 (iii) If P is a path in $\mathcal{A}_1$ from σ to τ, then there is a path Q in $\mathcal{A}_2$ from
        $f(\sigma)$ to $f(\tau)$ such that $\mathrm{Sg}(Q) = \mathrm{Sg}(P)$.

Proposition 7.1. *If there is an expanding morphism f from $\mathcal{A}_1$ to $\mathcal{A}_2$,
then $L(\mathcal{A}_1) \subseteq L(\mathcal{A}_2)$.*

Proof. Suppose that W is in $L(\mathcal{A}_1)$. There is a path P in $\mathcal{A}_1$ from α in
A_1 to ω in Ω_1 such that $W = \mathrm{Sg}(P)$. By the definition of an expanding
morphism, there is a path Q in $\mathcal{A}_2$ from $f(\alpha)$ to $f(\omega)$ such that $W = \mathrm{Sg}(Q)$.
Since $f(\alpha)$ is in A_2 and $f(\omega)$ is in Ω_2, this means that W is in $L(\mathcal{A}_2)$. $\square$

Condition (iii) in the definition of expanding morphism refers to all paths.
However, it suffices to consider only paths of length 1. (See Exercise 7.2.)
Thus if $\mathcal{A}_1$ and $\mathcal{A}_2$ are finite, it is possible to decide whether a given map
from Σ_1 to Σ_2 is an expanding morphism.

Proposition 7.2. *If $\mathcal{A}_1$ is accessible, then there is at most one expanding
morphism from $\mathcal{A}_1$ to $\mathcal{A}_2$.*

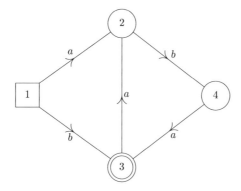

Figure 3.7.1

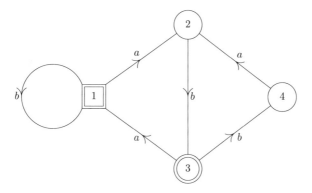

Figure 3.7.2

Proof. If $A_1 = \emptyset$, then $\Sigma_1 = \emptyset$ and the empty map is the unique expanding morphism from $\mathcal{A}_1$ to $\mathcal{A}_2$. Thus we may assume that $A_1 = \{\alpha\}$. Suppose there is an expanding morphism f from $\mathcal{A}_1$ to $\mathcal{A}_2$. Then $f(\alpha)$ is the unique element of A_2. For any σ in Σ_1 there is a path P in $\mathcal{A}_1$ from α to σ. Therefore $f(\sigma) = f(\alpha)^W$, where $W = \mathrm{Sg}(P)$. Thus f is unique. $\square$

Example 7.1. Let $X = \{a, b\}$ and let $\mathcal{A}_1$ and $\mathcal{A}_2$ be as shown in Figures 3.7.1 and 3.7.2. If there is an expanding morphism f from $\mathcal{A}_1$ to $\mathcal{A}_2$, then condition (i) forces f to map 1 to 1. Since $2 = 1^a$ in $\mathcal{A}_1$, we must have $f(2) = f(1)^a = 1^a = 2$ in $\mathcal{A}_2$. Similarly $f(3) = f(1)^b = 1^b = 1$ and $f(4) = f(2)^b = 2^b = 3$. It is not hard to check that the map f so defined is an expanding morphism.

The following are the most important results about the existence of expanding morphisms.

Proposition 7.3. *Let* $\mathcal{A} = (\Sigma, X, E, A, \Omega)$ *be a complete, accessible automaton. There is a unique expanding morphism f from $\mathcal{A}$ to $A(\mathcal{L})$, where $\mathcal{L} = L(\mathcal{A})$. Moreover, f is surjective as a map on states. If f is injective, then f is an isomorphism from $\mathcal{A}$ to $A(\mathcal{L})$.*

Proof. The uniqueness of f follows from Proposition 7.2. Let $A = \{\alpha\}$ and let σ be any element of Σ. Since $\mathcal{A}$ is accessible, there is a path P in $\mathcal{A}$ from α to σ.

Lemma 7.4. $C(\mathcal{L}, \mathrm{Sg}(P))$ *depends only on σ, not on the choice of P.*

Proof. A word W is in $C(\mathcal{L}, \mathrm{Sg}(P))$ if and only if $\mathrm{Sg}(P)W$ is in $\mathcal{L}$. Since $\mathcal{A}$ recognizes $\mathcal{L}$, this is the case if and only if there is a path Q in $\mathcal{A}$ from α to Ω such that $\mathrm{Sg}(Q) = \mathrm{Sg}(P)W$. If such a Q exists, then P is an initial segment of Q. Thus W is in $C(\mathcal{L}, \mathrm{Sg}(P))$ if and only if there is a path from σ to Ω having W as its signature. Therefore $C(\mathcal{L}, \mathrm{Sg}(P))$ depends only on σ. $\square$

By Lemma 7.4, we can define a map $f : \Sigma \to C(\mathcal{L})$ by $f(\sigma) = C(\mathcal{L}, \mathrm{Sg}(P))$. Given U in X^*, there is a σ in Σ such that $\sigma = \alpha^U$, since $\mathcal{A}$ is complete. The image $f(\sigma)$ is $C(\mathcal{L}, U)$, and hence f is surjective. Now $f(\alpha) = C(\mathcal{L}, \varepsilon) = \mathcal{L}$, the initial state of $A(\mathcal{L})$. Also, α^U is in Ω if and only if U is in $\mathcal{L}$, which holds if and only if $C(\mathcal{L}, U)$ is a terminal state of $A(\mathcal{L})$. Thus f maps Ω onto the set of terminal states in $A(\mathcal{L})$. If Q is a path in $\mathcal{A}$ from σ to τ, then concatenating P and Q yields a path from α to τ. Thus $f(\tau) = C(\mathcal{L}, \mathrm{Sg}(P)\,\mathrm{Sg}(Q))$. By the definition of $A(\mathcal{L})$, there is a path in $A(\mathcal{L})$ from $f(\sigma)$ to $f(\tau)$ having signature $\mathrm{Sg}(Q)$. Therefore f is an expanding morphism.

Suppose that $e = (\sigma, x, \tau)$ is in E. The preceding argument shows that $(f(\sigma), x, f(\tau))$ is an edge of $A(\mathcal{L})$. Since $\mathcal{A}$ is complete, f maps E onto the set of edges of $A(\mathcal{L})$. Thus if f is injective, then f is an isomorphism. $\square$

Corollary 7.5. *If $\mathcal{A}$ is a finite, complete automaton recognizing $\mathcal{L}$, then $\mathcal{A}$ has at least $|C(\mathcal{L})|$ states. If $\mathcal{A}$ has exactly $|C(\mathcal{L})|$ states, then $\mathcal{A}$ is isomorphic to $A(\mathcal{L})$.*

Proof. Replacing $\mathcal{A}$ by its accessible part can only decrease the number of states. Thus we may assume that $\mathcal{A}$ is accessible and apply Proposition 7.3. $\square$

In view of Corollary 7.5, it is reasonable to consider $|C(\mathcal{L})|$ to be a measure of the complexity of $\mathcal{L}$. Relatively minor changes in the proofs of Proposition 7.3 and Corollary 7.5 yield the following results.

Proposition 7.6. *Let* $\mathcal{A} = (\Sigma, X, E, A, \Omega)$ *be a trim, deterministic automaton recognizing a language* $\mathcal{L}$. *There is a unique expanding morphism* f *from* $\mathcal{A}$ *to* $A_t(\mathcal{L})$. *Moreover,* f *is surjective as a map of states. If* f *is injective, then* f *is an isomorphism of* $\mathcal{A}$ *with* $A_t(\mathcal{L})$.

Corollary 7.7. *If* $\mathcal{A}$ *is a finite, trim, deterministic automaton recognizing* $\mathcal{L}$, *then* $\mathcal{A}$ *has at least* $|C_t(\mathcal{L})|$ *states. If* $\mathcal{A}$ *has exactly* $|C_t(\mathcal{L})|$ *states, then* $\mathcal{A}$ *is isomorphic to* $A_t(\mathcal{L})$.

Corollary 7.7 makes it natural to consider the problem of determining an automaton isomorphic to $A_t(\mathcal{L})$, where $\mathcal{L}$ is given as $L(\mathcal{A})$ for some finite automaton $\mathcal{A}$. By applying the accessible subset construction and taking the trim part if necessary, we may assume that $\mathcal{A}$ is deterministic and trim. Our algorithm for computing $A_t(\mathcal{L})$ involves the possibly nondeterministic automaton $\mathcal{A}^\dagger = (\Sigma, X, E^\dagger, \Omega, A)$ defined in Exercise 2.8.

Proposition 7.8. *Let* $\mathcal{A} = (\Sigma, X, E, A, \Omega)$ *be a finite, trim, deterministic automaton, let* $f : \Sigma \to C_t(L(\mathcal{A}))$ *be the map of Proposition 7.6, and let* $\mathcal{B}$ *be the result of applying the accessible subset construction to* $\mathcal{A}^\dagger$. *If* σ *and* τ *are in* Σ, *then* $f(\sigma) = f(\tau)$ *if and only if, for each state* Γ *of* $\mathcal{B}$, *either both* σ *and* τ *are in* Γ *or neither is in* Γ.

Proof. For any word U in X^*, let $\Delta(U)$ be the set of states σ in Σ such that there is a path in $\mathcal{A}^\dagger$ from Ω to σ having signature U, or, equivalently, there is a path in $\mathcal{A}$ from σ to Ω having signature $U^\dagger$. Then the states in $\mathcal{B}$ are the possible sets $\Delta(U)$ as U ranges over X^*, and $U^\dagger$ is in $f(\sigma)$ if and only if σ is in $\Delta(U)$. Thus $f(\sigma)$ is the set of words $U^\dagger$ such that σ is in $\Delta(U)$. Therefore $f(\sigma) = f(\tau)$ if and only if σ and τ belong to exactly the same states of $\mathcal{B}$. □

Since the edges in $A_t(\mathcal{L})$ are images of the edges in $\mathcal{A}$, we can now construct an automaton isomorphic to $A_t(\mathcal{L})$.

Example 7.2. Let $X = \{a, b\}$, $\Sigma = \{1, 2, 3, 4, 5, 6, 7\}$, $A = \{1\}$, and $\Omega = \{3, 6\}$, and let E be given by Table 3.7.1. Then $\mathcal{A} = (\Sigma, X, E, A, \Omega)$ is trim. The edges of $\mathcal{A}^\dagger$ are given by Table 3.7.2. Here, for each state σ and each generator x the possible endpoints τ in edges (σ, x, τ) of $\mathcal{A}^\dagger$ are listed. Applying the accessible subset construction to $\mathcal{A}^\dagger$, we find that the accessible states in the resulting automaton correspond to the following subsets of Σ.

$$\{3, 6\}, \quad \{1, 4, 7\}, \quad \{2, 3, 5, 6\}, \quad \emptyset, \quad \{1, 2, 3, 4, 5, 6, 7\}.$$

By Proposition 7.8, $f(\sigma) = f(\tau)$ if and only if σ and τ are in the same block of the partition $\{\{1, 4, 7\}, \{2, 5\}, \{3, 6\}\}$ of Σ. Thus we may identify

Table 3.7.1

	a	b
1	3	2
2	4	6
3	1	6
4	3	5
5	7	3
6	1	3
7	6	2

Table 3.7.2

State	Label	Endpoints
1	a	3, 6
2	b	1, 7
3	a	1, 4
	b	5, 6
4	a	2
5	b	4
6	a	7
	b	2, 3
7	a	5

$C_t(L(\mathcal{A}))$ with $\{1, 2, 3\}$ in such a way that f maps 1, 4, and 7 to 1, f maps 2 and 5 to 2, and f maps 3 and 6 to 3. The edges of $A_t(L(\mathcal{A}))$ are the triples $(f(\sigma), x, f(\tau))$, where (σ, x, τ) ranges over the edges of $\mathcal{A}$. Therefore $A_t(L(\mathcal{A}))$ is isomorphic to $(\{1, 2, 3\}, X, E_1, \{1\}, \{3\})$, where E_1 corresponds to the following transition table:

	a	b
1	3	2
2	1	3
3	1	3

Exercises

7.1. Suppose that f is an expanding morphism from $\mathcal{A}_1$ to $\mathcal{A}_2$ and g is an expanding morphism from $\mathcal{A}_2$ to $\mathcal{A}_3$. Show that $f \circ g$ is an expanding morphism from $\mathcal{A}_1$ to $\mathcal{A}_3$.

7.2. Show that it is sufficient to assume condition (iii) in the definition of an expanding morphism for paths of length 1.

7.3. Let $\mathcal{A}$ be a trim, deterministic automaton, and assume that there is an expanding morphism from $A_t(L(\mathcal{A}))$ to $\mathcal{A}$. Prove that $\mathcal{A}$ and $A_t(L(\mathcal{A}))$ are isomorphic.

7.4. Construct an automaton isomorphic to $A_t(L(\mathcal{A}))$, where $\mathcal{A}$ is the automaton in Figure 3.7.3.

3.8 Standard automata

Given two finite automata, it is clearly a finite problem to decide whether the automata are isomorphic. In the case of accessible, deterministic automata it is possible to select representatives from the isomorphism classes in a natural way. This facilitates the determination of isomorphism. To define isomorphism class representatives of finite, accessible, deterministic automata with alphabet X, we fix a reduction ordering $\prec$ on X^*.

Let $\mathcal{A} = (\Sigma, X, E, A, \Omega)$ be a finite, accessible, deterministic automaton. If $A = \emptyset$, then $\mathcal{A}$ is the unique automaton over X with no states or edges. If $A \neq \emptyset$, then A contains a single state α, and for each σ in Σ there is at

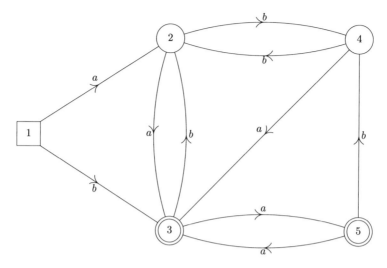

Figure 3.7.3

least one path in $\mathcal{A}$ from α to σ. Let $\mathcal{L}_\sigma$ be the language recognized by $(\Sigma, X, E, A, \{\sigma\})$. Thus $\mathcal{L}_\sigma$ is the set of signatures of paths in $\mathcal{A}$ from α to σ. Since $\mathcal{L}_\sigma \neq \emptyset$, there is a smallest element W_σ in $\mathcal{L}_\sigma$ with respect to $\prec$. If $\sigma \neq \tau$, then $W_\sigma \neq W_\tau$. Since ε is in $\mathcal{L}_\alpha$, it follows that $W_\alpha = \varepsilon$.

We shall say that $\mathcal{A}$ is *standard* with respect to $\prec$ if the following conditions hold:

(i) $\Sigma = \{1, \ldots, n\}$, where $n = |\Sigma|$.
(ii) $A \subseteq \{1\}$.
(iii) If σ and τ are in Σ and $\sigma < \tau$, then $W_\sigma \prec W_\tau$.

Under this definition, $(\emptyset, X, \emptyset, \emptyset, \emptyset)$ is standard. Condition (iii) can be replaced by

(iii') Suppose U is in X^*, $U \preceq W_\sigma$, and 1^U is defined. Then $1^U \leq \sigma$.
Note that whether or not $\mathcal{A}$ is standard does not depend on Ω. In the discussion that follows, we shall tend to ignore the sets of terminal states in our automata.

Proposition 8.1. *Every finite, accessible, deterministic automaton with alphabet X is isomorphic to a unique automaton which is standard with respect to $\prec$.*

Proof. Let $\mathcal{A} = (\Sigma, X, E, A, \Omega)$ be a finite, accessible, deterministic automaton. There is a unique way to number the elements of Σ as $\sigma_1, \ldots, \sigma_n$ such that $W_{\sigma_i} \prec W_{\sigma_j}$ if $i < j$. If $A = \{\alpha\}$, then $W_\alpha = \varepsilon$, so $\alpha = \sigma_1$. Let $\Sigma_1 = \{1, \ldots, n\}$, $A_1 = \{1\} \cap \Sigma_1$, $\Omega_1 = \{i \mid \sigma_i \in \Omega\}$, and

$E_1 = \{(i, x, j) \mid (\sigma_i, x, \sigma_j) \in E\}$. We leave it as an exercise to show that $\mathcal{A}_1 = (\Sigma_1, X, E_1, A_1, \Omega_1)$ is standard and $\mathcal{A}_1$ is the only standard automaton isomorphic to $\mathcal{A}$.

$\square$

Although we know that the words W_{σ_i} in the proof of Proposition 8.1 must exist, we have as yet no algorithm for computing them. Thus the proof of Proposition 8.1 does not lead immediately to an algorithm for *standardization*, that is, for finding $\mathcal{A}_1$. See Exercise 8.5.

It is possible to place the notion of a standard automaton into a broader context. Let $\mathcal{A} = (\Sigma, X, E, A, \Omega)$ be a nonempty, finite, accessible, deterministic automaton and let α be the unique initial state of $\mathcal{A}$. Set $\mathcal{W} = \mathcal{W}(\mathcal{A})$ equal to the set of words W in X^* such that α^W is defined, and let $f = f_{\mathcal{A}}$ be the function from $\mathcal{W}$ to Σ given by $f(W) = \alpha^W$. The pair $(\mathcal{W}, f)$ determines Σ, E, and A, but not Ω. Since $\mathcal{A}$ is accessible, f maps $\mathcal{W}$ onto Σ and $\alpha = f(\varepsilon)$. If σ and τ are in Σ and x is in X, then (σ, x, τ) is in E if and only if there is a word W in $\mathcal{W}$ such that Wx is also in $\mathcal{W}$, $f(W) = \sigma$ and $f(Wx) = \tau$.

Let us say that $\mathcal{A}$ is *numeric* if Σ is a set of positive integers. Our choice of a reduction ordering $\prec$ on X^* defines an ordering of finite, accessible, deterministic, numeric automata with alphabet X. (Strictly speaking, the ordering is on the triples (Σ, E, A) which arise in such automata, but as noted above we are ignoring the sets of terminal states.) The empty automaton $(\emptyset, X, \emptyset, \emptyset, \emptyset)$ comes first. Let $\mathcal{A}_i = (\Sigma_i, X, E_i, A_i, \Omega_i)$, $i = 1, 2$, be nonempty, finite, accessible, deterministic, numeric automata. Set $f_i = f_{\mathcal{A}_i}$. If $(\Sigma_1, E_1, A_1) \neq (\Sigma_2, E_2, A_2)$, then there is some word W such that $f_1(W)$ and $f_2(W)$ differ in the sense that one is defined and the other is not, or both are defined but they have different values. Let W be the first word at which f_1 and f_2 differ. Then we shall say that $\mathcal{A}_1$ *precedes* $\mathcal{A}_2$ and write $\mathcal{A}_1 \prec \mathcal{A}_2$ if either $f_1(W)$ is not defined and $f_2(W)$ is defined or both are defined and $f_1(W) < f_2(W)$. If neither automaton precedes the other, then they differ only in their sets of terminal states. If in addition the automata are isomorphic, then they must have the same terminal states as well, so $\mathcal{A}_1 = \mathcal{A}_2$. The precedence relation on finite, accessible, deterministic, numeric automata with alphabet X is transitive.

Proposition 8.2. *Let $\mathcal{A}$ be a finite, accessible, deterministic, numeric automaton with alphabet X. Then $\mathcal{A}$ is standard if and only if $\mathcal{A}$ precedes every other numeric automaton which is isomorphic to $\mathcal{A}$.*

Proof. Clearly we may assume that $\mathcal{A}$ is nonempty. Let $\mathcal{A} = (\Sigma, X, E, A, \Omega)$ and suppose first that $\mathcal{A}$ is standard. Let $\mathcal{A}' = (\Sigma', X, E', A', \Omega')$ be a numeric automaton which is isomorphic to $\mathcal{A}$ and different from $\mathcal{A}$. Then, as noted above, one of $\mathcal{A}$ and $\mathcal{A}'$ precedes the other. Suppose that

$\mathcal{A}'$ precedes $\mathcal{A}$. Since $\mathcal{A}$ and $\mathcal{A}'$ are isomorphic, $\mathcal{W}(\mathcal{A}) = \mathcal{W}(\mathcal{A}')$. Thus if W is the first word for which f_A and $f_{A'}$ differ, both $f_A(W)$ and $f_{A'}(W)$ are defined, and $f_{A'}(W) < f_A(W)$. Now $f_A(\varepsilon) = 1$ and $f_{A'}(\varepsilon)$ is positive. Therefore $W \neq \varepsilon$ and $A = A' = \{1\}$.

Let $\sigma = f_{A'}(W)$ and $\tau = f_A(W)$. Then $1 \leq \sigma < \tau$. Because $\mathcal{A}$ is standard, σ is a state of $\mathcal{A}$. Let V be the first word such that $1^V = \sigma$ in $\mathcal{A}$. Again because $\mathcal{A}$ is standard, $V \prec W$. By the definition of W, $\sigma = f_A(V) = f_{A'}(V)$. But this says that $1^V = 1^W$ in $\mathcal{A}'$ but $1^V \neq 1^W$ in $\mathcal{A}$. Since $\mathcal{A}$ and $\mathcal{A}'$ are isomorphic, this is impossible. Hence $\mathcal{A}$ precedes $\mathcal{A}'$.

Now suppose that $\mathcal{A}$ precedes all other numeric automata isomorphic to it. Then $\mathcal{A}$ is isomorphic to a standard automaton $\mathcal{A}'$, and by the previous argument $\mathcal{A}'$ precedes all numeric automata isomorphic to it. Thus if $\mathcal{A}$ and $\mathcal{A}'$ are different, then each precedes the other, which is impossible.   □

The orderings which will be used most frequently to define standard automata are length-plus-lexicographic orderings and basic wreath product orderings. It is important to develop efficient techniques for standardization and for deciding precedence of numeric automata with respect to these two classes of orderings. We shall begin with length-plus-lexicographic orderings.

Let $\mathcal{A} = (\Sigma, X, E, A, \Omega)$ be a finite, accessible, deterministic, numeric automaton. We can define an ordering on E, which we denote by $\ll$, as follows: $(\sigma, x, \tau) \ll (\varphi, y, \psi)$ if $\sigma < \varphi$ or $\sigma = \varphi$ and $x \prec y$. Since $\mathcal{A}$ is deterministic, the equalities $\sigma = \varphi$ and $x = y$ imply that $\tau = \psi$. If $A = \{\alpha\}$, then there may not be any edges going into α. However, since $\mathcal{A}$ is accessible, for every τ in $\Sigma - A$ there is an edge going into τ. Let e_τ be the first such edge with respect to $\ll$.

The automaton $\mathcal{A}$ is said to be *edge standard* relative to $\prec$ if the following conditions hold:

   (i) $\Sigma = \{1, \ldots, n\}$, where $n = |\Sigma|$.
   (ii) $A \subseteq \{1\}$.
   (iii) $e_2 \ll e_3 \ll \cdots \ll e_n$.

Assuming that conditions (i) and (ii) hold, condition (iii) is quite simple to check. We arrange the edges in order and look to see whether for $2 \leq \tau < n$ the first occurrence of τ as a third component comes before the first occurrence of $\tau + 1$.

Example 8.1. Let $X = \{a, b\}$ and let $\prec$ be the length-plus-lexicographic ordering on X^* with $a \prec b$. Suppose that $\mathcal{A} = (\Sigma, X, E, A, \Omega)$ is the automaton in which $\Sigma = \{1, 2, 3, 4\}$, $A = \{1\}$, and E is given by the following transition table:

	a	b
1	1	2
2	3	2
3	1	4
4	2	3

To examine the edges in order, we have only to go through entries in the table row by row. In doing so, we first encounter 2 as a third component in $e_2 = (1, b, 2)$. We first encounter 3 in $e_3 = (2, a, 3)$ and 4 in $e_4 = (3, b, 4)$. Since $e_2 \ll e_3 \ll e_4$, $\mathcal{A}$ is edge standard. In our previous notation, $W_1 = \varepsilon$, $W_2 = b$, $W_3 = ba$, and $W_4 = bab$. Thus $W_1 \prec W_2 \prec W_3 \prec W_4$. Hence $\mathcal{A}$ is also standard with respect to $\prec$. This is no accident.

Proposition 8.3. *Let $\mathcal{A}$ be a finite, accessible, deterministic, numeric automaton with alphabet X. If $\prec$ is a length-plus-lexicographic ordering of X^*, then $\mathcal{A}$ is edge standard with respect to $\prec$ if and only if $\mathcal{A}$ is standard with respect to $\prec$.*

Proof. We may clearly assume that $\mathcal{A}$ is not empty. Let $\mathcal{A} = (\Sigma, X, E, A, \Omega)$, where $\Sigma = \{1, \ldots, n\}$ and $A = \{1\}$. For σ in Σ let W_σ be the first word with respect to $\prec$ such that

$$1^{W_\sigma} = \sigma.$$

Lemma 8.4. *Suppose W_σ ends in an element x of X. Then $W_\sigma = (W_\tau)x$, where $\tau^x = \sigma$.*

Proof. Let $W_\sigma = Ux$ and let $\tau = 1^U$. Thus $\tau^x = \sigma$ and

$$1^{(W_\tau)x} = \sigma.$$

Therefore $(W_\tau)x \succeq W_\sigma = Ux$, so $W_\tau \succeq U$. But by the minimality of W_τ it follows that $U = W_\tau$. □

Suppose first that $\mathcal{A}$ is standard with respect to $\prec$.

Lemma 8.5. *Suppose that $2 \leq \tau \leq n$ and $e_\tau = (\sigma, x, \tau)$. Then $W_\tau = (W_\sigma)x$.*

Proof. By Lemma 8.4, W_τ has the form $(W_\rho)y$, where (ρ, y, τ) is an edge. By the definition of e_τ, $\sigma < \rho$ or $\sigma = \rho$ and $x \preceq y$. Since $\mathcal{A}$ is standard, $W_\sigma \preceq W_\rho$, so $(W_\sigma)x \preceq (W_\rho)y$. Since

$$1^{(W_\sigma)x} = \tau,$$

we have $W_\tau = (W_\sigma)x$. □

Now suppose that $2 \le \sigma < \tau \le n$. Let $e_\sigma = (\varphi, x, \sigma)$ and $e_\tau = (\psi, y, \tau)$. Then

$$(W_\varphi)x = W_\sigma \prec W_\tau = (W_\psi)y.$$

Therefore either $W_\varphi \prec W_\psi$ or $W_\varphi = W_\psi$ and $x \prec y$. But this means that either $\varphi < \psi$ or $\varphi = \psi$ and $x \prec y$. In either case, $e_\sigma \ll e_\tau$. Therefore $\mathcal{A}$ is edge standard with respect to $\prec$.

Now suppose that $\mathcal{A}$ is edge standard. We must prove that $\mathcal{A}$ is standard. Suppose $\mathcal{A}$ is not standard. Then there exist σ and τ in Σ such that $W_\sigma \prec W_\tau$ and $\tau < \sigma$. Among such pairs (σ, τ) choose one in which W_σ is as small as possible. Now $W_1 = \varepsilon$, so $1 < \tau < \sigma$. Thus we can write W_σ as $(W_\rho)x$. Then $W_\rho \prec W_\sigma$, so by the choice of σ we have $\rho < \sigma$. The triple (ρ, x, σ) is in E. Since $\mathcal{A}$ is edge standard, there is an edge (φ, y, τ) in E such that either $\varphi < \rho$ or $\varphi = \rho$ and $y \prec x$. But

$$(W_\rho)x = W_\sigma \prec W_\tau \preceq (W_\varphi)y.$$

If $\varphi = \rho$, then $(W_\rho)x \succ (W_\varphi)y$, which is not the case. Therefore $W_\rho \prec W_\varphi$ and $\varphi < \rho$. This contradicts the choice of σ. □

Here is a standardization procedure for the length-plus-lexicographic ordering of X^* determined by a given linear ordering of X. It is based on the definition of an edge standard automaton.

Procedure LENLEX_STND($\mathcal{A}, \prec; \mathcal{A}_1$);
 Input: $\mathcal{A}$: a nonempty, finite, accessible, deterministic
                    automaton $(\Sigma, X, E, A, \Omega)$;
          $\prec$: a linear ordering on X;
 Output: $\mathcal{A}_1$: the automaton which is isomorphic to $\mathcal{A}$ and is
                    standard with respect to the length-plus-lexicographic
                    ordering of X^* determined by $\prec$;
 Begin
   Let $A = \{\alpha\}$; $n := |\Sigma|$; $m := 1$; $\sigma_1 := \alpha$; $\Lambda := \{\sigma_1\}$;
   For $i := 1$ to n do
     For each edge (σ_i, x, τ) in E do
       (* The edges are considered in the $\prec$-order of their labels. *)
       If $\tau \notin \Lambda$ then begin $m := m + 1$; $\sigma_m := \tau$; $\Lambda := \Lambda \cup \{\tau\}$ end;
   $\Sigma_1 := \{1, \ldots, n\}$; $A_1 := \{1\}$; $\Omega_1 := \{i \mid \sigma_i \in \Omega\}$;
   $E_1 := \{(i, x, j) \mid (\sigma_i, x, \sigma_j) \in E\}$; $\mathcal{A}_1 := (\Sigma_1, X, E_1, A_1, \Omega_1)$
 End.

Table 3.8.1   Table 3.8.2

	x	y
1	6	
2		6
3	5	1
4	3	
5	3	2
6		4

	x	y
1	2	3
2	1	4
3	5	
4		5
5		6
6	1	

Proposition 8.6. *If $\mathcal{A}$ is a nonempty, finite, accessible, deterministic automaton with alphabet X and $\prec$ is a linear ordering of X, then* LENLEX_STND$(\mathcal{A}, \prec; \mathcal{A}_1)$ *returns $\mathcal{A}_1$ as the unique automaton which is isomorphic to $\mathcal{A}$ and standard with respect to the length-plus-lexicographic ordering of X^* determined by $\prec$.*

Proof. Exercise. $\square$

Example 8.2. Suppose $\mathcal{A} = (\Sigma, X, E, A, \Omega)$, where $\Sigma = \{1, 2, 3, 4, 5, 6\}$, $X = \{x, y\}$, $A = \{3\}$, and E is given by Table 3.8.1. If $x \prec y$, then in the call LENLEX_STND$(\mathcal{A}, \prec; \mathcal{A}_1)$ we obtain 3, 5, 1, 2, 6, 4 as the sequence $\sigma_1, \ldots, \sigma_6$. In $\mathcal{A}_1$ we have $A_1 = \{1\}$ and E_1 is given by Table 3.8.2. Note that the set of terminal states does not play a significant role in the standardization process.

It turns out in practice that we only need to decide the precedence of two automata when both are standard. Let $\mathcal{A} = (\Sigma, X, E, A, \Omega)$ be a nonempty automaton which is standard with respect to $\prec$. Suppose $|\Sigma| = n$ and $x_1, \ldots, x_r$ are the elements of X in increasing $\prec$-order. The *transition matrix* of $\mathcal{A}$ is the n-by-r matrix $T = T(\mathcal{A})$ such that $T[\sigma, j] = \tau$ if (σ, x_j, τ) is in E and $T[\sigma, j] = 0$ if σ^{x_j} is undefined in $\mathcal{A}$. Thus T is essentially the transition table with blanks filled in with zeros. We can put an ordering on the set of integer matrices by comparing two matrices lexicographically according to the sequences of their entries in row-major order.

Proposition 8.7. *Suppose that $\prec$ is a length-plus-lexicographic ordering of X^* and that $\mathcal{A}_1$ and $\mathcal{A}_2$ are finite, nonempty, accessible, deterministic automata which are standard with respect to $\prec$. Then $\mathcal{A}_1$ precedes $\mathcal{A}_2$ if and only if the transition matrix for $\mathcal{A}_1$ comes lexicographically before the transition matrix for $\mathcal{A}_2$.*

Proof. Set $f_i = f_{\mathcal{A}_i}$ and let T_i be the transition matrix for $\mathcal{A}_i$. Suppose that $\mathcal{A}_1$ precedes $\mathcal{A}_2$. Let W be the first word at which f_1 and f_2 differ. Then $W \neq \varepsilon$, since 1 is the initial state in both $\mathcal{A}_1$ and $\mathcal{A}_2$. Let $W = Ux_j$. Since 1^W is defined in $\mathcal{A}_2$, it follows that $\sigma = 1^U$ is defined in $\mathcal{A}_2$. Since

Table 3.8.3

	x	y
1	2	4
2	3	
3	1	6
4	5	
5		3
6	1	

$U \prec W$, we know that f_1 and f_2 agree at U. Thus $\sigma = 1^U$ in $\mathcal{A}_1$. Let $\tau = f_2(W) = \sigma^{x_j}$ in $\mathcal{A}_2$. Then either $f_1(W)$ is not defined or $f_1(W)$ is defined and $f_1(W) < \tau$. In the first case $T_1[\sigma, j] = 0$, and in the second case $0 < T_1[\sigma, j] < \tau$. In either case, $T_1[\sigma, j] < T_2[\sigma, j]$.

It remains to show that if the entries of T_1 and T_2 are compared in row-major order, then the first difference occurs at the (σ, j)-entries. If $1 \le k < j$, then $Ux_k \prec W$, so $f_1(Ux_k)$ and $f_2(Ux_k)$ agree. This means that the entries $T_1[\sigma, k]$ and $T_2[\sigma, k]$ are equal. Now suppose that $1 \le \rho < \sigma$. Let V be the first word such that $1^V = \rho$ in $\mathcal{A}_1$. Then $V \prec U$, so f_1 and f_2 agree at V. Therefore $1^V = \rho$ in $\mathcal{A}_2$. Let $1 \le k \le r$. Because $\prec$ is a length-plus-lexicographic ordering, $Vx_k \prec Ux_j = W$. Hence f_1 and f_2 agree at Vx_k. This means that the entries $T_1[\rho, k]$ and $T_2[\rho, k]$ are equal. Thus T_1 comes lexicographically before T_2.

Now suppose that T_1 is lexicographically earlier than T_2 and let the T_i differ first in their (σ, j)-entries. Let U be the first word such that $1^U = \sigma$ in $\mathcal{A}_1$. It is not hard to show that in computing 1^U we need only look at entries in T_1 which come before the (σ, j)-th entry in row-major order. This means that $1^U = \sigma$ in $\mathcal{A}_2$ as well. Since $T_1[\sigma, j] \ne T_2[\sigma, j]$, it follows that f_1 and f_2 do not agree at Ux_j. Let W be any word with $W \prec Ux_j$. Then to decide whether $f_1(W)$ is defined, we need only look at entries of T_1 which come before the (σ, j)-th. Thus f_1 and f_2 agree at W. Therefore $\mathcal{A}_1$ precedes $\mathcal{A}_2$. $\square$

In Section 5.6 we shall find it useful to consider automata which are standard with respect to a basic wreath product ordering $\prec$. Using wreath product orderings makes it more difficult to recognize standard automata.

Example 8.3. Let $X = \{x, y\}$ and let $\prec$ be the basic wreath product ordering of X^* in which $x \prec y$. Suppose $\mathcal{A} = (\Sigma, X, E, \{1\}, \{1\})$, where $\Sigma = \{1, \ldots, 6\}$ and E is given by Table 3.8.3. The earliest words with respect to $\prec$ are the powers of x. The only states of the form 1^W with W a power of x are 1, 2, and 3. It is easy to see that $W_1 = \varepsilon$, $W_2 = x$, and $W_3 = x^2$. Now we have to consider words which contain at least one y. The first such word is y, and y gets us to a new state, since $1^y = 4$. This means that $W_4 = y$.

Next we look at words which consist of y followed by a power of x. We see that $W_5 = yx$. We cannot reach 6 using a word yx^i, but we can reach 6 with x^2y. It is easy to see that $W_6 = x^2y$. Since $W_1 \prec W_2 \prec \cdots \prec W_6$, the automaton is standard with respect to $\prec$.

Here is a procedure which standardizes an automaton $\mathcal{A}$ with respect to a basic wreath product ordering:

Procedure WREATH_STND($\mathcal{A}, \prec; \mathcal{A}_1$);
Input: $\mathcal{A}$: a nonempty, finite, accessible, deterministic
                                automaton $(\Sigma, X, E, A, \Omega)$;
             $\prec$: a linear ordering on X;
Output: $\mathcal{A}_1$: the automaton which is isomorphic to $\mathcal{A}$ and is
                                standard with respect to the basic wreath product
                                ordering of X^* determined by $\prec$;
Begin
   Let $A = \{\alpha\}$; $n := |\Sigma|$, $m := 1$; $\sigma_1 := \alpha$; $\Lambda := \{\sigma_1\}$;
   For x in X do $j_x := 1$;
99: For x in X do   (* Elements of X are taken in $\prec$-order. *)
      While $j_x \leq m$ do
         If $(\sigma_{j_x})^x$ is not defined then $j_x := j_x + 1$
         Else begin
            $\tau := (\sigma_{j_x})^x$; $j_x := j_x + 1$;
            If τ is not in Λ then begin
               $m := m + 1$; $\sigma_m := \tau$; $\Lambda := \Lambda \cup \{\tau\}$; goto 99
            End
         End;
   $\Sigma_1 := \{1, \ldots, n\}$; $A_1 := \{1\}$; $\Omega_1 := \{i \mid \sigma_i \in \Omega\}$;
   $E_1 := \{(i, x, j) \mid (\sigma_i, x, \sigma_j) \in E\}$; $\mathcal{A}_1 := (\Sigma_1, X, E_1, A_1, \Omega_1)$
End.

It is left as an exercise to verify the correctness of WREATH_STND.

In Sections 4.9 and 5.6 we shall devise backtrack searches for various special classes of standard, accessible, deterministic automata. Although there is unlikely to be a strong need to produce *all* accessible, deterministic automata with a given alphabet X which are standard with respect to a fixed ordering $\prec$ of X^* and have a certain number n of states, the following discussion illustrates the ideas behind these searches. It turns out to be most natural to list all standard, accessible, deterministic automata having at most n states, where n is some specified positive integer.

Let X be a finite set and let $\prec$ be a reduction ordering on X^*. Define $\mathcal{S}(X, \prec, n)$ to be the set of nonempty, accessible, deterministic automata $\mathcal{A}$ with alphabet X and at most n states such that $\mathcal{A}$ is standard with respect to $\prec$. An element $\mathcal{A} = (\Sigma, X, E, \{1\}, \Omega)$ of $\mathcal{S}(X, \prec, n)$ is determined by E and Ω, since Σ consists of 1 and the set of endpoints of all edges in E. Given E, the set Ω can be any subset of Σ. We shall concentrate on describing a backtrack search for the edge sets E which can occur, or, equivalently, for the transition matrices corresponding to these edge sets.

Let $x_1, \ldots, x_r$ be the elements of X in $\prec$-order. The nodes of the search tree consist of "incomplete transition matrices", m-by-r matrices with $m \le n$ whose entries either are in $\{0, 1, \ldots, m\}$ or are '?', indicating that the entry is not yet determined. Suppose that $X = \{x, y\}$ and that $\prec$ is the length-plus-lexicographic ordering with $x \prec y$. The matrix

$$\begin{bmatrix} 0 & 2 \\ 3 & ? \\ 1 & ? \end{bmatrix}$$

describes those automata with three states which are standard with respect to $\prec$ and in which 1^x is definitely not defined, $1^y = 2$, $2^x = 3$, $3^x = 1$, and 2^y and 3^y are unspecified. For any word W in X^* we can ask whether 1^W is defined, undefined, or unspecified. In this example, 1^{x^2} and 1^{yx^3} are undefined, $1^{yx^2y} = 2$, and 1^{y^2} is unspecified.

The root of the search tree is the 1-by-r matrix all of whose entries are '?'. Let T be an m-by-r incomplete transition matrix which corresponds to a node in the search tree. If there are no words W such that 1^W is unspecified, then T is the transition matrix for an element in $\mathcal{S}(X, \prec, n)$. Otherwise, let W be the first word with respect to $\prec$ such that 1^W is unspecified. Then $W \ne \varepsilon$ and W has the form Ux_j, where $\sigma = 1^U$ is defined. Let s be the minimum of n and $m + 1$. For $k = 0, \ldots, s$ we get a descendant of T in the search tree as follows: Set $T[\sigma, j] = k$. If $k = m + 1$, then also add a k-th row to T consisting of all question marks.

Let us continue the example above and take $n \ge 3$. The root is $[? \; ?]$. The descendants of the root are

$$[0 \; ?], \quad [1 \; ?], \quad \begin{bmatrix} 2 & ? \\ ? & ? \end{bmatrix}$$

With each of these matrices, $W = y$. The descendants of $[0 \; ?]$ are

$$[0 \; 0], \quad [0 \; 1], \quad \begin{bmatrix} 0 & 2 \\ ? & ? \end{bmatrix},$$

The descendants of $[1 \; ?]$ are

$$[1 \; 0], \quad [1 \; 1], \quad \begin{bmatrix} 1 & 2 \\ ? & ? \end{bmatrix},$$

Table 3.8.4

	x	y
1	4	
2	5	3
3		2
4	3	5
5	1	

The descendants of $\begin{bmatrix} 2 & ? \\ ? & ? \end{bmatrix}$ are

$$\begin{bmatrix} 2 & 0 \\ ? & ? \end{bmatrix}, \quad \begin{bmatrix} 2 & 1 \\ ? & ? \end{bmatrix}, \quad \begin{bmatrix} 2 & 2 \\ ? & ? \end{bmatrix}, \quad \begin{bmatrix} 2 & 3 \\ ? & ? \\ ? & ? \end{bmatrix}.$$

Let us now change the reduction ordering on X and take $\prec$ to be the basic wreath product ordering with $x \prec y$. The root and its descendants are the same as before. For the first two of these descendants, $W = y$, and the nodes at depth 2 are the same. However, with $\begin{bmatrix} 2 & ? \\ ? & ? \end{bmatrix}$ we have $W = x^2$, and the descendants are

$$\begin{bmatrix} 2 & ? \\ 0 & ? \end{bmatrix}, \quad \begin{bmatrix} 2 & ? \\ 1 & ? \end{bmatrix}, \quad \begin{bmatrix} 2 & ? \\ 2 & ? \end{bmatrix}, \quad \begin{bmatrix} 2 & ? \\ 3 & ? \\ ? & ? \end{bmatrix}.$$

If the tree is traversed in what is known as a depth-first search, then the transition matrices are produced in order of precedence. In the current context, to traverse the tree in a depth-first search means to consider all descendants of a given node before considering another node at the same level and to consider the immediate descendants of a given node in increasing order of the number k described earlier. The number of transition matrices corresponding to elements of $S(X, \prec, n)$ does not depend on $\prec$. For example, if $|X| = 2$, then there are 49 transition matrices corresponding to elements of $S(X, \prec, 2)$. If one simply wants to count these transition matrices, one does not have to list them. See Exercises 8.6 and 8.7.

Exercises

8.1. Let $X = \{x, y\}$ and assume $x \prec y$. Standardize the automaton $\mathcal{A} = (\{1, 2, 3, 4, 5\}, \{x, y\}, E, \{2\}, \{4\})$ with respect to the length-plus-lexicographic ordering and the basic wreath product ordering determined by $\prec$. Here E is given by Table 3.8.4.

8.2. Prove that precedence of automata is a transitive relation.

8.3. Using backtrack, devise a depth-first search for standard, accessible, deterministic automata along the lines suggested in the text.

8.4. Modify the backtrack search in Exercise 8.3 so that only complete automata are produced.

8.5. Devise an algorithm for standardization with respect to an arbitrary reduction ordering $\prec$. (Hint: Assume in the notation of the proof of Proposition 8.1 that $\sigma_1, \ldots, \sigma_i$ have been found for some i with $1 \leq i < n$. Show that $W_{\sigma_{i+1}}$ is the first word with respect to $\prec$ of the form $(W_{\sigma_j})x$, where $1 \leq j \leq i$, and $(\sigma_j)^x$ is defined and is not in $\{\sigma_1, \ldots, \sigma_i\}$.)

8.6. Let $X = \{1, \ldots, r\}$ and $\Sigma = \{1, \ldots, n\}$. Fix a reduction ordering on X^*. Suppose S is an n-by-r matrix with entries in $\Sigma \cup \{0\}$. Set $\mathcal{B} = (\Sigma, X, E, \{1\}, \emptyset)$, where E is the set of triples (i, x, j) in $\Sigma \times X \times \Sigma$ such that $j = S[i, x]$. Let $\mathcal{A}$ be the standard automaton isomorphic to the accessible part of $\mathcal{B}$. Prove that if $\mathcal{A}$ has m states, then the number of choices for S which yield $\mathcal{A}$ is

$$(n-1)(n-2) \cdots (n-m+1)(n+1)^{r(n-m)}.$$

8.7. In the notation of Exercise 8.6, let $g(m, r)$ be the number of m-by-r matrices occurring as transition matrices for elements of $\mathcal{S}(X, \prec, m)$. Show that

$$(n+1)^{rn} = \sum_{m=1}^{n} g(m, r)(n-1) \cdots (n-m+1)(n+1)^{r(n-m)}.$$

Use this formula to compute $g(m, 2)$ for $m = 1, 2, 3$.

8.8. Suppose we wish to count transition matrices for complete automata in $\mathcal{S}(X, \prec, m)$. Derive results analogous to Exercises 8.6 and 8.7.

3.9 Additional constructions

We have already seen a number of constructions related to automata. Section 3.3 discussed the completion. In Section 3.4 we learned about the subset construction. In Section 3.6 we saw the construction of the trim part and the accessible subset construction. Section 3.8 added a standardization procedure. In this section we shall discuss several more automaton constructions.

Let X be a finite set. By Proposition 1.2, the set of rational languages over X is closed under the operations of union, intersection, complement, product, and submonoid generation. Suppose $\mathcal{A}_1 = (\Sigma_1, X, E_1, A_1, \Omega_1)$ and $\mathcal{A}_2 = (\Sigma_2, X, E_2, A_2, \Omega_2)$ are finite automata over X recognizing languages $\mathcal{L}_1$ and $\mathcal{L}_2$, respectively. We shall give constructions for finite automata recognizing each of the languages $\mathcal{L}_1 \cup \mathcal{L}_2, \mathcal{L}_1 \cap \mathcal{L}_2, X^* - \mathcal{L}_1, \mathcal{L}_1 \mathcal{L}_2$, and $(\mathcal{L}_1)^*$. Some of the constructions require the $\mathcal{A}_i$ to be deterministic or even complete, and some require Σ_1 and Σ_2 to be disjoint. By applying the accessible subset construction and forming the completion and perhaps taking an isomorphic copy, these conditions can always be satisfied. In the examples of the constructions given later in this section, we shall frequently use the automata $\mathcal{A}_1$ of Figure 3.9.1 and $\mathcal{A}_2$ of Figure 3.9.2.

If $\Sigma_1 \cap \Sigma_2 = \emptyset$, then the automaton

$$\mathcal{A}_3 = (\Sigma_1 \cup \Sigma_2, X, E_1 \cup E_2, A_1 \cup A_2, \Omega_1 \cup \Omega_2)$$

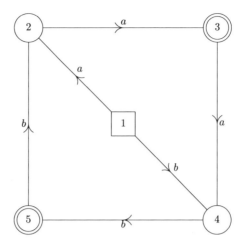

Figure 3.9.1

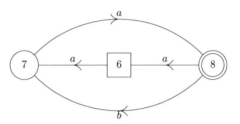

Figure 3.9.2

recognizes $\mathcal{L}_1 \cup \mathcal{L}_2$. Of course $\mathcal{A}_3$ will in general not be deterministic, even if both $\mathcal{A}_1$ and $\mathcal{A}_2$ are deterministic. A diagram for $\mathcal{A}_3$ is obtained simply by placing diagrams for $\mathcal{A}_1$ and $\mathcal{A}_2$ next to each other and considering the result as a single diagram. If we form $\mathcal{A}_3$ using our two examples, apply the accessible subset construction, and then take the trim part, we get a deterministic automaton with 11 states recognizing $\mathcal{L}_1 \cup \mathcal{L}_2$.

To get an automaton recognizing $\mathcal{L}_1 \cap \mathcal{L}_2$, we form $\Sigma_4 = \Sigma_1 \times \Sigma_2, A_4 = A_1 \times A_2$, and $\Omega_4 = \Omega_1 \times \Omega_2$. Let E_4 be the set of triples $((\sigma_1, \sigma_2), x, (\tau_1, \tau_2))$, where (σ_i, x, τ_i) is in E_i, $i = 1, 2$. Put $\mathcal{A}_4 = (\Sigma_4, X, E_4, A_4, \Omega_4)$. If P is a path in $\mathcal{A}_4$, then looking only at the first components of the states involved, we get a path P_1 in $\mathcal{A}_1$. Similarly, looking at the second components of the states, we get a path P_2 in $\mathcal{A}_2$. The signatures of P, P_1, and P_2 are the same word U. The path P goes from A_4 to Ω_4 if and only if P_i goes from A_i to Ω_i, $i = 1, 2$. Thus $\mathcal{A}_4$ recognizes $L(\mathcal{A}_1) \cap L(\mathcal{A}_2)$. In practice, we construct only the accessible part of $\mathcal{A}_4$ in a manner similar to the accessible subset construction. If $\mathcal{A}_1$ and $\mathcal{A}_2$ are deterministic, then so is $\mathcal{A}_4$.

Table 3.9.1

	a	b		a	b
1	2	3	7	7	10
2	4	5	8	4	10
3	2	6	9	11	9
4	7	5	10	12	9
5	8	6	11	7	10
6	2	9	12	7	10

Table 3.9.2

	a	b
1	2	3
2	4	5
3	2	6
4		5
5	8	6
6	2	
8	4	

Let us find an automaton recognizing $\mathcal{L}_1 \cap \mathcal{L}_2$ in the case of our two examples $\mathcal{A}_1$ and $\mathcal{A}_2$. The single initial state of $\mathcal{A}_4$ is $\beta_1 = (1,6)$. Now $1^a = 2$ in $\mathcal{A}_1$ and $6^a = 7$ in $\mathcal{A}_2$. Therefore $\mathcal{A}_4$ has an edge (β_1, a, β_2), where $\beta_2 = (2,7)$. In $\mathcal{A}_2$ the value of 6^b is undefined, so $(\beta_1)^b$ is undefined in $\mathcal{A}_4$. Continuing in this manner, we find that the accessible part of $\mathcal{A}_4$ has four states: β_1, β_2, $\beta_3 = (3,8)$, and $\beta_4 = (4,6)$. Of these, β_1 is initial and β_3 is terminal. The edges are given by the following table, where i has been written for β_i:

	a	b
1	2	
2	3	
3	4	
4		

Only states 1, 2, and 3 are trim. It is not hard to see that $L(\mathcal{A}_4)$ consists of the single word a^2.

To obtain an automaton recognizing $X^* - \mathcal{L}_1$, we first replace $\mathcal{A}_1$, if necessary, by a complete automaton recognizing the same language. Then $\mathcal{A}_5 = (\Sigma_1, X, E_1, A_1, \Sigma_1 - \Omega_1)$ recognizes $X^* - \mathcal{L}_1$. For given a word U in X^*, there is a unique path P in $\mathcal{A}_1$ such that P starts in A_1 and $U = \mathrm{Sg}(P)$. Let σ be the endpoint of P. Then U is in $\mathcal{L}_1$ if and only if σ is in Ω_1, and U is in $X^* - \mathcal{L}_1$ if and only if σ is in $\Sigma_1 - \Omega_1$.

Example 9.1. In Example 6.2 we found that the ideal $\mathcal{I}$ in $\{a,b\}^*$ generated by a^3, b^3, and $(ab)^2$ is recognized by the complete automaton $\mathcal{A} = (\Sigma, X, E, A, \Omega)$, where $\Sigma = \{1, 2, \ldots, 12\}$, $A = \{1\}$, $\Omega = \{7, 9, 10, 11, 12\}$, and E is given by Table 3.9.1. If we replace Ω by $\Sigma - \Omega$, we get an automaton $\mathcal{B}$ recognizing $X^* - \mathcal{I}$. Applying TRIM to $\mathcal{B}$, we find that the trim part of $\mathcal{B}$ has $\{1, 2, 3, 4, 5, 6, 8\}$ as the set of states and $\{1\}$ as the set of initial states. All states are terminal, and the edges are given by Table 3.9.2. This automaton has a circuit starting at 2 with signature $b^2 a$. By Proposition 6.4, $X^* - \mathcal{I}$ is infinite.

Because the set of canonical forms for a congruence is the complement of an ideal, the construction of an automaton for such a language is of considerable importance. If we have a finite, minimal generating set $\mathcal{U}$ for an ideal $\mathcal{I}$ of X^*, then we can construct an automaton recognizing $X^* - \mathcal{I}$ directly. Since $\mathcal{U}$ is minimal, no element of $\mathcal{U}$ contains another element of $\mathcal{U}$ as a subword. Let Σ be the set of words which are proper prefixes of elements of $\mathcal{U}$. If $\mathcal{U} = \emptyset$, then $\mathcal{I} = \emptyset$ and $X^* - \mathcal{I} = X^*$. If $\mathcal{U} = \{\varepsilon\}$, then $\mathcal{I} = X^*$ and $X^* - \mathcal{I} = \emptyset$. Let us assume that we are not in one of these cases, so $\Sigma \neq \emptyset$, and in particular ε is in Σ. Put $A = \{\varepsilon\}$ and $\Omega = \Sigma$. The set E of edges consists of the triples (U, x, V), where U is in Σ, x is in X, Ux does not have an element of $\mathcal{U}$ as a suffix, and V is the longest suffix of Ux which is in Σ.

Proposition 9.1. *The automaton $\mathcal{A} = (\Sigma, X, E, A, \Omega)$ recognizes $X^* - \mathcal{I}$.*

Proof. The proof is similar to that of Proposition 5.4 and is omitted. $\square$

Example 9.2. Let us apply Proposition 9.1 to the case $X = \{a, b\}$ and $\mathcal{U} = \{a^3, b^3, (ab)^2\}$. The elements of Σ are ε, a, b, a^2, ab, b^2, and aba. The longest suffix of $a^2 b$ which is in Σ is ab. Thus (a^2, b, ab) is an edge of $\mathcal{A}$. Since a^3 is in $\mathcal{U}$, there is no edge (a^2, a, U). If we number the elements of Σ as 1, 2, 3, 4, 5, 6, and 8, then $\mathcal{A}$ is exactly the automaton at the end of Example 9.1.

Now let us return to our discussion of operations on automata corresponding to the operations on rational languages of union, intersection, complement, product, and submonoid generation. Given finite automata $\mathcal{A}_1$ and $\mathcal{A}_2$, our next task is to construct an automaton recognizing $L(\mathcal{A}_1)L(\mathcal{A}_2)$. For this construction we need to assume that $\mathcal{A}_1$ and $\mathcal{A}_2$ have no states in common. Let

$$E_6 = E_1 \cup E_2 \cup \{(\omega, \varepsilon, \alpha) \mid \omega \in \Omega_1, \alpha \in A_2\}$$

and set $\mathcal{A}_6 = (\Sigma_1 \cup \Sigma_2, X, E_6, A_1, \Omega_2)$.

Proposition 9.2. $\mathcal{A}_6$ *recognizes* $L(\mathcal{A}_1)L(\mathcal{A}_2)$.

Proof. Suppose that U_i is in $\mathcal{A}_i$, $i = 1, 2$. For $i = 1, 2$, there is a path P_i in $\mathcal{A}_i$ from α_i in A_i to ω_i in Ω_i such that $U_i = \text{Sg}(P_i)$. All of the edges of P_1 and P_2 are in E_6, as is $(\omega_1, \varepsilon, \alpha_2)$. Let P be the path in $\mathcal{A}_6$ consisting of the edges of P_1, the edge $(\omega_1, \varepsilon, \alpha_2)$, and the edges of P_2. The signature of P is $U_1 U_2$, so $U_1 U_2$ is in $L(\mathcal{A}_6)$. Therefore $L(\mathcal{A}_1)L(\mathcal{A}_2) \subseteq L(\mathcal{A}_6)$. The proof of the reverse inclusion is left as an exercise. $\square$

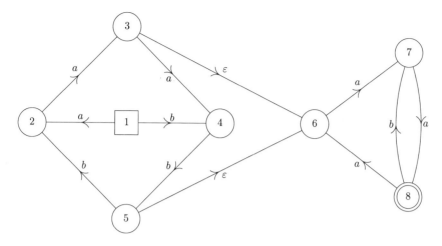

Figure 3.9.3

If we take $\mathcal{A}_1$ and $\mathcal{A}_2$ to be our standard examples, then $\mathcal{A}_6$ is given by Figure 3.9.3. This automaton is clearly not deterministic.

Finally, let us construct an automaton recognizing $(\mathcal{L}_1)^*$. Let ∞ be an object not in Σ_1. Set $\Sigma_7 = \Sigma_1 \cup \{\infty\}$,

$$E_7 = E_1 \cup \{(\infty, \varepsilon, \alpha) \mid \alpha \in A_1\} \cup \{(\omega, \varepsilon, \infty) \mid \omega \in \Omega_1\},$$

and $\mathcal{A}_7 = (\Sigma_7, X, E_7, \{\infty\}, \{\infty\})$.

Proposition 9.3. *The automaton $\mathcal{A}_7$ recognizes $(\mathcal{L}_1)^*$.*

Proof. Let $U_1, \ldots, U_m$ be in $\mathcal{L}_1 = L(\mathcal{A}_1)$. For $1 \leq i \leq m$ there are states α_i in A_1 and ω_i in Ω_1 and a path P_i in $\mathcal{A}_1$ from α_i to ω_i such that $U_i = \mathrm{Sg}(P_i)$. Let Q_i consist of $(\infty, \varepsilon, \alpha_i)$, P_i, and $(\omega_i, \varepsilon, \infty)$, and let P be the concatenation of $Q_1, \ldots, Q_m$. Then P is a path in $\mathcal{A}_7$, and $\mathrm{Sg}(p) = U_1 \ldots U_m$. Thus $(\mathcal{L}_1)^* \subseteq L(\mathcal{A}_7)$.

Conversely, let U be a word in $L(\mathcal{A}_7)$ and let P be a path in $\mathcal{A}_7$ from ∞ to ∞ such that $U = \mathrm{Sg}(P)$. We proceed by induction on the length of P to show that U is in $(\mathcal{L}_1)^*$. If P is empty, then $U = \varepsilon$ is in $(\mathcal{L}_1)^*$. Thus we may assume that P is not empty. The first edge e_1 of P has the form $(\infty, \varepsilon, \alpha)$, where α is in A_1. The next edge e_2 in P which involves ∞ must be of the form $(\omega, \varepsilon, \infty)$, where ω is in Ω_1. Let Q be the sequence of edges between e_1 and e_2. Then Q is a path in $\mathcal{A}_1$ from α to ω, and the signature V of Q is in $\mathcal{L}_1$. Let R consist of the edges of P after e_2. Then R goes from ∞ to ∞. By induction, the signature W of R is in $(\mathcal{L}_1)^*$. Therefore $U = VW$ is in $(\mathcal{L}_1)^*$. $\square$

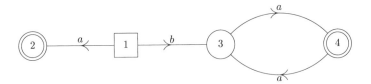

Figure 3.9.4

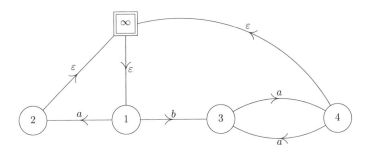

Figure 3.9.5

Example 9.3. Let $X = \{a, b\}$ and let $\mathcal{A}_1$ be the automaton in Figure 3.9.4. Then $\mathcal{A}_7$ is given by Figure 3.9.5.

Proposition 9.4. *Let $\mathcal{A}_1$ and $\mathcal{A}_2$ be finite automata with the same alphabet X. It is possible to decide whether $L(\mathcal{A}_1) \subseteq L(\mathcal{A}_2)$ and whether $L(\mathcal{A}_1) = L(\mathcal{A}_2)$.*

Proof. Let $\mathcal{L}_i = L(\mathcal{A}_i)$, $i = 1, 2$. Now $\mathcal{L}_1 \subseteq \mathcal{L}_2$ if and only if $\mathcal{L}_1 \cap (X^* - \mathcal{L}_2) = \emptyset$. We can construct a finite automaton $\mathcal{B}$ recognizing $\mathcal{L}_1 \cap (X^* - \mathcal{L}_2)$, and by Proposition 6.2 we can decide whether $L(\mathcal{B}) = \emptyset$. Since $\mathcal{L}_1 = \mathcal{L}_2$ if and only if $\mathcal{L}_1 \subseteq \mathcal{L}_2$ and $\mathcal{L}_2 \subseteq \mathcal{L}_1$, we can decide whether $\mathcal{L}_1$ and $\mathcal{L}_2$ are equal. $\qquad\square$

Exercises

9.1. Find a trim automaton recognizing $L(\mathcal{A}_1) \cap L(\mathcal{A}_2)$, where $\mathcal{A}_1$ and $\mathcal{A}_2$ are the automata defined at the beginning of this section.

9.2. Construct automata recognizing the set of canonical forms for the confluent rewriting systems in Examples 5.1, 5.3, and 5.5 in Chapter 2.

9.3. Let $\mathcal{A}$ be a finite automaton and let $\mathcal{L} = L(\mathcal{A})$. Give constructions for finite automata which recognize the following languages:

(a) The set of prefixes of elements of $\mathcal{L}$.
(b) The set $\mathcal{L}^\dagger$.
(c) The set of suffixes of elements of $\mathcal{L}$.
(d) The set of subwords of elements of $\mathcal{L}$.

9.4. Let $\mathcal{A}$ be a finite automaton with alphabet X and let W be a word in X^*. Show how to compute the longest prefix of W, if any, which is in $L(\mathcal{A})$.

9.5. Let $\mathcal{U}$ be a finite set of words in X^* such that no element of $\mathcal{U}$ is a prefix of another element of $\mathcal{U}$. Modify the construction of Proposition 9.1 to obtain an automaton recognizing the complement of the right ideal generated by $\mathcal{U}$.

3.10 More rewriting applications

In Section 3.5 we saw how to use automata as indexes to rewriting systems. These indexes speed up the processes of rewriting and of locating overlaps in the Knuth-Bendix procedure. This section describes some additional applications of automata to problems related to rewriting systems.

First we complete a discussion begun at the end of Section 2.5. Let $(X, \mathcal{R})$ be a finite presentation for a monoid M. If $\mathcal{R}$ is a confluent rewriting system with respect to some reduction ordering $\prec$ on X^*, then it is possible to decide whether M is finite. Let $\mathcal{U}$ be the complement of the ideal of X^* generated by the left sides of the rules in $\mathcal{R}$. The elements of $\mathcal{U}$ are canonical forms for the elements of M, so there is a one-to-one correspondence between $\mathcal{U}$ and M. Using the methods of Section 3.9, we can construct a finite automaton recognizing $\mathcal{U}$. Then by Corollary 6.5 we can decide whether $\mathcal{U}$ is finite or infinite.

Example 10.1. Let $X = \{a, b\}$. With respect to the length-plus-lexicographic ordering on X^* in which $a \prec b$, the set $\mathcal{R}$ of rules

$$a^3 \to \varepsilon, \quad b^3 \to \varepsilon, \quad baba \to a^2 b^2, \quad b^2 a^2 \to abab$$

is a rewriting system. This system is reduced and confluent. The proper prefixes of the left sides are

$$\varepsilon, \quad a, \quad b, \quad a^2, \quad ba, \quad b^2, \quad bab, \quad b^2 a.$$

Numbering these from 1 to 8 in the order given and applying the construction of Proposition 9.1, we obtain an automaton $\mathcal{A} = (\Sigma, X, E, \{1\}, \Sigma)$ recognizing $\mathcal{U}$, where $\Sigma = \{1, \ldots, 8\}$ and E is given Table 3.10.1. To decide whether $\mathcal{U} = L(\mathcal{A})$ is infinite, we use the approach of Exercise 6.3. Set $\mathcal{B}_1 = \mathcal{A}$. The set of states in $\mathcal{B}_1$ which have edges coming into them is $\Sigma_2 = \{2, \ldots, 8\}$. Let $\mathcal{B}_2$ be the restriction of $\mathcal{B}_1$ to Σ_2. In $\mathcal{B}_2$, the set of states with incoming edges is $\Sigma_3 = \{3, \ldots, 8\}$. Let $\mathcal{B}_3$ be the restriction of $\mathcal{B}_2$ to Σ_3. In $\mathcal{B}_3$ every state has an incoming edge. Since the labels on the edges of $\mathcal{B}_3$ are nonempty, $L(\mathcal{A})$ is infinite. The presentation $(X, \mathcal{R})$ is easily seen to define $M = \mathrm{Mon}\,\langle a, b \mid a^3 = b^3 = (ab)^3 = 1 \rangle$, which is a group.

In Section 2.8 we considered a right congruence $\sim$ on a free monoid X^* such that $\sim$ is finitely generated modulo a (two-sided) congruence $\cong$, where

Table 3.10.1

	a	b
1	2	3
2	4	3
3	5	6
4		3
5	4	7
6	8	
7		6
8		7

$\cong$ is finitely generated as a congruence. Let $\mathcal{R}$ and $\mathcal{S}$ be finite subsets of $X^* \times X^*$ such that $\mathcal{R}$ generates $\cong$ and $\sim$ is generated by $\cong$ and $\mathcal{S}$. To apply Knuth-Bendix techniques in this situation, we form $Y = X \cup \{\#\}$ and $\mathcal{T} = \{(\#P, \#Q) \mid (P, Q) \in \mathcal{S}\}$ and define $\equiv$ to be the congruence on Y^* generated by $\mathcal{R} \cup \mathcal{T}$. Here $\#$ denotes some object not in X. We then apply the Knuth-Bendix procedure to the presentation $(Y, \mathcal{R} \cup \mathcal{T})$.

It may happen that the Knuth-Bendix procedure does not terminate even when $\sim$ has only finitely many congruence classes. However, if the number of classes is finite, it is possible to determine the number of classes and to decide for any two words U and V whether $U \sim V$. To do so, we choose a reduction ordering $\prec$ which is consistent with length. Then we periodically interrupt the Knuth-Bendix procedure to apply an additional test.

At any point in the computation let $\mathcal{V}$ be the set of words in Y^* irreducible with respect to the current rewriting system and let $\mathcal{W} = \mathcal{V} \cap (\#X^*)$. By a simple extension of Proposition 8.6 in Chapter 2, if $\sim$ has only finitely many classes, then eventually $\mathcal{W}$ will be finite. When $\mathcal{W}$ is observed to be finite, we can use an automaton recognizing $\mathcal{W}$ to compute an upper bound n for the lengths of the words in $\mathcal{W}$. Let r be the maximum length of a left side in $\mathcal{R}$ and let t be the maximum length of a left side in $\mathcal{T}$. Set m equal to the larger of $n + r$ and t. We now throw away all rules with left sides of length greater than m. We continue the Knuth-Bendix procedure, forming only overlaps of length at most m. Since $\prec$ is consistent with length, no rules with left sides of length greater than m will result. Thus the process will terminate. When it does, the rules obtained are sufficient to rewrite any word $\#U$ in $\#X^*$ into the canonical form with respect to $\prec$ of its $\equiv$-class. Hence we may determine the canonical form of the $\sim$-class containing U.

The following is a sketch of a proof of the assertion just made. Let $\mathcal{V}$ and $\mathcal{W}$ be defined with respect to the final rewriting system. Since all elements of $\mathcal{W}$ have length at most n, the rules of length at most $n + 1$ are enough to rewrite any word in $\#X^*$ into an element of $\mathcal{W}$. Let U be a word in Y^* of length at most m. Then U can be rewritten into a unique element $\overline{U}$ of $\mathcal{V}$. For each x in X, we can define a map f_x from $\mathcal{W}$ into itself by $Wf_x = \overline{Wx}$, which is defined since $m \geq n+1$. Let φ be the homomorphism of X^* onto the monoid generated by the maps f_x under composition. For each W in $\mathcal{W}$ and each V in X^*, the image of W under $\varphi(V)$ is obtained by rewriting WV from the left. Let (L, R) be one of the original generators of $\cong$. Then during the execution of the Knuth-Bendix procedure there is always a word S derivable from both L and R. Thus this condition is satisfied by the final rewriting system. Let W be in $\mathcal{W}$. One way to rewrite WL is to rewrite WL into WS and then continue reducing. But WL has length at most m, so $W\varphi(L) = \overline{WL} = \overline{WS}$. By the same argument, $W\varphi(R) = \overline{WS}$. Therefore $\varphi(L) = \varphi(R)$. That is, the maps f_x satisfy the defining relations for M.

Now define a relation $\approx$ on X^* by $P \approx Q$ if and only if $\#\varphi(P) = \#\varphi(Q)$. Clearly $\approx$ is a right congruence. We have already seen that $\cong$ is contained in $\approx$. Suppose (P, Q) is in the original set $\mathcal{S}$. There is a word T derivable from both $\#P$ and $\#Q$ using the final rewriting system. Since $m \geq |\#P| \geq |\#Q|$, we have $\#\varphi(P) = \overline{\#P} = T = \overline{\#Q} = \#\varphi(Q)$. Therefore $P \approx Q$, so $\sim$ is contained in $\approx$. For any words U and V in X^*, if $\#V$ is derivable from $\#U$ using the final rewriting system, then $V \sim U$. This means that $\sim$ and $\approx$ are the same and that the set of words U in X^* such that $\#U$ is in $\mathcal{W}$ is the set of canonical forms for $\sim$ with respect to $\prec$.

Example 10.2. Let $X = \{x, y, y^{-1}\}$ and let $\prec$ be the length-plus-lexicographic ordering with $x \prec y \prec y^{-1}$. Set $Y = X \cup \{\#\}$ and extend $\prec$ to Y^* by making it the length-plus-lexicographic ordering with $\# \prec x$. Let $\mathcal{R}$ consist of the following rules:

$$x^2 \to \varepsilon, \quad yy^{-1} \to \varepsilon, \quad y^{-1}y \to \varepsilon, \quad y^3 \to \varepsilon, \quad yxyxyxy \to xy^{-1}xy^{-1}xy^{-1}x.$$

In the presence of the first three rules, the last rule is equivalent to $(xy)^7 \to \varepsilon$ but is more balanced in the sense of Section 2.7. The value of r is 7. The monoid G presented by $(X, \mathcal{R})$ is infinite, as is $\mathrm{RC}(X, \prec, \mathcal{R})$. Let $\cong$ be the congruence on X^* generated by $\mathcal{R}$ and let $\sim$ be the right congruence generated by $\cong$ and the set $\mathcal{S}$ made up of the following rules:

$$y \to x, \quad (yxy^{-1}x)^2 \to (xyxy^{-1})^2.$$

Given $\cong$, the second rule is equivalent to $(xyxy^{-1})^4 \to \varepsilon$.

The set $\mathcal{T}$ contains the rules

$$\#y \to \#x, \quad \#(yxy^{-1}x)^2 \to \#(xyxy^{-1})^2,$$

so $t = 9$. The set $\mathcal{W}$ is infinite at this point. For example, all words consisting of $\#$ followed by a power of y^{-1} are irreducible. The first step in applying KBS_2 to $(Y, \mathcal{R} \cup \mathcal{T})$ is to make the set of rules reduced. The only change occurs in the second rule in $\mathcal{T}$. We can rewrite the left side of this rule as follows:

$$\underline{\#y}xy^{-1}xyxy^{-1}x \rightarrow \underline{\#xx}y^{-1}xyxy^{-1}x \rightarrow \#y^{-1}xyxy^{-1}x.$$

Since the result is less than the original right side, we replace the rule by

$$\#xyxy^{-1}xyxy^{-1} \rightarrow \#y^{-1}xyxy^{-1}x.$$

If we now continue KBS_2, restricting overlaps to have length at most 9, we obtain the following rewriting system:

$$x^2 \rightarrow \varepsilon, \quad y^2 \rightarrow y^{-1}, \quad yy^{-1} \rightarrow \varepsilon, \quad y^{-1}y \rightarrow \varepsilon, \quad y^{-2} \rightarrow y,$$

$$yxyxyxy \rightarrow xy^{-1}xy^{-1}xy^{-1}x, \quad y^{-1}xy^{-1}xy^{-1}xy^{-1} \rightarrow xyxyxyx,$$

$$xyxyxyxy^{-1} \rightarrow y^{-1}xy^{-1}xy^{-1}xy, \quad xy^{-1}xy^{-1}xy^{-1}xy \rightarrow yxyxyxy^{-1},$$

$$yxy^{-1}xy^{-1}xy^{-1}x \rightarrow y^{-1}xyxyxy, \quad y^{-1}xyxyxyx \rightarrow yxy^{-1}xy^{-1}xy^{-1},$$

$$\#y \rightarrow \#x, \quad \#xy \rightarrow \#y^{-1}, \quad \#xy^{-1} \rightarrow \#,$$

$$\#y^{-1}xyxyxy \rightarrow \#y^{-1}xy^{-1}xy^{-1}x, \quad \#y^{-1}xyxy^{-1}xy \rightarrow \#y^{-1}xy^{-1}xyx,$$

$$\#y^{-1}xy^{-1}xyxy \rightarrow \#y^{-1}xyxy^{-1}xy^{-1}, \quad \#y^{-1}xy^{-1}xyxy^{-1} \rightarrow \#y^{-1}xyxy^{-1}x,$$

$$\#y^{-1}xy^{-1}xy^{-1}xy \rightarrow \#y^{-1}xyxyxy^{-1}.$$

The set $\mathcal{W}$ is still infinite.

Increasing the length of the overlaps allowed to 13 leads to a rewriting system with 26 rules for which $\mathcal{W}$ is finite and contains words of length at most 10. In this rewriting system, the rules with left sides of length at most 11 are the preceding 19 and the following three:

$$\#y^{-1}xyxyxy^{-1}xyx \rightarrow \#y^{-1}xyxyxy^{-1}xy^{-1},$$

$$\#y^{-1}xyxyxy^{-1}xy^{-1} \rightarrow \#y^{-1}xyxyxy^{-1}xy,$$

$$\#y^{-1}xyxy^{-1}xy^{-1}xyx \rightarrow \#y^{-1}xyxy^{-1}xy^{-1}xy^{-1}.$$

Continuing the Knuth-Bendix procedure until all overlaps of length 17 have been processed does not produce any rules with left sides of length less than 12 and hence does not change $\mathcal{W}$. At this point we are able to compute canonical forms for $\sim$. Further calculation shows that $|\mathcal{W}| = 24$.

For simple problems such as the one discussed in Example 10.2 it is easier to determine $\sim$ using a version of the procedure coset enumeration, which is presented in Chapters 4 and 5. However, Section 5.8 provides evidence that in situations where space is limited the Knuth-Bendix approach may sometimes be superior to coset enumeration.

4

Subgroups of free products of cyclic groups

In this chapter we shall discuss techniques for describing and manipulating finitely generated subgroups of a group F which is the free product of a finite number of cyclic groups. An important special case occurs when F is a free group, that is, when all of the cyclic free factors of F are infinite. Many questions about finitely generated subgroups of F can be answered with standard methods of automata theory. However, there is another approach called coset enumeration, which is normally more efficient. Coset enumeration is the next in our collection of major group-theoretic procedures. Only an introduction to coset enumeration is presented in this chapter. It is considered in more detail in Chapter 5.

One of the reasons that we can give a nice treatment of finitely generated subgroups of a free product of cyclic groups is that free products of cyclic groups make up a class of groups which have confluent rewriting systems of a particularly simple form. These rewriting systems are discussed in the first section.

4.1 Niladic rewriting systems

Let $(X, \mathcal{R})$ be a monoid presentation. We shall say that $\mathcal{R}$ is *niladic* if every element of $\mathcal{R}$ has the form (L, ε), where $|L| \geq 2$. The condition on the length of L is purely technical. If (x, ε) is in $\mathcal{R}$ for some x in X, then using a Tietze transformation we can delete x from X and from each element of $\mathcal{R}$ and obtain another presentation for the same monoid. If $\mathcal{R}$ is niladic, then $\mathcal{R}$ is a rewriting system with respect to every reduction ordering of X^*.

Example 1.1. Let $X = \{x, y, y^{-1}\}$ and let $\mathcal{R}$ consist of the rules

$$x^2 \to \varepsilon, \quad yy^{-1} \to \varepsilon, \quad y^{-1}y \to \varepsilon.$$

Then $\mathcal{R}$ is a confluent, reduced, niladic rewriting system on X^* and $\mathrm{Mon} \langle X \mid \mathcal{R} \rangle$ is isomorphic to the free product of $\mathbb{Z}_2$ and $\mathbb{Z}$.

Example 1.2. Let $X = \{x_1, \ldots, x_s\}$, let n be a positive integer, and let $\mathcal{R}$ consist of the pairs (L, ε), where L is a cyclic permutation of the word $(x_1 \ldots x_s)^n$. In Proposition 3.6 of Chapter 2 we saw that $M = \text{Mon} \langle X \mid \mathcal{R} \rangle$ is a group which is the free product of $\mathbb{Z}_n$ and $s - 1$ copies of $\mathbb{Z}$. Again $\mathcal{R}$ is a confluent, reduced, niladic rewriting system. It turns out that every confluent, reduced, niladic rewriting system which defines a group is built up from systems of this type.

The class of groups described by confluent, niladic rewriting systems is very restricted. These groups are determined in Proposition 1.2. Rewriting with respect to a confluent, niladic rewriting system has some special properties. The following proposition describes one such property.

Proposition 1.1. *Let $\mathcal{R}$ be a confluent, niladic rewriting system on X^* and let $\sim$ be the congruence generated by $\mathcal{R}$. Suppose that U, V, and W are words such that $W \sim UV$ and UV is irreducible with respect to $\mathcal{R}$. Then there are words A and B such that $W = AB$, $A \sim U$, and $B \sim V$.*

Proof. Since $\mathcal{R}$ is confluent, there is a sequence of words $W = W_0$, W_1, $\ldots$, $W_r = UV$ such that W_{i+1} is derivable in one step from W_i using $\mathcal{R}$. Set $A_r = U$ and $B_r = V$. Next assume that for $0 < i \leq r$ we have W_i written as $A_i B_i$, where $A_i \sim U$ and $B_i \sim V$. Now W_{i-1} has the form PLQ, where L is a left side in $\mathcal{R}$ and $PQ = W_i = A_i B_i$.

Suppose that P is a prefix of A_i. Then $A_i = PS$ for some word S and $Q = SB_i$. Let $A_{i-1} = PLS$ and $B_{i-1} = B_i$. Then $A_{i-1} \sim PS = A_i \sim U$ and $B_{i-1} \sim V$. Similarly, if $P = A_i S$ and $B_i = SQ$, then we may take $A_{i-1} = A_i$ and $B_{i-1} = SLQ$. Continuing in this manner, we have $W = W_0 = A_0 B_0$, where $A_0 \sim U$ and $B_0 \sim V$. $\square$

Example 1.3. Suppose that X and $\mathcal{R}$ are as in Example 1.1. Let $W = x^3 y^{-1} y x^2 y^{-1} y y^{-1} x y$, $U = xy^{-1}$, and $V = xy$. Then $W \sim UV$ and UV is irreducible with respect to $\mathcal{R}$. There are two pairs (A, B) such that $W = AB$, $A \sim U$, and $B \sim V$. These pairs are $(x^3 y^{-1}, yx^2 y^{-1} y y^{-1} xy)$ and $(x^3 y^{-1} y x^2 y^{-1} y y^{-1}, xy)$.

Example 1.4. Some condition on $\mathcal{R}$ is necessary in Proposition 1.1. Let $X = \{a, b\}$ and $\mathcal{R} = \{(ba, ab)\}$. Then $\mathcal{R}$ is confluent. Let $W = ba$, $U = a$, and $V = b$. Then $W \sim UV$ and UV is irreducible, but we cannot write W as AB, where $A \sim U$ and $B \sim V$.

The following result was first proved in (Cochet 1976).

Proposition 1.2. *Let X be a finite set and suppose that $\mathcal{R}$ is a confluent, reduced, niladic rewriting system on X^* such that $F = \text{Mon} \langle X \mid \mathcal{R} \rangle$ is a* group. *Then there are nonempty words* W_1, ..., W_r *and positive integers* n_1, ..., n_r *such that the following conditions hold:*

(a) *Each element x of X occurs in one of the W_i.*
(b) $|W_1| + |W_2| + \cdots + |W_r| = |X|.$
(c) *$\mathcal{R}$ consists of all pairs (L, ε), where L is a cyclic permutation of one of the words $W_i^{n_i}$.*

The group F is a free product of cyclic groups.

Proof. Let $\sim$ be the congruence generated by $\mathcal{R}$, and for U in X^* let $[U]$ be the $\sim$-class containing U. By assumption, the quotient F of X^* modulo $\sim$ is a group. If $X = \emptyset$, then $\mathcal{R} = \emptyset$ and we may take $r = 0$. Thus we may assume that $X \neq \emptyset$. For each x in X let U_x be the word in $[x]^{-1}$ which is irreducible with respect to $\mathcal{R}$. Now ε is derivable from xU_x using $\mathcal{R}$, so some left side in $\mathcal{R}$ must start with x. Suppose that (xV, ε) is in $\mathcal{R}$. Since $\mathcal{R}$ is reduced, V is irreducible with respect to $\mathcal{R}$ and $[x][V] = [xV] = [\varepsilon] = 1$. Therefore $V = U_x$. Thus $\mathcal{R} = \{(xU_x, \varepsilon) \mid x \in X\}$, so $\mathcal{R}$ is finite and $|\mathcal{R}| = |X|$. By essentially the same argument, $\mathcal{R} = \{(U_x x, \varepsilon) \mid x \in X\}$. Hence the set of left sides in $\mathcal{R}$ is closed under cyclic permutation.

Let (W, ε) be in $\mathcal{R}$ and assume that some generator occurs more than once in W. Among the generators in W, let x occur the maximum number of times. By taking a cyclic permutation of W if necessary, we may assume that W starts with x. Let $W = xA_1xA_2x\ldots xA_n$, where x does not occur in any of the A_i. The cyclic permutation $V = xA_2xA_3x\ldots xA_nxA_1$ is also a left side in $\mathcal{R}$. Since a left side is determined by its first term, we must have $V = W$. This means that $A_1 = A_2 = \cdots = A_n = A$, say, and that $W = (xA)^n$. By the choice of x, no generator occurs more than once in A. If (U, ε) is in $\mathcal{R}$ and $U = V^m$, where no generator occurs more than once in V, then either V and xA have no generator in common or they are cyclic permutations of each other and $m = n$. Let W_1, ..., W_r be representatives of the classes under cyclic permutation of words U such that no generator occurs more than once in U and (U^m, ε) is in $\mathcal{R}$ for some positive integer m. Then $|W_1| + \cdots + |W_r| = |X|$ and W_i^m is a left side in $\mathcal{R}$ for a unique integer $m = n_i$. Thus $\mathcal{R}$ is the set of pairs (L, ε) such that L is a cyclic permutation of one of the words $W_i^{n_i}$.

By Propositions 3.6 and 3.3 in Chapter 2, a presentation of this form is confluent and defines a free product of cyclic groups. $\square$

In most cases, the niladic rewriting systems to which coset enumeration has been applied satisfy the condition that all left sides have length exactly 2. Let us define a *classical* niladic rewriting system to be one which satisfies the conditions of Proposition 1.2 and in which all left sides have length 2. Thus $n_i|W_i| = 2$ for $1 \leq i \leq r$. For most of this chapter we shall

consider only classical niladic rewriting systems. Many of the results carry
over to the general case, which is discussed in Section 4.11.

Let X be a finite set and let $\mathcal{R}$ be a classical niladic rewriting system
on X^*. If x is in X, then there is a unique left side xy in $\mathcal{R}$ which starts
with x. If $y = x$, then the image of x in $F = \mathrm{Mon}\,\langle X \mid \mathcal{R}\rangle$ generates a cyclic
group of order 2. If $y \neq x$, then the images of x and y in F generate an
infinite cyclic group. Thus F is the free product of infinite cyclic groups
and cyclic groups of order 2. We shall denote y by x^{-1}, since y represents
the inverse of x in F.

Let $\sim$ be the congruence on X^* generated by $\mathcal{R}$. The set $\mathcal{C}$ of words
which are irreducible with respect to $\mathcal{R}$ is the set of canonical forms for $\sim$
with respect to any reduction ordering of X^*. Since $\mathcal{C}$ is the complement of
the ideal generated by the left sides in $\mathcal{R}$, the set $\mathcal{C}$ is a rational language.

Example 1.5. Let X and $\mathcal{R}$ be as in Example 1.1. The construction which
is the subject of Proposition 9.1 in Chapter 2 produces an automaton
$(\Sigma, X, E, A, \Omega)$ recognizing $\mathcal{C}$, where $\Sigma = \Omega = \{1, 2, 3, 4\}$, $A = \{1\}$, and E is
given by the following transition table:

	x	y	y^{-1}
1	2	3	4
2		3	4
3	2	3	
4	2		4

For any word U, let $[U]$ denote the $\sim$-class containing U and let $\overline{U}$ be the
unique element of $[U] \cap \mathcal{C}$. If $U = x_1 \ldots x_s$, then U^{-1} will denote $x_s^{-1} \ldots x_1^{-1}$,
which is in $[U]^{-1}$. It is easy to see that U is in $\mathcal{C}$ if and only if U^{-1} is in $\mathcal{C}$.
Let $g : X^* \to X^*$ be the homomorphism which maps x in X to x^{-1}. Then
U^{-1} can be described as $g(U^{\dagger})$.

Proposition 1.3. *If $\mathcal{W}$ is a rational language over X, then $\mathcal{W}^{-1} = \{W^{-1} \mid W \in \mathcal{W}\}$ is rational.*

Proof. This follows immediately from Exercise 3.3 and Exercise 1.5 or 2.8
in Chapter 3. □

Corollary 1.4. *If $\mathcal{W}$ is a rational language over X and $H = \mathrm{Grp}\langle [W] \mid W \in \mathcal{W}\rangle$, then there is a rational language $\mathcal{V}$ such that $H = \mathrm{Mon}\,\langle [V] \mid V \in \mathcal{V}\rangle$.*

Proof. We may take $\mathcal{V} = \mathcal{W} \cup \mathcal{W}^{-1}$. □

Proposition 1.5. *Suppose that U and V are words and U is in $\mathcal{C}$. Let
$W = \overline{UV}$. Then U and W have a common prefix of length at least $|U| - |V|$.*

Proof. Since U is irreducible, when a left side L in $\mathcal{R}$ is deleted in the rewriting of UV to obtain W, at least one generator from V must occur in L. Thus removing L deletes at least one generator from V and at most one generator from U. Therefore at most $|V|$ deletions are made and at least $|U| - |V|$ generators in U remain. $\square$

Proposition 1.6. *Suppose that $|X| \geq 2$ and U is in $\mathcal{C}$. There exists x in X such that Ux is in $\mathcal{C}$.*

Proof. If $U = \varepsilon$, then x may be any element of X. If U ends in y, then x may be any element of $X - \{y^{-1}\}$. $\square$

Corollary 1.7. *If $|X| \geq 2$ and U is in $\mathcal{C}$, then there exist arbitrarily long words V such that UV is in $\mathcal{C}$.*

Corollary 1.8. *The group F is infinite if and only if $|X| \geq 2$.*

Proof. If $|X| \geq 2$, then F is infinite by Corollary 1.7. If $|X| = 1$ and $X = \{x\}$, then $\mathcal{R} = \{(x^2, \varepsilon)\}$. Thus F is cyclic of order 2. If $|X| = 0$, then $|F| = 1$. $\square$

Later in this chapter we shall need some fairly detailed information about the multiplication in F. The following propositions are quite easy and are left as exercises.

Proposition 1.9. *Suppose that A and B are elements of $\mathcal{C}$. If A and B do not end in the same generator, then AB^{-1} is in $\mathcal{C}$. If A and B do not start with the same generator, then $B^{-1}A$ is in $\mathcal{C}$.*

Proposition 1.10. *Let A, B, and C be elements of $\mathcal{C}$ such that $B \neq \varepsilon$ and both AB and BC are in $\mathcal{C}$. Then ABC is in $\mathcal{C}$.*

Proposition 1.11. *Suppose that B, C, and W are words in $\mathcal{C}$ such that $W \neq \varepsilon$, B and W do not start with the same generator, and W and C do not end with the same generator. Then $B^{-1}WC^{-1}$ is in $\mathcal{C}$.*

We turn now to a study of the effect on rational languages over X of rewriting with respect to $\mathcal{R}$. Suppose that $\mathcal{L}$ is a rational language over X. Is it necessarily true that $\overline{\mathcal{L}} = \{\overline{U} \mid U \in \mathcal{L}\}$ is a rational language? We shall show that the answer is yes. However, if we drop the assumption that $\mathcal{R}$ is niladic, the answer may be no.

Example 1.6. Let $X = \{a, b\}$ and let $\mathcal{R} = \{(ba, ab)\}$, as in Example 1.4. The set $\mathcal{L} = \{(ab)^i \mid i \geq 0\}$ is rational. It is recognized by the automaton in

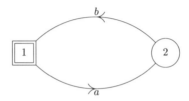

Figure 4.1.1

Figure 4.1.1. However, if $^-$ denotes rewriting with respect to $\mathcal{R}$, then $\overline{\mathcal{L}} = \{a^i b^i \mid i \geq 0\}$, and we saw in Example 1.5 of Chapter 3 that this set is not rational.

For any language $\mathcal{L}$ over X, let $\mathcal{L}'$ be the set of all words derivable from words in $\mathcal{L}$ using $\mathcal{R}$.

Proposition 1.12. *If $\mathcal{L}$ is a rational language over X, then $\mathcal{L}'$ is rational.*

Proof. Let $\mathcal{A} = (\Sigma, X, E, A, \Omega)$ be a finite automaton recognizing $\mathcal{L}$. Let us say that $\mathcal{A}$ is *compatible* with $\mathcal{R}$ if whenever there is a path P from σ to τ in $\mathcal{A}$ such that $\mathrm{Sg}(P)$ is a left side in $\mathcal{R}$, then either $\sigma = \tau$ or $(\sigma, \varepsilon, \tau)$ is in E. Let us define a sequence of automata $\mathcal{A} = \mathcal{A}_0, \mathcal{A}_1, \ldots, \mathcal{A}_m$, where $\mathcal{A}_i = (\Sigma, X, E_i, A, \Omega)$. If $\mathcal{A}_i$ is compatible with $\mathcal{R}$, then stop with $m = i$. Otherwise, there is a path P in $\mathcal{A}$ from σ to τ such that $\mathrm{Sg}(P)$ is a left side in $\mathcal{R}$, but $\sigma \neq \tau$ and $(\sigma, \varepsilon, \tau)$ is not in E_i. Set $E_{i+1} = E_i \cup \{(\sigma, \varepsilon, \tau)\}$. Since Σ is finite, there is only a finite number of edges of the form $(\sigma, \varepsilon, \tau)$, so this process must terminate with an automaton $\mathcal{A}_m$ which is compatible with $\mathcal{R}$. Let $\mathcal{L}_i = L(\mathcal{A}_i)$, $0 \leq i \leq m$. Clearly $\mathcal{L}_i \subseteq \mathcal{L}_{i+1}$.

Lemma 1.13. *Suppose that $0 \leq i < m$. If W is in $\mathcal{L}_{i+1}$, then W is derivable from a word in $\mathcal{L}_i$ using $\mathcal{R}$.*

Proof. Let $E_{i+1} = E_i \cup \{(\sigma, \varepsilon, \tau)\}$ and let P be a path in $\mathcal{A}_i$ from σ to τ such that $L = \mathrm{Sg}(P)$ is a left side in $\mathcal{R}$. Let Q be a path in $\mathcal{A}_{i+1}$ from A to Ω such that $W = \mathrm{Sg}(Q)$. If Q is a path in $\mathcal{A}_i$, then W is in $\mathcal{L}_i$. Suppose that Q is not a path in $\mathcal{A}_i$. Then Q involves the edge $(\sigma, \varepsilon, \tau)$ at least once. Let R be the path obtained from Q by replacing each occurrence of $(\sigma, \varepsilon, \tau)$ with P. Then R is a path in $\mathcal{A}_i$ from A to Ω, and W is derivable from $\mathrm{Sg}(R)$ using the rule (L, ε). $\square$

Lemma 1.14. $\mathcal{L}' = \mathcal{L}_m$.

Proof. By Lemma 1.13, we have $\mathcal{L} \subseteq \mathcal{L}_m \subseteq \mathcal{L}'$. Therefore it suffices to show that $(\mathcal{L}_m)' = \mathcal{L}_m$. Suppose that ULV is in $\mathcal{L}_m$ and (L, ε) is in $\mathcal{R}$. We want to show that UV is in $\mathcal{L}_m$. Let P be a path in $\mathcal{A}_m$ from A to Ω such

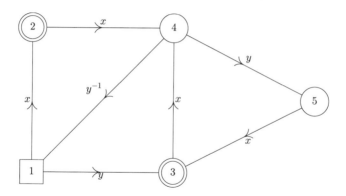

Figure 4.1.2

Figure 4.1.3

that $ULV = \mathrm{Sg}(P)$. Then P has a subpath Q with signature L. Let Q go from σ to τ. Since $\mathcal{A}_m$ is compatible with $\mathcal{R}$, either $\sigma = \tau$ or $(\sigma, \varepsilon, \tau)$ is in E_m. If $\sigma = \tau$, let R be the path obtained from P by deleting Q. If $\sigma \neq \tau$, then let R be the result of replacing Q by $(\sigma, \varepsilon, \tau)$ in P. In either case, R is a path in $\mathcal{A}_m$ from A to Ω such that $\mathrm{Sg}(R) = UV$. Therefore UV is in $\mathcal{L}_m$. □

By Lemma 1.14, $\mathcal{L}' = \mathcal{L}_m = L(\mathcal{A}_m)$ is rational. □

Corollary 1.15. *If $\mathcal{L}$ is a rational language over X, then $\overline{\mathcal{L}}$ is rational.*

Proof. By definition, $\overline{\mathcal{L}} = \mathcal{L}' \cap \mathcal{C}$. Since both $\mathcal{L}'$ and $\mathcal{C}$ are rational, so is $\overline{\mathcal{L}}$. □

We shall refer to the construction of $\mathcal{A}_m$ from $\mathcal{A}$ in Proposition 1.12 as the *derivation construction*. This construction can be modified to handle the case in which right sides in $\mathcal{R}$ have length 0 or 1. See Exercise 1.4.

Example 1.7. Let X and $\mathcal{R}$ be as in Examples 1.1, 1.3, and 1.5, and let $\mathcal{A}$ be the automaton in Figure 4.1.2. We shall construct automata recognizing $\mathcal{L}'$ and $\overline{\mathcal{L}}$, where $\mathcal{L} = L(\mathcal{A})$. Let us initialize a new automaton $\mathcal{A}'$ to $\mathcal{A}$. In $\mathcal{A}'$ we have the paths shown in Figure 4.1.3. Thus we add $(1, \varepsilon, 4)$ and $(4, \varepsilon, 3)$ to the edges of $\mathcal{A}'$. Now we have the paths shown in Figure 4.1.4. Therefore we add the edges $(2, \varepsilon, 4)$ and $(4, \varepsilon, 5)$. This creates the paths shown in Figure 4.1.5, so $(2, \varepsilon, 3)$ and $(2, \varepsilon, 5)$ are added to $\mathcal{A}'$. At this point $\mathcal{A}'$ is compatible with $\mathcal{R}$ and $\mathcal{L}' = L(\mathcal{A}')$.

Figure 4.1.4

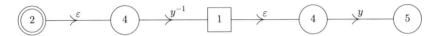

Figure 4.1.5

Table 4.1.1

	x	y	y^{-1}
1	2	3	4
2		5	4
3	6		
4	2		4
5	7		
6		5	4
7			

To compute an automaton recognizing $\overline{\mathcal{L}} = \mathcal{L}' \cap \mathcal{C}$, we need a deterministic automaton $\mathcal{A}_1$ recognizing $\mathcal{L}'$. Applying the accessible subset construction to $\mathcal{A}'$ and taking the trim part of the result gives us such an $\mathcal{A}_1$ with 6 states and 13 edges. An automaton recognizing $\mathcal{C}$ is given in Example 1.5. The accessible intersection procedure produces $\mathcal{A}_2 = (\Sigma_2, X, E_2, A_2, \Omega_2)$ recognizing $\overline{\mathcal{L}}$, where $\Sigma_2 = \{1, 2, 3, 4, 5, 6, 7\}$, $A_2 = \{1\}$, $\Omega_2 = \{1, 2, 3, 4, 6, 7\}$, and E_2 is given by Table 4.1.1.

Exercises

1.1. Let $\mathcal{R}$ be a classical niladic rewriting system on X^*. A word U in X^* is *cyclically reduced* if every cyclic permutation of U is in $\mathcal{C}$. Show that the set of all cyclically reduced words is a rational language over X.

1.2. Let X and $\mathcal{R}$ be as in Example 1.2. Show that conjugacy in $F = \mathrm{Mon}\,\langle X \mid \mathcal{R}\rangle$ can be decided.

1.3. In Proposition 1.2, suppose $X = \{a, b, c\}$, $r = 1$. $W_1 = abc$, and $n_1 = 2$. Compute an automaton isomorphic to $A_t(\mathcal{C})$.

1.4. Show that the derivation construction can be extended to work with confluent rewriting systems in which right sides have length 0 or 1.

1.5. Let $X = \{x, y\}$ and let $\mathcal{R}$ consist of the rules $x^2 \to \varepsilon$, $y^2 \to \varepsilon$. The group $G = \mathrm{Mon}\,\langle X \mid \mathcal{R}\rangle$ is called the *infinite dihedral group*. Show that the image of xy in G generates a normal infinite cyclic subgroup of G of index 2. Prove also that any subgroup of G of order larger than 2 has finite index in G.

4.2 Subgroups and their languages

In this section we continue the notation established in Section 4.1. We are studying a group $F = \mathrm{Mon}\,\langle X \mid \mathcal{R}\rangle$, where X is a finite set and $\mathcal{R}$ is a classical niladic rewriting system on X^*. The congruence generated by $\mathcal{R}$ is $\sim$, and the set of canonical forms for $\sim$ is $\mathcal{C}$. If U is a word in X^*, then $[U]$ is the $\sim$-class containing U and $\overline{U}$ is the element of $\mathcal{C}$ in $[U]$. We can compute $\overline{U}$ using $\mathcal{R}$.

Let us fix a subgroup H of F. We shall be studying the elements of H and the right cosets of H in F. In doing so, we shall associate various languages and automata with H. It is important to remember that multiplying a right coset of H on the right by an element of F yields a right coset. Thus if σ is a right coset of H and U is in X^*, then $\sigma[U]$ is a right coset of H.

Perhaps the most obvious language to associate with H is $\mathcal{L}_s = \mathcal{L}_s(H) = \{U \in X^* \mid [U] \in H\}$. Unfortunately, $\mathcal{L}_s$ need not be rational, even when H is finitely generated.

Example 2.1. Let $X = \{a, a^{-1}\}$ and let $\mathcal{R}$ consist of the rules

$$aa^{-1} \to \varepsilon, \quad a^{-1}a \to \varepsilon.$$

Set H equal to the trivial subgroup of F. Then $\mathcal{L}_s = \{U \in X^* \mid \overline{U} = \varepsilon\}$. More generally, for any integer k, let $\mathcal{D}_k = \{U \in X^* \mid \overline{U} = a^k\}$. Thus $\mathcal{L}_s = \mathcal{D}_0$. Then the cone $C(\mathcal{L}_s, a^k)$ is $\mathcal{D}_{-k}$, so $C(\mathcal{L}_s)$ is infinite.

Of course $\mathcal{L}_s$ can be rational. If $H = F$, then $\mathcal{L}_s = X^*$.

Let $\mathcal{A}_s = \mathcal{A}_s(H) = (\Sigma_s, X, E_s, \{H\}, \{H\})$ be the Schreier automaton of F relative to H and the map $x \mapsto [x]$. Thus $\Sigma_s = \Sigma_s(H)$ is the set of right cosets of H in F, and E_s is the set of triples (σ, x, τ), where σ and τ are in Σ_s, x is in X, and $\sigma[x] = \tau$. If φ and ψ are in Σ_s and U is in X^*, then $\varphi^U = \psi$ if and only if $\varphi[U] = \psi$.

Proposition 2.1. *The automaton* $\mathcal{A}_s$ *is complete and trim. In addition,* $\mathcal{L}_s = L(\mathcal{A}_s)$ *and* $\mathcal{A}_s$ *is a minimal complete automaton recognizing* $\mathcal{L}_s$

Proof. It is clear that $\mathcal{A}_s$ is complete. If σ is in Σ_s, then σ has the form $H[U]$ for some word U and $\sigma = H^U$. Therefore σ is accessible. Also $\sigma^{U^{-1}} = H[U][U^{-1}] = H$, so σ is coaccessible. Hence $\mathcal{A}_s$ is trim. A word W is in $\mathcal{L}_s$ if and only if $H[W] = H$. Therefore $\mathcal{L}_s = L(\mathcal{A}_s)$. Also, the cone $C(\mathcal{L}_s, W)$ is the set of words V such that

$$H = H[WV] = H[W][V],$$

so $C(\mathcal{L}_s, W) = \{V \in X^* \mid [V]^{-1} \in H[W]\}$. Thus $C(\mathcal{L}_s, W_1) = C(\mathcal{L}_s, W_2)$ if and only if $H[W_1] = H[W_2]$. It is easy to check that the map taking $C(\mathcal{L}_s, W)$ to $H[W]$ is an isomorphism from $A(\mathcal{L}_s)$ to $\mathcal{A}_s$. $\square$

Corollary 2.2. *The language* $\mathcal{L}_s$ *is rational if and only if the index of H in F is finite.*

For σ in Σ_s, let $\mathcal{L}_s(\sigma)$ be $\{U \in X^* \mid [U] \in \sigma\}$, the set of words defining elements of σ.

Corollary 2.3. *If H has finite index in F, then $\mathcal{L}_s(\sigma)$ is rational for all right cosets σ.*

Proof. The language $\mathcal{L}_s(\sigma)$ is recognized by $(\Sigma_s, X, E_s, \{H\}, \{\sigma\})$. $\square$

Proposition 2.4. *If $\mathcal{A} = (\Sigma, X, E, A, \Omega)$ is a trim, deterministic automaton such that $L(\mathcal{A}) \subseteq \mathcal{L}_s$, then there is an expanding morphism f from $\mathcal{A}$ to $\mathcal{A}_s$.*

Proof. Let σ be in Σ. Since $\mathcal{A}$ is trim, there is a path P in $\mathcal{A}$ from A to σ. Let $U = \mathrm{Sg}(P)$. We would like to map σ to $H[U]$, but we need to know that this is well defined. Suppose that Q is another path from A to σ and $V = \mathrm{Sg}(Q)$. Since σ is coaccessible, there is a path R from σ to Ω. Let $W = \mathrm{Sg}(R)$. Then UW and VW are both in $L(\mathcal{A})$. Therefore,

$$H[U] = H[U][W][W]^{-1} = H[UW][W]^{-1} = H[W]^{-1}.$$

Similarly, $H[V] = H[W]^{-1}$, so $H[U] = H[V]$. Thus we may define $f : \Sigma \to \Sigma_s(H)$ by $f(\sigma) = H[U]$.

If σ is in A, then we may take $U = \varepsilon$, and $f(\sigma) = H[\varepsilon] = H$. If σ is in Ω, then U is in $L(\mathcal{A})$. Hence $f(\sigma) = H[U] = H$. Suppose that (σ, x, τ) is in E. Then Ux is the signature of a path in $\mathcal{A}$ from A to τ. Thus $f(\tau) = H[Ux] = H[U][x] = f(\sigma)[x] = f(\sigma)^x$. Therefore f is an expanding morphism. $\square$

Corollary 2.5. *Let K be a subgroup of H. There is an expanding morphism from $\mathcal{A}_s(K)$ to $\mathcal{A}_s(H)$.*

Proof. Clearly $\mathcal{L}_s(K) \subseteq \mathcal{L}_s(H)$. Since $\mathcal{A}_s(K)$ is trim and deterministic and $\mathcal{A}_s(K)$ recognizes $\mathcal{L}_s(K)$, Proposition 2.4 applies. The morphism is the obvious map from $\Sigma_s(K)$ to $\Sigma_s(H)$ which takes $K[U]$ to $H[U]$ for any U in X^*. □

Another language associated with our subgroup H is $\mathcal{L}_c = \mathcal{L}_c(H) = \mathcal{L}_s(H) \cap \mathcal{C}$, the set of canonical forms for elements in H. If $\mathcal{L}_s$ is rational, then so is $\mathcal{L}_c$. However, $\mathcal{L}_c$ may be rational even when $\mathcal{L}_s$ is not.

Proposition 2.6. *If H is finitely generated, then $\mathcal{L}_c$ is a rational language.*

Proof. Since H is finitely generated as a group, by Corollary 1.4 H is finitely generated as a monoid. Let $\mathcal{U}$ be a finite set of words such that $\{[U] \mid U \in \mathcal{U}\}$ is a monoid generating set for H. Every element of H contains an element of $\mathcal{U}^*$, so $\mathcal{L}_c = \overline{\mathcal{U}^*}$. Since $\mathcal{U}$ is finite, $\mathcal{U}$ is rational. Therefore $\mathcal{U}^*$ is rational by Proposition 1.2 in Chapter 3. By Corollary 1.15, $\mathcal{L}_c$ is rational. □

For a right coset σ of H, put $\mathcal{L}_c(\sigma) = \mathcal{L}_s(\sigma) \cap \mathcal{C}$, the set of canonical forms for elements of σ.

Corollary 2.7. *If H is finitely generated, then $\mathcal{L}_c(\sigma)$ is rational for all σ in Σ_s.*

Proof. Let U be a word such that $[U]$ is in σ. Then $\mathcal{L}_c(\sigma) = \overline{\mathcal{L}_c(H)U}$, which is rational by Proposition 1.2 in Chapter 3 and Corollary 1.15. □

When H is given by a finite generating set, the proof of Proposition 2.6 provides us with a procedure for constructing an automaton recognizing $\mathcal{L}_c(H)$. To get started, we need an automaton recognizing $\mathcal{U}^*$. We can use the construction in Section 3.9. However, sometimes there is an easier way. See Exercises 2.1 and 2.2. Once we have our automaton, we apply the derivation construction, the accessible subset construction, and the accessible intersection construction. Finally we take the trim part. Details are left as an exercise. This procedure will be referred to as CANON.

Example 2.2. Let $X = \{x, y\}^{\pm}$, let $\mathcal{R}$ consist of the standard defining relations for the free group generated by $\{x, y\}$, and let H be the subgroup of F generated by $[xyx]$ and $[x^{-2}y]$. To get a monoid generating set for H, we add the inverses of the generators and take $\mathcal{U} = \{xyx, x^{-1}y^{-1}x^{-1}, x^{-2}y, y^{-1}x^2\}$. Applying CANON and standardizing the result with respect

Table 4.2.1

	x	x^{-1}	y	y^{-1}
1	2	3		4
2			5	
3		6		7
4	8			
5	9	6		
6			10	
7		11		
8	9			
9	2			4
10	2	3		
11				4

to the length-plus-lexicographic ordering with $x \prec x^{-1} \prec y \prec y^{-1}$, we get $\mathcal{A} = (\Sigma, X, E, A, \Omega)$, where $\Sigma = \{1, 2, \dots, 11\}$, $A = \{1\}$, $\Omega = \{1, 9, 10, 11\}$, and E is given by Table 4.2.1.

Proposition 2.8. *Suppose that $\mathcal{L}_c(H)$ is rational. Given $\mathcal{R}$ and a finite automaton $\mathcal{A}$ recognizing $\mathcal{L}_c(H)$, we can decide membership in H and decide equality of right cosets of H.*

Proof. To see if $[U]$ is in H, we compute $\overline{U}$ using $\mathcal{R}$ and check whether $\overline{U}$ is in $L(\mathcal{A})$. The right cosets $H[U]$ and $H[V]$ are equal if and only if $[U][V]^{-1} = [UV^{-1}]$ is in H. $\square$

Exercises

2.1. Let $\mathcal{U}$ be a finite set of words which is a monoid generating set for a subgroup H of F. Show that we may assume that the elements of $\mathcal{U}$ are nonempty and reduced and no element of $\mathcal{U}$ is a proper prefix of another element of $\mathcal{U}$.

2.2. Assume that $\mathcal{U}$ satisfies the conditions of Exercise 2.1. Let Σ be the set of proper prefixes of the elements of $\mathcal{U}$ and let E be the set of triples (U, x, V), where U and V are in Σ and either $V = Ux$ or Ux is in $\mathcal{U}$ and $V = \varepsilon$. Show that $(\Sigma, X, E, \{\varepsilon\}, \{\varepsilon\})$ recognizes $\mathcal{U}^*$.

2.3. Describe the procedure CANON in detail.

2.4. Let $\mathcal{L}$ be a language over the finite alphabet X and let $\mathcal{A}$ be an automaton with alphabet X such that $L(\mathcal{A}) \subseteq \mathcal{L}$. Show that there need not be an expanding morphism from $\mathcal{A}$ to $A(\mathcal{L})$. Thus the morphism in Proposition 2.4 is an exception.

4.3 Important cosets

We continue the notation of the previous sections in this chapter. We are investigating the elements and right cosets of a subgroup H of F. It turns out that when H is finitely generated, there is a finite subset Σ_I of Σ_s such that in most interesting computations involving right cosets of H, only elements of Σ_I occur.

A right coset σ of H will be called *important* if $\sigma = H[U]$, where U is a prefix of an element of $\mathcal{L}_c(H)$, the set of canonical forms for elements of H. If UV is in $\mathcal{L}_c(H)$ and $\sigma = H[U]$, then (U, V) will be called a *defining pair* for σ. The set of all important right cosets of H will be denoted $\Sigma_I = \Sigma_I(H)$. Since ε is in $\mathcal{L}_c(H)$, the pair $(\varepsilon, \varepsilon)$ is a defining pair for H. Thus H is in Σ_I.

Proposition 3.1. *Suppose that $\mathcal{A} = (\Sigma, X, E, A, \Omega)$ is a trim, deterministic automaton such that $\mathcal{L}_c(H) \subseteq L(\mathcal{A}) \subseteq \mathcal{L}_s(H)$. Let f be the expanding morphism from $\mathcal{A}$ to $\mathcal{A}_s(H)$. The image of Σ under f contains Σ_I.*

Proof. Suppose that σ is in Σ_I and (U, V) is a defining pair for σ. Then UV is in $\mathcal{L}_c(H)$, so there is a path P in $\mathcal{A}$ from A to Ω such that $UV = \mathrm{Sg}(P)$. There is an initial segment Q of P such that $U = \mathrm{Sg}(Q)$. Let Q end at φ. Then $f(\varphi) = \sigma$. $\square$

The *important-coset automaton* of H is the restriction $\mathcal{A}_I = \mathcal{A}_I(H)$ of $\mathcal{A}_s(H)$ to Σ_I. Thus $\mathcal{A}_I = (\Sigma_I, X, E_I, \{H\}, \{H\})$, where E_I is the set of triples (σ, x, τ) such that σ and τ are in Σ_I, x is in X, and $\sigma[x] = \tau$. We shall denote $L(\mathcal{A}_I)$ by $\mathcal{L}_I = \mathcal{L}_I(H)$.

Example 3.1. Suppose that H is the trivial subgroup of F. Then $\mathcal{L}_c(H) = \{\varepsilon\}$ and $\Sigma_I = \{H\}$. For no x in X is $H[x] = H$, so $\mathcal{A}_I = (\{H\}, X, \emptyset, \{H\}, \{H\})$.

Example 3.2. If $H = F$, then $\Sigma_I = \Sigma_s = \{H\}$ and $\mathcal{A}_I = \mathcal{A}_s$.

Proposition 3.2. *The automaton $\mathcal{A}_I(H)$ is trim and $\mathcal{L}_c(H) \subseteq \mathcal{L}_I(H) \subseteq \mathcal{L}_s(H)$.*

Proof. Since $\mathcal{A}_I$ is the restriction of $\mathcal{A}_s$ to Σ_I, it follows that $\mathcal{L}_I = L(\mathcal{A}_I) \subseteq L(\mathcal{A}_s) = \mathcal{L}_s$. Let $U = x_1 \ldots x_s$ be in $\mathcal{L}_c(H)$. By the definition of Σ_I, the cosets $\sigma_i = H[x_1 \ldots x_i]$, $0 \le i \le s$, are in Σ_I and the edges $(\sigma_{i-1}, x_i, \sigma_i)$ are in E_I. Therefore U is in $L(\mathcal{A}_I)$. Hence $\mathcal{L}_c(H) \subseteq \mathcal{L}_I(H)$. Each of the cosets σ_i is trim as a state in $\mathcal{A}_I(H)$. Since every coset in Σ_I arises as a σ_i for an appropriate choice of U, it follows that $\mathcal{A}_I(H)$ is trim. $\square$

Proposition 3.3. *Suppose that σ and τ are important right cosets of H and W is a word in $\mathcal{C}$ such that $\sigma[W] = \tau$. For all prefixes U of W the coset $\sigma[U]$ is important.*

Proof. The proposition states that if P is a path in $\mathcal{A}_s$ from Σ_I to Σ_I and $\mathrm{Sg}(P)$ is in $\mathcal{C}$, then P is a path in $\mathcal{A}_I$. If $W = \varepsilon$, then there is nothing to prove. We proceed by induction on $|W|$. Let $W = UV$. We wish to show that $\sigma[U]$ is important. We may assume that both U and V are nonempty. Let $U = xU_1$, where x is in X, and let (S,T) be a defining pair for σ. Suppose for the moment that T starts with x, so T has the form xT_1. Then (Sx, T_1) is a defining pair for $\varphi = \sigma[x]$, and $\tau = \varphi[U_1 V]$. By induction on $|W|$, the coset $\sigma[U] = \varphi[U_1]$ is important. Thus we may assume that T and W do not start with the same element of X.

Let (C, D) be a defining pair for τ. By an argument similar to the one in the previous paragraph, we may assume that W and C do not end with the same element of X. By Proposition 1.11, $T^{-1}WC^{-1}$ is in C. Now

$$H[T^{-1}WC^{-1}] = H[ST][T^{-1}WC^{-1}] = H[S][WC^{-1}] = \sigma[W][C^{-1}]$$
$$= \tau[C^{-1}] = H[C][C^{-1}] = H.$$

Therefore $T^{-1}WC^{-1} = T^{-1}UVC^{-1}$ is in $\mathcal{L}_c(H)$ and $\sigma[U] = H[T^{-1}U]$ is important. $\square$

Corollary 3.4. *If σ, τ, and W are as in Proposition 3.3, then there is a defining pair for σ of the form (M, WN).*

Proof. Let S, T, C, and D be as in the proof of Proposition 3.3. Then at least one of the words $SWD, SWC^{-1}, T^{-1}WD$, or $T^{-1}WC^{-1}$ is in C. The details are left as an exercise. $\square$

We turn now to a fundamental result due to Schreier.

Proposition 3.5. *A subgroup of finite index in a finitely generated group is finitely generated.*

Proof. Let G be a group and let G be generated as a monoid by a subset Z. Let H be a subgroup of G and let Σ denote the set of right cosets of H in G. For σ in Σ let u_σ be a representative for σ. We shall assume that $u_H = 1$. If σ is in Σ and z is in Z, set $y(\sigma, z) = u_\sigma z u_\tau^{-1}$, where $\tau = \sigma z$. Since $u_\sigma z$ is in τ, the element $y(\sigma, z)$ is in H.

Lemma 3.6. *The elements $y(\sigma, z)$ generate H as a monoid.*

Proof. Let h be an element of H. Since Z generates G, we can write $h = z_1 \ldots z_r$, where the z_i are in Z. Let $\sigma_i = Hz_1 \ldots z_i$, $0 \le i \le r$, and set $u_i = u_{\sigma_i}$ and $y_i = y(\sigma_{i-1}, z_i) = u_{i-1} z_i u_i^{-1}$. Then

$$h = 1h = u_0 z_1 \ldots z_r$$
$$= u_0 z_1 u_1^{-1} u_1 z_2 \ldots z_r$$
$$= y_1 u_1 z_2 \ldots z_r = y_1 u_1 z_2 u_2^{-1} u_2 z_3 \cdots z_r$$
$$= y_1 y_2 u_2 z_3 \cdots z_r = \cdots = y_1 y_2 \ldots y_r u_r.$$

But $\sigma_r = Hh = H$ and hence $u_r = 1$. Therefore $h = y_1 \ldots y_r$. □

Set $Y = \{y(\sigma, z) \mid \sigma \in \Sigma,\ z \in Z\}$. If H has finite index in G and Z is finite, then Y is finite and $|Y| \le |G : H||Z|$.

The same inequality holds if we consider group generators. If Z generates G as a group, then $Z \cup Z^{-1}$ generates G as a monoid. If z is in Z and σ is in Σ, then $y(\sigma, z^{-1}) = u_\sigma z^{-1} u_\tau^{-1} = (u_\tau z u_\sigma^{-1})^{-1}$, where $\tau = \sigma z^{-1}$. Since $\sigma = \tau z$, it follows that $y(\sigma, z^{-1}) = y(\tau, z)^{-1}$. Therefore $Y = \{y(\sigma, z) \mid \sigma \in \Sigma,\ z \in Z\}$ generates H as a group. □

The elements of Y in the proof of Proposition 3.5 are called the *Schreier generators* of H relative to Z and the choice of the elements u_σ. If $H = G$, then the Schreier generators are just the elements of Z. We can reduce slightly the bound on the number of elements needed to generate H.

Proposition 3.7. *If H is a subgroup of finite index n in a group G which is generated as a group by r elements, then H is generated as a group by $1 + n(r - 1)$ elements.*

Proof. Let Z be a set of r elements which generate G as a group and let $X = Z \cup Z^{-1}$. Fix a reduction ordering $\prec$ on X^*, and for each right coset σ of H in G let U_σ be the first word with respect to $\prec$ in X^* which defines an element of σ. Set u_σ equal to the element of G defined by U_σ. Then H is generated as a group by the Schreier generators $y(\sigma, z)$. Suppose $\sigma \ne H$ and $U_\sigma = x_1 \ldots x_s$, where the x_i are in X. Assume first that x_s is z^{-1} for some z in Z. If $\tau = \sigma z$, then U_τ must be $x_1 \ldots x_{s-1}$ and so $u_\tau = u_\sigma z$. Therefore $y(\sigma, z) = 1$. Now assume that $x_s = z$ is in Z. If $\tau = \sigma z^{-1}$, then $U_\tau = x_1 \ldots x_{s-1}$ and $y(\tau, z)$ is trivial. Thus $n - 1$ of the elements $y(\sigma, z)$ are trivial and hence H can be generated as a group by $nr - (n-1) = 1 + n(r-1)$ elements. □

In the case of subgroups of our group F, we can improve on Proposition 3.5 somewhat.

Proposition 3.8. *A subgroup H of F is finitely generated if and only if $\Sigma_I(H)$ is finite.*

Proof. Assume first that H is finitely generated. By Proposition 2.6, $\mathcal{L}_c(H)$ is rational. By Proposition 3.1, there is an expanding morphism f from $\mathcal{A} = A(\mathcal{L}_c(H))$ to $\mathcal{A}_s(H)$ and the image of f contains $\Sigma_I(H)$. Since $\mathcal{A}$ is finite, $\Sigma_I(H)$ is finite.

Now suppose that $\Sigma_I(H)$ is finite. For each σ in $\Sigma_I(H)$, let $(S(\sigma), T(\sigma))$ be a defining pair for σ. For each x in X such that σ^x is defined in $\mathcal{A}_I(H)$, let $Y(\sigma, x)$ be the word $S(\sigma) x S(\sigma^x)^{-1}$. Since $\sigma = H[S(\sigma)]$, it follows that

$\sigma^x = H[S(\sigma)x] = H[S(\sigma^x)]$. Therefore $[Y(\sigma, x)]$ is in H. It is convenient to denote H by α when we are considering H to be an element of Σ_I. We may choose $S(\alpha) = T(\alpha) = \varepsilon$.

Let U be an element of $\mathcal{L}_c(H)$ and let $U = x_1 \ldots x_t$, where each x_i is in X. For $0 \le i \le t$ let $\sigma_i = \alpha[x_1 \ldots x_i]$. Then $\sigma_0 = \alpha$ and $\sigma_t = \alpha[U] = \alpha$. Each σ_i is in Σ_I and $\sigma_i = (\sigma_{i-1})^{x_i}$, $0 < i \le t$. By arguments very similar to those in the proof of Lemma 3.6, we can prove that $x_1 \ldots x_i \sim Y(\sigma_0, x_1) Y(\sigma_1, x_2) \cdots Y(\sigma_{i-1}, x_i) S(\sigma_i)$, $0 \le i \le t$. Taking $i = t$, we find that $U \sim Y(\sigma_0, x_1) \cdots Y(\sigma_{t-1}, x_t)$, since $S(\sigma_t) = \varepsilon$. Every element of H is represented by an element of $\mathcal{L}_c(H)$. Therefore H is generated as a monoid by

$$\{[Y(\sigma, x)] \mid \sigma \text{ and } \sigma^x \text{ in } \Sigma_I\}.$$

Since Σ_I and X are finite, H is finitely generated. $\square$

Corollary 3.9. *The language $\mathcal{L}_c(H)$ is rational if and only if H is finitely generated.*

Proof. In the proof of Proposition 3.8 we saw that $\mathcal{L}_c(H)$ rational implies that Σ_I is finite and hence that H is finitely generated. In Proposition 2.6 we showed that $\mathcal{L}_c(H)$ is rational when H is finitely generated. $\square$

Corollary 3.10. *If $\mathcal{U}$ is a rational language over X, then $H = \mathrm{Grp}\langle [U] \mid U \in \mathcal{U}\rangle$ is finitely generated.*

Proof. Let $\mathcal{V} = \mathcal{U} \cup \mathcal{U}^{-1}$. Then $\mathcal{V}$ is rational and $H = \mathrm{Mon}\langle [V] \mid V \in \mathcal{V}\rangle$. Therefore $\mathcal{L}_c(H) = \overline{\mathcal{V}^*}$ is rational, and so H is finitely generated. $\square$

Example 3.3. Suppose that F is the free group on $\{x, y\}$, so $X = \{x, y\}^{\pm}$. Let $\mathcal{A} = (\Sigma, X, E, \{1\}, \{1\})$, where $\Sigma = \{1, 2, 3, 4\}$ and E is given by the following table:

	x	x^{-1}	y	y^{-1}
1	2		3	4
2	3	1	4	
3		2		1
4	4	4	1	2

In Section 4.4 a procedure is presented for verifying that $\mathcal{A}$ is isomorphic to $\mathcal{A}_I(H)$ for some subgroup H of F. Assuming this, let us use the method given in the proof of Proposition 3.8 to get generators for H. The words $S(\sigma)$ and $T(\sigma)$ can be chosen as follows:

$$S(1) = T(1) = \varepsilon, \quad S(2) = x, \quad T(2) = xy^{-1},$$

$$S(3) = y, \quad T(3) = x^{-2}, \quad S(4) = y^{-1}, \quad T(4) = xy.$$

In each case the word $S(\sigma)T(\sigma)$ is reduced and belongs to $L(\mathcal{A})$. Moreover, $1^{S(\sigma)} = \sigma$. There are 12 edges in E and hence 12 words $Y(\sigma, u)$ with u in X.

$$
\begin{aligned}
Y(1, x) &= xx^{-1}, & Y(3, x^{-1}) &= yx^{-2}, \\
Y(1, y) &= yy^{-1}, & Y(3, y^{-1}) &= yy^{-1}, \\
Y(1, y^{-1}) &= y^{-1}y, & Y(4, x) &= y^{-1}xy, \\
Y(2, x) &= x^2 y^{-1}, & Y(4, x^{-1}) &= y^{-1}x^{-1}y, \\
Y(2, x^{-1}) &= xx^{-1}, & Y(4, y) &= y^{-1}y, \\
Y(2, y) &= xy^2, & Y(4, y^{-1}) &= y^{-2}x.
\end{aligned}
$$

Half of these words are freely equivalent to the empty word. The remaining six words consist of three pairs, with the words in each pair representing inverse elements of F. Thus H is generated as a group by $[x^2 y^{-1}]$, $[xy^2]$, and $[y^{-1}xy]$.

For a subgroup H of F, the language $\mathcal{L}_I(H)$ is less easy to visualize than $\mathcal{L}_s(H)$ and $\mathcal{L}_c(H)$.

Proposition 3.11. *The language $\mathcal{L}_I(H)$ is the set of words W in $\mathcal{L}_s(H)$ such that for all prefixes U of W the coset $H[U]$ is important.*

Proof. Exercise. $\square$

Proposition 3.12. *The index of H in F is finite if and only if F is finite or $\mathcal{A}_I(H)$ is finite and complete.*

Proof. If F is finite, then $|X| = 1$ and F has order 2. If H is the trivial subgroup of F, then $\mathcal{A}_I(H)$ is finite but not complete.

Suppose that $\mathcal{A}_I(H)$ is finite and complete. Then for every important right coset σ and every x in X, the coset $\sigma[x]$ is important. Therefore $\Sigma_s(H) = \Sigma_I(H)$ and $\Sigma_s(H)$ is finite.

Now assume that F is infinite and $|F : H|$ is finite. Then $\Sigma_s(H)$ is finite, so $\Sigma_I(H)$ is finite. We must prove that $\Sigma_I = \Sigma_s$. Suppose U is a word in $\mathcal{C}$. We shall show that $H[U]$ is important.

For each σ in Σ_s choose a word W_σ such that $\sigma = H[W_\sigma]$. Let m be the maximum length of the words W_σ^{-1}. By Corollary 1.7, we can find a word V of arbitrary length such that UV is in $\mathcal{C}$. Choose V so that $|V| \geq m$. Let $\sigma = H[UV]$. Then $[UVW_\sigma^{-1}]$ is in H. Let $S = \overline{UVW_\sigma^{-1}}$. By Proposition 1.5, U is a prefix of S. Let $S = UT$. Then (U, T) is a defining pair for $H[U]$, so $H[U]$ is important. Thus every right coset of H is important and $\mathcal{A}_I(H)$ is complete. $\square$

Proposition 3.13. *Suppose that $\mathcal{A} = (\Sigma, X, E, A, \Omega)$ is a trim, deterministic automaton such that $L(\mathcal{A}) \subseteq \mathcal{L}_I(H)$. Let f be the expanding morphism from $\mathcal{A}$ to $\mathcal{A}_s(H)$. Then f is an expanding morphism from $\mathcal{A}$ to $\mathcal{A}_I(H)$.*

Proof. Let $f : \Sigma \to \Sigma_s(H)$ be the morphism of Proposition 2.4. Let σ be in Σ. There is a path P from A to σ and a path Q from σ to Ω. Let $U = \mathrm{Sg}(P)$ and $V = \mathrm{Sg}(Q)$. The word UV is in $L(\mathcal{A})$ and hence in $\mathcal{L}_I(H)$. By Proposition 3.11, the coset $H[U] = f(\sigma)$ is in $\Sigma_I(H)$. Since $\mathcal{A}_I(H)$ is the restriction of $\mathcal{A}_s(H)$ to $\Sigma_I(H)$ and f is an expanding morphism from $\mathcal{A}$ to $\mathcal{A}_s$ which maps Σ into Σ_I, it follows that f is an expanding morphism from $\mathcal{A}$ to $\mathcal{A}_I(H)$. $\square$

Corollary 3.14. *Suppose that K is a subgroup of H. There is an expanding morphism from $\mathcal{A}_I(K)$ to $\mathcal{A}_I(H)$.*

Proof. Suppose that σ is in $\Sigma_I(K)$ and let (S, T) be a defining pair for σ. Then $\sigma = K[S]$ and ST is in $\mathcal{L}_c(K) \subseteq \mathcal{L}_c(H)$. Therefore (S, T) is a defining pair for $H[S]$. Let V be in $\mathcal{L}_I(K)$ and let U be a prefix of V. Then $K[U]$ is in $\Sigma_I(K)$ and so $H[U]$ is in $\Sigma_I(H)$. By Proposition 3.11, V is in $\mathcal{L}_I(H)$. Therefore Proposition 3.13 applies. $\square$

Corollary 3.15. *The automaton $\mathcal{A}_I(H)$ is isomorphic to $A_t(\mathcal{L}_I(H))$.*

Proof. By Proposition 3.2, $\mathcal{A}_I(H)$ is trim. By Proposition 3.13 there is an expanding morphism from $A_t(\mathcal{L}_I(H))$ to $A_I(H)$. By Exercise 3.7.3, $\mathcal{A}_I(H)$ and $A_t(\mathcal{L}_I(H))$ are isomorphic. $\square$

If we have $\mathcal{R}$ and an automaton isomorphic to $\mathcal{A}_I(H)$, we can decide equality of cosets of H more easily than with the method described in Proposition 2.8. Suppose that U and V are words in X^* and we want to decide whether $H[U]$ and $H[V]$ are equal. Given $\mathcal{R}$, we can easily compute $\overline{U}$ and $\overline{V}$, so we may assume that U and V are in $\mathcal{C}$.

Proposition 3.16. *Suppose that U and V are elements of $\mathcal{C}$. Let $U = BR$ and $V = CS$, where B and C are the longest prefixes of U and V, respectively, such that $H[B]$ and $H[C]$ are in $\Sigma_I(H)$. Then $H[U] = H[V]$ if and only if $H[B] = H[C]$ and $R = S$.*

Proof. If $H[B] = H[C]$ and $R = S$, then clearly $H[U] = H[V]$. Assume that $H[U] = H[V]$, so $[UV^{-1}]$ and $[VU^{-1}]$ are in H. Suppose first that U and V do not end with the same element of X. By Proposition 1.9, the words UV^{-1} and VU^{-1} are in $\mathcal{C}$ and hence in $\mathcal{L}_c(H)$. Therefore $H[U]$ and $H[V]$ are in $\Sigma_I(H)$, so $B = U$, $C = V$, and $R = S = \varepsilon$.

Now suppose that U and V end in the same element x of X. Say $U = U_1 x$ and $V = V_1 x$. Then

$$H[U_1] = H[U][x]^{-1} = H[V][x]^{-1} = H[V_1].$$

Let B_1 and C_1 be the longest prefixes of U_1 and V_1, respectively, such that $H[B_1]$ and $H[C_1]$ are in $\Sigma_I(H)$, and let $U_1 = B_1 R_1$ and $V_1 = C_1 S_1$. By induction on $|U|$, we have $H[B_1] = H[C_1]$ and $R_1 = S_1$.

Lemma 3.17. *Either $B = B_1$, or $B_1 = U_1$ and $B = U$.*

Proof. Clearly B_1 is a prefix of B. Let $B = B_1 B_2$. By Proposition 3.3, for every prefix D of B, the coset $H[D]$ is in $\Sigma_I(H)$. But this means that B_1 is the longest common prefix of B and U_1. Thus either $B_2 = \varepsilon$, or $B_1 = U_1$ and $B_2 = x$. $\square$

Suppose that $B = U$. Then $C = V$ and $R = S = \varepsilon$. Thus we may assume that $B \neq U$ and $C \neq V$. By Lemma 3.17, $B = B_1$ and $C = C_1$. Therefore $R = R_1 x = S_1 x = S$. $\square$

Given a word U in $\mathcal{C}$ and an automaton $\mathcal{A} = (\Sigma, X, E, \{\alpha\}, \{\alpha\})$ isomorphic to $\mathcal{A}_I(H)$, we can easily compute the decomposition $U = BR$ of Proposition 3.16 using the procedure TRACE defined in Section 3.2. The word B is the longest prefix of U which is the signature of a path in $\mathcal{A}$ starting at α. Thus we perform the call $\mathrm{TRACE}(\mathcal{A}, \alpha, U; \sigma, B, R)$. The state σ corresponds to $H[B]$.

Example 3.4. There are several ways to construct an automaton isomorphic to $\mathcal{A}_I(H)$ when H is given by a finite generating set. The preferred method is normally coset enumeration. In Example 5.5, coset enumeration will be applied to the subgroup H of Example 2.2. Here F is the free group on $\{x, y\}$ and H is generated by $[xyx]$ and $[x^{-2}y]$. In this case, $\mathcal{A}_I(H)$ is isomorphic to $(\Sigma, X, E, A, \Omega)$, where $\Sigma = \{1, 2, 3, 4\}$, $A = \Omega = \{1\}$, and E is given by the following table:

	x	x^{-1}	y	y^{-1}
1	2	3		4
2		1	3	
3	1	4		2
4	3		1	

Here is an example of the use of Proposition 3.16. Let

$$U = xyx^{-1}yx^{-1}y^{-2}xy,$$
$$V = y^{-1}xy^{-1}x^{-3}yx^2y.$$

Both U and V are in C. The call TRACE($\mathcal{A}, 1, U; \sigma, B, R$) returns $\sigma = 2$, $B = xyx^{-1}yx^{-1}y^{-1}$, and $R = y^{-1}xy$. The call TRACE($\mathcal{A}, 1, V; \tau, C, S$) returns $\tau = 2$, $C = y^{-1}xy^{-1}x^{-3}yx$, and $S = xy$. Since $\sigma = \tau$, we have $H[B] = H[C]$, but $R \neq S$, so $H[U] \neq H[V]$.

We close this section with a theoretical result concerning the structure of F.

Proposition 3.18. *Suppose that H is finitely generated and has infinite index in F. Then H contains no nontrivial normal subgroup of F.*

Proof. We may assume that $|X| \geq 2$. The intersection N of the conjugates of H is the largest normal subgroup of F contained in H. An element u of F is contained in N if and only if $\sigma u = \sigma$ for every right coset σ of H in F. Suppose that U is a nonempty word in C such that $[U]$ is in N. Since H is finitely generated, $\Sigma_I(H)$ is finite. Let σ be a coset in $\Sigma_s(H) - \Sigma_I(H)$ and let W be a word in C such that $\sigma = H[W]$.

Lemma 3.19. *It is possible to choose σ and W so that WU is in C.*

Proof. Choose σ "far away" from $\Sigma_I(H)$ in the sense that $\sigma[V]$ is not in $\Sigma_I(H)$ for any short words V. Let W end in x, let U start with y, and suppose that WU is not in C. Then $y = x^{-1}$. If $Y = X - \{x, y\}$ is not empty, then we can choose z in Y and replace σ by $\sigma[z]$ and W by Wz. Now WU is in C.

Suppose that Y is empty. Since F is infinite, the following conditions must hold:

(i) $x \neq y$.
(ii) $X = \{x, y\}$.

Thus F is isomorphic to $\mathbb{Z}$, which has no nontrivial subgroups of infinite index. $\square$

By Lemma 3.19, we may assume that WU is in C. Let B be the longest prefix of W such that $H[B]$ is in $\Sigma_I(H)$ and let $W = BR$. Then $R \neq \varepsilon$. Therefore B is the longest prefix of WU defining an element of $\Sigma_I(H)$. Since $R \neq RU$, the cosets σ and $\sigma[U] = H[WU]$ are distinct by Proposition 3.16. $\square$

Exercises

3.1. Let $\mathcal{A}$ be a finite automaton with alphabet X. Prove that it is possible to decide whether $\{[U] \mid U \in L(\mathcal{A})\}$ is a subgroup of F.

Table 4.4.1

	x	x^{-1}	y	y^{-1}
1		2	3	
2	1		4	3
3		5	2	1
4	4	4		2
5	3			

3.2. Let $\prec$ be a reduction ordering on X^* and let H be a finitely generated subgroup of F. For each right coset σ of H in F let U_σ be the first word with respect to $\prec$ defining an element of σ. Define $\mathcal{L}_R(H)$ to be $\{U_\sigma \mid \sigma \in \Sigma_s(H)\}$ and let $\mathcal{M} = \{U_\sigma \mid \sigma \in \Sigma_I(H)\}$. Let $\mathcal{P}$ be the set of words Ux such that U is in $\mathcal{M}$, x is in X, $H[Ux]$ is in $\Sigma_I(H)$, but Ux is not in $\mathcal{M}$. Finally, let $\mathcal{J}$ be the right ideal of X^* generated by $\mathcal{P}$. Show that $\mathcal{L}_R(H) = \mathcal{C} \cap (X^* - \mathcal{J})$. Conclude that $\mathcal{L}_R(H)$ is a rational language.

3.3. Suppose in Exercise 3.2 that $\prec$ is a length-plus-lexicographic ordering. Let $\mathcal{A}$ be an automaton isomorphic to $\mathcal{A}_I(H)$. Show how to construct an automaton recognizing $\mathcal{L}_R(H)$. (Hint: Start by standardizing $\mathcal{A}$ with respect to $\prec$.)

4.4 Coset automata

In Sections 4.2 and 4.3, various languages and automata were associated with a subgroup of F. In this section, an axiomatic characterization is given of the automata which can occur. In particular, we characterize those automata which are isomorphic to $\mathcal{A}_I(H)$ for some finitely generated subgroup H of F.

Let $\mathcal{A} = (\Sigma, X, E, A, \Omega)$ be an automaton with alphabet X. We shall say that $\mathcal{A}$ is a *coset automaton* relative to $\mathcal{R}$ if the following conditions are satisfied:

(i) $\mathcal{A}$ is accessible and deterministic.

(ii) $A = \Omega \neq \emptyset$.

(iii) If (σ, x, τ) is in E, then (τ, x^{-1}, σ) is in E.

In the literature on coset enumeration, the term "coset table" occurs frequently. A *coset table* is the transition table for a coset automaton. In (iii), if $e = (\sigma, x, \tau)$, then (τ, x^{-1}, σ) will be denoted e' and called the edge *paired* with e. Clearly $e'' = e$.

Example 4.1. Suppose that $X = \{x, y\}^\pm$ and $\mathcal{R}$ consists of the rules

$$xx^{-1} \to \varepsilon, \quad x^{-1}x \to \varepsilon, \quad yy^{-1} \to \varepsilon, \quad y^{-1}y \to \varepsilon.$$

Let $\mathcal{A} = (\Sigma, X, E, \{1\}, \{1\})$, where $\Sigma = \{1, 2, 3, 4, 5\}$ and E is given by Table 4.4.1. Then $\mathcal{A}$ is a coset automaton relative to $\mathcal{R}$.

Proposition 4.1. *For every subgroup H of F the automata $\mathcal{A}_s(H)$ and $\mathcal{A}_I(H)$ are coset automata.*

Proof. Both $\mathcal{A}_s(H)$ and $\mathcal{A}_I(H)$ satisfy conditions (i) and (ii). Suppose that (σ, x, τ) is an edge of $\mathcal{A}_s(H)$. Then $\sigma[x] = \tau$, so $\sigma = \tau[x]^{-1} = \tau[x^{-1}]$. Thus (τ, x^{-1}, σ) is an edge of $\mathcal{A}_s(H)$. If σ and τ are in $\Sigma_I(H)$, then (τ, x^{-1}, σ) is an edge of $\mathcal{A}_I(H)$. $\square$

Proposition 4.2. *Let $\mathcal{A}$ be a coset automaton and suppose that (ρ, x, τ) and (σ, x, τ) are edges of $\mathcal{A}$. Then $\rho = \sigma$.*

Proof. By condition (iii), both (τ, x^{-1}, ρ) and (τ, x^{-1}, σ) are edges of $\mathcal{A}$. Since $\mathcal{A}$ is deterministic, $\rho = \sigma$. $\square$

Corollary 4.3. *Let $\mathcal{A}$ be a coset automaton and suppose that P and Q are paths in $\mathcal{A}$ with the same endpoint and the same signatures. Then P and Q are the same path.*

Proof. We use Proposition 4.2 and induction on the length of P and Q.

$\square$

Proposition 4.4. *Suppose that $\mathcal{A}$ is a coset automaton. If there is a path P in $\mathcal{A}$ from σ to τ, then there is a path Q in $\mathcal{A}$ from τ to σ such that $\mathrm{Sg}(Q) = \mathrm{Sg}(P)^{-1}$.*

Proof. Let P consist of the edges $(\sigma_{i-1}, x_i, \sigma_i)$, $1 \le i \le s$. By condition (iii), the sequence Q of edges $(\sigma_s, x_s^{-1}, \sigma_{s-1})$, ..., $(\sigma_1, x_1^{-1}, \sigma_0)$ is a path in $\mathcal{A}$, and $\mathrm{Sg}(Q) = \mathrm{Sg}(P)^{-1}$. $\square$

Corollary 4.5. *Coset automata are trim.*

Proof. Let $\mathcal{A} = (\Sigma, X, E, A, \Omega)$ be a coset automaton and let σ be in Σ. Since $\mathcal{A}$ is accessible, there is a path in $\mathcal{A}$ from A to σ. By Proposition 4.4, there is a path from σ to $A = \Omega$. Thus $\mathcal{A}$ is trim. $\square$

Proposition 4.6. *Let $\mathcal{A}$ be a coset automaton. If $\sigma^U = \tau$ in $\mathcal{A}$, then $\sigma^{\overline{U}} = \tau$.*

Proof. Let P be the path in $\mathcal{A}$ from σ to τ such that $\mathrm{Sg}(P) = U$. If U is not in $\mathcal{C}$, then $U = SxyT$, where xy is a left side in $\mathcal{R}$. Let the subpath of P having signature xy consist of the edges (ν, x, τ) and (τ, y, ρ). Then $y = x^{-1}$, and, by condition (iii) and the fact that $\mathcal{A}$ is deterministic, it follows that $\nu = \rho$. Thus we may delete these two edges and have a path from σ to τ with signature ST. Continuing in this way, we construct a path from σ to τ with signature $\overline{U}$. $\square$

Corollary 4.7. *Let $\mathcal{A}$ be a coset automaton and set $\mathcal{L} = L(\mathcal{A})$. Then $\mathcal{L}' = \mathcal{L}$.*

Proof. The proof of Proposition 4.6 actually shows that $\sigma^V = \tau$ for every word V derivable from U using $\mathcal{R}$. Taking both σ and τ to be the element of A, we see that $\mathcal{L}' = \mathcal{L}$. $\square$

For any automaton $\mathcal{A}$ with alphabet X, let $K(\mathcal{A})$ be $\{[U] \mid U \in L(\mathcal{A})\}$, the image of $L(\mathcal{A})$ under the natural map from X^* to F.

Proposition 4.8. *Suppose that $\mathcal{A}$ is a coset automaton relative to $\mathcal{R}$. Then $K(\mathcal{A})$ is a subgroup of F. If $\mathcal{A}$ is finite, then $K(\mathcal{A})$ is finitely generated.*

Proof. Let $\mathcal{A} = (\Sigma, X, E, A, \Omega)$ and set $\mathcal{L} = L(\mathcal{A})$. Since $A = \Omega \neq \emptyset$, the empty word ε is in $\mathcal{L}$ and $\mathcal{L}\mathcal{L} \subseteq \mathcal{L}$. Therefore $K(\mathcal{A})$ is a submonoid of F. By Proposition 4.4, $K(\mathcal{A})$ is a subgroup. If $\mathcal{A}$ is finite, then $\mathcal{L}$ is rational. Therefore $\mathcal{L}_c(K(\mathcal{A})) = \mathcal{L}' \cap \mathcal{C} = \mathcal{L} \cap \mathcal{C}$ is rational. Thus $K(\mathcal{A})$ is finitely generated by Corollary 3.9. $\square$

If $\mathcal{A}$ is a coset automaton, then $\mathcal{A}$ is trim and $L(\mathcal{A})$ is contained in $L(\mathcal{A}_s(K(\mathcal{A})))$. By Proposition 2.4, there is an expanding morphism from $\mathcal{A}$ to $\mathcal{A}_s(K(\mathcal{A}))$.

Proposition 4.9. *Suppose that $\mathcal{A}$ is a coset automaton and f is the expanding morphism from $\mathcal{A}$ to $\mathcal{A}_s(K(\mathcal{A}))$. Then f is one-to-one as a map on states, and the image of f contains $\Sigma_I(K(\mathcal{A}))$. Every edge of $\mathcal{A}_I(K(\mathcal{A}))$ has the form $(f(\sigma), x, f(\tau))$ for some edge (σ, x, τ) of $\mathcal{A}$.*

Proof. Let $\mathcal{A} = (\Sigma, X, E, \{\alpha\}, \{\alpha\})$ and set $K = K(\mathcal{A})$. The image of f contains $\Sigma_I(K)$ by Proposition 3.1. Suppose that σ and τ are in Σ and $f(\sigma) = f(\tau)$. Let $\sigma = \alpha^U$ and let $\tau = \alpha^V$. By Proposition 4.6, we may assume that U and V are in $\mathcal{C}$. Since $K[U] = f(\sigma) = f(\tau) = K[V]$, it follows that $[UV^{-1}]$ is in K.

Suppose first that UV^{-1} is in $\mathcal{C}$. Then UV^{-1} is in $\mathcal{L}_c(K) \subseteq L(\mathcal{A})$. Therefore there is a path P in $\mathcal{A}$ from α to α such that $\mathrm{Sg}(P) = UV^{-1}$. Thus there is a path Q in $\mathcal{A}$ from σ to α such that $\mathrm{Sg}(Q) = V^{-1}$. But then, by Propositions 4.4 and 4.6, there is a path R from α to σ such that $\mathrm{Sg}(R) = V$. Therefore $\sigma = \tau$.

Now suppose that UV^{-1} is not in $\mathcal{C}$. Then U and V end with the same element of X. Let $U = Cx$ and $V = Dx$ and set $\mu = \alpha^C$ and $\nu = \alpha^D$. Then $f(\mu) = K[C] = K[U][x]^{-1} = K[V][x]^{-1} = K[D] = f(\nu)$. By induction on $|U|$, $\mu = \nu$. Therefore $\sigma = \mu^x = \nu^x = \tau$.

Finally, let (φ, x, ψ) be an edge of $\mathcal{A}_I(K(\mathcal{A}))$. By Corollary 3.4, there is a defining pair (U, V) of φ such that V starts with x. Then $\alpha^{UV} = \alpha$ in $\mathcal{A}$, so $\sigma = \alpha^U$ and $\tau = \sigma^x$ are defined. Thus (σ, x, τ) is an edge of $\mathcal{A}$, and $f(\sigma) = \varphi$ and $f(\tau) = \psi$. $\square$

Now we are in a position to characterize the automata $\mathcal{A}_I(H)$, where H is a finitely generated subgroup of F. Let $\mathcal{A} = (\Sigma, X, E, A, \Omega)$ be a finite coset automaton. We shall say that $\mathcal{A}$ is *reduced* if for all states σ in $\Sigma - A$ there are at least two edges in E leaving σ.

Proposition 4.10. *Let $\mathcal{A}$ be a finite coset automaton. Then $\mathcal{A}$ is isomorphic to $\mathcal{A}_I(K(\mathcal{A}))$ if and only if $\mathcal{A}$ is reduced.*

Proof. Set $K = K(\mathcal{A})$. We shall first show that $\mathcal{A}_I(K)$ is reduced. Let σ be an important coset of K different from $\alpha = K$ and let (S, T) be a defining pair for σ. Since $\sigma \neq \alpha$, the words S and T are not empty. Let $S = Ux$ and $T = yV$, with x and y in X, and set $\rho = \alpha^U$ and $\tau = \sigma^y$. Then (ρ, x, σ) and (σ, y, τ) are edges in $\mathcal{A}_I(K)$. But (σ, x^{-1}, ρ) is also an edge and $y \neq x^{-1}$, since $ST = UxyV$ is in $\mathcal{L}_c(K)$. Therefore there are at least two edges leaving σ.

Now assume that $\mathcal{A} = (\Sigma, X, E, \{\beta\}, \{\beta\})$ is reduced and let f be the expanding morphism from $\mathcal{A}$ to $\mathcal{A}_s(K)$. Suppose that σ is in Σ and $\sigma = \beta^U$. If $\sigma = \beta$, then $f(\sigma) = K$ is important. Suppose $\sigma \neq \beta$. By Proposition 4.6, we may assume that U is in $\mathcal{C}$. Since there are at least two edges in E leaving σ, there is an element x of X such that Ux is reduced and $\tau = \sigma^x$ is defined. If $\tau = \beta$, then (U, x) is a defining pair for $f(\sigma) = K[U]$ and $f(\sigma)$ is important. If $\tau \neq \beta$, then we can repeat this process and find a y in X such that Uxy is reduced and τ^y is defined. Continuing in this manner, either we find a word V such that UV is reduced and $\sigma^V = \beta$ or we produce arbitrarily long words V such that UV is reduced and σ^V is defined in $\mathcal{A}$. In the first case, (U, V) is a defining pair for $f(\sigma)$, so $f(\sigma)$ is important. In the second case, we use the trick in the proof of Proposition 3.12. We choose V long enough so that for every ρ in Σ there is a word W of length less than $|V|$ so that $\rho^W = \beta$. Taking ρ to be σ^V, we see that U is a prefix of $C = \overline{UVW}$. If $C = UD$, then (U, D) is a defining pair for $f(\sigma)$. Thus in all cases $f(\sigma)$ is important, so f maps Σ into $\Sigma_I(K)$. By Proposition 4.9, f is an isomorphism of $\mathcal{A}$ onto $\mathcal{A}_I(K)$. $\square$

The automaton $\mathcal{A}$ of Example 4.1 is not reduced since there is only one edge leaving 5. If we remove the edges $(3, x^{-1}, 5)$ and $(5, x, 3)$, the result $\mathcal{B}$ is again a coset automaton. Since no path in $\mathcal{A}$ having a reduced signature involves either of the deleted edges, $\mathcal{L}_c(K(\mathcal{A})) = \mathcal{L}_c(K(\mathcal{B}))$. Therefore $K(\mathcal{B}) = K(\mathcal{A})$. The automaton $\mathcal{B}$ is reduced. Hence $\mathcal{A}_I(K(\mathcal{A}))$ is isomorphic to $\mathcal{B}$. This reduction process can be applied to any finite coset automaton to produce a reduced coset automaton defining the same subgroup of F.

Procedure REDUCE($\mathcal{A}; \mathcal{B}$);
Input: $\mathcal{A}$: a finite coset automaton;

Table 4.4.2

	x	x^{-1}	y	y^{-1}
1		2	3	4
2	1	5		6
3	4	7	8	1
4		3	1	
5	2			
6			2	
7	3	8	7	7
8	7			3

Output: $\mathcal{B}$: a finite, reduced coset automaton with $K(\mathcal{B}) = K(\mathcal{A})$;
Begin
    $\mathcal{B} := \mathcal{A}$;
    While there is a noninitial state σ in $\mathcal{B}$ such that there is only one
        edge (σ, x, τ) in $\mathcal{B}$ do begin
    Delete the edges (σ, x, τ) and (τ, x^{-1}, σ) from $\mathcal{B}$;
    Delete the state σ from $\mathcal{B}$
    End
End.

For any given iteration of the While-loop in REDUCE there may be several choices for σ. However, the final automaton $\mathcal{B}$ consists of those states and edges of $\mathcal{A}$ which are involved in paths P from the initial state to itself such that $\mathrm{Sg}(P)$ is in $\mathcal{C}$.

Exercise

4.1. Assume that X and $\mathcal{R}$ are as in Example 4.1. Let $\mathcal{A} = (\Sigma, X, E, \{1\}, \{1\})$, where $\Sigma = \{1, \ldots, 8\}$ and E is given by Table 4.4.2. Show that $\mathcal{A}$ is a coset automaton. Apply the procedure REDUCE to $\mathcal{A}$ to produce an automaton isomorphic to $\mathcal{A}_I(K(\mathcal{A}))$. Use this automaton to construct generators for $K(\mathcal{A})$.

4.5 Basic coset enumeration

In its most general meaning, the term "coset enumeration" refers to the computation of $\mathcal{A}_I(H)$ for some subgroup H of our free product $F = \mathrm{Mon}\langle X \mid \mathcal{R} \rangle$ of cyclic groups. If F is a free group and H has finite index in F, then every right coset of H is important, so finding $\mathcal{A}_I(H)$ really does involve enumerating, or counting, the right cosets of H.

The first coset enumeration procedure was described in (Todd & Coxeter 1936). This section is concerned only with a special case of the Todd-Coxeter procedure. Various versions of the full procedure are discussed in Chapter 5. Alternatives to the Todd-Coxeter approach are considered in Section 4.10.

Suppose that $\mathcal{U} = \{U_1, \ldots, U_m\}$ is a finite subset of X^* and $H = \mathrm{Grp}\langle [U] \mid U \in \mathcal{U}\rangle$. For $0 \le i \le m$, let $H_i = \mathrm{Grp}\langle [U_j] \mid 1 \le j \le i\rangle$. Thus H_0 is trivial and $H_m = H$. To construct an automaton isomorphic to $\mathcal{A}_I(H)$ using the Todd-Coxeter method, we construct coset automata $\mathcal{A}_0, \ldots, \mathcal{A}_m$ such that $H_i = K(\mathcal{A}_i)$ and then apply the procedure REDUCE, defined at the end of the previous section, to $\mathcal{A}_m$. We start with $\mathcal{A}_0 = (\{1\}, X, \emptyset, \{1\}, \{1\})$.

This approach requires us to solve the following problem: Given a finite coset automaton $\mathcal{A}$ with respect to $\mathcal{R}$ and a word U in X^*, construct a coset automaton $\mathcal{B}$ such that $K(\mathcal{B}) = \mathrm{Grp}\langle K(\mathcal{A}), [U]\rangle$. For reasons of efficiency, it is desirable that $\mathcal{A}$ be modified "in place" to obtain $\mathcal{B}$, so that the overhead of making a copy of $\mathcal{A}$ is avoided. We shall solve our problem with the aid of several auxiliary procedures. In all of them, $\mathcal{R}$ and $\mathcal{A} = (\Sigma, X, E, \{1\}, \{1\})$ will be available as global variables. States will be positive integers, and n will be a global variable giving the largest integer used so far for a state. In order not to have to write out revised versions of these procedures later, a few statements have been included in some of the procedures which should be ignored for the time being. Most of the statements in question have the form

If *save* then push (σ, x) onto the deduction stack;

The variable *save* is a global boolean variable, which should be assumed to be false for now. In the context of coset enumeration, a *deduction* is a pair (σ, x) of a state and a generator such that σ^x has recently been defined or changed. The use of *save* and the deduction stack is explained in Section 5.3. In the procedure DEFINE below, there is a reference to a global variable p, a vector indexed by positive integers. The significance of p is described in the next section.

The procedure DEFINE adds new edges to $\mathcal{A}$ without changing the subgroup $K(\mathcal{A})$.

Procedure DEFINE(σ, x);
Input: σ : a state in Σ;
     x : an element of X such that σ^x is not defined in $\mathcal{A}$;
($*$ Adds two edges to $\mathcal{A}$ so that σ^x is defined. $*$)
Begin
   $n := n + 1$; $p[n] := n$; Add n to Σ;
   Add (σ, x, n) and (n, x^{-1}, σ) to E;
   If *save* then push (σ, x) and (n, x^{-1}) onto the deduction stack
End.

The procedure DEFINE acts somewhat like an inverse to REDUCE. After DEFINE is executed, any path P in $\mathcal{A}$ from 1 to 1 involving n has (σ, x, n) and (n, x^{-1}, σ) as consecutive edges. Thus $\mathrm{Sg}(P)$ is not in $\mathcal{C}$. Therefore $L(\mathcal{A}) \cap \mathcal{C}$ is unchanged, and hence $K(\mathcal{A})$ remains the same.

The next procedure does change $K(\mathcal{A})$.

Procedure JOIN(σ, x, τ);
Input: σ, τ : states in Σ;
         x : an element of X such that σ^x and $\tau^{x^{-1}}$ are not
                defined;
($*$ Defines σ^x to be τ. $*$)
Begin
   Add (σ, x, τ) to E;
   If *save* then push (σ, x) onto the deduction stack;
   If $(\sigma, x) \neq (\tau, x^{-1})$ then begin
     Add (τ, x^{-1}, σ) to E;
     If *save* then push (τ, x^{-1}) onto the deduction stack
   End
End.

Before the execution of JOIN, let $H = K(\mathcal{A})$ and let U and V be words in X^* such that $1^U = \sigma$ and $\tau^V = 1$.

Proposition 5.1. *After the execution of* JOIN(σ, x, τ), $K(\mathcal{A}) = $ Grp $\langle H, [UxV] \rangle$.

Proof. After execution of JOIN, $1^{UxV} = \sigma^{xV} = \tau^V = 1$. Thus UxV is in $L(\mathcal{A})$, and $K(\mathcal{A})$ contains Grp $\langle H, [UxV] \rangle$. Now suppose that $1^W = 1$ and let P be the corresponding path. If none of the edges in P is an edge added by JOIN, then $[W]$ is in H. We proceed by induction on the number of new edges contained in P. Suppose P is the concatenation of P_1, (σ, x, τ), and P_2, where P_1 is a path from 1 to σ and P_2 is a path from τ to 1. Let $W_j = \mathrm{Sg}(P_j)$, $j = 1, 2$. Then $1^{W_1 U^{-1}} = \sigma^{U^{-1}} = 1$ and $1^{V^{-1}W_2} = \tau^{W_2} = 1$. By induction, $[W_1 U^{-1}]$ and $[V^{-1}W_2]$ are in Grp $\langle H, [UxV] \rangle$. But then

$$[W] = [W_1 x W_2] = [W_1 U^{-1}][UxV][V^{-1}W_2]$$

is in Grp $\langle H, [UxV] \rangle$. A similar argument handles the case in which P contains the edge (τ, x^{-1}, σ). $\square$

Example 5.1. Let $X = \{x, y, y^{-1}\}$, let $\mathcal{R}$ consist of the rules

$$x^2 \to \varepsilon, \quad yy^{-1} \to \varepsilon, \quad y^{-1}y \to \varepsilon,$$

and let $\mathcal{A} = (\Sigma, X, E, \{1\}, \{1\})$, where $\Sigma = \{1, 2, 3\}$ and E is given by the following coset table:

	x	y	y^{-1}
1	2		3
2	1		
3		1	

If $n = 3$, then the call DEFINE$(3, y^{-1})$ adds 4 to Σ and changes the table to

	x	y	y^{-1}
1	2		3
2	1		
3		1	4
4		3	

Note that $1^x = 2$ and $4^{y^2} = 1$. If at this point JOIN$(2, y, 4)$ is executed, then the table becomes

	x	y	y^{-1}
1	2		3
2	1	4	
3		1	4
4		3	2

and $[xy^3]$ is added to $K(\mathcal{A})$. The call JOIN$(4, x, 4)$ yields the table

	x	y	y^{-1}
1	2		3
2	1	4	
3		1	4
4	4	3	2

and adds $[xyxy^{-1}x]$ to $K(\mathcal{A})$.

The third procedure for modifying a coset table is more complicated than the first two. Before describing it, we must introduce the notion of a congruence on a coset automaton. Let $\mathcal{A} = (\Sigma, X, E, \{\alpha\}, \{\alpha\})$ be a coset automaton. A *congruence* on $\mathcal{A}$ is an equivalence relation $\approx$ on Σ such that whenever (σ, x, φ) and (τ, x, ψ) are in E and $\sigma \approx \tau$, then $\varphi \approx \psi$. The intersection of congruences is again a congruence. Thus given a subset S of $\Sigma \times \Sigma$, we can speak of the congruence generated by S, the smallest congruence $\approx$ for which $\sigma \approx \tau$ for all (σ, τ) in S.

Suppose that $\approx$ is a congruence on $\mathcal{A}$. Let Λ be the set of $\approx$-classes. For σ in Σ let $[\sigma]$ be the $\approx$-class containing σ. Put $D = \{([\sigma], x, [\tau]) \mid (\sigma, x, \tau) \in E\}$, $B = \{[\alpha]\}$, and $\mathcal{B} = (\Lambda, X, D, B, B)$.

Table 4.5.1

	x	x^{-1}	y	y^{-1}
1	2	3	4	4
2		1	5	5
3	1	6		
4		5	1	1
5	4		2	2
6	3		6	6

Proposition 5.2. *Under the preceding hypothesis, $\mathcal{B}$ is a coset automaton and the map $\sigma \mapsto [\sigma]$ is an expanding morphism from $\mathcal{A}$ to $\mathcal{B}$.*

Proof. If (σ, x, φ) and (τ, x, ψ) are in E and $[\sigma] = [\tau]$, then $[\varphi] = [\psi]$. Therefore there is at most one edge with a given label leaving $[\sigma]$. The remaining parts of the definition of a deterministic automaton are easily checked. If (σ, x, φ) is in E, then so is $(\varphi, x^{-1}, \sigma)$. Thus $([\varphi], x^{-1}, [\sigma])$ is in D. If there is a path P in $\mathcal{A}$ from σ to τ, then there is a path Q in $\mathcal{B}$ from $[\sigma]$ to $[\tau]$ such that $\mathrm{Sg}(Q) = \mathrm{Sg}(P)$. Thus $\mathcal{B}$ is a coset automaton and $\sigma \mapsto [\sigma]$ is an expanding morphism. $\square$

We shall call $\mathcal{B}$ the *quotient* of $\mathcal{A}$ modulo $\approx$.

Example 5.2. Let $X = \{x, y\}^{\pm}$ and let $\mathcal{R}$ be the set of rules defining the free group on $\{x, y\}$. Suppose that $\mathcal{A} = (\Sigma, X, E, \{1\}, \{1\})$, where $\Sigma = \{1, 2, 3, 4, 5, 6\}$ and E is given by Table 4.5.1. The sets $\Lambda_1 = \{1, 4, 6\}$ and $\Lambda_2 = \{2, 3, 5\}$ are the equivalence classes of a congruence $\approx$ on $\mathcal{A}$. The edges of the quotient of $\mathcal{A}$ modulo $\approx$ are described by the table

	x	x^{-1}	y	y^{-1}
1	2	2	1	1
2	1	1	2	2

where we have written i for Λ_i.

The third procedure of this section replaces $\mathcal{A}$ by an automaton isomorphic to the quotient of $\mathcal{A}$ by the congruence generated by a single pair (σ, τ) of states.

Procedure COINCIDENCE(σ, τ);
Input: σ, τ : states in Σ;
($*$ This is a preliminary version. See Section 4.6. $*$)
Begin
    Let $\approx$ be the congruence on $\mathcal{A}$ generated by (σ, τ);

(* For φ in Σ let $[\varphi]$ denote the $\approx$-class containing φ. *)
Let $\mathcal{B}$ be the quotient of $\mathcal{A}$ modulo $\approx$;
Let Π be the set of states in Σ which are the smallest elements of
    their $\approx$-class;
Replace E by the set of triples (φ, x, ψ), where φ and ψ are in Π
    and $([\varphi], x, [\psi])$ is an edge of $\mathcal{B}$;
Replace Σ by Π
End.

The preceding listing is only an outline. The implementation of the coincidence procedure is the most challenging part of writing a coset enumeration program. The details depend heavily on the data structure used to store the edges in E and to describe partitions of Σ. To avoid getting bogged down in these implementation details now, we defer to Section 4.6 a more complete discussion of the coincidence procedure. It is important to note here that the speed of the Todd-Coxeter approach to coset enumeration comes largely from the fact that COINCIDENCE can be implemented quite efficiently.

Example 5.3. Let X, $\mathcal{R}$, and $\mathcal{A}$ be as in Example 5.2 and let $\sigma = 4$ and $\tau = 6$. Here is one way to determine the set Λ of equivalence classes of $\approx$ during the execution of COINCIDENCE(4,6). Initially, let Λ consist of the following sets:

$$\{1\}, \quad \{2\}, \quad \{3\}, \quad \{4,6\}, \quad \{5\}.$$

Comparing the edges in $\mathcal{A}$ leaving 4 and 6, we find $(4, y, 1)$ and $(6, y, 6)$. Therefore we merge the first and fourth sets together to get:

$$\{1,4,6\}, \quad \{2\}, \quad \{3\}, \quad \{5\}.$$

Looking at the edges leaving 1 and 4, we find $(1, x^{-1}, 3)$ and $(4, x^{-1}, 5)$. This causes us to merge the last two sets, forming

$$\{1,4,6\}, \quad \{2\}, \quad \{3,5\}.$$

Finally, because of the edges $(1, x, 2)$ and $(6, x, 3)$, the blocks containing 2 and 3 are merged. At this point Λ consists of

$$\{1,4,6\}, \quad \{2,3,5\}.$$

As we saw in Example 5.2, these are the blocks of a congruence on $\mathcal{A}$.

Before the call COINCIDENCE(σ, τ), let $H = K(\mathcal{A})$ and let S and T be words such that $1^S = \sigma$ and $\tau^T = 1$.

Proposition 5.3. *After the call* COINCIDENCE(σ, τ) *we have* $K(\mathcal{A}) =$ Grp $\langle H, [ST] \rangle$.

Proof. Let $M = \text{Grp} \langle H, [ST] \rangle$ and let $\mathcal{A}_0$ denote the automaton $\mathcal{A}$ before the call to COINCIDENCE. After the call, $\mathcal{A}$ is isomorphic to the quotient $\mathcal{B}$ of $\mathcal{A}_0$ modulo $\approx$. By Proposition 5.2, the map taking each state φ of $\mathcal{A}_0$ to $[\varphi]$ is an expanding morphism from $\mathcal{A}_0$ to $\mathcal{B}$. Thus $K(\mathcal{B})$ contains H. In $\mathcal{A}_0$, $1^S = \sigma$ and $\tau^T = 1$. Then in $\mathcal{B}$ we have $[1]^S = [\sigma]$ and $[\tau]^T = [1]$. Since $[\sigma] = [\tau]$, it follows that $[1]^{ST} = [1]$ and $[ST]$ is contained in $K(\mathcal{B})$. Thus $K(\mathcal{B}) \supseteq M$.

By Proposition 2.4, there is an expanding morphism g from $\mathcal{A}_0$ to $\mathcal{A}_s(M)$. For states φ and ψ of $\mathcal{A}_0$, let $\varphi \equiv \psi$ if $g(\varphi) = g(\psi)$. The relation $\equiv$ is a congruence on $\mathcal{A}_0$. Moreover, $g(\sigma) = M[S] = M[ST]^{-1}[S] = M[T^{-1}S^{-1}S] = M[T^{-1}] = g(\tau)$, so $\sigma \equiv \tau$. By definition, $\approx$ is the intersection of all congruences on $\mathcal{A}_0$ in which σ and τ are in the same equivalence class. Therefore, if $\varphi \approx \psi$, then $\varphi \equiv \psi$. Thus there is a map h from the set of states of $\mathcal{B}$ to the set of right cosets of M such that $h([\varphi]) = g(\varphi)$ for all states φ of $\mathcal{A}_0$. It is not hard to check that h is an expanding morphism from $\mathcal{B}$ to $\mathcal{A}_s(M)$. Thus $K(\mathcal{B}) \subseteq M$. Hence $K(\mathcal{B}) = M$, and, since $\mathcal{A}$ is isomorphic to $\mathcal{B}$, the proposition follows. $\square$

The procedure TRACE of Section 3.2 relies on the fact that in a deterministic automaton a path is determined by its starting point and its signature. Because of Corollary 4.3, in a coset automaton we can trace backward, starting at the endpoint of the path.

```
Procedure BACK_TRACE(A, σ, U; τ, B, D);
    Input:    A        : a coset automaton (Σ, X, E, A, A);
              σ        : a state in Σ;
              U        : a word in X*;
    Output:   τ        : a state in Σ;
              B, D     : words in X* such that U = BD, τ^D = σ, and |D| is
                         as large as possible;
Begin
    τ := σ;  B := U;  D := ε;  done := false;

    While B ≠ ε and not done do begin
        Let B = Sx, where x is in X;

        If there is no edge with label x coming into τ then done := true
        Else begin
            Let (ρ, x, τ) be in E;  τ := ρ;  B := S;  D := xD
        End
    End
End.
```

At the heart of the Todd-Coxeter approach to coset enumeration is the concept of a two-sided trace. There are two types of two-sided traces, partial and full. Given a state σ in $\mathcal{A}$ and a word W in $\mathcal{C}$, the object of a two-sided trace is to modify $\mathcal{A}$, if necessary, so that $\sigma^W = \sigma$. More precisely, we want to modify $\mathcal{A}$ in such a way that there is an expanding morphism f from the old $\mathcal{A}$ to the new $\mathcal{A}$ such that $f(\sigma)^W = f(\sigma)$. If the trace is partial, then f must be surjective as a map on states.

Both types of two-sided traces start out the same way. We find words S, T, and U and states μ and ν such that $W = STU$, $\sigma^S = \mu$, $\nu^U = \sigma$, $|S|$ is as large as possible, and, subject to this, $|U|$ is as large as possible. If $T = \varepsilon$ and $\mu = \nu$, then $S = W$ and $\sigma^W = \sigma$ already. If $T = \varepsilon$ and $\mu \neq \nu$, then we identify μ and ν using COINCIDENCE. Suppose $|T| = 1$, so $T = x$ is an element of X. Then we use JOIN to make $\mu^x = \nu$. If $|T| > 1$, then a partial trace terminates unsuccessfully, without modifying $\mathcal{A}$. If $|T| > 1$ and the trace is full, then DEFINE is used to add a new edge so the trace can be carried further.

Here is the complete description of a two-sided trace. All of the global variables defined earlier continue to be available.

Procedure TWO_SIDED_TRACE($\sigma, W, full$);
Input: σ : a state in Σ;
       W : a word in $\mathcal{C}$;
       $full$: a boolean variable, with true indicating a full trace
              and false indicating a partial trace;
Begin
   TRACE($\mathcal{A}, \sigma, W; \mu, S, C$);   BACK_TRACE($\mathcal{A}, \sigma, C; \nu, T, U$);

   If $full$ then
     While $|T| > 1$ do begin
       Let T start with x;   DEFINE(μ, x);
        TRACE($\mathcal{A}, \sigma, W; \mu, S, C$);   BACK_TRACE($\mathcal{A}, \sigma, C; \nu, T, U$)
     End;

   If $T = \varepsilon$ and $\mu \neq \nu$ then COINCIDENCE(μ, ν)
   Else if $|T| = 1$ then begin
     Let T be the element x of X;   JOIN(μ, x, ν)
   End
End.

In a careful implementation of TWO_SIDED_TRACE, the forward and backward traces in the While-loop would not be repeated from the beginning but continued from the point at which they last stopped.

Before the call TWO_SIDED_TRACE($\sigma, W, full$), let $H = K(\mathcal{A})$ and let L and M be words such that $1^L = \sigma$ and $\sigma^M = 1$.

Table 4.5.2

	x	x^{-1}	y	y^{-1}
1	2	3	3	4
2	4	1	2	2
3	1		5	1
4	6	2	1	6
5		6		3
6	5	4	4	

Proposition 5.4. *After the call* TWO_SIDED_TRACE$(\sigma, W, full)$, *either* $K(\mathcal{A}) = \mathrm{Grp}\,\langle H, [LWM] \rangle$ *or the trace was partial and terminated unsuccessfully.*

Proof. Calls to DEFINE do not change $K(\mathcal{A})$ and are made only in full traces. If the call to COINCIDENCE is made, then just before the call we have $1^{LS} = \mu$ and $\nu^{UM} = 1$. By Proposition 5.3, $K(\mathcal{A}) = \mathrm{Grp}\,\langle H, [LWM] \rangle$ immediately after the call to COINCIDENCE, since LSUM = LWM as $T = \varepsilon$. If the call to JOIN is made, then just after the call $K(\mathcal{A}) = \mathrm{Grp}\,\langle H, [LWM] \rangle$ by Proposition 5.1. $\square$

Proposition 5.5. *If* $\mathcal{A}$ *is isomorphic to* $\mathcal{A}_I(K(\mathcal{A}))$ *before a call to* TWO_SIDED_TRACE, *then* $\mathcal{A}$ *is isomorphic to* $\mathcal{A}_I(K(\mathcal{A}))$ *after the call.*

Proof. Exercise. (Show that the statement is false if W is not required to be in $\mathcal{C}$.) $\square$

Example 5.4. Let us perform a few two-sided traces with X and $\mathcal{R}$ as in Example 5.2 and $\mathcal{A}$ given by Table 4.5.2. First let us carry out a partial trace of $x^{-1}yxyxy^3$ at 1. When tracing by hand, it is convenient to write the word without the use of exponents and place the starting state at the beginning and the end. In this case, we have

$$x^{-1} \; y \; x \; y \; x \; y \; y \; y$$
$$1 \qquad\qquad\qquad\qquad\qquad 1.$$

Now we work from left to right as far as we can and then work from right to left. Since $1^{x^{-1}} = 3$, $3^y = 5$, and 5^x is not defined, the results of the forward trace can be summarized as follows:

$$x^{-1} \; y \; x \; y \; x \; y \; y \; y$$
$$1 \quad 3 \; 5 \qquad\qquad\qquad 1.$$

To work from the right, we must find the state σ such that $\sigma^y = 1$. We could look down the column headed by y until we found 1, but there is an easier way. Since $\mathcal{A}$ is a coset automaton, $\sigma^y = 1$ if and only if $1^{y^{-1}} = \sigma$. Thus σ must be $1^{y^{-1}} = 4$. Similarly, $4^{y^{-1}} = 6$, and $6^{y^{-1}}$ is not defined. Thus, after the forward and backward traces, we have reached the following configuration.

$$x^{-1} \; y \; x \; y \; x \; y \; y \; y$$
$$1 \quad\;\; 3 \; 5 \qquad\quad 6 \; 4 \; 1.$$

At this point we are "stuck". Since the trace is partial, we give up.

Now let us trace $xy^{-2}x^{-1}y$ at 2. We start with

$$x \; y^{-1} \; y^{-1} \; x^{-1} \; y$$
$$2 \qquad\qquad\qquad\quad 2.$$

The result of the forward trace is

$$x \; y^{-1} \; y^{-1} \; x^{-1} \; y$$
$$2 \; 4 \quad 6 \qquad\qquad 2,$$

so we must trace $y^{-1}x^{-1}y$ backward from 2. Since $2^{y^{-1}} = 2$, $2^x = 4$, and $4^y = 1$, we reach the state 1, which "clashes" with the 6. This situation is pictured as follows:

$$x \; y^{-1} \; y^{-1} \; x^{-1} \; y$$
$$2 \; 4 \quad 6 \quad\; 4 \quad 2 \; 2.$$
$$\qquad\qquad 1$$

This is an example of the case in which $T = \varepsilon$ and $\mu \neq \nu$. The next step would be to call COINCIDENCE(6,1), but we shall not go any further with this trace.

The third trace will be a full trace of $y^2xy^2x^{-2}y^2$ at 1. The first forward and backward traces leave a "gap" with $|T|$ of length 4.

$$y \; y \; x \; y \; y \; x^{-1} \; x^{-1} \; y \; y$$
$$1 \; 3 \; 5 \qquad\qquad 5 \quad\; 6 \; 4 \; 1.$$

Since $\mu = 5$ and T starts with x, we make the call DEFINE(5, x). This changes the coset table as shown in Table 4.5.3. Both the forward and backward traces can now go further. We have

$$y \; y \; x \; y \; y \; x^{-1} \; x^{-1} \; y \; y$$
$$1 \; 3 \; 5 \; 7 \qquad 7 \quad\; 5 \quad 6 \; 4 \; 1.$$

Here $T = y^2$, so we execute DEFINE(7, y), changing the table as shown in

Table 4.5.3

	x	x^{-1}	y	y^{-1}
1	2	3	3	4
2	4	1	2	2
3	1		5	1
4	6	2	1	6
5	7	6		3
6	5	4	4	
7		5		

Table 4.5.4

	x	x^{-1}	y	y^{-1}
1	2	3	3	4
2	4	1	2	2
3	1		5	1
4	6	2	1	6
5	7	6		3
6	5	4	4	
7		5	8	
8				7

Table 4.5.5

	x	x^{-1}	y	y^{-1}
1	2	3	3	4
2	4	1	2	2
3	1		5	1
4	6	2	1	6
5	7	6		3
6	5	4	4	
7		5	8	8
8			7	7

Table 4.5.4. At this point, the traces stop with $T = y$, which we signal with an underscore:

$$y \ y \ x \ y \ y \ x^{-1} \ x^{-1} \ y \ y$$
$$1 \ 3 \ 5 \ 7 \ \underline{8 \ 7} \quad 5 \quad 6 \ 4 \ 1.$$

We call JOIN$(8, y, 7)$, which changes the table as shown in Table 4.5.5. The two edges $(8, y, 7)$ and $(7, y^{-1}, 8)$ added by JOIN are said to be *deduced* from the trace. This is the origin of the use of the term "deduction" for a position in the table whose entry has recently been filled in.

Here is the basic coset enumeration procedure. As before, the reference to the vector p should be ignored for now.

Procedure BASIC_CE$(X, \mathcal{R}, \mathcal{U}; \mathcal{A})$
Input: X : a finite set;
           $\mathcal{R}$: a classical niladic rewriting system on X^*;
           $\mathcal{U}$: a finite subset of X^*;
Output: $\mathcal{A}$: a coset automaton isomorphic to $\mathcal{A}_I(H)$, where H is
                 the subgroup of $F = \text{Mon} \langle X \mid \mathcal{R} \rangle$ generated by
                 $\{[U] \mid U \in \mathcal{U}\}$;
Begin
   $n := 1; \ p[1] := 1; \ \mathcal{A} := (\{1\}, X, \emptyset, \{1\}, \{1\})$;
   For U in $\mathcal{U}$ do begin
     REWRITE$(\mathcal{R}, U; V)$;
     TWO_SIDED_TRACE$(1, V, \text{true})$ ($*$ A full trace $*$)
   End
End.

The second argument of TWO_SIDED_TRACE is supposed to be an irreducible word. To be sure, we rewrite the elements of $\mathcal{U}$. The correctness of BASIC_CE is a direct consequence of Propositions 5.4 and 5.5. Each call to TWO_SIDED_TRACE adds one of the generators $[U]$ of H to $K(\mathcal{A})$.

Example 5.5. Let us apply BASIC_CE to the group of Example 2.2, which was also discussed in Example 3.4. Here $X = \{x,y\}^{\pm}$, $\mathcal{R}$ defines the free group on $\{x,y\}$, and $\mathcal{U} = \{xyx, x^{-2}y\}$. Initially, the coset table for $\mathcal{A}$ is

	x	x^{-1}	y	y^{-1}
1				

After the first trace, the states are 1, 2, and 3, and the edges are given by

	x	x^{-1}	y	y^{-1}
1	2	3		
2		1	3	
3	1			2

After the second trace, we have the table

	x	x^{-1}	y	y^{-1}
1	2	3		4
2		1	3	
3	1	4		2
4	3		1	

This is the table quoted in Example 3.4. Suppose we add $xy^2x^{-1}y$ as a third element of $\mathcal{U}$. Performing the two-sided trace of $xy^2x^{-1}y$ at 1 results in the call COINCIDENCE(2,3). The resulting automaton has states 1 and 2 and the following coset table:

	x	x^{-1}	y	y^{-1}
1	2	2	1	1
2	1	1	2	2

The corresponding subgroup has index 2 in F.

The output of BASIC_CE is determined only up to isomorphism since the order in which the elements of $\mathcal{U}$ are processed is not specified. In general, it is a good idea to standardize the output of BASIC_CE to remove the ambiguity. Section 4.7 describes how to standardize a finite coset automaton in place.

Exercises

5.1. Let X and $\mathcal{R}$ be as in Example 5.1 and let $\mathcal{A}$ be the automaton $(\Sigma, X, E, \{1\}, \{1\})$, where $\Sigma = \{1, \ldots, 8\}$ and E is given by Table 4.5.6. Determine the congruence classes of the congruence on $\mathcal{A}$ generated by $(8,6)$. How would $\mathcal{A}$ be changed by the call COINCIDENCE(8,6)?

Table 4.5.6

	x	y	y^{-1}
1	2	3	4
2	1	5	6
3		7	1
4		1	8
5	7		2
6	8	2	
7	5		3
8	6	4	

5.2. Let X and $\mathcal{R}$ be as in Example 5.5. Starting with $\mathcal{A} = (\{1\}, X, \emptyset, \{1\}, \{1\})$ and $n = 1$, carry out the following sequence of operations:

$$\text{DEFINE}(1, x), \quad \text{DEFINE}(1, y^{-1}), \quad \text{JOIN}(2, x, 3).$$

Now let $U = x^{-1}yxy$ and $V = x^{-1}y^{-1}x^{-1}yx$ and continue with the traces

$$\text{TWO\_SIDED\_TRACE}(1, U, \text{true}), \quad \text{TWO\_SIDED\_TRACE}(3, V, \text{true}).$$

5.3. Apply BASIC_CE to the following input data:

(a) X and $\mathcal{R}$ as in Example 5.1 and $\mathcal{U} = \{yxy^{-2}x, \, yxy^{-1}, \, y^2xy^2\}$.
(b) X and $\mathcal{R}$ as in Example 5.5 and $\mathcal{U} = \{(x^{-1}yxy^{-1})^2, \, (y^2x)^2, \, x^{-1}yxy^3x\}$.

4.6 The coincidence procedure

The version of the coincidence procedure given in the previous section was very sketchy. The procedure can make substantial changes in a coset automaton, and it is not clear from the outline presented how time-consuming calls to COINCIDENCE can be. As the first step in studying these issues, we shall consider some data structures and procedures designed to facilitate the manipulation of partitions. These data structures and procedures are referred to collectively as the *union-find procedure*. Proofs will be omitted. They can be found in [Aho et al. 1974].

Let n be a positive integer and let $\Sigma = \{1, \ldots, n\}$. A partition Π of Σ is defined conveniently by a vector p, indexed by the elements of Σ. The value of $p[\sigma]$ is set equal to the smallest element of the block of Π which contains σ. Thus if $n = 6$ and Π consists of the blocks

$$\{1, 3, 4\}, \quad \{2, 6\}, \quad \{5\},$$

then p is the vector $(1, 2, 1, 1, 5, 2)$. Partitions of Σ are in one-to-one correspondence with integer vectors p indexed by Σ such that $1 \le p[\sigma] \le \sigma$ and $p[p[\sigma]] = p[\sigma]$ for all σ in Σ.

It is useful to weaken the conditions on the vector p, although doing so destroys the one-to-one correspondence. The following conditions allow for a computationally effective description of a partition Π:

(i) $p[\sigma]$ is some element of the block of Π containing σ.
(ii) $p[\sigma] = \sigma$ for exactly one element σ of each block.
(iii) For any σ in Σ, the sequence $\sigma, p[\sigma], p[p[\sigma]], \ldots$ eventually becomes constant.

When (iii) holds, we shall denote by $p^{\infty}[\sigma]$ the value at which the sequence $\sigma, p[\sigma], p[p[\sigma]], \ldots$ stabilizes. If all three conditions hold, then two elements σ and τ of Σ are in the same block of Π if and only if $p^{\infty}[\sigma] = p^{\infty}[\tau]$. The preceding partition Π is described by the vector $p = (3, 6, 3, 1, 5, 6)$.

To compute $p^{\infty}[\sigma]$, we compute the sequence $\sigma_1, \ldots, \sigma_s$, where $\sigma_1 = \sigma$, $\sigma_{i+1} = p[\sigma_i]$, and s is the first index such that $p[\sigma_s] = \sigma_s$. If $s \geq 3$, then we may save time later if we modify p by redefining $p[\sigma_i]$ to be σ_s for $1 \leq i \leq s - 2$. This does not change the partition described by p. For example, with $p = (3, 6, 3, 1, 5, 6)$ and $\sigma = 4$, we have $\sigma_1 = 4$, $\sigma_2 = 1$, and $\sigma_3 = 3$. Thus we change $p[4]$ to 3. This process is called *path compression*.

Suppose that we have to compute $p^{\infty}[\sigma]$ for n not necessarily distinct elements σ of Σ. Without path compression, the time required can be quadratic in n. For example, suppose that $p[1] = 1$ and $p[\sigma] = \sigma - 1$ for $2 \leq \sigma \leq n$. Then computing $p^{\infty}[\sigma]$ takes time proportional to σ. With path compression, the time to perform n computations $p^{\infty}[\sigma]$ is linear in n.

Now suppose that we start with the trivial partition Π whose blocks are the sets $\{\sigma\}$ with σ in Σ. Next we are given pairs (σ, τ) of elements of Σ and asked to do the following: If σ and τ are in different blocks of Π, then combine those blocks into a single block. This is precisely what was done in Example 5.3 as we worked through the operation of COINCIDENCE.

Assume that there are going to be roughly n of the pairs (σ, τ) and that one pair must be processed before the next pair is received. The best-known procedure for carrying out this process maintains two vectors p and m. The vector p describes the current partition Π. For each σ in Σ such that $p[\sigma] = \sigma$, the value of $m[\sigma]$ is the size of the block containing σ. We initialize $p[\sigma]$ to σ and $m[\sigma]$ to 1 for each σ in Σ. Then for each pair (σ, τ) the following procedure is executed:

Procedure UNION(σ, τ);
Begin
    $\mu := p^{\infty}[\sigma]$; Perform path compression at σ;
    $\nu := p^{\infty}[\tau]$; Perform path compression at τ;
    If $\mu \neq \nu$ then
        If $m[\nu] \leq m[\mu]$ then
            Begin $p[\nu] := \mu$; $m[\mu] := m[\mu] + m[\nu]$ end

Else
        Begin $p[\mu] := \nu$; $m[\nu] := m[\mu] + m[\nu]$ end
End.

If Γ and Δ are the blocks of Π containing σ and τ, respectively, and $\Gamma \neq \Delta$, then we choose the representative for the new block $\Gamma \cup \Delta$ to be the representative for the larger of Γ and Δ. In this way the computation of $p^\infty[\rho]$ remains unchanged for ρ in the larger of Γ and Δ.

The total time needed to perform n calls to UNION is known not to be linear in n in the worst case. However, the time is very nearly linear. Let F be the function defined for nonnegative integers such that $F(0) = 1$ and

$$F(n) = 2^{F(n-1)}, \qquad n \geq 1.$$

Thus $F(1) = 2$, $F(2) = 4$, $F(3) = 16$, $F(4) = 2^{16} = 65536$, and $F(5) = 2^{65536}$. Let G be the inverse function to F, so that $G(n)$ is the smallest integer k such that $F(k) \geq n$. Although G goes to infinity as n goes to infinity, it does so very slowly, for $G(n) \leq 5$ if $n \leq 2^{65536}$. The total time required for n calls to UNION is at most a constant times $nG(n)$, which is linear in n for all practical purposes.

It is possible to dispense with the vector m, although there is a modest penalty in reduced execution speed. The test

If $m[\nu] \leq m[\mu]$ then

in UNION is replaced by

If $\mu < \nu$ then

In this way we choose the smaller of μ and ν to be the representative for the merged block. Now the total execution time for n calls to UNION is at worst a constant times $n \log(n)$.

Traditionally the vector m has not been used in coset enumeration coincidence procedures. There are several reasons for this:

• Space is usually at a premium in coset enumerations.
• The worst-case performance of $n \log(n)$ is rarely encountered in practice.
• It is more difficult to ensure termination of the coset enumeration procedures discussed in Chapter 5 if $p[\sigma]$ may be larger than σ.

To be consistent with current practice, the vector m will not be used here.

The version of COINCIDENCE presented in this section uses two subroutines. The function ACTIVE computes $p^\infty[\sigma]$ and performs path compression.

Function ACTIVE(σ): integer;
Input: σ : a positive integer not exceeding n;
(* Returns the representative for the equivalence class containing σ. *)
Begin
    (* Find $p^\infty[\sigma]$. *)
    $\tau := \sigma$; $\rho := p[\tau]$;

    While $\rho \neq \tau$ do begin $\tau := \rho$; $\rho := p[\tau]$ end;
    ACTIVE $:= \tau$; $\mu := \sigma$; $\rho := p[\mu]$;
    (* Perform path compression. *)
    While $\rho \neq \tau$ do begin $p[\mu] := \tau$; $\mu := \rho$; $\rho := p[\mu]$ end
End.

Within COINCIDENCE, we maintain a list $\eta_1, \ldots, \eta_k$ of integers which
have been deleted from Σ. This list is accessible to MERGE, which is a
modified version of UNION.

Procedure MERGE(σ, τ);
Input: σ, τ : positive integers not exceeding n;
Begin
    $\varphi :=$ ACTIVE(σ); $\psi :=$ ACTIVE(τ);

    If $\varphi \neq \psi$ then begin
        $\mu := \min(\varphi, \psi)$; $\nu := \max(\varphi, \psi)$; $p[\nu] := \mu$;
        $k := k + 1$; $\eta_k := \nu$
    End
End.

Here is the complete description of COINCIDENCE:

Procedure COINCIDENCE(σ, τ);
Input: σ, τ : elements of Σ;
Begin
    $k := 0$; MERGE(σ, τ); $i := 1$;

    While $i \leq k$ do begin
        $\nu := \eta_i$; $E_1 := \emptyset$; $i := i + 1$;

        (* Remove all edges involving ν. *)

        For x in X do
            If there is an edge (ν, x, φ) in E then begin
                Delete (ν, x, φ) from E and add it to E_1;
                If $(\varphi, x^{-1}) \neq (\nu, x)$ then delete (φ, x^{-1}, ν) from E
            End;

        (* Make sure that for each edge e in E_1 there is an edge in E
            equivalent to e. *)

For (ν, x, φ) in E_1 do begin
  $\mu := \text{ACTIVE}(\nu)$; $\psi := \text{ACTIVE}(\varphi)$;
  If μ^x is defined then $\text{MERGE}(\psi, \mu^x)$
  Else if $\psi^{x^{-1}}$ is defined then $\text{MERGE}(\mu, \psi^{x^{-1}})$
  Else begin
    Add (μ, x, ψ) to E;
    If *save* then push (μ, x) onto the deduction stack;
    If $(\psi, x^{-1}) \neq (\mu, x)$ then begin
      Add (ψ, x^{-1}, μ) to E;
      If *save* then push (ψ, x^{-1}) onto the deduction stack
    End
  End
 End
End
End.

As discussed in the previous section, references to the boolean variable *save* and to the deduction stack should be ignored for the time being.

In one of the comments in COINCIDENCE there is a reference to equivalent edges. If $\equiv$ is an equivalence relation on $\mathcal{A}$, we shall say that edges $e = (\nu, x, \varphi)$ and $f = (\mu, y, \psi)$ are *equivalent* if $x = y$, $\nu \equiv \mu$, and $\varphi \equiv \psi$. When this is the case, then the paired edges e' and f' are also equivalent.

We must show that COINCIDENCE does what it is supposed to do. Let $\mathcal{A}_0$ denote the current automaton just before a call to COINCIDENCE and let $\mathcal{A}$ denote the current automaton at the termination of the call. When an edge $e = (\nu, x, \varphi)$ is deleted from E, the edge e' is also deleted. When an edge $f = (\mu, x, \psi)$ is added to E, then f' is added also. Moreover, f is not added unless both μ^x and $\psi^{x^{-1}}$ are not defined. Thus $\mathcal{A}$ is deterministic and its edge set is closed under the pairing operation.

For states β and γ of $\mathcal{A}_0$, write $\beta \equiv \gamma$ if $\text{ACTIVE}(\beta) = \text{ACTIVE}(\gamma)$, where ACTIVE uses the vector p as it is at the termination of COINCIDENCE. Clearly $\equiv$ is an equivalence relation on the states of $\mathcal{A}_0$. Let $\approx$ be the congruence on $\mathcal{A}_0$ generated by (σ, τ). Whenever MERGE is called, its arguments are congruent with respect to $\approx$. Therefore $\beta \equiv \gamma$ implies $\beta \approx \gamma$.

If $e = (\nu, x, \varphi)$ is deleted, then the second For-loop makes sure that there is an edge e_1 such that e is equivalent to e_1 with respect to $\equiv$. If e_1 is later deleted, then there will be an edge e_2 such that e_1 is equivalent to e_2 with respect to $\equiv$. But then e is equivalent to e_2. Continuing in this way, we see that there is an edge f in $\mathcal{A}$ such that e is equivalent to f. Since no distinct states of $\mathcal{A}$ are equivalent with respect to $\equiv$, the edge f must be $(\text{ACTIVE}(\nu), x, \text{ACTIVE}(\varphi))$. Let us denote f by $\overline{e}$. Every edge of $\mathcal{A}$ is $\overline{e}$ for some edge e of $\mathcal{A}_0$.

Suppose now that β and γ are states of $\mathcal{A}_0$ such that $\beta \equiv \gamma$, and (β, x, φ) and (γ, x, ψ) are edges of $\mathcal{A}_0$. Then $(\text{ACTIVE}(\beta), x, \text{ACTIVE}(\varphi))$

and $(\text{ACTIVE}(\gamma), x, \text{ACTIVE}(\psi))$ are edges of $\mathcal{A}$. Since $\text{ACTIVE}(\beta) = \text{ACTIVE}(\gamma)$ and $\mathcal{A}$ is deterministic, we have $\text{ACTIVE}(\varphi) = \text{ACTIVE}(\psi)$ and hence $\varphi \equiv \psi$. Therefore $\equiv$ is a congruence on $\mathcal{A}_0$. Since $\sigma \equiv \tau$, it follows that $\equiv$ and $\approx$ are the same. Therefore $\mathcal{A}$ is isomorphic to the quotient of $\mathcal{A}_0$ by $\approx$.

As has been mentioned several times already, space is often a problem in coset enumeration. The vector p used by COINCIDENCE takes a substantial amount of space. In (Beetham 1984) it was observed that with some data structures for the edge set E no extra space is needed for p or the sequence of η_j. If $|X| = 2$, then during the execution of COINCI-DENCE the quantity $k - i$ is bounded. (See Exercise 6.2.) Thus only a fixed amount of space is needed to store the sequence $\eta_i, \ldots, \eta_k$ and the entries in p for these states. Once any edges involving η_i have been removed and processed, there is room to store $p[\eta_i]$ and to link η_i to η_{i+1} in the space formerly used to store edges leaving η_i. In the general case, we choose a subset Y of X such that $|Y| = 2$ and Y is closed under the map $x \mapsto x^{-1}$. Whenever MERGE finds two states which are equivalent, those two states are processed first by a special coincidence routine which looks only at edges with labels in Y. In the space released there is room to store both the values of p and a linked list of the η_j. Then the procedure returns to processing the edges which involve the current η_i and have labels in $X - Y$, as in COINCIDENCE. See the Appendix for more details.

Exercises

6.1. Work through the coincidence of Exercise 5.1 by hand using the procedure COINCI-DENCE of this section.

6.2. Suppose that $|X| = 2$. Show that $k - i \leq 1$ at all times during the While-loop in COINCIDENCE.

6.3. Let the vector p describe the partition of $\{1, \ldots, 10\}$ into 10 one-point blocks. Suppose UNION is called with the arguments $(10, i)$, with $i = 9, 8, \ldots, 1$, in that order. How many references to p are made? What is the answer if no path compression is used?

4.7 Standardization in place

The output of BASIC_CE is determined only up to isomorphism. To remove this ambiguity, it is useful to standardize the output. This section describes how to standardize the coset automaton of BASIC_CE in place. The techniques involved will be used again in Section 5.4. The global variables n, p, and $\mathcal{A} = (\Sigma, X, E, \{1\}, \{1\})$ are assumed to be available.

Standardization is accomplished through a sequence of transformations called switches. Fix two distinct integers σ and τ greater than 1 and let f be the permutation of the set $\mathbb{Z}^+$ of positive integers which interchanges σ and τ and leaves every other element of $\mathbb{Z}^+$ fixed. To *switch* σ and τ in $\mathcal{A}$

means to replace $\mathcal{A}$ by $f(\mathcal{A}) = (f(\Sigma), X, f(E), \{1\}, \{1\})$, where

$$f(E) = \{(f(\varphi), x, f(\psi)) \mid (\varphi, x, \psi) \in E\}.$$

Clearly the restriction of f to Σ is an isomorphism from $\mathcal{A}$ to $f(\mathcal{A})$. If $\{\sigma, \tau\} \cap \Sigma = \emptyset$, then $f(\mathcal{A}) = \mathcal{A}$.

Switches will be performed by the procedure SWITCH described next. The function FF defined within SWITCH implements the function f.

Procedure SWITCH(σ, τ);
Input: σ, τ : distinct integers between 2 and n;

  Function $FF(\rho)$: positive integer;
  Input: ρ: a positive integer;
  Begin
    If $\rho = \sigma$ then $FF := \tau$
    Else if $\rho = \tau$ then $FF := \sigma$
    Else $FF := \rho$
  End;

Begin
  Let E_1 be the set of edges (φ, x, ψ) in E such that $\{\varphi, \psi\} \cap \{\sigma, \tau\} \neq \emptyset$;
  $E := E - E_1$;
  For (φ, x, ψ) in E_1 do begin
    $\mu := FF(\varphi)$; $\nu := FF(\psi)$;   Add (μ, x, ν) to E
  End;
  ($*$ Make sure $p[\rho] = \rho$ if and only if ρ is in Σ. $*$)
  $\varphi := p[\sigma]$; $\psi := p[\tau]$;
  If $\varphi = \sigma$ then $p[\tau] := \tau$
  Else $p[\tau] := 0$;
  If $\psi = \tau$ then $p[\sigma] := \sigma$
  Else $p[\sigma] := 0$
End.

We continue to leave open the manner in which E is represented. The set E_1 has at most $4|X|$ elements. Normally this is quite small compared to $|E|$. In the following discussion we shall use the length-plus-lexicographic ordering of X^* based on the given order of the generators. Other orderings could be used. The reader is encouraged to consider what changes would be necessary if a basic wreath-product ordering were used.

Example 7.1. Suppose that $X = \{x, y, y^{-1}\}$ and $\mathcal{R}$ consists of the rules

$$x^2 \to \varepsilon, \quad yy^{-1} \to \varepsilon, \quad y^{-1}y \to \varepsilon.$$

Let $\mathcal{A} = (\Sigma, X, E, \{1\}, \{1\})$, where $\Sigma = \{1, 2, 3, 5, 7\}$ and E is given by

	x	y	y^{-1}
1	3		5
2	5		3
3	1	2	7
5	2	1	
7	7	3	

If we switch 2 and 3 in $\mathcal{A}$, then Σ is unchanged and E is now given by

	x	y	y^{-1}
1	2		5
2	1	3	7
3	5		2
5	3	1	
7	7	2	

After switching 3 and 5, the table is

	x	y	y^{-1}
1	2		3
2	1	5	7
3	5	1	
5	3		2
7	7	2	

Switching 4 and 5 changes Σ to $\{1, 2, 3, 4, 7\}$ and produces the table

	x	y	y^{-1}
1	2		3
2	1	4	7
3	4	1	
4	3		2
7	7	2	

After the switch of 7 and 5 we have $\Sigma = \{1, 2, 3, 4, 5\}$ and the table

	x	y	y^{-1}
1	2		3
2	1	4	5
3	4	1	
4	3		2
5	5	2	

At this point $\mathcal{A}$ is standard.

Table 4.7.1

	x	x^{-1}	y	y^{-1}
1	4	3	5	
2	5			4
3	1			7
4		1	2	
5	7	2		1
7		5	3	

Here is a formal description of the in-place standardization procedure illustrated in Example 7.1. The procedure is based on LENLEX_STND of Section 3.8 and is designed to be used as part of BASIC_CE as given in Section 4.5.

Procedure STANDARDIZE;
($*$ Standardize the current automaton with respect to a
    length-plus-lexicographic ordering. $*$)
Begin
   $\tau := 1$; $\sigma := 1$;
  While $\sigma \leq \tau$ do begin
    For x in X do   ($*$ Elements of X are taken in $\prec$-order. $*$)
     If σ^x is defined then begin
      $\rho := \sigma^x$;
      If $\rho > \tau$ then begin
       $\tau := \tau + 1$;
       If $\rho > \tau$ then SWITCH(τ, ρ)
      End
     End;
    $\sigma := \sigma + 1$
  End
 End.

At the termination of STANDARDIZE, the automaton $\mathcal{A}$ is standard with respect to the length-plus-lexicographic ordering of X^* determined by the given ordering of X. The switches in Example 7.1 are the ones that STANDARDIZE performs as it standardizes the original automaton in that example, assuming that $x \prec y \prec y^{-1}$.

Example 7.2. Let $X = \{x, y\}^{\pm}$, let $\mathcal{R}$ define the free group on $\{x, y\}$, and let $\mathcal{A} = (\Sigma, X, E, \{1\}, \{1\})$, where $\Sigma = \{1, 2, 3, 4, 5, 7\}$ and E corresponds to Table 4.7.1. If we assume that $x \prec x^{-1} \prec y \prec y^{-1}$, then STANDARDIZE performs the following switches in standardizing $\mathcal{A}$. With $\sigma = 1$, there is a switch of 2 and 4 followed by a switch of 4 and 5. When σ is 2, there

Table 4.7.2

	x	x^{-1}	y	y^{-1}
1	2	3	4	
2		1	5	
3				6
4	6	5		1
5	4			2
6			4	3

Table 4.7.3

	x	x^{-1}	y	y^{-1}
1	9			6
2		7	4	
4	7		6	2
6		9	1	4
7	2	4		
9	6	1	9	9

are no switches. With $\sigma = 3$, a switch of 6 and 7 occurs. At this point the automaton is standard. No further switches are carried out. The final coset table is Table 4.7.2.

The number of calls to SWITCH in STANDARDIZE is at most $|\Sigma| - 1$, and, with any reasonable representation for E, the execution time for a single call to SWITCH should be at worst proportional to $|X|$. Thus the time to standardize should be at most a constant times $|X||\Sigma|$.

Exercises

7.1. Show that $\sigma = p[\sigma]$ each time the For-loop in STANDARDIZE is executed. Thus σ is always in Σ until $\sigma > \tau$.

7.2. Assume that X and $\mathcal{R}$ are as in Example 7.2. Standardize the coset automaton $(\Sigma, X, E, \{1\}, \{1\})$, where E is given by Table 4.7.3.

7.3. Devise an in-place standardization procedure which uses the basic wreath-product ordering determined by a given ordering of the generators.

4.8 Computation with subgroups

Let H be a finitely generated subgroup of our free product $F = \text{Mon}\,\langle X \mid \mathcal{R}\rangle$. We have associated with H three automata, $\mathcal{A}_s(H)$, $\mathcal{A}_c(H)$, and $\mathcal{A}_I(H)$, and the languages they recognize, $\mathcal{L}_s(H)$, $\mathcal{L}_c(H)$, and $\mathcal{L}_I(H)$. Each automaton is a minimal trim, deterministic automaton recognizing its language. The automaton $\mathcal{A}_s(H)$ is finite if and only if H has finite index in F. The other automata are always finite (when H is finitely generated).

An automaton isomorphic to $\mathcal{A}_I(H)$ is the preferred description of H. However, any finite coset automaton $\mathcal{A}$ such that $H = K(\mathcal{A})$ is almost as good. Given $\mathcal{A}$ and $\mathcal{R}$, we can decide membership in H, decide equality of cosets in H, and obtain a finite set $\mathcal{U}$ of words such that $\{[U] \mid U \in \mathcal{U}\}$ is a generating set for H. Conversely, given a finite generating set, we can construct an automaton isomorphic to $\mathcal{A}_I(H)$ using BASIC_CE.

We shall now look at some additional computations with subgroups of F. Suppose that we are given finite coset automata $\mathcal{A}_1$ and $\mathcal{A}_2$. Let $H_i = K(\mathcal{A}_i)$, $i = 1, 2$. We shall first construct finite coset automata defining $\text{Grp}\,\langle H_1, H_2\rangle$ and $H_1 \cap H_2$.

One way to construct a coset automaton defining $M = \text{Grp}\langle H_1, H_2 \rangle$ is to use $\mathcal{A}_1$ and $\mathcal{A}_2$ to obtain finite sets $\mathcal{U}_1$ and $\mathcal{U}_2$ of words describing generators for H_1 and H_2, respectively, and then to use BASIC_CE with the set $\mathcal{U}_1 \cup \mathcal{U}_2$, which describes a set of generators for M. However, it suffices to construct just $\mathcal{U}_1$ and then to perform two-sided traces of the elements of $\mathcal{U}_1$ at the initial state of $\mathcal{A}_2$. See Exercise 8.5 for an alternative approach.

The procedure for finding a coset automaton corresponding to $H_1 \cap H_2$ is based on the following proposition.

Proposition 8.1. *Let H_1 and H_2 be subgroups of F and set $L = H_1 \cap H_2$. Then*

(a) $\mathcal{L}_s(L) = \mathcal{L}_s(H_1) \cap \mathcal{L}_s(H_2)$.
(b) $\mathcal{L}_c(L) = \mathcal{L}_c(H_1) \cap \mathcal{L}_c(H_2)$.

Proof. The first equality says that $[U]$ is in L if and only if $[U]$ is in H_1 and in H_2. The second equality says the same thing for freely reduced words. $\square$

As a corollary, we get a result first proved in (Howson 1954).

Corollary 8.2 (Howson's theorem). *If H_1 and H_2 are finitely generated subgroups of F, then $H_1 \cap H_2$ is finitely generated.*

Proof. This follows from Proposition 8.1 and Corollary 3.9. $\square$

Proposition 8.3. *Let $\mathcal{A}_1$ and $\mathcal{A}_2$ be finite coset automata and let $\mathcal{A}$ be the result of applying the accessible intersection procedure to $\mathcal{A}_1$ and $\mathcal{A}_2$. Then $\mathcal{A}$ is a coset automaton and $K(\mathcal{A}) = K(\mathcal{A}_1) \cap K(\mathcal{A}_2)$.*

Proof. Let $\mathcal{A}_i = (\Sigma_i, X, E_i, A_i, A_i)$, $i = 1, 2$. The full intersection procedure forms $\mathcal{A}_3 = (\Sigma_3, X, E_3, A_3, A_3)$, where $\Sigma_3 = \Sigma_1 \times \Sigma_2$, $A_3 = A_1 \times A_2$, and there is an edge in E_3 from (σ_1, σ_2) to (τ_1, τ_2) with label x if and only if (σ_i, x, τ_i) is in E_i for $i = 1, 2$. If this is the case, then for $i = 1, 2$ the triple $(\tau_i, x^{-1}, \sigma_i)$ is in E_i, and hence there is an edge labeled x^{-1} from (τ_1, τ_2) to (σ_1, σ_2). Therefore $\mathcal{A}_3$ satisfies conditions (ii) and (iii) of the definition of a coset automaton. The automaton $\mathcal{A}$ is the accessible part of $\mathcal{A}_3$, so $\mathcal{A}$ satisfies condition (i) as well.

Now $L(\mathcal{A}) = L(\mathcal{A}_1) \cap L(\mathcal{A}_2)$. Thus

$$
\begin{aligned}
\mathcal{L}_c(K(\mathcal{A})) &= L(\mathcal{A}) \cap \mathcal{C} \\
&= (L(\mathcal{A}_1) \cap \mathcal{C}) \cap (L(\mathcal{A}_2) \cap \mathcal{C}) \\
&= \mathcal{L}_c(K(\mathcal{A}_1)) \cap \mathcal{L}_c(K(\mathcal{A}_2)) \\
&= \mathcal{L}_c(K(\mathcal{A}_1) \cap K(\mathcal{A}_2)).
\end{aligned}
$$

Table 4.8.1

	x	x^{-1}	y	y^{-1}
$(1,1) \leftrightarrow 1$		2	3	4
$(3,2) \leftrightarrow 2$	1			3
$(2,3) \leftrightarrow 3$		5	2	1
$(4,2) \leftrightarrow 4$	6		1	
$(1,3) \leftrightarrow 5$	3	7	8	6
$(4,1) \leftrightarrow 6$		4	5	
$(3,3) \leftrightarrow 7$	5			9
$(2,2) \leftrightarrow 8$			10	5
$(2,1) \leftrightarrow 9$		11	7	11
$(3,1) \leftrightarrow 10$				8
$(1,2) \leftrightarrow 11$	9		9	12
$(4,3) \leftrightarrow 12$	12	12	11	

Therefore $K(\mathcal{A}) = K(\mathcal{A}_1) \cap K(\mathcal{A}_2)$. $\square$

Example 8.1. Let $X = \{x, y\}^{\pm}$ and let $\mathcal{R}$ be the set of defining rules for the free group on $\{x, y\}$. For $i = 1, 2$ let $\mathcal{A}_i = (\Sigma_i, X, E_i, \{1\}, \{1\})$, where $\Sigma_1 = \{1, 2, 3, 4\}$, $\Sigma_2 = \{1, 2, 3\}$, and E_1 and E_2 are given, respectively, by the following coset tables:

	x	x^{-1}	y	y^{-1}
1	2	3	2	4
2		1	3	1
3	1			2
4	4	4	4	1

	x	x^{-1}	y	y^{-1}
1		2	3	2
2	1		1	3
3	3	3	2	1

Applying the accessible intersection construction to these automata, we get $\mathcal{A} = (\Sigma, X, E, \{1\}, \{1\})$, where $\Sigma = \{1, 2, \ldots, 12\}$ and E is given by Table 4.8.1, which indicates the correspondence between Σ and $\Sigma_1 \times \Sigma_2$. Applying the procedure REDUCE described at the end of Section 4.4, we remove states 8 and 10 from $\mathcal{A}$ and find that $\mathcal{A}_I(K(\mathcal{A}))$ is isomorphic to the restriction of $\mathcal{A}$ to $\Sigma - \{8, 10\}$.

Now let $\mathcal{A} = (\Sigma, X, E, \{\alpha\}, \{\alpha\})$ be any finite coset automaton and let $H = K(\mathcal{A})$. Suppose U is a word in X^*. The conjugate $M = [U]^{-1}H[U]$ is finitely generated and so can be described by a finite coset automaton. To find such an automaton, one could determine a generating set of H, conjugate that generating set by $[U]$, and then construct $\mathcal{A}_I(M)$ using BA-SIC_CE. However, there is an easier way, particularly if α^U is defined in $\mathcal{A}$.

Proposition 8.4. *Let* $\mathcal{A} = (\Sigma, X, E, \{\alpha\}, \{\alpha\})$ *be a coset automaton, let* U *be a word in* X^*, *and suppose that* $\alpha^U = \sigma$ *in* $\mathcal{A}$. *Then* $\mathcal{B} = (\Sigma, X, E, \{\sigma\}, \{\sigma\})$ *is a coset automaton and* $K(\mathcal{B}) = [U]^{-1} K(\mathcal{A})[U]$.

Proof. Clearly $\mathcal{B}$ is deterministic and satisfies conditions (ii) and (iii) of the definition of a coset automaton. We must prove that $\mathcal{B}$ is accessible. By Proposition 4.4, there is a word V such that $\sigma^V = \alpha$ and $[U]^{-1} = [V]$. Since $\mathcal{A}$ is accessible, we can get from α to any state by a path. Therefore we can get from σ to any state and $\mathcal{B}$ is accessible.

Now suppose that $\sigma^W = \sigma$. Then $\alpha^{UWV} = \alpha$, so $[UWV] = [U][W][U]^{-1}$ is in $K(\mathcal{A})$, or $[W]$ is in $[U]^{-1} K(\mathcal{A})[U]$. Therefore $K(\mathcal{B}) \subseteq [U]^{-1} K(\mathcal{A})[U]$. Conversely, if $\alpha^W = \alpha$, then $\sigma^{VWU} = \sigma$, from which it follows that $K(\mathcal{B}) \supseteq [U]^{-1} K(\mathcal{A})[U]$. $\square$

Example 8.2. Suppose that $X = \{x, y, y^{-1}\}$ and $\mathcal{R}$ consists of the rules

$$x^2 \to \varepsilon, \quad yy^{-1} \to \varepsilon, \quad y^{-1}y \to \varepsilon.$$

Let $\mathcal{A} = (\Sigma, X, E, \{1\}, \{1\})$, where $\Sigma = \{1, 2, 3\}$ and E is given by the following coset table:

	x	y	y^{-1}
1	1	2	2
2	3	1	1
3	2	3	3

Then $\mathcal{A}$ is a complete coset automaton, which is standard with respect to the length-plus-lexicographic ordering with $x \prec y \prec y^{-1}$. The subgroup $H = K(\mathcal{A})$ has index 3 in F. The conjugates of H are H and the groups $K(\mathcal{A}_\sigma)$, $\sigma = 2, 3$, where $\mathcal{A}_\sigma = (\Sigma, X, E, \{\sigma\}, \{\sigma\})$. The coset tables for the standard automata isomorphic to $\mathcal{A}_2$ and $\mathcal{A}_3$ are, respectively,

	x	y	y^{-1}
1	2	3	3
2	1	2	2
3	3	1	1

and

	x	y	y^{-1}
1	2	1	1
2	1	3	3
3	3	2	2

In general, new states and edges must be added as in DEFINE before Proposition 8.4 can be applied. Here is the general procedure for computing conjugates of finitely generated subgroups of F.

Procedure CONJUGATE($\mathcal{A}, U; \mathcal{D}$);
Input: $\mathcal{A}$: a coset automaton $(\Sigma, X, E, \{\alpha\}, \{\alpha\})$;
           U : a word in X^*;

Output: $\mathcal{D}$　　: a coset automaton such that $K(\mathcal{D}) = [U]^{-1}K(\mathcal{A})[U]$;
Begin
　$\mathcal{B} := \mathcal{A}$; REWRITE$(X, \mathcal{R}, U; V)$; TRACE$(\mathcal{B}, \alpha, V; \sigma, B, D)$;
　Let $D = x_1 \ldots x_s$;

　For $i := 1$ to s do begin
　　Let τ be an object which is not a state of $\mathcal{B}$;
　　Add τ to the set of states of $\mathcal{B}$ and add (σ, x_i, τ) and
　　　(τ, x_i^{-1}, σ) to the set of edges of $\mathcal{B}$;
　　$\sigma := \tau$
　End;

　Change the initial and terminal state of $\mathcal{B}$ to σ;
　REDUCE$(\mathcal{B}; \mathcal{D})$
End.

Now let us turn to the problem of computing normalizers. The *normalizer* in F of a subgroup H of F is the subgroup $N_F(H)$ of elements u in F such that $u^{-1}Hu = H$. Since $H = u^{-1}Hu$ if and only if $A_I(H)$ is isomorphic to $A_I(u^{-1}Hu)$, we can decide membership in $N_F(H)$. But we can do more. If $H = 1$, then $N_F(H) = F$. This special case can be handled trivially. Therefore let us assume that $H \neq 1$. Since $N_F(u^{-1}Hu) = u^{-1}N_F(H)u$, we may replace H temporarily by a conjugate if we wish.

Suppose that $\mathcal{A} = (\Sigma, X, E, \{\alpha\}, \{\alpha\})$ is isomorphic to $A_I(H)$. For σ in Σ, let $\mathcal{A}_\sigma = (\Sigma, X, E, \{\sigma\}, \{\sigma\})$ and $H_\sigma = K(\mathcal{A}_\sigma)$. By Proposition 8.4, H_σ is a conjugate of H. Also, $|\Sigma_I(H_\sigma)| \leq |\Sigma|$. If $|\Sigma_I(H_\sigma)| < |\Sigma|$, then let us replace H by H_σ. Continuing in this way, we may assume that $|\Sigma_I(H_\sigma)| = |\Sigma|$ for all σ in Σ. When this condition holds, we shall say that $\mathcal{A}$ is *cyclically reduced*.

Proposition 8.5. *Suppose that $\mathcal{A} = (\Sigma, X, E, \{\alpha\}, \{\alpha\})$ is a finite, cyclically reduced coset automaton such that $H = K(\mathcal{A})$ is nontrivial. If U is an element of $\mathcal{C}$ such that $[U]$ is in $N_F(H)$, then α^U is defined in $\mathcal{A}$.*

Proof. Suppose that α^U is not defined. Perform the operation CONJUGATE$(\mathcal{A}, U; \mathcal{B})$. Let $\mathcal{B} = (\Lambda, X, Q, \{\sigma\}, \{\sigma\})$. The state σ was added in the last iteration of the For-loop of CONJUGATE. At the termination of CONJUGATE, there was only one edge leaving σ. The procedure REDUCE only removes edges, so there is at most one edge leaving σ in $\mathcal{B}$. However, $K(\mathcal{B}) = K(\mathcal{A})$ is nontrivial, so there is exactly one edge (σ, x, τ) in $\mathcal{B}$. Moreover, $\tau \neq \sigma$ and (τ, x^{-1}, σ) is the only edge coming into σ. Now $\mathcal{B}$ must be isomorphic to $\mathcal{A}$, so there are edges (α, x, β) and (β, x^{-1}, α) in $\mathcal{A}$, where $\beta \neq \alpha$, and these are the only two edges in $\mathcal{A}$ involving α. But in $\mathcal{A}_\beta = (\Sigma, X, E, \{\beta\}, \{\beta\})$ the state α is nonterminal and has only one edge

Table 4.8.2

	x	x^{-1}	y	y^{-1}
1	2	3	4	5
2		1	6	
3	1			7
4	7			1
5		6	1	
6	5	8	9	2
7	10	4	3	11
8	6			12
9	12			6
10		7	12	
11		12	7	
12	11		9	8	10

leaving it. Therefore $\mathcal{A}_\beta$ is not reduced, contradicting the assumption that $\mathcal{A}$ is cyclically reduced. Hence α^U is defined in $\mathcal{A}$. □

Corollary 8.6. *If H is a nontrivial, finitely generated subgroup of F, then H has finite index in $N_F(H)$.*

Proof. We may assume that $H = K(\mathcal{A})$, where $\mathcal{A}$ is a finite, cyclically reduced coset automaton. By Proposition 8.5, every right coset of H in $N_F(H)$ is important, and H has only finitely many important cosets. □

Example 8.3. Let X and $\mathcal{R}$ be as in Example 8.1, and let $\mathcal{A} = (\Sigma, X, E, \{1\}, \{1\})$, where $\Sigma = \{1, 2, \ldots, 12\}$ and E is given by Table 4.8.2. Since two edges leave every state, it follows that $\mathcal{A}_I(H_\sigma)$ is isomorphic to $\mathcal{A}_\sigma$ for every σ in Σ. Thus if $H = K(\mathcal{A})$, then $|N_F(H) : H|$ is the number of states σ in Σ such that $\mathcal{A}_\sigma$ is isomorphic to $\mathcal{A}$. Now four edges leave 1, so if $\mathcal{A}$ and $\mathcal{A}_\sigma$ are isomorphic, then four edges must leave σ. Thus σ must be one of 1, 6, 7, and 12. Standardizing $\mathcal{A}_6$, $\mathcal{A}_7$, and $\mathcal{A}_{12}$ with respect to the order $x \prec x^{-1} \prec y \prec y^{-1}$ gives $\mathcal{A}$ in each case. Therefore $|N_F(H) : H| = 4$. Since $1^{xy} = 6$, $1^{x^{-1}y^{-1}} = 7$, and $1^{xyx^{-1}y^{-1}} = 12$, the elements $1 = [\varepsilon]$, [x y], $[x^{-1}y^{-1}]$, and $[xyx^{-1}y^{-1}]$ are coset representatives for H in $N_F(H)$. In fact $N_F(H) = \mathrm{Grp}\,\langle H, [xy], [x^{-1}y^{-1}] \rangle$.

Exercises

8.1. Explain why it is useful in CONJUGATE to rewrite U before executing TRACE.

8.2. Suppose that $\mathcal{A} = (\Sigma, X, E, \{\alpha\}, \{\alpha\})$ is a cyclically reduced coset automaton with at least one edge. Show that there is a nonempty, cyclically reduced word U such that $\alpha^U = \alpha$.

8.3. Let X and $\mathcal{R}$ be as in Example 8.1. Suppose $\mathcal{A}_1$ and $\mathcal{A}_2$ are the standard coset automata defined by the following tables:

Table 4.8.3

	x	x^{-1}	y	y^{-1}
1	2			
2	3	1	4	
3		2	3	3
4	5	5		2
5	4	4		6
6	7		5	
7		6	7	7

	x	x^{-1}	y	y^{-1}
1	2	3	4	
2	3	1		4
3	1	2		
4			2	1

	x	x^{-1}	y	y^{-1}
1	2	3	4	2
2		1	1	4
3	1	4		
4	3		2	1

Determine the important-coset automaton for $K(\mathcal{A}_1) \cap K(\mathcal{A}_2)$.

8.4. Continuing with X and $\mathcal{R}$ as in Example 8.1, let $\mathcal{A} = (\Sigma, X, E, \{1\}, \{1\})$, where $\Sigma = \{1, \ldots, 7\}$ and E is given by Table 4.8.3. Set $H = K(\mathcal{A})$. Determine $|N_F(H) : H|$ and find a word U such that $N_F(H) = \text{Grp}\,\langle H, [U] \rangle$.

8.5. Let X be a finite set and let $\mathcal{R}$ be a classical niladic rewriting system on X^*. For $i = 1, 2$ let $\mathcal{A}_i$ be a finite coset automaton with respect to $\mathcal{R}$. Show that the procedure CA_JOIN described below returns a coset automaton $\mathcal{A}$ such that $K(\mathcal{A}) = \text{Grp}\,\langle K(\mathcal{A}_1), K(\mathcal{A}_2) \rangle$. A function f from states of $\mathcal{A}_1$ to states of $\mathcal{A}$ is built up. The subroutine FORCE, which is a version of the coincidence procedure, assumes that $\mathcal{A}$ and f are global variable and takes two states from $\mathcal{A}$ as arguments.

Procedure FORCE(μ, ν);
Begin
    Let $\mathcal{A}'$ be the quotient of $\mathcal{A}$ modulo the congruence generated
        by (μ, ν) and let g be the expanding morphism from $\mathcal{A}$ to $\mathcal{A}'$;
    Replace $\mathcal{A}$ by $\mathcal{A}'$ and f by $f \circ g$
End.

Procedure CA_JOIN($\mathcal{A}_1, \mathcal{A}_2; \mathcal{A}$);
Begin
    $\mathcal{A} := \mathcal{A}_2$;
    Let γ_1 be the initial state of $\mathcal{A}_1$;
    Define $f(\gamma_1)$ to be the initial state of $\mathcal{A}$;
    $s := 1$; $\Gamma := \{\gamma_1\}$; $i := 1$;

    While $i \le s$ do
      For x in X do
        If γ_i^x is defined in $\mathcal{A}_1$ then begin

          $\varphi := \gamma_i^x$; $\sigma := f(\gamma_i)$;

          If $\varphi \in \Gamma$ then begin
            $\psi := f(\varphi)$;

If σ^x is defined in $\mathcal{A}_1$ then
    FORCE(σ^x, ψ)
Else if $\psi^{x^{-1}}$ is defined in $\mathcal{A}$ then
    FORCE$(\psi^{x^{-1}}, \sigma)$
Else add the edges (σ, x, ψ) and (ψ, x^{-1}, σ)
      to $\mathcal{A}$
End

Else begin
  $s := s + 1$; $\gamma_s := \varphi$; Add φ to Γ;
  If σ^x is defined in $\mathcal{A}$ then
    $\tau := \sigma^x$
  Else add to $\mathcal{A}$ a new state τ and edges (σ, x, τ)
    and (τ, x^{-1}, σ);
  $f(\varphi) := \tau$
End
    End
End.

4.9 Standard coset tables

Section 3.8 sketched a procedure which can be used to list representatives for the isomorphism classes of accessible, deterministic automata with alphabet X and at most a given number n of states. The process involves choosing a reduction ordering $\prec$ on X^* and constructing transition matrices $T(\mathcal{A})$ corresponding to the accessible, deterministic automata $\mathcal{A}$ which are standard with respect to $\prec$. The matrices are produced in the precedence order defined by $\prec$.

Among the finite, accessible, deterministic automata with alphabet X are the finite coset automata with respect to $\mathcal{R}$. A standard coset automaton $\mathcal{A}$ is determined by $T(\mathcal{A})$. We can use the precedence ordering of these matrices to define an ordering of finitely generated subgroups of F. Given a finitely generated subgroup H, we associate with H the standard automaton $\mathcal{B}(H)$ isomorphic to $\mathcal{A}_I(H)$. If K is another finitely generated subgroup of F, then H comes before K if $T(\mathcal{B}(H))$ comes before $T(\mathcal{B}(K))$.

In this section we shall consider the following problems:

(1) List all standard coset tables with at most n rows.
(2) List all standard, reduced coset tables with at most n rows.
(3) List all complete, standard coset tables with at most n rows.
(4) List the complete, standard coset tables which correspond to subgroups H of F such that $|F : H| \leq n$ and H is first in its conjugacy class.

Problem (2) is equivalent to finding all subgroups of F with at most n important cosets, problem (3) is equivalent to finding all subgroups of F with index at most n, and problem (4) is equivalent to finding representatives

for the conjugacy classes of such subgroups. The techniques developed here
will be incorporated into the low-index subgroup algorithm of Section 5.6.

Perhaps the most obvious way to solve problem (1) is to use the procedure
of Section 3.8 to list all transition matrices with at most n rows and delete
those matrices which do not correspond to coset automata with respect to
$\mathcal{R}$. However, only a very small fraction of transition matrices define coset
automata with respect to $\mathcal{R}$, so it is more efficient to build the definition
of a coset automaton into the backtrack search.

Let $x_1, \ldots, x_r$ be the elements of X in $\prec$-order. In the search procedure of
Section 3.8, nodes of the search tree correspond to certain incomplete transition matrices. In our search for standard coset automata, the nodes will
correspond to incomplete transition matrices T which satisfy the following
condition: If $T[\sigma, i]$ is a positive integer τ and $x_i^{-1} = x_j$, then $T[\tau, j] = \sigma$.

The root of the tree is again the 1-by-r matrix all of whose entries are
'?'. Suppose that T is an m-by-r transition matrix which is a node in the
tree. If there is no word W such that 1^W is unspecified, then T is one of
the desired transition matrices. Otherwise, let W be the first word with
respect to $\prec$ such that 1^W is unspecified. Then W has the form Ux_i, where
$\sigma = 1^U$ is defined. Let $x_i^{-1} = x_j$ and let C be the set of indices τ such that
$T[\tau, j]$ is unspecified. Let D consist of 0, the elements in C, and, if $m < n$,
also $m+1$. For each k in D we get a descendant of T by the following steps:

$T[\sigma, i] := k$;
If $k = m + 1$ then add a k-th row of question marks;
if $k \neq 0$ then $T[k, j] := \sigma$;

As in Section 3.8, if the tree is examined in a depth-first search, then the
transition matrices are produced in order of precedence.

Example 9.1. Let $X = \{x, y\}^{\pm}$, let $\mathcal{R}$ define the free group on $\{x, y\}$, and let
$\prec$ be the length-plus-lexicographic ordering of X^* with $x \prec x^{-1} \prec y \prec y^{-1}$.
Assume that $n \geq 3$. The root of the search tree is $[?\ ?\ ?\ ?]$ and the
descendants of the root are

$$[0\ ?\ ?\ ?], \quad [1\ 1\ ?\ ?], \quad \begin{bmatrix} 2 & ? & ? & ? \\ ? & 1 & ? & ? \end{bmatrix}.$$

With the first and third of these matrices, $W = x^{-1}$, while with the second,
$W = y$. The descendants of $[0\ ?\ ?\ ?]$ are

$$[0\ 0\ ?\ ?], \quad \begin{bmatrix} 0 & 2 & ? & ? \\ 1 & ? & ? & ? \end{bmatrix}.$$

The descendants of $[1\ 1\ ?\ ?]$ are

$$[1\ 1\ 0\ ?], \quad [1\ 1\ 1\ 1], \quad \begin{bmatrix} 1 & 1 & 2 & ? \\ ? & ? & ? & 1 \end{bmatrix}.$$

The descendants of $\begin{bmatrix} 2 & ? & ? & ? \\ ? & 1 & ? & ? \end{bmatrix}$ are

$$\begin{bmatrix} 2 & 0 & ? & ? \\ ? & 1 & ? & ? \end{bmatrix}, \quad \begin{bmatrix} 2 & 2 & ? & ? \\ 1 & 1 & ? & ? \end{bmatrix}, \quad \begin{bmatrix} 2 & 3 & ? & ? \\ ? & 1 & ? & ? \\ 1 & ? & ? & ? \end{bmatrix}.$$

If we take $\prec$ to be the basic wreath-product ordering with $x \prec x^{-1} \prec y \prec y^{-1}$, then the root and its descendants are the same. However, with $\begin{bmatrix} 2 & ? & ? & ? \\ ? & 1 & ? & ? \end{bmatrix}$, now $W = x^2$, and the descendants are

$$\begin{bmatrix} 2 & ? & ? & ? \\ 0 & 1 & ? & ? \end{bmatrix}, \quad \begin{bmatrix} 2 & 2 & ? & ? \\ 1 & 1 & ? & ? \end{bmatrix}, \quad \begin{bmatrix} 2 & ? & ? & ? \\ 3 & 1 & ? & ? \\ ? & 2 & ? & ? \end{bmatrix}.$$

Example 9.2. Let $X = \{x, y, y^{-1}\}$, let $\prec$ be the length-plus-lexicographic ordering of X^* with $x \prec y \prec y^{-1}$, and let $\mathcal{R}$ consist of

$$x^2 \to \varepsilon, \quad yy^{-1} \to \varepsilon, \quad y^{-1}y \to \varepsilon.$$

The root of the search tree is $[?\ ?\ ?]$ and its descendants are

$$[0\ ?\ ?], \quad [1\ ?\ ?], \quad \begin{bmatrix} 2 & ? & ? \\ 1 & ? & ? \end{bmatrix}.$$

If $n = 2$, then a total of 23 coset automata are produced.

The 23 automata of Example 9.2 are not all reduced and so do not correspond to 23 different subgroups of F. For example, four of the automata correspond to the trivial subgroup. To produce only reduced automata, we must modify our search criteria slightly. If $\sigma > 1$, $T[\sigma, i]$ is the only '?' in the σ-th row, and this row contains only one positive entry, then 0 is not allowed to be in D. If in Example 9.2 we make this change, then 17 standard, reduced automata are produced.

In a search for standard, complete coset automata, 0 is never allowed to be in D. Let X and $\mathcal{R}$ be as in Example 9.2, let $\prec$ be the length-plus-lexicographic ordering with $x \prec y \prec y^{-1}$, and take $n = 3$. Then 11 complete, standard automata are produced. The transition matrices are

$$
[1\ 1\ 1], \quad
\begin{bmatrix} 1 & 2 & 2 \\ 2 & 1 & 1 \end{bmatrix}, \quad
\begin{bmatrix} 1 & 2 & 2 \\ 3 & 1 & 1 \\ 2 & 3 & 3 \end{bmatrix}, \quad
\begin{bmatrix} 1 & 2 & 3 \\ 2 & 3 & 1 \\ 3 & 1 & 2 \end{bmatrix},
$$

$$
\begin{bmatrix} 1 & 2 & 3 \\ 3 & 3 & 1 \\ 2 & 1 & 2 \end{bmatrix}, \quad
\begin{bmatrix} 2 & 1 & 1 \\ 1 & 2 & 2 \end{bmatrix}, \quad
\begin{bmatrix} 2 & 1 & 1 \\ 1 & 3 & 3 \\ 3 & 2 & 2 \end{bmatrix}, \quad
\begin{bmatrix} 2 & 2 & 2 \\ 1 & 1 & 1 \end{bmatrix},
$$

$$
\begin{bmatrix} 2 & 2 & 3 \\ 1 & 3 & 1 \\ 3 & 1 & 2 \end{bmatrix}, \quad
\begin{bmatrix} 2 & 3 & 2 \\ 1 & 1 & 3 \\ 3 & 2 & 1 \end{bmatrix}, \quad
\begin{bmatrix} 2 & 3 & 3 \\ 1 & 2 & 2 \\ 3 & 1 & 1 \end{bmatrix}.
$$

From this list of matrices we see that the free product F of $\mathbb{Z}_2$ and $\mathbb{Z}$ has one subgroup of index 1, three subgroups of index 2, and seven subgroups of index 3.

If we only want to know the number of subgroups of F of index at most n, then we can proceed in a manner analogous to Exercises 8.6 to 8.8 in Chapter 3. The columns of a complete standard coset table with m rows are permutations of $\{1, \dots, m\}$, and the columns headed x and x^{-1} are inverses of each other. Thus these columns define a homomorphism f of F into the symmetric group $\mathrm{Sym}(m)$. The first thing to do is to count these homomorphisms. If $x \neq x^{-1}$, then we may choose the image $f(x)$ arbitrarily and map x^{-1} to $f(x)^{-1}$. If x^2 is a left side in $\mathcal{R}$, then we must map x to an element of order 1 or 2 in $\mathrm{Sym}(m)$.

Proposition 9.1. *Given positive integers e and n, let $g(e, n)$ be the number of elements u in $\mathrm{Sym}(n)$ such that $u^e = 1$. Then*

$$
g(e, n) = \sum_{d \mid e} g(e, n - d)(n - 1) \cdots (n - d + 1).
$$

Proof. Exercise. □

Using Proposition 9.1, we can compute as many values of $g(2, n)$ as we need. For example, the values for $n = 1, 2, \dots, 5$ are 1, 2, 4, 10, and 26, respectively. Let X have s elements x such that $x^{-1} = x$, and write $|X|$ as $s + 2t$. Then the number of homomorphisms f of F into $\mathrm{Sym}(n)$ is

$$
(n!)^t g(2, n)^s.
$$

Given f, we can form the automaton $\mathcal{B} = (\Sigma, X, E, \{1\}, \{1\})$, where $\Sigma = \{1, 2, \ldots, n\}$ and E consists of the triples $(\sigma, x, \sigma^{f(x)})$. We can then define $\mathcal{A}$ to be the standard automaton isomorphic to the accessible part of $\mathcal{B}$. It is easy to check that $\mathcal{A}$ is a complete coset automaton.

Proposition 9.2. *If $\mathcal{A}$ has m states, then the number of homomorphisms f which produce $\mathcal{A}$ is*

$$(n-1)\ldots(n-m+1)[(n-m)!]^t g(2, n-m)^s.$$

Proof. Exercise. $\square$

Let $h(n)$ denote the number of complete, standard coset tables with n rows relative to $\mathcal{R}$.

Proposition 9.3. *The values of h are given by the following recursive formula:*

$$h(n) = \frac{1}{(n-1)!}\left[(n!)^t g(2, n)^s \right.$$

$$\left. - \sum_{m=1}^{n-1} h(m)(n-1)\ldots(n-m+1)[(n-m)!]^t g(2, n-m)^s \right].$$

Proof. As in Exercise 8.7 of Chapter 3, the equality

$$(n!)^t g(2, n)^s = \sum_{m=1}^{n} h(m)(n-1)\ldots(n-m+1)[(n-m)!]^t g(2, n-m)^s$$

counts the homomorphisms of F into $\mathrm{Sym}(n)$ in two ways. We have only to solve for $h(n)$. $\square$

In Example 9.2 we have $s = t = 1$. In this case, the formula of Proposition 9.3 gives the following values: $h(1) = 1$, $h(2) = 3$, $h(3) = 7$, $h(4) = 23$, $h(5) = 71$.

Solving the fourth problem stated at the beginning of this section involves a more complicated use of backtracking. The following example shows the reason.

Example 9.3. Assume that X, $\prec$, and $\mathcal{R}$ are as in Example 9.2. Suppose that $n \geq 4$ and that the first two rows of $T(\mathcal{A})$ are

$$\begin{bmatrix} 2 & 3 & 4 \\ 1 & 2 & 2 \end{bmatrix}$$

No standard coset automaton with $T(\mathcal{A})$ of this form corresponds to a subgroup of finite index in F which is first in its conjugacy class, for if we change the initial and terminal state to 2 and standardize, then the resulting automaton $\mathcal{B}$ corresponds to a conjugate of $K(\mathcal{A})$, and the first row of $T(\mathcal{B})$ will be $[2 \ 1 \ 1]$. Any such matrix comes lexicographically before one whose first row is $[2 \ 3 \ 4]$.

Suppose in our search for standard, complete automata we are considering a transition matrix T in the tree and its possible extensions. We may be able to use the ideas of Example 9.3 to show that no standard, complete coset automaton associated with a descendant of T can correspond to a subgroup of finite index in F which is first in its conjugacy class. Let $\mathcal{A}$ be the coset automaton with initial state 1 obtained by taking as edges the triples (σ, x_i, τ) with $T[\sigma, i] = \tau \neq$ '?' and let m be the number of rows of T. For $2 \leq i \leq m$, let $\mathcal{A}_i$ be the result of changing the initial and terminal state in $\mathcal{A}$ to i and standardizing the result. The matrices T and $T(\mathcal{A}_i)$ will in general have some entries which are '?'. Let W be the first word at which $f_{\mathcal{A}}$ and $f_{\mathcal{A}_i}$ differ. Suppose that $f_{\mathcal{A}_i}(W)$ and $f_{\mathcal{A}}(W)$ are both defined, that $f_{\mathcal{A}_i}(W) < f_{\mathcal{A}}(W)$, and that 1^V is determined in both $\mathcal{A}$ and $\mathcal{A}_i$ for all words $V \prec W$. Then no descendant of T can correspond to a subgroup of finite index in F which is first in its conjugacy class.

Implementing the test just described is somewhat involved. However, the test plays an important part in the low-index subgroup algorithm of Section 5.6, so we give a detailed version of it for the case of a length-plus-lexicographic ordering. In computing $\mathcal{A}_i$, we change the initial and terminal state to i and then arrange the states $1, \ldots, m$ as $i = \sigma_1, \sigma_2, \ldots$, ordered according to the first word which is the signature of a path in $\mathcal{A}$ from i to σ_j. In FIRST_CONJ, listed below, σ_j is stored as new$[j]$, and old$[$new$[j]]$ is set equal to j. The arrays new and old allow us to go back and forth between $\mathcal{A}$ and $\mathcal{A}_i$. There is an edge (σ, x, τ) in $\mathcal{A}_i$ if and only if (new$[\sigma], x,$ new$[\tau]$) is an edge of $\mathcal{A}$. The value returned by FIRST_CONJ is false if the test shows that $\mathcal{A}$ cannot be extended to a complete automaton describing a subgroup first in its conjugacy. Otherwise, FIRST_CONJ returns true.

Function FIRST_CONJ(T): boolean;
($*$ This procedure assumes that $\prec$ is a length-plus-lexicographic
   ordering. $*$)
Begin
  For $i := 1$ to m do new$[i] := 0$;
  For $i := 2$ to m do begin
    $s := 1$; old$[1] := i$; new$[i] := 1$;
    For $u := 1$ to m do
      For $y := 1$ to r do begin

a := $T[u, y]$; b := $T[\text{old}[u], y]$;
If $a = \text{`?'}$ or $b = \text{`?'}$ then goto 98; (∗ Test at i is
    inconclusive. ∗)
If new$[b] = 0$ then begin
    $s := s + 1$; old$[s] := b$; new$[b] := s$
End;

$c := \text{new}[b]$;

If $c < a$ then
    Begin FIRST_CONJ := $false$; goto 99 end
Else if $c > a$ then goto 98
End;

(∗ Reset new for the next iteration of the loop on i. ∗)
98: For $t := 1$ to s do new[old[t]] := 0
End;

FIRST_CONJ := $true$;
99:End.

One way to solve problem (4) stated at the beginning of this section
is to incorporate FIRST_CONJ into our search criteria. That is, in the
new search tree, only transition matrices on which FIRST_CONJ returns
true are permitted as nodes. If this is done, then with X and $\mathcal{R}$ as in
Example 9.2 and $n = 3$ the following transition matrices are produced:

$$
\begin{bmatrix} 1 & 1 & 1 \end{bmatrix}, \quad
\begin{bmatrix} 1 & 2 & 2 \\ 2 & 1 & 1 \end{bmatrix}, \quad
\begin{bmatrix} 1 & 2 & 2 \\ 3 & 1 & 1 \\ 2 & 3 & 3 \end{bmatrix}, \quad
\begin{bmatrix} 1 & 2 & 3 \\ 2 & 3 & 1 \\ 3 & 1 & 2 \end{bmatrix},
$$

$$
\begin{bmatrix} 1 & 2 & 3 \\ 3 & 3 & 1 \\ 2 & 1 & 2 \end{bmatrix}, \quad
\begin{bmatrix} 2 & 1 & 1 \\ 1 & 2 & 2 \end{bmatrix}, \quad
\begin{bmatrix} 2 & 2 & 2 \\ 1 & 1 & 1 \end{bmatrix}.
$$

Thus F has one conjugacy class of subgroups of index 1, three classes of
subgroups of index 2, and three classes of subgroups of index 3.

Exercises

9.1. Determine the answers to problems (1) to (4) stated at the beginning of this section in
the following cases:

(a) $X = \{x, y\}$, $\mathcal{R} = \{x^2 \to \varepsilon, y^2 \to \varepsilon\}$, $n = 5$.
(b) $X = \{x, y\}^{\pm}$, $\mathcal{R} = \text{FGRel}(\{x, y\})$, $n = 3$.

9.2. Determine the number of subgroups of index n, $1 \leq n \leq 5$, in a free group of rank 2.

9.3. Devise a version of FIRST_CONJ which uses a basic wreath-product ordering. Use the procedure WREATH_STND of Section 3.8 as a guide.

9.4. Using the techniques of Section 1.8, construct a procedure for estimating the size of the search tree associated with the solution to problem (4) stated at the beginning of the section.

*4.10 Other methods

Let H be a subgroup of F given by a finite generating set. Using BASIC_CE, we can construct a coset automaton isomorphic to $\mathcal{A}_I(H)$. However, there are at least two other methods for constructing such an automaton. One method is based on more or less standard techniques in automata theory, and the other is based on Nielsen reduction, an application of the Knuth-Bendix procedure discussed in Section 2.8. The purpose of this section is to sketch these other approaches to determining $\mathcal{A}_I(H)$. By comparing these three methods, one can gain insight into the relationships among coset enumeration, the Knuth-Bendix procedure for strings, and constructions of automata theory. Proofs will be omitted.

Let $\mathcal{U}$ be a finite set of words defining generators for H. The procedure CANON outlined at the end of Section 4.2 produces a trim, deterministic automaton $\mathcal{A}_1 = (\Sigma_1, X, E_1, A_1, \Omega_1)$ recognizing $\mathcal{L}_c(H)$. The expanding morphism f from $\mathcal{A}_1$ to $\mathcal{A}_I(H)$ of Proposition 3.13 actually maps Σ_1 onto $\Sigma_I(H)$ and E_1 onto $E_i(H)$. Thus all we need to do is decide when $f(\sigma) = f(\tau)$ for σ and τ in Σ_1. Let $\alpha^U = \sigma$ and $\alpha^V = \tau$, where $A_1 = \{\alpha\}$. Then $f(\sigma) = f(\tau)$ if and only if $H[U] = H[V]$. By Proposition 2.8, this can be decided. Let us write $\sigma \approx \tau$ if $f(\sigma) = f(\tau)$ and denote the equivalence classes of $\approx$ by $\Lambda_1, \ldots, \Lambda_s$, where α is in Λ_1. Let $\Sigma_2 = \{1, \ldots, s\}$ and define $g : \Sigma_1 \to \Sigma_2$ so that σ is in $\Lambda_{g(\sigma)}$. Set E_2 equal to the set of triples $(g(\sigma), x, g(\tau))$ such that (σ, x, τ) is in E_1. Then $(\Sigma_2, X, E_2, \{1\}, \{1\})$ is isomorphic to $\mathcal{A}_I(H)$.

Example 10.1. Let us apply the preceding construction to the group H of Example 2.2. Here F is the free group on $\{x, y\}$ and $\mathcal{A}_1$ has $\Sigma_1 = \{1, 2, \ldots, 11\}$, $A_1 = \{1\}$, $\Omega_1 = \{1, 9, 10, 11\}$, and E_1 given by Table 4.10.1. Since states 1, 9, 10, and 11 are terminal states, they must all map to H under f. Now $2 = 1^x$, $7 = 1^{x^{-1}y^{-1}}$, and $1^{x^{-1}y^{-1}x^{-1}} = 7^{x^{-1}} = 11$ is terminal. Thus $[x^{-1}y^{-1}x^{-1}]$ is in H, so $H[x] = H[x^{-1}y^{-1}]$. Therefore $f(2) = f(7)$. In this way we find that the $\approx$-classes are

$$\Lambda_1 = \{1, 9, 10, 11\}, \quad \Lambda_2 = \{2, 7\}, \quad \Lambda_3 = \{3, 5, 8\}, \quad \Lambda_4 = \{4, 6\}.$$

Therefore $\mathcal{A}_I(H)$ is isomorphic to $(\{1, 2, 3, 4\}, X, E_2, \{1\}, \{1\})$, where E_2 is given by

	x	x^{-1}	y	y^{-1}
1	2	3		4
2		1	3	
3	1	4		2
4	3		1	

This agrees with the result of Example 5.2.

The problem with this approach to constructing $\mathcal{A}_I(H)$ is that we might have to look at all possible pairs (σ, τ) in order to determine $\approx$. There are ways to speed up the computation of the sets Λ_i, but these techniques tend to make the method look like coset enumeration. The best way to handle partitions is with the union-find procedure, which is central to the coincidence procedure.

Now let us turn to the Knuth-Bendix procedure as a means of constructing $\mathcal{A}_I(H)$. As described in Section 2.8, we choose an object $\#$ not in X, set $Y = X \cup \{\#\}$, and select a reduction ordering $\prec$ on Y^*. We also set $\mathcal{T} = \{(\#U, \#) \mid U \in \mathcal{U}\}$. Given the input Y, $\prec$, and $\mathcal{T} \cup \mathcal{R}$, the Knuth-Bendix procedure for strings terminates. Let $\mathcal{V}$ be the reduced, confluent rewriting system thus obtained. The canonical forms for elements in $\#X^*$ correspond to the right cosets of H in F. Given words V and W in X^*, the words $\#V$ and $\#W$ can be reduced to the same canonical form using $\mathcal{V}$ if and only if $H[V] = H[W]$. Let $\mathcal{W}$ be the set of pairs (B, C) such that $(\#B, \#C)$ is in $\mathcal{V}$ and let $\mathcal{P}$ be the set of proper prefixes of the left sides in $\mathcal{W}$. Then $\mathcal{P}$ gives a set of representatives for the important right cosets of H. To get the edges of $\mathcal{A}_I(H)$, we take each P in $\mathcal{P}$ and each x in X and rewrite $\#Px$ using $\mathcal{V}$. Let $\#Q$ be the result. If Q is in $\mathcal{P}$, then form the edge (P, x, Q). If E is the set of edges thus constructed, then $(\mathcal{P}, X, E, \{\varepsilon\}, \{\varepsilon\})$ is isomorphic to $\mathcal{A}_I(H)$.

Example 10.2. We shall apply the Knuth-Bendix approach to the group of Example 10.1. The set $\mathcal{U}$ is $\{xyx, x^{-2}y\}$. Let us use the length-plus-lexicographic ordering of Y^* in which $\# \prec x \prec x^{-1} \prec y \prec y^{-1}$. The input set of rules for the Knuth-Bendix procedure is

$$xx^{-1} \to \varepsilon, \quad x^{-1}x \to \varepsilon, \quad yy^{-1} \to \varepsilon, \quad y^{-1}y \to \varepsilon,$$

$$\#xyx \to \#, \quad \#x^{-2}y \to \#.$$

The reduced, confluent rewriting system $\mathcal{V}$ returned consists of the rules in $\mathcal{R}$ together with

$$\#xy \to \#x^{-1}, \quad \#x^{-2} \to \#y^{-1}, \quad \#x^{-1}y^{-1} \to \#x, \quad \#y^{-1}x \to \#x^{-1}.$$

Table 4.10.1

	x	x^{-1}	y	y^{-1}
1	2	3		4
2			5	
3		6		7
4	8			
5	9	6		
6			10	
7		11		
8	9			
9	2			4
10	2	3		
11				4

Thus $\mathcal{P}$ consists of $S_1 = \varepsilon$, $S_2 = x$, $S_3 = x^{-1}$, and $S_4 = y^{-1}$. Rewriting $\#S_2 y$ gives $\#S_3$, so we form the edge (S_2, y, S_3). Writing i for S_i, the edges turn out to be exactly the same as those obtained in Example 10.1.

*4.11 General niladic systems

In Proposition 1.2 we classified all confluent, reduced, niladic rewriting systems which define groups. However, since that time we have considered only classical niladic rewriting systems. Most of the main results about subgroups of groups defined by classical niladic rewriting systems carry over to subgroups of groups defined by general confluent, niladic rewriting systems, although the details are somewhat more complicated. There is not sufficient space here to give the proper generalizations of all the results in Sections 4.1 to 4.10 for groups defined by general confluent, niladic rewriting systems. This section presents some of the main ideas. With a little effort, interested readers can work out the rest for themselves.

Let us fix a confluent, reduced, niladic rewriting system $\mathcal{R}$ on X^* defining a group F. Then F is a free product of cyclic groups. By changing the generating set, we could assume that all left sides in $\mathcal{R}$ were of the form x^m or xy with $x \neq y$. However, we shall not make this assumption. As before, $\sim$ will denote the congruence on X^* generated by $\mathcal{R}$, $[U]$ will be the $\sim$-class containing U, $\overline{U}$ will denote the irreducible word in $[U]$, and $\mathcal{C}$ will be the set of all irreducible words.

We begin by reviewing the main results of Section 4.1. For each x in X, there are unique elements y and z of X such that xy and zx are subwords of a left side in $\mathcal{R}$. Let us denote y by $\dot{x}$ and z by $\ddot{x}$. Note that $x = \ddot{y} = \dot{z}$. If U_x is the irreducible word in $[x]^{-1}$, then U_x starts with $\dot{x}$ and ends with $\ddot{x}$. For any word $W = x_1 \ldots x_s$ in X^*, define W^{-1} to be $U_{x_s} \ldots U_{x_1}$ and $W^{\diamond}$ to be $\overline{W^{-1}}$. Normally we shall only consider $W^{\diamond}$ when W is in $\mathcal{C}$. If $\mathcal{R}$ is classical and W is in $\mathcal{C}$, then $W^{\diamond} = W^{-1}$. Let A and B be in $\mathcal{C}$. We shall say that (A, B) is *susceptible to cancellation* if both A and B are nonempty and, for some x in X, the word A ends with x and B starts with $\dot{x}$. If (A, B) is not susceptible to cancellation, then AB is in $\mathcal{C}$.

Every word U in X^* can be written as $U_1 \ldots U_s$, where each U_i is a nonempty subword of a left side in $\mathcal{R}$. If s is as small as possible, then U_1, $\ldots$, U_s is a *minimal decomposition* of U. An important result is that an element U of $\mathcal{C}$ has a unique minimal decomposition U_1, $\ldots$, U_s and $U_s^\Diamond$, $\ldots$, $U_1^\Diamond$ is the minimal decomposition of $U^\Diamond$. The number s will be denoted $\|U\|$. Thus $\|U\| = 1$ if and only if U is a proper, nonempty subword of a left side.

Proposition 1.3 and Corollary 1.4 remain true, but Proposition 1.5 must be modified to say that U and W have a common prefix of length at least $|U| - (k-1)|V|$, where k is the maximum length of a left side in $\mathcal{R}$. Proposition 1.6 and its corollaries continue to hold. Proposition 1.9 must be changed slightly. The words AB^{-1} and $B^{-1}A$ must be replaced by $AB^\Diamond$ and $B^\Diamond A$. Proposition 1.10 requires that both (A, B) and (B, C) not be susceptible to cancellation. The derivation construction is still valid. In proving results about coset automata in the general case, we need a fairly technical lemma describing the circumstances under which one can have words U and V in $\mathcal{C}$ such that U is a prefix of $V^\Diamond$ and V is a suffix of $U^\Diamond$. This lemma is stated as Exercise 11.1.

The definitions of $\mathcal{L}_s(H)$, $\mathcal{A}_s(H)$, and $\mathcal{L}_c(H)$ are unchanged, and the main results of Section 4.2 carry over essentially without modification. The definitions of important coset, $\mathcal{L}_I(H)$, and $\mathcal{A}_I(H)$ remain the same. The results of Section 4.3 are still correct, but the proofs of some, such as Proposition 3.3, are more complicated.

When we come to Section 4.4, some definitions must be changed. Condition (iii) of the definition of a coset automaton relative to $\mathcal{R}$ should now read as follows:

(iii) If (σ, x, τ) is in E, then there is a path P in $\mathcal{A}$ from τ to σ such that $\text{Sg}(P) = U_x$.

The definition of a reduced coset automaton must also be modified if Proposition 4.10 is to continue to be true. Let $\mathcal{A}$ be a coset automaton and let P be a path in $\mathcal{A}$ such that $\text{Sg}(P)$ is a left side in $\mathcal{R}$. Then P begins and ends at the same state and so has the form

$$(\sigma_0, x_1, \sigma_1), (\sigma_1, x_2, \sigma_2), \ldots, (\sigma_{s-1}, x_s, \sigma_0).$$

The automaton $\mathcal{A}$ is *reduced* if for every such path P one of the following holds:

(1) $|\{\sigma_0, \ldots, \sigma_{s-1}\}| < s$.

(2) At least two distinct σ_i have two edges leaving them.
(3) Some σ_i is the initial state of $\mathcal{A}$ and some other σ_j has at least two edges leaving it.

In the revised version of REDUCE, whenever a path P is found in $\mathcal{A}$ such that $\mathrm{Sg}(P)$ is a left side in $\mathcal{R}$ and none of these three conditions is satisfied, then all the edges of P are deleted along with any states that are no longer accessible.

Basic coset enumeration is a little more involved. Let $x_1 \ldots x_s$ be a left side in $\mathcal{R}$ with $x_1 = x$. The call $\mathrm{DEFINE}(\sigma, x)$ must add $s - 1$ states $\tau_1, \ldots, \tau_{s-1}$ and edges $(\tau_{i-1}, x_i, \tau_i)$, $1 \leq i \leq s$, where $\tau_0 = \tau_s = \sigma$. The procedure JOIN becomes much more complicated. First of all, the second argument can be any nonempty, proper subword of a left side in $\mathcal{R}$. The call $\mathrm{JOIN}(\sigma, U, \tau)$ still modifies the current automaton so that $\sigma^U = \tau$, but some interesting things can happen. Suppose that $(ab)^{10}$ is a left side in $\mathcal{R}$, $U = (ab)^4$, and $\sigma = \tau$. Then $\sigma^{(ab)^4}$ and $\sigma^{(ab)^{10}}$ must both equal σ. This implies that $\sigma^{(ab)^2} = \sigma$. Here is the new definition of JOIN:

Procedure $\mathrm{JOIN}(\sigma, U, \tau)$;
Input: σ, τ : states in Σ;
      U : a nonempty, proper subword of a left side in $\mathcal{R}$ such that there is no edge leaving σ with label x and no edge coming into τ with label y, where U starts with x and ends with y;
Begin
  Let UV be a left side in $\mathcal{R}$;
  Let $U = x_1 \ldots x_s$ and $V = y_1 \ldots y_t$;
  If $\sigma = \tau$ and $x_1 = y_1$ then begin
    $s := \gcd(s, t)$; $t := 0$
  End;
  $\rho := \sigma$;
  For $i := 1$ to $s - 1$ do begin
    $n := n + 1$; $p[n] := n$; Add (ρ, x_i, n) to E; $\rho := n$
  End;
  Add (ρ, x_s, τ) to E;
  If $t > 0$ then begin
    $\rho := \tau$;
    For $i := 1$ to $t - 1$ do begin
      $n := n + 1$; $p[n] := n$; Add (ρ, y_i, n) to E; $\rho := n$
    End;
    Add (ρ, y_t, σ) to E
  End
End.

The sketch of the coincidence procedure given in Section 4.5 is still valid. However, the implementation details change significantly.

Procedure COINCIDENCE(σ, τ);
Input: σ, τ : elements of Σ;
Begin
  $k := 0$; MERGE(σ, τ); $i := 1$;
  While $i \leq k$ do begin
    $\nu := \eta_i$;
    For x in X do
      If ν^x is defined then begin
        Let $L = x_1 \ldots x_s$ be the left side in $\mathcal{R}$ with $x_1 = x$;
        Let $L = W^r$ where W is not a proper power;
        $\nu_0 := \nu$; $\mu_0 :=$ ACTIVE(ν_0);
        For $j := 1$ to s do begin
          $\nu_j := \nu_{j-1}^{x_j}$; $\mu_j :=$ ACTIVE(ν_j)
        End;
        Delete all edges (ν_{j-1}, x_j, ν_j), $1 \leq j \leq s$, from E;
        If $\mu_{j-1}^{x_j}$ is defined for some j then begin
          Choose one such j;
          Let $\varphi_0, \ldots, \varphi_s$ be states such that $\varphi_{j-1} = \mu_{j-1}$ and
            $\varphi_{m-1}^{x_m} = \varphi_m$, $1 \leq m \leq s$;
          For $m := 1$ to s do MERGE(μ_m, φ_m)
        End
        Else begin
          Let $\equiv$ be the smallest equivalence relation on $\{\varphi_0, \ldots,$
            $\varphi_{s-1}\}$ such that whenever $\varphi_{a-1} \equiv \varphi_{b-1}$ and $x_a = x_b$ then
            $\varphi_a \equiv \varphi_b$;
          Let t be the smallest multiple of $|W|$ such that $\varphi_0 \equiv \varphi_t$;
          For $m := t$ to s do MERGE(φ_{m-t}, φ_m);
          For $m := 0$ to t do $\psi_m :=$ ACTIVE(φ_m);
          For $m := 1$ to t do add $(\psi_{m-1}, x_m, \psi_m)$ to E
        End
      End
    End
  End
End.

It is assumed that the left sides in $\mathcal{R}$ are not long, so the variable s will have a small value. Thus the equivalence relation $\equiv$ will not be difficult to determine.

The procedure BACK_TRACE remains the same. It is more difficult to decide whether a two-sided trace can be done without defining new states. The "clause" in TWO_SIDED_TRACE beginning

Else if $|T| = 1$ then

is changed to the following:

Else if T is a subword of a left side in $\mathcal{R}$ then begin
  Let TV be a left side in $\mathcal{R}$;
  (∗ Since W is in $\mathcal{C}$, the word V is nonempty. ∗)
  If *full* or
    $(|T| = |V| = 1)$ or
    $((\mu = \nu)$ and $(TV$ has the form $x^m)$ and
    $(\gcd(|T|, |V|) = 1))$
    then JOIN(μ, T, ν)
End

The condition

$(|T| = |V| = 1)$ or
$((\mu = \nu)$ and $(TV$ has the form $x^m)$ and
  $(\gcd(|T|, |V|) = 1))$

describe the situation in which JOIN does not add any new states.

It is still possible to do all the computations with subgroups of F described in Section 4.8.

Exercise

11.1. Suppose that U and V are words in $\mathcal{C}$ such that U is a prefix of $V^\diamond$ and V is a suffix of $U^\diamond$. Show that $\|U\| = \|V\|$. Let $U_1, \ldots, U_s$ and $V_1, \ldots, V_s$ be the minimal decompositions of U and V, respectively. Prove that $V_i = U_j^\diamond$, where $2 \leq i \leq s$ and $j = s + 1 - i$, and that $V_1 U_s$ is a subword of a left side.

5

Coset enumeration

Chapter 4 dealt with finitely generated subgroups of a free product F of cyclic groups. The basic coset enumeration procedure presented in Sections 4.5 and 4.6 allows us to compute the important-coset automaton $\mathcal{A}_I(H)$ for such a subgroup H described by a finite generating set. In general, coset enumeration is a procedure for trying to find $\mathcal{A}_I(H)$, where H is a subgroup of F which is finitely generated modulo a normal subgroup N of F. When N is the normal closure of a finite set and H has finite index in F, we can compute $\mathcal{A}_I(H)$, and as we shall see in Chapter 6, we can find a finite presentation for H/N.

The notation established in Sections 4.1 to 4.10 remains in effect. Thus X is a finite set and $\mathcal{R}$ is a classical niladic rewriting system on X^*. The congruence on X^* generated by $\mathcal{R}$ is $\sim$, and $\mathcal{C}$ is the set of canonical forms for $\sim$. If U is in X^*, then $[U]$ is the $\sim$-class containing U and $\overline{U}$ is the unique element of $[U] \cap \mathcal{C}$. If x is in X, then x^{-1} is the element of X representing the inverse of $[x]$. If $U = x_1 \ldots x_s$, then $U^{-1} = x_s^{-1} \ldots x_1^{-1}$. If $|X| = 1$, then F is cyclic of order 2 and we have no trouble studying its subgroups. Therefore we shall assume that $|X| \geq 2$.

5.1 The general case

Let N be a normal subgroup of F and let K be a subgroup of F/N. The inverse image H of K under the natural homomorphism from F to F/N is a subgroup of F. The map $K \mapsto H$ is a one-to-one correspondence between the subgroups of F/N and the subgroups of F which contain N. The index of H in F is the same as the index of K in F/N. If K is finitely generated, then H may not be finitely generated, but there is a finite subset Y of F such that $H = \mathrm{Grp}\langle Y, N \rangle$. If in addition N is the normal closure of a finite subset Z of F, then the pair (Y, Z) provides a finite description of H. If H has finite index in F, then H is finitely generated by Proposition 3.5 in Chapter 4. If $N \neq 1$ and H has infinite index in F, then H is not finitely generated by Proposition 3.18 in Chapter 4.

In the general case of coset enumeration, we are given two finite subsets $\mathcal{U}$ and $\mathcal{V}$ of X^*. Let H be the subgroup of F generated by $\{[U] \mid U \in \mathcal{U}\}$ and the normal closure N of $\{[V] \mid V \in \mathcal{V}\}$. Thus H is generated by $\{[W] \mid W \in \mathcal{W}\}$, where

$$\mathcal{W} = \mathcal{U} \cup \{SVS^{-1} \mid S \in X^*, V \in \mathcal{V}\}.$$

To simplify the exposition, we shall sometimes assume that $\mathcal{U}$ is a subset of $\mathcal{C}$ and that the elements of $\mathcal{V}$ are cyclically reduced. That is, all cyclic permutations of the elements of $\mathcal{V}$ are in $\mathcal{C}$. Given any finite subsets $\mathcal{U}$ and $\mathcal{V}$, reducing the elements of $\mathcal{U}$ and cyclically reducing the elements of $\mathcal{V}$ does not change H.

We would like to be able to decide whether H has finite index in F, and if so, to determine that index. Unfortunately, there is no algorithmic solution to this problem. If there were an algorithm which could decide whether H has finite index in F, then we could apply that algorithm to the case in which $\mathcal{U} = \emptyset$. Then $H = N$ would be normal in F, and we would be able to decide whether the finitely presented group F/N was finite. As discussed in Section 1.5, this is one of the questions known not to be answerable algorithmically.

Although we cannot *decide* whether H has finite index in F, if the index is finite, then we can *verify* that fact and determine the index. The first observation in this regard is that we can decide whether a subgroup of finite index in F contains N. Let us say that a complete coset automaton $\mathcal{A}$ is *compatible with* $\mathcal{V}$ if $\sigma^V = \sigma$ in $\mathcal{A}$ for all states σ and all V in $\mathcal{V}$.

Proposition 1.1. *Let L be a subgroup of F. Then $N \subseteq L$ if and only if $\sigma[V] = \sigma$ for each right coset σ of L and each word V in $\mathcal{V}$.*

Proof. We shall have $N \subseteq L$ if and only if L contains all conjugates of the elements $[V]$ with V in $\mathcal{V}$. Let V be in $\mathcal{V}$ and let W be in X^*. Then $[W][V][W]^{-1}$ is in L if and only if $L[W][V][W]^{-1} = L$ or, equivalently, $L[W][V] = L[W]$. $\qquad\square$

Proposition 1.1 can be restated as follows: $N \subseteq L$ if and only if $\mathcal{A}_s(L)$ is compatible with $\mathcal{V}$.

Example 1.1. Let $X = \{a, b\}^{\pm}$, let $\mathcal{R}$ define the free group F on $\{a, b\}$, and let $\mathcal{V} = \{(ab)^2\}$. Suppose $\mathcal{A} = (\Sigma, X, E, \{1\}, \{1\})$, where $\Sigma = \{1, 2, 3, 4, 5, 6\}$ and E is given by Table 5.1.1. Then $\mathcal{A}$ is a coset automaton, $L = K(\mathcal{A})$ has index 6 in F, and $\mathcal{A}$ is isomorphic to $\mathcal{A}_s(L) = \mathcal{A}_I(L)$. Let $V = (ab)^2$. Then $1^V = 1$ and $2^V = 2$ in $\mathcal{A}$, but $3^V = 6$. Therefore L does not contain the normal closure of $[V]$ in F.

Table 5.1.1

	a	a^{-1}	b	b^{-1}
1	1	1	2	3
2	3	4	3	1
3	4	2	1	2
4	2	3	5	6
5	5	5	6	4
6	6	6	4	5

Let us return to our subgroup H generated by $\{[W] \mid W \in \mathcal{W}\}$. Assume that H has finite index in F. The automata $\mathcal{A}_I(H)$ and $\mathcal{A}_s(H)$ are the same. Since H is finitely generated, by Proposition 3.2 in Chapter 1 there is a finite subset $\mathcal{W}'$ of $\mathcal{W}$ such that $H = \operatorname{Grp} \langle [W] \mid W \in \mathcal{W}' \rangle$. Suppose that we have chosen a sequence $\mathcal{W}_0, \mathcal{W}_1, \ldots$ of finite subsets of $\mathcal{W}$ such that

(i) $\mathcal{U} \subseteq \mathcal{W}_i \subseteq \mathcal{W}_{i+1}$, $i \geq 0$.
(ii) $\bigcup_{i=0}^{\infty} \mathcal{W}_i = \mathcal{W}$.

Let $L_i = \operatorname{Grp} \langle [W] \mid W \in \mathcal{W}_i \rangle$. Then $\mathcal{W}' \subseteq \mathcal{W}_i$ for some i and $H = L_i$. Suppose that we begin computing $\mathcal{A}_I(L_1), \mathcal{A}_I(L_2), \ldots$ using BASIC_CE. At some point $\mathcal{A}_I(L_i)$ will be complete and compatible with $\mathcal{V}$. When this happens, we shall know that $H = L_i$. Of course, if H does not have finite index, then we shall go on computing coset automata forever.

There are many ways to choose a sequence of subsets $\mathcal{W}_i$. We could take, for example,

$$\mathcal{U} \cup \{SVS^{-1} \mid S \in X^*, V \in \mathcal{V}, |S| \leq i\},$$
$$\mathcal{U} \cup \{SVS^{-1} \mid S \in X^*, V \in \mathcal{V}, |SV| \leq i\},$$

or

$$\mathcal{U} \cup \{SVS^{-1} \mid S \in X^*, V \in \mathcal{V}, |SVS^{-1}| \leq i\}.$$

The following procedure uses the first of these choices.

Procedure GENERAL_CE($X, \mathcal{R}, \mathcal{U}, \mathcal{V}; \mathcal{A}$);
Input: X : a finite set with at least two elements;
           $\mathcal{R}$: a classical niladic rewriting system on X^*;
           $\mathcal{U}, \mathcal{V}$: finite subsets of X^*;
Output: $\mathcal{A}$: a coset automaton isomorphic to $\mathcal{A}_I(H)$, where H is
                       the subgroup of Mon $\langle X \mid \mathcal{R} \rangle$ generated by
                       $\{[U] \mid U \in \mathcal{U}\}$ and the normal closure of
                       $\{[V] \mid V \in \mathcal{V}\}$;

(∗ WARNING – TERMINATION DOES NOT OCCUR IF H DOES
NOT HAVE FINITE INDEX. ∗)

Begin
   $i := 0$; $done :=$ false;
   While not $done$ do begin
     $\mathcal{T} = \mathcal{U} \cup \{SVS^{-1} \mid S \in X^*, V \in \mathcal{V}, |S| \leq i\}$;
     BASIC_CE$(X, \mathcal{R}, \mathcal{T}; \mathcal{A})$
     If $\mathcal{A}$ is complete and $\mathcal{A}$ is compatible with $\mathcal{V}$ then $done :=$ true;
     $i := i + 1$
   End.
End.

The output of GENERAL_CE is determined only up to isomorphism,
since we do not specify the order in which the words in $\mathcal{T}$ are to be processed
by BASIC_CE. In the following examples, we shall assume that BASIC_CE
standardizes its output automaton using STANDARDIZE, with the elements of X taken in the order they are listed. The value of the automaton
$\mathcal{A}$ at the end of the While-loop for a particular value of i will be denoted $\mathcal{A}_i$.

Example 1.2. Let $X = \{x, y, y^{-1}\}$, let $\mathcal{R}$ consist of the rules

$$x^2 \to \varepsilon, \quad yy^{-1} \to \varepsilon, \quad y^{-1}y \to \varepsilon,$$

and let $\mathcal{U} = \{x\}$ and $\mathcal{V} = \{y^3, (xy)^3\}$. The coset table for $\mathcal{A}_0$ is

	x	y	y^{-1}
1	1	2	3
2	4	3	1
3	5	1	2
4	2	5	
5	3		4

The coset table of $\mathcal{A}_1$ is

	x	y	y^{-1}
1	1	2	3
2	4	3	1
3	5	1	2
4	2	5	6
5	3	7	4
6	7	4	
7	6		5

For $\mathcal{A}_2$ we get

Table 5.1.2

	x	x^{-1}	y	y^{-1}
1	2	3	4	2
2	3	1	1	4
3	1	2		
4	5	6	2	1
5		4		6
6	4		5	

Table 5.1.3

	x	x^{-1}	y	y^{-1}
1	2	3	4	2
2	3	1	1	4
3	1	2	5	6
4	7	8	2	1
5	9	6	6	3
6	5	10	3	5
7	8	4	10	8
8	4	7	7	9
9		5	8	
10	6			7

Table 5.1.4

	x	x^{-1}	y	y^{-1}
1	2	3	4	2
2	3	1	1	4
3	1	2	5	6
4	7	8	2	1
5	9	6	6	3
6	5	9	3	5
7	8	4	9	8
8	4	7	7	9
9	6	5	8	7

	x	y	y^{-1}
1	1	2	3
2	4	3	1
3	5	1	2
4	2	5	6
5	3	6	4
6	6	4	5

The procedure terminates with $\mathcal{A} = \mathcal{A}_2$.

Let us interpret what we have done as a result about the group $G = \mathrm{Grp} \langle x, y \mid x^2 = y^3 = (xy)^3 = 1 \rangle$. If F is the free product defined by $\mathcal{R}$ and N is the normal closure of (the image of) $\mathcal{V}$ in F, then G is isomorphic to F/N. Let K be the subgroup of G generated by x. Then $|G : K| = |F : H| = 6$. Because of the relation $x^2 = 1$, the order of K is at most 2. The columns of the final coset table satisfy the defining relations of G. Therefore there is a homomorphism of G onto the group M generated by these columns. The column headed x is the permutation $(1)(2,4)(3,5)(6)$, which has order 2. Thus the order of K must be 2. Hence the order of G is $6 \cdot 2 = 12$.

Example 1.3. Let $X = \{x, y\}^{\pm}$, let $\mathcal{R}$ define the free group on $\{x, y\}$, let $\mathcal{U} = \{xy\}$, and let $\mathcal{V} = \{x^3, y^3, (xy)^3, (xy^{-1})^3\}$. Tables 5.1.2 to 5.1.4 are the coset tables for $\mathcal{A}_0$, $\mathcal{A}_1$, and $\mathcal{A}_2$, respectively. The automaton $\mathcal{A}_2$ is compatible with $\mathcal{V}$, so termination occurs with $\mathcal{A} = \mathcal{A}_2$.

Let $G = \mathrm{Grp} \langle x, y \mid x^3 = y^3 = (xy)^3 = (xy^{-1})^3 = 1 \rangle$ and let K be the subgroup of G generated by xy. Our enumeration shows that $|G : K| = 9$, and the relation $(xy)^3 = 1$ says that K has order at most 3. The product of the permutations given by the columns headed x and y in the final coset table is $(1)(2,5,8)(3,4,9)(6)(7)$, which has order 3. Therefore $|K| = 3$ and $|G| = 9 \cdot 3 = 27$.

Table 5.1.5

	x	x^{-1}	y	y^{-1}
1	1	1	2	3
2	2	2	4	1
3	3	3	1	5
4	4	4	6	2
5	5	5	2	7
⋮	⋮	⋮	⋮	⋮
$2i-2$	$2i-2$	$2i-2$	$2i$	$2i-3$
$2i-1$	$2i-1$	$2i-1$	$2i-3$	$2i+1$
$2i$	$2i$	$2i$		$2i-2$
$2i+1$	$2i+1$	$2i+1$	$2i-2$	

Example 1.4. Let X and $\mathcal{R}$ be as in Example 1.3, and let $\mathcal{U} = \emptyset$ and $\mathcal{V} = \{x\}$. Then in GENERAL_CE the coset tables for $\mathcal{A}_0$, $\mathcal{A}_1$, and $\mathcal{A}_2$ are, respectively,

	x	x^{-1}	y	y^{-1}
1	1	1		

	x	x^{-1}	y	y^{-1}
1	1	1	2	3
2	2	2		1
3	3	3	1	

	x	x^{-1}	y	y^{-1}
1	1	1	2	3
2	2	2	4	1
3	3	3	1	5
4	4	4		2
5	5	5	3	

In general, the coset table for $\mathcal{A}_i$, $i \geq 1$, looks like Table 5.1.5. Thus GENERAL_CE loops forever.

The procedure GENERAL_CE has a number of weaknesses. Two of the main ones are the following:

• The size of the set $\mathcal{T}$ grows exponentially with i.
• During a given iteration of the Repeat-loop, all work done on previous iterations is ignored.

Let us see how we can address these problems.

Suppose that $\mathcal{A} = (\Sigma, X, E, \{\alpha\}, \{\alpha\})$ is a coset automaton. A sufficient condition for $K(\mathcal{A})$ to contain $[SVS^{-1}]$ is that $\alpha^S = \sigma$ for some state σ and that $\sigma^V = \sigma$. (This condition is not necessary since SVS^{-1} may not be in $\mathcal{C}$.) Thus for $K(\mathcal{A})$ to contain all elements $[SVS^{-1}]$ with V in $\mathcal{V}$ and $|S| \leq i$, the following conditions are sufficient:

(i) α^S is defined in $\mathcal{A}$ for all S in X^* such that $|S| \leq i$.
(ii) $\sigma^V = \sigma$ for every state σ which can be reached from α by a path of length at most i.

Let us define a class of procedures which we shall call *coset enumeration procedures*. The arguments are to be the same as in GENERAL_CE, except that we shall assume that the words in $\mathcal{U}$ are reduced and the words in $\mathcal{V}$ are cyclically reduced. The same global variables as in BASIC_CE are maintained and the procedures DEFINE, JOIN, COINCIDENCE, TRACE, BACK_TRACE, and TWO_SIDED_TRACE are used. The first four statements of any coset enumeration procedure are essentially the same as those of BASIC_CE:

$n := 1;\ p[1] := 1;\ E := \emptyset;$
For U in $\mathcal{U}$ do TWO_SIDED_TRACE$(1, U, \text{true})$

At this point, the procedure carries out a sequence of operations of the following types:

(1) Choose σ in Σ and x in X such that σ^x is not defined, and execute DEFINE(σ, x).
(2) Choose σ in Σ and V in $\mathcal{V}$ and call TWO_SIDED_TRACE$(\sigma, V, full)$, where *full* is true or false. Thus either full or partial traces can be performed.
(3) If $\mathcal{A}$ is complete and compatible with $\mathcal{V}$, then stop.

The sequence of operations is subject to two conditions: First, if termination is possible, then it must occur. Second, for every integer σ which is ever in Σ, one of the following holds:

(a) At some time $p[\sigma]$ becomes less than σ, so σ is deleted from Σ.
(b) Eventually σ^x is defined for all x in X, and $\sigma^V = \sigma$ for all V in $\mathcal{V}$.

Let X, $\mathcal{R}$, $\mathcal{U}$, and $\mathcal{V}$ be as in GENERAL_CE, let H be the subgroup of F generated by the image of $\mathcal{U}$ and the normal closure N of the image of $\mathcal{V}$, and let CE be a coset enumeration procedure.

Proposition 1.2. *If the call* CE$(X, \mathcal{R}, \mathcal{U}, \mathcal{V}; \mathcal{A})$ *terminates, then* $H = K(\mathcal{A})$.

Proof. After the first four statements, $K(\mathcal{A}) = \text{Grp}\langle [U] \mid U \in \mathcal{U}\rangle$. Calls to DEFINE do not change $K(\mathcal{A})$ and by Proposition 4.5.4, calls of the form TWO_SIDED_TRACE$(\sigma, V, full)$ either leave $\mathcal{A}$ unchanged or add $[SVS^{-1}]$ to $K(\mathcal{A})$, where $1^S = \sigma$. Thus at all times $K(\mathcal{A}) \subseteq H$. If CE terminates, then $\mathcal{A}$ is compatible with $\mathcal{V}$, so $K(\mathcal{A}) \supseteq N$ by Proposition 1. Hence $K(\mathcal{A}) = H$. $\square$

Proposition 1.3. *If H has finite index in F, then* CE$(X, \mathcal{R}, \mathcal{U}, \mathcal{V}; \mathcal{A})$ *terminates.*

Proof. Suppose that the procedure does not terminate.

Lemma 1.4. *For every word S in X^*, eventually 1^S is defined in $\mathcal{A}$.*

Proof. We proceed by induction on $|S|$. Note that 1 is always in Σ. Thus 1^ε is always defined and is equal to 1. Suppose that $S = Tx$ with x in X. By induction, 1^T is eventually defined. Once it is defined, it remains defined, but the value may change due to the action of COINCIDENCE. However, if 1^T changes, then the new value is positive and less than the old value. Therefore eventually $\sigma = 1^T$ remains constant. Therefore $p[\sigma]$ remains equal to σ, so by (b) above, $\sigma^x = (1^T)^x = 1^S$ eventually is defined. $\square$

Lemma 1.5. *For every S in X^* and every V in $\mathcal{V}$, eventually $K(\mathcal{A})$ contains $[SVS^{-1}]$.*

Proof. By Lemma 1.4 and its proof, $1^S = \sigma$ eventually is defined and remains constant. By (b), eventually $\sigma^V = \sigma$, so $K(\mathcal{A})$ contains $[SVS^{-1}]$. $\square$

Since the index of H in F is finite, H is generated by the image of the elements of $\mathcal{U}$ and a finite number of conjugates of the images of elements of $\mathcal{V}$. Thus eventually $K(\mathcal{A}) = H$. But then $\mathcal{A}$ is complete and compatible with $\mathcal{V}$, so termination is forced. $\square$

There is a great deal of freedom in designing a coset enumeration procedure. In the next few sections we shall discuss several strategies which have been proposed. However, we shall not have much to say about doing coset enumeration by hand, although for nearly 20 years, before the first computer implementation, long enumerations were carried out by hand, sometimes with subgroups of index several hundred. The reader can find extensive discussions of coset enumeration by hand in [Coxeter & Moser 1980], (Leech 1984), and (Neubüser 1982). No computer implementation of coset enumeration quite captures these methods.

We close this section with a slight digression. In most of our discussions, we have not considered the possibility that memory may be limited. In the real world this is very often the case. When a coset enumeration program is run, there is a fixed amount of memory available. When that memory is filled, the program must stop without producing a result. This raises the question of how long a coset enumeration can run as a function of the memory available to it. We do not have a good answer to this question. However, there is some evidence that the worst-case running time may not be bounded by a polynomial function of memory size.

The evidence referred to comes from looking at a rather artificial version of the problem. We consider sequences of coset automata $\mathcal{A}_0, \mathcal{A}_1, \ldots, \mathcal{A}_s$

such that $\mathcal{A}_0 = (\{1\}, X, \emptyset, \{1\}, \{1\})$ and for $1 \leq i < s$ the automaton $\mathcal{A}_{i+1}$ is obtained from $\mathcal{A}_i$ by an operation of one of the following two types:

(i) Execution of DEFINE(σ, x), where σ is a state in $\mathcal{A}_i$, x is in X, and σ^x is not defined in $\mathcal{A}_i$.
(ii) Execution of COINCIDENCE(σ, τ), where σ and τ are distinct states of $\mathcal{A}_i$.

Assume that X and $\mathcal{R}$ are fixed. Given a positive integer M, what is the longest such sequence in which each $\mathcal{A}_i$ has at most M states? We shall sketch a proof that the answer is not bounded above by a polynomial function of M as long as F is not abelian.

Let $X = \{x, y\}^{\pm}$ and let $\mathcal{R}$ define the free group on $\{x, y\}$. For $r \geq 1$, let $\mathcal{B}_r$ be the coset automaton $(\Sigma_r, X, E_r, \{1\}, \{1\})$, where $\Sigma_r = \{1, 2, \ldots, 2r\}$ and E_r consists of the edges $(2i-1, y, 2i), (2i-1, y^{-1}, 2i), (2i, y, 2i-1)$, and $(2i, y^{-1}, 2i-1)$, $1 \leq i \leq r$, and the edges $(2i, x, 2i+1)$, and $(2i+1, x^{-1}, 2i)$, $1 \leq i < r$. The coset tables for $\mathcal{B}_1, \mathcal{B}_2$, and $\mathcal{B}_3$ are, respectively,

	x	x^{-1}	y	y^{-1}
1			2	2
2			1	1

	x	x^{-1}	y	y^{-1}
1			2	2
2	3		1	1
3		2	4	4
4			3	3

	x	x^{-1}	y	y^{-1}
1			2	2
2	3		1	1
3		2	4	4
4	5		3	3
5		4	6	6
6			5	5

Suppose that r and M are positive integers. Let $h(r, M) = s$, where $\mathcal{A}_0, \mathcal{A}_1, \ldots, \mathcal{A}_s$ is a sequence of coset automata of maximal length such that $\mathcal{A}_0$ is trivial, $\mathcal{A}_s$ is isomorphic to $\mathcal{B}_r$, and $\mathcal{A}_{i+1}$ is obtained from $\mathcal{A}_i$ by an operation of type (i) or (ii), $1 \leq i < s$, and none of the automata has more than M states. If no such sequence exists, set $h(r, M) = 0$. If $M \geq 3$, then the following statements construct $\mathcal{B}_1$ using only operations of types (i) and (ii) in such a manner that no more than M states are ever active at one time.

$n := 1; \; p[1] := 1; \; E := \emptyset;$
For $i := 1$ to $M - 1$ do DEFINE(i, y);
COINCIDENCE($1, 3$)

This construction uses $M-1$ operations of type (i) and one operation of type (ii). Thus $h(1, M) \geq M$ if $M \geq 3$. It is easy to show that $h(1, 1) = h(1, 2) = 0$ and $h(1, 3) = 3$.

We shall now present a procedure BUILD which constructs $\mathcal{B}_r$ using a long sequence of operations of types (i) and (ii) such that there are never more than M active states. The basic idea of this construction arose out of conversations between the author and C. M. Hoffmann. The bulk of the work is done by the subroutine ATTACH, which "attaches" a copy of $\mathcal{B}_r$ to the current automaton. The standard global variables n and p are used.

Procedure ATTACH(r, M, ρ);
Input: r, M : positive integers;
      ρ : an active state such that ρ^y and $\rho^{y^{-1}}$ are not defined;
($*$ A copy of $\mathcal{B}_r$ is attached to the current automaton at ρ. No existing
     state is removed and no more than $M - 1$ new states are ever
     active at one time. $*$)
Begin
   $m := 1;\ \sigma := \rho$;
   If $r = 1$ then $k := 1$
   Else $k := 2r$;
   While $M - m \geq k$ do begin
     DEFINE(σ, y);
     If $r > 1$ then begin
       $\tau := n;$ DEFINE$(\tau, x);\ \varphi := n;$ ATTACH$(r - 1, M - m - 1, \varphi)$;
       DEFINE$(\tau, y);\ m := m + 2r$
     End
     Else $m := m + 1$;
     $\sigma := n$
   End;
   $\sigma := \rho^{y^2};$ COINCIDENCE(ρ, σ)
End.

The procedure BUILD simply initializes the automaton and calls AT-
TACH.

Procedure BUILD(r, M);
Input: r, M : positive integers;
($*$ Builds a copy of $\mathcal{B}_r$ in such a way that no more than M states are
     active at one time. $*$)
Begin
   $n := 1;\ p[1] := 1;\ E := \emptyset$;
   ATTACH$(r, M, 1)$
End.

Let N_d and N_c denote the number of calls to DEFINE and COINCI-
DENCE, respectively, during one execution of BUILD. Table 5.1.6 shows
some sample results. It is left to the reader to experiment further with
BUILD, to analyze its performance, and to prove the following proposition.

Proposition 1.6. *For each positive integer r there is a positive constant c_r
such that $h(r, M) \geq c_r M^r$ if $M \geq 2r + 1$.*

Table 5.1.6

r	M	N_d	N_c
2	10	16	3
2	20	56	5
2	40	216	10
3	50	1028	57
3	100	7576	217

Exercises

1.1. Let $\mathcal{R}$ be a classical niladic rewriting system on X^* and suppose that t is an object not in X. Set $Y = X \cup \{t\}$ and $\mathcal{S} = \mathcal{R} \cup \{t^2 \to \varepsilon\}$ and let $\mathcal{T}$ be the set of words of the form $txtytx^{-1}ty^{-1}$ with x and y in X. Suppose CE is a coset enumeration procedure and $\mathcal{V}$ is a finite subset of X^*. Let $\mathcal{W}$ consist of the elements of $\mathcal{V}$ together with t and all words of the form xtx with x in X. Prove that the call $\text{CE}(X, \mathcal{R}, \emptyset, \mathcal{V}; \mathcal{A})$ terminates if and only if the call $\text{CE}(Y, \mathcal{S}, \mathcal{W}, \mathcal{T}; \mathcal{B})$ terminates. Given $\mathcal{B}$, show how to get $\mathcal{A}$. Conclude that for a specific pair $(X, \mathcal{R})$ we may build a machine for enumerating cosets of the trivial subgroup in which the relations are "hardwired". (Hint: The group $\text{Mon} \langle Y \mid \mathcal{S} \rangle$ acts faithfully as a group of permutations of the elements of $F = \text{Mon} \langle X \mid \mathcal{R} \rangle$, where an element x of X acts as right multiplication by $[x]$ and t acts as the inverse map. See also Exercise 3.9 in Chapter 1.)

1.2. Suppose that an enumeration of the right cosets of a subgroup H in a finitely presented group G produces a coset automaton $(\Sigma, X, E, \{\alpha\}, \{\alpha\})$. Let σ be a state in Σ and assume that $\sigma^U = \sigma$ for all words U defining generators of H. Suppose that W is a word such that $\alpha^W = \sigma$. Show that W defines an element g of G such that $g^{-1} H g = H$.

1.3. Coset enumeration may sometimes be effective in studying groups which are not finitely presented. The most common examples are groups satisfying a finite exponent condition. These are groups in which the n-th power of every element is known to be trivial for some positive integer n. Let G be a group defined by a finite number of relators and the condition that G has exponent dividing n. Let H be a subgroup of G generated by the images of a finite set $\mathcal{U}$ of words. Our desire is to prove that $|G : H|$ is finite. We take a finite set of words containing the ordinary relators of G and some number of n-th powers. We then invoke $\text{CE}(X, \mathcal{R}, \mathcal{U}, \mathcal{V}; \mathcal{A})$ for some coset enumeration procedure CE. Suppose during the enumeration we find a state $\sigma \neq 1$ such that $\sigma^U = \sigma$ for all U in $\mathcal{U}$. Prove that we may force $\sigma = 1$ by a call to the coincidence procedure without affecting the eventual termination of the enumeration.

5.2 The HLT strategy

The HLT strategy for coset enumeration described in this section grew out of work by Haselgrove, Leech, and Trotter. It is discussed at length in (Cannon et al. 1973). The idea of the HLT strategy is easy to state. It consists of executing the following statements:

$\sigma := 1$;

While $\sigma \leq n$ do begin

    For each V in $\mathcal{V}$ perform a full two-sided trace of V at σ;

    For each x in X do if σ^x is not defined then $\text{DEFINE}(\sigma, x)$;

$$\sigma := \sigma + 1$$
End;

Recall that n is the largest integer used so far as a state. This simple description does not allow for the possibility that states may be removed by COINCIDENCE. The full procedure must consider this eventuality. A positive integer σ not exceeding n is currently a state if and only if $p[\sigma] = \sigma$.

Procedure HLT$(X, \mathcal{R}, \mathcal{U}, \mathcal{V}; \mathcal{A})$;
(* A coset enumeration procedure. *)
Begin
    $n := 1$; $p[1] := 1$; $E := \emptyset$;
    For U in $\mathcal{U}$ do TWO_SIDED_TRACE$(1, U, \text{true})$;
    $\sigma := 1$;
    While $\sigma \leq n$ do begin
      If $p[\sigma] = \sigma$ then begin
        For V in $\mathcal{V}$ do begin
          TWO_SIDED_TRACE(σ, V, true);
          If $p[\sigma] \neq \sigma$ then break
        End;
        If $p[\sigma] = \sigma$ then for x in X do if σ^x is not defined then
            DEFINE(σ, x)
      End;
      $\sigma := \sigma + 1$
    End;
    STANDARDIZE
End.

The standardization step is not logically necessary, but it facilitates the comparison of the output from different coset enumeration procedures with the same input data.

The procedure HLT clearly satisfies the definition of a coset enumeration procedure. Thus Propositions 1.3 and 1.4 apply to HLT. The operation of HLT is dependent on the order in which U, V, and x are taken in the three For-loops. In the following examples, the order given in the descriptions of $\mathcal{U}$, $\mathcal{V}$, and X will be used.

Example 2.1. Let $X = \{a, b\}^{\pm}$ and let $\mathcal{R}$ define the free group F on $\{a, b\}$. Suppose that $\mathcal{U} = \{U\}$ and $\mathcal{V} = \{V_1, V_2, V_3\}$, where $U = b$, $V_1 = a^3$, $V_2 = b^4$, and $V_3 = (ab)^2$. After the trace of U at 1, we have the following coset table:

	a	a^{-1}	b	b^{-1}
1			1	1

Next we trace V_1 at 1, which yields

	a	a^{-1}	b	b^{-1}
1	2	3	1	1
2	3	1		
3	1	2		

Tracing V_2 at 1 produces no changes. After V_3 is traced at 1, we have

	a	a^{-1}	b	b^{-1}
1	2	3	1	1
2	3	1	3	
3	1	2		2

Since 1^x is defined for all x in X, we begin tracing the elements of $\mathcal{V}$ at 2. Tracing V_1 does not give anything new, but the trace of V_2 produces

	a	a^{-1}	b	b^{-1}
1	2	3	1	1
2	3	1	3	5
3	1	2	4	2
4			5	3
5			2	4

After V_3 is traced, we have

	a	a^{-1}	b	b^{-1}
1	2	3	1	1
2	3	1	3	5
3	1	2	4	2
4	5		5	3
5		4	2	4

No more changes occur until we begin to trace the elements of $\mathcal{V}$ at 4. When V_1 is traced at 4, the table becomes

	a	a^{-1}	b	b^{-1}
1	2	3	1	1
2	3	1	3	5
3	1	2	4	2
4	5	6	5	3
5	6	4	2	4
6	4	5		

The next change occurs when V_3 is traced at 5. This gives

	a	a^{-1}	b	b^{-1}
1	2	3	1	1
2	3	1	3	5
3	1	2	4	2
4	5	6	5	3
5	6	4	2	4
6	4	5	6	6

The table is now complete and compatible with $\mathcal{V}$, so no more changes are made and $|F : H|$ is 6. The final standardization produces the following table:

	a	a^{-1}	b	b^{-1}
1	2	3	1	1
2	3	1	3	4
3	1	2	5	2
4	6	5	2	5
5	4	6	4	3
6	5	4	6	6

If $G = \mathrm{Grp}\,\langle a, b \mid a^3 = b^4 = (ab)^2 = 1\rangle$, then the subgroup K of G generated by b has index 6 in G. Clearly $|K|$ is at most 4. Since the column headed b in the final coset table defines the permutation $(1)(2, 3, 5, 4)(6)$, which has order 4, the order of G is 24.

Example 2.2. Let $X = \{x, y, y^{-1}\}$ and let $\mathcal{R}$ consist of the rules

$$x^2 \to \varepsilon, \quad yy^{-1} \to \varepsilon, \quad y^{-1}y \to \varepsilon.$$

Suppose that $\mathcal{U} = \emptyset$ and $\mathcal{V} = \{V_1, V_2, V_3\}$, where $V_1 = y^3$, $V_2 = xyxy^{-1}$, and $V_3 = (xy)^7$. Tracing V_1 at 1, we have

	x	y	y^{-1}
1		2	3
2		3	1
3		1	2

After the trace of V_2 at 1, the table is

	x	y	y^{-1}
1	4	2	3
2	5	3	1
3		1	2
4	1	5	
5	2		4

Now tracing V_3 at 1 gives

	x	y	y^{-1}			x	y	y^{-1}
1	4	2	3		7	8		6
2	5	3	1		8	7	9	
3	6	1	2		9	10		8
4	1	5			10	9	11	
5	2		4		11	12		10
6	3	7	12		12	11	6	

Tracing V_1 at 2 produces no change. Tracing V_2 at 2 causes states 5 and 12 to be identified. After the call to COINCIDENCE, states 9, 10, 11, and 12 are no longer in Σ and the coset table is

	x	y	y^{-1}			x	y	y^{-1}
1	4	2	3		5	2	6	4
2	5	3	1		6	3	7	5
3	6	1	2		7	8		6
4	1	5	8		8	7	4	

Tracing V_3 at 2 and V_1 at 3 produces no change. Tracing V_2 at 3 causes states 8 and 1 to be identified. At this point the table collapses to

	x	y	y^{-1}
1	1	1	1

Therefore the group $\mathrm{Grp}\,\langle x, y \mid x^2 = y^3 = xyxy^{-1} = (xy)^7 = 1 \rangle$ is trivial.

It has been found that the HLT strategy as described does not perform as well as one would like when memory is restricted. We shall now consider an extension to the HLT strategy which makes it work significantly better with limited memory.

One memory issue which arises in connection with all coset enumeration procedures is the possibility of reusing integers for labeling elements of Σ. In the procedure DEFINE of Section 4.5, whenever a new state is needed, the variable n is incremented and the new value of n is used as the new state. This is done even when there are positive integers less than n which have been deleted from Σ and could be used. With some data structures for storing E, there is considerable space savings to be realized if these inactive state labels are reused.

Even when this is done, there can still be problems with the HLT approach. To address these problems, a "lookahead phase" may be added to the HLT strategy. Suppose at some point during the execution of the While-loop in HLT there is no more space available to process a full trace at

σ or to execute a call DEFINE(σ, x). In its simplest version, the lookahead phase consists of carrying out the following loop:

$\tau := \sigma;$
While $\tau \leq n$ do begin
  If $p[\tau] = \tau$ then
    For V in $\mathcal{V}$ do begin
      TWO_SIDED_TRACE$(\tau, V, \text{false});$ (∗ A partial trace. ∗)
      If $p[\tau] \neq \tau$ then break
    End
  $\tau := \tau + 1$
End;

Here partial two-sided traces are performed in the hope that COINCI-DENCE will be called and will free up space for new states. A number of variations of lookahead have been suggested. Several may be found in (Cannon et al. 1973) and (Leon 1980b).

Even though enough space is released by a lookahead pass to resume the primary or "defining" phase of the HLT strategy, it may be necessary to invoke lookahead again if memory runs low. Repeated lookahead passes can become time-consuming and result in only small amounts of space being made available. Typically, the enumeration is abandoned when the space released falls below some preset threshold.

Exercises

2.1. Let X, $\mathcal{R}$, and $\mathcal{V}$ be as in Example 2.1 and set $\mathcal{U} = \{a\}$. Execute HLT$(X, \mathcal{R}, \mathcal{U}, \mathcal{V}; \mathcal{A})$.
2.2. Redo the enumeration in Example 2.2 with $V_2 = (xyxy^{-1})^2$.
2.3. Find the index of Grp $\langle ab \rangle$ in Grp $\langle a, b \mid a^2 = b^3 = (ab)^5 = 1 \rangle$.

5.3 The Felsch strategy

This section describes a strategy for coset enumeration first suggested in (Felsch 1961). It differs substantially from the HLT strategy. In outline, the Felsch strategy is as follows:

(1) Use only partial traces with elements of $\mathcal{V}$.
(2) Do not invoke DEFINE while it is possible to make a change in the current automaton by a partial two-sided trace of an element of $\mathcal{V}$.
(3) When a call DEFINE(σ, x) is made, choose σ as small as possible, and for that σ choose x as early as possible in some fixed ordering of X.

The goal of the Felsch strategy is to limit the size of Σ by obtaining as much information as possible before defining new states. Although the strategy is often successful, there are cases where other coset enumeration procedures perform better.

The main difficulty in implementing the Felsch strategy is finding an efficient way to fulfill the second statement in the outline. The obvious approach would be to run the following loop repeatedly until no change in the automaton occurs:

$\sigma := 1$;
While $\sigma \leq n$ do begin
   If $p[\sigma] = \sigma$ then
      For V in $\mathcal{V}$ do begin
         TWO_SIDED_TRACE$(\sigma, V, \text{false})$;
         If $p[\sigma] \neq \sigma$ then break
      End;
   $\sigma := \sigma + 1$
End;

However, this would be very time-consuming.

Suppose that we have reached a point where the preceding loop produces no changes in the current automaton $\mathcal{A}$. If we now make a call DEFINE(σ, x), then which calls TWO_SIDED_TRACE(ρ, V, false) could cause a change in $\mathcal{A}$? Clearly the trace must involve one or both of the new edges (σ, x, τ) and (τ, x^{-1}, σ) created by DEFINE. Thus either V has the form AxB, where $\rho^A = \sigma$, or V has the form $Cx^{-1}D$, where $\sigma^D = \rho$. If V has the first form, then the call TWO_SIDED_TRACE$(\sigma, xBA, \text{false})$ will produce the same result as the call TWO_SIDED_TRACE(ρ, V, false). If V has the second form, then calling TWO_SIDED_TRACE$(\tau, x^{-1}DC, \text{false})$ accomplishes the same thing as calling TWO_SIDED_TRACE(ρ, V, false).

Let $\mathcal{W}$ denote the set of cyclic permutations of the elements of $\mathcal{V}$. The previous discussion can be summarized as follows: After a call DEFINE(σ, x) we only need to carry out traces of the form TWO_SIDED_TRACE$(\sigma, W, \text{false})$ or TWO_SIDED_TRACE(τ, W, false), where W is in $\mathcal{W}$ and in the first case W starts with x and in the second case W starts with x^{-1} and $\tau = \sigma^x$.

It is quite possible that one of these calls to TWO_SIDED_TRACE will cause a change in $\mathcal{A}$, which will make it necessary to perform other traces. Changes in $\mathcal{A}$ resulting from a call to TWO_SIDED_TRACE occur either in JOIN or COINCIDENCE. Suppose the call JOIN(σ, x, τ) now makes it possible for TWO_SIDED_TRACE(ρ, V, false) to change $\mathcal{A}$. Then the trace of V at ρ must involve an edge introduced by JOIN. If COINCIDENCE deletes states and edges from $\mathcal{A}$, then we shall need to try all traces TWO_SIDED_TRACE$(\varphi, W, \text{false})$, where W is in $\mathcal{W}$, W starts with y, and φ^y was defined or changed by COINCIDENCE. Recall that a deduction is a pair (ρ, x) for which ρ^x has recently been defined. When COINCIDENCE changes the value of ρ^x, the original edge (ρ, x, φ) is removed, so ρ^x is temporarily undefined. Then a new edge (ρ, x, ψ) is added and so (ρ, x) becomes a deduction. To satisfy the second point in the outline of the Felsch strategy it suffices to save deductions and to trace cyclic permutations of the elements of $\mathcal{V}$ whenever the first edge of the forward trace corresponds to a deduction.

Most implementations of the Felsch strategy maintain a list of deductions. When a pair (σ, x) is placed on the list, then the program performs all calls TWO_SIDED_TRACE$(\sigma, W, \text{false})$, where W is in $\mathcal{W}$ and W starts with x, unless at some point before or during the processing of this deduction σ is removed from Σ. Pairs get put on the list in DEFINE, JOIN, and COINCIDENCE. Instructions to accomplish this have already been included in the listings of these procedures presented in Sections 4.5 and 4.6. In these procedures, the list is maintained as a stack, referred to as the deduction stack. (A queue or other structure would also work.) When the global variable *save* is set to true, the necessary pairs are pushed onto the deduction stack.

The following procedure removes pairs from the deduction stack and performs the appropriate traces. For each x in X the set $\mathcal{W}_x$ is the set of words in $\mathcal{W}$ which start with x.

Procedure PROCESS_DEDUCTIONS;
Begin
   While the deduction stack is not empty do begin
    Pop (σ, x) off the deduction stack;

    If $p[\sigma] = \sigma$ then
     For W in $\mathcal{W}_x$ do begin
      TWO_SIDED_TRACE$(\sigma, W, \text{false})$;
      If $p[\sigma] \neq \sigma$ then break
    End
   End
End.

It is possible for one pair (σ, x) to be on the deduction stack more than once. When (σ, x) is popped off the stack, it would be valid to delete immediately all other occurrences of (σ, x) on the stack. However, the stack data structure does not make such an operation easy. In practice, processing redundant deductions does not seem to cause significant problems. To save space, whenever (σ, x) is pushed on the stack, the pair (σ^x, x^{-1}) is usually assumed to be pushed as well, but not actually stored. The procedure PROCESS_DEDUCTIONS must be changed accordingly.

The complete Felsch strategy is described by the following procedure:

Procedure FELSCH$(X, \mathcal{R}, \mathcal{U}, \mathcal{V}; \mathcal{A})$;
(∗ A coset enumeration procedure. ∗)
Begin
   Let $\mathcal{W}$ be the set of cyclic permutations of the elements of $\mathcal{V}$;
   For x in X do let $\mathcal{W}_x$ be the set of elements of $\mathcal{W}$ which start with x;
   $n := 1$; $p[1] := 1$; $E := \emptyset$; $save :=$ true; Clear the deduction stack;

   For U in $\mathcal{U}$ do begin
      TWO_SIDED_TRACE$(1, U,$ true$)$; PROCESS_DEDUCTIONS
   End;

   $\sigma := 1$;

   While $\sigma \leq n$ do begin
     If $p[\sigma] = \sigma$ then
       For x in X do
         If σ^x is not defined then begin
           DEFINE(σ, x); PROCESS_DEDUCTIONS;
           If $p[\sigma] \neq \sigma$ then break

     End;

     $\sigma := \sigma + 1$
   End;

   STANDARDIZE
End.

To completely specify the operation of FELSCH, the order of the elements in each of the sets $\mathcal{W}_x$ must be given. Unless there is a statement to the contrary, the order used in the examples is obtained as follows: The elements of $\mathcal{V}$ are taken in the order given. For each V in $\mathcal{V}$ the cyclic permutations of V are formed by cyclicly permuting V to the left one step at a time. This gives an order for $\mathcal{W}$, and the order for $\mathcal{W}_x$ is the one inherited from $\mathcal{W}$.

Let us follow FELSCH through the enumerations of Examples 2.1 and 2.2.

Example 3.1. Let $X = \{a, b\}^{\pm}$ and let $\mathcal{R}$ define the free group on $\{a, b\}$. Set $\mathcal{U} = \{U\}$ and $\mathcal{V} = \{V_1, V_2, V_3\}$, where $U = b$, $V_1 = a^3$, $V_2 = b^4$, and $V_3 = (ab)^2$. Then $\mathcal{W} = \{W_1, W_2, W_3, W_4\}$, where $W_1 = V_1$, $W_2 = V_2$, $W_3 = V_3$, and $W_4 = (ba)^2$. The first step in the enumeration is to make a full trace of U at 1. This gives

	a	a^{-1}	b	b^{-1}
1			1	1

Now we call DEFINE$(1, a)$, which produces

	a	a^{-1}	b	b^{-1}
1	2			1
2		1		

The deductions $(1, a)$ and $(2, a^{-1})$ are pushed onto the stack but no changes are made as the stack is cleared.

Next comes the call DEFINE$(1, a^{-1})$. This makes the table

	a	a^{-1}	b	b^{-1}
1	2	3	1	1
2		1		
3	1			

The deductions $(1, a^{-1})$ and $(3, a)$ are pushed onto the stack, and then $(3, a)$ is popped off. The call TWO_SIDED_TRACE$(3, W_1)$ results in JOIN$(2, a, 3)$. Tracing W_3 at 3 leads to JOIN$(2, b, 3)$. No further changes occur as the stack is cleared. The table is now

	a	a^{-1}	b	b^{-1}
1	2	3	1	1
2	3	1	3	
3	1	2		2

Now DEFINE$(2, b^{-1})$ extends the table to

	a	a^{-1}	b	b^{-1}
1	2	3	1	1
2	3	1	3	4
3	1	2		2
4			2	

This time PROCESS_DEDUCTIONS does not add any more edges, so DEFINE$(3, b)$ is called. This gives

	a	a^{-1}	b	b^{-1}
1	2	3	1	1
2	3	1	3	4
3	1	2	5	2
4			2	
5				3

Tracing W_2 at 3 leads to JOIN$(5, b, 4)$ and tracing W_4 at 3 results in JOIN$(5, a, 4)$. Clearing the stack does not produce any more changes. At this point the table is

	a	a^{-1}	b	b^{-1}
1	2	3	1	1
2	3	1	3	4
3	1	2	5	2
4		5	2	5
5	4		4	3

The next definition is DEFINE$(4, a)$, which gives us

	a	a^{-1}	b	b^{-1}
1	2	3	1	1
2	3	1	3	4
3	1	2	5	2
4	6	5	2	5
5	4		4	3
6		4		

Tracing W_1 at 4 causes JOIN$(6, a, 5)$. Then tracing W_3 at 4 results in JOIN$(6, b, 6)$. The table is now complete, and no coincidences are produced as the remaining deductions on the stack are processed. The current table is

	a	a^{-1}	b	b^{-1}
1	2	3	1	1
2	3	1	3	4
3	1	2	5	2
4	6	5	2	5
5	4	6	4	3
6	5	4	6	6

This table is standard, so the final standardization step does not modify the table. This table is the same one obtained in Example 2.1 using HLT.

Example 3.2. Let $X = \{x, y, y^{-1}\}$ and let $\mathcal{R}$ consist of the rules

$$x^2 \to \varepsilon, \quad yy^{-1} \to \varepsilon, \quad y^{-1}y \to \varepsilon.$$

Suppose that $\mathcal{U} = \emptyset$ and $\mathcal{V} = \{V_1, V_2, V_3\}$, where $V_1 = y^3$, $V_2 = xyxy^{-1}$, and $V_3 = (xy)^7$. Then $\mathcal{W} = \{W_1, \ldots, W_7\}$, where $W_1 = V_1$, $W_2 = V_2$, $W_3 = yxy^{-1}x$, $W_4 = xy^{-1}xy$, $W_5 = y^{-1}xyx$, $W_6 = V_3$, and $W_7 = (yx)^7$. After the first two calls to DEFINE, no modifications to the table result from the calls to PROCESS_DEDUCTIONS. At this point the table is

	x	y	y^{-1}
1	2	3	
2	1		
3			1

The next definition is made by DEFINE$(1, y^{-1})$, which sets $1^{y^{-1}} = 4$ and $4^y = 1$ and pushes $(1, y^{-1})$ and $(4, y)$ onto the stack. Then $(4, y)$ is popped off the stack. The call TWO_SIDED_TRACE$(4, W_1, \text{false})$ leads to JOIN$(3, y, 4)$, which sets $3^y = 4$ and $4^{y^{-1}} = 3$. No further changes occur as the stack is cleared. The table is now

	x	y	y^{-1}
1	2	3	4
2	1		
3		4	1
4		1	3

Next DEFINE$(2, y)$ makes $2^y = 5$ and $5^{y^{-1}} = 2$. Then TWO_SIDED_TRACE$(5, W_5, \text{false})$ results in JOIN$(3, x, 5)$, which sets $3^x = 5$ and $5^x = 3$. These actions produce the following table:

	x	y	y^{-1}
1	2	3	4
2	1	5	
3	5	4	1
4		1	3
5	3		2

At this point DEFINE$(2, y^{-1})$ defines $2^{y^{-1}} = 6$ and $6^y = 2$. Then TWO_SIDED_TRACE$(6, W_1, \text{false})$ results in JOIN$(5, y, 6)$, which makes $5^y = 6$ and $6^{y^{-1}} = 5$. After this, TWO_SIDED_TRACE$(6, W_3, \text{false})$ leads to JOIN$(4, x, 6)$, which sets $4^x = 6$ and $6^x = 4$. The table is now complete.

	x	y	y^{-1}
1	2	3	4
2	1	5	6
3	5	4	1
4	6	1	3
5	3	6	2
6	4	2	5

The stack is not yet clear. The elements of $\mathcal{W}_y$ are still being traced at 6. The call TWO_SIDED_TRACE$(6, W_7, \text{false})$ causes 1 and 6 to be identified. The result is

	x	y	y^{-1}
1	1	1	1

which is consistent with Example 2.2.

In designing an implementation of the Felsch strategy, one should carefully consider the amount of memory to be reserved for the deduction stack. Experimental evidence suggests that the average size of the stack is quite small, although it can be large, especially after a massive collapse produced by COINCIDENCE. The idea of having the stack is that at any given moment there are usually only a few places where two-sided traces could lead to change, and these places should be recorded to save looking for them. However, if the stack has become very large, then there is really no need for the stack, and it may be more efficient to trace each element of $\mathcal{V}$ at each element of Σ rather than invoke PROCESS_DEDUCTIONS to clear the stack.

The HLT strategy has its lookahead variation, and it should come as no surprise that the Felsch strategy has variations too. One proposed by G. Havas is called the *small-gap strategy*. Suppose we are executing the procedure FELSCH and TWO_SIDED_TRACE(σ, W, false) has been called by PROCESS_DEDUCTIONS. After the call to BACK_TRACE, the word T measures the "gap" in the tracing of W at σ. If T is short in some sense, then it is reasonable to ask that the arguments in the next call to DEFINE be chosen so that this gap is shortened. Thus, as the stack is being cleared, we keep track of short gaps which are encountered. When the stack is empty and it is time to call DEFINE, then this is done in a way that shortens one or more of the listed gaps. Care should be exercised in implementing this approach. Followed strictly, it does not define a coset enumeration procedure in the sense of Section 5.1.

Exercises

3.1. Does the procedure FELSCH work correctly if $\mathcal{V}$ contains elements of X?

3.2. Carry out the enumerations in Exercises 2.1 to 2.3 using FELSCH instead of HLT.

3.3. Suppose that memory for the coset table is limited. Prove that adding extra or redundant relations in a Felsch enumeration cannot hurt, in the sense that if the enumeration completed before the addition, it will complete after the addition. Show by example that adding extra relators can hurt in an HLT enumeration.

5.4 Standardizing strategies

In both HLT and FELSCH, the last step in the enumeration is standardization, the primary purpose of which is to make the output independent of the order chosen for the subgroup generators in $\mathcal{U}$ and the relators in $\mathcal{V}$. However, there are coset enumeration strategies in which partial standardization plays a central role and is performed throughout the procedure. This has the effect of making the operation of the procedure as a whole, not just the final output, less dependent on the order used for $\mathcal{U}$ and $\mathcal{V}$. It also sometimes results in better performance.

Standardization with respect to a length-plus-lexicographic ordering can
be incorporated easily into HLT and FELSCH. Both procedures involve a
loop of the following form:

$\sigma := 1$;
While $\sigma \leq n$ do begin
  If $p[\sigma] = \sigma$ then begin
    $\vdots$
  End;
  $\sigma := \sigma + 1$
End;

Suppose that during some iteration of this loop $p[\sigma] = \sigma$, so σ is in Σ.
Let P be a word of minimal length such that $1^P = \sigma$, and suppose that
Q is a word such that $|Q| < |P|$. It may happen that 1^Q is not defined or
that $1^Q > 1^P$. In tracing an element V of $\mathcal{V}$ at σ, we are trying to make
sure that $[PVP^{-1}]$ is in the subgroup H defined by the current automaton.
It is plausible that it would be a good idea to add such elements with $|P|$
small before adding those with $|P|$ large. This can be arranged very easily
by performing a partial standardization of the coset automaton within the
basic loop. The corresponding strategies will be called the *standardizing*
HLT and Felsch strategies, respectively.

The main loop of STANDARDIZE is also a loop over the states. To
implement the standardizing strategies, we combine the main loop of STAN-
DARDIZE with the main loop in HLT or FELSCH. The result has the
following form:

$\sigma := 1$; $\tau := 1$;
While $\sigma \leq n$ do Begin
  If $p[\sigma] = \sigma$ then begin
    $\vdots$
    If $p[\sigma] = \sigma$ then
      For x in X do begin
        $\rho := \sigma^x$;
        If $\rho > \tau$ then begin
          $\tau := \tau + 1$;
          If $\rho > \tau$ then SWITCH(ρ, τ)
        End
      End
  End;
  $\sigma := \sigma + 1$
End;

Table 5.4.1

	a	a^{-1}	b	b^{-1}
1	5	4	2	7
2	3		6	1
3		2		4
4	1	5	3	
5	4	1	8	
6			7	2
7		8	1	6
8	7			5

Table 5.4.2

	a	a^{-1}	b	b^{-1}
1	2	4	5	7
2	4	1		8
3		5		4
4	1	2	3	
5	3		6	1
6			7	5
7		8	1	6
8	7			2

Table 5.4.3

	a	a^{-1}	b	b^{-1}
1	2	3	4	5
2	3	1	8	
3	1	2	7	
4	7		6	1
5		8	1	6
6		5		4
7	4			3
8	5			2

The statements indicated by dots are the same as the statements in the body of the main loop of HLT or FELSCH.

Let us apply the standardizing HLT and Felsch strategies to a sample problem. Suppose that $X = \{a,b\}^{\pm}$, $\mathcal{R}$ defines the free group on $\{a,b\}$, $\mathcal{U} = \{U\}$, and $\mathcal{V} = \{V_1, V_2, V_3\}$, where $U = bab^{-1}a$, $V_1 = a^3$, $V_2 = b^4$, and $V_3 = (ab)^2$. The tracing of the elements of $\mathcal{U}$ is the same for all coset enumeration procedures. After the full trace of U at 1, the coset table is

	a	a^{-1}	b	b^{-1}
1		4	2	
2	3			1
3		2		4
4	1		3	

At this point, the HLT and Felsch strategies diverge. Let us follow the standardizing HLT strategy first. After full traces of V_1, V_2, and V_3 at 1, we have Table 5.4.1. Now we standardize the first row. First 2 and 5 are switched, giving Table 5.4.2. Then, in turn, the pairs 3 and 4, 4 and 5, and 5 and 7 are switched. We now have Table 5.4.3.

Two-sided traces of the elements of $\mathcal{V}$ at 2 produce Table 5.4.4. To standardize the second row, we first switch 6 and 8, and then 7 and 10. The result is Table 5.4.5.

Tracing the elements of $\mathcal{V}$ at 3 introduces two new states, 11 and 12, but they are found to be coincident with previous states by the trace of V_3. To standardize row 3 we switch 8 and 10. We now have Table 5.4.6.

Tracing the elements of $\mathcal{V}$ at 4 and standardizing the fourth row gives Table 5.4.7.

The traces at 5 introduce two new states. In many implementations, the state labels 11 and 12 would be reused. However, the version of DEFINE in Section 4.5 gives them new numbers, 13 and 14. The switch of 10 and 13 is needed to standardize the fifth row. The coset table is now Table 5.4.8.

Table 5.4.4

	a	a^{-1}	b	b^{-1}
1	2	3	4	5
2	3	1	8	10
3	1	2	7	
4	7		6	1
5		8	1	6
6			5	4
7	10	4		3
8	5		9	2
9			10	8
10		7	2	9

Table 5.4.5

	a	a^{-1}	b	b^{-1}
1	2	3	4	5
2	3	1	6	7
3	1	2	10	
4	10		8	1
5		6	1	8
6	5		9	2
7		10	2	9
8			5	4
9			7	6
10	7	4		3

Table 5.4.6

	a	a^{-1}	b	b^{-1}
1	2	3	4	5
2	3	1	6	7
3	1	2	8	8
4	8		10	1
5		6	1	10
6	5		9	2
7		8	2	9
8	7	4	3	3
9			7	6
10			5	4

Table 5.4.7

	a	a^{-1}	b	b^{-1}
1	2	3	4	5
2	3	1	6	7
3	1	2	8	8
4	8	7	9	1
5		6	1	9
6	5		10	2
7	4	8	2	10
8	7	4	3	3
9			5	4
10			7	6

Table 5.4.8

	a	a^{-1}	b	b^{-1}
1	2	3	4	5
2	3	1	6	7
3	1	2	8	8
4	8	7	9	1
5	10	6	1	9
6	5	10	13	2
7	4	8	2	13
8	7	4	3	3
9		14	5	4
10	6	5	14	
13			7	6
14	9			10

Table 5.4.9

	a	a^{-1}	b	b^{-1}
1	2	3	4	5
2	3	1	6	7
3	1	2	8	8
4	8	7	9	1
5	10	6	1	9
6	5	10	11	2
7	4	8	2	11
8	7	4	3	3
9		14	5	4
10	6	5	14	
11			7	6
14	9			10

Table 5.4.10   Table 5.4.11

	a	a^{-1}	b	b^{-1}
1	2	3	4	5
2	3	1	6	7
3	1	2	8	8
4	8	7	9	1
5	10	6	1	9
6	5	10	11	2
7	4	8	2	11
8	7	4	3	3
9	11	12	5	4
10	6	5	12	
11	12	9	7	6
12	9	11		10

	a	a^{-1}	b	b^{-1}
1	2	3	4	5
2	3	1	6	7
3	1	2	8	8
4	8	7	9	1
5	10	6	1	9
6	5	10	11	2
7	4	8	2	11
8	7	4	3	3
9	11	12	5	4
10	6	5	12	12
11	12	9	7	6
12	9	11	10	10

Traces at 6 produce no changes. To standardize the sixth row, we switch 11 and 13. This is done even though prior to the switch 11 is not in Σ. At this point we have Table 5.4.9.

Traces at 7 add two more edges. No further changes occur until the traces at 9, which add two edges. Standardization of row 9 requires that 12 and 14 be switched. This produces Table 5.4.10.

Traces at 10 introduce two new states labeled 15 and 16, but they are immediately found to be coincident with 10 and 12, respectively. The table is now Table 5.4.11. No further changes occur.

Let us return to the point immediately after the trace of U at 1 and follow the standardizing Felsch strategy through this enumeration. The trace of U placed eight deductions on the stack. As these deductions are processed, two more edges are added. The table is now

	a	a^{-1}	b	b^{-1}
1		4	2	
2	3			1
3		2	4	4
4	1		3	3

After the definition of 5 as 1^a, we find that $5^a = 4$. Defining $1^{b^{-1}} = 6$ does not lead to any new edges. Currently, we have Table 5.4.12. Standardizing the first row produces Table 5.4.13.

Defining 2^b to be 7 leads to two more edges as the stack is cleared. The definition of 8 as $2^{b^{-1}}$ yields four more edges. After the second row is standardized, we have Table 5.4.14.

The next definition is $4^b = 9$ and this yields another pair of edges. The definition $10 = 5^a$ leads to $10^a = 6$. No switches are needed to standardize either row 4 or row 5. At this point we have Table 5.4.15.

Table 5.4.12

	a	a^{-1}	b	b^{-1}
1	5	4	2	6
2	3			1
3		2	4	4
4	1	5	3	3
5	4	1		
6				1

Table 5.4.13

	a	a^{-1}	b	b^{-1}
1	2	3	4	5
2	3	1		
3	1	2	6	6
4	6			1
5				1
6		4	3	3

Table 5.4.14

	a	a^{-1}	b	b^{-1}
1	2	3	4	5
2	3	1	6	7
3	1	2	8	8
4	8	7		1
5			6	1
6	5			2
7	4	8	2	
8	7	4	3	3

Table 5.4.15

	a	a^{-1}	b	b^{-1}
1	2	3	4	5
2	3	1	6	7
3	1	2	8	8
4	8	7	9	1
5	10	6	1	9
6	5	10		2
7	4	8	2	
8	7	4	3	3
9			5	4
10	6	5		

Table 5.4.16

	a	a^{-1}	b	b^{-1}
1	2	3	4	5
2	3	1	6	7
3	1	2	8	8
4	8	7	9	1
5	10	6	1	9
6	5	10	11	2
7	4	8	2	11
8	7	4	3	3
9	11	12	5	4
10	6	5	12	12
11	12	9	7	6
12	9	11	10	10

The definitions $11 = 6^b$ and $12 = 9^{a^{-1}}$ produce enough additional edges to complete the table. No coincidences are found and the enumeration is finished. The final automaton is the same as the one obtained earlier, namely, Table 5.4.16.

In general, the table produced by one of the standardizing strategies is not completely standard. The states σ are in order according to the first word W such that $1^W = \sigma$, but Σ may not consist of the integers from 1 to $|\Sigma|$. Thus a final standardizing pass may still be necessary. More examples using standardizing strategies are discussed in the next section.

Exercises

4.1. Carry out the enumerations in Exercises 2.1 to 2.3 using the standardizing HLT and Felsch strategies.

4.2. Under what circumstances might it be useful to use a standardizing strategy based on a basic wreath-product ordering? Develop such strategies for both the HLT and Felsch approaches.

Table 5.5.1

Number	Type	*save*	Standardize
1	HLT	false	No
2	HLT	false	Yes
3	HLT	true	No
4	HLT	true	Yes
5	CHLT	false	No
6	CHLT	false	Yes
7	CHLT	true	No
8	CHLT	true	Yes
9	Felsch	true	No
10	Felsch	true	Yes

5.5 Ten versions

This section describes a procedure TEN_CE which implements ten different coset enumeration strategies. These ten have been chosen because they fit together nicely in a single procedure. There are many other strategies which have been investigated and found useful. However, space does not permit giving an exhaustive catalog. Comparing the ten versions of this section will provide a good introduction to the issues involved in choosing an enumeration procedure for a particular application.

A strategy is selected within TEN_CE by specifying three things: its type, the value of the boolean variable *save*, and whether or not standardization is to be used. The type can be Felsch, HLT, or CHLT. The acronym CHLT stands for "cyclic HLT". In strategies of this type, full two-sided traces of all cyclic permutations of the elements of V are performed at the states in Σ. The set W of these cyclic permutations must be available for the Felsch strategies, so there is very little extra work needed to include the strategies of type CHLT.

The use of the deduction stack was introduced as part of the Felsch strategy. However, it can be used with HLT and CHLT strategies as well. Such combinations may be considered as alternatives to lookahead variations of the HLT strategy. The use of lookahead is controlled by the amount of memory available. An HLT strategy using the deduction stack will perform the same way with any amount of memory above the minimum needed for the enumeration to complete. This has considerable theoretical advantages as we attempt to analyze running times. Also, there can be practical advantages in enumerations which would require repeated lookahead phases.

There are three possible types and two possible values for *save*. However, with enumeration of the Felsch type, *save* must be true, so there are only five valid combinations. With each of these five, standardization may or may not be used. The ten strategies available in TEN_CE will be numbered from 1 to 10 as in Table 5.5.1.

In addition to the normal arguments of a coset enumeration procedure, TEN_CE has an input argument m which specifies the strategy number.

Procedure TEN_CE($X, \mathcal{R}, \mathcal{U}, \mathcal{V}, m; \mathcal{A}$);
(∗ Ten coset enumeration procedures. The strategy used is determined
    by the value of m. ∗)
Begin
    Let hlt be true if and only if m is in $\{1, 2, 3, 4\}$;
    Let $chlt$ be true if and only if m is in $\{5, 6, 7, 8\}$;
    Let $save$ be true if and only if m is in $\{3, 4, 7, 8, 9, 10\}$;
    Let $stnd$ be true if and only if m is in $\{2, 4, 6, 8, 10\}$;

    Let $\mathcal{W}$ be the set of cyclic permutations of the elements of $\mathcal{V}$;
    For x in X do let $\mathcal{W}_x$ be the set of elements of $\mathcal{W}$ which start with x;
    $n := 1$; $p[1] := 1$; $E := \emptyset$;
    If $save$ then clear the deduction stack;

    For U in $\mathcal{U}$ do begin
        TWO_SIDED_TRACE($1, U$, true);
        If $save$ then PROCESS_DEDUCTIONS
    End;

    If hlt then $\mathcal{D} := \mathcal{V}$
    Else if $chlt$ then $\mathcal{D} := \mathcal{W}$;

    $\sigma := 1$;
    If $stnd$ then $\tau := 1$;

    While $\sigma \leq n$ do begin
        If $p[\sigma] = \sigma$ then begin
            If hlt or $chlt$ then
                For D in $\mathcal{D}$ do begin
                    TWO_SIDED_TRACE(σ, D, true);
                    If $save$ then PROCESS_DEDUCTIONS;
                    If $p[\sigma] \neq \sigma$ then goto 99
                End;

            For x in X do
                If σ^x is not defined then begin
                    DEFINE(σ, x);
                    If $save$ then PROCESS_DEDUCTIONS;
                    If $p[\sigma] \neq \sigma$ then goto 99
                End;

            If $stnd$ then
                For x in X do begin
                    $\rho := \sigma^x$;
                    If $\rho > \tau$ then begin
                        $\tau := \tau + 1$;

If $\rho > \tau$ then $\mathrm{SWITCH}(\rho, \tau)$
  End
End
End;   (* If $p[\sigma] = \sigma$ *)

99: $\sigma := \sigma + 1$
End;   (* While $\sigma \leq n$*)

STANDARDIZE
End.

To completely determine the operation of TEN_CE, one must specify the order of the elements in X, $\mathcal{U}$, $\mathcal{V}$, and $\mathcal{W}$.

It is not a simple matter to decide how the ten strategies in TEN_CE should be compared and evaluated. Since we have not specified how the edge set E is to be stored, we cannot talk in detail about running times. Moreover, giving a worst-case or average run-time analysis is currently beyond our ability. In fact, there is no common agreement on what a "random presentation" is, so we cannot even define the average running time.

About all we can do is to look at specific examples. But how should these examples be chosen? They could be generated randomly using some scheme for choosing sets of words. However, such presentations tend not to look like the ones that arise "naturally" in mathematics. In group presentations that arise in the course of research in group theory, topology, and other fields of mathematics, powers and commutators occur much more frequently than they do in randomly generated presentations. (See Section 9.1 for the definition of commutators.)

If we choose as our test problems various presentations which have appeared in the literature, then what will our comparisons mean? Do we have any confidence that observations made about these problems will be applicable to problems researchers may encounter in the future?

If an enumeration terminates, then we can record the total number of states used, which is the value of n at the end, and the maximum number of states active at any one time. If these numbers are small, this is a favorable outcome. But suppose an enumeration does not terminate, either because the subgroup has infinite index or memory is exhausted before the index can be determined. Does it make any sense to say one strategy is better than another in such cases? Comparisons with strategies involving lookahead are even more difficult, since performance depends on the amount of memory available.

The preceding comments should make it clear that we shall not be able to study enough examples here to come to any firm conclusions about the relative performance of the ten strategies in TEN_CE. Nevertheless, a few examples will provide some useful insight. The performance quoted is for

Table 5.5.2

Strategy	Total	Max. Active
1	66	64
2	66	62
3	60	60
4	60	60
5	78	66
6	66	62
7	60	60
8	60	60
9	60	60
10	60	60

Table 5.5.3

Strategy	Total	Max. Active
1	1550	1502
2	1548	1458
3	673	673
4	393	393
5	1864	1808
6	1473	1413
7	620	620
8	592	590
9	588	588
10	588	588

one particular implementation of TEN_CE. Some variation is possible, depending on the order used for $X, \mathcal{U}, \mathcal{V}$, and $\mathcal{W}$. Most of the sample problems are taken from (Cannon et al. 1973). The first few have $\mathcal{U} = \emptyset$.

Example 5.1. Let us start with a simple enumeration. Suppose that $X = \{x, y, y^{-1}\}$ and $\mathcal{R}$ consists of the rules

$$x^2 \to \varepsilon, \quad yy^{-1} \to \varepsilon, \quad y^{-1}y \to \varepsilon.$$

Let $\mathcal{U} = \emptyset$ and let $\mathcal{V} = \{y^3, (xy)^5\}$. If N is the normal closure of the image of $\mathcal{V}$ in F, then $G = F/N$ has order 60, and the map $x \mapsto (1,2)(3,4)$ and $y \mapsto (1)(2,3,5)(4)$ defines an isomorphism of G with the alternating group Alt(5). Table 5.5.2 summarizes the results of determining the order of G using each of the ten strategies in TEN_CE. For each strategy the total number of states used and the maximum active at any one time are tabulated.

Example 5.2. Suppose that $X = \{r, s, t\}^{\pm}$ and $\mathcal{R}$ defines the free group on $\{r, s, t\}$. Let $\mathcal{U} = \emptyset$ and let $\mathcal{V} = \{t^{-1}rtr^{-2}, r^{-1}srs^{-2}, s^{-1}tst^{-2}\}$. In this case, $G = F/N$ is trivial. Table 5.5.3 shows the results of verifying this using TEN_CE.

Example 5.3. The group $G_k = \text{Grp}\langle x, y \mid x^2 = y^3 = (xy)^7 = (xyxy^{-1})^k = 1\rangle$ is finite for $k \leq 8$. The orders of $G_1, \ldots, G_8$ are 1, 1, 1, 168, 1, 1092, 1092, and 10752, respectively. Table 5.5.4 summarizes the performance of TEN_CE in determining the orders of G_k, $4 \leq k \leq 8$. Here X and $\mathcal{R}$ are as in Example 5.1, $\mathcal{U} = \emptyset$, and $\mathcal{V} = \{y^3, (xy)^7, (xyxy^{-1})^k\}$. The two entries in the table for each strategy and each value of k are the total number of states and the maximum number of active states.

Now let us look at some enumerations with $\mathcal{U} \neq \emptyset$.

Table 5.5.4

Strategy	\multicolumn{5}{c}{k}				
	4	5	6	7	8
1	292/208	698/490	2650/1514	8746/6258	128562/87254
2	274/230	730/562	2234/1474	4418/3674	110902/80744
3	168/168	266/266	1122/1092	1960/1934	32320/31678
4	168/168	310/310	1120/1092	2170/2150	39528/38942
5	324/228	874/558	3334/1808	15326/7792	178620/99632
6	274/230	730/562	2234/1474	4418/3675	110902/80744
7	168/168	296/296	1172/1092	2748/2684	31365/30108
8	168/168	316/316	1120/1092	2170/2150	39534/38948
9	168/168	336/336	1092/1092	1644/1490	39745/39745
10	168/168	336/336	1092/1092	1644/1490	39745/39745

Table 5.5.5

Strategy	Total	Max. Active
1	2635	2174
2	2662	2136
3	1212	1199
4	1492	1454
5	2619	2213
6	2027	1773
7	1284	1258
8	1518	1488
9	1306	1302
10	1306	1302

Example 5.4. The group $G = \text{Grp} \langle a, b \mid a^8 = b^7 = (ab)^2 = (a^{-1}b)^3 = 1 \rangle$ has order 10752 and the subgroup K generated by the images of a^2 and $a^{-1}b$ has index 448 in G. Table 5.5.5 shows the results of computing $|G : K|$. Here $X = \{a, b\}^{\pm}$, $\mathcal{R}$ defines the free group on $\{a, b\}$, $\mathcal{U} = \{a^2, a^{-1}b\}$, and $\mathcal{V} = \{a^8, b^7, (ab)^2, (a^{-1}b)^3\}$.

Example 5.5. Let n be an integer greater than 1 and let $B_n = \text{Grp}\langle x, y \mid x^n = y, y^n = x \rangle$. In B_n,

$$x^{n^2} = (x^n)^n = y^n = x.$$

Clearly B_n is generated by x, and therefore B_n is cyclic of order dividing $n^2 - 1$. For u in $\mathbb{Z}$, let $[u]$ denote the congruence class of u modulo $n^2 - 1$. The map $x \mapsto [1]$ and $y \mapsto [n]$ defines a homomorphism of B_n onto the additive group of integers modulo $n^2 - 1$. Therefore $|B_n| = n^2 - 1$.

Table 5.5.6

Strategy	n				
	10	12	14	15	16
1	170/163	252/243	350/339	405/393	464/451
2	166/135	247/198	344/273	401/316	457/360
3	99/99	143/143	195/195	224/224	255/255
4	99/99	143/143	195/195	224/224	255/255
5	99/99	143/143	195/195	224/224	255/255
6	99/99	143/143	195/195	224/224	255/255
7	99/99	143/143	195/195	224/224	255/255
8	99/99	143/143	195/159	224/224	255/255
9	406/243	1205/729	3597/2187	6393/4374	10768/6561
10	406/243	1205/729	3597/2187	6393/4374	10768/6561

Table 5.5.7

Strategy	Total	Max. Active
1	1204	575
2	72742	43242
3	272	257
4	288	255
5	255	255
6	255	255
7	255	255
8	255	255
9	10768	6561
10	10768	6561

Table 5.5.8

	x	x^{-1}	y	y^{-1}
1	10	6	7	2
2	3		1	
3	4	2		
4	5	3		
5	6	4		
6	1	5		
7			8	1
8			9	7
9			10	8
10		1		9

Using the strategies in TEN_CE to confirm the order for particular values of n provides some interesting results. Let $X = \{x, y\}^{\pm}$, let $\mathcal{R}$ define the free group on $\{x, y\}$, let $\mathcal{U} = \emptyset$, and let $\mathcal{V} = \{x^n y^{-1}, y^n x^{-1}\}$. Table 5.5.6 shows the statistics on total states used and maximum number of active states for selected values of n.

The statistics depend on the choice of the words in $\mathcal{V}$. If we replace the words by cyclic permutations, we get different results. For example, suppose that $n = 16$ and $\mathcal{V} = \{x^8 y^{-1} x^8, y^8 x^{-1} y^8\}$, then the statistics are as shown in Table 5.5.7.

All of the results tabulated here were obtained using the ordering on $\mathcal{W}$ described in Section 5.3. With different orderings of $\mathcal{W}$, some of the CHLT strategies can define redundant states. For example, suppose that $n = 4$ and we are following strategy 5. If we first trace $y^{-1} x^4$ at 1 and then trace $y^4 x^{-1}$, the coset table is as shown in Table 5.5.8. If we now trace $xy^{-1}x^3$ at 1, then states 4 and 9 are identified.

It is not hard to show that the number of states used by strategies 9 and 10 to find the order of B_n is exponential in n. The number of states used by strategies 6 and 8 is bounded above by a cubic polynomial in n, no matter what ordering $\mathcal{W}$ is chosen. The proofs of these statements and further investigation of the asymptotic behavior of strategies 1 to 10 on this family of examples are left as exercises.

Exercises

5.1. Suppose that TEN_CE is run with $\mathcal{U} = \emptyset$. Show that strategies 9 and 10 are identical.

5.2. Suppose that $|X| \geq 3$, $\mathcal{U} = \emptyset$, and every element of $\mathcal{V}$ has length at least r. Prove that the execution time for TEN_CE using either strategy 9 or 10 is bounded below by an exponential function of r. (If termination does not occur, then the execution time is considered to be infinite.) Conclude that Felsch strategies are not a good choice for presentations like B_n.

*5.3. Let $\mathcal{A}$ be a complete coset automaton with respect to $\mathcal{R}$ such that $K(\mathcal{A})$ is normal in $F = \mathrm{Mon} \langle X \mid \mathcal{R} \rangle$. Given two states σ and τ of $\mathcal{A}$ and a word W, the state σ^W is equal to σ if and only if τ^W is equal to τ. Suppose memory is limited and a coset enumeration with $\mathcal{U} = \emptyset$ terminates unsuccessfully, returning an incomplete coset automaton $\mathcal{B}$. Is it possible that there are states σ and τ of $\mathcal{B}$ and a word W such that $\sigma^W = \sigma$ but that a two-sided trace of W at τ would lead to a coincidence? How hard would it be to test for the existence of such a triple (σ, τ, W)?

5.6 Low-index subgroups

Let G be a group given by a finite presentation. One of the first things we would like to know about G is whether or not G is trivial. If G is trivial, then it is possible to verify this using coset enumeration or the Knuth-Bendix procedure. However, there is no algorithmic way to verify nontriviality. In particular cases, one can show that G is nontrivial by exhibiting a nontrivial homomorphic image of G. This is normally done by producing elements in some well understood group which satisfy the defining relations of G and are not all trivial. Chapter 11 discusses methods for finding nontrivial solvable quotients of G. (See Section 9.2 for the definition of a solvable group.) This section may be viewed as describing methods for finding nontrivial finite permutation groups which are homomorphic images of G.

If there is a nontrivial permutation group M on a finite set Ω such that M is a homomorphic image of G, then M has an orbit Γ of size greater than 1 and the permutation group induced on Γ by M is a nontrivial homomorphic image of M and hence also a nontrivial homomorphic image of G. Thus, if our goal is to show that G is nontrivial, we may assume that M is transitive on Ω. In this case, M is determined up to permutation isomorphism by the conjugacy class of the subgroups G_α, where α is in Ω and G_α is the set of elements in G whose images in M fix α. The index of G_α in G is $|\Omega|$. Thus the problem of deciding whether such an M exists with $|\Omega|$ not exceeding some bound n is equivalent to deciding whether G has any proper subgroups of index not exceeding n.

Let us assume that G is given as F/N, where F is the free product of cyclic groups defined by a classical niladic rewriting system $\mathcal{R}$ and N is the normal closure of the image in F of a finite set $\mathcal{V}$ of cyclically reduced words in X^*. Given a reduction ordering $\prec$ on X^*, subgroups of finite index in G are in one-to-one correspondence with the finite, standard, complete coset automata with respect to $\mathcal{R}$ which are compatible with $\mathcal{V}$. In Section 4.9 we learned how to generate the standard, complete coset automata which have at most a given number n of states. If we test each of these automata for compatibility with $\mathcal{V}$, then the automata which pass describe the subgroups K of G with $|G : K| \leq n$. It is likely that very few of the tests will be successful. Therefore it is reasonable to try to incorporate compatibility with $\mathcal{V}$ some way into the backtrack criteria. The set of subgroups of G with index at most n is closed under conjugation. Thus it suffices to find one representative from each conjugacy class of subgroups. In some applications we may not be interested in all subgroups of G with small index, but only those which contain a given finitely generated subgroup.

In this section we shall consider the following problem: Let $\mathcal{U}$ and $\mathcal{V}$ be finite subsets of X^* and let G be the group F/N, where N is the normal closure of the image of $\mathcal{V}$ in F. We shall assume that the elements of $\mathcal{U}$ are reduced and the elements of $\mathcal{V}$ are cyclically reduced. Given a positive integer n, find subgroups $H_1, \ldots, H_s$ of G such that each H_i contains the image of $\mathcal{U}$ in G, $|G : H_i| \leq n$, and if H is any subgroup of G containing the image of $\mathcal{U}$ and having index at most n, then H is conjugate to exactly one of the H_i.

The techniques currently available to solve this problem are limited. The best we can hope for in general is to produce a solution when n and $|X|$ are fairly small. For this reason, the methods of this section are usually referred to as the *low-index subgroup algorithm*. Continued research is needed and the procedures described here should be viewed as models for further investigation.

Fix a reduction ordering $\prec$ on X^*. We shall have a solution to our problem if we can list the standard, complete coset automata $\mathcal{A} = (\Sigma, X, E, \{1\}, \{1\})$ with respect to $\mathcal{R}$ satisfying the following conditions:

(i) $\mathcal{A}$ has at most n states.

(ii) $\mathcal{A}$ is compatible with $\mathcal{V}$.

(iii) $1^U = 1$ in $\mathcal{A}$ for all U in $\mathcal{U}$.

(iv) Let σ be a state in $\mathcal{A}$ such that $\sigma^U = \sigma$ for all U in $\mathcal{U}$ and let $\mathcal{A}_\sigma = (\Sigma, X, E, \{\sigma\}, \{\sigma\})$. Then the $\mathcal{A}$ is the same as or precedes the standard coset automaton isomorphic to $\mathcal{A}_\sigma$.

The subgroups of G corresponding to the subgroups $K(\mathcal{A})$ of F defined by the automata $\mathcal{A}$ satisfying conditions (i) to (iv) give us our solution. The low-index subgroup procedure lists the transition matrices for the coset automata satisfying these conditions.

The simplest low-index procedure is obtained by adding conditions (ii) to (iv) above to the backtrack criteria in the search described in Section 4.9 for representatives of the conjugacy classes in F of subgroups of index at most n. Whenever a pair of edges is added to the current automaton, a check is made to see whether there is a state σ and an element V of $\mathcal{V}$ such that σ^V is defined and unequal to σ, or whether there is a word U in $\mathcal{U}$ such that 1^U is defined and unequal to 1. Condition (iv) is tested by a modification of the technique described in Section 4.9. For example, if $\prec$ is a length-plus-lexicographic ordering, then the loop

For $i := 2$ to m do begin
$\qquad \vdots$
End;

in FIRST_CONJ is changed to

For $i := 2$ to m do
$\quad$ If $i^U = i$ for all U in $\mathcal{U}$ then begin
$\qquad \vdots$
$\quad$ End;

Unfortunately, this straightforward approach is not efficient enough to be of much use. To get a practical procedure, we need to incorporate ideas from coset enumeration, such as the deduction stack and two-sided traces. Also, if $\mathcal{V}$ contains a very long word V, then σ^V may almost never be defined except when the automaton is complete. To avoid wasting time in fruitless traces, it is probably a good idea to divide $\mathcal{V}$ into two sets, $\mathcal{V}_1$ and $\mathcal{V}_2$. The set $\mathcal{V}_1$ contains the shorter words, and compatibility with $\mathcal{V}_1$ is made part of the backtrack criteria. Compatibility with $\mathcal{V}_2$ is checked only when the table is complete as a filter on the output.

Suppose that W is a cyclic permutation of an element of $\mathcal{V}_1$ and σ is a state in the current automaton. It might happen that a partial two-sided trace of W at σ would lead to JOIN being called to add a new edge. If the current set of edges is part of a solution to our problem, then the new edge must be part of that same solution. Thus we should add the edge immediately. Adding this edge may make it possible to deduce further edges with other partial two-sided traces. If a two-sided trace would lead to a call to COINCIDENCE, then no way of extending the current edge set could produce an automaton which is compatible with $\mathcal{V}_1$. We can also perform partial two-sided traces of the elements of $\mathcal{U}$ at 1 to see if any further edges are forced or incompatibilities are found.

As in Section 4.9, let $x_1, \ldots, x_r$ be the elements of X in $\prec$-order, let $\mathcal{V}_1$, $\mathcal{V}_2$, and $\mathcal{U}$ be given, and let n be an upper bound for the index of subgroups being sought. The root of our search tree is the 1-by-r incomplete transition matrix all of whose entries are '?'. Suppose T is an m-by-r transition matrix which is a node in the tree. If all entries of T are positive integers, then we check whether the corresponding automaton $\mathcal{A}$ is compatible with $\mathcal{V}_2$. If $\mathcal{A}$ is compatible with $\mathcal{V}_2$, then T is one of the transition matrices we are looking for. If $\mathcal{A}$ is not compatible with $\mathcal{V}_2$, then T is not one of the desired matrices and T has no descendants in the search tree. Now suppose that T has at least one entry which is '?'. There is a word W such that 1^W is unspecified. Let W be the first such word. Then $W = Ux_i$ where $\sigma = 1^U$ is defined. Let x_i^{-1} be x_j and let D be the set of indices τ such that $T[\tau, j]$ is equal to '?'. If $m < n$, then add $m + 1$ to D. For each k in D we carry out the following steps:

$S := T$;
$S[\sigma, i] := k$; Push (σ, x_i) onto the deduction stack;
If $k = m + 1$ then add a k-th row to S consisting of all question marks;

If $(k, j) \neq (\sigma, i)$ then begin
    $S[k, j] := \sigma$; Push (k, x_j) onto the deduction stack
End;

While the deduction stack is not empty do begin
    Process deductions as in the Felsch strategy of coset enumeration by
        performing partial two-sided traces of elements of $\mathcal{V}_1$;
    If further deductions are found then make the corresponding entries
        in S;
    If at any time a coincidence is discovered then abandon processing of
        this value of k;
    Perform partial two-sided traces of the elements of $\mathcal{U}$ at 1 abandoning
        this value of k if a coincidence is found and adding any
        deductions discovered to the deduction stack
End;

Perform the first in conjugacy class check as discussed above;
If S passes the check then S is a descendant of T in the search tree;

The search procedure just outlined will be referred to as LOW_INDEX. The output from LOW_INDEX depends only on $\mathcal{V}_1 \cup \mathcal{V}_2$. However, the running time will depend on the division of the words between the two sets $\mathcal{V}_1$ and $\mathcal{V}_2$. In a given case, it is difficult to predict which division of the relators will yield the fastest execution time.

Example 6.1. Suppose we have the following data: $X = \{x, y\}^{\pm}$, $\prec$ is the length-plus-lexicographic ordering with $x \prec x^{-1} \prec y \prec y^{-1}$, $\mathcal{R}$ defines the free

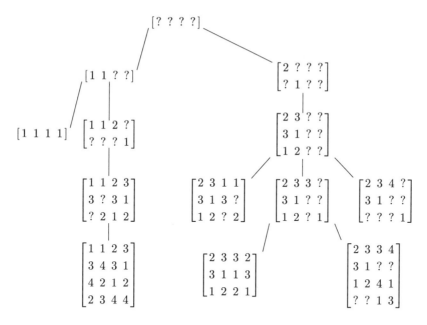

Figure 5.6.1

group on $\{x,y\}$, $\mathcal{V} = \mathcal{V}_1 = \{x^3, y^3, (xy)^2\}$, $\mathcal{V}_2 = \mathcal{U} = \emptyset$, and $n = 4$. Figure 5.6.1 shows the entire search tree.

Let us verify the determination of descendants at some of the nodes of this tree. If T is $\begin{bmatrix} 1 & 1 & 2 & ? \\ ? & ? & ? & 1 \end{bmatrix}$, then $W = y^{-1}$ and $D = \{2, 3\}$. If $k = 2$, then setting $S[1,4]$ equal to 2 leads to an incompatibility with the relator y^3. If $k = 3$, then a two-sided trace of y^3 at 1 leads to the deductions $2^y = 3$ and $3^{y^{-1}} = 2$. Then tracing $(xy)^2$ at 1 produces the deductions $2^x = 3$ and $3^{x^{-1}} = 2$. Thus T has exactly one descendant, as shown.

Now suppose that T is

$$\begin{bmatrix} 2 & 3 & ? & ? \\ 3 & 1 & ? & ? \\ 1 & 2 & ? & ? \end{bmatrix}.$$

In this case, $W = y$ and $D = \{1, 2, 3, 4\}$. With $k = 1$, a two-sided trace of $(xy)^2$ yields the deductions $2^y = 3$ and $3^{y^{-1}} = 2$. Taking $k = 2$ results in the deduction $3^y = 3$. But now the first in conjugacy class test fails, since changing the initial state to 3 and standardizing produces a transition matrix whose first row is $[2\ 3\ 1\ 1]$, and any such matrix precedes S. No further deductions are made when $k = 3$ or $k = 4$.

Finally, suppose that T is

$$\begin{bmatrix} 2 & 3 & 1 & 1 \\ 3 & 1 & 3 & ? \\ 1 & 2 & ? & 2 \end{bmatrix}.$$

Here $W = xy^{-1}$ and $D = \{3,4\}$. If $k = 3$, then an incompatibility with y^3 is found. When we take $k = 4$, we get the deductions $3^y = 4$ and $4^{y^{-1}} = 3$ from y^3 and then $4^x = 4$ from $(xy)^2$. But now the first in conjugacy-class test fails, since changing the initial state to 4 and standardizing leads to a matrix whose first row is $[1\ 1\ 2\ 3]$.

Example 6.2. Let $X = \{a, b, b^{-1}\}$, let $\mathcal{R}$ consist of the rules

$$a^2 \to \varepsilon, \quad bb^{-1} \to \varepsilon, \quad b^{-1}b \to \varepsilon,$$

and let $\prec$ be the length-plus-lexicographic ordering with $a \prec b \prec b^{-1}$. Set $\mathcal{U} = \{abab^{-1}\}$, $\mathcal{V} = \{b^3, (ab)^7\}$, and $n = 10$. The following transition matrices are produced by LOW_INDEX.

$$[1\ 1\ 1], \quad \begin{bmatrix} 1 & 2 & 3 \\ 2 & 3 & 1 \\ 4 & 1 & 2 \\ 3 & 5 & 6 \\ 5 & 6 & 4 \\ 7 & 4 & 5 \\ 6 & 7 & 7 \end{bmatrix}, \quad \begin{bmatrix} 1 & 2 & 3 \\ 2 & 3 & 1 \\ 4 & 1 & 2 \\ 3 & 5 & 6 \\ 7 & 6 & 4 \\ 6 & 4 & 5 \\ 5 & 7 & 7 \end{bmatrix}.$$

Therefore Grp $\langle a, b \mid a^2 = b^3 = (ab)^7 = 1 \rangle$ has only two conjugacy classes of proper subgroups of index at most 10 which contain a conjugate of the element $abab^{-1}$. With $\mathcal{V}_1 = \mathcal{V}$, the search tree has 24 nodes.

There are several ideas used in LOW_INDEX which come from coset enumeration. It is possible to use coset enumeration even more in an effort to prune the search tree further. Sometimes this is effective, but it may happen that the extra time required per node of the search tree is not offset by the fact that fewer nodes are examined.

Example 6.3. Suppose that we are asked to study the group $G =$ Grp $\langle r, s, t \mid t^{-1}rt = r^2, r^{-1}sr = s^2, s^{-1}ts = t^2 \rangle$ of Example 5.2, but we are not told that the group is trivial. A reasonable first step would be to look for subgroups of small index in G. Let $X = \{r, s, t\}^{\pm}$, let $\mathcal{R}$ define the free group on $\{r, s, t\}$, let $\prec$ be the length-plus-lexicographic ordering with $r \prec r^{-1} \prec s \prec s^{-1} \prec t \prec t^{-1}$, let $\mathcal{V}_1 = \{t^{-1}rtr^{-2}, r^{-1}srs^{-2}, s^{-1}tst^{-2}\}$, and let $\mathcal{V}_2 = \mathcal{U} = \emptyset$. With $n = 8$, LOW_INDEX examines a search tree with 548 nodes and reports finding only G. With $n = 12$, the tree has 8179 nodes. At this point the running times are beginning to be noticeable. A coset enumeration of the trivial subgroup in G shows that $|G| = 1$, so trying larger values of n is clearly a waste of time.

Coset enumeration can be used at any node of the search tree being examined by LOW_INDEX. One takes the current automaton to be the result of tracing the subgroup generators and proceeds with any enumeration strategy using $\mathcal{V}$ as the set of relators. Let Σ denote the set of states at the start of the enumeration. If the enumeration produces a coincidence among states in Σ, then the current node can be abandoned. If no coincidence is encountered, then the enumeration is interrupted at some point and the search resumed with the restriction to Σ of the coset automaton produced by the enumeration. Unfortunately, we do not have good heuristics on how long to let the enumeration run as we look for a coincidence among the states in Σ or for more edges connecting two of these states.

As implemented in LOW_INDEX, the conjugacy testing is not as efficient as it could be. Let $\mathcal{A}$ be an automaton representing a node in the search tree and let $\mathcal{B}$ be an immediate descendant of $\mathcal{A}$. Every state of $\mathcal{A}$ is a state of $\mathcal{B}$. Suppose that σ is a state in $\mathcal{A}$ such that $\sigma^U = \sigma$ for all U in $\mathcal{U}$. If the conjugacy testing in $\mathcal{A}$ showed that the standard coset table corresponding to $\mathcal{A}(\sigma)$ comes after $\mathcal{A}$, then the standard coset table corresponding to $\mathcal{B}(\sigma)$ comes after $\mathcal{B}$, so no conjugacy testing at σ is necessary in $\mathcal{B}$.

Even with all the extensions discussed so far, the low-index subgroup algorithm can be defeated fairly easily. In (Havas & Kovács 1984), some computations are described in which it was necessary to locate subgroups of index 13 in two groups, each of which was defined by a presentation on three generators with two relators of length roughly 50. Only by using detailed information about the groups available from other considerations was it possible to find the desired subgroups.

Up to now we have used length-plus-lexicographic orderings with LOW_INDEX. However, sometimes other orderings produce more efficient searches.

Example 6.4. Let $H = \text{Grp}\langle a, b \mid a^{11} = b^5 = bab^{-1}a^{-3} = 1 \rangle$. It is easy to see that H is finite and has order 55. By Sylow's theorem there are only two conjugacy classes of proper, nontrivial subgroups in H. Thus there are four classes altogether, consisting of subgroups of index 1, 5, 11, and 55. Let $X = \{a, b\}^{\pm}$, let $\mathcal{R}$ define the free group on $\{a, b\}$, let $\mathcal{V}_1 = \{a^{11}, b^5, bab^{-1}a^{-3}\}$, let

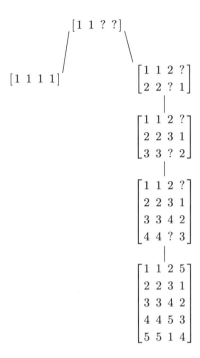

Figure 5.6.2

$\mathcal{V}_2 = \mathcal{U} = \emptyset$, and let $n = 11$. If we take $\prec$ to be the length-plus-lexicographic ordering with $a \prec a^{-1} \prec b \prec b^{-1}$, then the LOW_INDEX search tree has 244 nodes. However, if $\prec$ is the basic wreath product ordering with the same restriction to X, then the tree has only 18 nodes. Let us look at this tree. There are only two descendants of the root. They are

$$[1\ 1\ ?\ ?], \quad \begin{bmatrix} 2 & ? & ? & ? \\ ? & 1 & ? & ? \end{bmatrix}.$$

The subtree rooted at $[1\ 1\ ?\ ?]$ is shown in Figure 5.6.2.
    The only descendant of $\begin{bmatrix} 2 & ? & ? & ? \\ ? & 1 & ? & ? \end{bmatrix}$ is

$$\begin{bmatrix} 2 & ? & ? & ? \\ 3 & 1 & ? & ? \\ ? & 2 & ? & ? \end{bmatrix}.$$

The only descendant of this is

$$\begin{bmatrix} 2 & ? & ? & ? \\ 3 & 1 & ? & ? \\ 4 & 2 & ? & ? \\ ? & 3 & ? & ? \end{bmatrix}.$$

We continue in this way with each node having only one descendant until we reach

$$\begin{bmatrix} 2 & 11 & 1 & 1 \\ 3 & 1 & 5 & 4 \\ 4 & 2 & 9 & 7 \\ 5 & 3 & 2 & 10 \\ 6 & 4 & 6 & 2 \\ 7 & 5 & 10 & 5 \\ 8 & 6 & 3 & 8 \\ 9 & 7 & 7 & 11 \\ 10 & 8 & 11 & 3 \\ 11 & 9 & 4 & 6 \\ 1 & 10 & 8 & 9 \end{bmatrix}.$$

The reason that the wreath-product ordering produces a much smaller search tree is that with this ordering an orbit of $\text{Grp}\langle a \rangle$ is completed before the next image under b is considered. Since a has order 11, orbits of $\text{Grp}\langle a \rangle$ have length 1 or 11. Thus there are very few possibilities for the action of a. Given a on a union of orbits of $\text{Grp}\langle a \rangle$, the possibilities for b are very restricted.

The phenomenon observed in Example 6.4 is even more apparent in the following example.

Example 6.5. The group $\text{Grp}\langle a, b, c \mid a^{11} = b^5 = c^4 = (ac)^3 = c^{-1}bcb^{-2} = bab^{-1}a^{-3} = 1 \rangle$ is the first Mathieu simple group M_{11} of order 7920, which contains the group H of Example 6.4. In M_{11} there are three conjugacy classes of subgroups of index at most 12. These subgroups have index 1, 11, and 12. Let $X = \{a, b, c\}^{\pm}$, let $\mathcal{R}$ define the free group on $\{a, b, c\}$, let $\mathcal{V} = \mathcal{V}_1$ consist of the six relators for M_{11}, let $\mathcal{U} = \emptyset$, and let $n = 12$. If we take $\prec$ to be the basic wreath product ordering with $a \prec a^{-1} \prec b \prec b^{-1} \prec c \prec c^{-1}$, then the search tree has 54 nodes. However, if $\prec$ is the length-plus-lexicographic ordering with the same restriction to X, then the tree has 82928 nodes. (This number can be reduced using coset enumeration, as discussed earlier.)

The reason for the superiority of the wreath-product ordering is that it allows orbits of H to be built up completely before images under c are considered. Since H has very few permutation representations and when the procedure is working on an orbit of H it uses the efficient search of Example 6.4, the entire search tree is quite small.

Sometimes, when a systematic search for subgroups of small index in a finitely presented group fails to turn up any proper subgroups, the following probabilistic procedure is able to find such a subgroup. We start by enumerating as many cosets of the trivial subgroup as we can conveniently manage. Then two states are chosen at random, and the coincidence routine is called to force them to represent the same coset. This has the effect of adding a random element to the subgroup whose cosets are being enumerated. The enumeration is then resumed until either termination occurs or space is again used up. In the latter case, another random coincidence is made and the process repeated. It is likely that, when a subgroup of finite index is found, the subgroup is the whole group. To find proper subgroups of finite index, it may be necessary to repeat the entire procedure a number of times, making different choices for the coincidences in each run.

Exercises

6.1. Use LOW_INDEX to find the conjugacy classes of subgroups of index at most 4 in $\langle x, y \mid x^3 = y^3 = (xy)^3 = 1 \rangle$.

6.2. Find representatives for the conjugacy classes of subgroups H of index at most 4 in $\langle a, b \mid a^4 = b^4 = (ab)^3 = 1 \rangle$ such that H contains a conjugate of ab.

6.3. Let X, $\mathcal{R}$, $\mathcal{U}$, $\mathcal{V}$, and G be as defined at the beginning of this section, and let H be the subgroup of G generated by the image of $\mathcal{U}$. How should LOW_INDEX be modified to construct representatives for the conjugacy classes under the action of $N_G(H)$ of the subgroups of G which contain H and have index at most n?

6.4. Devise a way of using the technique of Knuth described in Section 1.8 to estimate the size of the search tree to be examined by LOW_INDEX.

5.7 Other applications

Coset enumeration and ideas related to it have a number of applications. The low-index subgroup procedure of Section 5.6 is one important application. This section sketches three more. In the first, we are given a finitely presented group G and a finite, complete coset automaton $\mathcal{A}$ which is compatible with the defining relations of G, so that $K(\mathcal{A})$ describes a subgroup H of G with finite index. We are asked to find a "small" generating set for H.

Example 7.1. Let $X = \{x, y\}^{\pm}$, let $\mathcal{R}$ be the classical niladic rewriting system defining the free group on $\{x, y\}$, and let $G = F/N$, where N is the normal

Table 5.7.1

	x	x^{-1}	y	y^{-1}
1	2	3	3	4
2	3	1	5	6
3	1	2	4	1
4	7	8	1	3
5	9	6	6	2
6	5	9	2	5
7	8	4	9	8
8	4	7	7	9
9	6	5	8	7

closure in F of the image of $\mathcal{V} = \{x^3, y^3, (xy)^3\}$. The coset automaton $\mathcal{A} = (\Sigma, X, E, \{1\}, \{1\})$ defined by Table 5.7.1 is compatible with $\mathcal{V}$. Thus $K(\mathcal{A})$ contains N and $K(\mathcal{A})/N$ is a subgroup H of G of index 9. By Proposition 3.7 in Chapter 4, $K(\mathcal{A})$ is generated as a group by $1 + 9(2-1) = 10$ Schreier generators, and the images of these generators in G generate H. However, it is plausible that H can be generated by considerably fewer than 10 elements.

In general, we are given X, a classical niladic rewriting system $\mathcal{R}$ on X^*, a finite set $\mathcal{V}$ of cyclically reduced words, and a finite, complete coset automaton $\mathcal{A}$ which is compatible with $\mathcal{V}$. It is useful to assume that $\mathcal{A}$ is standard. Let F be the free product of cyclic groups defined by X and $\mathcal{R}$ and let $G = F/N$, where N is the normal closure of the image of $\mathcal{V}$ in F. Our goal is to find as small a generating set as we can for $H = K(\mathcal{A})/N$.

There is no algorithm which can be guaranteed to find a generating set for H of minimal cardinality. We can always take the set of Schreier generators, but that set is usually too large to work with conveniently. If we work a little harder, often we can find smaller generating sets, but we must always consider the possibility that with just a little more effort we could reduce the number of generators even further. The procedure to be described returns a subset of the Schreier generators. It is quite possible that by considering more general elements one could find a smaller generating set.

Let CE be a standardizing Felsch coset enumeration procedure which has been modified to allow the user to specify a fifth input argument M, the maximum number of active states to be allowed. If more than M states are needed at some point to continue the enumeration, then the procedure halts, reports its failure, and returns the current coset table, which we shall assume is standard. Let n be the number of states in $\mathcal{A}$ and let M be any integer not less than n. Fix an ordering of X. Let $\mathcal{U}$ be the set of words found so far which define elements of H. Initially, $\mathcal{U}$ is the empty set. We call $CE(X, \mathcal{R}, \mathcal{U}, \mathcal{V}, M; \mathcal{B})$. Since $K(\mathcal{B})$ is a subgroup of $K(\mathcal{A})$, the number of states of $\mathcal{B}$ is at least n. If $\mathcal{B}$ is complete and has exactly n states, then

Table 5.7.2   Table 5.7.3

	x	x^{-1}	y	y^{-1}
1	2	3	4	5
2	3	1	6	7
3	1	2	8	9
4	10		5	1
5			1	4
6			7	2
7			2	6
8			9	3
9			3	8
10		4		

	x	x^{-1}	y	y^{-1}
1	2	3	3	4
2	3	1	5	6
3	1	2	4	1
4	7	8	1	3
5	9	10	6	2
6			2	5
7	8	4		
8	4	7		
9	10	5	8	
10	5	9		

$\mathcal{B}$ is $\mathcal{A}$ and $\mathcal{U}$ defines a set of generators for H. Otherwise, let (σ, x) be the lexicographically first pair such that either σ^x is not defined in $\mathcal{B}$ or σ^x is defined in $\mathcal{B}$ but has a different value from its value in $\mathcal{A}$. Let U be the first word in the length-plus-lexicographic ordering of X^* such that $1^U = \sigma$ in $\mathcal{A}$. Then $1^U = \sigma$ in $\mathcal{B}$. (Why?) Let V be the first word such that $1^V = \sigma^x$ in $\mathcal{A}$ and set $W = UxV^{-1}$. Add W to $\mathcal{U}$ and redo the enumeration.

We shall refer to the procedure just defined as GENERATORS. There are many questions which remain to be considered. Does GENERATORS always terminate? Do we have to use a Felsch coset enumeration procedure? Does it have to be a standardizing procedure? The elements of $\mathcal{U}$ are traced in CE in a particular order. Does it make any difference when the new element W is traced on the next iteration? Finding the answers to these questions will be left as an exercise. To make the task a little easier, we shall describe the operation of GENERATORS with the data of Example 7.1.

Let us take $M = 10$. Normally one would take M somewhat larger in comparison to n, but the coset tables would take too much space if we did that here. Initially $\mathcal{U}$ is empty. The enumeration produces an automaton $\mathcal{B}$ with the coset table as in Table 5.7.2. The first place that $\mathcal{A}$ and $\mathcal{B}$ differ is in the value of 1^y, which is 3 in $\mathcal{A}$ and 4 in $\mathcal{B}$. Since $3 = 1^{x^{-1}}$ in $\mathcal{A}$, the new generator W is $y(x^{-1})^{-1} = yx$.

The preceding description says to redo the enumeration with $\mathcal{U} = \{yx\}$. A careful implementation would simply force the coincidence of 4 and 3 in $\mathcal{B}$ and resume the enumeration. In either case, the new table for $\mathcal{B}$ is Table 5.7.3. Now the first place at which $\mathcal{A}$ and $\mathcal{B}$ differ is in the value of $5^{x^{-1}}$. Since in $\mathcal{A}$ we have $5^{x^{-1}} = 6$, $1^{xy} = 5$, and $1^{xy^{-1}} = 6$, our next generator is $W = xyx^{-1}yx^{-1}$, and we add W to $\mathcal{U}$. After the next enumeration, $\mathcal{B}$ agrees with $\mathcal{A}$. Thus H is generated by two elements, the images of yx and $xyx^{-1}yx^{-1}$.

The idea behind GENERATORS can be used to look for a small set of words whose normal closure is a given normal subgroup N of finite index in F. This time we assume that $\mathcal{A}$ is a finite, standard, complete coset automaton such that $N = K(\mathcal{A})$ is normal in F. To find a set $\mathcal{V}$ of words such that N is the normal closure of the image of $\mathcal{V}$, we initialize $\mathcal{V}$ to the empty set and choose an integer M greater than or equal to the index of N in F. We then iterate the following steps. First we call $\mathrm{CE}(X, \mathcal{R}, \emptyset, \mathcal{V}, M; \mathcal{B})$. If $\mathcal{B}$ agrees with $\mathcal{A}$, then we are done. If $\mathcal{B}$ is not the same as $\mathcal{A}$, then we find the first place where they differ and construct a word W as in GENERATORS. However, now before we add W to $\mathcal{V}$ and loop, we cyclically reduce W. Let us call this procedure NORMAL_GENS.

In Section 6.4, various techniques for simplifying presentations are discussed. One of these techniques is based on NORMAL_GENS. Other approaches use Tietze transformations or the Knuth-Bendix procedure.

The last application of coset enumeration will be to the problem of determining the order of a finite permutation group defined by a set of generators. The approach to be described here is called the *Schreier-Todd-Coxeter* method, and was first described in (Leon 1980b). Suppose that X is a set of permutations on $\Omega = \{1, \ldots, n\}$ and X is closed under taking inverses. Let G be the group generated by X. We shall also think of X as a set of "abstract generators" and consider the classical niladic rewriting system $\mathcal{R}$ consisting of all rules of the form $xx^{-1} \to \varepsilon$ with x in X. Let Y be the subset of X consisting of those permutations which fix 1 and let H be the subgroup of G generated by Y. We assume by induction on the degree that we know the order of H and a presentation $(Y, \mathcal{S})$ for H in terms of the generators in Y. Moreover, we assume that, given a permutation f of Ω which fixes 1, we can decide whether f is in H and, if it is, express f as a word in Y^*.

Now let Γ be the orbit of G containing 1. The order of G is $|\Gamma||G_1|$ and H is contained in G_1. Thus if we could show that $|G : H| = |\Gamma|$, we would know that $H = G_1$ and $|G| = |\Gamma||H|$. Let $\mathcal{A}_0$ be the coset automaton $(\Gamma, X, E, \{1\}, \{1\})$, where E is the set of triples (γ, x, γ^x) with γ in Γ and x in X, and let $\mathcal{A}$ be the standard automaton isomorphic to $\mathcal{A}_0$. Let $\mathcal{T}$ be the set of relators currently known to be satisfied by the permutations in X. Initially $\mathcal{T}$ is $\mathcal{S}$. Choose an integer $M \geq |\Gamma|$ and execute $\mathrm{CE}(X, \mathcal{R}, Y, \mathcal{T}, M; \mathcal{B})$. The number of states of $\mathcal{B}$ will be at least $|\Gamma|$. If $\mathcal{A}$ and $\mathcal{B}$ agree, then we know the order of G. If $\mathcal{A}$ and $\mathcal{B}$ do not agree, then we can find a word W as before which describes a Schreier generator of $K(\mathcal{A})$ which is not in $K(\mathcal{B})$. If we consider the terms of W to be permutations on Ω, then we can multiply these permutations together to get a permutation f which fixes 1. If f is in H, then we can express f as a word U in Y^*. In this case, we add WU^{-1} to $\mathcal{T}$ and repeat the enumeration. If f is not in H, then we must add f and f^{-1} to X and Y and then redetermine H. When we come back to the problem of deciding whether or not $|G : H|$ is $|\Gamma|$, we can retain the relations in $\mathcal{T}$. Eventually we shall have added enough Schreier generators to Y so that H is G_1, and we shall have found enough relators satisfied by the elements of X to prove by coset enumeration that $H = G_1$. At that point we shall know the order of G.

All of the applications discussed in this section involve performing a coset enumeration, which may be unsuccessful, examining the output of the enumeration, and then performing a closely related enumeration. To obtain efficiency, it is useful to have implementations of coset enumeration which can be stopped and then resumed with slightly modified data, such as an additional subgroup generator, an additional relator, or more memory for the coset automaton. The design of such a "restartable" coset enumeration procedure depends greatly on the strategy adopted and on the data structure used to represent edges.

Exercises

7.1. Let $G = \langle x, y \mid x^3 = y^3 = (xy)^3 = 1 \rangle$. The permutations $x = (1,2,3)(4,7,8)(5,6,9)$ and $y = (1)(2,4,5)(3,6,7)(8)(9)$ satisfy the defining relations for G. Thus there is a homomorphism from G onto the group generated by these permutations. Let H be the subgroup of G consisting of those elements whose images fix 1. Find a small set of words in $\{x, y\}^{\pm *}$ which defines a generating set for H.

7.2. The permutations $a = (1,2,6,3)(4,9,5,10)(7,11,8,12)(13,16,14,15)$ and $b = (1,4,6,5)$ $(2,7,3,8)(9,13,10,14)(11,15,12,16)$ generate a group P of order 16. Find a presentation for P in terms of a and b.

5.8 A comparison with the Knuth-Bendix procedure

So far in this chapter we have considered only the Todd-Coxeter approach to coset enumeration. However, in Section 4.10 it was pointed out that Knuth-Bendix techniques can also be used to compute important-coset automata. It is time to take a look at the possibility of using the Knuth-Bendix procedure to do coset enumeration. This section discusses a nontrivial example in which the Knuth-Bendix approach is superior to the other available coset enumeration methods.

In Examples 5.2 and 6.3 we studied the group $G = \mathrm{Grp}\langle r, s, t \mid t^{-1}rtr^{-2} = r^{-1}srs^{-2} = s^{-1}tst^{-2} = 1 \rangle$ and showed that G is trivial. This presentation was first studied in (Higman 1951). B. H. Neumann used this presentation to construct an infinite set of presentations for the trivial group. Let

$$R_1 = t^{-1}rtr^{-2}, \quad S_1 = r^{-1}srs^{-2}, \quad T_1 = s^{-1}tst^{-2}.$$

The second presentation in the set is obtained by substituting R_1, S_1, and T_1 for r, s, and t in R_1, S_1, and T_1. This gives

$$R_2 = T_1^{-1}R_1T_1R_1^{-2} = t^2s^{-1}t^{-1}st^{-1}rtr^{-2}s^{-1}tst^{-2}r^2t^{-1}r^{-1}tr^2t^{-1}r^{-1}t,$$

$$S_2 = R_1^{-1}S_1R_1S_1^{-2} = r^2t^{-1}r^{-1}tr^{-1}srs^{-2}t^{-1}rtr^{-2}s^2r^{-1}s^{-1}rs^2r^{-1}s^{-1}r,$$

$$T_2 = S_1^{-1}T_1S_1T_1^{-2} = s^2r^{-1}s^{-1}rs^{-1}tst^{-2}r^{-1}srs^{-2}t^2s^{-1}t^{-1}st^2s^{-1}t^{-1}s.$$

In $G_2 = \mathrm{Grp}\,\langle r, s, t \mid R_2 = S_2 = T_2 = 1 \rangle$, the elements defined by R_1, S_1, and T_1 are trivial by the computations in Example 5.2. But then by the same computations, G_2 is trivial. More complicated presentations of the trivial subgroup can be generated by repeating this construction.

In attempting to enumerate the cosets of the trivial subgroup in G_2, each of the coset enumeration procedures available in TEN_CE defines more than 100,000 states without finding a single coincidence. It is not known how much memory is necessary to allow one of these procedures to show that G_2 is trivial. However, a slightly modified version of KBS_2 is able to prove this result fairly easily. The modification referred to is the addition of a fourth input argument m, which is a positive integer, and the replacement of the statement

Push (AQ_j, Q_iC) onto the stack

in OVERLAP_2 by the statement

If $|A| + |B| + |C| \leq m$ then push (AQ_j, Q_iC) onto the stack

Thus m sets a bound on the length of the overlaps ABC which are to be processed.

Let $X = \{r, s, t\}^{\pm}$ and let $\mathcal{R}$ define the free group on $\{r, x, t\}$. Let $\prec$ be the length-plus-lexicographic ordering of X^* with $r \prec r^{-1} \prec s \prec s^{-1} \prec t \prec t^{-1}$ and let $\mathcal{S}$ consist of the rules in $\mathcal{R}$ together with the rules

$$R_2 \to \varepsilon, \quad S_2 \to \varepsilon, \quad T_2 \to \varepsilon.$$

After KBS_2$(X, \prec, \mathcal{S}, 26; \mathcal{T})$, the rewriting system $\mathcal{T}$ consists of the six rules

$$r \to \varepsilon, \quad r^{-1} \to \varepsilon, \quad s \to \varepsilon, \quad s^{-1} \to \varepsilon, \quad t \to \varepsilon, \quad t^{-1} \to \varepsilon.$$

Because overlaps are not processed if they have length greater than 26, a great deal of information is thrown away. Despite this, the procedure is able to conclude that G_2 is trivial. During the computation, 8388 rules are generated, with a maximum of 2522 active at one time. Since the left and right sides of these rules have lengths at most 26, the total length of the active rules never exceeds $2522 \cdot 52 = 131144$. Thus 150,000 bytes is enough to store the active rules. In order to get a good execution time, additional memory must be used for an index structure. Even with an index, the total amount of memory used is well under a megabyte. This is much better than any known traditional coset enumeration procedure.

In examples like this one, where the final answer is small but long words are involved at intermediate stages of the verification, the use of the Knuth-Bendix procedure should be considered seriously.

Exercises

8.1. As described in this chapter, the Todd-Coxeter approach to coset enumeration appears to be associated with length-plus-lexicographic orderings on free monoids. Are there useful Todd-Coxeter coset enumeration procedures which make use of other reduction orderings, such as wreath product orderings?

8.2. We have seen that the Knuth-Bendix procedure can be used to perform coset enumeration. It is also possible to use coset enumeration to construct confluent rewriting systems. Suppose that $\mathcal{A}$ is a finite, standard, complete coset automaton with respect to $\mathcal{R}$ such that $K(\mathcal{A})$ is normal in $F = \mathrm{Mon}\langle X \mid \mathcal{R}\rangle$. Show how to construct a reduced, confluent rewriting system which defines $F/K(\mathcal{A})$.

5.9 Historical notes

Coset enumeration as a formal procedure originated with (Todd & Coxeter 1936). However, there are arguments in Chapter XIII of [Dickson 1958], originally published in 1901, which have the feel of coset enumerations. The first efforts to produce a computer implementation of coset enumeration were by Haselgrove in 1953, but computer memories were too small at that time for useful results to be obtained. More information about early implementations can be found in (Leech 1963). Important papers in the development of the HLT strategy are (Trotter 1964), (Cannon et al. 1973), and (Leon 1980b). The Felsch strategy was proposed in (H. Felsch 1961). The paper (Atkinson 1981) was the first to suggest special treatment for generators of finite order greater than 2.

In connection with the study of finite simple groups one sometimes encounters a coset enumeration of a very special type. The group is $G = \mathrm{Grp}\langle X \mid \mathcal{R}\rangle$ and the subgroup whose cosets are to be enumerated is generated by $X - \{z\}$ for some element z of X. Moreover, the relations in $\mathcal{R}$ which do not involve z define a finite group H and we know how to compute efficiently in H with elements and subgroups, for example, because we know a faithful permutation representation of H of small degree. Finally, among the relations involving z are some which show that z normalizes a large subgroup of (the image of) H. Given all this information, it is possible to define a specialized coset enumeration procedure called *enumeration of double cosets*. One version of this procedure is described in (Linton 1991a).

An implementation of the low-index subgroup procedure was developed by the author during the period 1964–5. This program emphasized the use of coset enumeration in a manner similar to the procedure GENERATORS of Section 5.7. Independently, students of Joachim Neubüser developed a somewhat different approach. Conversations on this topic took place at the Oxford conference of 1967. The work of Neubüser's students reported in (Dietze & Schaps 1974) placed emphasis on the backtrack search approach.

Procedures analogous to coset enumeration can be defined whenever one has a class of algebraic structures for which there are free objects. If G is a group and K is a field, then the class of G-modules defined over K has free objects. An analogue of the low-index subgroup procedure can be used to determine, for a finitely presented group G, all G-modules of dimension not exceeding a prescribed bound over a given finite field. An implementation of this technique is discussed in (Linton 1991b).

6

The Reidemeister-Schreier procedure

In Proposition 3.5 of Chapter 4, we proved that subgroups of finite index in finitely generated groups are finitely generated. In this chapter we shall give a constructive proof that subgroups of finite index in finitely presented groups are finitely presented. The algorithm developed as part of the proof is the last of the major tools available for studying subgroups of finitely presented groups. This algorithm is usually referred to as the Reidemeister-Schreier procedure.

Let G be a group given by a presentation and let H be a subgroup of G. In order to obtain a presentation for H, we need to determine some additional information about H and the way H is embedded in G. Section 6.1 describes the data needed, demonstrates how to get the presentation from that data, and shows how to derive the data in a special case. Section 6.2 gives some examples showing how to derive the data in the general case using coset enumeration. Section 6.3 formalizes the procedure. The initial presentations of subgroups are often unpleasant. Section 6.4 discusses ways of simplifying these presentations.

6.1 Presentations of subgroups

For the time being we shall work with monoid presentations. Let $G = \operatorname{Mon} \langle X \mid S \rangle$ and assume that G is a group. The image in G of a word U in X^* will be denoted $[U]$, and $\equiv$ will be the congruence on X^* generated by S. Let H be a subgroup of G. Our goal is to find a set Y and a subset T of $Y^* \times Y^*$ such that H is isomorphic to $\operatorname{Mon} \langle Y \mid T \rangle$. Eventually we shall assume that X and S are finite and that H has finite index in G. In this case we shall want Y and T to be finite.

Suppose that Y is a set and $g : Y^* \to X^*$ is a monoid homomorphism. We can define a monoid homomorphism $h : Y^* \to G$ by $h(A) = [g(A)]$. Let us assume that h maps Y^* onto H. To save a little space as we construct a presentation for H relative to h we shall write $\langle A \rangle$ for $g(A)$ if A is in Y^*.

The cosets of H in G are described by the Schreier automaton $\mathcal{A}_s$ for G relative to H. However, we shall need more information than is contained in $\mathcal{A}_s$. Let $\mathcal{A} = (\Sigma, X, E, \{\alpha\}, \{\alpha\})$ be any automaton isomorphic to $\mathcal{A}_s$. Then $H = \{[U] \mid U \in X^*, \alpha^U = \alpha\}$. For each σ in Σ let W_σ be a word in X^* such that $\alpha^{W_\sigma} = \sigma$. The elements $[W_\sigma]$ with σ in Σ form a set of right coset representatives for H in G. We shall always assume that $W_\alpha = \varepsilon$. If (σ, x, τ) is in E, then $H[W_\sigma x] = H[W_\tau]$, so $[W_\sigma x][W_\tau]^{-1}$ is in H. By our assumption concerning g, there is a word P in Y^* such that $[\langle P\rangle] = [W_\sigma x][W_\tau]^{-1}$, or, equivalently, $W_\sigma x \equiv \langle P\rangle W_\tau$. The word P is not unique. For each edge (σ, x, τ) in E choose one such P and let D be the set of quadruples (σ, x, P, τ) so obtained. The quintuple $\mathcal{E} = (\Sigma, X, Y, D, \{\alpha\})$ will be called an *extended Schreier automaton* for G relative to g and the choice of the words W_σ. An element $e = (\sigma, x, P, \tau)$ in D will still be called an edge. The generator x is the *primary label* of e, and P is the *secondary label*. A path in $\mathcal{E}$ now has two signatures: the *primary signature*, which is the product of the primary labels on the edges of the path, and the *secondary signature*, which is the product of the secondary labels. For any word U in X^* and any σ in Σ, there is a unique path Q in $\mathcal{E}$ which starts at σ and has U as its primary signature. The secondary signature of Q will be denoted $P(\sigma, U)$. The endpoint of Q will continue to be written σ^U. If (σ, x, P, τ) is in D, then $P = P(\sigma, x)$.

Proposition 1.1. *Suppose that σ is in Σ and U and V are in X^*. Let $\tau = \sigma^U$. Then $P(\sigma, UV) = P(\sigma, U)P(\tau, V)$ and $W_\sigma U \equiv \langle P(\sigma, U)\rangle W_\tau$.*

Proof. The first equation simply says that the secondary signature of the concatenation of two paths is the product of the signatures. Suppose that $U = x_1 \ldots x_s$ with each x_i in X. Let the path from σ to τ with primary signature U consist of the edges $(\sigma_{i-1}, x_i, P_i, \sigma_i)$, $1 \le i \le s$. Thus $\sigma_0 = \sigma$, $\sigma_s = \tau$, and $P(\sigma, U) = P_1 \ldots P_s$. Then

$$W_\sigma U = W_{\sigma_0} x_1 \ldots x_s \equiv \langle P_1\rangle W_{\sigma_1} x_2 \ldots x_s \equiv \langle P_1\rangle \langle P_2\rangle W_{\sigma_2} x_3 \ldots x_s$$
$$\equiv \langle P_1\rangle \ldots \langle P_s\rangle W_{\sigma_s} = \langle P_1 \ldots P_s\rangle W_\tau = \langle P(\sigma, U)\rangle W_\tau. \quad \square$$

An extended Schreier automaton $\mathcal{E} = (\Sigma, X, Y, D, \{\alpha\})$ has an *extended transition table*. As with ordinary transition tables, the rows are indexed by Σ and the columns by X. The entry in row σ and column x contains $P(\sigma, x)$ followed by σ^x, normally written without separators or delimiters. If $P(\sigma, x) = \varepsilon$, then the entry consists of σ^x alone. (The term "extended transition table" is used rather than "extended coset table" because at this point we are not assuming that $\mathcal{S}$ contains a classical niladic rewriting system.)

Example 1.1. Probably the simplest possible extended tables occur when $H = G$, $Y = X$, and g is the identity map. If $\Sigma = \{1\}$, then for each x in X we can take $P(1, x)$ to be x. For example, if $X = \{x, y, z\}$, then the following is an extended transition table:

	x	y	z
1	$x1$	$y1$	$z1$

Example 1.2. Suppose that G is a finite cyclic group generated by an element x of order divisible by 3 and H is the unique subgroup of G of index 3. Let $X = \{x, x^{-1}\}$, $Y = \{y, y^{-1}\}$, $g(y) = x^3$, and $g(y^{-1}) = x^{-3}$. If $\Sigma = \{1, 2, 3\}$ and $\alpha = 1$, then an extended table for G relative to g and the choice $W_1 = \varepsilon$, $W_2 = x$, $W_3 = x^2$ is

	x	x^{-1}
1	2	$y^{-1}3$
2	3	1
3	$y1$	2

To check the $(3, x)$-entry, we note that $[W_3 x] = [x^2 x] = [x^3] = [\langle y \rangle W_1]$. If W_3 is changed to x^{-1}, then one possible extended transition table is

	x	x^{-1}
1	2	3
2	$y3$	1
3	1	$y^{-1}2$

Let $\mathcal{E} = (\Sigma, X, Y, D, \{\alpha\})$ be an extended Schreier automaton for G relative to g and a choice of coset representatives. Suppose that U and V are words in X^* such that $U \equiv V$ and suppose that σ is in Σ. Then $[U] = [V]$, so $\sigma^U = \sigma^V = \tau$, say. Then

$$\langle P(\sigma, U) \rangle W_\tau \equiv W_\sigma U \equiv W_\sigma V \equiv \langle P(\sigma, V) \rangle W_\tau.$$

Since G is a group, this implies that $\langle P(\sigma, U) \rangle \equiv \langle P(\sigma, V) \rangle$. Thus the relation $P(\sigma, U) = P(\sigma, V)$ holds in H with respect to the homomorphism h.

Now suppose that y is in Y. Then $\langle y \rangle$ defines an element of H, so $\alpha^{\langle y \rangle} = \alpha$. Therefore

$$\langle y \rangle = W_\alpha \langle y \rangle \equiv \langle P(\alpha, \langle y \rangle) \rangle W_\alpha = \langle P(\alpha, \langle y \rangle) \rangle.$$

Hence the relation $y = P(\alpha, \langle y \rangle)$ also holds in H with respect to h.

Let T_1 be the subset of $Y^* \times Y^*$ consisting of the pairs $(P(\sigma, S), P(\sigma, T))$, where σ ranges over Σ and (S, T) ranges over $\mathcal{S}$. Let T_2 be the set of pairs

$(y, P(\alpha, \langle y \rangle))$ with y in Y. Set $\mathcal{T} = \mathcal{T}_1 \cup \mathcal{T}_2$ and let $\approx$ be the congruence on Y^* generated by $\mathcal{T}$.

Proposition 1.2. *The pair $(Y, \mathcal{T})$ is a monoid presentation for H relative to h.*

Proof. Let Q and R be words in Y^*. We want to show that $[\langle Q \rangle] = [\langle R \rangle]$ if and only if $Q \approx R$. But $[\langle Q \rangle] = [\langle R \rangle]$ if and only if $\langle Q \rangle \equiv \langle R \rangle$. Thus we need to prove that $\langle Q \rangle \equiv \langle R \rangle$ if and only if $Q \approx R$. The preceding arguments show that $\langle B \rangle \equiv \langle C \rangle$ for all (B, C) in $\mathcal{T}$. The condition $\langle Q \rangle \equiv \langle R \rangle$ defines a congruence on Y^* which contains $\approx$. Therefore, if $Q \approx R$, then $\langle Q \rangle \equiv \langle R \rangle$. It remains to prove the reverse implication.

Lemma 1.3. *If U and V are words in X^* such that $U \equiv V$, then $P(\sigma, U) \approx P(\sigma, V)$ for all σ in Σ.*

Proof. Suppose that $U \equiv V$. There exists a sequence of words $U = U_0$, $U_1, \ldots, U_s = V$ such that U_{i+1} is obtained from U_i by replacing a subword S by a word T, where either (S, T) or (T, S) is in $\mathcal{S}$. A simple induction argument shows that we need only consider the case $s = 1$. Thus we may assume that $U = ASB$, $V = ATB$, and (S, T) is in $\mathcal{S}$. By Proposition 1.1,

$$P(\sigma, U) = P(\sigma, A)P(\rho, S)P(\tau, B),$$

where $\rho = \sigma^A$ and $\tau = \rho^S$. But $S \equiv T$, so $\tau = \rho^T$ as well. Therefore

$$P(\sigma, V) = P(\sigma, A)P(\rho, T)P(\tau, B).$$

Now $P(\rho, S) \approx P(\rho, T)$ by the definition of $\approx$. Since $\approx$ is a congruence, $P(\sigma, U) \approx P(\sigma, V)$. $\square$

Lemma 1.4. *If Q is in Y^*, then $Q \approx P(\alpha, \langle Q \rangle)$.*

Proof. Let $Q = y_1 \ldots y_s$ with each y_i in Y. Then $\langle Q \rangle = \langle y_1 \rangle \ldots \langle y_s \rangle$ and

$$P(\alpha, \langle Q \rangle) = P(\sigma_1, \langle y_1 \rangle)P(\sigma_2, \langle y_2 \rangle) \ldots P(\sigma_s, \langle y_s \rangle),$$

where $\sigma_1 = \alpha$ and $\sigma_{i+1} = \sigma_i^{\langle y_i \rangle}$, $1 \le i < s$. But $\langle y_i \rangle$ defines an element of H, so $\sigma_1 = \sigma_2 = \cdots = \sigma_s = \alpha$. Since $y_i \approx P(\alpha, \langle y_i \rangle)$ for each i, it follows that $Q \approx P(\alpha, \langle Q \rangle)$. $\square$

Now suppose that $\langle Q \rangle \equiv \langle R \rangle$ for two elements Q and R of Y^*. Then by Lemmas 1.3 and 1.4,

$$Q \approx P(\alpha, \langle Q \rangle) \approx P(\alpha, \langle R \rangle) \approx R. \quad \square$$

To apply Proposition 1.2, we need the presentation $(X, \mathcal{S})$, the homomorphism g, and an extended Schreier automaton $\mathcal{E} = (\Sigma, X, Y, D, \{\alpha\})$. The words W_σ do not play a role in the construction of $\mathcal{T}$, but their existence is part of the definition of an extended Schreier automaton. It is tempting to try to give an axiomatic treatment for objects one might call complete extended coset automata similar to the axiomatic treatment of coset automata given in Section 4.4. However, there are serious difficulties connected with such an attempt. Given X, $\mathcal{S}$, and $\mathcal{E}$, it is not obvious how to decide whether $\mathcal{E}$ really is an extended Schreier automaton for the subgroup of all elements $[U]$ of G such that $\alpha^U = \alpha$. In order to know that $\mathcal{E}$ is an extended Schreier automaton, it appears that we have to know how $\mathcal{E}$ was constructed.

It is useful to give a name to the procedure for constructing the set $\mathcal{T}$ of Proposition 1.2. Let us call the procedure RS_BASIC. It is invoked as follows: RS_BASIC$(X, \mathcal{S}, Y, g, \mathcal{E}; \mathcal{T})$. The input arguments are the presentation for G, the set Y, the homomorphism g, and the extended Schreier automaton. The output is the set of defining relations for H.

The easiest way to get data which can correctly be fed into RS_BASIC is to start with a monoid presentation $(X, \mathcal{S})$ which is known to define a group G and a homomorphism f of G into the symmetric group on a finite set Σ such that the image of G is transitive. If α is in Σ and H is the stabilizer of α in G, then we can get a presentation for H in terms of Schreier generators.

Procedure RS_SGEN$(X, \mathcal{S}, \Sigma, f, \alpha; Y, g, \mathcal{T})$;
Input: X : a finite set;
           $\mathcal{S}$: a finite subset of $X^* \times X^*$ such that $G = \mathrm{Mon} \langle X \mid \mathcal{S} \rangle$ is a group;
           Σ : a finite set;
           f : a map from X to $\mathrm{Sym}(\Sigma)$ such that the permutations $f(x)$ with x in X satisfy the relations in $\mathcal{S}$ and generate a transitive group;
           α : an element of Σ;
Output: Y : a finite set;
           g : a homomorphism from Y^* to X^*;
           $\mathcal{T}$: a finite subset of $Y^* \times Y^*$ such that $(Y, \mathcal{T})$ is a monoid presentation for the stabilizer H of α in G relative to the composition h of g and the map $U \mapsto [U]$;
Begin
   Let E be the set of triples (σ, x, τ) with σ and τ in Σ and x in X such that τ is the image of σ under $f(x)$;
   $\mathcal{A} := (\Sigma, X, E, \{\alpha\}, \{\alpha\})$;   (* $\mathcal{A}$ is a trim automaton. *)
   For σ in Σ do let W_σ be a word in X^* such that $\alpha^{W_\sigma} = \sigma$ and let V_σ be a word in X^* representing the inverse of $[W_\sigma]$ in G;

(* The words W_α and V_α are taken to be ε. *)
Let Y be a set in one-to-one correspondence with E;
$D := \emptyset$;

For σ in Σ do
  For x in X do begin
    Let y in Y correspond to (σ, x, τ) in E;
    $g(y) := W_\sigma x V_\tau$;
    Add (σ, x, y, τ) to D
  End;
  $\mathcal{E} := (\Sigma, X, Y, D, \{\alpha\})$; RS_BASIC$(X, \mathcal{S}, Y, g, \mathcal{E}; \mathcal{T})$
End.

Our requirement that G in RS_SGEN be a group includes the assumption that we can compute inverses. That is, given a word W in X^*, we can find a word V representing the inverse of $[W]$. Since G acts transitively on Σ, the right cosets of H in G correspond to the elements of Σ. Therefore $\mathcal{A}$ is isomorphic to the Schreier automaton of G relative to H. In Lemma 3.6 of Chapter 4 we proved that the Schreier generators $[W_\sigma x V_\tau]$ with (σ, x, τ) in E do in fact generate H. Thus g maps Y onto a set of words whose image in G generates H. Moreover, if (σ, x, y, τ) is in D, then

$$W_\sigma x \equiv W_\sigma x V_\tau W_\tau = \langle y \rangle\, W_\tau,$$

since $V_\tau W_\tau \equiv \varepsilon$. Therefore $\mathcal{E}$ is an extended Schreier automaton for G relative to g and the choice of the W_σ. Hence RS_BASIC returns a set of defining relations for H.

Example 1.3. The monoid $G = \text{Mon} \langle u, v \mid u^4 = v^4 = (uv)^4 = 1 \rangle$ is obviously a group, since $[u]^{-1}$ and $[v]^{-1}$ can be taken to be $[u^3]$ and $[v^3]$, respectively. Let $\Sigma = \{1, 2\}$ and $\alpha = 1$. Suppose that f takes u to $(1, 2)$ and v to the identity $(1)(2)$. The input assumptions of RS_SGEN are satisfied. The set E is defined by the following table:

	u	v
1	2	1
2	1	2

A set of coset representatives is given by $W_1 = \varepsilon$ and $W_2 = u$, and we can take V_1 and V_2 to be ε and u^3, respectively. Let $Y = \{a, b, c, d\}$, where the extended edges are given by

	u	v
1	a2	b1
2	c1	d2

Now $\langle a \rangle = g(a) = W_1 u V_2 = \varepsilon u u^3 = u^4$. Similarly, $g(b) = v$, $g(c) = u^2$, and $g(d) = u v u^3$. There are three relations in $\mathcal{S}$, so RS_BASIC produces six relations for H of the type $P(\sigma, S) = P(\sigma, T)$ with (S, T) in $\mathcal{S}$. These relations are

$$P(1, u^4) = (ac)^2 = 1,$$
$$P(2, u^4) = (ca)^2 = 1,$$
$$P(1, v^4) = b^4 = 1,$$
$$P(2, v^4) = d^4 = 1,$$
$$P(1, (uv)^4) = (adcb)^2 = 1,$$
$$P(2, (uv)^4) = (cbad)^2 = 1.$$

There are four elements of Y and hence there are four relations $y = P(1, \langle y \rangle)$. They are

$$a = P(1, u^4) = (ac)^2,$$
$$b = P(1, v) = b,$$
$$c = P(1, u^2) = ac,$$
$$d = P(1, uvu^3) = adcac.$$

Since $a = (ac)^2 = 1$, we can eliminate a from the presentation using a Tietze transformation. A few more simple transformations lead to the presentation $H \cong \mathrm{Mon} \langle b, c, d \mid b^4 = c^2 = d^4 = (dcb)^2 = 1 \rangle$.

The presentations obtained using RS_SGEN will usually be simpler if we make the following assumptions:

(i) $\mathcal{S}$ contains a classical niladic rewriting system $\mathcal{R}$ on X^*.
(ii) The word W_σ is the first word with respect to some fixed reduction ordering of X^* such that $\alpha^{W_\sigma} = \sigma$.
(iii) The word V_σ is W_σ^{-1}, which is defined using $\mathcal{R}$.

Let us assume that these conditions are satisfied. Suppose that (σ, x, a, τ) and $(\tau, x^{-1}, b, \sigma)$ are in D. Then the relations $P(\sigma, xx^{-1}) = 1$ and $P(\tau, x^{-1}x) = 1$ yield $ab = ba = 1$, so a and b define inverse elements of H. It is possible that $g(a) = W_\sigma x W_\tau^{-1}$ can be rewritten to ε using $\mathcal{R}$. Since W_σ and W_τ are irreducible with respect to $\mathcal{R}$, this can happen only when $W_\tau = W_\sigma x$ or $W_\sigma = W_\tau x^{-1}$. In this case, both a and b define the identity of H and can be removed from the presentation by a Tietze transformation. Thus we can take Y to be a set in one-to-one correspondence with the subset E_1 of E

consisting of those edges (σ, x, τ) such that $W_\sigma x W_\tau^{-1}$ cannot be rewritten to ε using $\mathcal{R}$, and we define D to consist of the quadruples (σ, x, a, τ), where a in Y corresponds to (σ, x, τ) in E_1, together with the quadruples $(\sigma, x, \varepsilon, \tau)$ with (σ, x, τ) in $E - E_1$. Let $\mathcal{Q}$ be the set of pairs (ab, ε), where a in Y corresponds to (σ, x, τ) in E_1 and b corresponds to (τ, x^{-1}, σ). Then $\mathcal{Q}$ is a classical niladic rewriting system on Y^*, the relations in $\mathcal{Q}$ are satisfied in H, and all relations $P(\sigma, xx^{-1}) = 1$ are consequences of $\mathcal{Q}$.

Suppose that σ is in Σ. By induction on the length of W_σ it is easy to show that $P(\alpha, W_\sigma) = \varepsilon = P(\sigma, W_\sigma^{-1})$. If a in Y corresponds to (σ, x, τ) in E_1, then

$$P(\alpha, \langle a \rangle) = P(\alpha, W_\sigma x W_\tau^{-1}) = P(\alpha, W_\sigma) P(\sigma, x) P(\tau, W_\tau^{-1}) = \varepsilon a \varepsilon = a.$$

Therefore the relation $a = P(\alpha, \langle a \rangle)$ reduces to the identity $a = a$ and can be ignored. Thus H is defined by the relations in $\mathcal{Q}$ together with the relations $P(\sigma, S) = P(\sigma, T)$, where (S, T) ranges over $\mathcal{S} - \mathcal{R}$. This version of RS_SGEN will be called RS_SGEN2.

The preceding discussion provides the proof of the following result.

Proposition 1.5. *Let H be a subgroup of index n in a group G defined by a group presentation with r generators and s relations. Then H has a group presentation with $1 + n(r - 1)$ generators and ns relations.*

Example 1.4. Let $M = \mathrm{Grp} \langle x, y \mid (xyx^{-1}y^{-1})^2 = 1 \rangle$. The map $x \mapsto (1, 2, 3)$ and $y \mapsto (2, 3, 4)$ defines a homomorphism from M to $\mathrm{Sym}(4)$, whose image is transitive. We shall find a presentation for the stabilizer H in M of $\alpha = 1$. To get a monoid presentation, we add generators x^{-1} and y^{-1} and the relations $xx^{-1} = x^{-1}x = yy^{-1} = y^{-1}y = 1$. The set E of edges in the automaton $\mathcal{A}$ of RS_SGEN is given by the following table.

	x	x^{-1}	y	y^{-1}
1	2	3	1	1
2	3	1	3	4
3	1	2	4	2
4	4	4	2	3

Using the length-plus-lexicographic ordering with $x \prec x^{-1} \prec y \prec y^{-1}$, we get the coset representatives $W_1 = \varepsilon$, $W_2 = x$, $W_3 = x^{-1}$, and $W_4 = xy^{-1}$. The set E_1 has cardinality 10. There are no pairs of the form (a^2, ε) in $\mathcal{Q}$. Thus it is reasonable to take $Y = \{a, b, c, d, e\}^\pm$ and to associate the elements of Y with Schreier generators so that the edges of $\mathcal{E}$ are given by the following extended table:

	x	x^{-1}	y	y^{-1}
1	2	3	$a1$	$a^{-1}1$
2	$b3$	1	$c3$	4
3	1	$b^{-1}2$	$d4$	$c^{-1}2$
4	$e4$	$e^{-1}4$	2	$d^{-1}3$

The map g is defined by

$$g(a) = y, \quad g(b) = x^3, \quad g(c) = xyx, \quad g(d) = x^{-1}y^2x^{-1}, \quad g(e) = xy^{-1}xyx^{-1},$$

and $g(a^{-1}) = g(a)^{-1}, \ldots, g(e^{-1}) = g(e)^{-1}$. The relations in $\mathcal{Q}$ define the free group on $\{a, b, c, d, e\}$. The subgroup H is defined by $\mathcal{Q}$ and the four relations $P(\sigma, (xyx^{-1}y^{-1})^2) = 1$. These relations are

$$cb^{-1}ea^{-1} = ea^{-1}cb^{-1} = bde^{-1}d^{-1}ac^{-1} = d^{-1}ac^{-1}bde^{-1} = 1.$$

The first two relators and the last two relators are cyclic permutations of each other. We may use the first relator to find that $e = bc^{-1}a$ and then apply a Tietze transformation to get $H = \mathrm{Grp}\langle a, b, c, d \mid bda^{-1}cb^{-1}d^{-1}ac^{-1} = 1\rangle$.

Exercises

1.1. Suppose that $\mathcal{R}$ is a classical niladic rewriting system on X^* and H is a subgroup of $F = \mathrm{Mon}\langle X \mid \mathcal{R}\rangle$. Let Σ be the set of right cosets of H in F and for x in X let $f(x)$ be the permutation of Σ induced by right multiplication by $[x]$. Show that it makes sense to invoke RS_SGEN2 even if Σ is infinite and that the set $\mathcal{T}$ of relations produced is a classical niladic rewriting system on Y^*. What does this say about the structure of subgroups of F? In particular, what does it say about subgroups of free groups?

1.2. Suppose in RS_SGEN2 that $\mathcal{S} = \mathcal{R} \cup \mathcal{V}$, where $\mathcal{R}$ is a classical niladic rewriting system on X^* and $\mathcal{R} \cap \mathcal{V} = \emptyset$. Let $\mathcal{Q}$ be the niladic rewriting system on Y^* consisting of the relations $P(\alpha, L) = 1$, where L is a left side in $\mathcal{R}$. Let σ be in Σ and let (V^m, ε) be in $\mathcal{V}$. Show that the words $P(\sigma, V^m)$ and $P(\sigma^V, V^m)$ are cyclic permutations of each other. Thus, in the presence of the relations in $\mathcal{Q}$, each of these two relators implies the other. Conclude that we only need to form the relators $P(\sigma, V^m)$ as σ ranges over a set of representatives for the cycles of V acting on Σ.

1.3. There is a homomorphism of $G = \mathrm{Grp}\langle x, y \mid x^3 = y^3 = (xy)^3 = 1\rangle$ into $\mathrm{Sym}(7)$ in which x maps to $(1)(2, 4, 5)(3, 6, 7)$ and y maps to $(1, 2, 3)(4, 7, 6)(5)$. Find a presentation for the stabilizer H of 1 in G in terms of Schreier generators.

6.2 Examples of extended coset enumeration

In Section 6.1 we learned how to find a finite presentation for a subgroup H of finite index in a finitely presented group G when the generators for H are taken to be Schreier generators. However, we may need to have a presentation for H in terms of other generators. As an important special case, we may have a second generating set for G, and we would like to have a presentation for G in terms of these generators.

In this section we shall look at some examples of a procedure called *extended coset enumeration*, which produces the extended Schreier automata needed to get presentations of subgroups of finite index in terms of specific sets of generators. (Other authors have referred to this procedure as modified or generalized coset enumeration.) In Section 6.3, a formal description will be given of an extended coset enumeration procedure based on the HLT strategy of coset enumeration. Any coset enumeration procedure can be modified to obtain an extended procedure. The HLT procedure will be used because it is the simplest to describe.

The input arguments X, $\mathcal{R}$, and $\mathcal{V}$ of a coset enumeration procedure are retained in the extended procedure. Thus X is a finite set, $\mathcal{R}$ is a classical niladic rewriting system on X^*, and $\mathcal{V}$ is a finite subset of X^* giving defining relators for G as a quotient of the group $F = \mathrm{Mon}\,\langle X \mid \mathcal{R}\rangle$. The set $\mathcal{U}$ of words defining group generators for H is replaced by three new input arguments: a finite set Y, a classical rewriting system $\mathcal{Q}$ on Y^*, and a homomorphism $g\colon Y^* \to X^*$ which defines a homomorphism of $\mathrm{Mon}\,\langle Y \mid \mathcal{Q}\rangle$ onto H. As before, we shall usually write $\langle Q\rangle$ for $g(Q)$.

Given the set $\mathcal{U}$, we can construct one triple Y, $\mathcal{Q}$, g as follows: Replace $\mathcal{U}$ by $\mathcal{U} \cup \mathcal{U}^{-1}$, if necessary, so that $\mathcal{U} = \mathcal{U}^{-1}$. Take Y to be a set with $|\mathcal{U}|$ elements, choose a bijection $g\colon Y \to \mathcal{U}$, and extend g to a homomorphism from Y^* to X^*. Let $\mathcal{Q}$ consist of the pairs (ab, ε) with a and b in Y and $g(b) = g(a)^{-1}$.

Example 2.1. Let $G = \mathrm{Grp}\,\langle x, y \mid x^2 = y^3 = (xy)^4 = 1\rangle$ and let H be the subgroup of G generated by the images of xyx and yxy^{-1}. The obvious choice is to take $X = \{x, y, y^{-1}\}$ and let $\mathcal{R}$ consist of the rules $x^2 \to \varepsilon$, $yy^{-1} \to \varepsilon$, $y^{-1}y \to \varepsilon$. A simple coset enumeration shows that $|G : H| = 4$ and produces the following coset table:

	x	y	y^{-1}
1	2	3	4
2	1	2	2
3	3	4	1
4	4	1	3

To do the corresponding extended enumeration, we must first choose Y, g, and $\mathcal{Q}$. Since $(yxy^{-1})^{-1} = yxy^{-1}$, we can take $Y = \{u, u^{-1}, v\}$, define $g(u) = xyx$, $g(u^{-1}) = xy^{-1}x$, and $g(v) = yxy^{-1}$, and let $\mathcal{Q}$ consist of the rules $uu^{-1} \to \varepsilon$, $u^{-1}u \to \varepsilon$, $v^2 \to \varepsilon$.

As we build the extended coset automaton $\mathcal{E} = (\Sigma, X, Y, D, \{1\})$, we must construct words W_σ which give coset representatives. Initially, $\Sigma = \{1\}$ and

$W_1 = \varepsilon$. When a new state τ is defined as σ^z for some z in X, then W_τ is defined to be $W_\sigma z$. Edges $(\sigma, z, \varepsilon, \tau)$ and $(\tau, z^{-1}, \varepsilon, \sigma)$ are added to D.

The first step in the enumeration is to perform full two-sided traces of the subgroup generators at 1. The first generator is $g(u) = xyx$. In the trace, we define 2 to be 1^x and W_2 to be x. This adds the edges $(1, x, \varepsilon, 2)$ and $(2, x, \varepsilon, 1)$ to D. Now we conclude that $2^y = 2$, so there is an edge $(2, y, Q, 2)$ in D for some word Q in Y^*. Among the defining relations to be constructed for H is the relation $u = P(1, \langle u \rangle) = P(1, xyx)$. Let us diagram the trace of xyx at 1, including the secondary labels on the edges:

$$\begin{array}{cccc} x & y & x \\ 1 & 2 & 2 & 1. \\ \varepsilon & Q & \varepsilon \end{array}$$

The primary labels are written above and between the states encountered, and the secondary labels are written below the corresponding primary labels. Thus $P(1, xyx) = \varepsilon Q \varepsilon = Q$. The relation $u = P(1, \langle u \rangle)$ means that the elements of H represented by u and Q are the same. Thus we should take Q to be u and add $(2, y, u, 2)$ to D. We must also add an edge $(2, y^{-1}, Q_1, 2)$ to D. Another relation for H is $P(1, yy^{-1}) = P(1, \varepsilon)$. The trace of yy^{-1} at 2 is

$$\begin{array}{ccc} y & y^{-1} \\ 2 & 2 & 2. \\ u & Q_1 \end{array}$$

Hence uQ_1 must represent the trivial element of H, so we take Q_1 to be u^{-1}. At this point D has four edges as described by the following extended table:

	x	y	y^{-1}
1	2		
2	1	$u2$	$u^{-1}2$

Now we trace yxy^{-1} at 1. After defining 1^y to be 3 and W_3 to be y, we conclude that 3^x is 3. To determine the secondary label Q on the $(3, x)$-entry in the table, we look again at the trace of yxy^{-1} at 1, this time including the secondary labels.

$$\begin{array}{cccc} y & x & y^{-1} \\ 1 & 3 & 3 & 1. \\ \varepsilon & Q & \varepsilon \end{array}$$

Since the relation $v = P(1, yxy^{-1}) = \varepsilon Q \varepsilon$ must hold in H, we take $Q = v$. The extended table is now

	x	y	y^{-1}
1	2	3	
2	1	$u2$	$u^{-1}2$
3	$v3$		2

At this point we start tracing the relators at the states in Σ, starting with y^3 at 1. The definition $4 = 3^y$ is made and then $4^y = 1$ is deduced. Therefore there is an edge $(4, y, Q, 1)$ in D. The trace of y^3 with both primary and secondary labels is

$$
\begin{array}{cccc}
y & y & y \\
1 & 3 & 4 & 1. \\
\varepsilon & Q & \varepsilon
\end{array}
$$

Since y^3 is trivial in G, it follows that $P(\sigma, y^3) = Q$ represents the identity element of H. Hence we take $Q = \varepsilon$. This gives the following table:

	x	y	y^{-1}
1	2	3	4
2	1	$u2$	$u^{-1}2$
3	$v3$	4	1
4		1	3

The trace of $(xy)^4$ at 1 leads to the deduction $4^x = 4$, but again we must determine the secondary label Q on the edge $(4, x, Q, 4)$. The trace is

$$
\begin{array}{cccccccc}
x & y & x & y & x & y & x & y \\
1 & 2 & 2 & 1 & 3 & 3 & 4 & 4 & 1. \\
& u & & & v & & Q
\end{array}
$$

This time empty secondary labels have been omitted. Since $P(1, (xy)^4)$ must represent the identity element of H, we take $Q = vu^{-1}$. Our extended table is now complete:

	x	y	y^{-1}
1	2	3	4
2	1	$u2$	$u^{-1}2$
3	$v3$	4	1
4	$vu^{-1}4$	1	3

The defining relations for H can be read off using the table. We know that H is a quotient group of $L = \text{Mon}\,\langle Y \mid \mathcal{Q} \rangle$, so we start with the relations in $\mathcal{Q}$. The relations $u = P(1, \langle u \rangle)$ and $v = P(1, \langle v \rangle)$ were used in the construction of the table. They give the identities $u = u$ and $v = v$. The only relation of the form $P(\sigma, x^2) = 1$ to yield anything new is obtained by taking $\sigma = 4$. This gives $(vu^{-1})^2 = 1$, which with the relation $v^2 = 1$ is equivalent to $(uv)^2 = 1$. The relations $P(\sigma, yy^{-1}) = 1$ and $P(\sigma, y^{-1}y) = 1$ provide no new information. The only nontrivial relation $P(\sigma, y^3) = 1$ is $P(2, y^3) = u^3 = 1$. The words $P(\sigma, (xy)^4)$ are all cyclic permutations of $uvvu^{-1}$, which are all trivial in L. Thus H is $\mathrm{Grp}\,\langle u, v \mid u^3 = v^2 = (uv)^2 = 1 \rangle$.

The extended enumeration of Example 2.1 is relatively easy since no coincidences were encountered. As should be expected, the handling of coincidences in an extended enumeration is fairly complicated.

Example 2.2. Let $G = \mathrm{Grp}\,\langle a, b \mid a^4 = b^4 = (a^2b)^2 = (ab)^4 = 1 \rangle$ and let H be the subgroup of G generated by the images of a and b^2. Here we take $X = \{a, b\}^{\pm}$ and let $\mathcal{R}$ define the free group on $\{a, b\}$. An ordinary coset enumeration shows that $|G : H| = 4$ and gives the following coset table:

	a	a^{-1}	b	b^{-1}
1	1	1	2	2
2	3	3	1	1
3	2	2	4	4
4	4	4	3	3

We can choose Y and $\mathcal{Q}$ to be the same as in Example 2.1, since $(b^2)^2$ is trivial in H. The homomorphism g maps u, u^{-1}, and v to a, a^{-1}, and b^2, respectively. As before, we start the extended enumeration with $\Sigma = \{1\}$ and $W_1 = \varepsilon$. The two-sided trace of the first subgroup generator $g(u) = a$ at 1 yields $1^a = 1$. Since $u = P(1, \langle u \rangle)$ in H, the secondary label for the $(1, a)$-entry in the extended table is taken to be u. The secondary label on the paired entry is u^{-1}. Thus at the end of the trace the extended table is

	a	a^{-1}	b	b^{-1}
1	$u1$	$u^{-1}1$		

The trace of b^2 at 1 first defines 2 as 1^b and sets $W_2 = b$. Next the deduction $2^b = 1$ is made. Since $P(1, b^2)$ and v must represent the same element in H, the secondary label on the $(2, b)$-entry should be v, and the secondary label on the $(1, b^{-1})$ entry should also be v. This gives

	a	a^{-1}	b	b^{-1}
1	$u1$	$u^{-1}1$	2	$v2$
2			$v1$	1

Now we start the traces of the relators of G. We already have $1^{a^4} = 1^{b^4} = 1$. We could get some relations for H by looking at the secondary signatures

Table 6.2.1

	a	a^{-1}	b	b^{-1}
1	u1	u^{-1}1	2	v2
2	3	vu^{-2}3	v1	1
3	$u^{-2}v$2	2	4	u^{-1}5
4	5			3
5			4	u3

of the corresponding paths, but this will be done later. To complete the trace of $(a^2b)^2$ at 1, we must define 2^a to be 3 and set $W_3 = ba$. Next the deduction $3^a = 2$ is made, so there is an edge of the form $(3, a, Q, 2)$ in D. The trace of $(a^2b)^2$ at 1 with secondary labels is

$$
\begin{array}{cccccc}
a & a & b & a & a & b \\
1 & 1 & 1 & 2 & 3 & 2 & 1. \\
u & u & & Q & v &
\end{array}
$$

Again empty labels have been omitted. Since $P(1, (a^2b)^2)$ must represent the identity element of H, we take $Q = u^{-2}v$ to make u^2Qv the identity in $L = \mathrm{Mon}\langle Y \mid Q \rangle$. The label on the paired edge is vu^2. At this point the table is

	a	a^{-1}	b	b^{-1}
1	u1	u^{-1}1	2	v2
2	3	$vu^2$3	v1	1
3	$u^{-2}v$2	2		

The trace of $(ab)^4$ at 1 causes 3^b to be defined to be 4 and 4^a to be defined to be 5. The coset representatives W_4 and W_5 are bab and $baba$, respectively. The deduction $5^b = 3$ is then made. If Q is the secondary label on the $(5, b)$-entry in the table, then the trace of $(ab)^4$ at 1 is

$$
\begin{array}{cccccccc}
a & b & a & b & a & b & a & b \\
1 & 1 & 2 & 3 & 4 & 5 & 3 & 2 & 1. \\
u & & & & & Q & u^{-2}v & v &
\end{array}
$$

Now $P(1, (ab)^4) = uQu^{-2}vv$ must be the identity element in H. If we take Q to be u, then $P(1, (ab)^4)$ is the identity element in L. The secondary label on the $(3, a^{-1})$-entry should be u^{-1}. When the new edges are entered, we have Table 6.2.1.

None of the traces at 2 produces any change in the table. Neither does the trace of a^4 at 3. However, the trace of b^4 at 3 results in the definitions $6 = 4^b$ and $W_6 = bab^2$, and the deduction $6^b = 5$. If Q is the secondary label

Table 6.2.2

	a	a^{-1}	b	b^{-1}
1	$u1$	$u^{-1}1$	2	$v2$
2	3	vu^23	$v1$	1
3	$u^{-2}v2$	2	4	$u^{-1}5$
4	5		6	3
5		4	$u3$	$u6$
6			$u^{-1}5$	4

on the $(6,b)$-entry, then the trace of b^4 at 3 is

$$
\begin{array}{cccc}
b & b & b & b \\
3\ 4\ 6 & 5 & & 3. \\
Q & u
\end{array}
$$

Since the product of the secondary labels must represent the identity in H, we take Q to be u^{-1}. The label on the paired entry is u. This gives Table 6.2.2.

Looking only at the primary labels, the result of tracing $(a^2b)^2$ at 3 is the following:

$$
\begin{array}{cccccc}
a & a & b & a & a & b \\
3\ 2\ 3\ 4\ 5\ 5 & 3. \\
& & & 4
\end{array}
$$

Thus we have found a coincidence. States 4 and 5 correspond to the same right coset of H. However, we must interpret the coincidence not merely as an equality of cosets but as an equality of elements. Our representative $W_5 = baba$ defines an element in the coset represented by $W_4 = bab$. Thus there must be a word Q in Y^* such that W_5 and $\langle Q \rangle W_4$ define the same element of H. It is necessary to find one such word Q. To do so, we look at the forward trace at 4 of the word $(aba)^2$, which is a cyclic permutation of $(a^2b)^2$ and so is the identity in G. This trace, including the secondary edges, is

$$
\begin{array}{cccccc}
a\ b & a & a\ b\ a \\
4\ 5\ 3 & & 2\ 3\ 4\ 5. \\
u\ u^{-2}v
\end{array}
$$

This means that

$$
W_4 \equiv W_4(aba)^2 \equiv \langle uu^{-2}v \rangle W_5 = \langle u^{-1}v \rangle W_5,
$$

or

$$W_5 \equiv \langle u^{-1}v \rangle^{-1} W_4 \equiv \langle vu \rangle W_4.$$

Now that we have interpreted the coincidence of 4 and 5 as an equality of elements, we must see if any further coincidences follow, and interpret them as equalities of elements. Looking at the entries in column b, we see that the coincidence of 4 and 5 implies the coincidence of 6 and 3. At the level of words, the $(4, b)$-entry says that $W_4 b \equiv W_6$. The $(5, b)$-entry says that $W_5 b \equiv \langle u \rangle W_3$. Using the preceding result about W_5, we have

$$W_6 \equiv W_4 b \equiv \langle u^{-1}v \rangle W_5 b \equiv \langle u^{-1}v \rangle \langle u \rangle W_3 = \langle u^{-1}vu \rangle W_3.$$

No further coincidences are found. The $(4, a^{-1})$-entry can now be deduced. Since $W_5 a^{-1} \equiv W_4$, we have $\langle vu \rangle W_4 a^{-1} \equiv W_4$, or $W_4 a^{-1} \equiv \langle u^{-1}v \rangle W_4$. After we insert this entry, replace all 5's in the table by $vu4$ and all 6's by $u^{-1}vu3$, and delete 5 and 6 as states, we have the following extended table:

	a	a^{-1}	b	b^{-1}
1	$u1$	$u^{-1}1$	2	$v2$
2	3	$vu^2 3$	$v1$	1
3	$u^{-2}v2$	2	4	$u^{-1}vu^4 4$
4	$vu4$	$u^{-1}v4$	$u^{-1}vu3$	3

The final step of applying RS_BASIC and obtaining the presentation for H is left as an exercise.

Exercises

2.1. Finish Example 2.2 by deriving the presentation for H.

2.2. Let X and $\mathcal{R}$ be given and let Y, $\mathcal{Q}$, and g be constructed as described in the text. Show that $\langle Q \rangle^{-1} \equiv \langle Q^{-1} \rangle$ for every word Q in Y^*, where the first inverse is with respect to $\mathcal{R}$ and the second is with respect to $\mathcal{Q}$.

2.3. Carry out an extended coset enumeration with the following input data: $X = \{x, y\}^{\pm}$, $\mathcal{R} = \text{FGRel}(\{x, y\})$, $\mathcal{V} = \{x^3, y^3, (xy)^3\}$, $Y = \{u, v\}^{\pm}$, $\mathcal{Q} = \text{FGRel}(\{u, v\})$, $g(u) = x$, $g(u^{-1}) = x^{-1}$, $g(v) = yxy^{-1}$, $g(v^{-1}) = yx^{-1}y^{-1}$.

6.3 An extended HLT enumeration procedure

In the previous section we worked through two examples of extended coset enumeration based on the HLT strategy. In this section, we shall give a formal description of an extended HLT coset enumeration procedure. The procedure has the following input arguments:

X : a finite set.

$\mathcal{R}$: a classical niladic rewriting system on X^*.

$\mathcal{V}$: a finite set of words in X^* defining relators for a group G as a quotient of $F = \text{Mon}\langle X \mid \mathcal{R} \rangle$.

Y : a finite set.

$\mathcal{Q}$: a classical niladic rewriting system on Y^*.

g : a homomorphism from Y^* to X^* which defines a homomorphism from $L = \text{Mon}\langle Y \mid \mathcal{Q} \rangle$ onto a subgroup H of finite index in G.

The output argument is an extended Schreier automaton $\mathcal{E} = (\Sigma, X, Y, D, \{1\})$ for G relative to g and a particular choice of coset representatives. The coset representatives W_σ are not explicitly constructed, although they are frequently referred to in the discussion which follows. They are defined implicitly as new states are defined.

For words in X^*, inversion and reduction are with respect to $\mathcal{R}$. For words in Y^*, these operations are with respect to $\mathcal{Q}$. If S and T are in X^*, then $S \equiv T$ means that S and T represent the same element of G. Thus $\equiv$ is the congruence generated by $\mathcal{R}$ and the pairs (V, ε) with V in $\mathcal{V}$. If Q is in Y^*, then $g(Q)$ is written $\langle Q \rangle$.

The global variable n will again denote the largest integer used so far to label a state in $\mathcal{E}$, and the vector p will be used to record the fact that a state has become coincident with an earlier state. However, since now coincidences must be described by equalities of elements of G, we shall need to store some additional information about coincidences. If $\sigma > p[\sigma] = \tau$, then S_σ will be a word in Y^* such that $W_\sigma \equiv \langle S_\sigma \rangle W_\tau$. To keep the description as simple as possible, we shall not use path compression when computing $p^\infty[\sigma]$. Because an HLT strategy is being used, there will be no references to the boolean variable *save* or to the deduction stack.

The procedure HLT of Section 5.2 calls directly or indirectly eight subprocedures: DEFINE, JOIN, ACTIVE, MERGE, COINCIDENCE, TRACE, BACK_TRACE, and TWO_SIDED_TRACE. The extended procedure, which will be called HLT_X, will need versions of all eight of the subprocedures. They will be discussed in the order given, and the names of the new procedures will be formed by adding _X to the original names.

The procedure DEFINE_X is very similar to DEFINE, which is discussed in Section 4.5.

Procedure DEFINE_X(σ, x);
Input: σ : a state;
       x : an element of X such that σ^x is not defined;
$(*$ A new state is defined as σ^x. $*)$
Begin
    $n := n + 1$; $p[n] := n$; Add $(\sigma, x, \varepsilon, n)$ and $(n, x^{-1}, \varepsilon, \sigma)$ to D
End.

However, JOIN_X is more complicated than JOIN, since JOIN_X must handle elements of G, not just cosets.

Procedure JOIN_X(A, σ, x, B, τ);
Input: A, B : words in Y^*;
    σ, τ : states;
    x : an element of X such that neither σ^x nor $\tau^{x^{-1}}$ is
        defined;
$(*$ Adds edges to D which imply that $\langle A \rangle\, W_\sigma x \equiv \langle B \rangle\, W_\tau.$ $*)$
Begin
  REWRITE$(Y, Q, A^{-1}B; Q)$; Add (σ, x, Q, τ) to D;
  If $(\sigma, x) \neq (\tau, x^{-1})$ then add $(\tau, x^{-1}, Q^{-1}, \sigma)$ to D
End.

The procedure ACTIVE_X not only computes $\tau = p^\infty[\sigma]$ but also finds a word A in Y^* such that $W_\sigma \equiv \langle A \rangle\, W_\tau$. Because two objects are returned, ACTIVE_X is not written as a function.

Procedure ACTIVE_X$(\sigma; A, \tau)$;
Input: σ : an integer between 1 and n representing a possibly
            inactive state;
Output: A : a word in Y^*;
     τ : an active state such that $W_\sigma \equiv \langle A \rangle\, W_\tau$;
Begin
  $\tau := \sigma$; $B := \varepsilon$; $\rho := p[\tau]$;
  If $\rho \neq \tau$ then begin
    $B := BS_\tau$; $\tau := \rho$; $\rho := p[\tau]$
  End;
  REWRITE$(Y, Q, B; A)$
End.

There is no logical necessity in JOIN_X and ACTIVE_X to rewrite words using Q. However, without any rewriting, the secondary labels on edges in D could get very long. The approach adopted here is to rewrite frequently. The procedure MERGE_X must "merge" two elements of G, not just two cosets of H. Recall that $\eta_1, \ldots, \eta_k$ are the states found to be coincident with earlier states so far during the current call to the coincidence procedure.

Procedure MERGE_X(A, σ, B, τ);
Input: A, B : words in Y^*;
    σ, τ : integers between 1 and n;
$(*$ Note the coincidence $\langle A \rangle\, W_\sigma \equiv \langle B \rangle\, W_\tau.$ $*)$
Begin

ACTIVE_X$(\sigma; P, \varphi)$; ACTIVE_X$(\tau; Q, \psi)$;
If $\varphi \neq \psi$ then begin
  $k := k + 1$;
  If $\varphi < \psi$ then begin
    $p[\psi] := \varphi$; $\eta_k := \psi$; REWRITE$(Y, \mathcal{Q}, Q^{-1} B^{-1} A P; S_\psi)$ end
  Else begin
    $p[\varphi] := \psi$; $\eta_k := \varphi$; REWRITE$(Y, \mathcal{Q}, P^{-1} A^{-1} B Q; S_\varphi)$
  End
End
End.

The coincidence procedure also processes coincidences of elements of G.

Procedure COINCIDENCE_X(A, σ, B, τ);
Input: A, B : words in Y^*;
    σ, τ : integers between 1 and n;
($*$ Process the coincidence $\langle A \rangle W_\sigma \equiv \langle B \rangle W_\tau$ and any consequences of
  it. $*$)
Begin
  $k := 0$; MERGE_X(A, σ, B, τ); $i := 1$;
  While $i \leq k$ do begin
  $\nu := \eta_i$; $D_1 := \emptyset$;
    For x in X do
      If there is an edge (ν, x, Q, φ) in D then begin
        Delete (ν, x, Q, φ) from D and add it to D_1;
        If $(\varphi, x^{-1}) \neq (\nu, x)$ then delete $(\varphi, x^{-1}, Q^{-1}, \nu)$ from D
      End;
    For (ν, x, Q, φ) in D_1 do begin
      ACTIVE_X$(\nu; T, \mu)$; ACTIVE_X$(\varphi; U, \psi)$;
      If there is an edge (μ, x, P, β) in D then
        MERGE_X(QU, ψ, TP, β)
      Else if there is an edge $(\psi, x^{-1}, R, \gamma)$ in D then
        MERGE_X$(Q^{-1} T, \mu, U R, \gamma)$
      Else begin
        REWRITE$(Y, \mathcal{Q}, T^{-1} Q U; P)$; Add (μ, x, P, ψ) to D;
        If $(\mu, x) \neq (\psi, x^{-1})$ then add $(\psi, x^{-1}, P^{-1}, \mu)$ to D
      End
    End;
    $i := i + 1$
  End
End.

The procedures TRACE of Section 3.2 and BACK_TRACE of Section 4.5 must be modified to compute secondary signatures. In both, the extended automaton $\mathcal{E}$ will be a global variable.

Procedure TRACE_X$(\sigma, U; \tau, B, C, P)$;

Input:	σ	: a state;
	U	: a word in X^*;
Output:	τ	: a state;
	B, C	: words in X^* such that $U = BC$, $\sigma^B = \tau$, and $|B|$ is as big as possible;
	P	: a word in Y^*, the secondary signature $P(\sigma, B)$;

Begin

$\quad \tau := \sigma$; $B := \varepsilon$; $C := U$; $Q := \varepsilon$; $done :=$ false;

$\quad$ While $C \neq \varepsilon$ and not $done$ do begin

$\quad\quad$ Let $C = xT$ with x in X;

$\quad\quad$ If τ^x is not defined then $done :=$ true

$\quad\quad$ Else begin

$\quad\quad\quad$ Let (τ, x, K, ρ) be in D; $B := Bx$; $C := T$; $Q := QK$; $\tau := \rho$

$\quad\quad$ End

$\quad$ End;

$\quad$ REWRITE$(Y, \mathcal{Q}, Q; P)$

End.

Procedure BACK_TRACE_X$(\sigma, U; \tau, B, C, P)$;

Input:	σ	: a state;
	U	: a word in X^*;
Output:	τ	: a state;
	B, C	: words in X^* such that $U = BC$, $\tau^C = \sigma$, and $|C|$ is as big as possible;
	P	: a word in Y^*, the secondary signature $P(\tau, C)$;

Begin

$\quad \tau := \sigma$; $B := U$; $C := \varepsilon$; $Q := \varepsilon$; $done :=$ false;

$\quad$ While $B \neq \varepsilon$ and not $done$ do begin

$\quad\quad$ Let $B = Tx$ with x in X;

$\quad\quad$ If $\tau^{x^{-1}}$ is not defined then $done :=$ true

$\quad\quad$ Else begin

$\quad\quad\quad$ Let (ρ, x, K, τ) be in D; $B := T$; $C := xC$; $Q := KQ$; $\tau := \rho$

$\quad\quad$ End

$\quad$ End;

$\quad$ REWRITE$(Y, \mathcal{Q}, Q; P)$

End.

Since we are using the HLT strategy, all two-sided traces are full traces. Thus we do not need the argument *full* of TWO_SIDED_TRACE. In its place, we put a word P in Y^*. The object of the trace now is to make sure that $W_\sigma V \equiv \langle P \rangle W_\sigma$. When V is a relator, P will be empty. If $\sigma = 1$ and V is the subgroup generator $g(y)$, then P will be y.

Procedure TWO_SIDED_TRACE_X(σ, V, P);
Input: σ : a state;
      V : a word in X^*;
      P : a word in Y^*;
Begin
   TRACE_X$(\sigma, V; \mu, S, C, Q)$; BACK_TRACE_X$(\sigma, C; \nu, T, U, R)$;
   While $|T| > 1$ do begin
     Let T start with x; DEFINE_X(μ, x);
     TRACE_X$(\sigma, V; \mu, S, C, Q)$; BACK_TRACE_X$(\sigma, C; \nu, T, U, R)$
   End;
   If $T = \varepsilon$ and $\mu \neq \nu$ then COINCIDENCE_X$(P^{-1}Q, \mu, R^{-1}, \nu)$
   Else begin
     Let T be the element x of X; JOIN_X$(P^{-1}Q, \mu, x, R^{-1}, \nu)$
   End
End.

Now we can define HLT_X.

Procedure HLT_X$(X, \mathcal{R}, \mathcal{V}, Y, \mathcal{Q}, g; \mathcal{E})$;
(∗ An extended HLT coset enumeration procedure. The arguments are
    explained at the beginning of this section. ∗)
Begin
   $n := 1$; $p[1] := 1$; $D := \emptyset$;
   For y in Y do TWO_SIDED_TRACE_X$(1, g(y), y)$;
   $\sigma := 1$;
   While $\sigma \leq n$ do begin
     If $p[\sigma] = \sigma$ then begin
       For V in $\mathcal{V}$ do begin
         TWO_SIDED_TRACE_X(σ, V, ε);
         If $p[\sigma] \neq \sigma$ then goto 99
       End;
       For x in X do if σ^x is not defined then DEFINE_X(σ, x)
     End;
     99: $\sigma := \sigma + 1$
   End;
   Let Σ be the set of integers σ with $1 \leq \sigma \leq n$ and $p[\sigma] = \sigma$;
   $\mathcal{E} := (\Sigma, X, Y, D, \{1\})$; Standardize $\mathcal{E}$
End.

An extended Schreier automaton is standard if the corresponding ordinary Schreier automaton is standard. To standardize $\mathcal{E}$ in HLT_X, only trivial modifications of the procedure STANDARDIZE of Section 4.7 are required.

Example 3.1. Let $G = \mathrm{Grp}\,\langle x, y \mid x^2 = y^3 = (xy)^7 = 1 \rangle$. The words xy and yx define generators of G. Let us find a presentation for G in terms of these generators. The first step is to call HLT_X with the following input arguments:

$$
\begin{array}{ll}
X & : \{x, y, y^{-1}\}, \\
\mathcal{R} & : \{x^2 \to \varepsilon, yy^{-1} \to \varepsilon, y^{-1}y \to \varepsilon\}, \\
\mathcal{V} & : \{y^3, (xy)^7\}, \\
Y & : \{u, v\}^{\pm}, \\
\mathcal{Q} & : \mathrm{FGRel}(\{u, v\}), \\
g & : \text{the homomorphism defined by } g(u) = xy,\ g(u^{-1}) = y^{-1}x, \\
& \quad g(v) = yx,\ g(v^{-1}) = xy^{-1}.
\end{array}
$$

Let us follow the operation of HLT_X. First the subgroup generators are traced at 1. The trace of $g(u)$ produces the following table:

	x	y	y^{-1}
1	2		$u^{-1}2$
2	1	$u1$	

After $g(v)$ is traced, we have

	x	y	y^{-1}
1	2	$v2$	$u^{-1}2$
2	1	$u1$	$v^{-1}1$

Now let us observe what happens in the call TWO_SIDED_TRACE_X$(1, y^3, \varepsilon)$. After TRACE_X$(1, y^3; \mu, S, C, Q)$ we have

$$\mu = 2, \quad S = y^3, \quad C = \varepsilon, \quad Q = vuv.$$

The call BACK_TRACE_X$(1, \varepsilon; \nu, T, U, R)$ produces

$$\nu = 1, \quad T = U = R = \varepsilon.$$

Thus we invoke COINCIDENCE_X$(vuv, 2, \varepsilon, 1)$. The call to MERGE_X makes $p[2] = 1$, $S_2 = v^{-1}u^{-1}v^{-1}$, and $\eta_1 = 2$. When η_1 is processed, all edges are removed from D and the three edges leaving 2 are put in D_1. When $(2, x, \varepsilon, 1)$ is processed, the edge $(1, x, vuv, 1)$ is put into D. When $(2, y, u, 1)$

is processed, the edges $(1, y, (vu)^2, 1)$ and $(1, y^{-1}, (u^{-1}v^{-1})^2, 1)$ are put in D. Thus the final extended coset table is

	x	y	y^{-1}
1	$vuv1$	$(vu)^2 1$	$(u^{-1}v^{-1})^2 1$

The relation $u = P(1, \langle u \rangle)$ is $u = vuvvuvu$, which is equivalent to $(uv^2)^2 = 1$. The relation $v = P(1, \langle v \rangle)$ is $v = vuvuvuv$, which is equivalent to $(uv)^3 = 1$. The relation $P(1, x^2) = 1$ is $(vuv)^2 = 1$, which is equivalent to $(uv^2)^2 = 1$ and can be dropped. The relations $P(1, yy^{-1}) = 1$ and $P(1, y^{-1}y) = 1$ are trivial. From $P(1, y^3) = 1$ we get $(vu)^6 = 1$, which is a consequence of $(uv)^3 = 1$ and can be omitted. Finally, from $P(1, (xy)^7) = 1$ we have $(vuvvuvu)^7 = 1$. Given $(uv)^3 = 1$, this is equivalent to $(vuvu^{-1}v^{-1})^7 = 1$, which in turn is equivalent to $v^7 = 1$. Thus a presentation of G on u and v is $(uv^2)^2 = (uv)^3 = v^7 = 1$.

Exercise

3.1. Let $P = \mathrm{Grp}\,\langle a, b \mid a^5 = b^5 = (ab)^5 = 1 \rangle$. Show that the normal closure N in P of the image of a has index 5 in P and is generated by $b^i ab^{-i}$, $0 \le i \le 3$. Find a presentation for N in terms of these generators.

6.4 Simplifying presentations

Presentations obtained using the Reidemeister-Schreier procedure frequently are difficult to work with. The presentations in terms of Schreier generators, described in Section 6.1, have a great many generators and relations. For example, if H is a subgroup of index 100 in a group defined by five relations on two generators, then the presentation for H will have 101 generators and 500 relations. If we know a small set $\mathcal{U}$ of generators for our subgroup H, then extended coset enumeration can be used to produce an extended Schreier automaton from which we can read off a set of defining relations for H in terms of the elements of $\mathcal{U}$. This can get the number of generators down to a reasonable level, but the number of relations is still large. Moreover, the secondary labels on the edges of the extended automaton can be quite long. If this is the case, then the relators obtained may be extremely long. In order to be useful, presentations constructed with the Reidemeister-Schreier procedure must be cleaned up or simplified in some manner. It may be enough just to remove relators which are cyclic permutations of other relators. However, usually more drastic methods must be employed.

There is no universal agreement as to when one presentation is simpler than another. Usually it is considered good if the number of generators can be reduced or the total length of the relators can be made smaller. If relators must be long, then high powers of short words are preferable. Eliminating a generator normally causes the total length of the relators to increase. A decrease in the total length often can be achieved by adding a generator to stand for a word which occurs frequently as a subword of the relators. The real challenge is to reduce both the number of generators and the total length of the presentation.

This section describes techniques which can be used to try to simplify any finite monoid presentation (X, S) which obviously defines a group. To simplify the exposition, we shall assume that for each x in X either there is a relation of the form $x^m = 1$, $m > 0$, or there are relations $xx^{-1} = x^{-1}x = 1$, where $x \neq x^{-1}$. The techniques used for simplification involve Tietze transformations from Section 1.4, the Knuth-Bendix procedure for strings, and coset enumeration methods similar to those discussed in Section 5.7. We start with Tietze transformations.

Suppose first that S contains a relation $S = 1$, where some generator x occurs exactly once in S, and x^{-1}, if it exists, does not occur at all in S. By taking a cyclic permutation of the relator S, if necessary, we may assume that $S = xA$. Using Tietze transformations, we can eliminate the generator x (and x^{-1} if it is present) by replacing x by A^{-1} and x^{-1} by A in all relations. Now suppose that S contains distinct relations $S = 1$ and $T = 1$, where some cyclic permutation of S has the form UV^{-1}, where $|U| > |V|$, and some cyclic permutation of T has the form AUB. Then $T = 1$ may be replaced by $AVB = 1$, thus decreasing the total length of the relations. These two steps can be iterated to achieve further simplifications.

It is possible to automate the use of Tietze transformations, but the best results frequently are obtained when the user controls the process interactively. This is particularly important if Tietze transformations are used to add generators. Such a step may temporarily make the presentation worse, but it can lead eventually to a much nicer presentation. An implementation of this approach is described in (Havas et al. 1984).

There is an approach to simplification based on the Knuth-Bendix procedure which may be viewed as a generalization of the technique just described. Now we think of our set S as a rewriting system and run the Knuth-Bendix procedure for a while. It is probably not realistic to hope that a finite, confluent rewriting system will be found. Our goal is to try to discover simple rules which are implied by our given ones. Our presentation will grow in size as the Knuth-Bendix procedure runs. At some point we run out of space or decide that we have tried long enough to find simple consequences of our original relations. Let T be the rewriting system produced by the Knuth-Bendix procedure. We now start the second phase of our process. We pick the simplest rule (S, T) in T, using whatever criterion of simplicity is appropriate for our group. Let $\mathcal{U} = \mathcal{V} = \{(S, T)\}$. We apply the

Knuth-Bendix procedure to the system $\mathcal{U}$ for a while to determine as many consequences of it as convenient. Let the rewriting system produced again be called $\mathcal{U}$. We now go back to $\mathcal{T}$. If all rules in $\mathcal{T}$ are consequences of $\mathcal{U}$, then we stop with $\mathcal{V}$ as our simplified presentation. Otherwise, we look for the simplest rule in $\mathcal{T}$ which is not implied by $\mathcal{U}$. This rule is added both to $\mathcal{U}$ and to $\mathcal{V}$, and then the Knuth-Bendix procedure is restarted on $\mathcal{U}$. The process continues until $\mathcal{V}$ consists of those rules (S, T) in $\mathcal{T}$ which are not obvious consequences of the rules in $\mathcal{T}$ which are simpler than (S, T).

If the goal of the simplification is to reduce the total length of the rules, then a length-plus-lexicographic ordering should be used. However, if the goal is to eliminate a particular generator x along with x^{-1}, if it is present, then one should use an ordering which gives preference to words with few occurrences of x and x^{-1}. One could order words first by the number of occurrences of x and x^{-1}, and then use some other ordering to break ties. An alternative would be to use a basic wreath-product ordering in which x and x^{-1} come last, or the wreath product of some reduction ordering on $(X - \{x, x^{-1}\})^*$ with the length-plus-lexicographic ordering of $\{x, x^{-1}\}^*$.

The third approach to simplification is a variation of the second, with coset enumeration substituted for the Knuth-Bendix procedure. It is very close to the procedure NORMAL_GENS of Section 5.7. We enumerate the cosets of the trivial subgroup in the group $\mathrm{Mon}\langle X \mid \mathcal{S} \rangle$ until we run out of space. It is probably best to use the Felsch strategy. Let the coset automaton produced be $\mathcal{A}$. Now we let $\mathcal{V}$ consist of just those relations in $\mathcal{S}$ which make the presentation obviously define a group. We start a second enumeration of the cosets of the trivial subgroup in $\mathrm{Mon}\langle X \mid \mathcal{V} \rangle$. After the enumeration has run for a while, we stop it. Let $\mathcal{B}$ be the coset automaton which results. We now find the first place in which $\mathcal{A}$ and $\mathcal{B}$ differ. This process is discussed in Section 5.7. It amounts to finding the lexicographically first pair (U, V) of reduced words such that U occurs later in the length-plus-lexicographic ordering of X^*, $1^U = 1^V$ in $\mathcal{A}$ but $1^U \neq 1^V$ in $\mathcal{B}$. Here, "reduced" means not containing any relators as subwords. The word UV^{-1} is cyclically reduced to give a word S, and the relation $S = 1$ is added to $\mathcal{V}$. The enumeration using $\mathcal{B}$ is restarted with this new relation, and the process is repeated until no more pairs (U, V) are found. The output presentation is $\mathcal{V}$, which consists of those relations apparent in $\mathcal{A}$ which are not implied by simpler relations holding in $\mathcal{A}$.

Given a subgroup H of finite index in a finitely presented group G, we have at least two ways to go about obtaining a nice presentation for H. We could construct a presentation in terms of Schreier generators and then simplify that presentation using the techniques of this section. We could also use the methods of Section 5.7 to find a small generating set for H, then apply extended coset enumeration followed by the Reidemeister-Schreier procedure to get a presentation in terms of those generators. This presentation would then be simplified. In the latter case the simplification would not have to involve removing redundant generators. At this point, we do not have enough experience to be able to tell which if either approach is more likely to prove successful.

Example 4.1. This example is based on one in (Havas et al. 1984). The Mathieu group M_{11} of order 7920 is defined by the presentation

$$\mathrm{Grp} \langle a, b, c \mid a^{11} = b^5 = c^4 = (ac)^3 = b^2 c^{-1} b^{-1} c = a^3 b a^{-1} b^{-1} = 1 \rangle.$$

The subgroup H of M_{11} generated by a, b, and c^2 is isomorphic to the simple group $\mathrm{PSL}(2, 11)$ and has order 660. Thus $|M_{11} : H| = 12$. Using an extended coset enumeration procedure based on the Felsch strategy and the Reidemeister-Schreier procedure, we obtain a presentation for H on generators $u = a$, $v = b$, and $w = c^2$ with the following relations:

$$u^{11} = 1, \quad v^5 = 1, \quad v^{10} = 1, \quad w^2 = 1, \quad (uwv^2)^3 = 1,$$
$$u^3 v u^{-1} v^{-1} = 1, \quad v^4 w^{-1} v^{-1} w = 1, \quad (uwv^2 w^{-1} u^{-1} v^6 uwv^{-2})^2 = 1,$$
$$(uwv^2 w^{-1} u^{-1} v^4 w^{-1} u^{-1} v^{-4} uwv^{-2})^3 = 1,$$
$$(uwuwv^{-2} w^{-1} u^{-1} v^2 w^{-1} u^{-1} v^4)^3 = 1, \quad (uwv^2 w^{-1} u^{-1} v^{-4} uwv^{-2})^2 = 1,$$
$$uwv^4 w^{-1} u^{-1} v^{-4} uwv^{-2} uwv^2 w^{-1} u^{-1} v^4 w^{-1} u^{-1} v^2 w^{-1} u^{-1} v^2 = 1,$$
$$uwv^2 w^{-1} u^{-1} v^6 uwv^{-2} uwv^2 w^{-1} u^{-1} v^{-4} uwv^{-2} = 1.$$

Actually, this presentation has already been cleaned up somewhat since relators which are cyclic permutations of other relators or their inverses have been deleted. Because of the relation $w^2 = 1$, it seems reasonable to remove w^{-1} as a monoid generator. Thus we take as our starting point the monoid presentation on generators u, u^{-1}, v, v^{-1}, and w consisting of the relations

$$uu^{-1} = 1, \quad u^{-1}u = 1, \quad vv^{-1} = 1, \quad v^{-1}v = 1, \quad w^2 = 1,$$

together with the relations obtained by replacing w^{-1} by w in the preceding relations.

Let us apply the Knuth-Bendix approach to presentation simplification with two choices of orderings on words. In each case a rule (S, T) will be considered simpler than another rule (U, V) if ST^{-1} comes before UV^{-1} in the chosen ordering. First we take the length-plus-lexicographic ordering with $u \prec u^{-1} \prec v \prec v^{-1} \prec w$. Using a Knuth-Bendix implementation which keeps the set of rules reduced and balanced, employs the heuristics described in Section 2.7, and processes all overlaps of length at most 5, we obtain a rewriting system $\mathcal{T}$ with 353 rules, all of whose left sides have lengths at most 7. In trying to determine the rules which are not consequences of earlier rules, it is probably a good idea to push the Knuth-Bendix process a little harder than we did initially. Thus we form overlaps of length up to 7 in processing the set $\mathcal{U}$. In this case, the final set $\mathcal{V}$ consists of the five rules giving the inverses of the generators plus the following rules:

$$wv \to v^{-1}w, \quad v^3 \to v^{-2}, \quad u^3v \to vu, \quad u^4 \to v^{-1}uv, \quad u^2wu^{-1}v^{-1} \to wuwv.$$

This gives us the description of H as

$$\mathrm{Grp}\langle u, v, w \mid w^2 = (wv)^2 = v^5 = u^3vu^{-1}v^{-1} = u^4v^{-1}u^{-1}v$$
$$= u^2wu^{-1}v^{-2}wu^{-1}w = 1\rangle.$$

When we use the ordering which has the generators in the same order as before but which is the wreath product of length-plus-lexicographic orderings on $\{u, u^{-1}\}^*$, $\{v, v^{-1}\}^*$, and $\{w\}^*$, we get a quite different presentation. Excluding the rules giving the inverses of the generators, the set $\mathcal{V}$ consists of the following rules:

$$u^6 \to u^{-5}, \quad uv \to vu^4, \quad v^3 \to v^{-2}, \quad vw \to wv^{-1}, \quad wuw \to u^2v^2wu^2.$$

The corresponding group presentation of H can be written

$$\mathrm{Grp}\,\langle u, v, w \mid u^{11} = uvu^{-4}v^{-1} = v^5 = (vw)^2 = wuwu^{-2}wv^{-2}u^{-2} = 1\rangle.$$

With this ordering, the entire confluent rewriting system for H consists of only 45 rules.

If a relator is considered simple if it is a power of a short word, then we can experiment with sets of relators drawn from our various presentations for H to obtain the following nice presentation:

$$H = \mathrm{Grp}\,\langle u, v, w \mid u^{11} = v^5 = w^2 = (vw)^2 = (uwv^2)^3 = u^3vu^{-1}v^{-1} = 1\rangle.$$

Exercise

4.1. Find simple presentations for the following subgroups H:

(a) $H = \langle aba, bab^{-1}\rangle$ in $\mathrm{Grp}\,\langle a, b \mid a^2 = b^3 = (ab)^7 = (abab^{-1})^4 = 1\rangle$.
(b) $H = \langle a, b^2ab^{-1}\rangle$ in $\mathrm{Grp}\,\langle a, b \mid a^5 = b^5 = (ab)^2 = (a^{-1}b)^4 = 1\rangle$.

6.5 Historical notes

The original versions of the Reidemeister-Schreier procedure are described in (Reidemeister 1926) and (Schreier 1927). The presentations obtained are in terms of Schreier generators. Procedures for giving presentations of subgroups of finite index in terms of arbitrary generators were developed in (Mendelsohn 1970) and (McLain 1977). The first computer implementation of the Reidemeister-Schreier procedure is described in (Havas 1974a), which contains a discussion of presentation simplification. An alternative approach to managing the data in an extended coset enumeration is given in (Arrell & Robertson 1984).

Generalized automata

The discussion in this chapter is motivated by an observation concerning the nature of coset tables in the early stages of HLT enumerations. Suppose we are performing an HLT coset enumeration and Table 7.0.1 describes our current automaton. Assume that our next step is to perform a full two-sided trace of $(xy)^7$ at 1. After the initial forward and backward traces, we have the following situation:

$$x \ y \ x \ y \ x \ y \ x \ y \ x \ y \ x \ y \ x \ y$$
$$1 \ 2 \ 2 \ 4 \qquad\qquad\qquad 5 \ 3 \ 1.$$

To complete the trace, we shall define eight more states and enter 18 edges into the automaton. With most data structures for edges, we shall need to reserve space for 32 new edges, since eventually 4 edges will leave each of the newly created states. However, at the completion of the trace, only half of these edges are defined. If there were more generators, a much larger fraction of the reserved space would be unused. This approach seems wasteful. It is quite common for an HLT enumeration involving a substantial number of generators to terminate unsuccessfully with only 10% of the memory available to store the coset table or equivalent data structure actually used. In our example, we are simply trying to record the facts that $4^T = 5$ and $5^{T^{-1}} = 4$, where $T = yxyxyxyxy$. This information could be stored with no new states and only two new edges if the labels on edges could be words of arbitrary length. In this case we would like to add "edges"

$$(4, yxyxyxyxy, 5) \quad \text{and} \quad (5, y^{-1}x^{-1}y^{-1}x^{-1}y^{-1}x^{-1}y^{-1}x^{-1}y^{-1}, 4).$$

Clearly more space will be needed to store two edges like this than two ordinary edges, but it seems likely that these two edges can be stored in much less space than must be reserved for 32 ordinary edges.

Table 7.0.1

	x	x^{-1}	y	y^{-1}
1	2			3
2	4	1	2	2
3	5		1	
4		2		5
5		3	4	

This chapter pursues the idea of using generalized automata, automata whose edges are labeled by words of any length, to do coset enumeration, primarily using the HLT strategy. The techniques presented have not been subjected to extensive experimentation. They are presented as objects for further study in the hope of encouraging other suggestions for saving space in coset enumerations.

*7.1 Definitions

This section provides the appropriate generalizations of the definitions and results about automata from Chapter 3. A *generalized automaton* is a quintuple $\mathcal{A} = (\Sigma, X, E, A, \Omega)$, where Σ and X are sets, A and Ω are subsets of Σ, and E is a subset of $\Sigma \times X^* \times \Sigma$. An element $e = (\sigma, U, \tau)$ in E is called an *edge*, and U is the *label* of e. In this context, if all labels have length at most 1, so that $\mathcal{A}$ is an automaton according to the definition in Section 3.2, we shall say that $\mathcal{A}$ is an *ordinary automaton*. The generalized automaton $\mathcal{A}$ is *finite* if Σ, X, and E are all finite. A *path* in $\mathcal{A}$ is a sequence of edges in which the endpoint of one edge is the starting point of the next. The *signature* of a path is the product of the labels on its edges. The *language accepted by* $\mathcal{A}$ is the set $L(\mathcal{A})$ of signatures of paths from A to Ω. We shall say that $\mathcal{A}$ is *deterministic* if $|A| \leq 1$, all edge labels have length at least 1, and for each state σ in Σ and each x in X there is at most one edge (σ, U, τ) in E such that U starts with x. In a deterministic automaton $\mathcal{A}$, a path P is determined by its starting point σ and its signature W. The endpoint of P will again be denoted σ^W. The term "complete" will be used only for ordinary automata. The statement of Exercise 2.4 of Chapter 3 is still correct, although its proof is not quite a triviality.

Transition tables are not adequate to represent the set of edges in a finite, generalized automaton, even a deterministic one. One way to describe the edges is to list the states and for each state σ to list the labels and endpoints of the edges leaving σ. For example, Table 7.1.1 describes a set of 10 edges, among which are $(1, yxy, 1)$ and $(3, y^2, 4)$. This representation makes it relatively easy to locate edges leaving a particular state, but it does not facilitate finding the edges coming into a given state. If this is important, then other schemes must be used.

Table 7.1.1

State	Label	Endpoint	State	Label	Endpoint
1	x^2y	2	3	x	1
	yxy	1		y^2	4
	y^2x^2	3	4	x^5	2
2	xy	4		x^3y	1
	yx	3		y^3	2

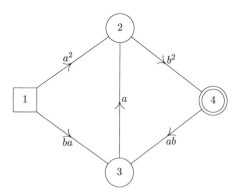

Figure 7.1.1

Using finite, generalized automata does not allow us to recognize a wider class of languages than using ordinary automata. If $\mathcal{A} = (\Sigma, X, E, A, \Omega)$ is a finite, generalized automaton, then we can construct an ordinary automaton recognizing the same language. Edges with labels of length at most 1 are retained. If $e = (\sigma, U, \tau)$ is an edge in E such that $U = x_1 \ldots x_s$ with $s > 1$, then we choose new states $\sigma_1, \ldots, \sigma_{s-1}$ and replace e by the edges $(\sigma_{i-1}, x_i, \sigma_i)$, $1 \leq i \leq s$, where $\sigma_0 = \sigma$ and $\sigma_s = \tau$. If Λ is the enlarged set of states and D is the enlarged set of edges, then $\mathcal{B} = (\Lambda, X, D, A, \Omega)$ is a finite, ordinary automaton recognizing the same language as $\mathcal{A}$. The automaton $\mathcal{B}$ is determined up to isomorphism and will be called an *expansion* of $\mathcal{A}$.

Example 1.1. If $\mathcal{A}$ is given by Figure 7.1.1, then an expansion of $\mathcal{A}$ is described by Figure 7.1.2.

Constructing an expansion reduces the length of edge labels at the expense of creating more states. We can sometimes go the other way, reducing the number of states and making labels longer. Suppose that $\mathcal{A} = (\Sigma, X, E, A, \Omega)$ is a finite, generalized automaton and assume that $e = (\rho, U, \sigma)$ and $f = (\sigma, V, \tau)$ are in E. If σ is not in Ω and f is the only edge leaving σ, then we may replace e by (ρ, UV, τ) without changing the language recognized. If at this point σ is not in A and there is no edge

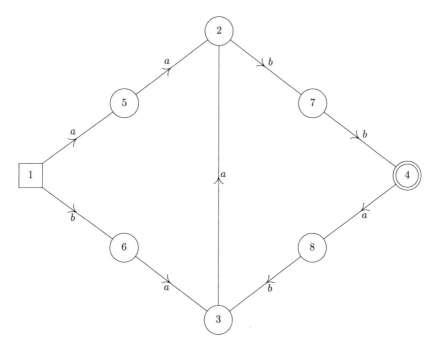

Figure 7.1.2

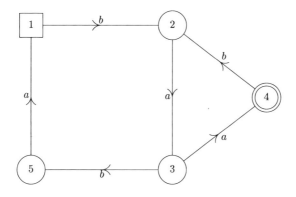

Figure 7.1.3

coming into σ, then we may also remove f from E and delete σ from Σ. If we iterate this process until no further changes are possible, then the resulting generalized automaton $\mathcal{B}$ is unique and is called the *contraction* of $\mathcal{A}$. If $\mathcal{A}$ is deterministic, then so is $\mathcal{B}$.

Example 1.2. Suppose $\mathcal{A}$ is given by Figure 7.1.3. Here is one way to construct the contraction of $\mathcal{A}$. First the edge $(1, b, 2)$ is replaced by

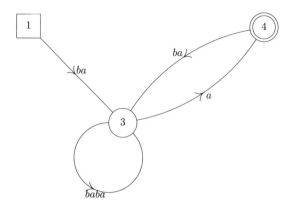

Figure 7.1.4

Figure 7.1.5

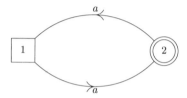

Figure 7.1.6

$(1, ba, 3)$. Then $(4, b, 2)$ is replaced by $(4, ba, 3)$. Next $(2, a, 3)$ is deleted and 2 is removed as a state. Now $(3, b, 5)$ is replaced by $(3, ba, 1)$, and both the edge $(5, a, 1)$ and the state 5 are deleted. Finally, $(3, ba, 1)$ is replaced by $(3, baba, 3)$. No further changes are possible and we are left with Figure 7.1.4.

The definitions of accessible, coaccessible, and trim states carry over without change. Thus we can refer to a trim generalized automaton, and if $\mathcal{A}$ is not trim, then we can take its trim part without changing the language recognized. It is not true that a deterministic generalized automaton which has the minimum number of states among all deterministic generalized automata recognizing a particular rational language $\mathcal{L}$ is uniquely determined by $\mathcal{L}$ up to isomorphism. For example, the automata $\mathcal{A}_1$ in Figure 7.1.5 and $\mathcal{A}_2$ in Figure 7.1.6 both recognize $\mathcal{L} = \{a^{2i+1} \mid i \geq 0\}$. It is easy to check that there is no generalized automaton, deterministic or not, which

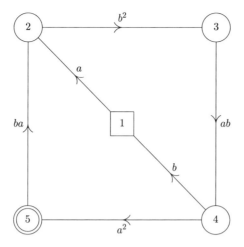

Figure 7.1.7

recognizes $\mathcal{L}$ and has fewer than two states. The automaton $\mathcal{A}_2$ has an additional minimality property. Suppose U is in $\mathcal{L}$. Then there is a path P in $\mathcal{A}_1$ and a path Q in $\mathcal{A}_2$ such that both paths go from 1 to 2 and $\mathrm{Sg}(P) = U = \mathrm{Sg}(Q)$. The length of Q is always less than or equal to the length of P. For example, if $U = a^3$, then P has length 3 and Q has length 2. If we add this extra minimality condition, then, for any rational language $\mathcal{L}$, a minimal generalized automaton recognizing $\mathcal{L}$ is isomorphic to the contraction of $\mathcal{A}_t(\mathcal{L})$.

Exercise

1.1. Let $\mathcal{A}$ be the generalized automaton described by Figure 7.1.7. Determine an expansion and the contraction of $\mathcal{A}$.

*7.2 Generalized coset automata

The previous section discussed analogues of results in Chapter 3 for generalized automata. This section does the same for the results in Sections 4.4 and 4.11. Let $\mathcal{R}$ be a reduced, confluent, niladic rewriting system on X^* which defines a group. We shall not assume that $\mathcal{R}$ is classical, and thus we shall draw heavily on Section 4.11. The congruence generated by $\mathcal{R}$ is denoted by $\sim$, the $\sim$-class containing a word U is $[U]$, and $\mathcal{C}$ denotes the set of words which are irreducible with respect to $\mathcal{R}$. The unique word in $[U] \cap \mathcal{C}$ is $\overline{U}$, and $U^{\Diamond}$ is the word in $\mathcal{C}$ such that $UU^{\Diamond} \sim \varepsilon$.

A generalized automaton $\mathcal{A} = (\Sigma, X, E, A, \Omega)$ will be called a *generalized coset automaton* relative to $\mathcal{R}$ if the following conditions hold:

Table 7.2.1

State	Label	Endpoint
1	x^2	2
	yxy	3
2	x	3
	y^2	2
	y^{-2}	2
3	x^2	1
	$y^{-1}x^4y^{-1}$	1

(i) $\mathcal{A}$ is accessible and deterministic.

(ii) $A = \Omega \neq \emptyset$.

(iii) If (σ, U, τ) is in E, then U is in $\mathcal{C}$ and there is a path P in $\mathcal{A}$ from τ to σ such that $\mathrm{Sg}(P) = U^{\diamond}$.

Conditions (i) and (ii) are the same as in the definition of an ordinary coset automaton.

Example 2.1. Let $X = \{x, y, y^{-1}\}$, let $\mathcal{R}$ consist of the rules

$$x^5 \to \varepsilon, \quad yy^{-1} \to \varepsilon, \quad y^{-1}y \to \varepsilon,$$

and let E be given by Table 7.2.1. The path P in condition (iii) for the first edge consists of the two edges $(2, x, 3)$ and $(3, x^2, 1)$. For $(1, yxy, 3)$, P is the single edge $(3, y^{-1}x^4y^{-1}, 1)$. The generalized automaton defined by this table satisfies conditions (i) to (iii).

Proposition 2.1. *Suppose that $\mathcal{A}$ is a generalized coset automaton. If there is a path P in $\mathcal{A}$ from σ to τ, then there is a path Q from τ to σ such that $[\mathrm{Sg}(Q)] = [\mathrm{Sg}(P)]^{-1}$.*

Proof. Let P consist of the edges $(\sigma_{i-1}, U_i, \sigma_i)$, $1 \leq i \leq s$. By condition (iii), there are paths Q_i from σ_i to σ_{i-1} such that $\mathrm{Sg}(Q_i) = U_i^{\diamond}$. Let Q be the concatenation of $Q_s, \ldots, Q_1$. Then Q is a path from $\sigma_s = \tau$ to $\sigma_0 = \sigma$, and $[\mathrm{Sg}(Q)] = [\mathrm{Sg}(P)]^{-1}$. $\square$

Corollary 2.2. *Generalized coset automata are trim.*

Recall the fact from Section 4.11 that a word U in $\mathcal{C}$ has a unique minimal decomposition as a product of nonempty subwords of left sides in $\mathcal{R}$. The number of terms in the product is denoted $\|U\|$.

Proposition 2.3. *Let $\mathcal{A}$ be a generalized coset automata and let (σ, U, τ) be an edge of $\mathcal{A}$. Either $\|U\| = 1$ or $(\tau, U^{\diamond}, \sigma)$ is an edge.*

Proof. There is a path P in $\mathcal{A}$ from τ to σ such that $\mathrm{Sg}(P) = U^\diamond$. Let (ρ, V, σ) be the last edge of P and let V end in x. Then U and $V^\diamond$ each start with $\dot{x}$. There is a path Q from σ to ρ such that $\mathrm{Sg}(Q) = V^\diamond$. Since $\mathcal{A}$ is deterministic, (σ, U, τ) is the first edge of Q. Therefore V is a suffix of $U^\diamond$ and U is a prefix of $V^\diamond$. By a remark in Section 4.11, this means that $\|U\| = \|V\|$ and $V^\diamond = UW$, where $\|W\| \le 1$. If $\|W\| = 0$, then $W = \varepsilon$ and $V = U^\diamond$. In this case, $\rho = \tau$ and $(\tau, U^\diamond, \sigma)$ is an edge. Suppose that $\|W\| = 1$. Let (φ, T, ρ) be the last edge in Q. Then T is a suffix of W, so $\|T\| = 1$. By the preceding argument, with U replaced by V and V by T, we have $\|V\| = \|T\|$. Therefore $\|U\| = \|V\| = 1$. $\square$

Corollary 4.3 in Chapter 4 holds for generalized coset automata as well as for ordinary ones.

Proposition 2.4. *Let $\mathcal{A} = (\Sigma, X, E, A, A)$ be a generalized coset automaton. Suppose that $e = (\rho, U, \tau)$ and that $f = (\sigma, V, \tau)$ are in E and that U and V end in the same element x of X. Then $e = f$.*

Proof. There is a path P in $\mathcal{A}$ from τ to ρ such that $\mathrm{Sg}(P) = U^\diamond$. Let $g = (\tau, W, \varphi)$ be the first edge of P. Thus g is the unique edge leaving τ whose label starts with $\dot{x}$. Therefore g is also the first edge in the path Q from τ to σ such that $\mathrm{Sg}(Q) = V^\diamond$. If $\|U\| > 1$, then $(\tau, U^\diamond, \rho)$ is an edge, so $W = U^\diamond$ and $\varphi = \rho$. Since $\|V\| = \|V^\diamond\| \ge \|W\| = \|U\|$, it follows that $\|V\| > 1$. Hence $W = V^\diamond$ and $\varphi = \sigma$, and therefore $e = f$.

Thus we may assume that $\|U\| = \|V\| = 1$. Therefore $\|W\| = 1$ and there are words S and T such that $L = WSU = WTV$ is a left side in $\mathcal{R}$, $\mathrm{Sg}(P) = WS$, and $\mathrm{Sg}(Q) = WT$. Thus the concatenation of P and e and the concatenation of Q and f are both paths from τ to τ and both have signature L. Therefore they are the same path and $e = f$. $\square$

Corollary 2.5. *In a generalized coset automaton, a path is determined by its endpoint and its signature.*

The main ideas behind the proofs of Propositions 2.3 and 2.4 are due to M. F. Newman.

Proposition 2.6. *Let $\mathcal{A}$ be a generalized coset automaton. If $\sigma^W = \tau$ in $\mathcal{A}$, then $\sigma^{\overline{W}} = \tau$.*

Proof. Let $\mathcal{A} = (\Sigma, X, E, A, A)$ and let P be a path in $\mathcal{A}$ from σ to τ such that $W = \mathrm{Sg}(P) = SLT$, where L is a left side in $\mathcal{R}$. It is not clear that there is a subpath of P whose signature is L. However, there is a minimal subpath Q such that L is a subword of $\mathrm{Sg}(Q)$. Let Q consist of the edges $(\sigma_{i-1}, U_i, \sigma_i)$, $1 \le i \le s$. Since edge labels are in $\mathcal{C}$, we have $s > 1$

and $L = CU_2 \ldots U_{s-1}D$, where C is a nonempty suffix of U_1 and D is a nonempty prefix of U_s. Therefore, by Proposition 2.3, for $2 \le i \le s$ the pair (U_{i-1}, U_i) is susceptible to cancellation. Thus either $U_{i-1} = U_i^\diamond$ and $\sigma_{i-2} = \sigma_i$ or $U_{i-1}U_i$ is a subword of a left side. In the first case, we can delete the edges $(\sigma_{i-2}, U_{i-1}, \sigma_{i-1})$ and $(\sigma_{i-1}, U_i, \sigma_i)$, and the result will still be a path R from σ to τ and $W = \mathrm{Sg}(R)$. By induction on the length of P, $\sigma^{\overline{W}} = \tau$.

We are left with the case in which, for $2 \le i \le s$, the word $U_{i-1}U_i$ is a subword of a left side in $\mathcal{R}$. In this situation, $V = U_1 \ldots U_s$ is a subword of a power of a left side and V contains a left side. This means that V has a prefix M which is a left side and U_1 is a proper prefix of M. Let $M = U_1V_1$. Then $V_1 = U_1^\diamond$ and there is a path R from σ_1 to σ_0 such that $\mathrm{Sg}(R) = V_1$. Since $\mathcal{A}$ is deterministic, R is a subpath of Q. The concatenation of $(\sigma_0, U_1, \sigma_1)$ and R is a path from σ_0 to σ_0 with signature M. As before, this subpath may be deleted from P without changing $[\mathrm{Sg}(P)]$. Again we are done by induction on the length of P. $\square$

Corollary 2.7. *Let $\mathcal{A}$ be a generalized coset automaton and set $\mathcal{L} = L(\mathcal{A})$. Then $\overline{\mathcal{L}} \subseteq \mathcal{L}$.*

Proof. Take σ and τ in Proposition 2.6 to be the initial state of $\mathcal{A}$. $\square$

As before, if $\mathcal{A}$ is a generalized coset automaton, then we set $K(\mathcal{A}) = \{[U] \mid U \in \mathcal{L}(\mathcal{A})\}$.

Proposition 2.8. *Let $\mathcal{A}$ be a generalized coset automaton. Then $H = K(\mathcal{A})$ is a subgroup of $F = \mathrm{Mon}\langle X \mid \mathcal{R}\rangle$.*

Proof. It is clear that ε is in $L(\mathcal{A})$ and that $L(\mathcal{A})$ is a submonoid of X^*. Therefore H is a submonoid of F. By Proposition 2.1, H is a subgroup of F. $\square$

Let $\mathcal{A}$ be a finite generalized coset automaton. It is useful to have a procedure for constructing an ordinary coset automaton $\mathcal{B}$ such that $K(\mathcal{B}) = K(\mathcal{A})$. The procedure may be thought of as roughly analogous to the expansion construction, although, in general, the expansion of a coset automaton is not a coset automaton. Let $\mathcal{A} = (\Sigma, X, E, A, A)$ and suppose that (σ, U, τ) is an edge in E such that $\|U\| > 1$. By Proposition 2.1, $(\tau, U^\diamond, \sigma)$ is also an edge. Let $U_1 \ldots U_s$ be the minimal decomposition of U as product of subwords of left sides in $\mathcal{R}$. Then $U_s^\diamond \ldots U_1^\diamond$ is the minimal decomposition of $U^\diamond$. Let $\mathcal{A}_1$ be the automaton obtained from $\mathcal{A}$ by replacing the edges (σ, U, τ) and $(\tau, U^\diamond, \sigma)$ with the edges $(\sigma_{i-1}, U_i, \sigma_i)$ and $(\sigma_i, U_i^\diamond, \sigma_{i-1})$, $1 \le i \le s$, where $\sigma_0 = \sigma$, $\sigma_s = \tau$, and $\sigma_1, \ldots, \sigma_{s-1}$ are distinct objects not in Σ.

Table 7.2.2

State	Label	Endpoint
1	a^2ba^3	2
	b^4a	3
2	$a^2b^4a^3$	1
	b^2	3
3	a^4b	1
	b^2	4
4	a	4
	b	2

Proposition 2.9. *Under these hypotheses, $\mathcal{A}_1$ is a generalized coset automaton and $K(\mathcal{A}_1) = K(\mathcal{A})$.*

Proof. Exercise. □

By repeated application of Proposition 2.9, we may assume that $\|U\| = 1$ for every edge (σ, U, τ) in E. Now we can take the expansion.

Proposition 2.10. *Let $\mathcal{A}$ be a generalized coset automaton such that the labels on edges are subwords of left sides of $\mathcal{R}$. The expansion $\mathcal{B}$ of $\mathcal{A}$ is a coset automaton and $K(\mathcal{B}) = K(\mathcal{A})$.*

Proof. Exercise. □

Example 2.2. Let us illustrate the constructions of Propositions 2.9 and 2.10. Suppose that $X = \{a, b\}$ and $\mathcal{R}$ consists of the rules $a^5 \to \varepsilon$ and $b^5 \to \varepsilon$. Let $\mathcal{A}_1$ be $(\Sigma, X, E, \{1\}, \{1\})$, where $\Sigma = \{1, 2, 3, 4\}$ and E is given by Table 7.2.2. In the first phase of the procedure, we split up those edges whose labels are not subwords of left sides. This involves adding three new states. To split the edge $(1, a^2ba^3, 2)$, let us add states 5 and 6 such that $1^{a^2} = 5$ and $5^b = 6$. To split $(1, b^4a, 3)$, we add a state 7 such that $1^{b^4} = 7$. The edges of the resulting automaton $\mathcal{A}_2$ are described by Table 7.2.3. One expansion $\mathcal{A}_3$ of this automaton has state set $\{1, \ldots, 24\}$ and edges given by Table 7.2.4. Although one must be careful about drawing conclusions about the size of computer data structures from the manner in which their information is displayed on the printed page, it would seem that $\mathcal{A}_1$ is a more concise description of $K(\mathcal{A}_1) = K(\mathcal{A}_3)$ than $\mathcal{A}_3$ is.

Just as we can form expansions of generalized coset automata, we can form contractions and find minimal coset automata defining a given finitely generated subgroup of F.

Let $\mathcal{A} = (\Sigma, X, E, \{\alpha\}, \{\alpha\})$ be a generalized coset automaton relative to $\mathcal{R}$, let $e = (\rho, U, \sigma)$ be in E, and suppose that $\sigma \neq \alpha$. There is at least one

Table 7.2.3

State	Label	Endpoint	State	Label	Endpoint
1	a^2	5	5	a^3	1
	b^4	7		b	6
2	a^2	6	6	a^3	2
	b^2	3		b^4	5
3	a^4	7	7	a	3
	b^2	4		b	1
4	a	4			
	b	2			

Table 7.2.4

	a	b		a	b		a	b
1	8	9	9		10	17		4
2	12	13	10		11	18	19	
3	14	17	11		7	19	1	
4	4	2	12	6		20	21	
5	18	6	13		3	21	2	
6	20	22	14	15		22		23
7	3	1	15	16		23		24
8	5		16	7		24		5

edge leaving σ, the first edge $f = (\sigma, V, \tau)$ of the path which goes from σ to ρ and has signature $U^\diamond$. Let us consider several cases.

Case 1. There is no edge $g = (\sigma, W, \varphi)$ leaving σ such that UW is in $\mathcal{C}$. This means that UV is not in $\mathcal{C}$, so by Proposition 2.3, $V = U^\diamond$, $\tau = \rho$, and f is the only edge leaving σ. Any path P which goes from α to α and involves either e or f must have e and f as consecutive edges. Thus $Sg(P)$ is not in $\mathcal{C}$. If we remove e and f from E and delete σ from Σ, then the result is still a generalized coset automaton describing $K(\mathcal{A})$.

Case 2. There is exactly one edge $g = (\sigma, W, \varphi)$ leaving σ such that UW is in $\mathcal{C}$.

Case 2.1. $f = g$. By Proposition 2.3, UV is a proper prefix of a left side in $\mathcal{R}$. The edges e and f may be replaced by the single edge (ρ, UV, τ).

Case 2.2. $f \neq g$ and there is an edge $h = (\varphi, W^\diamond, \sigma)$. Now $V = U^\diamond$ and $\tau = \rho$. The pair (U, W) is not susceptible to cancellation, so $(UW)^\diamond = W^\diamond U^\diamond = W^\diamond V$. The edges e, f, g, and h may be replaced by (ρ, UW, φ) and $(\varphi, W^\diamond V, \rho)$.

If we modify $\mathcal{A}$ as described until no more changes are possible, then the resulting automaton $\mathcal{B}$ is unique. We shall call $\mathcal{B}$ the *contraction of $\mathcal{A}$ as a coset automaton*.

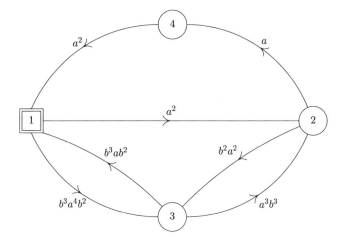

Figure 7.2.1

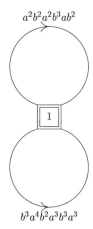

Figure 7.2.2

Example 2.3. Let X and $\mathcal{R}$ be as in Example 2.2 and let $\mathcal{A}$ be given by Figure 7.2.1. We may replace the edges $(2, a, 4)$ and $(4, a^2, 1)$ by the single edge $(2, a^3, 1)$ and delete 4 as a state. We may also replace the four edges coming into or going out of 3 by the pair of edges $(1, b^3 a^4 b^2 a^3 b^3, 2)$ and $(2, b^2 a^2 b^3 a b^2, 1)$ and then delete state 3. Finally, the four current edges may be replaced by $(1, b^3 a^4 b^2 a^3 b^3 a^3, 1)$ and $(1, a^2 b^2 a^2 b^3 a b^2, 1)$. Thus we obtain the automaton in Figure 7.2.2, as the contraction of $\mathcal{A}$ as a coset automaton.

Table 7.2.5

State	Label	Endpoint
1	ab^2	2
	ba^3b^3	3
2	a	4
	b^3a^4	1
3	a^2	2
	$b^2a^2b^4$	1
4	a^2	3
	b^2a	5
5	a^4b^3	4

Exercise

2.1. Let X and $\mathcal{R}$ be as in Examples 2.2 and 2.3. Suppose $\mathcal{A} = (\Sigma, X, E, A, A)$ is the generalized coset automaton with respect to $\mathcal{R}$ in which $A = \{1\}$ and E is given by Table 7.2.5. Determine an expansion and the contraction of $\mathcal{A}$ as a coset automaton.

*7.3 Basic operations

This section describes the basic operations on finite generalized coset automata needed to carry out coset enumeration using the HLT strategy. These operations include tracing, handling coincidences, and several new operations. In our examples, we shall assume that $X = \{x, y, z\}$, that $\mathcal{R}$ consists of the rules

$$(xy)^3 \to \varepsilon, \quad (yx)^3 \to \varepsilon, \quad z^5 \to \varepsilon,$$

and that $\mathcal{A}$ is a generalized coset automaton with respect to $\mathcal{R}$ which initially has the edges shown in Table 7.3.1.

In Section 4.11 we discussed the minimal decomposition of a word U in $\mathcal{C}$ as a product of subwords of left sides in $\mathcal{R}$. We shall now introduce another decomposition. There are words P and V such that $U = PVP^\diamond$ and $U^\diamond = PV^\diamond P^\diamond$. For example, we may take $P = \varepsilon$ and $V = U$. The pair (P, V) with $|P|$ maximal will be called the *cyclic decomposition* of U. Let $U = U_1 \ldots U_s$ be the minimal decomposition of U. If $U \neq U^\diamond$, then $P = U_1 \ldots U_i$, where i is the largest index such that $U_j = U_{s+1-j}^\diamond$ for $1 \leq j \leq i$. If $U = U^\diamond$, then $s = 2i+1$ is odd and $(U_{i+1})^2$ is a left side. Here $P = U_1 \ldots U_i$ and $V = U_{i+1}$. In our example, if $U = yxyz^2xyzxyz^3xyx$, then the cyclic decomposition of U is $(yxyz^2, xyzxy)$.

When we perform a forward trace of a word W in $\mathcal{A}$ starting at a state σ, we may not end at a state. That is, we may reach the following situation. We have $W = BC$, where $\tau = \sigma^B$ is defined and there is an edge (τ, U, φ) such that U and C have a nonempty common prefix. However, U is not a prefix of C. For example, in $\mathcal{A}$ let $\sigma = 1$ and $W = xyxyzyx$. The trace starts

Table 7.3.1

State	Label	Endpoint
1	$xyxz^2yx$	2
	z^2xyz	3
2	$yxyxz^3yxy$	1
	$zyxyxz^3x$	3
3	$yxyxyz^2yxz^4$	2
	z^4xyxyz^3	1

along the edge $(1, xyxz^2yx, 2)$, but leaves the edge in the middle. Suppose that we have a situation of this type. Let S be the longest common prefix of U and C, and let $U = ST$. In our example, $S = xyx$ and $T = z^2yx$. It will sometimes be necessary to split the edge (τ, U, φ) at S and introduce a new state ψ such that $\psi = \tau^S$. However, this splitting may actually require adding two or three new states, not just one. Let $U = U_1 \ldots U_t$ be the minimal decomposition of U as a product of nonempty subwords of left sides. The easiest case to handle is the one in which $t = 1$, for we simply add a new state ψ and replace (τ, U, φ) by (τ, S, ψ) and (ψ, T, φ).

Next let us consider the case in which $S = U_1 \ldots U_i$, where $1 \le i < t$. Here $U^\diamond = T^\diamond S^\diamond$. Usually we can replace (τ, U, φ) and $(\varphi, U^\diamond, \tau)$ by the four edges

$$(\tau, S, \psi), \quad (\psi, S^\diamond, \tau), \quad (\psi, T, \varphi), \quad (\varphi, T^\diamond, \psi),$$

where ψ is a new state. The one exception occurs when $\tau = \varphi$ and the first terms of S and $T^\diamond$ are the same. In this situation, we may have two edges leaving τ both of whose labels start with the same generator. Assume that we are in this exceptional situation. Since (τ, U, τ) and $(\tau, U^\diamond, \tau)$ are both edges, they must be the same edge. That is, $U = U^\diamond$. By replacing U by $U^\diamond$ and S by $T^\diamond$ if necessary, we may assume that $\|S\| \le \|T\|$. Let (P, V) be the cyclic decomposition of U. Then S is a prefix of P. Let $P = SQ$, let ψ be a new state, and replace the edge (τ, U, τ) by the three edges

$$(\tau, S, \psi), \quad (\psi, S^\diamond, \tau), \quad (\psi, QVQ^\diamond, \psi).$$

Finally, suppose that $S = U_1 \ldots U_{i-1}R$, where R is a proper, nonempty prefix of U_i. Using methods already described, we may split one or more edges so that $\rho = \tau^{U_1 \ldots U_{i-1}}$ and ρ^{U_i} are both defined. Now to insure that ρ^R is defined we are back to the case in which $t = 1$.

In our earlier example, to split $e = (1, xyxz^2yx, 2)$ at xyx, we delete e and $(2, yxyxz^3yxy, 1)$ and add $(1, xyx, 4)$, $(4, yxy, 1)$, $(4, z^2yx, 2)$, and $(2, yxyxz^3, 4)$. To split $f = (2, zyxyxz^3x, 3)$ at zy, we delete f and $(3, yxyxyz^2yxz^4, 2)$ and replace them by $(2, z, 5)$, $(5, z^4, 2)$, $(6, z^3x, 3)$,

Table 7.3.2

State	Label	Endpoint	State	Label	Endpoint
1	xyx	4	5	y	7
	z^2xyz	3		z^4	2
2	$yxyxz^3$	4	6	yx	5
	z	5		z^3x	3
3	$yxyxyz^2$	6	7	xyx	6
	z^4xyxyz^3	1			
4	yxy	1			
	z^2yx	2			

$(3, yxyxyz^2, 6)$, $(6, yx, 5)$, $(5, y, 7)$, and $(7, xyx, 6)$. At this point we have the edges shown in Table 7.3.2.

Now suppose that we have states σ and τ and a nonempty word U in $\mathcal{C}$. Let U start with x and end with y and assume that there is no edge leaving σ whose label starts with x and there is no edge coming into τ whose label ends with y. To *join* σ to τ by U means to add one or more edges so that $\sigma^U = \tau$. The obvious thing to try is to add edges (σ, U, τ) and $(\tau, U^\diamond, \sigma)$. This works fine unless $\sigma = \tau$, $x = \dot{y}$, and $U \neq U^\diamond$, for then there will be two edges leaving σ both of whose labels start with x. Suppose we have this special case and let (P, V) be the cyclic decomposition of U. If $P \neq \varepsilon$, then we introduce a new state ρ and join σ to ρ by P and ρ to ρ by V. This reduces us to the case in which $P = \varepsilon$. Now let $U = U_1 \ldots U_s$ be the minimal decomposition of U. Since $x = \dot{y}$, $P = \varepsilon$, and $U \neq U^\diamond$, either U_1 is a proper prefix of $U_s^\diamond$ or $U_s^\diamond$ is a proper prefix of U_1. By replacing U by $U^\diamond$ if necessary, we may assume that U_1 is a proper prefix of $U_s^\diamond$. Let $W = U_2 \ldots U_{s-1}$ and choose Q so that $U_1 Q U_s$ is a left side in $\mathcal{R}$. We choose two new states φ and ψ, add edges (σ, U_1, φ), (φ, Q, ψ), and (ψ, U_s, σ), and then join φ to ψ by W. For example, to join 7 to 7 by $yxyz^2xyzxyz^3xyx$, we must add three new states. One way to do this is to let the new states be 8, 9, and 10, and the new edges be $(7, yxyz^2, 8)$, $(8, z^3xyx, 7)$, $(8, xy, 9)$, $(9, xy, 10)$, $(10, xy, 8)$, $(9, z, 10)$, and $(10, z^4, 9)$. With these additions, our automaton $\mathcal{A}$ has the edges shown in Table 7.3.3.

The next operation is somewhat specialized. Suppose that $e = (\sigma, U, \tau)$ is an edge and U is a subword of a left side. Let $U = ST$, where both S and T are nonempty and T starts with x. Finally, let γ be a state such that there is no edge leaving γ whose label starts with x. To *insert* γ into e at S means to replace e by the pair of edges (σ, S, γ) and (γ, T, τ). For example, to insert 7 into $(5, z^4, 2)$ at z^2 in $\mathcal{A}$ changes the edge set as shown in Table 7.3.4.

The operations of two-sided tracing and processing coincidences become much more closely linked when generalized coset automata are used. We are familiar with the way two-sided traces can lead to coincidences. However, now a natural approach to coincidence processing leads to two-sided traces.

Table 7.3.3

State	Label	Endpoint	State	Label	Endpoint
1	xyx	4	6	yx	5
	z^2xyz	3		z^3x	3
2	$yxyxz^3$	4	7	xyx	6
	z	5		$yxyz^2$	8
3	$yxyxyz^2$	6	8	xy	9
	z^4xyxyz^3	1	9	xy	10
4	yxy	1		z	10
	z^2yx	2	10	xy	8
5	y	7		z^4	9
	z^4	2			

Table 7.3.4

State	Label	Endpoint	State	Label	Endpoint
1	xyx	4	6	yx	5
	z^2xyz	3		z^3x	3
2	$yxyxz^3$	4	7	xyx	6
	z	5		$yxyz^2$	8
3	$yxyxyz^2$	6		z^2	2
	z^4xyxyz^3	1	8	xy	9
4	yxy	1	9	xy	10
	z^2yx	2		z	10
5	y	7	10	xy	8
	z^2	7		z^4	9

Suppose that a state σ has been found to be coincident with an earlier state τ and $e = (\sigma, U, \rho)$ is an edge. Then we must remove e and make sure that $\tau^U = \rho$ if $\rho \neq \sigma$, or $\tau^U = \tau$ if $\rho = \sigma$. Doing this can be just as complicated as performing two-sided traces of subgroup generators or relators. Of course, when e is removed, we shall usually have to remove other edges so that the axioms for a generalized coset automaton continue to be satisfied. We no longer have only two-sided traces which start and end at the same state. We must allow for two-sided traces of a word in $\mathcal{C}$ from one state to some other state.

To facilitate both tracing and coincidence processing, we maintain a stack of *trace records*, which are triples (β, W, γ), where β and γ are possibly inactive states and W is a word. To process the trace record (β, W, γ), we first compute the active states δ and ϵ equivalent to β and γ, respectively, and then perform a two-sided trace of W from δ to ϵ. This trace may result in other trace records being pushed onto the stack. When an active state σ is found to be coincident with an earlier active state τ, we examine each edge $e = (\sigma, U, \rho)$ leaving σ. If U is not a subword of a left side, then e and $(\rho, U^\diamond, \sigma)$ are removed and e is pushed onto the trace-record stack.

However, if U is a subword of a left side, then let W^r be the left side of which U is a prefix, where W is not a proper power. Let W^t be the smallest power of W such that $\sigma^{W^t} = \sigma$. We remove all edges in the path from σ to σ having signature W^t and push these edges onto the trace record stack. Once all edges leaving σ have been removed, we record the fact that σ is equivalent to τ and go back to processing trace records.

Since we have moved most of the coincidence procedure into the procedure for performing two-sided traces, the latter procedure is now fairly complicated. Suppose that we wish to perform a full two-sided trace of a word U in $\mathcal{C}$ from β to γ. By performing forward and backward traces, we may assume that there is no edge leaving β whose label is a prefix of U and there is no edge coming into γ whose label is a suffix of U. If $U = \varepsilon$ and $\beta \neq \gamma$, then we have a coincidence. Assume $U \neq \varepsilon$. Let B be the longest prefix of U which is also a prefix of the label of an edge leaving β and let E be the longest suffix of U which is also a suffix of an edge coming into γ. Let $U = BC = DE$. Suppose B and C are both nonempty and let e be the edge leaving β whose label has B as a prefix. We split e at B and repeat the trace. In this way we may assume that $B = \varepsilon$ or $C = \varepsilon$. Similarly, by splitting an edge coming into γ, if necessary, we may assume that $D = \varepsilon$ or $E = \varepsilon$.

Suppose that $C = \varepsilon$, so $B = U$. Thus there is an edge (β, V, δ), where V has the form UP. If V is not a subword of a left side, then remove (β, V, δ) and $(\delta, V^\diamond, \beta)$ and push (γ, P, δ) onto the trace-record stack. In this way we arrange things so that either $B = \varepsilon$ or V is a subword of a left side. Similarly, we may assume that $E = \varepsilon$ or there is an edge (ϵ, QU, γ) such that QU is a power of a left side. If $B = E = \varepsilon$, then we can connect β to γ by U as described earlier. If $B = \varepsilon$ and $E = U$, then we insert β into (ϵ, QU, γ) at Q. If $B = U$ and $E = \varepsilon$, then we insert γ into (β, UP, δ) at U.

We are left with the case $B = E = U$, in which there are edges $e = (\beta, UP, \delta)$ and $f = (\epsilon, QU, \gamma)$ such that both UP and QU are subwords of left sides. At this point we must examine the triples e, f, and (β, U, γ), all the edges in the path from δ to β with signature $(UP)^\diamond$, and all the edges in the path from γ to ϵ with signature $(QU)^\diamond$. Let D be the set of these triples, let Δ be the set of states involved, let L be the left side in $\mathcal{R}$ which has U as a prefix, let Y be the set of generators occurring in L, and let $\mathcal{S}$ be the set of cyclic permutations of L. Then $\mathcal{S}$ is a confluent, niladic rewriting system on Y^* which defines a group. Our job is to find a generalized coset automaton $\mathcal{B} = (\Sigma_0, Y, E_0, \{\alpha\})$ with respect to $\mathcal{S}$ satisfying the following conditions:

(i) There is a path in $\mathcal{B}$ from α to α with signature L, and every edge in E_0 occurs at least once in this path.

(ii) There is a map g of Δ onto Σ_0 such that whenever (φ, R, ψ) is in D then $g(\varphi)^R = g(\psi)$ in $\mathcal{B}$.

(iii) $\mathcal{B}$ is as general as possible in the sense that it has the minimum possible number of states and the sum of the lengths of the edges is as large as possible.

Conditions (i) to (iii) determine $\mathcal{B}$ up to isomorphism. If $|L|$ is large, then constructing one such $\mathcal{B}$ involves a fair amount of work. However, assuming that $|L|$ is small, finding such a $\mathcal{B}$ is not hard. One approach involves using an ordinary coset automaton $\mathcal{B}'$. Let $L = y_1 \ldots y_t$. Initially let the state set of $\mathcal{B}'$ be $\{1, \ldots, t\}$ and let the edges of $\mathcal{B}'$ consist of the triples $(i, y_i, i+1)$, $1 \leq i < t$, and $(t, y_t, 1)$. For each φ in Δ there is a word W_φ which is a prefix of a power of L such that there is a path in the graph (Δ, D) from β to φ having signature W_φ. For each (φ, R, ψ) in D, perform a two-sided trace of $W_\varphi R(W_\psi)^{-1}$ in $\mathcal{B}'$. After this has been done, define the function g from Δ to the states of $\mathcal{B}'$ such that $1^{W_\varphi} = g(\varphi)$ in $\mathcal{B}'$. Take Σ_0 to be the image of g and take E_0 to be the set of triples (μ, S, ν), where μ and ν are in Σ_0, S is nonempty, $\mu^S = \nu$ in $\mathcal{B}'$, and for no proper, nonempty prefix T of S is μ^T defined.

Once $\mathcal{B}$ has been constructed, we make φ and ψ in Δ coincident if $g(\varphi) = g(\psi)$. Let Γ be the set of elements in Δ which are still active. Then g maps Γ bijectively onto Σ_0. Delete those edges in the original generalized coset automaton which are in D. Add the edges (φ, R, ψ) with φ and ψ in Γ and $(g(\varphi), R, g(\psi))$ in E_0. This completes the two-sided trace of U from β to γ.

One two-sided trace can result in many trace records being pushed onto the stack. Also states are created and removed during the execution of the procedure. Thus the issue of the termination of the two-sided trace procedure must be carefully examined. Termination does occur, but the proof will be omitted.

The construction of the auxiliary automaton $\mathcal{B}$ is somewhat involved. Two examples will be given to illustrate this part of the two-sided trace procedure. For the purpose of these examples, assume that $X = \{a, b, c\}$ and that $\mathcal{R}$ consists of the rules $L \to \varepsilon$, where L is a cyclic permutation of the word $(abc)^6$. In the first example, we shall assume that the following triples are edges of the current generalized coset automaton:

$$(4, abcab, 7), \quad (7, cabca, 10), \quad (10, bcab, 8), \quad (8, cabc, 4).$$

Now suppose that we must make a two-sided trace of $U = abca$ from 4 to 10. If $4^U = 10$, then $4^{Ubcabcabc} = 4$. That is, $4^{(abc)^4} = 4$. Since $4^{(abc)^6}$ is also 4, we must have $4^{(abc)^2} = 4$. It turns out that one possible choice for $\mathcal{B}$ has states 4, 7, 8, and 10, and the following edges:

$$(4, ab, 8), \quad (8, ca, 10), \quad (10, b, 7), \quad (7, c, 4).$$

No coincidences are involved here, and the four original edges are replaced by the four new edges.

In the second example, assume that the following edges are currently in the automaton:

$$(5, abca, 9), \quad (9, bc, 5), \quad (6, ab, 10), \quad (10, cab, 7), \quad (7, cabc, 6).$$

Now let us perform a two-sided trace of $U = ca$ from 10 to 9. Before the trace, $9^{(bca)^2} = 9$ and $10^{(cab)^3} = 10$. After the trace, if β denotes the active state equivalent to 10, we must have $\beta^{abc} = \beta$, since $\gcd(2,3) = 1$. It follows easily that the automaton $\mathcal{B}$ may be taken to have states 5, 7, and 9 and edges

$$(5, a, 9), \quad (9, b, 7), \quad (7, c, 5).$$

State 10 is coincident with state 7, and state 6 is coincident with state 5. The original five edges are removed, as are all edges involving 6 or 10. The three edges of $\mathcal{B}$ are added.

*7.4 Some examples

The previous section discussed the individual operations on generalized coset automata involved in doing coset enumeration. We shall not attempt to formalize an HLT enumeration procedure using generalized coset automata. However, we shall present some examples which illustrate how the techniques of Section 7.3 are used. As we work through the examples, many of the edges in the generalized coset automata will have labels of length 1, while some will have longer labels. To save space, the edges will be represented by hybrid tables. An ordinary coset table will be used to describe the edges with labels of length 1. The edges with labels of length greater than 1 are identified by pointers enclosed in brackets which refer to entries in an auxiliary table.

The first example is the enumeration of the cosets of the subgroup H generated by x and yxy^2 in the group $G = \mathrm{Mon}\langle x, y \mid x^2 = y^3 = (xy)^7 = (xyxy^2)^4 = 1\rangle$. We shall take $X = \{x, y\}$ and let $\mathcal{R}$ consist of the rules $x^2 \to \varepsilon$ and $y^3 \to \varepsilon$. Thus there are two subgroup generators and two relators for G as a quotient group of $F = \mathrm{Mon}\langle X \mid \mathcal{R}\rangle$. We initialize $\mathcal{A}$ to be the automaton with 1 as its only state and no edges. The traces of the subgroup generators are very easy. They result in the two edges $(1, x, 1)$ and $(1, yxy^2, 1)$. Note that $(yxy^2)^\lozenge = yxy^2$. Our hybrid tables take the following form:

	x	y
1	1	[1]

	Label	Endpoint
[1]	yxy^2	1

Table 7.4.1a

	x	y
1	1	2
2	2	3
3	4	1
4	3	5
5	[1]	6
6	[2]	4

Table 7.4.1b

	Label	Endpoint
[1]	$xyxyx$	6
[2]	xy^2xy^2x	5

Table 7.4.2a

	x	y
1	1	2
2	2	3
3	4	1
4	3	5
5	[1]	6
6	7	4
7	6	8
8	[2]	9
9	[3]	7

Table 7.4.2b

	Label	Endpoint
[1]	xyx	9
[2]	xy^2xyx	8
[3]	xy^2x	5

The trace of $(xy)^7$ from 1 to 1 requires the edge $(1, yxy^2, 1)$ to be split at yxy. This splitting introduces two new states and produces the following edges, all of whose labels have length 1:

	x	y
1	1	2
2	2	3
3		1

Now the trace of $(xy)^7$ from 1 to 1 leads to 3 being joined to 3 by $(xy)^4x$. This adds three new states and gives us the edges shown in Tables 7.4.1a and 7.4.1b.

At this point, $1^{(xyxy^2)^4} = 1$ and $2^{(xy)^7} = 2$. As we trace $(xyxy^2)^4$ from 2 to 2, we must split $(6, xy^2xy^2x, 5)$ at xy. Assuming that among the new edges introduced by the splitting are $(6, x, 7)$, $(7, y, 8)$, and $(8, y, 9)$, the trace adds the edge $(8, xy^2xyx, 8)$. Now we have Tables 7.4.2a and 7.4.2b.

The trace of $(xy)^7$ at 3 produces nothing new. When we trace $(xyxy^2)^4$ at 3, the edge $(5, xyx, 9)$ gets split into edges with labels of length 1. If we assume that among these edges are $(5, x, 11)$, $(11, y, 10)$, and $(10, y, 12)$, the trace adds the edge $(12, xyxy^2x, 12)$. The edge set now is shown in Tables 7.4.3a and 7.4.3b.

Table 7.4.3a

	x	y		x	y
1	1	2	7	6	8
2	2	3	8	[1]	9
3	4	1	9	10	7
4	3	5	10	9	12
5	11	6	11	5	10
6	7	4	12	[2]	11

Table 7.4.3b

	Label	Endpoint
[1]	xy^2xyx	8
[2]	$xyxy^2x$	12

Table 7.4.4a

	x	y		x	y
1	1	2	10	9	12
2	2	3	11	5	10
3	4	1	12	16	11
4	3	5	13	8	15
5	11	6	14	14	13
6	7	4	15	[1]	14
7	6	8	16	12	17
8	13	9	17	17	18
9	10	7	18	[2]	16

Table 7.4.4b

	Label	Endpoint
[1]	xy^2x	8
[2]	xyx	12

At this point $4^{(xy)^7} = 4$, $4^{(xyxy^2)^4} = 4$, and $5^{(xy)^7} = 5$. As we trace $(xyxy^2)^4$ at 5, both of the edges in the auxiliary table are split. Assuming that the edges created by the splitting include

$$(8, x, 13), \quad (13, y, 15), \quad (15, y, 14), \quad (12, x, 16), \quad (16, y, 17), \quad (17, y, 18),$$

the trace adds $(15, xy^2x, 18)$ and (18,x y x,15). Now we have Tables 7.4.4a and 7.4.4b.

The trace of $(xy)^7$ at 6 causes the two edges with labels of length greater than 1 to be split. Assuming that the edges added by the splitting include

$$(15, x, 19), \quad (19, y, 21), \quad (21, y, 20),$$

the trace adds $(21, x, 21)$. The automaton is now complete and compatible with the relators. The final coset table is Table 7.4.5.

In the first example, the use of generalized coset automata is hardly necessary. The second example provides somewhat better justification for their introduction. Suppose $G = \mathrm{Grp}\, \langle a, b \mid a^2 = b^3 = (ab)^7 = (abab^{-1})^8 = 1 \rangle$. We noted in Section 5.5 that the order of G is 10752. Let us compare what happens as we begin to enumerate the cosets of the trivial subgroup in G using both an ordinary coset automaton and a generalized coset automaton.

Table 7.4.5

	x	y		x	y		x	y
1	1	2	8	13	9	15	19	14
2	2	3	9	10	7	16	12	17
3	4	1	10	9	12	17	17	18
4	3	5	11	5	10	18	20	16
5	11	6	12	16	11	19	15	21
6	7	4	13	8	15	20	18	19
7	6	8	14	14	13	21	21	20

Table 7.4.6a

	a	b	b^{-1}
1	[1]	3	4
2	[2]	[3]	[4]
3	[5]	4	1
4	[6]	1	3

Table 7.4.6b

	Label	Endpoint
[1]	aba	2
[2]	$ab^{-1}a$	1
[3]	$(ba)^5$	4
[4]	$(b^{-1}aba)^7$	3
[5]	$(ab^{-1}ab)^7$	2
[6]	$(ab^{-1})^5$	2

We shall take $X = \{a, b, b^{-1}\}$ and let $\mathcal{R}$ consist of the rules

$$a^2 \to \varepsilon, \quad bb^{-1} \to \varepsilon, \quad b^{-1}b \to \varepsilon.$$

If we trace the three relators b^3, $(ab)^7$, and $(abab^{-1})^8$ at the first coset in an ordinary coset automaton, the result has 42 states. If we use a generalized coset automaton, then we need only 4 states. The edges are given by Tables 7.4.6a and 7.4.6b.

The next step in the enumeration is to perform two-sided traces of the relators at the state 1^a. In the case of the generalized automaton, edge [1] must be split in order for 1^a to be defined. After these traces are performed in the ordinary automaton, the total number of states defined is 52. In the generalized automaton, there are only 8. The generalized automaton is isomorphic to the automaton in Tables 7.4.7a and 7.4.7b. Note that the states other than 1 have been renumbered, so they may not correspond to the states with the same number used earlier.

If we now trace the relators at 1^b, then the ordinary automaton has 76 states. However, the generalized automaton has only 10 states and is described by Tables 7.4.8a and 7.4.8b..

In this example, the long relators are powers, and the edge labels could be written concisely. Thus we wrote $(b^{-1}aba)^6 b^{-1}ab$ instead of

$$b^{-1}abab^{-1}abab^{-1}abab^{-1}abab^{-1}abab^{-1}ab.$$

Table 7.4.7a

	a	b	b^{-1}
1	2	3	4
2	1	5	6
3	7	4	1
4	[1]	1	3
5	8	6	2
6	[2]	2	5
7	3	[3]	[4]
8	5	[5]	[6]

Table 7.4.7b

	Label	Endpoint
[1]	$(ab^{-1})^5$	8
[2]	$(ab{-}1)^5$	7
[3]	$(ba)^5$	6
[4]	$(b^{-1}aba)^6b^{-1}ab$	8
[5]	$(ba)^5$	4
[6]	$(b^{-1}bab)^6b^{-1}ab$	7

Table 7.4.8a

	a	b	b^{-1}
1	2	3	4
2	1	5	6
3	7	4	1
4	[1]	1	3
5	8	6	2
6	[2]	2	5
7	3	[3]	[4]
8	5	[5]	[6]
9	[7]	[8]	[9]
10	[10]	[11]	[12]

Table 7.4.8b

	Label	Endpoint
[1]	$ab^{-1}a$	10
[2]	$(ab^{-1})^4$	9
[3]	ba	9
[4]	$(b^{-1}aba)^6b^{-1}ab$	8
[5]	$(ba)^3b$	10
[6]	$(b^{-1}aba)^6b^{-1}ab$	7
[7]	ab^{-1}	7
[8]	$(ba)^4$	6
[9]	$(b^{-1}aba)^6b^{-1}ab$	10
[10]	aba	4
[11]	$(bab^{-1}a)^6b$	17
[12]	$(b^{-1}a)^3b^{-1}$	9

However, it can be shown that if $\mathcal{R}$ is classical, as it is in this example, then at any time during an HLT enumeration the labels on the edges of the generalized automaton are subwords of words in $\mathcal{U} \cup \mathcal{U}^{-1} \cup \mathcal{V} \cup \mathcal{V}^{-1}$, where $\mathcal{U}$ is the set of subgroup generators and $\mathcal{V}$ is the set of relators. Once $\mathcal{R}$, $\mathcal{U}$, and $\mathcal{V}$ have been stored, it takes only a constant amount of space to describe the edge labels, no matter how long the elements of $\mathcal{U}$ and $\mathcal{V}$ may be.

8

Abelian groups

With this chapter we begin a new topic in our study of finitely presented groups. Chapters 4 to 7 were primarily concerned with subgroups of finitely presented groups. We studied methods for trying to decide whether subgroups have finite index and for obtaining presentations for subgroups. In the special case of finitely generated subgroups of a free product of cyclic groups, we also considered the computation of intersections and normalizers and the determination of conjugacy. Now we begin the study of quotient groups of finitely presented groups. We shall try to answer questions like the following: Let $\mathcal{G}$ be a class of groups and let G be a group given by a finite presentation. Among the quotient groups of G which belong to the class $\mathcal{G}$, is there a largest one? If so, can we determine it?

The class $\mathcal{G}$ is usually chosen to contain groups in which we have a real chance of doing explicit computation. The hope is that we can identify large quotient groups of our finitely presented group G in which we can solve the word problem and compute with elements and subgroups. By studying these quotient groups, we may be able to determine properties of G. Although most questions about subgroups of finitely presented groups are in general undecidable, there are algorithms for determining the largest quotient group of G belonging to several important classes of groups.

Since our goal is to produce quotients in which we can compute, our first task is to look at some classes of groups and see what kinds of calculations can be carried out. We must examine carefully which computations are merely possible in principle, and which are actually practical for interesting groups. This chapter will be concerned with finitely generated abelian groups. Chapter 9 considers polycyclic groups, a class of groups in which a wide range of computations is possible. Additive notation will normally be used for abelian groups.

Up to now, using only elementary group theory, we have been able to develop quite powerful procedures. However, in this and later chapters, deeper results about groups and other algebraic objects, such as commutative rings and modules over them, will be needed. Limitations of space will

force us to draw more heavily on other sources for important facts about
these algebraic objects.

8.1 Free abelian groups

Free abelian groups play roughly the same role in the theory of abelian
groups that free groups play in the general theory of groups. This section
and the next two summarize the basic facts about finitely generated free
abelian groups, their subgroups, and their quotient groups.

Let $x_1, \ldots, x_r$ be elements of an abelian group G. An *integral linear
combination* of $x_1, \ldots, x_r$ is an element of G of the form $m_1 x_1 + \cdots + m_r x_r$,
where the m_i are integers. A *basis* for G is a generating set X of G such
that for any finite sequence $x_1, \ldots, x_r$ of distinct elements of X the only
way to express 0 as an integral linear combination of the x_i is the obvious
way, $0x_1 + 0x_2 + \cdots + 0x_r$. Not all abelian groups have bases. Those that
do are called *free abelian groups*.

The direct sum $\mathbb{Z} \oplus \cdots \oplus \mathbb{Z}$ of n copies of the additive group of the integers
is denoted $\mathbb{Z}^n$. Elements of $\mathbb{Z}^n$ are n-tuples or vectors $(a_1, \ldots, a_n)$ of integers
with addition and negation performed componentwise. For $1 \leq i \leq n$ let
x_i be the vector in $\mathbb{Z}^n$ with zeros in all positions except for the i-th, which
contains 1. Thus x_i is the i-th row of the n-by-n identity matrix. The linear
combination $m_1 x_1 + \cdots + m_n x_n$ is $(m_1, \ldots, m_n)$, which is the zero element
of $\mathbb{Z}^n$ if and only if $m_1 = \cdots = m_n = 0$. Therefore $\{x_1, \ldots, x_n\}$ is a basis
for $\mathbb{Z}^n$ and $\mathbb{Z}^n$ is free abelian. As just defined, bases are sets. However, we
shall usually assume that bases come with a linear ordering. The ordered
basis $x_1, \ldots, x_n$ is called the *standard basis* of $\mathbb{Z}^n$.

For any set X, the *free abelian group on* X is the set of all functions
$f: X \to \mathbb{Z}$ such that f maps all but a finite number of elements of X to 0.
The sum of two functions f and g is defined by $(f + g)(x) = f(x) + g(x)$
for all x in X. (See Exercise 1.1.) Since n-tuples of integers are functions
from $\{1, \ldots, n\}$ to $\mathbb{Z}$, the group $\mathbb{Z}^n$ is the free abelian group on $\{1, \ldots, n\}$.

Let G be a free abelian group with basis X and let K be any abelian
group. Every map from X to K has a unique extension to a homomorphism
of G into K. If the image of X generates K, then the homomorphism is
surjective and K is isomorphic to a quotient group of G. It follows that
any abelian group is isomorphic to a quotient group of a free abelian group.
The number of homomorphisms of G into the group $\mathbb{Z}_2$ of order 2 is $2^{|X|}$.
Hence if Y is any other basis, then $2^{|Y|} = 2^{|X|}$, so $|Y| = |X|$. Therefore any
two bases of G have the same cardinality, which is called the *rank* of G.
The rank of $\mathbb{Z}^n$ is n.

Any subgroup of $\mathbb{Z}^n$ can be generated by n or fewer elements. We shall not
prove this statement here, but a more general result is proved as Proposition 3.5 of Chapter 9. Since subgroups are finitely generated, the ascending

chain condition holds for subgroups of $\mathbb{Z}^n$. A subgroup H of $\mathbb{Z}^n$ can be specified by giving an m-by-n matrix A whose rows generate H. If this is the case, we shall write $H = S(A)$. The set $S(A)$ is the set of all integral linear combinations of the rows of A. These linear combinations are the vectors uA, where u ranges over $\mathbb{Z}^m$.

Our first important task is to develop algorithms for deciding membership in $S(A)$ for a given m-by-n integer matrix A. The first algorithm will be based on the concept of row equivalence of matrices. Two m-by-n integer matrices are said to be *row equivalent over* $\mathbb{Z}$ if one can be transformed into the other by a sequence of integer row operations. There are three types of *integer row operations*:

(1) Interchange two rows.
(2) Multiply a row by -1.
(3) Add an integral multiple of one row to another row.

Example 1.1. Let A be the following matrix:

$$\begin{bmatrix} 3 & -1 & 2 & 4 \\ 1 & 5 & -3 & 0 \\ -4 & 6 & 1 & 2 \end{bmatrix}.$$

Interchanging rows 1 and 2 of A gives

$$\begin{bmatrix} 1 & 5 & -3 & 0 \\ 3 & -1 & 2 & 4 \\ -4 & 6 & 1 & 2 \end{bmatrix}.$$

Now multiplying row 2 by -1 produces

$$\begin{bmatrix} 1 & 5 & -3 & 0 \\ -3 & 1 & -2 & -4 \\ -4 & 6 & 1 & 2 \end{bmatrix}.$$

Adding twice row 3 to row 1 yields

$$\begin{bmatrix} -7 & 17 & -1 & 4 \\ -3 & 1 & -2 & -4 \\ -4 & 6 & 1 & 2 \end{bmatrix}.$$

All four of these matrices are row equivalent.

The effect of one row operation can be undone by another row operation. In the case of row operations of types (1) and (2), applying the operation twice leaves the matrix unchanged. If q times the i-th row is added to the j-th row, then adding $-q$ times the i-th row to the j-th row returns the matrix to its original form. Thus if B can be derived from A by a sequence of integer row operations, then A can be derived from B. It follows easily that row equivalence over $\mathbb{Z}$ is an equivalence relation on integer matrices.

If B is obtained from A by an integer row operation, then the rows of B are in $S(A)$. Therefore $S(B) \subseteq S(A)$. Since A can be obtained from B by a row operation, we have $S(A) \subseteq S(B)$, and therefore $S(A) = S(B)$. Hence if A and B are row equivalent over $\mathbb{Z}$, then $S(A) = S(B)$.

It is possible to specify representatives for the row equivalence classes of integer matrices. Let A be an m-by-n matrix over $\mathbb{Z}$ and suppose that A has r nonzero rows. We say that A is *in row Hermite normal form* if the following conditions are satisfied:

(1) The first r rows of A are nonzero.
(2) For $1 \leq i \leq r$ let A_{ij_i} be the first nonzero entry in the i-th row of A. Then $j_1 < j_2 < \cdots < j_r$.
(3) $A_{ij_i} > 0$, $1 \leq i \leq r$.
(4) If $1 \leq k < i \leq r$, then $0 \leq A_{kj_i} < A_{ij_i}$.

The entries A_{ij_i} will be called the *corner entries* of A. Condition (2) states that successive corner entries occur in later columns, condition (3) says that corner entries are positive, and condition (4) asserts that an entry above a corner entry c is its own least nonnegative remainder under division by c. A matrix satisfying conditions (1) and (2) will be said to be in *row echelon form*. The reader is warned that there is not general agreement on the definition of the term "Hermite normal form". Also, Hermite's contributions to this subject are often described inaccurately. See the historical note at the end of this chapter.

Example 1.2. The matrix

$$\begin{bmatrix} 2 & 1 & 5 & 0 & -2 & 1 & -1 \\ 0 & 3 & -1 & 2 & 4 & 0 & 2 \\ 0 & 0 & 0 & 4 & 7 & 1 & 8 \\ 0 & 0 & 0 & 0 & 0 & 2 & 5 \\ 0 & 0 & 0 & 0 & 0 & 0 & 0 \end{bmatrix}$$

is in row Hermite normal form.

Every integer matrix B is row equivalent to a matrix A which is in row Hermite normal form. The procedure for determining A is called *integer row reduction*. Here is a generic row reduction procedure.

Procedure ROW_REDUCE($B; A$);
Input: B : an m-by-n integer matrix;
Output: A : the matrix in row Hermite normal form row
                equivalent to B;
Begin
  $A := B$; $i := 1$; $j := 1$;

  While $i \leq m$ and $j \leq n$ do
    If $A_{kj} = 0$ for $i \leq k \leq m$ then $j := j + 1$
    Else begin
        While there exist distinct indices k and ℓ between i and m with
        $0 < |A_{kj}| \leq |A_{\ell j}|$ do begin
        $q := A_{\ell j}$ div A_{kj};

          Subtract q times row k of A from row ℓ
        End;

        Let $A_{kj} \neq 0$ with $i \leq k \leq m$;   (* k is unique. *)
        If $k \neq i$ then interchange rows i and k of A;
        If $A_{ij} < 0$ then multiply row i of A by -1;

        For $\ell := 1$ to $i - 1$ do begin
          $q := A_{\ell j}$ div A_{ij};
          Subtract q times row i of A from row ℓ
        End;

        $i := i + 1$; $j := j + 1$
    End
End.

In ROW_REDUCE we have a great deal of freedom in choosing k and ℓ with $0 < |A_{kj}| \leq |A_{\ell j}|$. In the examples, we shall take $|A_{kj}|$ as small as possible and $|A_{\ell j}|$ as large as possible. Other strategies will be discussed later. The output matrix is described as *the* matrix in row Hermite normal form row equivalent to B. Uniqueness will be proved later in Corollary 1.2.

Example 1.3. Let us row reduce the following matrix:

$$\begin{bmatrix} 4 & -1 & 2 & -5 \\ -3 & 0 & 1 & -3 \\ 2 & 4 & -3 & 2 \\ 5 & -3 & 4 & 4 \end{bmatrix}.$$

Initially $i = j = 1$. On the first iteration of the inner While-loop, $k = 3$, $\ell = 4$, and $q = 5 \operatorname{div} 2 = 2$. Subtracting twice row 3 from row 4 gives

$$
\begin{bmatrix}
4 & -1 & 2 & -5 \\
-3 & 0 & 1 & -3 \\
2 & 4 & -3 & 2 \\
1 & -11 & 10 & 0
\end{bmatrix}.
$$

With $k = 4$, $\ell = 1$, and $q = 4$ we get

$$
\begin{bmatrix}
0 & 43 & -38 & -5 \\
-3 & 0 & 1 & -3 \\
2 & 4 & -3 & 2 \\
1 & -11 & 10 & 0
\end{bmatrix}.
$$

Two more iterations with $k = 4$ give

$$
\begin{bmatrix}
0 & 43 & -38 & -5 \\
0 & -33 & 31 & -3 \\
0 & 26 & -23 & 2 \\
1 & -11 & 10 & 0
\end{bmatrix}.
$$

Interchanging the first and fourth rows, we have

$$
\begin{bmatrix}
1 & -11 & 10 & 0 \\
0 & -33 & 31 & -3 \\
0 & 26 & -23 & 2 \\
0 & 43 & -38 & -5
\end{bmatrix},
$$

and we are finished with the first column.

Now $i = j = 2$ and we work on the entries in rows 2 to 4 of the second column. Taking $k = 3$, $\ell = 4$, and $q = 1$ gives

$$
\begin{bmatrix}
1 & -11 & 10 & 0 \\
0 & -33 & 31 & -3 \\
0 & 26 & -23 & 2 \\
0 & 17 & -15 & -7
\end{bmatrix}.
$$

Next $k = 4$, $\ell = 2$, and $q = -2$ produces

$$\begin{bmatrix} 1 & -11 & 10 & 0 \\ 0 & 1 & 1 & -17 \\ 0 & 26 & -23 & 2 \\ 0 & 17 & -15 & -7 \end{bmatrix}.$$

Two more row operations result in

$$\begin{bmatrix} 1 & -11 & 10 & 0 \\ 0 & 1 & 1 & -17 \\ 0 & 0 & -49 & 444 \\ 0 & 0 & -32 & 284 \end{bmatrix}.$$

Now we must adjust the (1,2)-entry so that condition (4) of the definition of row Hermite normal form is satisfied. Adding 11 times row 2 to row 1 yields

$$\begin{bmatrix} 1 & 0 & 21 & -187 \\ 0 & 1 & 1 & -17 \\ 0 & 0 & -49 & 444 \\ 0 & 0 & -32 & 284 \end{bmatrix}.$$

At this point $i = j = 3$. Five row operations of type (3) involving only the last two rows produce

$$\begin{bmatrix} 1 & 0 & 21 & -187 \\ 0 & 1 & 1 & -17 \\ 0 & 0 & 0 & -390 \\ 0 & 0 & 1 & 174 \end{bmatrix}.$$

We now interchange rows 3 and 4 and then adjust the (1,3)- and (2,3)-entries. This gives

$$\begin{bmatrix} 1 & 0 & 0 & -3841 \\ 0 & 1 & 0 & -191 \\ 0 & 0 & 1 & 174 \\ 0 & 0 & 0 & -390 \end{bmatrix}.$$

With $i = j = 4$, all we have to do is multiply the fourth row by -1 and adjust the entries above the fourth corner entry. Our final result is

$$
\begin{bmatrix}
1 & 0 & 0 & 59 \\
0 & 1 & 0 & 199 \\
0 & 0 & 1 & 174 \\
0 & 0 & 0 & 390
\end{bmatrix}.
$$

Note that the largest absolute value, 3841, of an entry occurring in the row reduction is nearly 10 times the largest entry in the final matrix. This is typical. Entries in the intermediate matrices may be very large, even when the initial matrix and the final result have only small entries.

Several authors suggest that using a strategy proposed in (Rosser 1952) may help to minimize the size of the entries in the intermediate matrices. In the Rosser strategy, we choose $|A_{\ell j}|$ in ROW_REDUCE to be as large as possible and $|A_{kj}|$ to be as large as possible with $k \neq \ell$. Delaying the enforcement of condition (4) of the definition of row Hermite normal form until A is in row echelon form may also be a good idea. If these strategies are used on the matrix of Example 1.3, then the largest absolute value of an entry in any intermediate matrix is 407.

If A is an integer matrix in row echelon form, then it is easy to decide membership in $S(A)$. The following example illustrates the method.

Example 1.4. Let us determine whether $u = (6, 0, 16, 6, 4, 13, 31)$ is in $S(A)$, where A is the matrix of Example 1.2. The last row of A is zero, so $S(A) = S(B)$, where

$$
B = \begin{bmatrix}
2 & 1 & 5 & 0 & -2 & 1 & -1 \\
0 & 3 & -1 & 2 & 4 & 0 & 2 \\
0 & 0 & 0 & 4 & 7 & 1 & 8 \\
0 & 0 & 0 & 0 & 0 & 2 & 5
\end{bmatrix}.
$$

We want to know whether there exist integers a, b, c, and d such that $xB = u$, where $x = (a, b, c, d)$. If we compute the entries of xB corresponding to the columns of B with corner entries, we see that a, b, c, and d must satisfy

$$
\begin{aligned}
2a &= 6, \\
a + 3b &= 0, \\
2b + 4c &= 6, \\
a + c + 2d &= 13.
\end{aligned} \tag{$*$}
$$

This system of equations can easily be solved. The first equation gives $a = 3$. Substituting this value into the second equation yields $b = -1$. From the third equation we get $c = 2$, and from the fourth equation $d = 4$.

Having solved the system $(*)$, we know that if u is in $S(B)$, then

$$u = (3, -1, 2, 4)B.$$

However, we must compare the third, fifth, and seventh components in these two 7-tuples, since those components were not computed in forming $(*)$. It is in fact true that $(3, -1, 2, 4)B = u$, so u is in $S(B)$.

Suppose that A is an m-by-n integer matrix which is in row echelon form and all rows of A are nonzero. Let u be an element of $\mathbb{Z}^n$. To decide whether u is in $S(A)$, we attempt to solve for a vector $x = (a_1, \ldots, a_m)$ such that $xA = u$. Looking only at the columns containing corner entries of A, we get m equations analogous to the system $(*)$ of Example 1.4. That is, the i-th equation involves only $a_1, \ldots, a_i$, and the coefficient of a_i is not zero. This system has a unique solution x with rational components. If x does not have integer components, then u is not in $S(A)$. If all components of x are integers, then we compute xA and compare it with u.

Taking $u = (0, \ldots, 0)$, we see that the only way to express u as an integral linear combination of the rows of A is as $(0, \ldots, 0)A$. Thus the rows of A form a basis for $S(A)$, and $S(A)$ is a free abelian group of rank at most n. For an integer matrix B, the *rank* of B is the rank of $S(B)$.

If A is in row echelon form, then we can rephrase our membership test for $S(A)$ as follows:

Procedure MEMBER$(A, u; flag, x)$;
Input: A : an m-by-n integer matrix in row echelon form;
           u : an element of $\mathbb{Z}^n$;
Output: $flag$: a boolean variable which is true if u is in $S(A)$ and false otherwise;
           x : if $flag$ is true, an element of $\mathbb{Z}^m$ such that $u = xA$;
Begin
    $v := u$; Let r be the number of nonzero rows in A;
    For $i := 1$ to r do begin
        Let A_{ij} be the first nonzero entry in the i-th row of A;
        If A_{ij} does not divide v_j then begin
           $flag :=$ false; Goto 99
        End;
        $x_i := v_j/A_{ij}$; Subtract x_i times the i-th row of A from v
    End;
    If v is not the zero vector then begin

$flag :=$ false; Goto 99
End;

For $i := r + 1$ to m do $x_i := 0$;
$x := (x_1, \ldots, x_m)$;
99: End.

Proposition 1.1. *Let H be a subgroup of $\mathbb{Z}^n$. There is a unique matrix A in row Hermite normal form such that $H = S(A)$ and A has no zero rows.*

Proof. Let A and B be matrices which are in row Hermite normal form such that $S(A) = H = S(B)$ and neither A nor B has any zero rows. Since the rows of A and of B are bases for H, the number m of rows is the same.

Suppose that $m = 1$. Since the single row of A is in $S(B)$, there is an integer a such that $A = aB$. Similarly there is an integer b such that $B = bA$. Thus $A = abA$. Since A has a nonzero entry, $ab = 1$. Since the first nonzero entries of A and B are both positive, this implies that $a = b = 1$. That is, $A = B$.

Now assume that $m > 1$. Let the corner entry a in row 1 of A occur in column j. Then all elements of $H = S(A)$ have zeros in components 1 to $j - 1$. Therefore the corner entry b in the first row of B must occur in column j or later. By symmetry, b occurs in column j. Let A_1 be the matrix obtained by deleting the first row of A, and let B_1 be similarly defined. Then $S(A_1)$ is the subgroup H_1 of H consisting of all elements with 0 as the j-th component. Again by symmetry, $S(B_1)$ is also H_1. Since A_1 and B_1 are in row Hermite normal form and have no zero rows, $A_1 = B_1$ by induction on m.

Let u be the first row of A and let v be the first row of B. The set D of j-th components of the elements of H is a subgroup of $\mathbb{Z}$ which is generated by a and also by b. Therefore $a = \pm b$. Since a and b are both positive, $a = b$. Thus $u - v$ is in H_1. Suppose that $u \neq v$. Let the first nonzero component of $u - v$ be c and let c be the k-th component of $u - v$. Then column k of A_1 contains a corner entry d and d divides c. But by condition (4) of the definition of row Hermite normal form, the k-th components of u and v are both between 0 and $d - 1$. Therefore their difference c has absolute value at most $d - 1$. This is a contradiction, so $u = v$. Hence $A = B$. $\square$

Corollary 1.2. *Each integer matrix B is row equivalent over $\mathbb{Z}$ to a unique matrix A which is in row Hermite normal form.*

Proof. We know that B is row equivalent to a matrix A which is in row Hermite normal form. The nonzero rows of A constitute the unique matrix C such that C is in row Hermite normal form, C has no nonzero rows, and $S(C) = S(A) = S(B)$. $\square$

The matrix A of Corollary 1.2 will be called the *row Hermite normal form* of B.

Exercises

1.1. Let X be a set and let F be the free abelian group on X. For x in X, define f_x in F by

$$f_x(y) = \begin{cases} 1, & x = y, \\ 0, & x \neq y. \end{cases}$$

Show that $\{f_x \mid x \in X\}$ is a basis of F.

1.2. Find the row Hermite normal form for each of the following matrices:

(a) $\begin{bmatrix} 4 & 3 \\ 6 & 7 \end{bmatrix}$, (b) $\begin{bmatrix} 3 & -1 & 2 \\ 4 & 0 & -3 \\ 5 & 1 & 4 \end{bmatrix}$, (c) $\begin{bmatrix} 4 & -2 & 5 & 3 & 4 \\ 2 & -1 & 7 & 4 & 3 \\ 0 & 0 & -6 & 0 & -3 \\ -6 & -3 & 0 & 3 & -7 \end{bmatrix}$.

1.3. Suppose that we define a fourth integer row operation, add or delete a row of zeros. Show now that two integer matrices A and B with n columns are row equivalent if and only if $S(A) = S(B)$.

1.4. Suppose that B is an m-by-n integer matrix. Let X be a set with n elements $x_1, \ldots, x_n$ and let $\mathcal{R}$ be the rewriting system on $X^{\pm *}$ containing the following rules:

$$x_i x_i^{-1} \to \varepsilon, \qquad 1 \leq i \leq n,$$
$$x_i^{-1} x_i \to \varepsilon, \qquad 1 \leq i \leq n,$$
$$x_j x_i \to x_i x_j, \qquad 1 \leq i < j \leq n,$$
$$x_1^{B_{i1}} \ldots x_n^{B_{in}} \to \varepsilon, \qquad 1 \leq i \leq m.$$

Let $\prec$ be the basic wreath product ordering of $X^{\pm *}$ with $x_n \prec x_n^{-1} \prec \cdots \prec x_1 \prec x_1^{-1}$. Prove that the Knuth-Bendix procedure for strings terminates with this input. Describe the confluent rewriting system constructed. Show how to read off the row Hermite normal form of B.

1.5. Show that any two free abelian groups of the same rank are isomorphic.

1.6. Let A be an m-by-n integer matrix in row echelon form and let the corner entries of A occur in columns $j_1 < \cdots < j_r$. Define a function c from $\mathbb{Z}^n$ to itself as follows: Given x in $\mathbb{Z}^n$, execute the statements

$y := x;$
For $i := 1$ to r do begin
$\quad q := y_{j_i} \text{ div } A_{ij_i};$
$\quad$ Subtract q times the i-th row of A from y
End;

and define $c(x)$ to be y. Show that x and x' are in the same coset of $S(A)$ in $\mathbb{Z}^n$ if and only if $c(x) = c(x')$.

1.7. Suppose that $\mathbb{Z}$ is ordered as follows: $0 \prec 1 \prec 2 \prec \cdots \prec -1 \prec -2 \prec \cdots$. Extend the ordering to $\mathbb{Z}^n$ lexicographically. If x is in $\mathbb{Z}^n$, prove that $c(x)$ in Exercise 1.6 is the first element of the coset $x + S(A)$.

1.8. Give appropriate definitions for integer column operations, column echelon form, and column Hermite normal form. Show that the column Hermite normal form of an integer matrix B is the transpose of the row Hermite normal form of the transpose of B.

1.9. Let A be an m-by-n integer matrix of rank r. Show that the set of x in $\mathbb{Z}^n$ such that $Ax = 0$ is a subgroup of rank $n - r$. (Hint: Prove that A may be assumed to be in row Hermite normal form.)

1.10. Let A and B be m-by-n integer matrices in row echelon form. Suppose that $S(B) \subseteq S(A)$ and that A and B have the same corner entries. Show that $S(B) = S(A)$.

1.11. Extend the notion of row equivalence to matrices with entries in a commutative ring R. In case $R = F[X]$, where F is a field, define row Hermite normal form and show that any matrix is row equivalent to a unique matrix in row Hermite normal form.

8.2 Elementary matrices

Let A be an m-by-n integer matrix and let I be the m-by-m identity matrix. Suppose that P is obtained by applying a row operation to I. Then PA is the result of applying the same row operation to A. For example, if

$$A = \begin{bmatrix} 4 & 6 & 1 \\ -3 & 5 & 2 \end{bmatrix},$$

we can add twice the first row of A to the second row by multiplying A on the left by

$$P = \begin{bmatrix} 1 & 0 \\ 2 & 1 \end{bmatrix}.$$

The proof of this result in general is easy and is left as an exercise.

An *elementary integer matrix* is a matrix obtained by applying one integer row operation to an identity matrix. Suppose that A and B are m-by-n integer matrices and A is obtained by applying a sequence of s row operations to B. If $P_1, \ldots, P_s$ is the corresponding sequence of m-by-m elementary matrices, then $A = P_s \ldots P_1 B$.

The group of units in the ring of m-by-m integer matrices is denoted $\mathrm{GL}(m, \mathbb{Z})$. The reader is assumed to be familiar with determinants and the fact that an m-by-m integer matrix Q is in $\mathrm{GL}(m, \mathbb{Z})$ if and only if $\det Q = \pm 1$. Such a matrix is said to be *unimodular*. If P is an m-by-m elementary matrix, then $\det P = -1$ if the corresponding row operation is of type (1) or (2), and $\det P = 1$ if the row operation is of type (3). Thus P is in $\mathrm{GL}(m, \mathbb{Z})$. Since every integer row operation has an inverse row operation, the inverses of elementary matrices are elementary. If P is the product $P_s \ldots P_1$ of the preceding elementary matrices, then P is in $\mathrm{GL}(m, \mathbb{Z})$ and $A = PB$. Thus we have the following result.

Proposition 2.1. *If A and B are row equivalent integer matrices with m rows, then there is an element P of $\mathrm{GL}(m, \mathbb{Z})$ such that $A = PB$.*

It is possible to modify the procedure ROW_REDUCE of Section 8.1 so that it returns an element P of $\mathrm{GL}(m,\mathbb{Z})$ with $A = PB$. The matrix P is initialized to be the m-by-m identity. Then whenever a row operation is applied to A, that same operation is applied to P. In this way the equality $A = PB$ is maintained at all times. If we follow the computation of Example 1.3, then P turns out to be

$$\begin{bmatrix} -7 & 0 & 2 & 5 \\ -24 & 1 & 7 & 17 \\ -21 & 1 & 6 & 15 \\ -47 & 1 & 13 & 33 \end{bmatrix}.$$

Suppose that B is an m-by-m integer matrix, A is the row Hermite normal form of B, and P is an element of $\mathrm{GL}(m,\mathbb{Z})$ such that $A = PB$. Then

$$\det A = (\det P)(\det B),$$

so $|\det A| = |\det B|$. Thus B is in $\mathrm{GL}(m,\mathbb{Z})$ if and only if A is in $\mathrm{GL}(m,\mathbb{Z})$. If $\det A = \pm 1$, then A cannot have a row of zeros. Thus A has m corner entries and $\det A$ is their product. Hence all corner entries in A are 1, A is the identity matrix, and $P = B^{-1}$. The procedure ROW_REDUCE modified as described here can be used to compute inverses for elements of $\mathrm{GL}(m,\mathbb{Z})$.

The following proposition summarizes the preceding discussion.

Proposition 2.2. *Let B be an m-by-m integer matrix. The following are equivalent:*

(a) $\det B = \pm 1$.
(b) B is in $\mathrm{GL}(m,\mathbb{Z})$.
(c) B is row equivalent to the m-by-m identity matrix.
(d) $S(B) = \mathbb{Z}^m$.
(e) B is a product of elementary matrices.

Corollary 2.3. *Two m-by-n integer matrices A and B are row equivalent if and only if there is an element P of $\mathrm{GL}(m,\mathbb{Z})$ such that $A = PB$.*

In Section 8.1, we learned how to decide whether a vector u in $\mathbb{Z}^n$ is in $S(B)$, where B is a given m-by-n integer matrix. We row reduce B to get the row Hermite normal form A of B. We can then determine a vector x in $\mathbb{Z}^m$ such that $u = xA$ or show that no such x exists. Suppose that u is in $S(B)$ and we would like to express u as yB for some y in $\mathbb{Z}^m$. If we have $A = PB$ with P in $\mathrm{GL}(m,\mathbb{Z})$, then $u = xA = xPB$, so we may take y to be xP.

Exercises

2.1. For each of the matrices B in Exercise 1.2 find a unimodular matrix P such that PB is in row Hermite normal form.

2.2. Let a and b be integers and set $d = \gcd(a,b)$. Prove that the row Hermite normal form of $B = \begin{bmatrix} a \\ b \end{bmatrix}$ is $A = \begin{bmatrix} d \\ 0 \end{bmatrix}$. Show that with just a minor change the procedure GCDX of Section 1.7 can be made to return a unimodular matrix P such that $A = PB$.

2.3. Find the inverses of the following matrices:

$$(a) \quad \begin{bmatrix} 5 & 4 \\ 9 & 7 \end{bmatrix}, \qquad (b) \quad \begin{bmatrix} 1 & 1 & -1 \\ -1 & 1 & 0 \\ -2 & 1 & 1 \end{bmatrix}.$$

2.4. Suppose that we modify the inner While-loop of ROW_REDUCE as follows:

> While there exist distinct indices $k < \ell$ between i and m such that both $a = A_{kj}$ and $b = A_{\ell j}$ are nonzero do begin
>
> $d := \gcd(a,b)$;
>
> Let P be an element of $\mathrm{GL}(2,\mathbb{Z})$ such that $P \begin{bmatrix} a \\ b \end{bmatrix} = \begin{bmatrix} d \\ 0 \end{bmatrix}$;
>
> $(* \text{ See Exercise 2.2. } *)$
>
> Multiply the submatrix of A consisting of rows k and ℓ by P
> End;

Is this likely to speed up the computation?

8.3 Finitely generated abelian groups

Let G be an abelian group written additively and suppose that G is generated by $g_1, \ldots, g_n$. The map f of $\mathbb{Z}^n$ onto G taking $(a_1, \ldots, a_n)$ to $a_1 g_1 + \cdots + a_n g_n$ is a homomorphism and G is isomorphic to $\mathbb{Z}^n/H$, where H is the kernel of f. As remarked in Section 8.1, H is finitely generated, so there is a matrix B such that $H = S(B)$. In this section we shall learn how to determine the isomorphism type of $\mathbb{Z}^n/S(B)$ for a given m-by-n integer matrix B.

Example 3.1. Let A be the following matrix:

$$\begin{bmatrix} 2 & 0 & 0 & 0 & 0 \\ 0 & 4 & 0 & 0 & 0 \\ 0 & 0 & 12 & 0 & 0 \end{bmatrix}.$$

There is a homomorphism f of $\mathbb{Z}^5$ onto $\mathbb{Z}_2 \oplus \mathbb{Z}_4 \oplus \mathbb{Z}_{12} \oplus \mathbb{Z}^2$ taking (a,b,c,d,e) to

$$([a]_2, [b]_4, [c]_{12}, d, e),$$

where $[u]_s$ denotes the congruence class modulo s containing u. The kernel H of f consists of all vectors (a, b, c, d, e) such that 2 divides a, 4 divides b, 12 divides c, and $d = e = 0$. Thus $H = S(A)$ and $\mathbb{Z}^5/S(A)$ is isomorphic to $\mathbb{Z}_2 \oplus \mathbb{Z}_4 \oplus \mathbb{Z}_{12} \oplus \mathbb{Z}^2$.

In general, it is difficult to see the structure of $\mathbb{Z}^n/S(B)$. However, it is possible to determine an m-by-n matrix A such that $\mathbb{Z}^n/S(B)$ and $\mathbb{Z}^n/S(A)$ are isomorphic and the structure of $\mathbb{Z}^n/S(A)$ can be read off easily as in Example 3.1. Constructing A involves the use of integer row and column operations.

The *integer column operations* are the obvious analogues of the integer row operations: interchange two columns, multiply a column by -1, and add an integral multiple of one column to another column. A column operation may be applied to a matrix B by applying the operation to the appropriate identity matrix and multiplying B on the right by the result. If P is obtained by applying an integer column operation to an identity matrix I, then P can also be obtained by applying an integer row operation to I. Thus P is elementary. (We do not have to distinguish between row elementary and column elementary.)

Two m-by-n integer matrices A and B are *equivalent* over $\mathbb{Z}$ if one can be obtained from the other by a sequence of integer row and column operations. This is the same as saying that there are matrices P and Q in $\mathrm{GL}(m, \mathbb{Z})$ and $\mathrm{GL}(n, \mathbb{Z})$, respectively, such that $A = PBQ$. Equivalence of matrices over $\mathbb{Z}$ is an equivalence relation.

Proposition 3.1. *Let A and B be m-by-n integer matrices which are equivalent over $\mathbb{Z}$. Then $\mathbb{Z}^n/S(A)$ and $\mathbb{Z}^n/S(B)$ are isomorphic.*

Proof. Let $A = PBQ$, where P is in $\mathrm{GL}(m, \mathbb{Z})$ and Q is in $\mathrm{GL}(n, \mathbb{Z})$. The map g taking x in $\mathbb{Z}^n$ to xQ is a homomorphism of $\mathbb{Z}^n$ into itself. Since $g(xQ^{-1}) = xQ^{-1}Q = x$, it follows that g is surjective. If $xQ = 0$, then $x = xQQ^{-1} = 0Q^{-1} = 0$. Thus the kernel of g is trivial and g is an automorphism of $\mathbb{Z}^n$. Now $S(BQ)$ is $g(S(B))$, for $S(B)$ is the set of yB with y in $\mathbb{Z}^m$ and $(yB)Q = y(BQ)$. By Exercise 3.6 in Chapter 1, $\mathbb{Z}^n/S(BQ)$ and $\mathbb{Z}^n/S(B)$ are isomorphic. Since $S(PBQ) = S(BQ)$, the result follows. $\square$

Just as the matrices in row Hermite normal form are representatives for the equivalence classes of integer matrices under integer row equivalence, there is an easily identified set of representatives under equivalence. An integer matrix A is in *Smith normal form* if for some $r \geq 0$ the entries $d_i = A_{ii}$, $1 \leq i \leq r$, are positive, A has no other nonzero entries, and d_i divides d_{i+1}, $1 \leq i < r$. The matrix A of Example 3.1 is in Smith normal form.

Every integer matrix B is equivalent to a matrix A in Smith normal form. Given B, the process of computing such an A is called *reduction*. If B is a zero matrix, then B is in Smith normal form already. Thus we may assume that B is not zero. Several reduction algorithms which use row and column operations try to reduce B first to the form

$$\begin{bmatrix} d & 0 & \dots & 0 \\ 0 & & & \\ \vdots & & C & \\ 0 & & & \end{bmatrix}, \tag{$*$}$$

where d is positive. Some algorithms arrange things so that d divides every entry of C at this point. Other algorithms attack C immediately and come back to the divisibility condition only at the end.

We look for a nonzero entry B_{ij} in B. It is usually convenient to choose $|B_{ij}|$ small, but this is not necessary. If B_{ij} divides all entries in row i and column j, then we interchange rows 1 and i and columns 1 and j, if necessary, to make $i = j = 1$. Then row operations of type (3) can be used to make $B_{k1} = 0$ for $k > 1$, and column operations of type (3) can be used to make $B_{1k} = 0$ for $k > 1$. Multiplying row 1 by -1, if necessary, insures that $B_{11} > 0$. Thus we have obtained the form $(*)$.

If B_{ij} does not divide B_{rj} for some r, then let $q = B_{rj} \operatorname{div} B_{ij}$, subtract q times row i from row r, and set (i, j) equal to (r, j). Similarly, if B_{ij} does not divide B_{is} for some s, then let $q = B_{is} \operatorname{div} B_{ij}$, subtract q times column j from column s, and set (i, j) equal to (i, s). In either case, the value of $|B_{ij}|$ is reduced, so eventually B_{ij} will divide each entry in row i and row j.

If, when we reach the form $(*)$, the entry $d = B_{11}$ does not divide some entry B_{rs} with $r \geq 2$ and $s \geq 2$, then we may proceed as follows: Let $q = B_{rs} \operatorname{div} d$. Add column s to column 1 and subtract q times row 1 from row r. The $(r, 1)$-entry is now positive and less than d. Set $(i, j) = (r, 1)$ and repeat the procedure of the previous two paragraphs.

Having obtained a matrix of the form $(*)$, if C is not zero, we iterate the procedure with C. In this way we produce a matrix

$$\begin{bmatrix} d_1 & & & & 0 \\ & d_2 & & & \\ & & \ddots & & \\ 0 & & & d_r & \\ & & & & 0 \end{bmatrix},$$

where the d_i are positive. If we always choose d in $(*)$ to divide the entries in C, then d_i divides d_{i+1}, $1 \leq i < r$. If this divisibility condition does not hold, then we use the following observation: Let a and b be positive integers and let $d = \gcd(a, b) = ua + vb$ and $\ell = \text{lcm}(a, b) = ab/d$. Set

$$
R = \begin{bmatrix} u & v \\ -\dfrac{b}{d} & \dfrac{a}{d} \end{bmatrix} \quad \text{and} \quad S = \begin{bmatrix} 1 & \dfrac{-vb}{d} \\ 1 & \dfrac{ua}{d} \end{bmatrix}.
$$

Then

$$
R \begin{bmatrix} a & 0 \\ 0 & b \end{bmatrix} S = \begin{bmatrix} d & 0 \\ 0 & \ell \end{bmatrix}.
$$

As long as there are indices i and j with $1 \le i < j \le r$ such that d_i does not divide d_j, we use this device to replace d_i and d_j by $\gcd(d_i, d_j)$ and $\text{lcm}(d_i, d_j)$, respectively. Eventually we reach the state in which d_i divides d_{i+1}, $1 \le i < r$.

Suppose that matrices P and Q are initialized at the beginning of the computation to the m-by-m and n-by-n identity matrices, respectively. If any row operations applied to B are also applied to P and any column operations applied to B are also applied to Q, then the final matrix A in Smith normal form is equal to PBQ, P is in $\text{GL}(m, \mathbb{Z})$, and Q is in $\text{GL}(n, \mathbb{Z})$.

Example 3.2. Let us compute a matrix in Smith normal form equivalent to the matrix

$$
B = \begin{bmatrix} 2 & 2 & 6 & 4 & 0 \\ 6 & 4 & 16 & 10 & -14 \\ 4 & 3 & 11 & 7 & 0 \\ 8 & 5 & 21 & 13 & -4 \end{bmatrix}.
$$

The entry in the $(1,1)$ position divides all of the entries in the first row and the first column. However, if we subtract the first row from the third row, we can get an entry of 1 in the $(3,2)$ position, and this is preferable. After the subtraction we have

$$
\begin{bmatrix} 2 & 2 & 6 & 4 & 0 \\ 6 & 4 & 16 & 10 & -14 \\ 2 & 1 & 5 & 3 & 0 \\ 8 & 5 & 21 & 13 & -4 \end{bmatrix}.
$$

Now we interchange rows 1 and 3 and columns 1 and 2. The result is

$$\begin{bmatrix} 1 & 2 & 5 & 3 & 0 \\ 4 & 6 & 16 & 10 & -4 \\ 2 & 2 & 6 & 4 & 0 \\ 5 & 8 & 21 & 13 & -4 \end{bmatrix}.$$

Three row operations of type (3) yield

$$\begin{bmatrix} 1 & 2 & 5 & 3 & 0 \\ 0 & -2 & -4 & -2 & -4 \\ 0 & -2 & -4 & -2 & -0 \\ 0 & -2 & -4 & -2 & -4 \end{bmatrix}.$$

Three column operations of type (3) give us

$$\begin{bmatrix} 1 & 0 & 0 & 0 & 0 \\ 0 & -2 & -4 & -2 & -4 \\ 0 & -2 & -4 & -2 & -0 \\ 0 & -2 & -4 & -2 & -4 \end{bmatrix}.$$

Now we can turn our attention to the 3-by-4 submatrix in the lower right corner. Multiplying row 2 by -1, we get

$$\begin{bmatrix} 1 & 0 & 0 & 0 & 0 \\ 0 & 2 & 4 & 2 & 4 \\ 0 & -2 & -4 & -2 & 0 \\ 0 & -2 & -4 & -2 & -4 \end{bmatrix}.$$

Two row operations of type (3) produce

$$\begin{bmatrix} 1 & 0 & 0 & 0 & 0 \\ 0 & 2 & 4 & 2 & 4 \\ 0 & 0 & 0 & 0 & 4 \\ 0 & 0 & 0 & 0 & 0 \end{bmatrix}.$$

and three column operations of type (3) yield

$$\begin{bmatrix} 1 & 0 & 0 & 0 & 0 \\ 0 & 2 & 0 & 0 & 0 \\ 0 & 0 & 0 & 0 & 4 \\ 0 & 0 & 0 & 0 & 0 \end{bmatrix}.$$

All we have to do now is interchange columns 3 and 5, and we obtain a matrix in Smith normal form:

$$\begin{bmatrix} 1 & 0 & 0 & 0 & 0 \\ 0 & 2 & 0 & 0 & 0 \\ 0 & 0 & 4 & 0 & 0 \\ 0 & 0 & 0 & 0 & 0 \end{bmatrix}.$$

Thus $G = \mathbb{Z}^5/S(B)$ is isomorphic to $\mathbb{Z}_1 \oplus \mathbb{Z}_2 \oplus \mathbb{Z}_4 \oplus \mathbb{Z}^2$. Since $\mathbb{Z}_1$ is a group of order 1, G is isomorphic to $\mathbb{Z}_2 \oplus \mathbb{Z}_4 \oplus \mathbb{Z}^2$.

Suppose that A is the matrix

$$\begin{bmatrix} d_1 & & & & 0 \\ & d_2 & & & \\ & & \ddots & & \\ & 0 & & d_r & \\ & & & & 0 \end{bmatrix},$$

where the d_i are positive integers, and n is the number of columns of A. Then

$$\mathbb{Z}^n/S(A) \cong \mathbb{Z}_{d_1} \oplus \cdots \oplus \mathbb{Z}_{d_r} \oplus \mathbb{Z}^s,$$

where $s = n - r$. It follows from this and the remarks at the beginning of the section that any finitely generated abelian group G is isomorphic to a direct sum

$$\mathbb{Z}_{d_1} \oplus \cdots \oplus \mathbb{Z}_{d_r} \oplus \mathbb{Z}^s.$$

If we assume that each $d_i > 1$ and d_i divides d_{i+1} for $1 \leq i < r$, then the integers r, s, and $d_1, \ldots, d_r$ are uniquely determined by G. This result is known as the fundamental theorem of finitely generated abelian groups. A proof of this theorem may be found in [Sims 1984] and in other books on algebra or group theory.

Proposition 3.2. *Every integer matrix B is equivalent over $\mathbb{Z}$ to a unique matrix A in Smith normal form.*

Proof. Let B have n columns. We know that B is equivalent to some matrix A in Smith normal form. Let $d_1, \ldots, d_r$ be the nonzero entries of A and assume that $d_1 = \cdots = d_t = 1$ and $d_i > 1$ for $i > t$. By the fundamental theorem, the numbers $r - t$, $n - r$, and $d_{t+1}, \ldots, d_r$ are uniquely determined by $\mathbb{Z}^n / S(B)$. Thus $t = n - (n - r) - (r - t)$ is determined. $\quad \square$

The nonzero entries of the matrix in Smith normal form equivalent to a given integer matrix B are called the *invariant factors* of B.

In Section 8.1 we referred to the fact that during row reduction the entries of the matrix can get very large. This phenomenon is even more pronounced in the case of the reduction algorithm sketched earlier. Even starting with a 10-by-10 matrix whose entries are 0 or ± 1, one can overflow 32-bit integers during reduction. In later sections we shall discuss better methods for computing the Smith normal form A of an integer matrix B and for finding unimodular matrices P and Q such that $A = PBQ$.

If B has n columns, then the invariant factors $d_1, \ldots, d_r$ of B determine the isomorphism type of $G = \mathbb{Z}^n / S(B)$. However, to determine an explicit isomorphism of G onto $\mathbb{Z}_{d_1} \oplus \cdots \oplus \mathbb{Z}_{d_r} \oplus \mathbb{Z}^{n-r}$ we need more information, such as the matrix Q.

Proposition 3.3. *Let B be an m-by-n integer matrix with invariant factors $d_1, \ldots, d_r$, and let P and Q be unimodular matrices such that $A = PBQ$ is in Smith normal form. Given x in $\mathbb{Z}^n$, let $xQ = y = (y_1, \ldots, y_n)$. The map f taking x to*

$$([y_1]_{d_1}, \ldots, [y_r]_{d_r}, y_{r+1}, \ldots, y_n)$$

is a homomorphism of $\mathbb{Z}^n$ onto $K = \mathbb{Z}_{d_1} \oplus \cdots \oplus \mathbb{Z}_{d_r} \oplus \mathbb{Z}^{n-r}$ with kernel $S(B)$.

Proof. The map π taking $(y_1, \ldots, y_n)$ to $([y_1]_{d_1}, \ldots, [y_r]_{d_r}, y_{r+1}, \ldots, y_n)$ is a homomorphism of $\mathbb{Z}^n$ onto K with kernel $S(A)$. The map α taking x to xQ is an automorphism of $\mathbb{Z}^n$ taking $S(B)$ to $S(A)$. Since f is the composition $\alpha \circ \pi$, the result follows. $\quad \square$

Corollary 3.4. *Let B be as in Proposition 3.3. If u_i is the i-th row of Q^{-1}, then the image $\overline{u}_i$ of u_i in $G = \mathbb{Z}^n / S(B)$ has order d_i if $1 \le i \le r$ and infinite order if $i > r$. Moreover, G is the internal direct sum of the subgroups $\mathrm{Grp}\langle \overline{u}_i \rangle$, $1 \le i \le n$.*

Proof. Under the map f of Proposition 3.3, u_i goes to the generator of the i-th summand of K. $\quad \square$

Since Q is the more important of P and Q, it seems reasonable to try to keep the entries in Q as small as possible. Giving preference to row operations may be a way to accomplish this.

Exercises

3.1. Computing unimodular matrices P and Q such that PBQ is in Smith normal form is not trivial even when B is already a diagonal matrix. Let B_n be the matrix

$$
\begin{bmatrix}
1 & & & & & \\
& 2 & & 0 & & \\
& & 3 & & & \\
& & & \ddots & & \\
& 0 & & & & \\
& & & & & n
\end{bmatrix},
$$

Describe the invariant factors of B_n. How many are greater than 1?

3.2. The text sketched several reduction procedures. Write out a formal description of one of them.

3.3. Suppose that A and Q are integer matrices such that AQ is defined and Q is unimodular. Show that A and AQ have the same rank. Conclude that equivalent matrices have the same rank.

3.4. Suppose that HNF is a function which returns the row Hermite normal form of an integer matrix. Define a new function F by

$$
F(B) = \mathrm{HNF}(\mathrm{HNF}(B)^t)^t,
$$

where A^t denotes the transpose of the matrix A. Show that $F(B)$ is equivalent to B. Let $B_1 = B$ and for $i \geq 1$ define B_{i+1} to be $F(B_i)$. Show that there is an index s such that $B_s = B_{s+1}$ and that B_s has the form

$$
\begin{bmatrix}
d_1 & & & & 0 \\
& d_2 & & & \\
& & \ddots & & \\
& & & d_r & \\
& 0 & & & \\
& & & & 0
\end{bmatrix},
$$

where the d_i are positive integers but d_i may not be divisible by d_j when $j < i$.

3.5. Let $d_1, \ldots, d_r$ be a sequence of positive integers. Show that no more than $r(r-1)/2$ steps replacing d_i and d_j by $\gcd(d_i, d_j)$ and $\mathrm{lcm}(d_i, d_j)$, respectively, are needed to modify the sequence so that d_i divides d_{i+1} for $1 \leq i < r$.

8.4 Modular techniques

Using the procedures of Section 8.3 to compute the invariant factors of an integer matrix having moderately large numbers of rows and columns requires arbitrary precision arithmetic. When the matrix has hundreds of rows and columns, these methods are not practical even with arbitrary precision. This section describes an alternative approach which has been useful with matrices having a thousand or more rows and columns. It draws heavily on (Havas & Sterling 1979). Similar results are also presented in

(Domich, Kannan, & Trotter 1987) and (Domich 1989). See also (Gerstein 1977). The new approach is based on a different characterization of invariant factors.

Given an m-by-n integer matrix B and an integer k with $0 \leq k \leq \min(m,n)$, let $D_k(B)$ be the gcd of the determinants of all k-by-k submatrices of B. (Here $D_0(B)$ is defined to be 1.) If S is a k-element subset of $\{1, \ldots, m\}$ and T is a k-element subset of $\{1, \ldots, n\}$, then $B(S,T)$ will denote the determinant of the k-by-k submatrix of B consisting of the entries in the rows indexed by S and the columns indexed by T.

Proposition 4.1. *Suppose that A and B are equivalent m-by-n integer matrices. Then $D_k(A) = D_k(B)$ for $0 \leq k \leq \min(m,n)$.*

Proof. We must show that integer row and column operations applied to A do not change $D_k(A)$. It clearly suffices to consider only row operations and to assume that $k \geq 1$. Let S and T be k-element subsets of row and column indices, respectively. Suppose that B is obtained from A by a single integer row operation. We need to see how $B(S,T)$ and $A(S,T)$ are related.

If row i of A is multiplied by -1, then $B(S,T) = -A(S,T)$ if i is in S and $B(S,T) = A(S,T)$ otherwise. Thus $D_k(B) = D_k(A)$ in this case. If rows i and j are interchanged, then $B(S,T) = \pm A(f(S),T)$, where f is the permutation of $\{1, \ldots, m\}$ which interchanges i and j and leaves all other points fixed. Again, up to signs, the sets of determinants of k-by-k submatrices of A and of B are the same and $D_k(A) = D_k(B)$.

Now suppose that q times row i is added to row j. If j is not in S, then $B(S,T) = A(S,T)$. Suppose that j is in S. Without loss of generality, we may assume that j is the first element of S. Thus

$$B(S,T) = \begin{vmatrix} a_{11} + qc_1 & \cdots & a_{1k} + qc_k \\ a_{21} & \cdots & a_{2k} \\ \vdots & & \vdots \\ a_{k1} & \cdots & q_{kk} \end{vmatrix},$$

where

$$A(S,T) = \begin{vmatrix} a_{11} & \cdots & q_{1k} \\ \vdots & & \vdots \\ a_{k1} & \cdots & a_{kk} \end{vmatrix},$$

and $c_1, \ldots, c_k$ are the entries of A in row i of the columns indexed by T. Expanding $B(S,T)$ by minors of the first row, we find that $B(S,T) -$

$A(S,T) + q(\det C)$, where

$$C = \begin{vmatrix} c_1 & \cdots & c_k \\ a_{21} & \cdots & a_{2k} \\ \vdots & & \vdots \\ a_{k1} & \cdots & a_{kk} \end{vmatrix}.$$

If i is in S, then C has two equal rows, so $\det C = 0$. If i is not in S, then up to sign $\det C = A(S',T)$, where S' is obtained from S by replacing j with i. In either case $B(S,T)$ is divisible by $D_k(A)$. Therefore $D_k(A)$ divides $D_k(B)$. But A can be obtained from B by adding $-q$ times row i to row j. Thus the same argument shows that $D_k(B)$ divides $D_k(A)$. Therefore $D_k(A) = D_k(B)$. □

Proposition 4.2. *If A is an m-by-n integer matrix and $0 \le k < \min(m,n)$, then $D_k(A)$ divides $D_{k+1}(A)$.*

Proof. Since $D_0(A) = 1$, we may assume that $k \ge 1$. Let S and T be $(k+1)$-element subsets of row and column indices, respectively. Expanding $A(S,T)$ by minors of one row, we see that $A(S,T)$ is an integral linear combination of determinants of k-by-k submatrices of A. Thus $D_k(A)$ divides $A(S,T)$, so $D_k(A)$ divides $D_{k+1}(A)$. □

Proposition 4.3. *If $d_1, \ldots, d_r$ are the nonzero diagonal entries of a matrix A in Smith normal form, then $D_k(A) = 0$ if $k > r$ and $D_k(A) = d_1 \ldots d_k$ if $k \le r$.*

Proof. Let S and T be k-element sets of row and column indices, respectively. Unless $k \le r$ and $S = T$, the submatrix indexed by S and T has a row of zeros and $A(S,T) = 0$. If $S = \{i_1, \ldots, i_k\}$, where $1 \le i_1 < i_2 < \cdots < i_k \le r$, then $A(S,S) = d_{i_1} \ldots d_{i_k}$. But d_1 divides d_{i_1}, d_2 divides d_{i_2}, and so on. Thus $d_1 \ldots d_k$ divides $A(S,S)$. Since $d_1 \ldots d_k = A(T,T)$, where $T = \{1, 2, \ldots, k\}$, it follows that $D_k(A) = d_1 \ldots d_k$. □

Corollary 4.4. *Let $d_1, \ldots, d_r$ be the invariant factors of an integer matrix B. Then r is the largest index such that $D_r(B) \ne 0$ and $d_k = D_k(B)/D_{k-1}(B)$, $1 \le k \le r$.*

By itself, Corollary 4.4 is not adequate to determine the invariant factors of B. The number of k-by-k submatrices is very large, so $D_k(B)$ cannot be computed from its definition. However, if we can find several k-by-k submatrices with nonzero determinants, then experience suggests that the gcd d of these determinants is likely to be a good approximation to $D_k(B)$ in the sense that $d/D_k(B)$ is "small".

The following steps outline the way we shall use these ideas to compute the invariant factors of an integer matrix B:

(1) Find the rank r of B.
(2) Find a small number of r-by-r submatrices of B which have nonzero determinants.
(3) Compute the gcd d of the determinants in (2).
(4) Determine the invariant factors of B modulo d.
(5) Recover the invariant factors of B from the invariant factors modulo d.

Step (4) explicitly mentions modular arithmetic. In fact, steps (1), (2), and (4) all use modular arithmetic. Let d be an integer greater than 1. In discussing the manipulation of matrices modulo d, we have two possible points of view. We can consider that we are working with matrices over the ring $\mathbb{Z}_d$ of integers modulo d or that our matrices have integer entries which are always reduced to their smallest nonnegative remainders modulo d. The points of view are equivalent, and we shall adopt the second. For an integer matrix A, let $\overline{A}$ be the result of reducing all entries of A modulo d.

The *row operations modulo d* on integer matrices are the following:

(1) Interchange two rows.
(2) Multiply a row by any integer relatively prime to d.
(3) Add an integral multiple of one row to another row.

In both (2) and (3), the modified entries are reduced modulo d.

Two integer matrices A and B are *row equivalent modulo d* if $\overline{A}$ can be transformed into $\overline{B}$ by a sequence of row operations modulo d. The matrix A is *in row Hermite normal form modulo d* if the following conditions are satisfied:

(i) $A = \overline{A}$.
(ii) A is in row Hermite normal form over $\mathbb{Z}$.
(iii) The corner entries of A divide d.

Proposition 4.5. *Every integer matrix B is row equivalent modulo d to a unique matrix A which is in row Hermite normal form modulo d.*

Proof. The proof is essentially the same as the proof of the corresponding result for row equivalence over $\mathbb{Z}$. The one extra point is the way operations of type (2) are used to insure that corner entries divide d. Suppose that a is the first nonzero entry in the i-th row. Then $0 < a < d$ and $a = uv$, where $u = \gcd(a,d)$ and $\gcd(v, d/u) = 1$. There is an integer w such that $wv \equiv 1$

$(\bmod d/u)$. It follows that $aw = uvw \equiv u \pmod d$. By Proposition 7.1 of Chapter 1 we may choose w so that $\gcd(w, d) = 1$. If we multiply row i by w and reduce modulo d, then the first nonzero entry of the i-th row is now u, which divides d. □

The number of nonzero rows of A in Proposition 4.5 will be called the *rank of B modulo d* or the *d-rank* of B.

Example 4.1. Let us row reduce the matrix

$$\begin{bmatrix} 10 & 11 & 3 & 9 \\ 8 & 7 & 7 & 5 \\ 6 & 0 & 4 & 2 \end{bmatrix}$$

modulo 12. The entries are already reduced modulo 12. Since $10 = 2 \cdot 5$ and 5 is its own multiplicative inverse modulo 6, we multiply row 1 by 5 modulo 12 to get

$$\begin{bmatrix} 2 & 7 & 3 & 9 \\ 8 & 7 & 7 & 5 \\ 6 & 0 & 4 & 2 \end{bmatrix}.$$

Now subtracting 4 times row 1 from row 2 and 3 times row 1 from row 3 and reducing modulo 12 gives

$$\begin{bmatrix} 2 & 7 & 3 & 9 \\ 0 & 3 & 7 & 5 \\ 0 & 3 & 7 & 11 \end{bmatrix}.$$

Subtracting twice row 2 from row 1 and row 2 from row 3 results in

$$\begin{bmatrix} 2 & 1 & 1 & 11 \\ 0 & 3 & 7 & 5 \\ 0 & 0 & 0 & 6 \end{bmatrix}.$$

Finally, subtracting row 3 from row 1 gives

$$\begin{bmatrix} 2 & 1 & 1 & 5 \\ 0 & 3 & 7 & 5 \\ 0 & 0 & 0 & 6 \end{bmatrix}.$$

Row reduction modulo d can be used to compute determinants modulo d. Let B be a square integer matrix. At the beginning of the row reduction, set $c = 1$. If two rows are interchanged, then replace c by $-c$ modulo d. If a row is multiplied by w and $wv \equiv 1 \pmod{d}$, then replace c by vc modulo d. If A is the final matrix which is in row Hermite normal form modulo d, then $\det B \equiv c(\det A) \pmod{d}$, and $\det A$ is the product of the diagonal entries of A.

Now let B be an m-by-n integer matrix. The d-rank of B is a lower bound for the rank of B. If d is chosen to be a random, moderately large prime, then the rank of B and the d-rank of B will almost certainly be the same. If we get the same rank r modulo several primes, then our confidence that r is the rank of B should be very high. However, to be absolutely sure, we must work a little harder.

The rank of B is the largest integer r such that B contains an r-by-r submatrix with nonzero determinant. We need an easily computable upper bound on the size of the determinant of any square submatrix of B. One such bound is the Hadamard bound given by the following proposition and its corollary.

Proposition 4.6. *Let A be an n-by-n real matrix. Then*

$$|\det A| \leq \prod_{i=1}^{n} \left(\sum_{j=1}^{n} A_{ij}^2 \right)^{1/2}.$$

Proof. See Exercise 15 of Section 4.6.1 of [Knuth 1973] and its solution. □

Corollary 4.7. *Let B be an m-by-n integer matrix, where $m \leq n$. The determinant of any square submatrix of B has absolute value at most the smaller of*

$$\prod_{i} \left(\sum_{j=1}^{n} B_{ij}^2 \right)^{1/2} \quad and \quad \left[\max_{j} \left(\sum_{i=1}^{m} B_{ij}^2 \right)^{1/2} \right]^m,$$

where the product is over the nonzero rows.

If B has more rows than columns, then we apply Corollary 4.7 to the transpose of B. If $m = n = 100$ and all entries in B have absolute value at most 1, then Corollary 4.7 says that the determinant of any square submatrix of B has absolute value at most 10^{100}.

Proposition 4.8. *Let B be an integer matrix and let $p_1, \ldots, p_s$ be a sequence of distinct primes such that $p_1 \ldots p_s$ exceeds the Hadamard bound for B. Then the rank of B is the maximum r of the p_i-rank of B, $1 \leq i \leq s$.*

Proof. Let b be the Hadamard bound for B. If C is a square submatrix of B with more than r rows, then $\det C \equiv 0 \pmod{p_i}$, $1 \le i \le s$. Therefore $\det C \equiv 0 \pmod{p_1 \dots p_s}$. Since $|\det C| \le b < p_1 \dots p_s$, this means $\det C = 0$. Since for some i there is an r-by-r submatrix C of B such that $\det C$ is not divisible by p_i, the rank of B is r. $\square$

When Proposition 4.8 is applied, the primes p_i are usually chosen to be as large as possible subject to p_i^2 being representable as a single precision integer.

Now that we can compute the rank r of a large integer matrix B, we need a way to find r-by-r submatrices of B with nonzero determinants and to compute those determinants. Assuming that Proposition 4.8 was used to find r, we know some prime p such that the p-rank of B is r. For $1 \le i \le m$ let B_i be the submatrix of B consisting of the first i rows. By a slight extension of the process of row reduction modulo p, we can determine the set S of row indices i such that the p-rank of B_i is greater than the p-rank of B_{i-1}. Let C be the submatrix of B consisting of the rows indexed by S and let T be the set of column indices corresponding to the corner entries in the row Hermite form modulo p of C. Then $B(S,T)$ is not congruent to 0 modulo p and hence is not 0.

Having found an r-by-r submatrix with nonzero determinant, we must compute that determinant. This again can be done with modular arithmetic.

Proposition 4.9. *Let E be a square matrix and let $p_1, \dots, p_s$ be a sequence of distinct primes such that $p_1 \dots p_s$ exceeds twice the Hadamard bound for E. Suppose that $\det E \equiv e_i \pmod{p_i}$, $1 \le i \le s$. Then $\det E$ is the integer e with smallest absolute value satisfying $e \equiv e_i \pmod{p_i}$, $1 \le i \le s$.*

Proof. Let b be the Hadamard bound for E. The numbers e_i, $1 \le i \le s$, determine $\det E$ modulo $p_1 \dots p_s$. Since $2b < p_1 \dots p_s$, no two integers between $-b$ and b are congruent modulo $p_1 \dots p_s$. $\square$

We can use row reduction modulo p to compute $\det E$ modulo p for sufficiently many primes p and use Proposition 4.9 to determine $\det E$. An efficient way of doing this is discussed in Exercise 4.1.

It may happen that B has only one r-by-r submatrix with nonzero determinant. However, it is probably worth spending a little more time looking for others. One way to do this is to apply the procedure used to find the first submatrix to the matrix obtained from B by randomly permuting the rows and the columns of B. If other r-by-r submatrices with nonzero determinants are found, then we compute the gcd of their determinants.

At this point we know the rank r of B and a multiple d of $D_r(B)$. The next step is to compute the elementary divisors of B modulo d. To define these, we need to introduce *column operations modulo d*, which are the obvious analogues of the row operations modulo d. Two integer matrices A and B are *equivalent modulo d* if $\overline{A}$ can be transformed into $\overline{B}$ by a sequence of row and column operations modulo d. The matrix A is in *Smith normal form modulo d* if the following conditions hold:

(i) $A = \overline{A}$.
(ii) A is in Smith normal form over $\mathbb{Z}$.
(iii) The nonzero entries of A divide d.

Proposition 4.10. *An integer matrix B is equivalent modulo d to a unique matrix A which is in Smith normal form modulo d.*

Proof. Exercise. □

If B and A are as in Proposition 4.10, then the nonzero entries of A will be called the *invariant factors of B modulo d*. The following result allows us to complete the final step in our modular approach to computing invariant factors.

Proposition 4.11. *Let B be an integer matrix with invariant factors $d_1, \ldots, d_r$ and let d be a positive multiple of d_r. Suppose the invariant factors of B modulo d are $c_1, \ldots, c_s$. Then $d_i = c_i$, $1 \le i \le s$, and $d_i = d$, $s < i \le r$.*

Proof. One way to compute the Smith normal form of B modulo d is to compute the Smith normal form over $\mathbb{Z}$ and reduce the entries modulo d. Since all the d_i divide d, the only effect of the reduction modulo d is to replace any d_i which is equal to d by 0. □

Example 4.2. Let us use the modular techniques we have developed to compute the invariant factors of the matrix

$$
B = \begin{bmatrix}
3 & 0 & 7 & -1 & 6 & -1 & -1 & 0 & -4 & -4 \\
14 & 14 & 8 & 3 & 4 & -4 & 4 & -6 & -6 & -4 \\
17 & 30 & 14 & -1 & 4 & -10 & 5 & -15 & -5 & -3 \\
4 & 2 & -4 & 1 & -4 & -3 & -1 & -3 & 0 & 4 \\
4 & 6 & -5 & 2 & -7 & 1 & 3 & 0 & 1 & 3 \\
10 & 16 & 11 & 1 & 7 & -7 & 2 & -9 & -1 & -3 \\
11 & 2 & -9 & 6 & 3 & 0 & 5 & -6 & 0 & 4 \\
-7 & 0 & 13 & 1 & 20 & -3 & 0 & -3 & 3 & -10
\end{bmatrix}.
$$

The 2-rank of B is 7, the 3-rank is 6, the 5-rank is 7, and the 7-rank is 8. Since the rank of B cannot exceed 8, the rank is 8. There are 45 8-by-8 submatrices of B. Let us consider first the submatrix C consisting of the first eight columns of B. The Hadamard bound for C is about 1.1×10^{11}. Assuming our computer can store eight-digit decimal integers in one word, we can compute the determinant of C modulo a prime near 1000 using single-precision calculations. The first four primes greater than 1000 are 1009, 1013, 1019, and 1021. It turns out that

$$\det C \equiv 14 \pmod{1009},$$
$$\equiv 944 \pmod{1013},$$
$$\equiv 331 \pmod{1019},$$
$$\equiv 810 \pmod{1021}.$$

Using Exercise 4.1 and Proposition 4.9, we find that $\det C = 276480$. Similar computations show that the determinants of the submatrices obtained by omitting the first two columns and by omitting columns 3 and 8 are -207360 and -172800, respectively. The gcd of 276480, 207360, and 172800 is 34560. The invariant factors of B modulo 34560 consist of six 1's, 3, and 11520. By Proposition 4.11, these are the invariant factors of B. Reduction of B modulo 34560 involves integers with more than eight digits. We may work modulo the prime power factors of 34560 if we choose. See Exercise 4.2.

Let B be an m-by-n integer matrix of rank r and let d be a multiple of the largest invariant factor d_r of B. The algorithm for reduction modulo d can be extended to produce square integer matrices P and Q such that PBQ is congruent modulo d to the Smith normal form of B, and P and Q are invertible modulo d, that is, their determinants are relatively prime to d. If $r = n$, then $G = \mathbb{Z}^n/S(B)$ is finite and is isomorphic to a quotient of $(\mathbb{Z}_d)^n$. Knowing the matrix Q, we can construct an explicit isomorphism of G with a direct sum of cyclic groups. However, if G is infinite, then modular techniques can tell us the isomorphism type of G, but they do not appear powerful enough to determine an explicit isomorphism of G with a direct sum of cyclic groups.

We close this section with some applications of the results obtained to row equivalence and row Hermite normal forms. Let B be an m-by-n integer matrix. For any subset T of $\{1, \ldots, n\}$ set $k = |T|$ and define $D_T(B)$ to be the gcd of the determinants of the k-by-k submatrices of B whose columns are indexed by T.

Proposition 4.12. *If A is row equivalent to B, then $D_T(A) = D_T(B)$ for all subsets T of column indices.*

Proof. See the proof of Proposition 4.1. $\square$

Let us order the set of nonempty subsets of $\{1, \ldots, n\}$ as follows: Given two nonempty subsets S and T, we compare the first element of S with the first element of T. If the first elements are equal, we compare the second elements, and so on. Suppose A is the row Hermite normal form of B and $j_1 < j_2 < \cdots < j_r$ are the indices of the columns of A containing the corner entries $c_1, \ldots, c_r$. Then for $1 \leq i \leq r$ the set $T_i = \{j_1, \ldots, j_i\}$ is characterized by the property that it is the first i-element subset T of $\{1, \ldots, n\}$ such that $D_T(B) \neq 0$. Moreover, $D_{T_i}(B) = c_1 \ldots c_i$. Thus each c_i is less than or equal to the Hadamard bound for B. Since all entries in the j_i-th column of A are between 0 and c_i, we have the following result:

Proposition 4.13. *Let B be an integer matrix, let b be the Hadamard bound for B, and let A be the row Hermite normal form of B. Then all entries of A in columns containing corner entries lie between 0 and b.*

Corollary 4.14. *Let B, b, and A be as in Proposition 4.13 and let B have m rows and n columns. Then no entry of A exceeds $\min(m, n)b^2$ in absolute value.*

Proof. For $1 \leq k \leq n$, let A_k and B_k be the k-th columns of A and B, respectively, and let the corner entries of A occur in columns $j_1 < \cdots < j_r$. If $k = j_i$ for some i, then no entry in A_k exceeds b in absolute value. Suppose that A_k does not contain a corner entry. Applying Exercise 4.5 to the submatrix of B consisting of columns k and $j_1, \ldots, j_r$, we obtain integers u and $v_1, \ldots, v_r$ such that

$$uB_k = \sum_{i=1}^{r} v_i B_{j_i},$$

$u \neq 0$, and none of u, $v_1, \ldots, v_r$ exceeds b in absolute value. The same relation holds on the columns of A. That is

$$uA_k = \sum_{i=1}^{r} v_i A_{j_i}.$$

Thus for $1 \leq p \leq m$ the entry A_{pk} has absolute value at most

$$\sum_{i=1}^{r} |v_i||A_{pj_i}| \leq rb^2 \leq \min(m, n)b^2. \qquad \square$$

Exercises

4.1. Suppose that $p_1, \ldots, p_s$ are distinct odd primes and $a_1, \ldots, a_s$ are integers with $|a_i| < p_i/2$. For $2 \leq i \leq s$ let q_i be the positive integer less than p_i such that $p_1 \ldots p_{i-1}q_i \equiv 1 \pmod{p_i}$. Show that after executing the statements

$x = a_1$;
For $i := 2$ to s do begin
    Choose y such that $y \equiv q_i(x - a_i) \pmod{p_i}$ and $|y| < p_i/2$;
    $x := x - y p_1 \ldots p_{i-1}$
End;

the value of x is the integer of smallest absolute value such that $x \equiv a_i \pmod{p_i}$, $1 \leq i \leq s$.

4.2. Assume that B is an integer matrix of rank r, p is a prime, and p^a is the highest power of p dividing the last invariant factor d_r of B. Suppose that $b \geq a$ and let the invariant factors of B modulo p^b be $c_1, \ldots, c_s$. Show that the highest power of p dividing the i-th invariant factor d_i of B is c_i if $1 \leq i \leq s$ and is p^b if $s < i \leq r$.

4.3. Is the converse of Proposition 4.12 true? That is, if A and B are integer matrices with the same numbers of rows and columns such that $D_T(A) = D_T(B)$ for all subsets T of column indices, is it necessarily true that A and B are row equivalent? (Hint: Consider 2-by-2 matrices which are in row Hermite normal form.)

4.4. Suppose that A is an $(n-1)$-by-n integer matrix of rank $n-1$. For $1 \leq i \leq n$ let A_i be the submatrix of A obtained by deleting the i-th column, and set $x_i = (-1)^{i+1} \det(A_i)$. Let $x = (x_1, \ldots, x_n)$. Show that $x \neq 0$ and $Ax = 0$.

4.5. Let A be an m-by-n integer matrix of rank less than n. Show that there is an $x \neq 0$ in $\mathbb{Z}^n$ such that $Ax = 0$ and no component of x exceeds the Hadamard bound for A in absolute value.

4.6. Let B be an m-by-n integer matrix and let $p_1, \ldots, p_s$ be a sequence of distinct primes whose product exceeds the Hadamard bound for B. Let A be the row Hermite normal form of B and for $1 \leq i \leq s$ let A_i be the row Hermite normal form modulo p_i of B. Suppose that the corner entries of A occur in columns $j_1 < \cdots < j_r$ and the corner entries of A_i occur in columns $j_1(i) < \cdots < j_{r_i}(i)$. Show that for $1 \leq k \leq r$ we have

$$j_k = \min_i j_k(i),$$

where the minimum is taken over those values of i for which $r_i \geq k$.

4.7. Let B be an m-by-n integer matrix of rank n and let d be a positive multiple of $D_n(B)$. Show how to recover the row Hermite normal form of B from the row Hermite normal form modulo d of B.

8.5 The Kannan-Bachem algorithm

In computer science, the traditional definition of tractability of an infinite class of computations is that the computations can be carried out in times which are a polynomial function of the sizes of the inputs. The size of an input to a computation is, roughly speaking, the number of bits in a binary encoding of that input. To make a careful statement about tractability, one must specify the model of computation being used. For the calculation of Hermite and Smith normal forms of integer matrices, a reasonable model is sequential computation in which arithmetic operations on integers of a fixed precision take unit time.

It is widely believed that the algorithms for row Hermite normal form in Section 8.1 and for Smith normal form in Section 8.3 are not polynomial-time algorithms. However, as this is written, no examples of nonpolynomial running times are known. The number of bits in the Hadamard bound for an integer matrix B is polynomial in the size of B. Corollaries 4.4 and 4.14 show that the sizes of the Smith normal form and the row Hermite normal form of B are polynomial in the size of B. The modular techniques of Section 8.4 provide a polynomial-time algorithm for computing invariant factors and hence for finding Smith normal forms. These modular techniques can be extended in a fairly straightforward way to provide an algorithm for computing the row Hermite normal form of an integer matrix whose rank is equal to the number of columns. Unfortunately, when applied to an integer matrix B, the methods of Section 8.4 do not produce unimodular matrices P and Q such that PBQ is in Smith normal form.

The first polynomial-time algorithm for finding Smith normal forms and associated unimodular matrices was given in (Kannan & Bachem 1979). This algorithm is based on an algorithm for Hermite normal forms. This section presents the main ideas of the Kannan-Bachem algorithm for Smith normal forms. The full Hermite normal form algorithm will not be given. (See Exercise 5.1.) To compute Smith normal forms, a slightly weaker procedure, called here KB_ROW, suffices.

The procedure KB_ROW takes as input an m-by-n integer matrix B and returns a matrix A and a permutation σ of $\{1, \ldots, n\}$ such that A is row equivalent to B and if C is defined by $C_{ij} = A_{i\sigma_j}$, then C is in row Hermite normal form. Here σ_j denotes the image of j under σ. It follows immediately that the nonzero rows of A form a basis for $S(B)$ and that membership in $S(B)$ can be decided easily using A and σ. Let $x = (x_1, \ldots, x_n)$ be in $\mathbb{Z}^n$ and define $y = (y_1, \ldots, y_n)$ by $y_j = x_{\sigma_j}$. Then x is in $S(B)$ if and only if y is in $S(C)$ and the latter can be decided using MEMBER from Section 8.1.

In KB_ROW there are calls to a subroutine ROD. The name is an abbreviation for the name of the corresponding subroutine in (Kannan & Bachem 1979), which is REDUCE OFF DIAGONAL. The purpose of ROD is to enforce conditions analogous to conditions (3) and (4) of the definition of row Hermite normal form. In ROD, the current matrix A and permutation σ are global variables.

Procedure ROD(j);
Input: j : an integer between 1 and m such that $A_{j\sigma_j} \neq 0$ and
            $A_{j\sigma_k} = 0$ for $1 \leq k < j$;
(* The entries $A_{i\sigma_j}$, $1 \leq i < j$, are reduced modulo $A_{j\sigma_j}$. *)
Begin
    If $A_{j\sigma_j} < 0$ then multiply row j of A by -1;
    For $i := 1$ to $j - 1$ do begin
    $q := A_{i\sigma_j}$ div $A_{j\sigma_j}$;
    Subtract q times row j of A from row i
    End
End.

Here is the weak version of the Kannan-Bachem row reduction algorithm:

Procedure KB_ROW$(B; A, \sigma)$;
Input: B : an m-by-n integer matrix;
Output: A : an integer matrix row equivalent to B;
          σ : a permutation of $\{1, \ldots, n\}$ such that the matrix C
                  defined by $C_{ij} = A_{i\sigma_j}$ is in row Hermite normal form;
Begin
    For $j := 1$ to n do $\sigma_j := j$;
    $A := B$; $s := m$; $i := 1$;
    ($*$ s will denote the index of the last row of A not known to be 0. $*$)

    While $i \leq s$ do begin
        For $k := 1$ to $i - 1$ do begin
            GCDX$(A_{k\sigma_k}, A_{i\sigma_k}; d, p, q)$;

            $U := \begin{bmatrix} p & q \\ -A_{i\sigma_k}/d & A_{k\sigma_k}/d \end{bmatrix}$;

            Multiply the submatrix of A consisting of rows k and i on the
                left by U;
            ROD(k)

        End;

        If the i-th row of A is now zero then begin
            If $i < s$ then interchange rows i and s of A;
            $s := s - 1$
        End

        Else begin
            Let j be the first index such that $A_{i\sigma_j} \neq 0$;
            If $j \neq i$ then interchange σ_i and σ_j;
            ROD(i); $i := i + 1$
        End
    End
End.

At first glance, the operation of KB_ROW may be a little confusing.
However, some examples should help to clarify matters.

Example 5.1. Let B be the matrix of Example 1.3, namely

$$\begin{bmatrix} 4 & -1 & 2 & -5 \\ -3 & 0 & 1 & -3 \\ 2 & 4 & -3 & 2 \\ 5 & -3 & 4 & 4 \end{bmatrix}.$$

Initially $\sigma = (1,2,3,4)$, the identity permutation. Note that we shall *not* use cycle notation for σ, but rather vector notation, in which σ is identified with the vector $(\sigma_1, \sigma_2, \ldots, \sigma_n)$. On the first pass through the While-loop no changes to A or σ are made. On the second pass, we have $i = 2$. In the For-loop with $k = 1$ we find $\gcd(4, -3) = 1 = 1 \cdot 4 + 1 \cdot (-3)$. Thus we multiply rows 1 and 2 of A by $\begin{bmatrix} 1 & 1 \\ 3 & 4 \end{bmatrix}$ to get

$$\begin{bmatrix} 1 & -1 & 3 & -8 \\ 0 & -3 & 10 & -27 \\ 2 & 4 & -3 & 2 \\ 5 & -3 & 4 & 4 \end{bmatrix}.$$

The call ROD(1) does nothing. The second row of A is nonzero and $j = 2$, so there is no change in σ. The call ROD(2) produces

$$\begin{bmatrix} 1 & 2 & -7 & 19 \\ 0 & 3 & -10 & 27 \\ 2 & 4 & -3 & 2 \\ 5 & -3 & 4 & 4 \end{bmatrix}.$$

On the third pass through the While-loop the value of i is 3. In the For-loop with $k = 1$ we multiply rows 1 and 3 by $\begin{bmatrix} 1 & 0 \\ -2 & 1 \end{bmatrix}$ to get

$$\begin{bmatrix} 1 & 2 & -7 & 19 \\ 0 & 3 & -10 & 27 \\ 0 & 0 & 11 & -36 \\ 5 & -3 & 4 & 4 \end{bmatrix},$$

and then call ROD(1), which does nothing. When $k = 2$, there is no change to A. Now $j = 3$ and the call ROD(3) results in

$$\begin{bmatrix} 1 & 2 & 4 & -17 \\ 0 & 3 & 1 & -9 \\ 0 & 0 & 11 & -36 \\ 5 & -3 & 4 & 4 \end{bmatrix}.$$

On the last pass through the While-loop $i = 4$. In the For-loop with $k = 1$, the matrix A is changed to

$$\begin{bmatrix} 1 & 2 & 4 & -17 \\ 0 & 3 & 1 & -9 \\ 0 & 0 & 11 & -36 \\ 0 & -13 & -16 & 89 \end{bmatrix}.$$

When $k = 2$, we get $1 = \gcd(3, -13) = -4 \cdot (3) - 1 \cdot (-13)$. Multiplying rows 2 and 4 of A by $\begin{bmatrix} -4 & -1 \\ 13 & 3 \end{bmatrix}$ gives

$$\begin{bmatrix} 1 & 2 & 4 & -17 \\ 0 & 1 & 12 & -53 \\ 0 & 0 & 11 & -36 \\ 0 & 0 & -35 & 150 \end{bmatrix},$$

and the call ROD(2) changes row 1 to produce

$$\begin{bmatrix} 1 & 0 & -20 & 89 \\ 0 & 1 & 12 & -53 \\ 0 & 0 & 11 & -36 \\ 0 & 0 & -35 & 150 \end{bmatrix}.$$

With $k = 3$, we multiply rows 3 and 4 by $\begin{bmatrix} 16 & 5 \\ 35 & 11 \end{bmatrix}$ to get

$$\begin{bmatrix} 1 & 0 & -20 & 89 \\ 0 & 1 & 12 & -53 \\ 0 & 0 & 1 & 174 \\ 0 & 0 & 0 & 390 \end{bmatrix}.$$

and then call ROD(3), which gives

$$\begin{bmatrix} 1 & 0 & 0 & 3569 \\ 0 & 1 & 0 & -2141 \\ 0 & 0 & 1 & 174 \\ 0 & 0 & 0 & 390 \end{bmatrix}.$$

The final call ROD(4) produces

$$\begin{bmatrix} 1 & 0 & 0 & 59 \\ 0 & 1 & 0 & 199 \\ 0 & 0 & 1 & 174 \\ 0 & 0 & 0 & 390 \end{bmatrix}.$$

The permutation σ is still (1,2,3,4), the identity, and the matrix returned is in row Hermite normal form.

Example 5.2. Now let us apply KB_ROW to the following matrix:

$$\begin{bmatrix} 3 & 1 & -3 & -3 & -5 \\ -1 & 5 & 5 & -1 & -3 \\ 2 & -2 & 3 & 3 & -1 \\ -2 & -3 & -5 & -3 & 5 \\ -3 & 4 & -4 & -4 & 1 \end{bmatrix}.$$

The first pass through the While-loop does not change A. Here are the values of A after each of the remaining iterations of the While-loop.

$$\begin{bmatrix} 1 & 11 & 7 & -5 & -11 \\ 0 & 16 & 12 & -6 & -14 \\ 2 & -2 & 3 & 3 & -1 \\ -2 & -3 & -5 & -3 & 5 \\ -3 & 4 & -4 & -4 & 1 \end{bmatrix}, \quad \begin{bmatrix} 1 & 3 & 8 & 2 & -4 \\ 0 & 8 & 13 & 1 & -7 \\ 0 & 0 & 14 & 8 & 0 \\ -2 & -3 & -5 & -3 & 5 \\ -3 & 4 & -4 & -4 & 1 \end{bmatrix},$$

$$\begin{bmatrix} 1 & 0 & 4 & 28 & 2 \\ 0 & 1 & 6 & 40 & 4 \\ 0 & 0 & 7 & 27 & 3 \\ 0 & 0 & 0 & 46 & 6 \\ -3 & 4 & -4 & -4 & 1 \end{bmatrix}, \quad \begin{bmatrix} 1 & 0 & 0 & 0 & 11 \\ 0 & 1 & 0 & 0 & 22 \\ 0 & 0 & 1 & 1 & 33 \\ 0 & 0 & 0 & 2 & 24 \\ 0 & 0 & 0 & 0 & 39 \end{bmatrix}.$$

Again σ is the identity, and the final matrix is in row Hermite normal form.

As long as σ does not change and no zero rows are encountered, after the i-th iteration of the While-loop the submatrix consisting of the first i rows of A is in row Hermite normal form. With any of the versions of ROW_REDUCE in Section 8.1, the submatrices consisting of the first i columns are put into row Hermite normal form for $i = 1, 2, \ldots$. Here is an example in which KB_ROW returns σ as a nonidentity permutation and encounters zero rows.

Example 5.3. Let B be the following matrix:

$$\begin{bmatrix} 0 & -11 & 0 & -10 & 5 \\ -6 & -8 & -6 & -5 & 0 \\ -4 & 4 & -4 & 6 & -4 \\ 2 & 6 & 2 & 3 & -2 \\ 8 & 2 & 8 & -8 & 2 \end{bmatrix}.$$

The values of the matrix A at the end of each iteration of the While-loop in KB_ROW are

$$\begin{bmatrix} 0 & 11 & 0 & 10 & -5 \\ -6 & -8 & -6 & -5 & 0 \\ -4 & 4 & -4 & 6 & -4 \\ 2 & 6 & 2 & 3 & -2 \\ 8 & 2 & 8 & -8 & 2 \end{bmatrix}, \quad \begin{bmatrix} 42 & 1 & 42 & -15 & 25 \\ 66 & 0 & 66 & -25 & 40 \\ -4 & 4 & -4 & 6 & -4 \\ 2 & 6 & 2 & 3 & -2 \\ 8 & 2 & 8 & -8 & 2 \end{bmatrix},$$

$$\begin{bmatrix} 0 & 1 & 0 & 6 & 1 \\ 2 & 0 & 2 & 23 & 8 \\ 0 & 0 & 0 & 28 & 8 \\ 2 & 6 & 2 & 3 & -2 \\ 8 & 2 & 8 & -8 & 2 \end{bmatrix}, \quad \begin{bmatrix} 0 & 1 & 0 & 6 & 1 \\ 2 & 0 & 2 & 23 & 8 \\ 0 & 0 & 0 & 28 & 8 \\ 8 & 2 & 8 & -8 & 2 \\ 0 & 0 & 0 & 0 & 0 \end{bmatrix}, \quad \begin{bmatrix} 0 & 1 & 0 & 6 & 1 \\ 2 & 0 & 2 & 23 & 8 \\ 0 & 0 & 0 & 28 & 8 \\ 0 & 0 & 0 & 0 & 0 \\ 0 & 0 & 0 & 0 & 0 \end{bmatrix}.$$

The value of σ returned is (2,1,4,3,5).

Just as we did with the algorithms of Section 8.1, if we initialize P to be the appropriate identity matrix and apply all row operations used in KB_ROW to P as well, then after the call KB_ROW$(B; A, \sigma)$ we have $A = PB$.

Proposition 5.1. *The procedure* KB_ROW, *modified to produce a unimodular matrix P such that $A = PB$, runs in polynomial time.*

Proof. In this proof, no effort will be made to obtain the best possible upper bound for the running time of KB_ROW. Also, some of the steps will only be sketched. The details are left to the reader as exercises. Clearly we may assume that B is not a matrix of zeros. Throughout this proof, "polynomial" as an adjective will mean "bounded by a polynomial in the size of B".

Lemma 5.2. *We may assume that the permutation σ returned by* KB_ROW *is the identity.*

Proof. Suppose the call KB_ROW$(B; A, \sigma)$ returns with σ not the identity. Define C and D by $C_{ij} = A_{i\sigma_j}$ and $D_{ij} = B_{i\sigma_j}$. Then the call KB_ROW$(D; E, \tau)$ returns with τ equal to the identity and $E = C$. Moreover, exactly the same row operations are used in the two calls to KB_ROW. □

The unimodular matrix P returned by KB_ROW is the product of the following matrices: the matrices U in the For-loop on k (expanded to m-by-m matrices), the elementary matrices corresponding to row operations performed in ROD, and the elementary matrices corresponding to certain interchanges of rows. It is easy to check that the total number of these factors is bounded by a cubic polynomial in m, and hence is certainly polynomial. It is not hard to show that the product of a polynomial number of matrices all of the same polynomial size is again of polynomial size. Thus, it suffices to show that each factor of P has polynomial size, for then, since $A = PB$ at all times during the computation, A has polynomial size throughout.

We are left with showing that the entries in the matrix U and the values of q in ROD have at all times sizes bounded by a fixed polynomial in the size of B. Let b be the Hadamard bound for B and define

$$b_1 = mb^2, \quad b_2 = (nb_1^2)^{m/2}, \quad b_3 = 2b_1 b_2, \quad b_4 = b_3(1 + b_3)^{m-1}.$$

Then $b \leq b_1 \leq b_2 \leq b_3 \leq b_4$. Even for matrices B of moderate size, the number b_4 is very large, but it is always of polynomial size.

Let E denote the upper left i-by-i submatrix of A at the start of some iteration of the loop "While $i \leq s \ldots$". Set $E^{(0)} = E$ and for $1 \leq k < i$ let $E^{(k)}$ be the upper left i-by-i submatrix of A after the completion of the k-th iteration of the For-loop during the current iteration of the While-loop.

Lemma 5.3. *The following inequalities hold for $1 \leq k < i$:*

(a) $|E_{it}^{(k)}| \leq b_2, \quad 1 \leq t \leq i.$
(b) $|E_{rt}^{(k)}| \leq b_1, \quad k < r < i, 1 \leq t \leq i.$
(c) $|E_{kt}^{(k)}| \leq b_3, \quad 1 \leq t \leq i.$
(d) $|E_{rt}^{(k)}| \leq b_3(1 + b_3)^{k-1}, \quad 1 \leq r < k, 1 \leq t \leq i.$

Proof. The first $i - 1$ rows of E are the nonzero rows in the row Hermite normal form of a submatrix of B, and entries in the i-th row of E are entries of B. By Corollary 4.14, all entries of E are bounded by b_1. Thus the Hadamard bound for E does not exceed b_2. The row operations performed during the first k iterations of the For-loop affect only the rows of E indexed by $S = \{1, 2, \ldots, k, i\}$. Therefore inequality (b) holds.

Let us prove (a). If $t \le k$, then $E_{it}^{(k)} = 0$. Thus we may assume that $k < t \le i$. Let $T = \{1, 2, \ldots, k, t\}$. Then, in the notation introduced at the beginning of Section 8.4,

$$|E^{(k)}(S, T)| = |E(S, T)| \le b_2$$

and

$$E^{(k)}(S, T) = E_{11}^{(k)} \ldots E_{kk}^{(k)} E_{it}^{(k)}.$$

Since $|E_{11}^{(k)} \ldots E_{kk}^{(k)}| \ge 1$, it follows that $|E_{it}^{(k)}| \le b_2$.

In the call to GCDX during the k-th iteration of the For-loop, we have $A_{kk} = E_{kk}^{(k-1)}$ and $A_{ik} = E_{1k}^{(k-1)}$. By (a), (b), and Exercise 7.4 in Chapter 1, $|p| \le b_2$ and $|q| \le b_1$. Let $1 \le t \le i$. After the multiplication by U,

$$|A_{kt}| = |E_{kt}^{(k)}| < |p| b_1 + |q| b_2 \le 2 b_1 b_2 = b_3.$$

Thus (c) holds. Now suppose that $1 \le r < k$. In the call ROD(k), we subtract $q = A_{rk}$ div A_{kk} times row k of A from row r. By induction, $|q| \le b_3 (1 + b_3)^{k-2} \le b_4$. Hence

$$|E_{rt}^{(k)}| = |E_{rt}^{(k-1)} - q E_{kt}^{(k)}| \le b_3 (1 + b_3)^{k-2} + b_3 (1 + b_3)^{k-2} b_3 = b_3 (1 + b_3)^{k-1}.$$

This proves (d). $\square$

The last iteration of the For-loop occurs with $k = i - 1$. If the value of $E_{ii}^{(i-1)}$ is 0, then the i-th row of A is 0, since we are assuming that σ is returned as the identity. If $E_{ii}^{(i-1)}$ is not 0, then proof of Lemma 5.3(d) shows that the values of q in the call ROD(i) have absolute value at most b_4. Thus all entries in the matrix U have absolute value at most b_2 and all values of q in ROD have absolute values at most b_4. This completes the proof of Proposition 5.1. $\square$

Using KB_ROW and the idea of Exercise 3.4, we can construct a polynomial-time algorithm for finding a diagonal matrix equivalent to a given integer matrix. To simplify the description, let us define a procedure KB_COL as follows:

Procedure KB_COL($B; A, \sigma$);

Input: B : an m-by-n integer matrix;
Output: A : an integer matrix column equivalent to B;
         σ : a permutation of $\{1, \ldots, m\}$ such that the matrix C
              defined by $C_{ij} = A_{\sigma_i j}$ is in column Hermite normal
              form;
Begin
   KB_ROW(B^t; A, σ); $A := A^t$ ($*$ A^t denotes the transpose of A. $*$)
End.

Here is the Kannan-Bachem diagonalization procedure:

Procedure KB_DIAG(B; A);
Input: B : an integer matrix;
Output: A : a diagonal integer matrix equivalent to B;
Begin
   KB_ROW(B; A, σ); Define C by $C_{ij} = A_{i\sigma_j}$;
   KB_COL(C; A, σ);

   While A is not diagonal do begin
      KB_ROW(A; C, σ); KB_COL(C; A, σ)
   End
End.

Let r be the rank of B in KB_DIAG. After the first line of KB_DIAG
has been executed, the matrix C is in row Hermite normal form and the i-th
corner entry of C is in column i. After all succeeding calls to KB_ROW and
KB_COL, the permutation σ returned is the identity and can be ignored.
After the second line of KB_DIAG has been executed, A has the form

$$\begin{bmatrix} E & 0 \\ 0 & 0 \end{bmatrix},$$

where E is an r-by-r matrix in column Hermite normal form and $|\det E|$
is $D_r(B)$. All entries in E are bounded by the Hadamard bound for B. At
the start of some iteration of the While-loop, let i be minimal such that the
i-th column of E contains at least two nonzero entries. At the end of the
iteration of the While-loop, the entries E_{jj} with $j < i$ are unchanged and
E_{ii} has been reduced by a factor of at least 2. It follows that the number
of iterations of the While-loop is polynomial in the size of B. Since the
entries in A remain bounded and KB_ROW runs in polynomial time, it
follows that KB_DIAG runs in polynomial time. To go from a diagonal
matrix to a matrix in Smith normal form, one can use the method discussed in Exercise 3.5. Since this computation is also polynomial, we have

a polynomial-time algorithm for finding Smith normal forms and associated unimodular matrices.

Example 5.4. Suppose B is the following matrix:

$$\begin{bmatrix} 1 & -1 & 1 & -1 & 1 & -1 & 1 & 1 \\ 0 & 1 & 1 & -1 & 0 & 1 & -1 & -1 \\ 0 & 0 & 0 & 0 & 1 & 0 & -1 & -1 \\ 0 & -1 & 0 & 0 & 0 & 0 & 1 & 1 \\ 1 & 0 & 0 & 0 & -1 & 1 & 1 & 0 \\ 0 & 1 & 0 & -1 & 0 & 0 & -1 & -1 \\ 1 & -1 & 0 & 0 & 1 & 1 & 0 & 1 \\ 1 & -1 & 1 & 1 & 0 & -1 & 1 & -1 \end{bmatrix}.$$

Let us follow the computation during the execution of KB_DIAG($B; A$). After the first line has been executed, C has the following value:

$$\begin{bmatrix} 1 & 0 & 0 & 0 & 0 & 2 & 0 & 0 \\ 0 & 1 & 0 & 0 & 0 & 0 & 0 & 0 \\ 0 & 0 & 1 & 0 & 0 & 0 & 0 & 0 \\ 0 & 0 & 0 & 1 & 0 & 0 & 0 & 0 \\ 0 & 0 & 0 & 0 & 1 & 1 & 0 & 0 \\ 0 & 0 & 0 & 0 & 0 & 3 & 0 & 1 \\ 0 & 0 & 0 & 0 & 0 & 0 & 1 & 1 \\ 0 & 0 & 0 & 0 & 0 & 0 & 0 & 2 \end{bmatrix}.$$

After the second line has been executed, A equals

$$\begin{bmatrix} 1 & 0 & 0 & 0 & 0 & 0 & 0 & 0 \\ 0 & 1 & 0 & 0 & 0 & 0 & 0 & 0 \\ 0 & 0 & 1 & 0 & 0 & 0 & 0 & 0 \\ 0 & 0 & 0 & 1 & 0 & 0 & 0 & 0 \\ 0 & 0 & 0 & 0 & 1 & 0 & 0 & 0 \\ 0 & 0 & 0 & 0 & 0 & 1 & 0 & 0 \\ 0 & 0 & 0 & 0 & 0 & 0 & 1 & 0 \\ 0 & 0 & 0 & 0 & 0 & 2 & 0 & 6 \end{bmatrix}.$$

Since A is not diagonal, the body of the While-loop is executed. After the call to KB_ROW, the value of C is

$$
\begin{bmatrix}
1 & 0 & 0 & 0 & 0 & 0 & 0 & 0 \\
0 & 1 & 0 & 0 & 0 & 0 & 0 & 0 \\
0 & 0 & 1 & 0 & 0 & 0 & 0 & 0 \\
0 & 0 & 0 & 1 & 0 & 0 & 0 & 0 \\
0 & 0 & 0 & 0 & 1 & 0 & 0 & 0 \\
0 & 0 & 0 & 0 & 0 & 1 & 0 & 0 \\
0 & 0 & 0 & 0 & 0 & 0 & 1 & 0 \\
0 & 0 & 0 & 0 & 0 & 0 & 0 & 6
\end{bmatrix}.
$$

The call to KB_COL just sets A equal to C. At this point the computation is complete.

Exercises

5.1. Let B be an m-by-n integer matrix of rank r. Suppose the call KB_ROW$(B; A, \sigma)$ returns with $\sigma_1 < \sigma_2 < \cdots < \sigma_r$. Show that A is the row Hermite normal form of B.

5.2. Let B be an m-by-n integer matrix of rank r. Show that there is a permutation τ of $\{1, \ldots, m\}$ such that, if C is defined by $C_{ij} = B_{\tau_i j}$, then the call KB_ROW$(C; A, \sigma)$ returns with $\sigma_1 < \sigma_2 < \cdots < \sigma_r$. Prove that τ can be found in polynomial time.

8.6 Lattice reduction

Up to now, our methods for recognizing that elements of $\mathbb{Z}^n$ are linearly independent over $\mathbb{Z}$ have been based ultimately on the observation that the nonzero rows of a matrix in row Hermite normal form are linearly independent. There is another way to recognize linear independence in $\mathbb{Z}^n$. It involves viewing $\mathbb{Z}^n$ as a subset of $\mathbb{R}^n$ and using geometric techniques in this Euclidean space.

Before exploring the new approach, let us review the basic facts about geometry in $\mathbb{R}^n$. Notions of length and angle are defined using an inner product. For $u = (u_1, \ldots, u_n)$ and $v = (v_1, \ldots, v_n)$ in $\mathbb{R}^n$, the inner product (u, v) of u and v is $u_1 v_1 + \cdots + u_n v_n$. This inner product is bilinear, symmetric, and positive definite. That is, for all u, v, and w in $\mathbb{R}^n$ and all r in $\mathbb{R}$,

$$
\begin{aligned}
(u + v, w) &= (u, w) + (v, w), \\
(u, v + w) &= (u, v) + (u, w), \\
(ru, v) &= r(u, v) = (u, rv), \\
(u, v) &= (v, u), \\
(u, u) &\geq 0,
\end{aligned}
$$

and $(u, u) = 0$ if and only if $u = 0$. The length of u is $\|u\| = \sqrt{(u, u)}$. If u and v are nonzero, then the angle α between u and v is defined by

$$\cos \alpha = \frac{(u, v)}{\|u\| \|v\|}, \quad 0 \le \alpha \le \pi.$$

In particular, u and v are perpendicular or orthogonal if $(u, v) = 0$.

Proposition 6.1. *Let B be a k-by-n matrix with real entries, and set $C = BB^t$. Then the rows of B are linearly independent over $\mathbb{R}$ if and only if C is nonsingular.*

Proof. Note that C_{ij} is the inner product of the i-th row and the j-th row of B. We shall actually prove that the rows of B are linearly dependent if and only if C is singular. Suppose first that the rows of B are linearly dependent. Then the rank of B is less than k. The rank of $C = BB^t$ is at most the rank of B, so C has rank less than k. Therefore C is singular.

Now suppose that C is singular. Then there is a nonzero element u of $\mathbb{R}^k$ such that $uC = 0$. But then $0 = (uC, u) = (uB, uB)$. Since $(,)$ is positive definite, this means that $uB = 0$. Therefore the rows of B are linearly dependent. $\square$

Corollary 6.2. *If $b_1, \ldots, b_k$ are nonzero elements of $\mathbb{R}^n$ which are pairwise orthogonal, then $b_1, \ldots, b_k$ are linearly independent over $\mathbb{R}$.*

Proof. Let B be the k-by-n matrix whose i-th row is b_i. Since $(b_i, b_j) = 0$ if $i \ne j$, the matrix $C = BB^t$ is diagonal and

$$\det C = \prod_{i=1}^{k} (b_i, b_i) \ne 0.$$

Therefore C is nonsingular. $\square$

In $\mathbb{R}^n$ we have the Gram-Schmidt orthogonalization procedure:

Procedure G_S$(b_1, \ldots, b_k; \mathit{flag}, b_1^*, \ldots, b_k^*)$;
Input: $b_1, \ldots, b_k$: elements of $\mathbb{R}^n$;
Output: flag : a boolean variable which is true if the b_i are linearly independent and false otherwise;
        $b_1^*, \ldots, b_k^*$: if flag is true, elements of $\mathbb{R}^n$ which form an orthogonal basis for the subspace of $\mathbb{R}^n$ spanned by the b_i;
Begin
    $b_1^* := b_1$; $\mathit{flag} := (b_1^* \ne 0)$; $i := 2$;

While $i \leq k$ and *flag* do begin
    $b_i^* := b_i$;
    For $j := 1$ to $i - 1$ do begin
        $\mu_{ij} := (b_i, b_j^*)/(b_j^*, b_j^*)$;
        $b_i^* := b_i^* - \mu_{ij} b_j^*$
    End;
    flag $:= (b_i^* \neq 0)$
End
End.

If in G_S the vectors $b_1, \ldots, b_k$ are linearly independent, then at the conclusion of the procedure

$$b_i^* = b_i - \sum_{j=1}^{i-1} \mu_{ij} b_i^*, \quad 1 \leq i \leq k.$$

Thus $b_1, \ldots, b_i$ and $b_1^*, \ldots, b_i^*$ span the same subspace V_i, and b_i^* is the component of b_i orthogonal to V_{i-1}. If the b_i have rational components, then the μ_{ij} are rational, as are the components of the b_i^*.

Example 6.1. Let $n = 3$ and

$$b_1 = (1, 1, -1), \quad b_2 = (2, 0, 1), \quad b_3 = (1, -1, -1).$$

Then

$$b_1^* = b_1, \quad \mu_{21} = \tfrac{1}{3}, \quad b_2^* = b_2 - \tfrac{1}{3} b_1^* = \left(\tfrac{5}{3}, -\tfrac{1}{3}, \tfrac{4}{3}\right),$$
$$\mu_{31} = \tfrac{1}{3}, \quad \mu_{32} = \tfrac{1}{7}, \quad b_3^* = b_3 - \tfrac{1}{3} b_1^* - \tfrac{1}{7} b_2^* = \left(\tfrac{3}{7}, -\tfrac{9}{7}, -\tfrac{6}{7}\right).$$

Continuing the notation used in G_S, let B and B^* be the matrices whose i-th rows are b_i and b_i^*, respectively.

Proposition 6.3. *If $b_1, \ldots, b_k$ are linearly independent in $\mathbb{R}^n$, then*

$$\det BB^t = \prod_{i=1}^{k} (b_i^*, b_i^*).$$

Proof. The matrix B^* is obtained from B by real row operations of type (3). Therefore $B^* = PB$, where P is a k-by-k real matrix and $\det P = 1$. Thus

$$\prod_{i=1}^{k} (b_i^*, b_i^*) = \det B^*(B^*)^t = \det PBB^t P^t = \det BB^t. \qquad \square$$

Suppose that the b_i are as in Proposition 6.3. For $1 \le i \le k$ define C_i to be the i-by-i submatrix of BB^t in the upper left corner and set $d_i = \det C_i$. By Proposition 6.3,

$$d_i = \det B_i B_i^t = \prod_{j=1}^{i} \|b_j^*\|^2,$$

where B_i is the matrix consisting of the first i rows of B. Now assume that the b_i are in $\mathbb{Z}^n$. Then each d_i is a positive integer. Since $b_i^* - b_i$ is in the $\mathbb{R}$-subspace spanned by $b_1, \ldots, b_{i-1}$, there exist real numbers λ_{ij}, $1 \le j < i \le k$, such that

$$b_i^* = b_i - \sum_{j=1}^{i-1} \lambda_{ij} b_j.$$

Since $(b_i^*, b_m) = 0$ for $1 \le m \le i-1$, we have

$$(b_i, b_m) = \sum_{j=1}^{i-1} \lambda_{ij}(b_j, b_m), \quad 1 \le m < i.$$

For a fixed value of i, the determinant of this system of equations for the λ_{ij} is $\det C_{i-1} = d_{i-1}$. By Cramer's rule, $d_{i-1}\lambda_{ij}$ is an integer, $1 \le j < i$. Now

$$d_{i-1}b_i^* = d_{i-1}b_i - \sum_{j=1}^{i-1} d_{i-1}\lambda_{ij} b_j$$

is in $\mathbb{Z}^n$. Finally, since $(b_j^*, b_j^*) = d_j/d_{j-1}$,

$$d_j \mu_{ij} = d_j \frac{(b_i, b_j^*)}{(b_j^*, b_j^*)} = d_{j-1}(b_i, b_j^*) = (b_i, d_{j-1}b_j^*)$$

is in $\mathbb{Z}$. Thus, if the b_i are in $\mathbb{Z}^n$, then the sizes of the denominators of all fractions occurring in the orthogonalization procedure are polynomial in the size of the b_i. It can be shown that the orthogonalization procedure runs in polynomial time.

A *lattice* in $\mathbb{R}^n$ is an additive subgroup generated by a set of elements which are linearly independent over $\mathbb{R}$. Thus $\mathbb{Z}^n$ is a lattice in $\mathbb{R}^n$. Every lattice is a free abelian group of rank at most n. Subgroups of lattices are lattices. However, not every free abelian subgroup of $\mathbb{R}^n$ is a lattice. For example, in $\mathbb{R}$ the numbers 1 and $\sqrt{2}$ additively generate a free abelian group H of rank 2, but H is not a lattice in $\mathbb{R}$. Basic facts about lattices can be found in [Cassels 1971], [Lovász 1986], and [Pohst & Zassenhaus 1989].

Let B be a k-by-n real matrix whose rows are linearly independent over $\mathbb{R}$, and let L be the lattice in $\mathbb{R}^n$ generated by rows of B. Any other basis of L as a free abelian group arises as the set of rows of a matrix $B' = PB$, where P is in $\mathrm{GL}(k, \mathbb{Z})$. Since

$$\det B'(B')^t = \det PBB^t P^t = \det BB^t,$$

it follows that $\det(BB^t)$ depends only on L. By Proposition 6.3, $\det BB^t$ is positive. The positive square root of $\det BB^t$ is called the *discriminant* of L and is written $d(L)$. By Proposition 6.3,

$$d(L) = \prod_{i=1}^{k} \|b_i^*\|,$$

where $b_1^*, \ldots, b_k^*$ is the orthogonalization of the rows of B. If $k = n$, then $d(L) = |\det B|$.

The proof of the following result can be found in the references cited earlier.

Proposition 6.4. *Let L be a lattice in $\mathbb{R}^n$. Any bounded region of $\mathbb{R}^n$ contains at most a finite number of elements of L. In particular, L contains shortest nonzero elements and the number of such elements is finite.*

If a lattice L is generated by a sequence $b_1, \ldots, b_k$ of nonzero, pairwise orthogonal vectors in $\mathbb{R}^n$, then it is easy to see that the shortest length of a nonzero vector in L is

$$\min_i \|b_i\|.$$

However, in general L may have nonzero elements which are much shorter than any of the vectors in a particular basis of L.

Proposition 6.4 is a special case of the following result.

Proposition 6.5. *If L is a lattice of rank k, then for $1 \leq i \leq k$ there is a positive lower bound on the discriminants of sublattices of L of rank i.*

It is possible, in principle, to find a shortest nonzero vector in a lattice L. However, no polynomial-time algorithm is known. In (Lenstra, Lenstra, & Lovász 1982) a polynomial-time algorithm was presented which takes a basis $b_1, \ldots, b_k$ of L as input and produces another basis of L containing a vector which is almost shortest in the sense that its length is at most $2^{(k-1)/2}$ times the length of a shortest nonzero vector. The first application of this algorithm, which is now called the LLL lattice reduction algorithm, was to prove that polynomials in $\mathbb{Q}[X]$ can be factored into irreducible factors in polynomial time. However, the LLL algorithm has many other important applications, including applications to finding Hermite and Smith normal forms. We shall be primarily interested in a modified version of the algorithm due to Pohst, but it will be useful to see the original algorithm as well.

Let $b_1, \ldots, b_k$ be a linearly independent sequence of vectors in $\mathbb{R}^n$, let $b_1^*, \ldots, b_k^*$ be its orthogonalization, and for $1 \leq i \leq k$ let

$$b_i^* = b_i - \sum_{j=1}^{i-1} \mu_{ij} b_j^* \quad \text{and} \quad d_i = \prod_{j=1}^{i} \|b_j^*\|^2.$$

Then $b_1, \ldots, b_k$ is LLL-*reduced* if the following conditions hold:

(1) $|\mu_{ij}| \leq \frac{1}{2}, \quad 1 \leq j < i \leq k$
(2) $\|b_i^* + \mu_{ii-1} b_{i-1}^*\|^2 \geq \frac{3}{4} \|b_{i-1}^*\|^2, \quad 1 < i \leq k.$

Let V_i be the $\mathbb{R}$-subspace spanned by $b_1, \ldots, b_i$ (and also by $b_1^*, \ldots, b_i^*$). Then $b_i^* + \mu_{ii-1} b_{i-1}^*$ is the component of b_i orthogonal to V_{i-2}, and condition (2) says that this component is not much shorter than the component of b_{i-1} orthogonal to V_{i-2}. The fraction $\frac{3}{4}$ is usually used in the definition, but any fixed number in the open interval $(\frac{1}{4}, 1)$ may be substituted. By Proposition 6.5 there is a positive lower bound for d_i which depends only on L. Thus $D = d_1 \ldots d_{k-1}$ is bounded away from 0. Of course, if L is contained in $\mathbb{Z}^n$, then each d_i is a positive integer, so clearly $D \geq 1$.

If $k = 2$, we can illustrate the definition of an LLL-reduced basis with a picture drawn in the plane determined by b_1 and b_2. Let b_1 and b_2 be drawn as directed line segments leaving the origin. For condition (1) to hold, b_2 must end in the strip of width $\|b_1\|$ which runs perpendicular to b_1 and has the origin on its central line. The vector $b_2^* + \mu_{21} b_1^*$ is just b_2. Thus condition (2) says that b_2 does not end inside the circle of radius $\frac{\sqrt{3}}{2} \|b_1\|$ with center at the origin. Hence, if both conditions hold, then b_2 ends in the shaded region in Figure 8.6.1. If condition (1) fails, we can add an integral multiple of b_1 to b_2 to make b_2 end in the perpendicular strip. This leaves b_1^* and b_2^* unchanged. If condition (2) fails, then interchanging b_1 and b_2 multiplies $D = d_1 = \|b_1^*\|^2$ by a factor less than $\frac{3}{4}$. Since D is bounded away from zero, such changes cannot be continued indefinitely. Thus eventually an LLL-reduced basis is reached.

Now suppose that $k = 3$ and that the sequence b_1, b_2 is already LLL-reduced. The condition that $|\mu_{3j}|$ not exceed $\frac{1}{2}$ for $j = 1, 2$ says that the

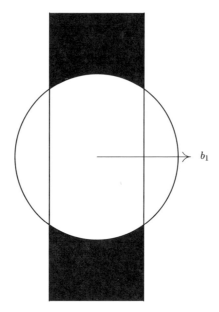

Figure 8.6.1

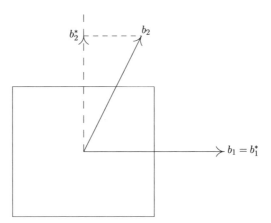

Figure 8.6.2

orthogonal projection u of b_3 into the plane determined by b_1 and b_2 lies in the $\|b_1^*\|$-by-$\|b_2^*\|$ rectangle which is centered at the origin and has its sides parallel to $b_1^* = b_1$ and b_2^*, as shown in Figure 8.6.2. If u does not lie in this rectangle, then we can add an integral multiple of b_2 to b_3 so that u lies in the correct strip parallel to b_1 and then add an integral multiple of b_1 to b_3 to get u into the desired rectangle.

The condition $\|b_3^* + \mu_{32} b_2^*\|^2 \geq \frac{3}{4} \|b_2^*\|^2$ says that the length of the projection v of b_3 into the plane determined by b_2^* and b_3^* has length at least $\frac{\sqrt{3}}{2} \|b_2^*\|$. If this condition fails, we interchange b_2 and b_3. It can be shown that this leaves b_1^* unchanged, replaces b_2^* by v, and replaces b_3^* by a vector of length $\|b_3^*\| \|b_2^*\| / \|v\|$. It follows that $D = d_1 d_2 = \|b_1^*\|^4 \|b_2^*\|^2$ is multiplied by a factor less than $\frac{3}{4}$.

These low-dimensional examples give a fairly good insight into the operation of the LLL lattice reduction algorithm. The algorithm proceeds by making changes to the current basis, either by replacing b_m by $b_m - r b_p$, where r is an integer and $1 \leq p < m$, or by interchanging two consecutive basis elements. The algorithm keeps track of the following data: the current basis $b_1, \ldots, b_k$, the numbers μ_{ij} with $1 \leq j < i \leq k$, and the numbers $B_i = (b_i^*, b_i^*)$, $1 \leq i \leq k$. Condition (2) in the definition of LLL-reduced can be written

$$B_i + \mu_{i\,i-1}^2 B_{i-1} \geq \tfrac{3}{4} B_{i-1}.$$

The effect of replacing b_m by $b_m - r b_p$ is to replace μ_{mp} by $\mu_{mp} - r$ and to replace μ_{mj} by $\mu_{mj} - r \mu_{pj}$ for $1 \leq j < p$. No other μ_{ij} is affected. Since no b_i^* is changed, the B_i and the d_i remain the same. The effect of interchanging b_{m-1} and b_m is more complicated. The values of B_{m-1} and B_m change, but their product remains the same. The μ_{ij} with $\{i,j\} \cap \{m-1, m\} \neq \emptyset$ are also modified. The value of d_{m-1} changes, while the d_i with $i \neq m-1$ are unaffected. The most important fact is that b_{m-1} and b_m are interchanged in the LLL algorithm only when the result of that action is to multiply d_{m-1} by a positive factor less than $\frac{3}{4}$. The value of $D = d_1 \ldots d_{k-1}$ is multiplied by the same factor. Since D is bounded away from 0, the number of times two basis vectors are interchanged is bounded. The details are given in the description of the procedure LLL, which follows. This description is based on the one in [Pohst & Zassenhaus 1989] and on the implementation distributed with Maple V. The subroutine ADJUST_MU enforces the condition that $|\mu_{mp}| \leq \frac{1}{2}$.

Procedure ADJUST_MU(m, p);
Input: m, p : integers with $n \geq m > p \geq 1$;
Begin
   If $|\mu_{mp}| > \frac{1}{2}$ then begin
      Let r be the nearest integer to μ_{mp};
      $b_m := b_m - r b_p$; $\mu_{mp} := \mu_{mp} - r$;
      For $j := 1$ to $p - 1$ do $\mu_{mj} := \mu_{mj} - r \mu_{pj}$
   End
End.

Procedure LLL$(c_1, \ldots, c_k; b_1, \ldots, b_k)$;

Input: $c_1, \ldots, c_k$: linearly independent elements of $\mathbb{R}^n$;
Output: $b_1, \ldots, b_k$: an LLL-reduced basis for the lattice generated
                               by the c_i;
Begin
  For $i := 1$ to k do begin
    $b_i := c_i$; $b_i^* := b_i$;

    For $j := 1$ to $i - 1$ do begin
      $\mu_{ij} := (b_i, b_j^*)/B_j$;
      $b_i^* := b_i^* - \mu_{ij} b_j^*$
    End;

    $B_i := (b_i^*, b_i^*)$
  End;   (* The vectors b_i^* may now be discarded. *)

  $m := 2$;

  While $m \leq k$ do begin
    ADJUST_MU$(m, m - 1)$;
    $\nu := \mu_{mm-1}$; $C := B_m + \nu^2 B_{m-1}$;
    If $C \geq \frac{3}{4} B_{m-1}$ then begin
      For $p := m - 2$ downto 1 do ADJUST_MU(m, p);
      $m := m + 1$
    End

    Else begin
      $\mu_{mm-1} := \nu B_{m-1}/C$; $B_m := B_{m-1} B_m/C$; $B_{m-1} := C$;
      Interchange b_{m-1} and b_m;
      For $j := 1$ to $m - 2$ do interchange μ_{m-1j} and μ_{mj};
      For $j := m + 1$ to k do
        $\begin{bmatrix} \mu_{jm-1} \\ \mu_{jm} \end{bmatrix} := \begin{bmatrix} 1 & \mu_{mm-1} \\ 0 & 1 \end{bmatrix} \begin{bmatrix} 0 & 1 \\ 1 & -\nu \end{bmatrix} \begin{bmatrix} \mu_{jm-1} \\ \mu_{jm} \end{bmatrix}$;
      If $m > 2$ then $m := m - 1$
    End
  End
End.

We shall be interested in using LLL only when the input vectors are in $\mathbb{Z}^n$. The numbers μ_{ij} and B_i used in the procedure will be rational, but in general not integers. Earlier we showed how to bound the size of the denominators needed by polynomials in the size of the b_i. In (Lenstra et al. 1982) it is shown that LLL runs in polynomial time. More precisely, the following result is proved.

Proposition 6.6. Let $c_1, \ldots, c_k$ be linearly independent vectors in $\mathbb{Z}^n$ and let M be a real number such that $M \geq 2$ and $\|c_i\|^2 \leq M$, $1 \leq i \leq k$. Then the number of arithmetic operations on integers carried out by LLL with *input* $c_1, \ldots, c_k$ *is* $O(n^4 \log M)$ *and the number of digits in the integers used is* $O(n \log M)$.

If in LLL we think of $b_1, \ldots, b_k$ as the rows of a matrix, then the transformations to that matrix correspond to row operations of types (1) and (3). If at the beginning of LLL we initialize P to be the k-by-k identity matrix and apply all row operations used to P, then the final value of P is unimodular and satisfies

$$b_i = \sum_{j=1}^{k} P_{ij} c_j, \quad 1 \le i \le k.$$

It is difficult to carry out LLL by hand, even on small matrices. Here are summaries of two computations with LLL.

Example 6.2. Let W be the following matrix:

$$\begin{bmatrix} -1 & -1 & 3 & -4 \\ 0 & 2 & -2 & 4 \\ 4 & 4 & 2 & -1 \end{bmatrix}.$$

Here are the changes made to W when LLL is applied to its rows. First row 1 is added to row 2 to give

$$\begin{bmatrix} -1 & -1 & 3 & -4 \\ -1 & 1 & 1 & 0 \\ 4 & 4 & 2 & -1 \end{bmatrix}.$$

Then rows 1 and 2 are interchanged:

$$\begin{bmatrix} -1 & 1 & 1 & 0 \\ -1 & -1 & 3 & -4 \\ 4 & 4 & 2 & -1 \end{bmatrix}.$$

Now row 1 is subtracted from row 2:

$$\begin{bmatrix} -1 & 1 & 1 & 0 \\ 0 & -2 & 2 & -4 \\ 4 & 4 & 2 & -1 \end{bmatrix}.$$

Finally, row 1 is subtracted from row 3. The final matrix is

$$
\begin{bmatrix}
-1 & 1 & 1 & 0 \\
0 & -2 & 2 & -4 \\
5 & 3 & 1 & -1
\end{bmatrix}.
$$

Example 6.3. When LLL is applied to the rows of

$$
\begin{bmatrix}
1 & 19 & -18 & -1 & -16 \\
20 & 2 & 8 & -15 & -1 \\
-18 & -2 & 17 & -2 & 15 \\
20 & 2 & -15 & 14 & -13
\end{bmatrix},
$$

ten row operations are used to produce the following matrix:

$$
\begin{bmatrix}
2 & 0 & 2 & 12 & 2 \\
22 & 2 & 10 & -3 & 1 \\
5 & 19 & 9 & -6 & 0 \\
18 & 2 & -17 & 2 & -15
\end{bmatrix}.
$$

The matrix has not been changed dramatically, but the product of the lengths of the rows has been reduced from $741846.33\ldots$ to $199242.53\ldots$.

The procedure LLL uses exact computation with real numbers. If the input vectors are in $\mathbb{Z}^n$, then only rational arithmetic is needed. In this case, experimental evidence described in (Lagarias & Odlyzko 1985) suggests that results can sometimes be obtained more quickly if floating-point computations are used to calculate the b_i^*, B_i, and μ_{ij}. A number of individuals have suggested variants of LLL in an effort to make the algorithm use only integer arithmetic or to reduce the precision needed for floating-point approximations. See (Schönhage 1984), (Schnorr 1988), (Schnorr & Euchner 1991), and Section 3.5 of [Weger 1989].

Let L be a subgroup of $\mathbb{Z}^n$. If we are given a basis $b_1, \ldots, b_k$ for L and the orthogonalization $b_1^*, \ldots, b_k^*$ of that basis, then we can decide membership in L. Let $x_1, \ldots, x_k$ be integers and set $u = x_1 b_1 + \cdots + x_k b_k$. We can write $u = y_1 b_1^* + \cdots + y_k b_k^*$. In general the y_i are rational numbers, but $y_k = x_k$ is an integer since $b_k - b_k^*$ is in the $\mathbb{R}$-subspace spanned by $b_1, \ldots, b_{k-1}$. Thus x_k is equal to

$$
\frac{(u, b_k^*)}{(b_k^*, b_k^*)}.
$$

This leads to the following test for deciding whether an element v of $\mathbb{Z}^n$ is in L:

$w := v$;

For $j := k$ downto 1 do begin
  $x := (w, b_j^*)/(b_j^*, b_j^*)$;
  If x is not an integer then v is not in L
  Else $w := w - xb_j$
End;

v is in L if and only if w is 0;

This membership test can be modified to produce almost-shortest elements in cosets of L in $\mathbb{Z}^n$. Suppose $b_1, \ldots, b_k$ is LLL-reduced and let v be in $\mathbb{Z}^n$. If we execute the following statements,

$w := v$;

For $j := k$ downto 1 do begin
  Let r be the nearest integer to $(w, b_j^*)/(b_j^*, b_j^*)$;
  $w := w - rb_j$
End;

then $L + w = L + v$ and the length of the projection of w onto the subspace spanned by L is within a factor of $2^{k/2}$ of being as small as possible. See (Babai 1986).

Another variation of this same technique can be used to find integral dependencies among elements of $\mathbb{Z}^n$. Suppose that $b_1, \ldots, b_k$ are as before and that v in $\mathbb{Z}^n$ is a rational linear combination of $b_1, \ldots, b_k$. We can find integers y and $x_1, \ldots, x_k$ such that $yv = x_1 b_1 + \cdots + x_k b_k$ as follows:

$w := v$; $y := 1$;

For $j := k$ downto 1 do begin
  Let $(w, b_j^*)/(b_j^*, b_j^*) = p/q$, where p and q are relatively prime integers;
  Multiply w, y, and $x_{j+1}, \ldots, x_k$ by q;
  $x_j := p$; $w := w - pb_j$
End;

Exercises

6.1. Derive the formulas for the changes to the B_i and the μ_{ij} in LLL.

6.2. Let L be a lattice. Show that a nonzero vector in L of minimal length is part of a $\mathbb{Z}$-basis of L.

6.3. Find LLL-reduced bases for the subgroups generated by the rows of the following matrices:

$$\text{(a)} \quad \begin{bmatrix} 4 & 6 \\ 5 & 7 \end{bmatrix}, \quad \text{(b)} \quad \begin{bmatrix} 2 & 1 & -1 \\ 5 & 3 & -2 \end{bmatrix}, \quad \text{(c)} \quad \begin{bmatrix} 6 & 5 & 4 & 1 \\ 5 & 2 & -1 & 3 \\ 4 & -1 & 3 & 2 \end{bmatrix}.$$

8.7 The modified LLL algorithm

A major drawback of the LLL algorithm described in Section 8.6 is the fact
the input vectors must be known in advance to be linearly independent. In
(Pohst 1987) a modified version of the algorithm was presented which takes
as input any generating set for a lattice L in $\mathbb{R}^n$ and returns an LLL-reduced
basis for L. A similar algorithm was sketched in (Hastad et al. 1986). The
discussion here follows (Pohst 1987) and [Pohst & Zassenhaus 1987]. In
general, it is difficult to determine whether a given sequence of real vectors
generates a lattice. However, if all the vectors have integer components,
then the additive subgroup generated is contained in $\mathbb{Z}^n$ and so is a lattice.

Suppose that b_1 and b_2 are linearly dependent vectors in $\mathbb{R}^n$ which generate a nonzero lattice L. Since the rank of L is the same as the dimension
of the $\mathbb{R}$-subspace which it spans, the rank of L must be 1. Thus L is
generated by a single element c and there are integers x_1 and x_2 such that
$b_1 = x_1 c$ and $b_2 = x_2 c$. Since L is generated by b_1 and b_2, the gcd of x_1 and
x_2 is 1. Now $x_2 b_1 = x_1 x_2 c = x_1 b_2$, so $x_2 b_1 - x_1 b_2 = 0$.

More generally, suppose that $b_1, \ldots, b_k$ are linearly dependent vectors
which generate a lattice L. Then there are integers $x_1, \ldots, x_k$ not all 0 such
that $x_1 b_1 + \cdots + x_k b_k = 0$. Such an equation will be called a $\mathbb{Z}$-*relation* on
the b_i. We may assume that $\gcd(x_1, \ldots, x_k) = 1$. If some x_i is ± 1, then
b_i is an integral linear combination of the other b's and b_i may be deleted
from the generating set. If $|x_i| \neq 1$ for all i, then it may happen that no
proper subset of $\{b_1, \ldots, b_k\}$ generates L. However, it is possible to find
$k - 1$ vectors which generate L.

Let $x = (x_1, \ldots, x_k)$ and let B be the k-by-n matrix whose i-th row is b_i.
Then $xB = 0$. By thinking of x as a 1-by-k matrix, we can find an element
P of $\mathrm{GL}(k, \mathbb{Z})$ such that $xP = (1, 0, \ldots, 0)$. Let $C = P^{-1}B$. Then the rows
of C generate L. But

$$(1, 0, \ldots, 0)C = xPP^{-1}B = xB = 0.$$

Therefore the first row of C is 0 and L is generated by rows 2 through k
of C. In practice, we need only find P in $\mathrm{GL}(k, \mathbb{Z})$ such that xP has a
component equal to 1.

Example 7.1. Suppose that b_1, b_2, and b_3 are vectors and we know that
$6b_1 + 10b_2 + 15b_3 = 0$. Let $x = (6, 10, 15)$ and write

$$B = \begin{bmatrix} b_1 \\ b_2 \\ b_3 \end{bmatrix},$$

so $xB = 0$. If we subtract twice the first "column" of x from the third "column", then x is changed to $(6,10,3)$. The corresponding elementary matrix is

$$P = \begin{bmatrix} 1 & 0 & -2 \\ 0 & 1 & 0 \\ 0 & 0 & 1 \end{bmatrix},$$

and P^{-1} is

$$\begin{bmatrix} 1 & 0 & 2 \\ 0 & 1 & 0 \\ 0 & 0 & 1 \end{bmatrix}.$$

Let us replace B by

$$P^{-1}B = \begin{bmatrix} b_1 + 2b_3 \\ b_2 \\ b_3 \end{bmatrix}.$$

Then it is still true that $xB = 0$. Now subtracting three times the third "column" of x from the second "column" gives $x = (6,1,3)$. In this case

$$P = \begin{bmatrix} 1 & 0 & 0 \\ 0 & 1 & 0 \\ 0 & -3 & 1 \end{bmatrix} \quad \text{and} \quad P^{-1} = \begin{bmatrix} 1 & 0 & 0 \\ 0 & 1 & 0 \\ 0 & 3 & 1 \end{bmatrix}.$$

Multiplying B on the left by P^{-1} gives

$$B = \begin{bmatrix} b_1 + 2b_3 \\ b_2 \\ 3b_2 + b_3 \end{bmatrix}.$$

The rows of B generate the same abelian group A as b_1, b_2, and b_3. However, since now $(6,1,3)B = 0$, the second row of B is an integral linear combination of the first and third rows. Thus A is generated by $b_1 + 2b_3$ and $3b_2 + b_3$.

Suppose that B is a k-by-n integer matrix of rank less than k. One way to find a nontrivial $\mathbb{Z}$-relation $xB = 0$ satisfied by the rows of B is with modular techniques. Another method makes use of the observation in the

last paragraph of the previous section. Once x is known, the method of
Example 7.1 can be applied to reduce the number of generators.

The modified LLL algorithm accepts input vectors which need not have
integral components. Thus modular techniques are not appropriate. A
different method is used to find implicitly a relation $xB = 0$ and to produce
a smaller generating set. To get an idea of how this algorithm works, let
us examine what happens when the original LLL algorithm is applied to
a sequence $b_1, \ldots, b_k$ of linearly dependent vectors. Let t be minimal such
that $b_1, \ldots, b_t$ are linearly dependent. Then $b_t^* = 0$ and, if $t < k$, the attempt
to compute μ_{it} with $i > t$ will lead to division by 0. However, $b_1, \ldots, b_t$ form
a redundant generating set for the lattice N they generate. Thus we may
temporarily ignore $b_{t+1}, \ldots, b_k$ and assume that $t = k$. If $t = 1$, then $b_1 = 0$
and we may drop b_1 from the generating set. Suppose that $t > 1$. By
Proposition 6.5, the numbers $d_i = B_1 \ldots B_i$, $1 \le i < t$, are bounded below
by a positive number depending only on N. Thus $D = d_1 \ldots d_{t-1}$ is bounded
away from 0. Each time b_m and b_{m-1} are interchanged during the execution
of LLL, the value of D is multiplied by a factor less than $\frac{3}{4}$.

Suppose that $b_1, \ldots, b_{t-1}$ never become linearly dependent. Each time
the body of the While-loop is executed with $m = t$, we shall always have
$B_m = 0$ and $\nu^2 \le \frac{1}{4}$. Thus C will be less than $3B_{m-1}/4$ and b_m and b_{m-1} will
be interchanged. Thus the procedure never terminates and D approaches 0
as a limit. As remarked earlier, this cannot happen. Therefore $b_1, \ldots, b_{t-1}$
become linearly dependent at some point. That is, t decreases. Eventually
some b_i becomes 0. Since it is not possible to compute all the b_i^* initially
and it is not possible to keep track of the changes in the numbers μ_{ij}, we
must retain the b_i^* during the computation.

The following version of the modified LLL algorithm is based on [Pohst
& Zassenhaus 1989]. The integer s is such that $b_{s+1}, \ldots, b_k$ are known to
be 0. At the start of an iteration of the loop "While $i \le s$ do", $b_1, \ldots, b_{i-1}$
are known to be linearly independent and LLL-reduced. Throughout this
iteration, either $t = i$ or $t < i$ and $b_t^* = 0$, which means that $b_1, \ldots, b_t$ are
linearly dependent. After the initialization, there are five places in MLLL
at which one or more of the b_i are changed. These places are marked with
comments, such as "$(* \ P1 \ *)$", to facilitate the discussion of examples.

Procedure MLLL($c_1, \ldots, c_k; b_1, \ldots, b_k$);
Input: $c_1, \ldots, c_k$: vectors in $\mathbb{R}^n$ which generate a lattice L;
Output: $b_1, \ldots, b_k$: a sequence of vectors whose nonzero terms form
                                an LLL-reduced basis for L;
Begin
    For $i := 1$ to k do $b_i := c_i$; $s := k$; $i := 1$;
99: While $i \le s$ do
        If $b_i = 0$ then begin
            If $i < s$ then interchange b_i and b_s; $(* \ P1 \ *)$

$s := s - 1$
End
Else begin $(* \ b_i \neq 0 \ *)$
 $\quad b_i^* := b_i;$
 $\quad$For $j := 1$ to $i - 1$ do begin
 $\quad\quad \mu_{ij} := (b_i, b_j^*)/B_j; \ b_i^* := b_i^* - \mu_{ij} b_j^*$
 $\quad$End;
 $\quad B_i := (b_i^*, b_i^*);$
 $\quad$If $i = 1$ then $i := 2$
 $\quad$Else begin $(* \ i > 1 \ *)$
 $\quad\quad t := i; \ m := i;$
 $\quad\quad$While $m \leq t$ do begin
 $\quad\quad\quad$ADJUST_MU$(m, m - 1);$ $\hspace{3cm}$ $(* \ P2 \ *)$
 $\quad\quad\quad \nu := \mu_{mm-1}; \ C := B_m + \nu^2 B_{m-1};$
 $\quad\quad\quad$If $C \geq \frac{3}{4} B_{m-1}$ then begin
 $\quad\quad\quad\quad$For $p := m - 2$ downto 1 do ADJUST_MU$(m, p);$ $(* \ P3 \ *)$
 $\quad\quad\quad\quad m := m + 1$
 $\quad\quad\quad$End
 $\quad\quad\quad$Else begin
 $\quad\quad\quad\quad$If $b_m = 0$ then begin
 $\quad\quad\quad\quad\quad$If $m < s$ then interchange b_m and $b_s;$ $\hspace{1.2cm}$ $(* \ P4 \ *)$
 $\quad\quad\quad\quad\quad s := s - 1; \ i := m;$
 $\quad\quad\quad\quad\quad$Goto 99
 $\quad\quad\quad\quad$End;
 $\quad\quad\quad\quad$If $C \neq 0$ then begin
 $\quad\quad\quad\quad\quad \mu_{mm-1} := \nu B_{m-1}/C; \ B_m := B_{m-1} B_m / C;$
 $\quad\quad\quad\quad\quad$For $j := m + 1$ to t do
 $\quad\quad\quad\quad\quad\quad \begin{bmatrix} \mu_{jm-1} \\ \mu_{jm} \end{bmatrix} := \begin{bmatrix} 1 & \mu_{mm-1} \\ 0 & 1 \end{bmatrix} \begin{bmatrix} 0 & 1 \\ 1 & -\nu \end{bmatrix} \begin{bmatrix} \mu_{jm-1} \\ \mu_{jm} \end{bmatrix}$
 $\quad\quad\quad\quad$End
 $\quad\quad\quad\quad B_{m-1} := C;$
 $\quad\quad\quad\quad$Interchange b_{m-1} and $b_m;$ $\hspace{3.5cm}$ $(* \ P5 \ *)$
 $\quad\quad\quad\quad$If $B_{m-1} = 0$ then $t := m - 1$ $\quad$ $(* \ b_1, \ldots, b_{m-1}$ are
 $\quad\quad\quad\quad\quad$dependent. $*);$
 $\quad\quad\quad\quad$For $j := 1$ to $m - 2$ do interchange μ_{m-1j} and $\mu_{mj};$
 $\quad\quad\quad\quad b_{m-1}^* := b_{m-1};$
 $\quad\quad\quad\quad$For $j := 1$ to $m - 2$ do $b_{m-1}^* := b_{m-1}^* - \mu_{m-1j} b_j^*;$
 $\quad\quad\quad\quad$If $m \leq t$ then begin
 $\quad\quad\quad\quad\quad b_m^* := b_m;$
 $\quad\quad\quad\quad\quad$For $j := 1$ to $m - 1$ do $b_m^* := b_m^* - \mu_{mj} b_j^*$
 $\quad\quad\quad\quad$End;
 $\quad\quad\quad\quad$If $m > 2$ then $m := m - 1$

Table 8.7.1

b_1	b_2	b_3	Place
36	84	100	
36	12	100	2
12	36	100	5
12	0	100	2
12	100	0	4
12	4	0	2
4	12	0	5
4	0	0	4

$$\text{End} \quad (* \; C < \tfrac{3}{4}B_{m-1} \; *)$$
$$\text{End;} \quad (* \; \text{While } m \le t \; *)$$
$$i := i + 1$$
$$\text{End} \; (* \; i > 1 \; *)$$
$$\text{End} \; (* \; b_i \ne 0 \; *)$$
$$\text{End.}$$

As with the procedure LLL of Section 8.6, we can modify MLLL to produce a k-by-k unimodular matrix P such that

$$b_i = \sum_j P_{ij} c_j.$$

Example 7.2. If $n = 1$ in MLLL, then MLLL performs the version of the Euclidean algorithm in which remainders under division are chosen to have minimum absolute value. For example, suppose that $c_1 = 36$, $c_2 = 84$, and $c_3 = 100$. Table 8.7.1 summarizes the changes in the values of b_1, b_2, and b_3 during the execution of MLLL$(c_1, c_2, c_3; b_1, b_2, b_3)$. The fourth column indicates the "places" in the procedure at which the changes occur.

Example 7.3. Now let us consider an example in $\mathbb{Z}^2$. Let $c_1 = (4, -1)$, $c_2 = (5, 4)$, and $c_3 = (-2, -4)$. Table 8.7.2 shows the changes in the values of b_1, b_2, and b_3 in the execution of MLLL.

Example 7.4. Our last example is in $\mathbb{Z}^3$. Suppose the input to MLLL consists of the rows of the matrix

$$\begin{bmatrix} 48 & -124 & 292 \\ 171 & -142 & 141 \\ -291 & 254 & -277 \end{bmatrix}.$$

Then the computation proceeds as shown in Table 8.7.3.

Table 8.7.2

b_1	b_2	b_3	Place
$(4,-1)$	$(5,4)$	$(-2,-4)$	
$(4,-1)$	$(1,5)$	$(-2,-4)$	2
$(4,-1)$	$(1,5)$	$(-1,1)$	2
$(4,-1)$	$(-1,1)$	$(1,5)$	5
$(-1,1)$	$(4,-1)$	$(1,5)$	5
$(-1,1)$	$(1,2)$	$(1,5)$	2
$(-1,1)$	$(1,2)$	$(-1,1)$	2
$(-1,1)$	$(-1,1)$	$(1,2)$	5
$(-1,1)$	$(0,0)$	$(1,2)$	2
$(-1,1)$	$(1,2)$	$(0,0)$	4

Table 8.7.3

b_1	b_2	b_3	Place
$(48,-124,292)$	$(171,-142,141)$	$(-291,254,-277)$	
$(48,-124,292)$	$(123,-18,-151)$	$(-291,254,-277)$	2
$(123,-18,-151)$	$(48,-124,292)$	$(-291,254,-277)$	5
$(123,-18,-151)$	$(171,-142,141)$	$(-291,254,-277)$	2
$(123,-18,-151)$	$(171,-142,141)$	$(51,-30,5)$	2
$(123,-18,-151)$	$(51,-30,5)$	$(171,-142,141)$	5
$(51,-30,5)$	$(123,-18,-151)$	$(171,-142,141)$	5
$(51,-30,5)$	$(21,42,-161)$	$(171,-142,141)$	2
$(51,-30,5)$	$(21,42,-161)$	$(192,-100,-20)$	2
$(51,-30,5)$	$(192,-100,-20)$	$(21,42,-161)$	5
$(51,-30,5)$	$(-12,20,-40)$	$(21,42,-161)$	2
$(-12,20,-40)$	$(51,-30,5)$	$(21,42,-161)$	5
$(-12,20,-40)$	$(39,-10,-35)$	$(21,42,-161)$	2
$(-12,20,-40)$	$(39,-10,-35)$	$(-18,52,-126)$	2
$(-12,20,-40)$	$(-18,52,-126)$	$(39,-10,-35)$	5
$(-12,20,-40)$	$(18,-8,-6)$	$(39,-10,-35)$	2
$(18,-8,-6)$	$(-12,20,-40)$	$(39,-10,-35)$	5
$(18,-8,-6)$	$(39,-10,-35)$	$(-12,20,-40)$	5
$(18,-8,-6)$	$(3,6,-23)$	$(-12,20,-40)$	2
$(18,-8,-6)$	$(3,6,-23)$	$(-18,8,6)$	2
$(18,-8,-6)$	$(-18,8,6)$	$(3,6,-23)$	5
$(18,-8,-6)$	$(0,0,0)$	$(3,6,-23)$	2
$(18,-8,-6)$	$(3,6,-23)$	$(0,0,0)$	4

Exercises

7.1. Suppose that B is an integer matrix with four rows such that $(105, -140, 126, 150)B = 0$. Find a 3-by-4 integer matrix Q such that $S(QB) = S(B)$.

7.2. Apply MLLL to the rows of the following matrices:

$$
(a) \quad
\begin{bmatrix}
-18 & 4 \\
-1 & -33 \\
90 & 72
\end{bmatrix}, \quad
(b) \quad
\begin{bmatrix}
286 & -111 & -127 \\
-70 & 213 & -185 \\
234 & -169 & -13
\end{bmatrix}
$$

8.8 A comparison

Given a description of an abelian group G as $\mathbb{Z}^n/S(B)$, where B is an m-by-n integer matrix, frequently we would like to obtain an explicit isomorphism of G with a direct sum of cyclic groups. If B has rank n, then modular techniques suffice. However, if the rank r of B is less than n, then modular methods do not appear adequate. Suppose we can find a unimodular matrix Q such that BQ has the form $[C\ 0]$, where C is an m-by-r matrix and 0 denotes an m-by-$(n-r)$ matrix of zeros. Then G is isomorphic to $H = (\mathbb{Z}^r/S(C)) \times \mathbb{Z}^{n-r}$, and, using Q, we can construct an explicit isomorphism of G with H. Since C has rank equal to the number of its columns, modular techniques can be applied to find an isomorphism of $\mathbb{Z}^r/S(C)$ with a direct sum of cyclic groups. Putting these isomorphisms together gives an isomorphism of G with a direct sum of cyclic groups.

In this chapter we have emphasized row operations. Multiplying on the right by a unimodular matrix corresponds to applying column operations. By taking transposes, we see that we can produce an isomorphism of our group G with a direct sum of cyclic groups provided we can solve the following problem: Given an integer matrix B of rank r, find a unimodular matrix P such that $A = PB$ has exactly r nonzero rows or, equivalently, such that the nonzero rows of A form a basis for $S(B)$. Ideally the entries in both P and A should be small.

We have studied three rather different algorithms for solving this problem. The first was the procedure ROW_REDUCE of Section 8.1, using, for example, the Rosser strategy. The second was the Kannan-Bachem algorithm embodied in KB_ROW of Section 8.5. Finally, we saw the modified LLL algorithm in Section 8.7. We know that KB_ROW and MLLL are polynomial-time algorithms and we suspect that the Rosser strategy is not polynomial. It would be useful to have some information about the average running times of these procedures. A rigorous analysis is currently not possible, and extensive computer experiments have yet to be carried out. However, looking at one moderately large example may give some insight.

Matrix 8.8.1

$$
\begin{bmatrix}
13 & -1 & 14 & -6 & -10 & -11 & -1 & 2 & -1 & -4 & -9 & -1 & 2 & 8 & 4 & 3 & -6 & -8 \\
5 & 12 & 15 & 2 & -6 & -6 & 5 & -4 & -10 & -10 & -3 & 7 & 1 & -7 & 9 & 4 & 5 & -3 \\
-10 & -3 & 5 & -2 & -5 & 15 & 1 & -11 & -8 & -2 & -13 & -9 & 9 & 1 & 5 & 5 & -4 & -16 \\
-4 & 3 & 1 & 6 & -11 & 11 & 3 & -8 & -8 & 6 & -5 & 6 & 15 & -2 & -1 & -1 & -6 & 6 \\
-2 & -3 & 13 & -3 & 2 & -9 & -2 & -4 & -11 & -9 & 2 & -4 & -2 & 1 & -7 & -12 & 2 & 8 \\
-4 & -3 & -10 & 9 & 10 & 5 & -6 & -2 & 4 & -1 & 0 & -5 & -10 & -2 & 3 & -1 & 9 & 8 \\
0 & 1 & 7 & 12 & -9 & -3 & 7 & 5 & -4 & -15 & 1 & 2 & -2 & -10 & -5 & 2 & 2 & 3 \\
8 & -4 & -18 & -7 & 7 & -1 & -4 & -2 & 21 & 6 & 10 & 3 & -7 & -3 & 6 & 13 & 8 & -19 \\
7 & -3 & 3 & -6 & -2 & -8 & -6 & -1 & 9 & -4 & 0 & -5 & -12 & 7 & 11 & -4 & 8 & -10 \\
12 & 10 & -5 & -4 & 21 & -18 & 1 & 3 & 9 & 10 & 5 & -9 & -21 & 2 & 17 & 2 & 11 & 11 \\
6 & -12 & -2 & -5 & 11 & 1 & -4 & -11 & -1 & 6 & 8 & -1 & -7 & -9 & 7 & -14 & 8 & -4 \\
5 & 9 & 3 & -2 & 16 & -13 & 7 & 5 & 0 & 1 & 11 & -5 & -16 & -7 & 10 & -4 & 9 & 6 \\
2 & 2 & -2 & -3 & 6 & 0 & 8 & -1 & 3 & 3 & -5 & -9 & -7 & -7 & 10 & 9 & 6 & -12 \\
0 & 2 & -5 & 5 & 0 & 3 & -7 & -5 & 3 & -8 & 4 & 17 & 1 & -13 & 4 & 7 & 6 & -13 \\
14 & -1 & -5 & 2 & 2 & 0 & -1 & -1 & 7 & 1 & 2 & 1 & -2 & 3 & 8 & 14 & -1 & -1 \\
-6 & -3 & 7 & 0 & -19 & 4 & -4 & -3 & -6 & 1 & -13 & 3 & 13 & 9 & -9 & -10 & -6 & 4 \\
5 & -10 & 1 & 1 & 7 & 4 & -6 & -1 & -5 & -6 & 12 & 4 & 2 & 3 & -3 & -2 & -10 & 6 \\
-2 & 9 & -3 & -18 & -3 & -12 & 6 & -5 & 8 & 11 & -3 & -4 & -4 & 5 & 4 & 0 & 10 & -6
\end{bmatrix}
$$

Example 8.1. Let B be Matrix 8.8.1. This matrix was obtained by forming a product UV, where U was an 18-by-14 matrix, V was a 14-by-18 matrix, and the entries of U and V were chosen randomly from the set $\{-2, -1, 0, 1, 2\}$. The rank of B is 14.

Two popular computer algebra systems are Maple and Mathematica. At the time this was written, Version V of Maple and Version 1.2 of Mathematica were available. Both systems included implementations of LLL, although neither provided the option of obtaining the unimodular matrix describing the new basis in terms of the old. Neither system had an implementation of MLLL.

Implementations of the Rosser strategy, KB_ROW, and MLLL were written in the Maple and Mathematica languages and applied to B using a Sun 3/60 workstation. Two variants of MLLL were tried. One used exact rational arithmetic throughout, and the other used floating-point computations for the b_i^*, B_i, and μ_{ij}. The precision of the floating-point numbers could be specified by the user.

The Rosser strategy produced the row Hermite normal form A of B given in Matrix 8.8.2 in about 90 seconds on each system. The entries in the unimodular matrix P returned were very large, the largest having absolute value of roughly 10^{50}. The procedure KB_ROW obtained the same row Hermite normal form in about 110 seconds on each system. In this case the entries in P were smaller. The largest absolute value was about 10^{16}.

The rational-arithmetic implementations of MLLL produced a basis of short vectors and a much better unimodular matrix at the expense of

Matrix 8.8.2

$$
\begin{bmatrix}
1 & 0 & 0 & 0 & 0 & 0 & 1 & 0 & 0 & 0 & 0 & 2 & 0 & 378896 & 1610460 & 518760 & -1994370 & -409551 \\
0 & 1 & 0 & 0 & 0 & 0 & 1 & 0 & 0 & 0 & 0 & 2 & 0 & 296918 & 1262024 & 406522 & -1562871 & -320942 \\
0 & 0 & 1 & 0 & 0 & 0 & 0 & 0 & 0 & 0 & 0 & 0 & 0 & 332069 & 1411420 & 454644 & -1747882 & -358934 \\
0 & 0 & 0 & 1 & 0 & 0 & 0 & 0 & 0 & 0 & 0 & 1 & 0 & 95135 & 404368 & 130254 & -500763 & -102834 \\
0 & 0 & 0 & 0 & 1 & 0 & 1 & 0 & 0 & 0 & 0 & 1 & 0 & 370118 & 1573153 & 506743 & -1948170 & -400065 \\
0 & 0 & 0 & 0 & 0 & 1 & 1 & 0 & 0 & 0 & 0 & 0 & 0 & 69826 & 296793 & 95603 & -367543 & -75478 \\
0 & 0 & 0 & 0 & 0 & 0 & 2 & 0 & 0 & 0 & 0 & 2 & 0 & 41797 & 177679 & 57234 & -220033 & -45190 \\
0 & 0 & 0 & 0 & 0 & 0 & 0 & 1 & 0 & 0 & 0 & 0 & 0 & 409937 & 1742375 & 561252 & -2157735 & -443094 \\
0 & 0 & 0 & 0 & 0 & 0 & 0 & 0 & 1 & 0 & 0 & 0 & 0 & 313155 & 1331029 & 428750 & -1648327 & -338491 \\
0 & 0 & 0 & 0 & 0 & 0 & 0 & 0 & 0 & 1 & 0 & 2 & 0 & 445463 & 1893391 & 609895 & -2344748 & -481501 \\
0 & 0 & 0 & 0 & 0 & 0 & 0 & 0 & 0 & 0 & 1 & 0 & 0 & 159065 & 676081 & 217778 & -837250 & -171930 \\
0 & 0 & 0 & 0 & 0 & 0 & 0 & 0 & 0 & 0 & 0 & 3 & 0 & 227370 & 966423 & 311302 & -1196803 & -245770 \\
0 & 0 & 0 & 0 & 0 & 0 & 0 & 0 & 0 & 0 & 0 & 0 & 1 & 471748 & 2005097 & 645881 & -2483087 & -509907 \\
0 & 0 & 0 & 0 & 0 & 0 & 0 & 0 & 0 & 0 & 0 & 0 & 0 & 489867 & 2082110 & 670687 & -2578457 & -529491 \\
0 & 0 & 0 & 0 & 0 & 0 & 0 & 0 & 0 & 0 & 0 & 0 & 0 & 0 & 0 & 0 & 0 & 0 \\
0 & 0 & 0 & 0 & 0 & 0 & 0 & 0 & 0 & 0 & 0 & 0 & 0 & 0 & 0 & 0 & 0 & 0 \\
0 & 0 & 0 & 0 & 0 & 0 & 0 & 0 & 0 & 0 & 0 & 0 & 0 & 0 & 0 & 0 & 0 & 0 \\
0 & 0 & 0 & 0 & 0 & 0 & 0 & 0 & 0 & 0 & 0 & 0 & 0 & 0 & 0 & 0 & 0 & 0
\end{bmatrix}
$$

Matrix 8.8.3

$$
\begin{bmatrix}
0 & -1 & 0 & 0 & -1 & -2 & 0 & 1 & 1 & 0 & 0 & 1 & 0 & -2 & -1 & -1 & -1 & -2 \\
0 & 1 & 1 & 0 & -3 & -2 & 0 & -1 & 1 & 1 & -2 & -1 & 1 & 1 & 1 & 0 & -1 & -1 \\
-2 & 1 & 0 & 0 & 2 & 1 & 2 & 0 & -2 & 2 & -1 & -2 & 0 & -1 & 0 & -1 & 0 & 2 \\
-2 & 2 & -1 & 1 & -2 & 1 & -1 & 0 & 1 & 1 & -1 & 1 & 1 & 1 & -1 & -1 & 2 & 1 \\
-1 & -2 & -3 & 2 & 0 & 0 & -1 & 0 & 2 & -1 & -1 & -1 & -1 & 0 & -1 & 2 & -1 & 1 \\
0 & -2 & -1 & 0 & 0 & 2 & 0 & 1 & 0 & -1 & -1 & 0 & -2 & 1 & 3 & 0 & -2 & 0 \\
-1 & -2 & 0 & 2 & -1 & 2 & 0 & 1 & -1 & -2 & -1 & -1 & 0 & 1 & -1 & 0 & -2 & 2 \\
-1 & 3 & 0 & 0 & -1 & 0 & 1 & 2 & 0 & 0 & 0 & 3 & -1 & -1 & 2 & -1 & 1 & 0 \\
-2 & 1 & 1 & 0 & 0 & 1 & 0 & -2 & -2 & -2 & 0 & 0 & 1 & 0 & 0 & 1 & 0 & 0 \\
1 & -2 & -3 & 2 & 0 & 0 & -1 & -1 & 2 & 2 & 1 & 1 & 1 & -2 & -2 & 0 & 0 & 1 \\
0 & 1 & 0 & 2 & 1 & -2 & -2 & 0 & 1 & 1 & 0 & -1 & 0 & 1 & 2 & 0 & -3 & 1 \\
1 & -1 & -2 & 1 & 2 & 2 & -2 & 0 & 1 & 0 & -2 & 0 & 0 & 0 & -1 & 3 & 2 & -1 \\
-1 & 1 & 0 & 1 & 0 & -1 & -1 & -1 & -1 & -1 & 0 & 2 & -1 & -1 & 0 & -2 & 2 & 3 \\
0 & 0 & 0 & -2 & -2 & -1 & 1 & 0 & -1 & 1 & -1 & 2 & 2 & 1 & -2 & 0 & -2 & 2 \\
0 & 0 & 0 & 0 & 0 & 0 & 0 & 0 & 0 & 0 & 0 & 0 & 0 & 0 & 0 & 0 & 0 & 0 \\
0 & 0 & 0 & 0 & 0 & 0 & 0 & 0 & 0 & 0 & 0 & 0 & 0 & 0 & 0 & 0 & 0 & 0 \\
0 & 0 & 0 & 0 & 0 & 0 & 0 & 0 & 0 & 0 & 0 & 0 & 0 & 0 & 0 & 0 & 0 & 0 \\
0 & 0 & 0 & 0 & 0 & 0 & 0 & 0 & 0 & 0 & 0 & 0 & 0 & 0 & 0 & 0 & 0 & 0
\end{bmatrix}
$$

considerably longer execution times, about 23 minutes for Maple and 74 minutes for Mathematica. The matrix A_1 returned is given by Matrix 8.8.3. The largest absolute value of an entry in P was slightly more than 10^8.

The main problem in devising a floating-point implementation of MLLL is deciding when the number $C = B_m + \mu_{mm-1}^2 B_{m-1}$ is 0. As noted in Section 8.6, $d_{m-1}B_{m-1}$ and $d_{m-1}\mu_{mm-1}$ are integers. The number d_{m-1} was computed as the nearest integer to $B_1 \ldots B_{m-1}$ and C was considered to be 0 if the floating-point approximations to B_m and μ_{mm-1} had absolute values less than $1/(2d_{m-1})$.

Using 15 or more decimal digits, the floating-point versions of MLLL produced bases for the lattice $S(B)$ generated by the rows of B. The bases depended on the precision chosen, but they were all very similar to the basis in A_1. In all cases, the largest entries in P had absolute values roughly 10^8. Since Maple and Mathematica have different ways of handling floating-point numbers, the Maple and Mathematica implementations sometimes produced different bases using the same precision. With 100 digits, Maple returned the matrix A_1 and Mathematica returned the matrix obtained by negating the 14-th row of A_1. In these computations Mathematica was faster, with times ranging from 6 to 10 minutes, depending on the precision. The corresponding Maple times were 16 to 29 minutes.

The rational-arithmetic version of MLLL for Maple took about 6 minutes to produce an LLL-reduced basis for $S(B)$ starting from the row Hermite normal form given earlier. Even including the time used by the Rosser strategy or KB_ROW, this way of obtaining an LLL-reduced basis was about three times as fast as letting the rational-arithmetic version of MLLL act directly on B. Of course the unimodular matrix produced had entries of absolute value much larger than 10^8.

The reader should not draw firm conclusions from this single example. A great deal of work remains to be done. For example, the number $\frac{3}{4}$ used in MLLL can be varied between $\frac{1}{4}$ and 1. If this value is changed, what are the effects on execution speed, the lengths of the basis vectors produced, and the size of the entries in the unimodular matrix? There are other ways of using the LLL algorithm. For example, suppose B is an m-by-n integer matrix and we have found by some method a unimodular matrix P such that PB has the form $\begin{bmatrix} C \\ 0 \end{bmatrix}$, where the rows of C form a basis for $S(B)$. Let r be the rank of B and write P as $\begin{bmatrix} Q \\ R \end{bmatrix}$, where Q and R have r and $n - r$ rows, respectively. Thus $QB = C$ and $RB = 0$. In fact, $S(R)$ is the set of all elements v in $\mathbb{Z}^m$ such that $vB = 0$. We can apply any row operations to R and we can add multiples of rows of R to rows of Q without affecting the product PB. Thus we can apply LLL to the rows of R and then replace each row u of Q by an almost-shortest element of the coset $S(R) + u$, as described at the end of Section 8.6. In this way we can hope to reduce the entries in P to manageable size.

8.9 Historical notes

Excluding the Euclidean algorithm, the oldest of the computational procedures described in this book are the techniques for manipulating integer matrices discussed in Sections 8.1 to 8.3. The papers (Hermite 1851) and

(Smith 1861) are usually cited as establishing the existence of normal forms
for integer matrices under row equivalence and equivalence, respectively.
However, both Hermite and Smith considered only square matrices with
nonzero determinant. The extensions to arbitrary integer matrices came
later. Moreover, Hermite includes no proof and gives the impression he
is stating a result which is common knowledge to his contemporaries. In
(Kronecker 1870), finite abelian groups were shown to be direct sums of
cyclic groups. The uniqueness of the orders of these cyclic groups under
the assumptions that they are greater than 1 and that each order divides
all larger ones was proved in (Frobenius & Stickelberger 1879). This paper established the connection between finite abelian groups and the work
of Smith. Results close to the modern statement of the fundamental theorem of finitely generated abelian groups are in (Tietze 1908). The first
polynomial-time algorithm for computing column Hermite normal form was
given in (Frumkin 1977), which considered only matrices whose rank is
equal to the number of columns. This is also the case which was considered in (Kannan & Bachem 1979), where it was shown that the entries in
the associated unimodular matrices have polynomial size. These results,
together with the modular techniques in (Havas & Sterling 1979), the LLL
lattice reduction algorithm in (Lenstra et al. 1982), and the modification
of the LLL algorithm in (Pohst 1987) were all important advances in our
efforts to study finitely presented abelian groups computationally.

9

Polycyclic groups

Polycyclic groups form a broad class of finitely presented groups in which extensive computation is possible. This chapter discusses the basic structure and properties of polycyclic groups and presents algorithms for computing with elements and subgroups of polycyclic groups. All finite solvable groups are polycyclic. The literature on algorithms for computing with finite solvable groups is too extensive to cover in detail here. See (Laue, Neubüser, & Shoenwaelder 1984), (Mecky & Neubüser 1989), and (Glasby & Slattery 1990). Emphasis will be placed on those algorithms which apply to infinite as well as finite polycyclic groups. Various computations with polycyclic groups have been shown to be possible in principle but have not yet been shown to be practical for interesting groups. Some of these algorithms will be mentioned, but details will not be given. By combining the rewriting techniques of Chapter 2 with the methods developed in this chapter, one obtains algorithms for solving a wide range of problems in *polycyclic-by-finite* groups. These are groups which have a polycyclic subgroup of finite index. The polycyclic-by-finite groups make up the largest class of finitely presented groups in which most computational problems concerning elements and subgroups have algorithmic solutions.

A word of caution is in order. In coset enumeration, right cosets of subgroups are used almost exclusively. However, in studying polycyclic groups it is traditional to use left cosets. It would be possible to present both subjects using the same type of cosets, but it seems best to remain consistent with other authors.

This chapter assumes that the reader has had a good introduction to the theory of groups. The material required is summarized, but for the most part proofs are omitted. This is particularly true for Sections 9.1 and 9.2, which review results about commutator subgroups and elementary properties of solvable and nilpotent groups. Details can be found in most standard texts on group theory.

9.1 Commutator subgroups

Let h and k be elements of a group G. The *commutator* of h and k is the element $[h,k] = h^{-1}k^{-1}hk$. The conjugate $h^k = k^{-1}hk$ of h by k is $h[h,k]$, and $hk = kh[h,k]$. Thus h and k commute if and only if their commutator is trivial. Suppose H and K are subgroups of G. The *commutator subgroup* of H and K is the group

$$[H,K] = \mathrm{Grp}\,\langle [h,k] \mid h \in H, k \in K \rangle.$$

Since $[k,h] = [h,k]^{-1}$, it follows that $[K,H] = [H,K]$. The *commutator subgroup* of G is $[G,G]$, which is also called the *derived subgroup* of G and denoted G'. The derived subgroup of G is trivial if and only if G is abelian.

Proposition 1.1. *The derived subgroup G' is normal in G and the quotient G/G' is abelian. If N is any normal subgroup of G such that G/N is abelian, then $N \supseteq G'$.*

Thus G/G' is the largest abelian quotient group of G. The following propositions present some more basic facts about commutator subgroups.

Proposition 1.2. *Suppose that H_1, K_1, H_2, and K_2 are subgroups of G such that $H_1 \subseteq H_2$ and $K_1 \subseteq K_2$. Then $[H_1, K_1] \subseteq [H_2, K_2]$.*

Proposition 1.3. *If $f\colon G \to Q$ is a homomorphism of groups, then $[f(H), f(K)] = f([H,K])$ for all subgroups H and K of G.*

Corollary 1.4. *Suppose that N is a normal subgroup of G and $^-$ is the natural homomorphism from G to G/N. Then $[\overline{H}, \overline{K}] = \overline{[H,K]}$.*

Proposition 1.5. *If H and K are normal subgroups of G, then $[H,K]$ is normal in G and $[H,K] \subseteq H \cap K$.*

If x, y, and z are elements of G, then $[x,y,z]$ is defined to be $[[x,y],z]$. In general, $[x_1, x_2, \ldots, x_n]$ is defined recursively to be $[[x_1, x_2, \ldots, x_{n-1}], x_n]$. Such commutators are said to be *left-normed*. If $H_1, \ldots, H_n$ are subgroups of G, then $[H_1, \ldots, H_n] = [[H_1, \ldots, H_{n-1}], H_n]$.

Proposition 1.6. *For any elements x, y, and z of G, the following hold:*

(a) $[xy, z] = [x, z][x, z, y][y, z]$.
(b) $[x, yz] = [x, z][x, y][x, y, z]$.
(c) $[x, y^{-1}, z]^y [y, z^{-1}, x]^z [z, x^{-1}, y]^x = 1$.

Proposition 1.7. *If* H, K, *and* L *are subgroups of* G *and* N *is a normal subgroup of* G *containing* $[K, L, H]$ *and* $[L, H, K]$, *then* N *contains* $[H, K, L]$.

Proof. In view of Corollary 1.4, we may pass to the quotient G/N and assume that N is trivial. Thus it suffices to prove that $[H, K, L] = 1$ whenever $[K, L, H] = [L, H, K] = 1$.

To show that $[H, K, L] = 1$, we must prove that each element of $[H, K]$ commutes with each element of L. For this it suffices to prove that each element in a generating set for $[H, K]$ commutes with each element of L. Thus it is enough to show that $[x, y, z] = 1$ for all x in H, y in K, and z in L. By Proposition 1.6(c),

$$[x, y, z]^{y^{-1}} [y^{-1}, z^{-1}, x]^z [z, x^{-1}, y^{-1}]^x = 1.$$

But $[y^{-1}, z^{-1}, x]$ is in $[K, L, H]$ and $[z, x^{-1}, y^{-1}]$ is in $[L, H, K]$. Therefore

$$[y^{-1}, z^{-1}, x]^z [z, x^{-1}, y^{-1}]^x = 1,$$

and hence $[x, y, z]^{y^{-1}} = 1$. Conjugating by y, we obtain $[x, y, z] = 1$. $\square$

Proposition 1.7 is sometimes called the three subgroups lemma.

There are many ways in which one could form "higher commutator subgroups" in G. Examples of such subgroups are $(G')' = [G', G'] = [[G, G], [G, G]]$ and $[G', G] = [G, G, G]$. Two sequences of these higher commutator subgroups have been found to be particularly useful. The *derived series* of G is obtained by taking successive derived subgroups. Thus we have $G^{(0)} = G$, $G^{(1)} = G'$, $G^{(2)} = (G')'$, and, in general,

$$G^{(i+1)} = (G^{(i)})' = [G^{(i)}, G^{(i)}].$$

The *lower central series* of G is defined by taking successive commutator subgroups with G. Here $\gamma_1(G) = G$, $\gamma_2(G) = G' = G^{(1)}$, and, in general,

$$\gamma_{i+1}(G) = [\gamma_i(G), G].$$

By Proposition 1.5, all of the terms in the derived series and the lower central series are normal in G. In addition, $G^{(0)} \supseteq G^{(1)} \supseteq G^{(2)} \supseteq \cdots$ and $\gamma_1(G) \supseteq \gamma_2(G) \supseteq \gamma_3(G) \supseteq \cdots$.

Proposition 1.8. *Suppose that* $f: G \to Q$ *is a homomorphism of groups. Then* $f(G)^{(i)} = f(G^{(i)})$ *and* $\gamma_i(f(G)) = f(\gamma_i(G))$.

Proof. Induction and Proposition 1.3. $\square$

Proposition 1.9. *If H is a subgroup of G, then $H^{(i)} \subseteq G^{(i)}$ and $\gamma_i(H) \subseteq \gamma_i(G)$.*

Proof. Induction and Proposition 1.2. □

Proposition 1.10. *For all $i \geq 1$ and $j \geq 1$, $[\gamma_i(G), \gamma_j(G)] \subseteq \gamma_{i+j}(G)$.*

Proof. For $j = 1$ we have the definition of $\gamma_{i+1}(G)$. We proceed by induction on j. Since

$$[\gamma_i(G), \gamma_j(G)] = [\gamma_i(G), [\gamma_{j-1}(G), G]] = [\gamma_{j-1}(G), G, \gamma_i(G)],$$

by Proposition 1.7 it suffices to prove that $[G, \gamma_i(G), \gamma_{j-1}(G)]$ and $[\gamma_i(G), \gamma_{j-1}(G), G]$ are contained in $\gamma_{i+j}(G)$. But $[G, \gamma_i(G), \gamma_{j-1}(G)] = [\gamma_{i+1}(G), \gamma_{j-1}(G)]$ is contained in $\gamma_{i+j}(G)$ by induction. Also by induction, $[\gamma_i(G), \gamma_{j-1}(G)] \subseteq \gamma_{i+j-1}(G)$. Therefore

$$[\gamma_i(G), \gamma_{j-1}(G), G] \subseteq [\gamma_{i+j-1}(G), G] = \gamma_{i+j}(G). \qquad \square$$

Corollary 1.11. $G^{(i)} \subseteq \gamma_{2^i}(G)$ *for $i \geq 0$.*

Proof. For $i = 0$ we have $G^{(0)} = G = \gamma_1(G)$. If $i > 1$, then

$$G^{(i)} = [G^{(i-1)}, G^{(i-1)}] \subseteq [\gamma_{2^{i-1}}(G), \gamma_{2^{i-1}}(G)] \subseteq \gamma_{2^{i-1}+2^{i-1}}(G) = \gamma_{2^i}(G). \quad \square$$

Exercises

1.1. Suppose that G is a group generated by a set X. Show that $G' = \gamma_2(G)$ is the normal closure in G of the set $\{[x, y] \mid x, y \in X\}$.

1.2. Generalize Exercise 1.1 by showing that $\gamma_s(G)$ is the normal closure in G of the set $\{[x_1, \ldots, x_s] \mid x_i \in X, 1 \leq i \leq s\}$ for $s \geq 2$.

1.3. Let F be the free group on $X = \{a, b\}$, $a \neq b$. Show that every commutator $[[x_1, x_2], [x_3, x_4]]$ with each x_i in X is trivial but that $F^{(2)}$ is not trivial.

9.2 Solvable and nilpotent groups

Let G be a group. We say that G is *solvable* if $G^{(i)}$ is trivial for some $i \geq 0$. If G is solvable, then the smallest value of i for which $G^{(i)} = 1$ is called the *derived length* of G. Groups of order 1 have derived length 0. Nontrivial abelian groups have derived length 1. The derived length of G is 2 if and only if G' is nontrivial and abelian. A group with derived length at most 2 is called *metabelian*.

Our group G is *nilpotent* if some term in the lower central series is trivial. In this case, the smallest integer c such that $\gamma_{c+1}(G) = 1$ is called the *nilpotency class*, or simply the *class*, of G. Trivial groups have class 0 and nontrivial abelian groups have class 1.

Proposition 2.1. *Subgroups and quotient groups of solvable groups are solvable. Subgroups and quotient groups of nilpotent groups are nilpotent.*

Proof. Let H and N be subgroups of G with N normal. Suppose $G^{(i)} = 1$. Then $H^{(i)} = 1$ by Proposition 1.9 and $(G/N)^{(i)} = 1$ by Proposition 1.8. Similarly, if $\gamma_j(G) = 1$, then $\gamma_j(H) = 1$ and $\gamma_j(G/N) = 1$. □

Proposition 2.2. *The group $G/G^{(i)}$ is solvable with derived length at most i. The group $G/\gamma_j(G)$ is nilpotent of class at most $j - 1$.*

Proof. By Proposition 1.9, $(G/G^{(i)})^{(i)} = G^{(i)}/G^{(i)} = 1$. Similarly $\gamma_j(G/\gamma_j(G)) = \gamma_j(G)/\gamma_j(G) = 1$. □

Proposition 2.3. *If N is a normal subgroup of G and both N and G/N are solvable, then G is solvable.*

Proof. Since G/N is solvable, there is an integer i such that $(G/N)^{(i)} = (G^{(i)}N)/N = 1$. This means that $G^{(i)} \subseteq N$. There is an integer j such that $N^{(j)} = 1$. Hence $G^{(i+j)} = (G^{(i)})^{(j)} \subseteq N^{(j)} = 1$. □

At this point nilpotent groups differ from solvable groups. Proposition 2.3 is false if "nilpotent" is substituted for "solvable". Let G be the symmetric group Sym(3), which has order 6, and let N be the alternating subgroup of G. Then N is generated by the 3-cycle (1,2,3). Both N and G/N are abelian and hence nilpotent. Now $[(1,2,3),(1,2)] = (1,2,3)$. Therefore $N = G' = [N,G]$. It follows easily that $N = \gamma_i(G)$ for all $i \geq 2$. Hence G is not nilpotent.

Proposition 2.4. *Nilpotent groups are solvable.*

Proof. Suppose $\gamma_j(G) = 1$. By Corollary 1.11, $G^{(i)} = 1$ provided $2^i \geq j$.

□

The symmetric group of degree 3 is an example of a solvable group which is not nilpotent.

Example 2.1. Let R be a commutative ring with $1 \neq 0$ and let n be a positive integer. For $r \geq 1$ let $U_n^{(r)}(R)$ consist of those n-by-n matrices A over R such that $A_{ij} = 0$ if $j < i+r$. Thus A is in $U_n^{(r)}(R)$ if and only if all entries on or below the main diagonal are 0 and all entries on the first $r-1$ diagonals above the main diagonal are also 0. Thus an element of $U_4^{(2)}(R)$ has the form

$$
\begin{bmatrix}
0 & 0 & * & * \\
0 & 0 & 0 & * \\
0 & 0 & 0 & 0 \\
0 & 0 & 0 & 0
\end{bmatrix},
$$

where $*$ denotes any element of R. If $r \geq n$, then $U_n^{(r)}(R)$ contains only the 0 matrix. Suppose that A is in $U_n^{(r)}(R)$ and B is in $U_n^{(s)}(R)$. A simple argument shows that AB is in $U_n^{(r+s)}(R)$. In particular, $A^n = 0$. Let $D_n^{(r)}(R)$ consist of all matrices $I + A$, where I is the n-by-n identity matrix and A is in $U_n^{(r)}(R)$. Since

$$
(I + A)(I + B) = I + A + B + AB
$$

and

$$
(I + A)^{-1} = I - A + A^2 - \cdots + (-A)^{n-1},
$$

the sets $D_n^{(r)}(R)$ are subgroups of $\mathrm{GL}(n, R)$. If A is in $U_n^{(r)}(R)$ and B is in $U_n^{(s)}(R)$, then the commutator $[I + A, I + B]$ is

$$
(I - A + \cdots + (-A)^{n-1})(I - B + \cdots + (-B)^{n-1})(I + A)(I + B).
$$

Direct computation shows that this matrix has the form $I + C$, where C is in $U_n^{(r+s)}$. Thus $[D_n^{(r)}(R), D_n^{(s)}(R)]$ is contained in $D_n^{(r+s)}(R)$. Therefore, if $D = D_n^{(1)}(R)$, then $\gamma_r(D) \subseteq D_n^{(r)}(R)$. In particular, $\gamma_n(D)$ is trivial, so D is nilpotent of class at most $n - 1$.

The *quotients* of the derived series of a group G are the groups $G^{(i)}/G^{(i+1)}$, $i = 0, 1, \ldots$. They are all abelian groups. The quotients of the lower central series of G are the groups $\gamma_i(G)/\gamma_{i+1}(G), i = 1, 2, \ldots$. These groups are also abelian. In fact $\gamma_i(G)/\gamma_{i+1}(G)$ is in the center of $G/\gamma_{i+1}(G)$.

If G is finitely generated, then $G/G' = G/\gamma_2(G)$ is finitely generated. However, the remaining quotients in the derived series may not be finitely generated, as the following example illustrates.

Example 2.2. Let x be the permutation of the integers which maps i to $i+2$ for all i. Thus x has "cycles" $(\ldots, -3, -1, 1, 3, 5, \ldots)(\ldots, -4, -2, 0, 2, 4, \ldots)$. Let $y = (0, 1)$ and let G be the group of permutations of the integers generated by x and y. The conjugates of y by powers of x are the 2-cycles $y_i = (2i, 2i+1), i = 0, \pm 1, \ldots$. The subgroup N generated by the y_i is abelian and normal in G. The commutator $[y, x]$ is $z = (0, 1)(2, 3)$. The conjugates of z by powers of x are the elements $z_i = (2i, 2i+1)(2i+2, 2i+3) = y_i y_{i+1}$. The subgroup M generated by the z_i is normal. By Exercise 1.1, $M = G'$. Each element of M moves at most a finite number of integers. Therefore any finitely generated subgroup of M moves only finitely many integers. However, M moves all the integers, so M is not finitely generated. Since N is abelian, so is M. Therefore $G'' = 1$ and G'/G'' is not finitely generated.

Again we have a difference between solvable and nilpotent groups.

Proposition 2.5. *If G is generated modulo G' by $x_1, \ldots, x_n$, then $\gamma_2(G)/\gamma_3(G)$ is generated by the images of $[x_j, x_i]$ with $1 \le i < j \le n$.*

Proof. We may work in $G/\gamma_3(G)$ and so we may assume $\gamma_3(G) = 1$. Thus elements of G' are in the center of G. By Proposition 1.6, $[xy, z] = [x, z][y, z]$ and $[x, yz] = [x, y][x, z]$. By induction one concludes that

$$[y_1 \cdots y_s, z_1 \cdots z_t] = \prod_{i,j} [y_i, z_j].$$

Now $1 = [xx^{-1}, y] = [x, y][x^{-1}, y]$, so $[x^{-1}, y] = [x, y]^{-1}$. Similarly, $[x, y^{-1}] = [x, y]^{-1}$ and $[x^{-1}, y^{-1}] = [x, y]$. It follows that the commutator of any two elements of $H = \mathrm{Grp}\langle x_1, \ldots, x_n \rangle$ is in the subgroup generated by the commutators $[x_j, x_i]$. Since $[x_i, x_i] = 1$ and $[x_i, x_j] = [x_j, x_i]^{-1}$, we may assume that $i < j$. By assumption, any element of G can be written as uz, where u is in H and z is in G'. If v is also in H and w is in G', then $[uz, vw] = [u, v]$. Therefore $G' = H'$. □

Proposition 2.6. *Suppose G is a group generated modulo G' by a set X and Y is a subset of $\gamma_i(G)$, $i \ge 2$, whose image in $\gamma_i(G)/\gamma_{i+1}(G)$ generates that group. Then $\gamma_{i+1}(G)/\gamma_{i+2}(G)$ is generated by the image of $Z = \{[y, x] \mid y \in Y, x \in X\}$.*

Proof. We may assume that $\gamma_{i+2}(G) = 1$. All of the elements of Z are in $\gamma_{i+1}(G)$. If u is in $\gamma_i(G)$ and v and w are in G, then by Proposition 1.6, $[u, vw] = [u, w][u, v][u, v, w]$. But $[u, v, w]$ is in $\gamma_{i+2}(G)$ and hence $[u, v, w]$ is trivial. Also $[[u, w], [u, v]]$ is in $\gamma_{2i+2}(G)$ and so is trivial. Therefore $[u, vw] = [u, v][u, w]$. If w is in G', then $[u, w]$ is in $\gamma_{i+2}(G)$ and so $[u, vw] = [u, v]$. By a similar argument one shows that $[uv, w] = [u, w][v, w]$ whenever u and v are in $\gamma_i(G)$ and w is in G. If v is in $\gamma_{i+1}(G)$, then $[uv, w] = [u, w]$. In addition $[u^{-1}, v] = [u, v]^{-1} = [u, v^{-1}]$ if u is in $\gamma_i(G)$ and v is in G. By an argument similar to the one in the proof of Proposition 2.5, $\gamma_{i+1}(G)$ is generated by the commutators $[u, v]$, where u ranges over a set of generators of $\gamma_i(G)$ modulo $\gamma_{i+1}(G)$ and v ranges over a set of generators of G modulo G'. □

Corollary 2.7. *If a group G is generated by n elements and $i \geq 2$, then $\gamma_i(G)/\gamma_{i+1}(G)$ is generated by $(n-1)n^{i-1}/2$ elements.*

Proof. The case $i = 2$ follows from Proposition 2.5. Now induction on i and Proposition 2.6 complete the proof. □

The bound in Corollary 2.7 can be improved somewhat. See Section 9.9.

Proposition 2.8. *If N is a subgroup of the center of G and G/N is nilpotent, then G is nilpotent.*

Proof. By assumption, $\gamma_i(G)$ is contained in N for some i. Since N is central, $[N, G] = 1$. But $\gamma_{i+1}(G) = [\gamma_i(G), G]$ is contained in $[N, G] = 1$, so $\gamma_{i+1}(G) = 1$. □

Exercises

2.1. Suppose that G is a finitely generated nilpotent group. Prove that all of the terms in the lower central series of G are finitely generated.

2.2. Let G be a nilpotent group and let X be a subset of G whose image in G/G' generates G/G'. Show that X generates G.

2.3. Let $D_n^{(r)}(R)$ be as in Example 2.1. Show that $[D_n^{(r)}(R), D_n^{(s)}(R)] = D_n^{(r+s)}(R)$.

9.3 Polycyclic groups

The class of polycyclic groups contains the class of finitely generated nilpotent groups and is contained in the class of finitely generated solvable groups. Polycyclic groups have finite presentations of a form which makes many types of computations practical.

Let G be a group. A *polycyclic series* of length n for G is a sequence of subgroups

$$G = G_1 \supseteq G_2 \supseteq \cdots \supseteq G_n \supseteq G_{n+1} = 1$$

such that for $1 \leq i \leq n$ the subgroup G_{i+1} is normal in G_i and G_i/G_{i+1} is cyclic. Note that we do not require that each G_i be normal in G. A group is *polycyclic* if it has a polycyclic series.

Proposition 3.1. *Polycyclic groups are solvable.*

Proof. Let $G = G_1 \supseteq G_2 \supseteq \cdots \supseteq G_n \supseteq G_{n+1} = 1$ be a polycyclic series for a group G. We proceed by induction on n. If $n = 0$, then G is trivial and hence solvable. Assume that $n > 0$. Then G_2 is normal in G and G/G_2 is cyclic and therefore solvable. The sequence $G_2 \supseteq \cdots \supseteq G_n \supseteq G_{n+1} = 1$ is a polycyclic series for G_2 of length $n - 1$. By induction, G_2 is solvable. Thus G is solvable by Proposition 2.3. □

Proposition 3.2. *Finitely generated abelian groups are polycyclic.*

Proof. Let G be an abelian group generated by $a_1, \ldots, a_n$. Set $G_i = \mathrm{Grp}\langle a_i, \ldots, a_n \rangle$, $1 \leq i \leq n+1$. Then G_{i+1} is contained in G_i and G_{i+1} is normal in G, so G_{i+1} is certainly normal in G_i. The quotient G_i/G_{i+1} is generated by the coset $a_i G_{i+1}$. Therefore G_i/G_{i+1} is cyclic and the G_i form a polycyclic series for G. $\square$

Proposition 3.3. *If N is a normal subgroup of a group G and both N and G/N are polycyclic, then G is polycyclic.*

Proof. A subgroup of G/N has the form H/N, where H is a subgroup of G containing N. If K is another subgroup of G containing N, then $K/N \subseteq H/N$ if and only if $K \subseteq H$, and K/N is normal in H/N if and only if K is normal in H. In this case, $(H/N)/(K/N)$ is isomorphic to H/K. Thus we can pull back a polycyclic series of G/N to a sequence of subgroups of G from G to N such that each subgroup is normal in the preceding one and the quotients are cyclic. Following this by a polycyclic series for N, we get a polycyclic series for G. $\square$

Proposition 3.4. *Finitely generated nilpotent groups are polycyclic.*

Proof. Let G be a finitely generated nilpotent group of class c. If $c \leq 1$, then G is abelian, and hence G is polycyclic by Proposition 3.2. Assume that $c > 1$. The last nontrivial term in the lower central series of G is $\gamma_c(G)$. By Proposition 1.10 and Corollary 2.7, $\gamma_c(G)$ is abelian and finitely generated. Therefore $\gamma_c(G)$ is polycyclic. The quotient $G/\gamma_c(G)$ is nilpotent of class $c-1$. By induction on c, $G/\gamma_c(G)$ is polycyclic. Thus G is polycyclic by Proposition 3.3. $\square$

Proposition 3.5. *If G is a group with a polycyclic series of length n, then G can be generated by n elements.*

Proof. Let $G = G_1 \supseteq \cdots \supseteq G_{n+1} = 1$ be a polycyclic series. For $1 \leq i \leq n$, let a_i be an element of G_i such that $a_i G_{i+1}$ generates G_i/G_{i+1}. Then every coset of G_{i+1} in G_i contains a power of a_i. Thus if g is in G, then $g = a_1^{\alpha_1} g_2$, where g_2 is in G_2. But $g_2 = a_2^{\alpha_2} g_3$, where g_3 is in G_3. Therefore $g = a_1^{\alpha_1} a_2^{\alpha_2} g_3$. Continuing in this manner, we find that $g = a_1^{\alpha_1} \ldots a_n^{\alpha_n} g_{n+1}$, where g_{n+1} is in G_{n+1}. But $G_{n+1} = 1$, so $g = a_1^{\alpha_1} \ldots a_n^{\alpha_n}$. Hence G is generated by $a_1, \ldots, a_n$. $\square$

Proposition 3.6. *Quotient groups of polycyclic groups are polycyclic.*

Proof. Let $G = G_1 \supseteq \cdots \supseteq G_{n+1} = 1$ be a polycyclic series for a group G and let N be a normal subgroup of G. For $1 \leq i \leq n+1$, the product $G_i N$

is a subgroup of G and $G_{i+1}N$ is normal in G_iN if $i \leq n$. Also

$$G_iN/G_{i+1}N \cong G_i/G_{i+1}(G_i \cap N) \cong (G_i/G_{i+1})/(G_{i+1}(G_i \cap N)/G_{i+1})$$

is a quotient of the cyclic group G_i/G_{i+1}. Thus $G_iN/G_{i+1}N$ is cyclic.

Define H_i to be G_iN/N. Then $G/N = H_1 \supseteq \cdots \supseteq H_{n+1} = N/N = 1$. Moreover, H_{i+1} is normal in H_i and $H_i/H_{i+1} \cong G_iN/G_{i+1}N$ is cyclic. Therefore G/N is polycyclic. $\square$

Proposition 3.7. *Subgroups of polycyclic groups are polycyclic.*

Proof. Let $G = G_1 \supseteq \cdots \supseteq G_{n+1} = 1$ be a polycyclic series for G and let H be a subgroup of G. Set $H_i = G_i \cap H$. It is easy to check that H_{i+1} is a normal subgroup of H_i. By the second isomorphism theorem,

$$H_i/H_{i+1} \cong H_i/(H_i \cap G_{i+1}) \cong (H_iG_{i+1})/G_{i+1},$$

which is a subgroup of the cyclic group G_i/G_{i+1}. Therefore $H_1 \supseteq \cdots \supseteq H_{n+1} = 1$ is a polycyclic series for H. $\square$

Corollary 3.8. *If G has a polycyclic series of length n, then every subgroup of G can be generated by n or fewer elements.*

Proof. The proof of Proposition 3.7 showed that every subgroup of G has a polycyclic series of length n. Thus Proposition 3.5 applies. $\square$

The following characterization of polycyclic groups is one of the main reasons that extensive computation in polycyclic groups is possible.

Proposition 3.9. *A group is polycyclic if and only if it is solvable and all subgroups are finitely generated.*

Proof. Let G be a group. We have already shown that G polycyclic implies that G is solvable and that subgroups are finitely generated. Now suppose that G is solvable and all subgroups of G are finitely generated. Let G have derived length k. If $k \leq 1$, then G is abelian and finitely generated. Therefore G is polycyclic by Proposition 3.2. Suppose $k > 1$. Then $G^{(k-1)}$ is a finitely generated abelian normal subgroup of G and $G/G^{(k-1)}$ has derived length $k - 1$. Subgroups of $G/G^{(k-1)}$ are images of subgroups of G and hence are finitely generated. By induction on k, $G/G^{(k-1)}$ is polycyclic. Therefore G is polycyclic by Proposition 3.3. $\square$

Example 3.1. Let D be the group $D_4^{(1)}(\mathbb{Z})$ defined in Example 2.1. Thus D is the subgroup of $\mathrm{GL}(4, \mathbb{Z})$ consisting of all matrices

$$A = \begin{bmatrix} 1 & x_1 & x_4 & x_6 \\ 0 & 1 & x_2 & x_5 \\ 0 & 0 & 1 & x_3 \\ 0 & 0 & 0 & 1 \end{bmatrix},$$

where the x_i are integers. If

$$B = \begin{bmatrix} 1 & y_1 & y_4 & y_6 \\ 0 & 1 & y_2 & y_5 \\ 0 & 0 & 1 & y_3 \\ 0 & 0 & 0 & 1 \end{bmatrix},$$

is another element of D, then AB has the form

$$\begin{bmatrix} 1 & x_1 + y_1 & * & * \\ 0 & 1 & x_2 + y_2 & * \\ 0 & 0 & 1 & x_3 + y_3 \\ 0 & 0 & 0 & 1 \end{bmatrix}.$$

Thus the map from D to $\mathbb{Z}^3$ taking A to (x_1, x_2, x_3) is a surjective homomorphism with kernel $D_4^{(2)}(\mathbb{Z})$. Similarly, if A is in $D_4^{(2)}(\mathbb{Z})$, then mapping A to (x_4, x_5) defines a homomorphism of $D_4^{(2)}(\mathbb{Z})$ onto $\mathbb{Z}^2$ with kernel $D_4^{(3)}(\mathbb{Z})$. Finally, $D_4^{(3)}(\mathbb{Z})$ is generated by

$$\begin{bmatrix} 1 & 0 & 0 & 1 \\ 0 & 1 & 0 & 0 \\ 0 & 0 & 1 & 0 \\ 0 & 0 & 0 & 1 \end{bmatrix}$$

and hence is cyclic. By Proposition 3.3, $D_4^{(2)}(\mathbb{Z})$ is polycyclic. Using the same proposition again, we see that D is polycyclic. If E_i consists of the matrices A given earlier with $x_1 = \cdots = x_{i-1} = 0$, then $D = E_1 \supset E_2 \supset \cdots \supset E_7 = 1$ is a polycyclic series for D.

A group G is *hopfian* if whenever $f: G \to G$ is a surjective homomorphism, then f is an isomorphism. This is equivalent to saying that if N is a normal subgroup of G such that G/N is isomorphic to G, then $N = 1$.

Proposition 3.10. *If G is a group which satisfies the ascending chain condition on subgroups, then G is hopfian.*

Proof. Let $f : G \to G$ be a surjective homomorphism such that the kernel N of f is nontrivial. For $i \geq 1$, the i-fold composition f^i of f with itself is a homomorphism of G onto itself. Set N_i equal to the kernel of f^i. Then N_i is the inverse image of 1 under f^i and N_{i+1} is the inverse image of N under f^i. Since N is nontrivial and f^i is surjective, N_{i+1} properly contains N_i. Therefore $N_1 \subset N_2 \subset \dots$ is a strictly increasing, infinite sequence of subgroups of G. This is impossible by the ascending chain condition. Therefore G is hopfian. $\square$

Corollary 3.11. *Polycyclic groups are hopfian.*

Proof. By Corollary 3.8, subgroups of polycyclic groups are finitely generated, so the ascending chain condition holds.

$\square$

Proposition 3.12. *Suppose that H and K are subgroups of a polycyclic group G. If $H \subseteq K$ and H is conjugate to K in G, then $H = K$.*

Proof. Suppose that $H \subset K = H^g$, where g is in G. Conjugating repeatedly by g, we see that the sequence $H \subset H^g \subset H^{g^2} \subset \cdots$ is strictly increasing. This cannot happen in a polycyclic group. Thus $H = K$. $\square$

Let G be a polycyclic group. Not all polycyclic series for G have the same length. However, the number of infinite quotients in a polycyclic series is the same for all series. This number is called the *Hirsch number* of G. It is possible to choose the polycyclic series so that all the infinite factors come after the finite factors. See Proposition 2 in Chapter 1 of [Segal 1983].

Exercises

3.1. Show that the order of a finite subgroup of a polycyclic group G divides the product of the orders of the finite quotients in any polycyclic series for G.

3.2. Generalize the discussion in Example 3.1 to $D_n^{(1)}(\mathbb{Z})$ for any $n > 1$.

9.4 Polycyclic presentations

Let $G = G_1 \supseteq \cdots \supseteq G_{n+1} = 1$ be a polycyclic series for a group G. For $1 \leq i \leq n$, let a_i be an element of G_i whose image in G_i/G_{i+1} generates that group. The sequence $a_1, \dots, a_n$ will be called a *polycyclic generating sequence* for G. Note that the order is important. (For finite solvable groups, the term *AG-system* was introduced in (Jürgensen 1970).) By the proof of Proposition 3.5, $G_i = \mathrm{Grp}\,\langle a_i, \ldots, a_n \rangle$ and every element g of G_i can be expressed in the form $a_i^{\alpha_i} \ldots a_n^{\alpha_n}$, where the exponents α_j are integers. Let $I = I(a_1, \ldots, a_n)$ denote the set of subscripts i such that G_i/G_{i+1} is finite, and let $m_i = |G_i : G_{i+1}|$, the order of a_i relative to G_{i+1}, if i is in I. We shall normally assume that the generating sequence is not redundant in the sense that no a_i is in G_{i+1}. Thus $m_i > 1$ for each i in I. We shall say that the expression for g is a *collected word* if $0 \le \alpha_j < m_j$ for j in I. Each element of G can be described by a unique collected word in the generators $a_1, \ldots, a_n$. If $a_1^{\alpha_1} \ldots a_n^{\alpha_n}$ is the collected word representing a nontrivial element g of G and α_i is the first nonzero exponent, then $a_i^{\alpha_i}$ and α_i will be called the *leading term* and the *leading exponent* of g, respectively.

Suppose i is in I. Then $a_i^{m_i}$ is in G_{i+1} and can be expressed as a collected word in the generators $a_{i+1}, \ldots, a_n$. The collected word representing a_i^{-1} has the form $a_i^{m_i - 1} u$, where u is in G_{i+1}. Thus a_i^{-1} can be eliminated from any word representing an element of G. Now suppose that $1 \le i < j \le n$. Then a_j is in G_{i+1}, which is normal in G_i. Thus the conjugate $a_i^{-1} a_j a_i$ is in G_{i+1}, so $a_j a_i$ can be expressed as the product of a_i and a collected word involving $a_{i+1}, \ldots, a_n$. Similarly, $a_j^{-1} a_i$ can be expressed in this form as well, although this is necessary only if j is not in I. Thus there are unique relations

$$
\begin{aligned}
a_j a_i &= a_i a_{i+1}^{\alpha_{iji+1}} \ldots a_n^{\alpha_{ijn}}, & j &> i, \\
a_j^{-1} a_i &= a_i a_{i+1}^{\beta_{iji+1}} \ldots a_n^{\beta_{ijn}}, & j &> i,\ j \notin I, \\
a_j a_i^{-1} &= a_i^{-1} a_{i+1}^{\gamma_{iji+1}} \ldots a_n^{\gamma_{ijn}}, & j &> i,\ i \notin I, \\
a_j^{-1} a_i^{-1} &= a_i^{-1} a_{i+1}^{\delta_{iji+1}} \ldots a_n^{\delta_{ijn}}, & j &> i,\ i,j \notin I, \\
a_i^{m_i} &= a_{i+1}^{\mu_{ii+1}} \ldots a_n^{\mu_{in}}, & i &\in I, \\
a_i^{-1} &= a_i^{m_i - 1} a_{i+1}^{\nu_{ii+1}} \ldots a_n^{\nu_{in}}, & i &\in I,
\end{aligned}
\tag{$*$}
$$

where the right sides are collected words. The relations $(*)$ constitute a group presentation for G, the *standard polycyclic presentation* relative to $a_1, \ldots, a_n$. If i is in I, then a_i^{-1} occurs only in one relation, the one giving the collected form of a_i^{-1}. We can get a monoid presentation for G in terms of the a_i and the a_i^{-1} with i not in I by adding the relations $a_i a_i^{-1} = a_i^{-1} a_i = 1$ for i not in I and deleting the relation with left side a_i^{-1} if i is in I. The result will be called the *standard monoid polycyclic presentation*. However, keeping all of the a_i^{-1} is often useful since it facilitates the computation of inverses of elements in G.

Let $X = \{a_1, \ldots, a_n\}$. Interpreted as pairs of words in $X^{\pm *}$, the relations $(*)$ are rewriting rules with respect to the basic wreath-product ordering with $a_n \prec a_n^{-1} \prec \cdots \prec a_1 \prec a_1^{-1}$. Let $\mathcal{R}$ be the set of these rules, together with the monoid rules $a_i a_i^{-1} \to \varepsilon$ and $a_i^{-1} a_i \to \varepsilon$ with i not in I. Using

$\mathcal{R}$, we can rewrite any word in $X^{\pm *}$ into collected form. Since collected forms are unique, the rewriting system is confluent. Since a finite, confluent rewriting system exists, if we start with any monoid presentation for G on the monoid generators in $X^{\pm}$, then, using the ordering $\prec$, the Knuth-Bendix procedure for strings will construct $\mathcal{R}$, which will be called the *standard polycyclic rewriting system* for G relative to the polycyclic generating sequence $a_1, \ldots, a_n$.

Any group presentation of the form $(*)$ defines a polycyclic group G. However, the order of G_i/G_{i+1} may be finite even if i is not in I, and the order of G_i/G_{i+1} may be less than m_i when i is in I. If a presentation $(*)$ is the standard polycyclic presentation for the group it defines, then the presentation is said to be *consistent*. In this context, "consistent" and "confluent" are essentially synonyms.

Example 4.1. Let D be the group $D_4^{(1)}(\mathbb{Z})$ studied in Example 3.1. Define $a_1, \ldots, a_6$ as follows:

$$
a_1 = \begin{bmatrix} 1 & 1 & 0 & 0 \\ 0 & 1 & 0 & 0 \\ 0 & 0 & 1 & 0 \\ 0 & 0 & 0 & 1 \end{bmatrix}, \quad
a_2 = \begin{bmatrix} 1 & 0 & 0 & 0 \\ 0 & 1 & 1 & 0 \\ 0 & 0 & 1 & 0 \\ 0 & 0 & 0 & 1 \end{bmatrix}, \quad
a_3 = \begin{bmatrix} 1 & 0 & 0 & 0 \\ 0 & 1 & 0 & 0 \\ 0 & 0 & 1 & 1 \\ 0 & 0 & 0 & 1 \end{bmatrix},
$$

$$
a_4 = \begin{bmatrix} 1 & 0 & 1 & 0 \\ 0 & 1 & 0 & 0 \\ 0 & 0 & 1 & 0 \\ 0 & 0 & 0 & 1 \end{bmatrix}, \quad
a_5 = \begin{bmatrix} 1 & 0 & 0 & 0 \\ 0 & 1 & 0 & 1 \\ 0 & 0 & 1 & 0 \\ 0 & 0 & 0 & 1 \end{bmatrix}, \quad
a_6 = \begin{bmatrix} 1 & 0 & 0 & 1 \\ 0 & 1 & 0 & 0 \\ 0 & 0 & 1 & 0 \\ 0 & 0 & 0 & 1 \end{bmatrix}.
$$

It is easy to check that $a_1, \ldots, a_6$ is a polycyclic generating sequence for D. The set I is empty.

Given an element of D, it is not difficult to determine the collected word $a_1^{\alpha_1} \ldots a_6^{\alpha_6}$ which represents the element. For example, let

$$
u = \begin{bmatrix} 1 & 2 & 0 & -2 \\ 0 & 1 & -1 & 3 \\ 0 & 0 & 1 & 1 \\ 0 & 0 & 0 & 1 \end{bmatrix}.
$$

Then α_1, α_2, and α_3 are the entries of u just above the main diagonal. That is, $\alpha_1 = 2$, $\alpha_2 = -1$, and $\alpha_3 = 1$. Multiplying u on the left by $(a_1^2 a_2^{-1} a_3)^{-1}$

gives

$$\begin{bmatrix} 1 & 0 & 2 & -8 \\ 0 & 1 & 0 & 4 \\ 0 & 0 & 1 & 0 \\ 0 & 0 & 0 & 1 \end{bmatrix}.$$

The entries on the second diagonal above the main diagonal give $\alpha_4 = 2$ and $\alpha_5 = 4$. Multiplying on the left by $(a_4^2 a_5^4)^{-1}$ yields

$$\begin{bmatrix} 1 & 0 & 0 & -8 \\ 0 & 1 & 0 & 0 \\ 0 & 0 & 1 & 0 \\ 0 & 0 & 0 & 1 \end{bmatrix},$$

from which we see that $\alpha_6 = -8$. Thus $u = a_1^2 a_2^{-1} a_3 a_4^2 a_5^4 a_6^{-8}$.

Using this technique, we can determine the standard polycyclic presentation for D with respect to $a_1, \ldots, a_6$. It consists of 60 relations. In the following description, α and β range independently over $\{1, -1\}$.

$$a_6^{\alpha} a_i^{\beta} = a_i^{\beta} a_6^{\alpha}, \quad 1 \le i \le 5,$$
$$a_5^{\alpha} a_1^{\beta} = a_1^{\beta} a_5^{\alpha} a_6^{-\alpha\beta},$$
$$a_5^{\alpha} a_i^{\beta} = a_i^{\beta} a_5^{\alpha}, \quad 2 \le i \le 4,$$
$$a_4^{\alpha} a_3^{\beta} = a_3^{\beta} a_4^{\alpha} a_6^{\alpha\beta},$$
$$a_4^{\alpha} a_i^{\beta} = a_i^{\beta} a_4^{\alpha}, \quad 1 \le i \le 2,$$
$$a_3^{\alpha} a_2^{\beta} = a_2^{\beta} a_3^{\alpha} a_5^{-\alpha\beta},$$
$$a_3^{\alpha} a_1^{\beta} = a_1^{\beta} a_3^{\alpha},$$
$$a_2^{\alpha} a_1^{\beta} = a_1^{\beta} a_2^{\alpha} a_4^{-\alpha\beta}.$$

In order to recognize that $a_1, \ldots, a_n$ is a polycyclic generating sequence for a group G, we do not have to be given the entire standard polycyclic presentation for G. It would be nice to be able to define a general polycyclic presentation on generators $a_1, \ldots, a_n$ to be any presentation which makes it obvious that the group G defined is polycyclic and that $a_1, \ldots, a_n$ is a polycyclic generating sequence for G. Unfortunately, "obvious" cannot be defined precisely, so we must resort to a somewhat more cumbersome definition, one which describes several cases where the group is obviously polycyclic. The reader should be aware that the terminology related to these presentations is not standard.

Let $a_1, \ldots, a_n$ be a sequence of generators and set $X_i = \{a_i, \ldots, a_n\}^{\pm}$. A *polycyclic presentation* on $a_1, \ldots, a_n$ is a group presentation on these generators such that the following conditions hold:

(i) For $1 \leq i < j \leq n$ there is a relation $a_j a_i = a_i S_{ij}$, where S_{ij} is in X_{i+1}^*.

(ii) One of the following holds:

    (a) All of the words S_{ij} in (i) have the form $a_j A_{ij}$, where A_{ij} is in X_{j+1}^*.

    (b) For $1 \leq i < n$, either there is a relation $a_i^{m_i} = W_i$, where $m_i > 0$ and W_i is in X_{i+1}^*, or for each j with $i < j \leq n$ there is a relation $a_j a_i^{-1} = a_i^{-1} U_{ij}$, where U_{ij} is in X_{i+1}^*.

A relation $a_j a_i = a_i S_{ij}$, $i \neq j$, will be called a *commutation relation*. It is trivial if $S_{ij} = a_j$. A relation $a_i^{m_i} = W_i$ will be called a *power relation*.

Condition (i) in the definition of a polycyclic presentation does not, by itself, imply that the group is polycyclic, as the following example shows.

Example 4.2. Let G be the group generated by a and b subject to the single defining relation $ba = ab^2$. With respect to the basic wreath-product ordering of $\{a, b\}^{\pm}$ in which $b \prec b^{-1} \prec a \prec a^{-1}$ the group G has the following confluent rewriting system:

$$aa^{-1} \to \varepsilon, \quad a^{-1}a \to \varepsilon, \quad bb^{-1} \to \varepsilon, \quad b^{-1}b \to \varepsilon,$$
$$b^2 a^{-1} \to a^{-1}b, \quad ba \to ab^2, \quad b^{-1}a \to ab^{-2}, \quad b^{-1}a^{-1} \to ba^{-1}b^{-1}.$$

The words b^i and aba^{-1} are all irreducible with respect to this system. Thus aba^{-1} is not equal to a power of b, so $H = \mathrm{Grp}\langle b \rangle$ is not equal to aHa^{-1}. Since $(aba^{-1})^2 = ab^2 a^{-1} = b$, we have $H \subset aHa^{-1}$. By Proposition 3.12, G is not polycyclic.

Proposition 4.1. *Suppose that G is a group defined by a polycyclic presentation on generators $a_1, \ldots, a_n$. Then G is polycyclic and $a_1, \ldots, a_n$ is a polycyclic generating sequence for G.*

Proof. If condition (iia) holds, then we can prove that G is nilpotent. Clearly this is always the case if $n = 1$. In general, taking $j = n$ in (iia), we see that $a_n a_i = a_i a_n$ for all i. Thus $N = \mathrm{Grp}\langle a_n \rangle$ is in the center of G and hence N is normal in G. A presentation for G/N is obtained by setting a_n equal to 1 in the relations for G. This presentation also satisfies conditions (i) and (iia). Therefore, by induction on n, G/N is nilpotent, and G is nilpotent by Proposition 2.8. By Proposition 3.4, G is polycyclic.

Now suppose that condition (iib) holds. For $1 \leq i \leq n$, let G_i be the subgroup of G generated by $a_i, \ldots, a_n$. We must show that G_{i+1} is normal

in G_i, $1 \le i < n$. It suffices to prove that $a_i^{-1} G_{i+1} a_i = G_{i+1}$. By condition
(i), $a_i^{-1} G_{i+1} a_i \subseteq G_{i+1}$. If $a_i^{m_i}$ is in G_{i+1} for some positive integer m_i, then

$$G_{i+1} \supseteq a_i^{-1} G_{i+1} a_i \supseteq a_i^{-2} G_{i+1} a_i^2 \supseteq \cdots \supseteq a_i^{-m_i} G_{i+1} a_i^{m_i} = G_{i+1}.$$

Thus $G_{i+1} = a_i^{-1} G_{i+1} a_i$.

If there is no relation $a_i^{m_i} = W_i$ with W_i in X_{i+1}^*, then condition (iib) says
that $a_i G_{i+1} a_i^{-1} \subseteq G_{i+1}$, or $G_{i+1} \subseteq a_i^{-1} G_{i+1} a_i$. Thus $G_{i+1} = a_i^{-1} G_{i+1} a_i$ in this
case too. $\square$

By pushing the analysis in the proof of Proposition 4.1 a little further,
one can show that the exponents β_{ijk}, δ_{ijk}, and ν_{ijk} in $(*)$ are determined
by the α_{ijk}, γ_{ijk}, and μ_{ik}. See Exercise 4.2. If $(*)$ is consistent, then
the γ_{ijk} are actually determined by the α_{ijk} and μ_{ik}. See Proposition 8.2.
Given a polycyclic presentation for a group, the corresponding standard
polycyclic presentation can be obtained using the Knuth-Bendix procedure
for strings. It can also be constructed using more specialized techniques.
(See Section 9.8.)

Example 4.3. The presentation

$$a_2 a_1 = a_1 a_2^3, \quad a_2^{-1} a_1 = a_1 a_2^{-3}, \quad a_2 a_1^{-1} = a_1^{-1} a_2^4, \quad a_2^{-1} a_1^{-1} = a_1^{-1} a_2^{-4}$$

has the form $(*)$ with $I = \emptyset$. As in the Knuth-Bendix procedure, we can
rewrite the overlap $a_2 a_1 a_1^{-1}$ in two ways:

$$\underline{a_2 a_1 a_1^{-1}} = a_2$$

and

$$\underline{a_2 a_1} a_1^{-1} = a_1 a_2^3 a_1^{-1} = a_1 a_2^2 a_1^{-1} a_2^4 = a_1 a_2 a_1^{-1} a_2^8 = a_1 a_1^{-1} a_2^{12} = a_2^{12}.$$

Thus $a_2 = a_2^{12}$. Since we are in a group, this implies that $a_2^{11} = 1$. If this
relation and $a_2^{-1} = a_2^{10}$ are added, the resulting presentation is consistent.

A polycyclic presentation for which condition (iia) holds will be called a
nilpotent presentation. A presentation

$$\begin{aligned}
a_i^{m_i} &= W_i, \quad 1 \le i \le n, \\
a_j a_i &= a_i S_{ij}, \quad 1 \le i < j \le n
\end{aligned} \tag{$**$}$$

where the words W_i and S_{ij} are in $\{a_{i+1}, \ldots, a_n\}^*$, is called a *power-conjugate* presentation and defines a finite solvable group G with order dividing $m_1 \ldots m_m$. The presentation $(**)$ is actually a monoid presentation for G. It is also a rewriting system with respect to the basic wreath-product ordering with $a_n \prec \cdots \prec a_1$. Nilpotent power-conjugate presentations are sometimes called *power-commutator presentations*, although this term is often reserved for the *prime-exponent* case, in which all of the exponents m_i are equal to a fixed prime.

The abbreviation "pc-presentation" is frequently used. However, "pc" could reasonably stand for "polycyclic", "power-conjugate", or "power-commutator". To avoid confusion, the abbreviation "pc" will not be used in this book.

Example 4.4. The following consistent power-conjugate presentation on generators a, b, c, d, e, f, g is a modification of one in (Felsch 1976):

$$g^2 = 1,$$
$$f^4 = 1, \quad gf = fg,$$
$$e^2 = 1, \quad ge = ef^2g, \quad fe = ef^3,$$
$$d^6 = f^2, \quad gd = def^3, \quad fd = def^2g, \quad ed = df^2g,$$
$$c^3 = 1, \quad gc = cd^3f^2g, \quad fc = cf, \quad ec = ced^3, \quad dc = cd^4f^3g,$$
$$b^2 = 1, \quad gb = bg, \quad fb = bf^3, \quad eb = bef^3, \quad db = bd^5ef, \quad cb = bc^2,$$
$$a^2 = 1, \quad ga = ag, \quad fa = afg, \quad ea = ad^3f, \quad da = acd^3ef^2,$$
$$ca = ad^4ef^2g, \quad ba = ab.$$

The group defined has order $2 \cdot 4 \cdot 2 \cdot 6 \cdot 3 \cdot 2 \cdot 2 = 1152$.

In Section 9.9 and in Chapter 11 we shall be working with a class of nilpotent polycyclic presentations which have additional structure. A γ-*weighted* presentation for a group G is a nilpotent polycyclic presentation $\mathcal{R}$ on generators $a_1, \ldots, a_n$ such that the following hold:

(a) Each a_i has associated with it a positive integer weight w_i such that $w_1 = 1$ and $w_i \leq w_{i+1}$ for $1 \leq i < n$.

(b) If there is a power relation $a_i^{m_i} = W_i$, then the generators occurring in W_i all have weight at least $w_i + 1$.

(c) For each commutation relation $a_j a_i = a_i a_j A_{ij}$, the generators occurring in A_{ij} all have weight at least $w_i + w_j$.

(d) If $w_k = e > 1$, then there are integers i and j with $w_i = 1$ and $w_j = e - 1$ such that $A_{ij} = a_k$. One such pair is fixed, and the relation $a_j a_i = a_i a_j a_k$ is called the *definition* of a_k.

Suppose that $\mathcal{R}$ is a γ-weighted presentation for G on $a_1, \ldots, a_n$. For $e \geq 1$, set $G(e)$ equal to the subgroup of G generated by the a_i with $w_i \geq e$.

By (c), $G(e)$ is normal in G and $G(e-1)/G(e)$ is central in $G/G(e)$. Thus $\gamma_e(G) \subseteq G(e)$. Let $c = w_n$. Condition (d) implies that $G(c) = [G(c-1), G]$. Now $[G(c-2), G]$ contains $[G(c-1), G]$ and by (d) $G(c-1)$ is generated by $[G(c-2), G]$ modulo $G(c)$. Therefore $G(c-1) = [G(c-2), G]$. Continuing in this way, we find that $G(e+1) = [G(e), G]$, $1 \le e < c$. Thus $\gamma_e(G) = G(e)$. Consistency is more easily checked for γ-weighted presentations than it is for general nilpotent presentations. (See Section 9.8.)

Example 4.5. Let us consider the nilpotent presentation on generators $a_1, \ldots, a_7$ with weights $w_1 = w_2 = w_3 = 1$, $w_4 = w_5 = 2$, and $w_6 = w_7 = 3$ in which $a_j a_i = a_i a_j$ when $w_i + w_j > 3$ and

$$a_2 a_1 = a_1 a_2 a_4^4 a_5^2 a_6^3, \quad a_3 a_1 = a_1 a_3 a_5, \quad a_3 a_2 = a_2 a_3 a_4, \quad a_4 a_1 = a_1 a_4 a_6,$$
$$a_4 a_2 = a_2 a_4 a_6^3 a_7^2, \quad a_4 a_3 = a_3 a_4 a_6^4, \quad a_5 a_1 = a_1 a_5 a_7^6,$$
$$a_5 a_2 = a_2 a_5 a_7, \quad a_5 a_3 = a_3 a_5 a_6^{-2} a_7^4.$$

This presentation is γ-weighted. The definitions of a_4 to a_7 are the relations

$$a_3 a_2 = a_2 a_3 a_4, \quad a_3 a_1 = a_1 a_3 a_5, \quad a_4 a_1 = a_1 a_4 a_6, \quad a_5 a_2 = a_2 a_5 a_7,$$

respectively.

Rewriting with respect to a standard polycyclic rewriting system or a power-conjugate system is now called *collection*. The term "collection" was originally introduced in (P. Hall 1934) and referred there to computation in free nilpotent groups as discussed in Section 9.10. A great deal of effort has gone into devising efficient collection strategies. Suppose the generators are $a_1, \ldots, a_n$. Hall used a strategy in which all occurrences of a_1 are moved left to the beginning of the word. Next, all occurrences of a_2 are moved left until they are adjacent to the a_1's. Then the a_3's are moved left, and so on. This collection strategy is called *collection to the left*. It has properties which make it useful in the proofs of various formulas for the collected form of special words, but it is usually not efficient for computation. For some time, a consensus favored *collection from the right*, in which the left side of a rule occurring nearest the end of the word is selected for replacement. However, evidence in (Leedham-Green & Soicher 1990) and (Vaughan-Lee 1990) suggests that in a substantial number of cases *collection from the left* is superior. In this strategy, the left side nearest the beginning of the word is chosen.

Example 4.6. Let us compare these three strategies as applied to collecting the word $a_1 a_2 a_3 a_4 a_1 a_2 a_3 a_4$ using the rewriting system of Example 4.1. Collection to the left uses 22 applications of the rules:

$$a_1 a_2 a_3 \underline{a_4 a_1} a_2 a_3 a_4 = a_1 a_2 \underline{a_3 a_1} a_4 a_2 a_3 a_4$$

$$= a_1 \underline{a_2 a_1} a_3 a_4 a_2 a_3 a_4$$

$$= a_1 a_1 a_2 a_4^{-1} a_3 \underline{a_4 a_2} a_3 a_4$$

$$= a_1 a_1 a_2 a_4^{-1} \underline{a_3 a_2} a_4 a_3 a_4$$

$$= a_1 a_1 a_2 \underline{a_4^{-1} a_2} a_3 a_5^{-1} a_4 a_3 a_4$$

$$= a_1 a_1 a_2 a_2 \underline{a_4^{-1}} a_3 a_5^{-1} a_4 a_3 a_4$$

$$= a_1 a_1 a_2 a_2 a_3 a_4^{-1} a_6^{-1} a_5^{-1} \underline{a_4 a_3} a_4$$

$$= a_1 a_1 a_2 a_2 a_3 a_4^{-1} a_6^{-1} \underline{a_5^{-1} a_3} a_4 a_6 a_4$$

$$= a_1 a_1 a_2 a_2 a_3 a_4^{-1} \underline{a_6^{-1} a_3} a_5^{-1} a_4 a_6 a_4$$

$$= a_1 a_1 a_2 a_2 a_3 \underline{a_4^{-1} a_3} a_6^{-1} a_5^{-1} a_4 a_6 a_4$$

$$= a_1 a_1 a_2 a_2 a_3 a_3 a_4^{-1} a_6^{-1} a_6^{-1} \underline{a_5^{-1} a_4} a_6 a_4$$

$$= a_1 a_1 a_2 a_2 a_3 a_3 a_4^{-1} a_6^{-1} \underline{a_6^{-1} a_4} a_5^{-1} a_6 a_4$$

$$= a_1 a_1 a_2 a_2 a_3 a_3 a_4^{-1} \underline{a_6^{-1} a_4} a_6^{-1} a_5^{-1} a_6 a_4$$

$$= a_1 a_1 a_2 a_2 a_3 a_3 \underline{a_4^{-1} a_4} a_6^{-1} a_6^{-1} a_5^{-1} a_6 a_4$$

$$= a_1 a_1 a_2 a_2 a_3 a_3 a_6^{-1} a_6^{-1} a_5^{-1} \underline{a_6 a_4}$$

$$= a_1 a_1 a_2 a_2 a_3 a_3 a_6^{-1} a_6^{-1} \underline{a_5^{-1} a_4} a_6$$

$$= a_1 a_1 a_2 a_2 a_3 a_3 a_6^{-1} \underline{a_6^{-1} a_4} a_5^{-1} a_6$$

$$= a_1 a_1 a_2 a_2 a_3 a_3 \underline{a_6^{-1} a_4} a_6^{-1} a_5^{-1} a_6$$

$$= a_1 a_1 a_2 a_2 a_3 a_3 a_4 a_6^{-1} \underline{a_6^{-1} a_5^{-1}} a_6$$

$$= a_1 a_1 a_2 a_2 a_3 a_3 a_4 a_6^{-1} \underline{a_5^{-1}} a_6^{-1} a_6$$

$$= a_1 a_1 a_2 a_2 a_3 a_3 a_4 a_5^{-1} a_6^{-1} \underline{a_6^{-1} a_6}$$

$$= a_1 a_1 a_2 a_2 a_3 a_3 a_4 a_5^{-1} a_6^{-1}.$$

Collection from the right does not do any better:

$$a_1 a_2 a_3 \underline{a_4 a_1} a_2 a_3 a_4 = a_1 a_2 a_3 a_1 \underline{a_4 a_2} a_3 a_4$$

$$= a_1 a_2 a_3 a_1 a_2 \underline{a_4 a_3} a_4$$

$$= a_1 a_2 a_3 a_1 a_2 a_3 a_4 \underline{a_6 a_4}$$

$$= a_1 a_2 \underline{a_3 a_1} a_2 a_3 a_4 a_4 a_6$$

$$= a_1 a_2 a_1 \underline{a_3 a_2} a_3 a_4 a_4 a_6$$

$$= a_1 a_2 a_1 a_2 a_3 \underline{a_5^{-1}} \underline{a_3} a_4 a_4 a_6$$

$$= a_1 a_2 a_1 a_2 a_3 a_3 \underline{a_5^{-1}} \underline{a_4} a_4 a_6$$

$$= a_1 a_2 a_1 a_2 a_3 a_3 a_4 \underline{a_5^{-1}} \underline{a_4} a_6$$

$$= a_1 \underline{a_2} \underline{a_1} a_2 a_3 a_3 a_4 a_4 a_5^{-1} a_6$$

$$= a_1 a_1 a_2 \underline{a_4^{-1}} \underline{a_2} a_3 a_3 a_4 a_4 a_5^{-1} a_6$$

$$= a_1 a_1 a_2 a_2 \underline{a_4^{-1}} \underline{a_3} a_3 a_4 a_4 a_5^{-1} a_6$$

$$= a_1 a_1 a_2 a_2 a_3 a_4^{-1} \underline{a_6^{-1}} \underline{a_3} a_4 a_4 a_5^{-1} a_6$$

$$= a_1 a_1 a_2 a_2 a_3 a_4^{-1} a_3 \underline{a_6^{-1}} \underline{a_4} a_4 a_5^{-1} a_6$$

$$= a_1 a_1 a_2 a_2 a_3 a_4^{-1} a_3 a_4 \underline{a_6^{-1}} \underline{a_4} a_5^{-1} a_6$$

$$= a_1 a_1 a_2 a_2 a_3 a_4^{-1} a_3 a_4 a_4 \underline{a_6^{-1}} \underline{a_5^{-1}} a_6$$

$$= a_1 a_1 a_2 a_2 a_3 a_4^{-1} a_3 a_4 a_4 a_5^{-1} \underline{a_6^{-1}} \underline{a_6}$$

$$= a_1 a_1 a_2 a_2 a_3 \underline{a_4^{-1}} a_3 a_4 a_4 a_5^{-1}$$

$$= a_1 a_1 a_2 a_2 a_3 a_3 a_4^{-1} \underline{a_6^{-1}} \underline{a_4} a_4 a_5^{-1}$$

$$= a_1 a_1 a_2 a_2 a_3 a_3 a_4^{-1} a_4 \underline{a_6^{-1}} \underline{a_4} a_5^{-1}$$

$$= a_1 a_1 a_2 a_2 a_3 a_3 a_4^{-1} a_4 a_4 \underline{a_6^{-1}} \underline{a_5^{-1}}$$

$$= a_1 a_1 a_2 a_2 a_3 a_3 \underline{a_4^{-1}} \underline{a_4} a_4 a_5^{-1} a_6^{-1}$$

$$= a_1 a_1 a_2 a_2 a_3 a_3 a_4 a_5^{-1} a_6^{-1}.$$

However, collection from the left requires only 12 applications of the rules:

$$a_1 a_2 a_3 \underline{a_4} \underline{a_1} a_2 a_3 a_4 = a_1 a_2 \underline{a_3} \underline{a_1} a_4 a_2 a_3 a_4$$

$$= a_1 \underline{a_2} \underline{a_1} a_3 a_4 a_2 a_3 a_4$$

$$= a_1 a_1 a_2 \underline{a_4^{-1}} a_3 a_4 a_2 a_3 a_4$$

$$= a_1 a_1 a_2 a_3 a_4^{-1} \underline{a_6^{-1}} \underline{a_4} a_2 a_3 a_4$$

$$= a_1 a_1 a_2 a_3 \underline{a_4^{-1}} \underline{a_4} a_6^{-1} a_2 a_3 a_4$$

$$= a_1 a_1 a_2 a_3 \underline{a_6^{-1}} \underline{a_2} a_3 a_4$$

$$= a_1 a_1 a_2 \underline{a_3} \underline{a_2} a_6^{-1} a_3 a_4$$

$$= a_1 a_1 a_2 a_2 a_3 a_5^{-1} \underline{a_6^{-1}} \underline{a_3} a_4$$

$$= a_1 a_1 a_2 a_2 a_3 \underline{a_5^{-1}} \underline{a_3} a_6^{-1} a_4$$

$$= a_1 a_1 a_2 a_2 a_3 a_3 a_5^{-1} \underline{a_6^{-1}} \underline{a_4}$$

$$= a_1 a_1 a_2 a_2 a_3 a_3 \underline{a_5^{-1}} a_4 a_6^{-1}$$
$$= a_1 a_1 a_2 a_2 a_3 a_3 a_4 a_5^{-1} a_6^{-1}.$$

In its basic form, neither collection from the right nor collection from the left is adequate. There are several ways of speeding up the process further. With either strategy, there is a collected part and an uncollected part of the current word. In collection from the right, the collected part is a suffix, while in collection from the left it is a prefix. The collected part $a_1^{\alpha_1} \ldots a_n^{\alpha_n}$ can be represented by its exponent vector $(\alpha_1, \ldots, \alpha_n)$. Let r be the smallest index such that $G_r = \mathrm{Grp}\,\langle a_r, \ldots, a_n \rangle$ is abelian. In Example 4.1, $r = 4$. If $i \geq r - 1$, then in collection from the left we have

$$(\alpha_1, \ldots, \alpha_n) a_i^\beta = (\alpha_1, \ldots, \alpha_i + \beta, \alpha_{i+1}, \ldots, \alpha_n).$$

Note that, if i is in I and $\alpha_i + \beta \geq m_i$ or $\alpha_i + \beta < 0$, then the vector on the right does not represent a collected word. In collection from the right, one has

$$a_i^\beta (0, \ldots, 0, \alpha_i, \ldots, \alpha_n) = (0, \ldots, 0, \alpha_i + \beta, \alpha_{i+1}, \ldots, \alpha_n)$$

for any i, and

$$a_i^\beta (0, \ldots, 0, a_r, \ldots, a_n) = (0, \ldots, 0, a_r, \ldots, a_{i-1}, a_i + \beta, a_{i+1}, \ldots, a_n)$$

provided $i \geq r$. When collecting in nilpotent groups, it is important to be able to develop formulas for the collected form of various families of words. The simplest formula is $a_j^\alpha a_i^\beta = a_i^\beta a_j^\alpha$ when $i < j$ and a_i and a_j commute. If $\mathrm{Grp}\,\langle a_i, a_j \rangle$ is nilpotent of class 2, then $a_j^\alpha a_i^\beta = a_i^\beta a_j^\alpha [a_j, a_i]^{\alpha\beta}$. The formulas defining the rules in Example 4.1 are valid for all integers α and β, not just in the case $|\alpha| = |\beta| = 1$. The use of more complicated formulas in an approach called *combinatorial collection* is described in (Havas & Nicholson 1976).

When formulas cannot easily be derived and memory is not a problem, then one can store the collected form for additional products $a_j^\alpha a_i^\beta$. In a power-conjugate system, this could mean storing up to

$$\sum_{i<j} (m_j - 1)(m_i - 1)$$

conjugation rules. With the presentation of Example 4.4, adding the 56 extra rules greatly reduces the time needed to collect words. In (Felsch 1976), a novel data structure is described which encodes a power-conjugate presentation as a family of subroutines, which are executed to carry out collection.

Extensive research on collection is in progress as of this writing. Most experimental evidence comes from collection within finite solvable groups. Complicating the situation is the observation that the words which arise in various important algorithms connected with polycyclic groups do not appear to be random. To get the best performance on these algorithms, the collection procedure must be tuned to the words which occur most frequently. In view of the inconclusive results available at the present time, no recommendation will be made here concerning the choice of collection procedures.

Although we may not know the best collection strategy, we certainly can solve the word problem in a group G given by polycyclic presentations. Probably the next problem about elements of G to consider is the conjugacy problem. Conjugacy of elements can be decided in principle, but practical algorithms for infinite polycyclic groups have not yet been developed. If two elements g and h of G are not conjugate, then there is a finite quotient group of G in which the images of g and h are not conjugate. A proof of this result may be found in [Segal 1983]. This leads to the following "algorithm" for deciding whether two elements g and h are conjugate. We start two computers running. The first computer systematically forms conjugates of g in G. The second computer systematically examines the finite quotients of G. The conjugacy problem in a finite group is clearly solvable in principle. Thus either the first computer will find an element u of G such that $u^{-1}gu = h$ or the second computer will find a finite quotient group of G in which the images of g and h are not conjugate. We simply wait to see which computer stops. Useful conjugacy algorithms for finite solvable groups have been developed. See (Mecky & Neubüser 1989). Conjugacy in nilpotent groups is discussed in Section 9.7.

The next two sections discuss computation in subgroups and quotient groups of polycyclic groups.

Exercises

4.1. Suppose that $a_1, \ldots, a_n$ is a polycyclic generating sequence for a group G. Let $a_1^{\alpha_1} \ldots a_n^{\alpha_n}$ be the collected word defining an element g of G, let $a_i^{\beta_i}$ be the leading term of an element h, and let $a_1^{\gamma_1} \ldots a_n^{\gamma_n}$ be the collected word defining gh. Show that $\gamma_j = \alpha_j$, $1 \leq j < i$, and that $\gamma_i = \alpha_i + \beta_i$ if i is not in I, while $\gamma_i = (\alpha_i + \beta_i) \bmod m_i$ if i is in I. Here $I = I(a_1, \ldots, a_n)$ and m_i is the relative order of a_i modulo $\mathrm{Grp}\langle a_{i+1}, \ldots, a_n \rangle$. Describe the leading terms of gh and h^{-1}.

4.2. Show that the exponents β_{ijk}, δ_{ijk}, and ν_{ik} in a standard polycyclic presentation $(*)$ are determined by the α_{ijk}, γ_{ijk}, and μ_{ik}.

4.3. Let $(X, \mathcal{R})$ be a nilpotent presentation satisfying conditions (a), (b), and (c) of the definition of a γ-weighted presentation. For $e \geq 1$, let $Q(e)$ be the abelian group generated by the a_i of weight e subject to the relations $a_i^{m_i} = 1$ for i in I and $w_i = e$. Given U in $X^{\pm *}$, let $\overline{U}$ denote the word obtained from U by deleting all generators of weight different from e. We shall say that $(X, \mathcal{R})$ is *weakly γ-weighted* if for $e \geq 2$ the group $Q(e)$ is generated by the images of the words $\overline{A_{ij}}$, where $w_i = 1$ and $w_j = e - 1$. Show that a γ-weighted presentation is weakly γ-weighted and that in a group G defined by a weakly γ-weighted presentation the group $\gamma_e(G)$ is generated by the a_i with $w_i \geq e$.

9.5 Subgroups

In Section 8.1 we discussed how to compute with subgroups of $\mathbb{Z}^n$. Subgroups were represented by integer matrices with n columns, and integer row operations were used to manipulate these matrices. This section uses ideas of M. F. Newman described in (Laue et al. 1984) to generalize the techniques of Section 8.1 to study the subgroups of a group G given by a polycyclic presentation on generators $a_1, \ldots, a_n$. The group Grp $\langle a_i, \ldots, a_n \rangle$ will be denoted G_i. We shall assume that the presentation is consistent and that we know the set I of indices i such that G_i/G_{i+1} is finite and the relative order m_i of a_i modulo G_{i+1} for i in I.

A subgroup H of G will be described by a sequence $U = (g_1, \ldots, g_s)$ of generating elements. Let the collected form of g_i be $a_1^{\alpha_{i1}} \ldots a_n^{\alpha_{in}}$. The s-by-n matrix A of integers α_{ij} will be used to represent U and will be called the *associated exponent matrix*. Corresponding to the elementary row operations of Section 8.1, we have the following *elementary operations* on U:

(1) Interchange g_i and g_j if $i \neq j$.
(2) Replace g_i by g_i^{-1}.
(3) Replace g_i by $g_i g_j^{\beta}$, where β is an integer and $i \neq j$.
(4) Add as a new component g_{s+1} any element of Grp $\langle g_1, \ldots, g_s \rangle$.
(5) Delete g_s, if $g_s = 1$.

Notice that the length of U may increase or decrease. This was not necessary in Section 8.1, since any finite sequence of generators of an abelian group is a polycyclic generating sequence. In general, a polycyclic group generated by a small set may require long polycyclic generating sequences.

Two sequences $U = (g_1, \ldots, g_s)$ and $V = (h_1, \ldots, h_t)$ are *equivalent under elementary operations* if one can be transformed into the other by a sequence of these operations. To see that this is an equivalence relation, we must show that the effect of an elementary operation can be undone by a sequence of one or more operations. Applying operations of types (1) and (2) twice to U leaves U unchanged. If an operation of type (3) is applied, then replacing g_i by $g_i g_j^{-\beta}$ restores U to its original form. After an operation of type (4), g_{s+1} is a product of powers of the g_i with $1 \leq i \leq s$. A sequence of operations of type (3) can make g_{s+1} the identity element, and then an operation of type (5) deletes g_{s+1}. Finally, if an operation of type (5) is performed, then undoing it is a special case of an operation of type (4).

Let us say that a sequence $U = (g_1, \ldots, g_s)$ of elements of G is in *standard form* if the associated exponent matrix A satisfies the following conditions:

(i) All rows of A are nonzero (i.e., no g_i is the identity).
(ii) A is row reduced over $\mathbb{Z}$.
(iii) If A_{ij} is a corner entry and j is in I, then A_{ij} divides m_j.

Suppose that U is in standard form. An *admissible sequence of exponents* for U is a sequence $(\beta_1, \ldots, \beta_s)$ of integers such that, if A_{ij} is a corner entry and j is in I, then $0 \leq \beta_i < m_j/A_{ij}$. Let $E(U)$ be the set of admissible sequences of exponents for U and let $S(U)$ be the set of products $g_1^{\beta_1} \ldots g_s^{\beta_s}$, where $(\beta_1, \ldots, \beta_s)$ ranges over $E(U)$.

Example 5.1. Suppose G is $\mathbb{Z}^n$ and $a_1, \ldots, a_n$ is the standard basis. Then a sequence $U = (g_1, \ldots, g_s)$ of elements of G is in standard form if and only if the associated matrix A is row reduced and has rank s. In this case, all s-tuples of integers are admissible for U. The set $S(U)$ is simply the subgroup $S(A)$ of Section 8.1. The components of U form a basis of $S(U)$.

In general, $S(U)$ is not a subgroup, but there is a one-to-one correspondence between elements of $S(U)$ and elements of $E(U)$.

Proposition 5.1. *Suppose that $U = (g_1, \ldots, g_s)$ is a sequence of elements of G in standard form and $(\beta_1, \ldots, \beta_s)$ and $(\gamma_1, \ldots, \gamma_s)$ are in $E(U)$. If $g_1^{\beta_1} \ldots g_s^{\beta_s} = g_1^{\gamma_1} \ldots g_s^{\gamma_s}$, then $\beta_i = \gamma_i$, $1 \leq i \leq s$.*

Proof. Let A be the matrix of exponents associated with U, and let $g = g_1^{\beta_1} \ldots g_s^{\beta_s} = g_1^{\gamma_1} \ldots g_s^{\gamma_s}$. Suppose A_{1j} is the corner entry in the first row of A. If $a_1^{\delta_1} \ldots a_n^{\delta_n}$ is the collected word defining g, then $\delta_k = 0$, $1 \leq k < j$. We have two cases depending on whether j is in I. Assume first that j is not in I. Then G_j/G_{j+1} is isomorphic to $\mathbb{Z}$ and $\delta_j = A_{1j}\beta_1 = A_{1j}\gamma_1$. Since $A_{1j} \neq 0$, this means that $\beta_1 = \gamma_1$. Now assume that j is in I. Then G_j/G_{j+1} is isomorphic to $\mathbb{Z}_{m_j}$ and $A_{1j}\beta_1 \equiv \delta_j \equiv A_{1j}\gamma_1 \pmod{m_j}$. But both β_1 and γ_1 are nonnegative and less than m_j/A_{1j}. Therefore $A_{1j}\beta_1$ and $A_{1j}\gamma_1$ are nonnegative and less than m_j. Thus $A_{1j}\beta_1 = A_{1j}\gamma_1$. Hence $\beta_1 = \gamma_1$ in this case too. Thus we may multiply g on the left by $g_1^{-\beta_1}$ and conclude that $g_2^{\beta_2} \ldots g_s^{\beta_s} = g_2^{\gamma_2} \ldots g_s^{\gamma_s}$. By induction applied to the $(s-1)$-tuple $(g_2, \ldots, g_s)$, we have $\beta_i = \gamma_i$, $2 \leq i \leq s$. $\square$

The proof of Proposition 5.1 gives us an algorithm for deciding membership in $S(U)$ for a sequence U of elements of G in standard form. It is a straightforward generalization of the algorithm in Section 8.1 for deciding membership in a subgroup of $\mathbb{Z}^n$ given as $S(B)$, where B is a row reduced integer matrix.

Function POLY_MEMBER(U, g): boolean;

Input: U : a sequence $(g_1, \ldots, g_s)$ of elements of G in standard form;

      g : an element of G;

(* The value true is returned if g is in $S(U)$, and false is returned otherwise. *)

Begin

   Let A be the exponent matrix associated with U; $h := g$;

   (* At all times $a_1^{\gamma_1} \ldots a_n^{\gamma_n}$ will be the collected word representing h. *)

   $i := 1$; $done := $ false;

   While $i \leq s$ and not $done$ do begin

      Let A_{ij} be the corner entry of A in the i-th row;

      If some $\gamma_k \neq 0$ for $1 \leq k < j$ then $done := $ true

      Else if A_{ij} does not divide γ_j then $done := $ true

      Else begin

         $q := \gamma_j / A_{ij}$; $h := g_i^{-q} h$;

      End;

      $i := i + 1$

   End;

   POLY_MEMBER $:= (h = 1)$

End.

Example 5.2. Let $D = D_4^{(1)}(\mathbb{Z})$ as given by the presentation on $a_1, \ldots, a_6$ derived in Example 4.1. If

$$g_1 = a_1^2 a_2^{-1} a_4,$$
$$g_2 = a_3^3 a_4 a_6,$$
$$g_3 = a_4^2 a_5 a_6,$$

then $U = (g_1, g_2, g_3)$ is in standard form. The associated exponent matrix is

$$\begin{bmatrix} 2 & -1 & 0 & 1 & 0 & 0 \\ 0 & 0 & 3 & 1 & 0 & 1 \\ 0 & 0 & 0 & 2 & 1 & 1 \end{bmatrix}.$$

Let us decide whether $u = a_1^{-6} a_3^3 a_3^6 a_4^{21} a_5^5 a_6^{64}$ is in $S(U)$. Since $I = \emptyset$, we simply want to know whether u can be expressed as $g_1^{\beta_1} g_2^{\beta_2} g_3^{\beta_3}$. The leading terms of u and g_1 are a_1^{-6} and a_1^2, respectively. Thus if u is in $S(U)$, then $\beta_1 = (-6)/2 = -3$. The product $g_1^3 u$ is $v = a_3^6 a_4^{12} a_5^5 a_6^{10}$. Therefore β_2 must be $6/3 = 2$. Multiplying g_2^{-2} and v yields $w = a_4^{10} a_5^5 a_6^5$. Now $\beta_3 = 10/2 = 5$. The product $g_3^{-5} w$ is 1, so $u = g_1^{-3} g_2^2 g_3^5$ is in $S(U)$.

Given a sequence $U = (g_1, \ldots, g_s)$ of elements of G in standard form, we can decide whether $S(U)$ is a subgroup of G. Let us say that U is *full* if the following conditions hold:

(i) For $1 \leq i < j \leq s$ the set $S(U)$ contains $g_i^{-1}g_jg_i$.
(ii) If A_{ij} is a corner entry of the matrix A of exponents associated with U and j is in I, then $S(U)$ contains g_i^q, where $q = m_j/A_{ij}$.

Example 5.3. The triple $U = (g_1, g_2, g_3)$ in Example 5.2 is not full. Let $u = g_1^{-1}g_2g_1 = a_3^3a_4a_5^3a_6^{-2}$. If u is in $S(U)$ and $u = g_1^{\beta_1}g_2^{\beta_2}g_3^{\beta_3}$, then $\beta_1 = 0$ and $\beta_2 = 1$. But $g_2^{-1}u = a_5^3a_6^{-3}$. Since no g_i has a power of a_5 as its leading term, u is not in $S(U)$.

Proposition 5.2. *If $U = (g_1, \ldots, g_s)$ is a sequence of elements of G in standard form, then $S(U)$ is a subgroup of G if and only if U is full. If U is full, then $g_1, \ldots, g_s$ is a polycyclic generating sequence for $S(U)$.*

Proof. The elements g_i are in $S(U)$. Thus, if $S(U)$ is a subgroup of G, then U is full. Assume now that U is full. If $s = 0$, then $S(U) = \{1\}$ is a subgroup. We proceed by induction on s and suppose that $s > 0$. Let A_{1k} be the corner entry in the first row of the matrix associated with U. Then $g_2, \ldots, g_s$ are contained in G_{k+1} and $S(U) \cap G_{k+1}$ is $S(V)$, where $V = (g_2, \ldots, g_s)$. If $2 \leq i < j \leq s$, then $g_i^{-1}g_jg_i$ is in $S(U)$ and is in G_{k+1}. Therefore $g_i^{-1}g_jg_i$ is in $S(V)$. By a similar argument, V satisfies condition (ii) of the definition of full. Therefore V is full and, by induction, $H = S(V)$ is a subgroup of G. If $2 \leq j \leq s$, then $g_1^{-1}g_jg_1$ is in G_{k+1} by Exercise 4.1. Since U is full, $g_1^{-1}g_jg_1$ is also in $S(U)$, and hence $g_1^{-1}g_jg_1$ is in H. Thus $g_1^{-1}Hg_1 \subseteq H$. By Proposition 3.12, $g_1^{-1}Hg_1 = H$. This implies that $g_1Hg_1^{-1} = H$. Thus H is normal in $K = \mathrm{Grp}\langle g_1, \ldots, g_s\rangle$ and every element of K can be written in the form $g_1^\alpha h$, where h is in H. If k is not in I, then every element $g_1^\alpha h$ is in $S(U)$, so $K = S(U)$. Suppose that k is in I and $q = m_k/A_{1k}$. Then $u = g_1^q$ is in H, so α can always be chosen so that $0 \leq \alpha < q$. Thus $K = S(U)$ in this case too. $\square$

The following result generalizes Proposition 1.1 in Chapter 8.

Proposition 5.3. *Let H be a subgroup of G. There is a unique sequence $U = (g_1, \ldots, g_s)$ in standard form such that $H = S(U)$.*

Proof. Let k be the largest index such that $H \subseteq G_k$. If $k = n+1$, then H is trivial and the empty sequence $U = ()$ is the only sequence in standard form such that $H = S(U)$. Suppose that $k \leq n$. Let $^-$ denote the canonical homomorphism from G_k to G_k/G_{k+1}. The image $\overline{H}$ is nontrivial. Let α be the least positive integer such that $(\overline{a_k})^\alpha$ is in $\overline{H}$. Then $\overline{H}$ is generated by $(\overline{a_k})^\alpha$, and, if k is in I, then α divides m_k. Let g_1 be an element of H such that $\overline{g_1} = (\overline{a_k})^\alpha$. Then the leading term of g_1 is a_k^α. By induction on $n - k$, there is a sequence $W = (g_2, \ldots, g_s)$ in standard form such that $H \cap G_{k+1} = S(W)$. The sequence $U = (g_1, g_2, \ldots, g_s)$ satisfies conditions (i) and (iii) of the definition of standard form, but it may not satisfy condition (ii) because the entries in row 1 of the matrix A associated with U which lie above corner entries of A may not be reduced modulo those corner entries. Thus it may be necessary to modify g_1. To do so, we execute the following instructions. At all times, $g_1 = a_1^{\alpha_1} \ldots a_n^{\alpha_n}$.

> For $i := 2$ to s do begin
>   Let a_j^β be the leading term of g_i; $q := \alpha_j$ div β; $g_1 := g_1 g_i^{-q}$
> End.

By Exercise 4.1, the entries in row 1 of A which lie above corner entries are now reduced modulo those entries. Hence U is in standard form. Every element of H has the form $g_1^\gamma u$, where u is in $H \cap G_{k+1}$ and $0 \le \gamma < m_k/\alpha$ if k is in I. Since $H \cap G_{k+1} = S(W)$, it follows that $H = S(U)$. Thus we have proved the existence part of the proposition.

To prove uniqueness, suppose that $H = S(U) = S(V)$, where both $U = (g_1, \ldots, g_s)$ and $V = (h_1, \ldots, h_t)$ are in standard form. The leading term of g_1 is a_k^α, where k is as defined earlier and $\alpha > 0$. If k is in I, then α divides m_k. The group $\overline{H}$ is generated by $\overline{g_1} = (\overline{a_k})^\alpha$, and $\overline{H}$ has order m_k/α if k is in I. These conditions uniquely determine α. By symmetry, the leading term of h_1 is also a_k^α. Now $H \cap G_{k+1} = S(g_2, \ldots, g_s) = S(h_2, \ldots, h_t)$. By induction on $n - k$, we have $s = t$ and $g_i = h_i$, $2 \le i \le s$. Let $u = g_1^{-1} h_1$. Then u is in $H \cap G_{k+1}$, so $u = g_2^{\beta_2} \ldots g_s^{\beta_s}$, where $(\beta_2, \ldots, \beta_s)$ is in $E(g_2, \ldots, g_s)$. If $u = 1$, then $g_1 = h_1$ and we are done. Suppose that $u \ne 1$ and let i be minimal such that $\beta_i \ne 0$. Let a_j^δ be the leading term of g_i and let the collected forms of g_1 and h_1 be $a_1^{\mu_1} \ldots a_n^{\mu_n}$ and $a_1^{\nu_1} \ldots a_1^{\nu_1}$, respectively. Assume that j is not in I. Since $h_1 = g_1 u$, we have $\nu_j = \mu_j + \beta_i \delta$. But this is not possible, since both μ_j and ν_j are reduced modulo δ. A similar argument takes care of the case in which j is in I. $\square$

Suppose $V = (h_1, \ldots, h_r)$ is a sequence of elements of G and H is the subgroup generated by the h_i. If we knew the full sequence U in standard form such that $H = S(U)$, then we could decide membership in H using POLY_MEMBER. It is in fact possible to transform V into U using elementary operations. The procedure is a relatively straightforward generalization of the row reduction procedure of Section 8.1.

Initially set U equal to V. The first observation is that we may apply elementary operations in such a way that the matrix A associated with U is in row echelon form. Suppose that g_i and g_j have leading terms a_k^β and a_k^γ with $|\beta| \ge |\gamma|$. Let $q = \beta$ div γ. Replacing g_i by $g_i g_j^{-q}$ sets the exponent of a_k in g_i equal to β mod γ. Repeating this step until it is no longer possible produces a sequence in which no two leading terms involve the same generator. By permuting the elements in the sequence U and deleting elements which are the identity, we can assume that A is in row echelon form and has no zero rows. By replacing g_i with g_i^{-1} if necessary, we can make all corner entries positive.

Now suppose that there is a corner entry $\alpha = A_{ik}$ such that k is in I and A_{ik} does not divide m_k. Let $\beta = \gcd(\alpha, m_k) = p\alpha + qm_k$. The leading term of g_i^p is a_k^β. Add g_i^p as a new member of the sequence U and repeat the procedure in the previous paragraph to put A back into row echelon form. Since this iteration either introduces a new column containing a corner entry or replaces a corner entry with a proper divisor, the process stops eventually with $U = (g_1, \ldots, g_s)$ such that the associated matrix A is in row echelon form and all rows are nonzero, corner entries are positive, and any corner entry A_{ik} with k in I divides m_k. To get U into standard form, we have only to execute the following statements:

For $i := 2$ to s do begin
  Let a_k^α be the leading term of g_i;

  For $j := 1$ to $i - 1$ do begin
    Let β be the exponent on a_k in the collected word representing g_j;
    $q := \beta \operatorname{div} \alpha$; $g_j := g_j g_i^{-q}$
  End
End.

Now that U is standard, we begin checking whether U is full. The test for fullness requires that various elements be in $S(U)$. Suppose that u is one of those elements and u is not in $S(U)$. Thus POLY_MEMBER$(U; u)$ returns false. Let v be the last value assigned to h within POLY_MEMBER. Add v as a new member of the sequence U and repeat the entire process. When U has again been put into standard form, either there will be a new column in A containing a corner entry or some corner entry will have been reduced. Thus this iteration must also stop. When it does, U will be full and $S(U)$ will be $H = \operatorname{Grp}\langle h_1, \ldots, h_r \rangle$.

Example 5.4. Let us continue Examples 5.2 and 5.3 and determine the subgroup of D generated by g_1, g_2, and g_3. The sequence U is already in standard form. However, $u = g_1^{-1} g_2 g_1$ is not in $S(U)$. This becomes clear when we compute $v = g_2^{-1} u = a_5^3 a_6^{-3}$. Thus we define g_4 to be v. Now A is

$$
\begin{bmatrix}
2 & -1 & 0 & 1 & 0 & 0 \\
0 & 0 & 3 & 1 & 0 & 1 \\
0 & 0 & 0 & 2 & 1 & 1 \\
0 & 0 & 0 & 0 & 3 & -3
\end{bmatrix},
$$

and U is still in standard form. Now let $u = g_1^{-1} g_3 g_1 = a_4^2 a_5 a_6^{-1}$. To check membership of u in $S(U)$, we compute $v = g_3^{-1} u = a_6^{-2}$. Clearly v is not in $S(U)$. In order to have a positive corner entry, we define g_5 to be $v^{-1} = a_6^2$. To put U in standard form, we have only to replace g_4 by $g_4 g_5^2 = a_5^3 a_6$. This gives

$$A = \begin{bmatrix} 2 & -1 & 0 & 1 & 0 & 0 \\ 0 & 0 & 3 & 1 & 0 & 1 \\ 0 & 0 & 0 & 2 & 1 & 1 \\ 0 & 0 & 0 & 0 & 3 & 1 \\ 0 & 0 & 0 & 0 & 0 & 2 \end{bmatrix}.$$

At this point $U = (g_1, g_2, g_3, g_4, g_5)$ is full.

Example 5.5. Let us now determine the subgroup generated by $h_1 = ad^3 eg$ and $h_2 = bf$ in the group of order 1152 in Example 4.4. Initially set $g_1 = h_1$, $g_2 = h_2$, and $U = (g_1, g_2)$. The associated matrix A is

$$\begin{bmatrix} 1 & 0 & 0 & 3 & 1 & 0 & 1 \\ 0 & 1 & 0 & 0 & 0 & 1 & 0 \end{bmatrix},$$

and U is in standard form. If $u = g_1^{-1} g_2 g_1$, then $u = bf^3$ and $g_2^{-1} u = f^2$. We define g_3 to be f^2. Now A is

$$\begin{bmatrix} 1 & 0 & 0 & 3 & 1 & 0 & 1 \\ 0 & 1 & 0 & 0 & 0 & 1 & 0 \\ 0 & 0 & 0 & 0 & 0 & 2 & 0 \end{bmatrix},$$

and U is still in standard form. The square of g_1 is g, and we define g_4 to be g. In order to get U into standard form, we replace g_1 by $g_1 g_4^{-1} = ad^3 e$. This gives

$$A = \begin{bmatrix} 1 & 0 & 0 & 3 & 1 & 0 & 0 \\ 0 & 1 & 0 & 0 & 0 & 1 & 0 \\ 0 & 0 & 0 & 0 & 0 & 2 & 0 \\ 0 & 0 & 0 & 0 & 0 & 0 & 1 \end{bmatrix}.$$

Now U is full. The order of H can now be seen to be 16.

Let us give the name POLY_SUBGROUP to the procedure for determining a full sequence U of generators for a subgroup of a polycyclic group described by generating elements. In order to formalize POLY_SUBGROUP, we must choose an order in which to test the elements $g_i^{-1}g_jg_i$ and g_i^q in the definition of a full sequence. It is also useful to try to arrange the computation so that, when a new generator is added to U, we can avoid repeating all of the tests made previously. There is inadequate experience at this point on which to base a firm recommendation. The reader is encouraged to experiment with various ways of spelling out the details of POLY_SUBGROUP.

Given a finite subset Y of G, we can compute the normal closure N of Y in G using POLY_SUBGROUP. There are at least two ways to organize the computation. In the most straightforward approach, we first find the full sequence U in standard form such that $\mathrm{Grp}\langle Y \rangle = S(U)$. For each element y of Y and each generator x of G we compute $u = x^{-1}yx$ and check whether u is in $S(U)$. If u is not in $S(U)$, then we add u to Y and recompute U. Since the ascending chain condition holds for subgroups of G, this process eventually stops.

The second method of computing normal closures works with the group $G \times G$. This group is polycyclic and we can easily get a polycyclic presentation for it. Let $b_1, \ldots, b_n$ be a sequence of generators distinct from $a_1, \ldots, a_n$. For each relation in the standard polycyclic presentation for G on the a_i's add the corresponding relation on the b_i's, so $\mathrm{Grp}\langle b_1, \ldots, b_n \rangle$ is isomorphic to G. Now add relations $b_i^\alpha a_j^\beta = a_j^\beta b_i^\alpha$ for all i and j and all α and β in $\{1, -1\}$. The group M generated by $a_1, \ldots, a_n$, $b_1, \ldots, b_n$ subject to these relations is isomorphic to $G \times G$. The subgroup $D = \mathrm{Grp}\langle a_1 b_1, \ldots, a_n b_n \rangle$ corresponds to the diagonal subgroup of $G \times G$. There are two obvious polycyclic generating sequences for M. They are $a_1, \ldots, a_n$, $b_1, \ldots, b_n$ and $b_1, \ldots, b_n$, $a_1, \ldots, a_n$. The only difference between the standard polycyclic presentations for these two sequences is the ordering of the left and right sides in the relations $b_i^\alpha a_j^\beta = a_j^\beta b_i^\alpha$.

According to Exercise 3.9 in Chapter 1, there is a one-to-one correspondence between the set of normal subgroups of G and the set of subgroups of M containing D. In this correspondence, a normal subgroup N of G corresponds to DN. Since $N = (DN) \cap G$, the following result is easily proved.

Proposition 5.4. *Let Y be a subset of G and let H be the subgroup of M generated by Y and $a_1 b_1, \ldots, a_n b_n$. Then the normal closure N of Y in G is $H \cap G$.*

Suppose that we compute in M using the polycyclic generating sequence $b_1, \ldots, b_n$, $a_1, \ldots, a_n$ and determine the full sequence $U = (h_1, \ldots, h_s)$ such that $H = S(U)$. Let h_r be the first component of U which is in G, that is, whose collected form involves only a_i's. Then $N = S(h_r, \ldots, h_s)$. Conceptually, this approach to normal closures is very appealing. We do not have to write a new routine. However, the presentation for M is more than twice as large as the presentation for G and it is not clear that this method is any faster than the first method.

Once normal closures can be computed, it is possible to compute commutator subgroups. If X and Y are finite generating sets for subgroups H and K of G, then $[H, K]$ is the normal closure in $\mathrm{Grp}\langle X, Y \rangle$ of the set of commutators $[x, y]$ with x in X and y in Y. With the ability to find commutator subgroups, we can determine the derived series and the lower central series of G.

Exercises

5.1. Show that POLY_MEMBER works correctly even if U does not satisfy condition (ii) of the definition of standard form.

5.2. Determine the order of the subgroup of the group in Example 4.4 generated by $abdf$ and ceg.

5.3. Let D and g_1, g_2, and g_3 be as in Examples 5.2 to 5.4 Determine the normal closure of $\{g_1, g_2, g_3\}$ in D.

9.6 Homomorphisms

In this section, various techniques are discussed for working with a homomorphism $f: G \to H$ from one polycyclic group to another. We shall assume that we know the standard polycyclic presentation for G relative to a polycyclic generating sequence $a_1, \ldots, a_n$. We shall want to do such things as describe the image of f, determine the kernel of f, and compute inverse images of elements and subgroups. As usual, G_i will be $\mathrm{Grp}\langle a_i, \ldots, a_n \rangle$.

Let us start with the case in which H is a quotient G/N, where N is a normal subgroup of G. Here $f: G \to H$ is the natural homomorphism. Suppose we have a full sequence $U = (g_1, \ldots, g_s)$ such that $N = S(U)$. For $1 \le i \le n$ let $b_i = f(a_i)$. Then $b_1, \ldots, b_n$ is a polycyclic generating sequence for H. Set $H_i = \mathrm{Grp}\langle b_i, \ldots, b_n \rangle$. An obvious problem is to find a polycyclic presentation for H in terms of $b_1, \ldots, b_n$. The commutation relations present no problems. We just replace each a_i by b_i in the commutation relations defining G. The only question concerns the power relations for H.

Suppose that $1 \le i \le n$. If no g_j has a leading term which is a power of a_i, then the order of b_i modulo H_{i+1} is the same as the order of a_i modulo G_{i+1}. If there is a power relation $a_i^{m_i} = W_i$ for G, then the power relation for b_i is obtained by replacing a's by the corresponding b's in this relation. If some $g_j = a_i^{\alpha_i} \ldots a_n^{\alpha_n}$ with $\alpha_i > 0$, then α_i is the order of b_i modulo H_{i+1} and $b_i^{\alpha_i} = b_n^{-\alpha_n} \ldots b_{i+1}^{-\alpha_{i+1}}$ holds in H.

Example 6.1. The presentation

$$c^\alpha a^\beta = a^\beta c^\alpha, \quad c^\alpha b^\beta = b^\beta c^\alpha, \quad b^\alpha a^\beta = a^\beta b^\alpha c^{\alpha\beta},$$

where α and β range over $\{1, -1\}$, is consistent. Let G be the group defined. If $g_1 = a^3b^3$, $g_2 = b^6c$, and $g_3 = c^3$, then $U = (g_1, g_2, g_3)$ is in standard form and is full. Moreover, $N = S(U)$ is normal in G. If $u = aN$, $v = bN$, and $w = cN$, then in $H = G/N$ we have

$$u^3 = v^{-3}, \quad v^6 = w^{-1}, \quad w^3 = 1.$$

Reworking these relations slightly and transferring the commutation relations from G to H yields the following consistent power-commutator presentation for H:

$$wu = uw, \quad wv = vw, \quad vu = uvw,$$
$$w^3 = 1, \quad v^6 = w^2, \quad u^3 = v^3w.$$

If our normal subgroup N is not given as $S(U)$ but as the normal closure of a finite subset T of G, then we have two possible courses of action. We could obtain a description of N as $S(U)$ using the techniques at the end of the previous section. We could also add the relations $t = 1$ for t in T as new defining relations. Here we are thinking of T as a set of collected words. The Knuth-Bendix procedure for strings can now be used to get the standard polycyclic presentation for H.

Now let us assume that H is a second polycyclic group described by a consistent polycyclic presentation on generators $b_1, \ldots, b_m$. A homomorphism $f : G \to H$ is determined by the images $u_i = f(a_i)$, $1 \le i \le n$. A map $a_i \mapsto u_i$ of the generators of G into H defines a homomorphism if and only if the u_i satisfy the defining relations for G. This can be checked, since we can compute collected words in H. The image K of f is generated by the u_i and thus can be determined. We also need to compute the kernel of f, and, for an element k of K, we want to be able to find g in G such that $f(g) = k$.

Essentially all the information we need to compute with f can be obtained with one invocation of POLY_SUBGROUP in $M = H \times G$. This approach is similar to the second method for finding normal closures described at the end of Section 9.5. Assuming the generating sets for G and H are disjoint, we get a presentation for M on the a_i's and the b_j's by combining the relations for G and H and adding relations which say that each a_i commutes with each b_j. All computation in M will be done using the polycyclic generating sequence $b_1, \ldots, b_m, a_1, \ldots, a_n$. Let L be the subgroup of M generated by the elements $u_1a_1, \ldots, u_na_n$ and let $W = (w_1, \ldots, w_s)$ be the full sequence such that $L = S(W)$. Each w_i can be written uniquely as h_ig_i, where h_i is in H and g_i is in G. Let r be the largest index such that h_r is not trivial.

Proposition 6.1. *The group L consists of all elements of M of the form $f(g)g$ with g in G. The sequence $U = (h_1, \ldots, h_r)$ is full and $K = S(U)$. The sequence $V = (g_{r+1}, \ldots, g_s)$ is full and $S(V)$ is the kernel of f. If k is in K and $k = h_1^{\beta_1} \ldots h_r^{\beta_r}$, then $f(g) = k$, where $g = g_1^{\beta_1} \ldots g_r^{\beta_r}$.*

Proof. The set P of products $f(g)g$ with g in G is easily checked to be a subgroup of M and the elements $u_i a_i$ are all in P. Thus L is contained in P. Given $g = a_1^{\alpha_1} \ldots a_n^{\alpha_n}$ in G, the element $(u_1 a_1)^{\alpha_1} \cdots (u_n a_n)^{\alpha_n} = u_1^{\alpha_1} \ldots u_n^{\alpha_n} a_1^{\alpha_1} \ldots a_n^{\alpha_n} = f(g)g$. Therefore $L = P$. If $1 \le i \le r$, then the leading term of $h_i g_i$ is the leading term of h_i. If $r + 1 \le i \le s$, then the leading term of $h_i g_i$ is the leading term of g_i. Since W is in standard form, it is easy to check that U and V are in standard form and $E(W)$ is the set of s-tuples $(\beta_1, \ldots, \beta_r, \alpha_{r+1}, \ldots, \alpha_s)$, where $(\beta_1, \ldots, \beta_r)$ is in $E(U)$ and $(\alpha_{r+1}, \ldots, \alpha_s)$ is in $E(V)$. Clearly $S(U) \subseteq K$. However, for any g in G the element $f(g)g$ is in $S(W)$. Thus $f(g)g = (h_1 g_1)^{\beta_1} \cdots (h_r g_r)^{\beta_r} g_{r+1}^{\alpha_{r+1}} \ldots g_s^{\alpha_s}$, where $(\beta_1, \ldots, \beta_r, \alpha_{r+1}, \ldots, \alpha_s)$ is in $E(W)$. Therefore $f(g) = h_1^{\beta_1} \ldots h_r^{\beta_r}$ and $f(g)$ is in $S(U)$. Hence $K = S(U)$. If $f(g) = 1$, then $\beta_1 = \cdots = \beta_r = 0$. Thus g is in $S(V)$. Since each g_i with $i > r$ is in the kernel of f, the kernel is equal to $S(V)$. Finally, if $k = h_1^{\beta_1} \ldots h_r^{\beta_r}$, then $(h_1 g_1)^{\beta_r} \cdots (h_r g_r)^{\beta_r} = kg$ is in L, where $g = g_1^{\beta_1} \ldots g_r^{\beta_r}$. Therefore $f(g) = k$. $\square$

Example 6.2. Let G be the group defined by the following power-conjugate presentation on the generators a, b, c:

$$c^8 = 1,$$

$$b^8 = 1, \quad cb = bc,$$

$$a^2 = b^2 c^2, \quad ba = ac, \quad ca = ac.$$

Let H be the group on generators u and v defined by the presentation

$$v^8 = 1,$$

$$u^4 = 1, \quad vu = uv.$$

The map $a \mapsto u^2 v^4$, $b \mapsto v^2$, $c \mapsto v^6$ defines a homomorphism f of G into H. To determine the kernel of f using Proposition 6.1, we apply POLY_SUBGROUP to $W = (u^2 v^4 a, v^2 b, v^6 c)$ in $M = H \times G$ using the polycyclic generating sequence u, v, a, b, c. The matrix associated with W is

$$\begin{bmatrix} 2 & 4 & 1 & 0 & 0 \\ 0 & 2 & 0 & 1 & 0 \\ 0 & 6 & 0 & 0 & 1 \end{bmatrix}.$$

The resulting full sequence has the following associated matrix:

$$
\begin{bmatrix}
2 & 0 & 1 & 0 & 2 \\
0 & 2 & 0 & 0 & 7 \\
0 & 0 & 0 & 1 & 1 \\
0 & 0 & 0 & 0 & 4
\end{bmatrix}.
$$

Thus the kernel of f is generated by bc and c^4.

Exercises

6.1. Use the ideas in the discussion following Example 6.1 to devise a third algorithm for computing normal closures in polycyclic groups based on the Knuth-Bendix procedure.

6.2. Let G be the group generated by a, b, c, d subject to the relations

$$ba = ac, \quad ca = ad, \quad da = ab, \quad cb = bc, \quad db = bd, \quad dc = cd,$$
$$a^3 = b^4 = c^4 = d^4 = 1,$$

and let H be a cyclic group of order 12 generated by an element u. The map

$$a \mapsto u^4, \quad b \mapsto u^3, \quad c \mapsto u^3, \quad d \mapsto u^3$$

extends to a homomorphism f of G onto H. Find the kernel of f.

9.7 Conjugacy in nilpotent groups

As noted at the end of Section 9.4, the conjugacy problem in polycyclic groups is solvable in principle, but the available algorithms are not practical for most infinite polycyclic groups. However, for finitely generated nilpotent groups the situation is much better. This section describes the computation of centralizers and the determination of conjugacy in nilpotent polycyclic groups.

Let G be a finitely generated nilpotent group. We shall assume that G is given by a standard polycyclic presentation on generators $a_1, \ldots, a_n$ and that this presentation is nilpotent as defined in Section 9.4. If the presentation is not nilpotent, then we can compute the lower central series of G, determine a polycyclic series which refines the lower central series, make a corresponding choice of a polycyclic generating sequence, and determine the standard polycyclic presentation of G with respect to the new generators.

The first problem to be considered is the computation of centralizers. Let g be an element of G. The following recursive procedure may be used to compute the centralizer $C_G(g)$ of g in G. The subgroup $N = \mathrm{Grp}\,\langle a_n \rangle$ is contained in the center of G and hence is normal. By induction on n, we can compute the centralizer K of gN in G/N and find the inverse image L

of K in G. The subgroup L is the set of all elements u of G such that $u^{-1}gu$ is in gN, or, equivalently, $f(u) = [g, u] = g^{-1}u^{-1}gu$ is in N. Restricted to L, the function f is a homomorphism into N, for if u_1 and u_2 are in L, then $f(u_1u_2) = [g, u_1u_2]$, which by Proposition 1.6(b) is $[g, u_2][g, u_1][g, u_1, u_2]$. But $[g, u_1]$ and $[g, u_2]$ are in N, so they commute. Also, $[g, u_1, u_2] = 1$. Thus $f(u_1u_2) = [g, u_1][g, u_2]$. The centralizer $C_G(g)$ is the kernel of f. Since N is a cyclic group, the computation of the kernel of f is an easy application of the methods of Section 9.6.

Example 7.1. Let us consider the group $D = D_4^{(1)}(\mathbb{Z})$ of Examples 2.1, 3.1, 4.1, 4.5, 5.2, 5.3, and 5.4. We shall find the centralizer of $g = a_1a_3^2$ in D. For $1 \leq i \leq 7$, let $N_i = \operatorname{Grp}\langle a_i, \ldots, a_6 \rangle$. Our approach is to compute the centralizer C_i of gN_i in D/N_i, $1 \leq i \leq 7$. The group we really want is C_7. For $1 \leq i \leq 4$, the group D/N_i is abelian, so $C_i = D/N_i$. The inverse image L_5 in D/N_5 of C_4 is generated by the images of a_1, a_2, a_3, and a_4. Working modulo N_5, we have

$$g^{-1}a_1^{-1}ga_1 \equiv 1, \quad g^{-1}a_2^{-1}ga_2 \equiv a_4, \quad g^{-1}a_3^{-1}ga_3 \equiv 1, \quad g^{-1}a_4^{-1}ga_4 \equiv 1.$$

The kernel C_5 of the homomorphism from L_5 into N_4/N_5 is generated by the images of a_1, a_3, and a_4. The inverse image L_6 in D/N_6 of C_5 is generated by the images of a_1, a_3, a_4, and a_5.

Modulo N_6, we have

$$g^{-1}a_1^{-1}ga_1 \equiv 1, \quad g^{-1}a_3^{-1}ga_3 \equiv 1, \quad g^{-1}a_4^{-1}ga_4 \equiv 1, \quad g^{-1}a_5^{-1}ga_5 \equiv 1.$$

Thus $C_6 = L_6$ and its inverse image L_7 in $D = D/N_7$ is generated by a_1, a_3, a_4, a_5, and a_6. In D,

$$g^{-1}a_1^{-1}ga_1 = 1, \quad g^{-1}a_3^{-1}ga_3 = 1, \quad g^{-1}a_4^{-1}ga_4 = a_6^{-2},$$
$$g^{-1}a_5^{-1}ga_5 = a_6, \quad g^{-1}a_6^{-1}ga_6 = 1.$$

The kernel C_7 of the homomorphism from L_7 to N_6 is $\operatorname{Grp}\langle a_1, a_3, a_4a_5^2, a_6 \rangle$.

The determination of conjugacy involves only a slight extension of the algorithm for computing centralizers. Now we have two elements g and h of G and we want to decide whether g and h are conjugate. We consider gN and hN in G/N, where $N = \operatorname{Grp}\langle a_n \rangle$. If gN and hN are not conjugate, then g and h are not conjugate in G. If gN and hN are conjugate, then we can find an element u of G such that $u^{-1}(hN)u = u^{-1}huN$ is gN. Replacing h by $u^{-1}hu$, we may assume that $gN = hN$. In this case, if g and h are conjugate, the conjugating element lies in the inverse image L of the centralizer in G/N of gN. We know how to compute L. Let f be the homomorphism from L to N mapping v to $[g, v]$. The conjugates of g by elements of L are the elements of $gf(L)$. Let $w = g^{-1}h$. Then h is conjugate to g if and only if w is in $f(L)$. We can decide this and, if w is in $f(L)$, we can find an element v of L such that $f(v) = w$. In this case $h = v^{-1}gv$.

Exercise

7.1. In the group D of Example 7.1 determine the centralizers of the elements $a_1 a_4 a_5^3$ and $a_2^2 a_4^4 a_6$.

9.8 Cyclic extensions

A group G is a *cyclic extension* of a group N if N is a normal subgroup of G and G/N is cyclic. In this section we shall take the point of view that N is given and we wish to construct cyclic extensions of N. A polycyclic group is a group which can be built up by starting with the trivial group, constructing a cyclic extension, then making a cyclic extension of that group, and continuing a finite number of times. The theory of cyclic extensions is well understood. The exposition here is based in large part on Sections III.7 and III.8 of [Zassenhaus 1958].

We shall look first at the case in which G/N is infinite cyclic. Let x be an element of G such that xN generates G/N. Since N is normal, the map σ taking an element v of N to $x^{-1}vx$ is an automorphism of N. The image of v under σ will be written v^σ. The group G is determined up to isomorphism by N and σ, since any g in G can be represented uniquely as $x^i v$ with v in N and

$$(x^i v)(x^j w) = x^i x^j x^{-j} v x^j w = x^{i+j} v^{\sigma^j} w.$$

Moreover, any automorphism of N can occur this way, for if σ is an automorphism of N, then the binary operation

$$(i, v)(j, w) = (i + j, v^{\sigma^j} w)$$

defines a group structure on $G = \mathbb{Z} \times N$. The identity element of G is $(0, 1_N)$. The inverse of (i, v) is $(-i, (v^{-1})^{\sigma^{-i}})$. The associative law is proved as follows.

$$(i, a)[(j, b)(k, c)] = (i, a)(j + k, b^{\sigma^k} c) = (i + j + k, a^{\sigma^{j+k}} b^{\sigma^k} c),$$

$$[(i, a)(j, b)](k, c) = (i + j, a^{\sigma^j} b)(k, c) = (i + j + k, (a^{\sigma^j} b)^{\sigma^k} c).$$

Since σ is an automorphism,

$$\left(a^{\sigma^j} b\right)^{\sigma^k} = \left(a^{\sigma^j}\right)^{\sigma^k} b^{\sigma^k} = a^{\sigma^{j+k}} b^{\sigma^k}.$$

If σ is the identity automorphism of N, then G is the ordinary direct product of $\mathbb{Z}$ and N. For any σ, if we identify $\{0\} \times N$ with N, then N is a normal subgroup of G and G/N is isomorphic to $\mathbb{Z}$.

There is an alternative way to see that any automorphism of N leads to a cyclic extension G of N such that G/N is infinite. Let x and x^{-1} be objects not in N and set $Y = N \cup \{x, x^{-1}\}$. Let $\mathcal{R}$ be the rewriting system on Y^* consisting of the following rules: the semigroup relations from the multiplication table of N as described in Section 2.3, the rules $xx^{-1} \to \varepsilon$ and $x^{-1}x \to \varepsilon$, and the rules $vx \to xv^{\sigma}$ and $vx^{-1} \to x^{-1}v^{\sigma^{-1}}$ for v in N. Without too much difficulty, it is possible to show that $\mathcal{R}$ is confluent. If S is the ideal of Y^* generated by N, then $(Y, \mathcal{R}, S)$ is a restricted presentation for the group G.

Now let us turn to the case in which G is a cyclic extension of N and N has finite index n in G. Again choose an element x such that xN generates G/N and let σ be the automorphism of N induced by x. Since the coset xN has order n in G/N, it follows that $(xN)^n = x^nN = N$. Therefore $u = x^n$ is an element of N. Knowing N, σ, n, and u determines G up to isomorphism. Every element of G can be expressed uniquely as x^iv, where v is in N and $0 \le i < n$. Also

$$(x^iv)(x^jw) = \begin{cases} x^{i+j}v^{\sigma^j}w, & i+j < n, \\ x^{i+j-n}uv^{\sigma^j}w, & i+j \ge n. \end{cases}$$

Not all pairs (σ, u) can occur. Since $u = x^n$, it follows that $u^{\sigma} = x^{-1}x^nx = x^n = u$. Thus σ fixes u. Also, for any v in N we have $v^{\sigma^n} = x^{-n}vx^n = u^{-1}vu$. Therefore σ^n is the inner automorphism of N induced by u. These two conditions are sufficient for the pair (σ, u) to arise in a cyclic extension G of N with G/N of order n. Again there are two ways to see this. We can define a binary operation on $\{0, 1, \ldots, n-1\} \times N$ by the formula

$$(i, v)(j, w) = \begin{cases} (i+j, v^{\sigma^j}w), & i+j < n, \\ (i+j-n, uv^{\sigma^j}w), & i+j \ge n, \end{cases}$$

and prove the structure defined is a group. On the other hand, we can choose an object x not in N, define $Y = N \cup \{x\}$, and form the rewriting system $\mathcal{R}$ consisting of the multiplication-table rules for N, the rule $x^n \to u$, and the rules $vx \to xv^{\sigma}$ for v in N. We then must prove that $\mathcal{R}$ is confluent and that $(Y, \mathcal{R}, S)$ is a restricted presentation for a group, where S is the ideal of Y^* generated by N. The amounts of work involved in the two approaches are roughly the same, and the choice reduces to a matter of taste.

The point of view of cyclic extensions gives us a new perspective on the test for consistency of a presentation with the form of a standard polycyclic presentation. In such a presentation we have generators $a_1, \ldots, a_n$, a subset I of $\{1, \ldots, n\}$, and for each i in I an integer m_i greater than 1. In this discussion we shall use the monoid version of the standard polycyclic presentation in which there is a generator a_i^{-1} only when i is not in I. The set $\mathcal{R}$ of monoid relations has the form

$$a_i a_1^{-1} = 1, \quad a_1^{-1} a_i = 1, \quad i \notin I,$$
$$a_j a_i = a_i a_{i+1}^{\alpha_{iji+1}} \ldots a_n^{\alpha_{ijn}}, \quad j > i,$$
$$a_j^{-1} a_i = a_i a_{i+1}^{\beta_{iji+1}} \ldots a_n^{\beta_{ijn}}, \quad j > i, \ j \notin I,$$
$$a_j a_i^{-1} = a_i^{-1} a_{i+1}^{\gamma_{iji+1}} \ldots a_n^{\gamma_{ijn}}, \quad j > i, \ j \notin I,$$
$$a_j^{-1} a_i^{-1} = a_i^{-1} a_{i+1}^{\delta_{iji+1}} \ldots a_n^{\delta_{ijn}}, \quad j > i, \ i,j \notin I,$$
$$a_i^{m_i} = a_{i+1}^{\mu_{iji+1}} \ldots a_n^{\mu_{in}}, \quad i \in I,$$

where the right sides are collected in the sense that the integers α_{ijk}, β_{ijk}, γ_{ijk}, δ_{ijk}, and μ_{ik} are between 0 and $m_k - 1$ when k is in I. Let Y be the set of generators occurring in $\mathcal{R}$. The monoid G defined by $(Y, \mathcal{R})$ is a group. If $(Y, \mathcal{R})$ is consistent, then many of the relations in $\mathcal{R}$ are redundant. To prove this, we shall use a criterion equivalent to consistency. Let $Y_i = Y \cap \{a_i, \ldots, a_n\}^{\pm}$, let $\mathcal{R}_i$ consist of the relations in $\mathcal{R}$ which involve only generators in Y_i, and let G_i be the subgroup of G generated by Y_i.

Proposition 8.1. *The presentation $(Y, \mathcal{R})$ is consistent if and only if for $1 \leq i \leq n$ any relation in G of the form $U = V$ with U and V in Y_i^* is a consequence of the relations in $\mathcal{R}_i$.*

Proof. Suppose that $(Y, \mathcal{R})$ is consistent. Then any word in Y_i^* can be rewritten into collected form using the relations in $\mathcal{R}_i$, interpreted as rewriting rules. If the relation $U = V$ holds in G and both U and V are in Y_i^*, then U and V have the same collected form W, and the relations $U = W$ and $V = W$ are consequences of $\mathcal{R}_i$. Thus $U = V$ is a consequence of $\mathcal{R}_i$.

Now suppose that any relation $U = V$ in G with U and V in Y_i^* is a consequence of $\mathcal{R}_i$. If $(Y, \mathcal{R})$ is not consistent, then for some i, $1 \leq i \leq n$, there is an integer $m > 0$ such that a_i^m is in G_{i+1} and either i is not in I or i is in I and $m < m_i$. The relations in $\mathcal{R}_i$ are satisfied if we set a_j equal to 1 for $i < j \leq n$. If this is done, then the group defined is infinite cyclic if i is not in I or cyclic of order m_i if i is in I. Therefore in the group generated by Y_i and defined by $\mathcal{R}_i$ no relation $a_i^m = W$ with W in Y_{i+1}^* can hold. Therefore $(Y, \mathcal{R})$ is consistent. $\square$

Proposition 8.2. *If $(Y, \mathcal{R})$ is consistent, then the relations in $\mathcal{R}$ with positive left sides define G as a group.*

Proof. Assume that $(Y, \mathcal{R})$ is consistent and let $\mathcal{S}_i$ denote the set of relations in $\mathcal{R}$ with left sides $a_k a_j$ or $a_j^{m_j}$, where $i \leq j$. Thus $\mathcal{S}_n$ is empty if n is not in I and $\mathcal{S}_n$ consists of the single relation $a_n^{m_n} = 1$ if n is in I. In either case, $\mathcal{S}_n$ defines G_n as a group. Now suppose that we know $\mathcal{S}_{i+1}$ defines G_{i+1}. To show that $\mathcal{S}_i$ defines G_i, we must prove that all the relations in $\mathcal{R}_i$ are consequences of $\mathcal{S}_i$. (Since we are considering $\mathcal{S}_i$ to be a set of group relations, the relations $a_i a_i^{-1} = a_i^{-1} a_i = 1$ come for free.) By assumption, the relations in $\mathcal{R}_{i+1}$ are consequences of $\mathcal{S}_{i+1}$ and hence of $\mathcal{S}_i$. Suppose that $j > i$ and j is not in I. The relation

$$a_j^{-1} a_i = a_i a_{i+1}^{\beta_{iji+1}} \ldots a_n^{\beta_{ijn}}$$

is equivalent to

$$a_i = a_j a_i a_{i+1}^{\beta_{iji+1}} \ldots a_n^{\beta_{ijn}},$$

which is equivalent modulo the relations in $\mathcal{S}_i$ to

$$a_i = a_i a_{i+1}^{\alpha_{iji+1}} \ldots a_n^{\alpha_{ijn}} a_{i+1}^{\beta_{iji+1}} \ldots a_n^{\beta_{ijn}},$$

which in turn is equivalent to

$$1 = a_{i+1}^{\alpha_{iji+1}} \ldots a_n^{\alpha_{ijn}} a_{i+1}^{\beta_{iji+1}} \ldots a_n^{\beta_{ijn}}.$$

This is a relation in G_{i+1} and hence is a consequence of $\mathcal{R}_{i+1}$ and thus of $\mathcal{S}_i$. Suppose that i is not in I and $j > i$. The relation

$$a_j a_i^{-1} = a_i^{-1} a_{i+1}^{\gamma_{iji+1}} \ldots a_n^{\gamma_{ijn}}$$

is equivalent to

$$a_j = a_i^{-1} a_{i+1}^{\gamma_{iji+1}} \ldots a_n^{\gamma_{ijn}} a_i.$$

Using only relations with left sides $a_k^\eta a_i$, $\eta = \pm 1$, and $a_i^{-1} a_i = 1$, one can rewrite the right side of this relation into a word W in Y_{i+1}^*. The relation $a_j = W$ is a consequence of $\mathcal{R}_{i+1}$ and hence of $\mathcal{S}_i$. Therefore the relation

$$a_j a_i^{-1} = a_i^{-1} a_{i+1}^{\gamma_{iji+1}} \ldots a_n^{\gamma_{ijn}}$$

is a consequence of $\mathcal{S}_i$. The relations with left sides $a_j^{-1} a_i^{-1}$ are handled like those with left sides $a_j^{-1} a_i$. By induction on $n - i$, the relations in $\mathcal{S}_1$ define $G_1 = G$. □

Suppose now that we have a presentation of the form $(Y, \mathcal{R})$ as before and we want to decide whether it is consistent. If $n = 1$, then either there are the two relations $a_1 a_1^{-1} = a_1^{-1} a_1 = 1$ or there is the single relation $a_1^{m_1} = 1$. In either case the presentation is consistent. Let us assume that $n > 1$. The group G defined by $(Y, \mathcal{R})$ is a cyclic extension of the subgroup G_2 generated by $a_2, \ldots, a_n$. By induction on n, we may assume that we have already checked that the relations in $(Y, \mathcal{R})$ which involve only $a_2, \ldots, a_n$ and their inverses form a consistent presentation of a group K. We can compute products and test equality of elements in K. There is an obvious homomorphism from K onto G_2. By Proposition 8.1, $(Y, \mathcal{R})$ is consistent if and only if this homomorphism is an isomorphism. This can be decided by checking that the conditions for a cyclic extension are satisfied.

We first test whether the map of generators

$$a_j \mapsto a_2^{\alpha_{1j2}} \ldots a_n^{\alpha_{1jn}}, \quad 2 \le j \le n,$$

extends to a homomorphism σ of K into itself. This is done by checking whether the images of $a_2, \ldots, a_n$ satisfy a set of defining relations for K. Proposition 8.2 can be used to reduce the number of relations which must be checked. Assuming that σ is defined, we next need to decide whether σ is surjective. This could be done by showing that $a_2^\sigma, \ldots, a_n^\sigma$ generate K. However, if 1 is not in I, then it is quicker to test whether

$$(a_2^{\gamma_{1j2}} \ldots a_n^{\gamma_{1jn}})^\sigma = a_j, \quad 2 \le j \le n.$$

If 1 is in I, then we shall have to test whether σ^{m_1} is the inner automorphism of K induced by $a_1^{m_1}$, and a positive result implies that σ maps K onto itself. If σ is surjective, then σ is an automorphism of K, since K is hopfian by Corollary 3.11. To complete the first phase of our consistency check, we test whether

$$a_2^{\beta_{1j2}} \ldots a_n^{\beta_{1jn}} = (a_2^{\alpha_{1j2}} \ldots a_n^{\alpha_{1jn}})^{-1}$$

in K if j is not in I and whether

$$a_2^{\delta_{1j2}} \ldots a_n^{\delta_{1jn}} = (a_2^{\gamma_{1j2}} \ldots a_n^{\gamma_{1jn}})^{-1}$$

if 1 and j are not in I.

Suppose that all the tests so far have been successful. Then the conditions for an infinite cyclic extension of K are satisfied. If 1 is not in I, then G_2 is isomorphic to K and $(Y, \mathcal{R})$ is consistent. However, if 1 is in I, then there is more work to do. Let u be the element $a_2^{\mu_{12}} \ldots a_n^{\mu_{1n}}$ of K. We must check whether $u^\sigma = u$ and whether $a_j^{\sigma^{m_1}} = u^{-1} a_j u$, $2 \le j \le n$. If these conditions are satisfied, then the pair (σ, u) defines a cyclic extension of K with quotient of order m_1.

We can rephrase the previous discussion in terms of rewriting rules. The following approach is inspired in part by (Vaughan-Lee 1984, 1985). Let us interpret the relations in $\mathcal{R}$ as rewriting rules. Among the overlaps of left sides in $\mathcal{R}$ are the following:

$$
\begin{aligned}
& a_k a_j a_i, \quad k > j > i, \\
& a_j^{m_j} a_i, \quad j \in I, \quad j > i, \\
& a_j a_i^{m_i}, \quad i \in I, \quad j > i, \\
& a_j a_i^{-1} a_i, \quad i \notin I, \quad j > i, \qquad (*) \\
& a_i^{m_i+1}, \quad i \in I, \\
& a_j^{-1} a_j a_i, \quad j \notin I, \quad j > i, \\
& a_j^{-1} a_j a_i^{-1}, \quad i, j \notin I, \, j > i.
\end{aligned}
$$

Proposition 8.3. *If local confluence holds at the overlaps* $(*)$*, then* $\mathcal{R}$ *is confluent.*

Proof. Let $\prec$ be the basic wreath-product ordering of Y^* with $a_n \prec \cdots \prec a_1$, $a_1 \prec a_1^{-1}$ if 1 is not in I, and $a_i \prec a_i^{-1} \prec a_{i-1}$ if $i > 1$ and i is not in I. To simplify the exposition, let us introduce the following notation:

$$
\begin{aligned}
S_{ij} &= a_{i+1}^{\alpha_{iji+1}} \ldots a_n^{\alpha_{ijn}}, \\
T_{ij} &= a_{i+1}^{\beta_{iji+1}} \ldots a_n^{\beta_{ijn}}, \\
U_{ij} &= a_{i+1}^{\gamma_{iji+1}} \ldots a_n^{\gamma_{ijn}}, \\
V_{ij} &= a_{i+1}^{\delta_{iji+1}} \ldots a_n^{\delta_{ijn}}, \\
W_i &= a_{i+1}^{\mu_{ij+1}} \ldots a_n^{\mu_{in}}.
\end{aligned}
$$

Suppose that $\mathcal{R}$ is not consistent. By induction on n, we may assume that the relations not involving a_1 or a_1^{-1} form a consistent presentation for a group K. This means that, if $k > j > 1$ and the indicated words are defined, then S_{jk} and T_{jk} represent inverse elements in K, as do U_{jk} and V_{jk}. Also, $U_{jk} a_j$ and $a_j a_k$ represent the same element of K. The notation $Q \rightarrow R$ will be used to indicate that R is obtained from Q by the application of one rewriting rule in $\mathcal{R}$, while $Q \overset{*}{\longrightarrow} R$ will signal that zero or more rules have been used. $\square$

Lemma 8.4. *If* $j > 1$ *and* j *is not in* I*, then* S_{1j} *and* T_{1j} *represent inverse elements of* K*.*

Proof. By assumption, there is local confluence at the word $a_j^{-1}a_ja_1$. The first few steps in processing this overlap are uniquely determined. They are

$$a_j^{-1}a_ja_1 \to a_1$$

and

$$a_j^{-1}a_ja_1 \to a_j^{-1}a_1S_{1j} \to a_1T_{1j}S_{1j}.$$

The word a_1 is irreducible with respect to $\mathcal{R}$. Since local confluence holds, there is a reduction of $T_{1j}S_{1j}$ to the empty word using $\mathcal{R}$. This means that S_{1j} and T_{1j} represent inverse elements of K. □

Let P be the first word with respect to $\prec$ at which confluence fails. By Proposition 7.1 in Chapter 2, P is an overlap containing exactly two left sides. Local confluence fails at P and P contains a_1 or a_1^{-1}. Since P is not one of the overlaps $(*)$, P must have one of the following forms:

(1) $a_k^{-1}a_ja_1$,	(9) $a_ka_j^{-1}a_1^{-1}$,
(2) $a_ka_j^{-1}a_1$,	(10) $a_k^{-1}a_j^{-1}a_1^{-1}$,
(3) $a_k^{-1}a_j^{-1}a_1$,	(11) $a_j^{-1}a_1^{-1}a_1$,
(4) $a_ja_j^{-1}a_1$,	(12) $a_ja_j^{-1}a_1^{-1}$,
(5) $a_j^{-1}a_1^{m_1}$,	(13) $a_ja_1a_1^{-1}$,
(6) $a_ka_ja_1^{-1}$,	(14) $a_j^{-1}a_1a_1^{-1}$,
(7) $a_k^{-1}a_ja_1^{-1}$,	(15) $a_1a_1^{-1}a_1$,
(8) $a_j^{m_j}a_1^{-1}$,	(16) $a_1^{-1}a_1a_1^{-1}$.

For each of these forms there are assumptions that one or more of the indices involved does or does not belong to I. For example, in (5) j is not in I and 1 is in I. Local confluence clearly holds in cases (15) and (16). There are many similarities in the consideration of the other 14 cases. Only a few cases will be discussed in detail. The remaining ones are left as exercises.

Case (1). Suppose that P has the form (1) for some indices j and k with $1 < j < k \le n$ and k not in I. Let y and z be the elements of K defined by S_{1j} and S_{1k}, respectively. Then T_{1k} defines z^{-1}. The word $Q = S_{jk}T_{jk}$ is in Y_{j+1}^*, so Qa_1 precedes P with respect to $\prec$. Therefore confluence holds at Qa_1. Let M and N be any words such that $S_{jk}a_1 \xrightarrow{*} a_1M$ and $T_{jk}a_1 \xrightarrow{*} a_1N$. Then we have the reductions

$$S_{jk}T_{jk}a_1 \xrightarrow{*} S_{jk}a_1N \xrightarrow{*} a_1MN$$

and

$$S_{jk}T_{jk}a_1 \xrightarrow{*} a_1,$$

since S_{jk} and T_{jk} represent inverse elements of K. Therefore MN must reduce to the empty word. Hence, if M represents the element u of K, then N represents u^{-1}.

By assumption, local confluence holds at $a_k a_j a_1$. The initial reductions here are

$$a_k a_j a_1 \to a_k a_1 S_{1j} \to a_1 S_{1k} S_{1j}$$

and

$$a_k a_j a_1 \to a_j S_{jk} a_1 \xrightarrow{*} a_j a_1 M \to a_1 S_{1j} M.$$

Hence $S_{1k}S_{1j}$ and $S_{1j}M$ define the same element of K. Therefore $zy = yu$, so $yu^{-1} = z^{-1}y$.

The initial reductions in processing the overlap $a_k^{-1} a_j a_1$ are

$$a_k^{-1} a_j a_1 \to a_k^{-1} a_1 S_{1j} \to a_1 T_{1k} S_{1j}$$

and

$$a_k^{-1} a_j a_1 \to a_j T_{jk} a_1 \xrightarrow{*} a_j a_1 N \to a_1 S_{1j} N.$$

Now $T_{1k}S_{1j}$ defines $z^{-1}y$ and $S_{1j}N$ defines yu^{-1}. By the previous remark, if R is the reduced word representing $z^{-1}y$, then $a_1 R$ is derivable from both $a_1 T_{1k} S_{1j}$ and $a_1 S_{1j} N$. Thus local confluence holds at $a_k^{-1} a_j a_1$ after all.

Case (2). Suppose that P has the form (2) for some j and k with $1 < j < k \le n$ and j not in I. Let y and z be as in case (1). Then T_{1j} represents y^{-1}. Let L be any word such that $U_{jk}a_1 \xrightarrow{*} a_1 L$ and let v be the element of K represented by L. Since $U_{jk}a_j$ and $a_j a_k$ represent the same element of K and $a_j a_k$ is reduced, we have $U_{jk}a_j \xrightarrow{*} a_j a_k$. The word $U_{jk}a_j a_1$ precedes P with respect to $\prec$, so confluence holds at $U_{jk}a_j a_1$. Now

$$U_{jk}a_j a_1 \to U_{jk}a_1 S_{1j} \xrightarrow{*} a_1 L S_{1j}$$

and

$$U_{jk}a_j a_1 \xrightarrow{*} a_j a_k a_1 \to a_j a_1 S_{1k} \to a_1 S_{1j} S_{1k}.$$

By confluence, $vy = yz$, so $zy^{-1} = y^{-1}v$.

Processing the overlap P, we have

$$a_k a_j^{-1} a_1 \to a_k a_1 T_{1j} \to a_1 S_{1k} T_{1j}$$

and

$$a_k a_j^{-1} a_1 \to a_j^{-1} U_{jk} a_1 \xrightarrow{*} a_j^{-1} a_1 L \to a_1 T_{1j} L.$$

If R is the reduced word defining zy^{-1}, then $a_1 R$ is derivable from both $a_1 S_{1k} T_{1j}$ and $a_1 T_{1j} L$. Therefore local confluence holds at P.

Case (5). Suppose that P has the form (5) with j not in I and 1 in I. Let M and N be words such that $S_{1j} a_1^{m_1-1} \xrightarrow{*} a_1^{m_1-1} M$ and $T_{1j} a_1^{m_1-1} \xrightarrow{*} a_1^{m_1-1} N$, and let u be the element of K represented by M. The word $S_{1j} T_{1j} a_1^{m_1-1}$ precedes P with respect to $\prec$. By essentially the same argument as used in case (1), N represents u^{-1}.

By assumption, local confluence holds at $a_j a_1^{m_1-1}$. The initial reductions at that word are

$$a_j a_1^{m_1} \to a_j W_1$$

and

$$a_j a_1^{m_1} \to a_1 S_{ij} a_1^{m_1-1} \xrightarrow{*} a_1^{m_1} M \to W_1 M.$$

Therefore $a_j W_1$ and $W_1 M$ represent the same element of K. From this it follows that $a_j^{-1} W_1$ and $W_1 N$ represent the same element.

The initial reductions at P are

$$a_j^{-1} a_1^{m_1} \to a_j^{-1} W_1$$

and

$$a_j^{-1} a_1^{m_1} \to a_1 T_{1j} a_1^{m_1-1} \xrightarrow{*} a_1^{m_1} N \to W_1 N.$$

By the remark above, local confluence holds.

If P is not of the first five forms, then confluence holds at all words not involving a_1^{-1}. Thus we may assume that 1 is not in I. We can define a map $^-$ from Y_2^* to itself as follows: Given Q in Y_2^*, define $\overline{Q}$ by $Q a_1 \xrightarrow{*} a_1 \overline{Q}$ and $a_1 \overline{Q}$ is irreducible with respect to $\mathcal{R}$. Confluence at $Q a_1$ implies that $\overline{Q}$ is well defined. If $Q \xrightarrow{*} R$, then one of the ways to reduce $Q a_1$ is

$$Q a_1 \xrightarrow{*} R a_1 \xrightarrow{*} a_1 \overline{R},$$

so $\overline{Q} = \overline{R}$. This means that we can define a map $\sigma \colon K \to K$ which takes the element represented by Q to the element represented by $\overline{Q}$.

It is easy to check that σ is a homomorphism. If $j > 1$, then we have local confluence at $a_j a_1^{-1} a_1$. The two reductions of this word are

$$a_j a_1^{-1} a_1 \to a_j$$

and

$$a_j a_1^{-1} a_1 \to a_1^{-1} U_{jk} a_1 \xrightarrow{*} a_1^{-1} a_1 M \to M \xrightarrow{*} \overline{U_{jk}},$$

where M is some word such that $U_{jk} a_1 \xrightarrow{*} a_1 M$. Thus $\overline{U_{jk}} = a_j$. Since the image of σ is a subgroup of K, that image is all of K. Therefore, by Corollary 3.11, σ is an automorphism of K. This mean that if Q and R are in Y_2^* and $\overline{Q} = \overline{R}$, then Q and R represent the same element of K.

Case (6). Suppose that P has the form (6) for some j and k with $1 < j < k \leq n$. Let L be any word such that $S_{jk} a_1^{-1} \xrightarrow{*} a_1^{-1} L$. Consider the reductions

$$S_{jk} a_1^{-1} a_1 \to S_{jk}$$

and

$$S_{jk} a_1^{-1} a_1 \xrightarrow{*} a_1^{-1} L a_1 \xrightarrow{*} a_1^{-1} a_1 \overline{L} \to \overline{L}.$$

Since $S_{jk} a_1^{-1} a_1$ precedes P, confluence holds and $\overline{L} = S_{jk}$.

Processing the overlap P leads to the following reductions:

$$a_k a_j a_1^{-1} \to a_k a_1^{-1} U_{1j} \to a_1^{-1} U_{1k} U_{1j}$$

and

$$a_k a_j a_1^{-1} \to a_j S_{jk} a_1^{-1} \xrightarrow{*} a_j a_1^{-1} L \to a_1^{-1} U_{1j} L.$$

Now

$$U_{1k} U_{1j} a_1 \xrightarrow{*} U_{1k} a_1 a_j \xrightarrow{*} a_1 a_k a_j \to a_1 a_j S_{jk}$$

and

$$U_{1j} L a_1 \xrightarrow{*} U_{1j} a_1 S_{jk} \xrightarrow{*} a_1 a_j S_{jk}.$$

Hence

$$\overline{U_{1j} L} = a_j S_{jk} = \overline{U_{1k} U_{1j}},$$

so $U_{1j}L$ and $U_{1k}U_{1j}$ represent the same element of K. This means that local confluence holds at P.

Corollary 8.5. *If $(Y, \mathcal{R})$ is not consistent, then the first word with respect to $\prec$ at which confluence fails is one of the overlaps* $(*)$.

If $I = \{1, \ldots, n\}$, Proposition 8.3 says nothing new. However, if $I = \emptyset$, then for large n Proposition 8.3 reduces the amount of work needed to check consistency by a factor of 8. The number of overlaps between left sides of commutation relations is cubic in n, while the number of all other overlaps is quadratic in n. Without Proposition 8.3, we would have to check local consistency at all overlaps $a_k^\epsilon a_j^\nu a_i^\chi$, where $k > j > i$ and ϵ, η, and χ range independently over $\{1, -1\}$. With Proposition 8.3, we need only consider the case $\epsilon = \eta = \chi = 1$.

It is possible to convert the consistency test of Proposition 8.3 into a procedure for producing a consistent presentation for the group given by a presentation $(Y, \mathcal{R})$ which is not consistent. The following examples illustrate the ideas. The details are left as an exercise to the reader.

Example 8.1. Let G be the group defined by the following power-conjugate presentation on generators x, y, z, in that order:

$$zy = yz, \quad yx = xz, \quad zx = xy, \quad z^{15} = 1, \quad y^9 = 1, \quad x^2 = 1.$$

The relations involving only y and z form a consistent presentation for $\mathbb{Z}_9 \times \mathbb{Z}_{15}$. Conjugation by x takes y to z and z to y. In order for this map to define an automorphism of Grp $\langle y, z \rangle$, y^{15} and z^9 must be trivial. This implies the power relations $y^3 = 1$ and $z^3 = 1$. With these new power relations the presentation is confluent.

Example 8.2. Let H be the group defined by the nilpotent presentation on generators a, b, c, d, e, f, g, h, in that order, in which the nontrivial commutation relations are

$$ba = abc, \quad ca = acd, \quad cb = bce,$$

and the power relations are

$$a^2 = fh, \quad b^2 = gh, \quad c^2 = h^2.$$

The relations on c, d, e, f, g, and h form a consistent presentation of an abelian group. Conjugation by b maps c to ce and fixes d, e, f, g, and h. In order for this map to extend to an automorphism σ, the relation $c^2 = h$ must be preserved. That is, $(ce)^2$ must equal h, so $e^2 = 1$. Let us add this relation to the presentation. Since b^2 is in Grp $\langle c, d, e, f, g, h \rangle$, which is abelian, σ^2 must be the identity. Now $\sigma^2(c) = \sigma(ce) = ce^2 = c$, so σ^2 is 1. The condition that σ must preserve $b^2 = gh$ is also satisfied. Thus with the addition of $e^2 = 1$ we have a consistent presentation on b, c, d, e, f, g, and h. Now let τ be the automorphism induced on Grp $\langle b, c, d, e, f, g, h \rangle$ by a. The condition that τ preserves the relations $c^2 = h^2$ and $b^2 = gh$ leads to the relations $d^2 = 1$ and $e = h^2$, respectively. The existing relation $e^2 = 1$ now implies that $h^4 = 1$. The condition that τ^2 is the identity and hence fixes a produces $d = h^2$. The original presentation together with the relations of $d = h^2$, $e = h^2$, and $h^4 = 1$ is consistent.

Exercises

8.1. Use the ideas of this section to determine a consistent polycyclic presentation for the group generated by a, b, c subject to the relations

$$ ba = ab^{-1}, \quad ba^{-1} = a^{-1}b^{-1}, \quad ca = abc, \quad ca^{-1} = a^{-1}c^{-1}, \quad cb = bc^{-1}, \quad cb^{-1} = b^{-1}c^{-1}. $$

8.2. Complete the case analysis in the proof of Proposition 8.3.

9.9 Consistency, the nilpotent case

This section continues the discussion of Section 9.8. The goal remains to reduce as much as possible the amount of work needed to check the consistency of a presentation $(Y, \mathcal{R})$ which has the form of a standard monoid polycyclic presentation. The notation established in the previous section is still in effect.

The consistency criterion of Proposition 8.3 can be strengthened if $\mathcal{R}$ is a γ-weighted presentation as defined in Section 9.4. In this case the generators $a_1, \ldots, a_n$ have positive integer weights $1 = w_1 \leq \cdots \leq w_n$. (We assign weight w_i to a_i^{-1} if i is not in I.) All generators in the word W_i have weight at least $w_i + 1$ and S_{ij} has the form $a_j A_{ij}$, where every generator in A_{ij} has weight at least $w_i + w_j$. For each index k with $w_k > 1$ there are indices i and j with $i < j$, $w_i = 1$, $w_j = w_k - 1$, and $A_{ij} = a_k$. We choose one such pair (i, j) and call the relation $a_j a_i = a_i a_j a_k$ the *definition* of a_k. For each $e \geq 1$, the subgroup $G(e)$ generated by the a_k with $w_k \geq e$ is normal in G. Thus a_i and a_j commute modulo $G(w_i + w_j)$. Therefore, if $(Y, \mathcal{R})$ is consistent, then

$$ S_{ij} = a_j A_{ij}, $$
$$ T_{ij} = a_j^{-1} B_{ij}, $$
$$ U_{ij} = a_j C_{ij}, $$
$$ V_{ij} = a_j^{-1} D_{ij}, $$

where generators occurring in A_{ij}, B_{ij}, C_{ij}, or D_{ij} have weight at least $w_i + w_j$. We shall make this additional assumption.

Let $c = w_n$ and consider the following set of overlaps:

$$
\begin{aligned}
a_k a_j a_i, \quad & k > j > i,\; w_i + w_j + w_k \le c,\\
a_j^{m_j} a_i, \quad & j \in I, \qquad j > i,\; w_i + w_j < c,\\
a_j a_i^{m_i}, \quad & i \in I, \qquad j > i,\; w_i + w_j < c,\\
a_i^{m_i+1}, \quad & i \in I, \qquad\qquad\quad 2w_i < c, \qquad\qquad (**)\\
a_j a_i^{-1} a_i, \quad & i \notin I, \qquad j > i,\; w_i + w_j \le c,\\
a_j^{-1} a_j a_i, \quad & i \notin I, \qquad j > i,\; w_i + w_j \le c,\\
a_j^{-1} a_j a_i^{-1}, \quad & i, j \notin I,\; j > i,\; w_i + w_j \le c.
\end{aligned}
$$

Proposition 9.1. *Suppose that $(Y, \mathcal{R})$ is γ-weighted and satisfies the additional assumption. If $(Y, \mathcal{R})$ is not confluent, then the first word P at which confluence fails is one of the overlaps $(**)$.*

Proof. By Corollary 8.5, P is one of the overlaps $(*)$. Suppose first that $P = a_k a_j a_i$ with $k > j > i$ and $w_i + w_j + w_k > c$. The initial reductions at P are

$$ a_k a_j a_i \to a_k a_i a_j A_{ij} \to a_i a_k A_{ik} a_j A_{ij} $$

and

$$ a_k a_j a_i \to a_j a_k A_{jk} a_i. $$

Now any a_r^α, $\alpha = \pm 1$, occurring in A_{jk} satisfies $r > i$ and $w_r \ge w_j + w_k$. Therefore $w_i + w_r > c$ and $a_r^\alpha a_i \to a_i a_r^\alpha$ is in $\mathcal{R}$. Thus

$$ a_j a_k A_{jk} a_i \xrightarrow{\;*\;} a_j a_k a_i A_{jk} \to a_j a_i a_k A_{ik} A_{jk} \to a_i a_j A_{ij} a_k A_{ik} A_{jk}. $$

By the minimality of P, $(Y_{i+1}, \mathcal{R}_{i+1})$ is consistent and defines a group K. Any monoid generator a_r^α occurring in A_{ij} has weight at least $w_i + w_j$. Hence a_k and A_{ij} define commuting elements of K. By the same argument, a_j and A_{ik} define commuting elements, as do A_{ij}, A_{ik}, and A_{jk}. Therefore, if u is the element of K defined by $a_k A_{ik} a_j A_{ij}$, then u is also defined by the following words:

$$ a_k a_j A_{ik} A_{ij}, \quad a_j a_k A_{jk} A_{ik} A_{ij}, \quad a_j A_{ij} a_k A_{ik} A_{jk}. $$

Therefore, if R is the collected word in Y_{i+1}^* defining u, then $a_i R$ is derivable from both $a_i a_k A_{ik} a_j A_{ij}$ and $a_i a_j A_{ij} a_k A_{ik} A_{jk}$. Therefore local confluence holds at P.

Now suppose that $P = a_j^{m_j} a_i$ with $j > i$ and $w_i + w_j > c$. Since A_{ij} is empty, the initial reductions at P are

$$a_j^{m_j} a_i \to W_j a_i$$

and

$$a_j^{m_j} a_i \to a_j^{m_j-1} a_i a_j \xrightarrow{*} a_i a_j^{m_j} \to a_i W_j.$$

Every generator in W_j has weight at least $w_j + 1$. Therefore $W_j a_i \xrightarrow{*} a_i W_j$, so local confluence holds at P.

If $P = a_j a_i^{m_i}$ with $j > i$ and $w_i + w_j > c$, then the initial reductions at P are

$$a_j a_i^{m_i} \to a_j W_i$$

and

$$a_j a_i^{m_i} \to a_i a_j a_i^{m_i-1} \xrightarrow{*} a_i^{m_i} a_j \to W_i a_j.$$

Each generator occurring in W_i commutes with a_j, so $a_j W_i$ and $W_i a_j$ define the same element of the group $K = \mathrm{Mon}\langle Y_{i+1} \mid \mathcal{R}_{i+1}\rangle$. Thus confluence holds at P.

Suppose $P = a_i^{m_i+1}$ with $2w_i \geq c$. The reductions at P are

$$a_i^{m_i+1} \to a_i W_i$$

and

$$a_i^{m_i+1} \to W_i a_i \xrightarrow{*} a_i W_i,$$

since each a_r^{α}, $\alpha = \pm 1$, which occurs in W_i has weight $w_r \geq w_i + 1$. Therefore $w_r + w_i > c$ and $a_r^{\alpha} a_i \to a_i a_r^{\alpha}$ is in $\mathcal{R}$.

Finally, if $j > i$ and $w_i + w_j > c$, then it is easy to check that local confluence holds at $a_j a_i^{-1} a_i$, $a_i^{-1} a_j a_i$, and $a_i^{-1} a_j a_i^{-1}$. $\square$

One can improve Proposition 9.1 using ideas from (Vaughan-Lee 1984). For any i and j with $1 \leq i < j \leq n$ the set $\mathcal{W} = Y_j^* a_i Y_j^*$ is closed under rewriting. Suppose that confluence holds on $\mathcal{W}$. If P is in Y_j^*, then $P \xrightarrow{*} Q$ if and only if $a_i P \xrightarrow{*} a_i Q$. Therefore confluence holds on Y_j^* and $(Y_j, \mathcal{R}_j)$ is a consistent presentation for a group K_j. Let u be an element of K_j, let P in Y_j^* represent u, and let $P a_i \xrightarrow{*} a_i Q$. (Such a Q always exists.) An argument in the proof of Proposition 8.3 shows that the element v of K_j defined by Q depends only on u and that the map $u \mapsto v$ is a homomorphism

σ_i of K_j into itself. Since $a_k a_i \rightarrow a_i a_k A_{ik}$ for $j \leq k \leq n$, it is easy to see that σ_i is surjective. Therefore σ_i is an automorphism of K_j. (In the proof of Proposition 8.3, we could be sure that W was closed under rewriting only when $j = i + 1$ and we needed local confluence at the overlaps $a_k a_i^{-1} a_i$ or $a_k a_i^{m_i}$ to know that σ_i was surjective.)

Now suppose that $1 < j \leq n$ and let $L = b_1 \ldots b_r$ be a word in $\{a_1, \ldots, a_{j-1}\}^*$ which is irreducible with respect to $\mathcal{R}$. Set $\mathcal{W}(j, L) = Y_j^* b_1 Y_j^* \ldots Y_j^* b_r Y_j^*$.

Proposition 9.2. *Assume that j and L are as described and that confluence holds on $\mathcal{W}(j, b_i) = Y_j^* b_i Y_j^*$, $1 \leq i \leq r$. Then confluence holds on $\mathcal{W}(j, L)$. If P in Y_j^* represents an element u of K_j and $PL \xrightarrow{*} LQ$, then Q represents u^τ, where $\tau = \tau_1 \cdots \tau_r$ and τ_i is the automorphism of K_j induced by b_i.*

Proof. If $r \leq 1$, then there is nothing to prove, so we may assume that $r \geq 2$ and that confluence does not hold on $\mathcal{W}(j, L)$. Let $\mathcal{V}$ be the union over all subwords M of L of the sets $\mathcal{W}(j, M)$. The set $\mathcal{V}$ is closed under rewriting and under taking subwords. Let R be the first word in $\mathcal{V}$ with respect to $\prec$ at which confluence fails. By Exercise 7.2 in Chapter 2, R is an overlap of precisely two left sides. By assumption, R must involve at least two consecutive factors from L. Since L is irreducible, R must contain elements of Y_j as well. All elements of Y_j come before the factors of L in the ordering $\prec$. A simple case analysis shows that no such R can exist.

Now let P be any word in Y_j^* and let P represent u in K_j. We can rewrite PL as follows:

$$PL = Pb_1 \ldots b_r \xrightarrow{*} b_1 P_1 b_2 \ldots b_r \xrightarrow{*} b_1 b_2 P_2 b_3 \ldots b_r \xrightarrow{*} \cdots \xrightarrow{*} b_1 \ldots b_r P_r$$
$$= LP_r,$$

where each P_i is in K_j and represents the image u_i of u under $\tau_1 \cdots \tau_i$. We may assume that LP_r is irreducible. Now let Q be any word in K_j such that $PL \xrightarrow{*} LQ$. Since we have confluence on $\mathcal{W}(j, L)$, it follows that $LQ \xrightarrow{*} LP_r$, which means that $Q \xrightarrow{*} P_r$. Therefore Q represents u_r too. $\square$

Proposition 9.3. *Assume that $(Y, \mathcal{R})$ is as in Proposition 9.1. In testing local confluence at the overlaps $(**)$, we may assume that $w_i = 1$ when considering $a_k a_j a_i$ with $k > j > i$ and $a_j^{m_j} a_i$ with $j > i$.*

Proof. Suppose that local confluence holds at all the overlaps $(**)$, except perhaps at some of the words $a_k a_j a_i$ or $a_j^{m_j} a_i$ with $w_i > 1$. The presentation $(Y_n, \mathcal{R}_n)$ is consistent and confluence holds on $\mathcal{W}(n, a_k)$ for all $k < n$. The automorphisms of K_n induced by $a_1, \ldots, a_{n-1}$ are all trivial. Suppose r is such that $(Y_r, \mathcal{R}_r)$ is consistent and confluence holds on $\mathcal{W}(r, a_k)$ for all k

with $1 \le k < r$. Each of the generators a_k induces an automorphism σ_k on K_r. Choose r minimal such that if $k < r$ and $w_k > 1$, then $\sigma_j \sigma_i = \sigma_i \sigma_j \sigma_k$, where $a_j a_i = a_i a_j a_k$ is the definition of a_k. As noted, $r \le n$.

If $r = 1$, then $(Y, \mathcal{R})$ is consistent. Suppose $r > 1$.

Lemma 9.4. *The presentation* $(Y_{r-1}, \mathcal{R}_{r-1})$ *is consistent.*

Proof. By assumption, a_{r-1} induces an automorphism σ_{r-1} on K_r. If $r-1$ is not in I, then σ_{r-1} defines a cyclic extension H of K_r with H/K_r infinite. If $r - 1$ is in I, then local confluence at the words $a_s a_{r-1}^{m_{r-1}}$ with $s \ge r - 1$ shows that σ_{r-1} and W_{r-1} define a cyclic extension H of K_r with H/K_r of order m_{r-1}. In either case, local confluence at the last three types of overlaps in $(**)$ with $i = r - 1$ implies that H is a homomorphic image of Mon $\langle Y_{r-1}, \mathcal{R}_{r-1} \rangle$. Therefore $(Y_{r-1}, \mathcal{R}_{r-1})$ is consistent. □

Suppose that confluence holds on all $\mathcal{W}(r - 1, a_k)$ with $k < r - 1$. Then each a_k defines an automorphism σ_k of K_{r-1}. Let $a_j a_i = a_i a_j a_k$ be the definition of some a_k with $k < r - 1$ and $w_k > 1$. The automorphisms $\sigma_j \sigma_i$ and $\sigma_i \sigma_j \sigma_k$ agree on K_r. To see if they agree on K_{r-1}, it suffices to check whether they agree on the element u represented by a_{r-1}. Since $w_i = 1$, we know that local confluence holds at $a_{r-1} a_j a_i$. The reductions confirming local confluence begin as follows:

$$a_{r-1} a_j a_i \rightarrow a_j a_{r-1} A_{jr-1} a_i,$$
$$a_{r-1} a_j a_i \rightarrow a_{r-1} a_i a_j a_k.$$

Any word derivable from both $a_j a_{r-1} A_{jr-1} a_i$ and $a_{r-1} a_i a_j a_k$ must have a_i occurring earlier in the word than a_j. Thus the first reduction must continue with

$$a_j a_{r-1} A_{jr-1} a_i \xrightarrow{*} a_j a_i M \rightarrow a_i a_j a_k M.$$

This means that the second reduction must involve

$$a_{r-1} a_i a_j a_k \xrightarrow{*} a_i a_j a_k N.$$

Local confluence means that M and N define the same element of K_{r-1}. But M represents the image of u under $\sigma_j \sigma_i$ and, by Proposition 9.2, N represents the image of u under $\sigma_i \sigma_j \sigma_k$. This means that $\sigma_j \sigma_i$ and $\sigma_i \sigma_j \sigma_k$ agree on u and therefore they agree on K_{r-1}.

By the choice of r, we conclude that confluence must fail on some $\mathcal{W}(r - 1, a_k)$ with $k < r - 1$. Choose k minimal. By Exercise 7.2 in Chapter 2, the first word R in $\mathcal{W}(r - 1, a_k)$ at which confluence fails is the overlap of precisely two left sides, local confluence fails at R, and R

involves a_k. By arguments similar to those in Proposition 8.3, R has the form $a_t a_s a_k$ with $t > s > k$ or $a_s^{m_s} a_k$. Since confluence holds on $\mathcal{W}(r, a_k)$, in either case $s = r - 1$. If $w_k = 1$, then local confluence is known to hold at R. Therefore $w_k > 1$. Let $a_j a_i = a_i a_j a_k$ be the definition of a_k. Then a_i and a_j define automorphisms σ_i and σ_j, respectively, of K_{r-1} and a_k defines an automorphism σ_k of K_r. On K_r, we have $\sigma_j \sigma_i = \sigma_i \sigma_j \sigma_k$.

As before, local confluence holds at $a_{r-1} a_j a_i$ and one of the reductions confirming this starts out

$$a_{r-1} a_j a_i \rightarrow a_j a_{r-1} A_{jr-1} a_i \xrightarrow{*} a_j a_i M \rightarrow a_i a_j a_k M.$$

We must analyze the other reduction a little more closely. It begins

$$a_{r-1} a_j a_i \rightarrow a_{r-1} a_i a_j a_k \rightarrow a_i a_{r-1} A_{ir-1} a_j a_k \xrightarrow{*} a_i a_{r-1} a_j P a_k Q,$$

where it is possible to write A_{ir-1} as EF and $F a_j a_k \xrightarrow{*} a_j a_k Q$ and $E a_j \xrightarrow{*} a_j P$. The reduction continues

$$a_i a_{r-1} a_j P a_k Q \rightarrow a_i a_j a_{r-1} A_{jr-1} P a_k Q \xrightarrow{*} a_i a_j a_{r-1} a_k L \rightarrow a_i a_j a_k a_{r-1} A_{kr-1} L.$$

Let u and v be the elements of K_{r-1} represented by a_{r-1} and $a_{r-1} A_{kr-1}$, respectively. Then A_{ir-1} and A_{jr-1} represent $u^{-1} u^{\sigma_i}$ and $u^{-1} u^{\sigma_j}$, respectively. The word L represents

$$(u^{-1} u^{\sigma_j})^{\sigma_k} (u^{-1} u^{\sigma_i})^{\sigma_j \sigma_k} = [u^{-1} u^{\sigma_j} (u^{-1} u^{\sigma_i})^{\sigma_j}]^{\sigma_k} = [u^{-1} u^{\sigma_i \sigma_j}]^{\sigma_k}.$$

The element $u^{-1} u^{\sigma_i \sigma_j}$ is in K_r and on K_r we know that σ_k is the same as $\sigma_j^{-1} \sigma_i^{-1} \sigma_j \sigma_i$. Therefore L represents

$$(u^{\sigma_j^{-1} \sigma_i^{-1} \sigma_j \sigma_i})^{-1} u^{\sigma_j \sigma_i}.$$

As before, M represents $u^{\sigma_j \sigma_i}$. Local confluence at $a_{r-1} a_j a_i$ implies that

$$v (u^{\sigma_j^{-1} \sigma_i^{-1} \sigma_j \sigma_i})^{-1} u^{\sigma_j \sigma_i} = u^{\sigma_j \sigma_i}$$

or

$$v = u^{\sigma_j^{-1} \sigma_i^{-1} \sigma_j \sigma_i}.$$

This means that all relations in $\mathcal{R}$ which can be used to rewrite elements of $\mathcal{W}(r - 1, a_k)$ are satisfied in the cyclic extension of K_{r-1} defined by $\sigma_j^{-1} \sigma_j^{-1} \sigma_j \sigma_i$. Therefore confluence must hold on $\mathcal{W}(r - 1, a_k)$ after all. This contradiction completes the proof of Proposition 9.3. $\square$

9.10 Free nilpotent groups

Let F be the free group on a finite set X with $|X| = r$ and let e be a positive integer. Any group isomorphic to $G = F/\gamma_{e+1}(F)$ is called a *free nilpotent group* of rank r and class e. The term "collection" was first used in (P. Hall 1934) in connection with computing in free nilpotent groups. In this section we shall describe polycyclic generating sequences for G. Consistent polycyclic presentations are given for G in terms of some of these sequences, but the proof of consistency requires considerably more space than is available here. Details can be found in [Hall 1959] and [Magnus et al. 1976].

Since G is finitely generated and nilpotent, it is polycyclic by Proposition 3.4. Suppose $X = \{x_1, \ldots, x_r\}$. By Propositions 2.5 and 2.6, for $k \geq 2$ the commutators $[x_{i_1}, \ldots, x_{i_k}]$ with $i_1 > i_2$ generate $\gamma_k(F)$ modulo $\gamma_{k+1}(F)$. Thus the images in G of these elements with $k \leq e$ form a polycyclic generating sequence for G when arranged in increasing order of k. A presentation for G is obtained by setting all commutators $[x_{i_1}, \ldots, x_{i_{k+1}}]$ with $i_1 > i_2$ equal to 1. The Knuth-Bendix procedure can be used to find the corresponding standard polycyclic presentation.

Example 10.1. Let $X = \{x_1, x_2\}$ and take $e = 5$. One polycyclic generating sequence for G is $x_1, \ldots, x_{17}$, where $x_3, \ldots, x_{17}$ are defined as follows:

$$x_3 = [x_2, x_1],$$
$$x_4 = [x_3, x_1] = [x_2, x_1, x_1],$$
$$x_5 = [x_3, x_2] = [x_2, x_1, x_2],$$
$$x_6 = [x_4, x_1] = [x_2, x_1, x_1, x_1],$$
$$x_7 = [x_4, x_2] = [x_2, x_1, x_1, x_2],$$
$$x_8 = [x_5, x_1] = [x_2, x_1, x_2, x_1],$$
$$x_9 = [x_5, x_2] = [x_2, x_1, x_2, x_2],$$
$$x_{10} = [x_6, x_1] = [x_2, x_1, x_1, x_1, x_1],$$
$$x_{11} = [x_6, x_2] = [x_2, x_1, x_1, x_1, x_2],$$
$$x_{12} = [x_7, x_1] = [x_2, x_1, x_1, x_2, x_1],$$
$$x_{13} = [x_7, x_2] = [x_2, x_1, x_1, x_2, x_2],$$
$$x_{14} = [x_8, x_1] = [x_2, x_1, x_2, x_1, x_1],$$
$$x_{15} = [x_8, x_2] = [x_2, x_1, x_2, x_1, x_2],$$
$$x_{16} = [x_9, x_1] = [x_2, x_1, x_2, x_2, x_1],$$
$$x_{17} = [x_9, x_2] = [x_2, x_1, x_2, x_2, x_2].$$

A presentation for G on these generators is $[x_j, x_i] = 1$, $10 \leq j \leq 17$, $1 \leq i \leq$ 2. The Knuth-Bendix procedure using the basic wreath-product ordering with $x_{17} \prec x_{17}^{-1} \prec x_{16} \prec \cdots \prec x_1 \prec x_1^{-1}$ will produce the standard polycyclic presentation for G on the x_i. The Knuth-Bendix procedure can be helped along by including some redundant relations $[x_j, x_i] = 1$, where $[x_j, x_i]$ is clearly in $\gamma_6(G) = 1$. Examples of such relations are $[x_6, x_3] = 1$ and $[x_j, x_i] = 1$ for $4 \leq i < j \leq 17$.

The standard presentation obtained in this case is too large to list here, but it is instructive to look at the power relations which occur. They are

$$x_7 = x_8 x_{11} x_{14}^{-1} x_{15} x_{16}^{-1},$$

$$x_{12} = x_{14},$$

$$x_{13} = x_{15}.$$

Thus $\gamma_4(F)/\gamma_5(F) \cong \gamma_4(G)/\gamma_5(G)$ is the abelian group generated by x_6, x_7, x_8, and x_9 subject to the single relation $x_7 = x_8$. Therefore $\gamma_4(G)/\gamma_5(G)$ is free abelian of rank 3. Similarly, $\gamma_5(F)/\gamma_6(F) \cong \gamma_5(G)$ is free abelian of rank 6.

Computations like those in Example 10.1 suggest that the quotients $\gamma_k(F)/\gamma_{k+1}(F)$ are always free abelian groups and that for large k the ranks of these groups are substantially smaller than the upper bound of

$$\frac{r^k - r^{k-1}}{2}$$

given by Propositions 2.5 and 2.6. The rank of $\gamma_k(F)/\gamma_{k+1}(F)$ is known and bases for these groups have been determined. To describe these bases we need to introduce the concept of a basic sequence of commutators.

A *basic sequence of commutators* in F is an infinite sequence $c_1, c_2, \ldots$ of elements of F, where each c_i has associated with it a positive integer w_i called its *weight*. The c_i must satisfy several conditions, which will now be described. The c_i are ordered by weight. That is, if $j > i$, then $w_j \geq w_i$. The commutators of weight 1 are $c_1, \ldots, c_r$, which are the elements of X arranged in some order. If $w_k > 1$, then c_k is described explicitly as the commutator $[c_j, c_i]$, where $j > i$ and $w_k = w_i + w_j$. If $w_j > 1$, so that c_j is described as $[c_q, c_p]$ with $q > p$, then $p \leq i$. Finally, for each $j > i$ such that either $w_j = 1$ or $w_j > 1$ and c_j is described as $[c_q, c_p]$ with $p \leq i$, there is a unique index k such that c_k is described as $[c_j, c_i]$. The phrase "sequence of basic commutators" is used by most authors, but being basic is a property of the sequence, not of the individual terms in the sequence. We shall say that a commutator u is basic only when a basic sequence of commutators has previously been specified and u is a term in that sequence.

Example 10.2. Suppose that $X = \{a, b\}$. The terms of weight at most 6 in one basic sequence are

$$
\begin{array}{ll}
c_1 = a, & c_{13} = [c_7, c_2], \\
c_2 = b, & c_{14} = [c_8, c_2], \\
c_3 = [c_2, c_1], & c_{15} = [c_5, c_4], \\
c_4 = [c_3, c_1], & c_{16} = [c_6, c_3], \\
c_5 = [c_3, c_2], & c_{17} = [c_7, c_3], \\
c_6 = [c_4, c_1], & c_{18} = [c_8, c_3], \\
c_7 = [c_4, c_2], & c_{19} = [c_{11}, c_1], \\
c_8 = [c_5, c_2], & c_{20} = [c_{11}, c_2], \\
c_9 = [c_4, c_3], & c_{21} = [c_{12}, c_2], \\
c_{10} = [c_5, c_3], & c_{22} = [c_{13}, c_2], \\
c_{11} = [c_6, c_1], & c_{23} = [c_{14}, c_2]. \\
c_{12} = [c_6, c_2], &
\end{array}
$$

Here c_1 and c_2 have weight 1, c_3 has weight 2, c_4 and c_5 have weight 3, c_6, c_7, and c_8 have weight 4, $c_9, \ldots, c_{14}$ have weight 5, and $c_{15}, \ldots, c_{23}$ have weight 6.

As defined here, a basic sequence of commutators includes the description of each term of weight greater than 1 as the commutator of earlier terms in the sequence. The reason for this is at this stage it is conceivable that $[c_j, c_i] = [c_q, c_p]$ even though $(j, i) \neq (q, p)$ and that some of the c_k could be trivial. This cannot occur, but the proof is somewhat involved. From now on we shall write $c_k = [c_j, c_i]$ for the assertion that c_k is described as $[c_j, c_i]$.

Let us fix a basic sequence $c_1, c_2, \ldots$ of commutators in F. The commutators of weight 1 are in $F = \gamma_1(F)$. By Proposition 1.10 and a simple induction argument, it is easy to prove that c_i is in $\gamma_{w_i}(F)$ for all i. Suppose that $c_1, \ldots, c_t$ are the commutators in the sequence with weight at most e, and for $1 \leq i \leq t$ let a_i be the image of c_i in $G = F/\gamma_{e+1}(F)$. We shall prove that the a_i form a polycyclic generating sequence for G. In the proof, the following proposition will be needed.

Proposition 10.1. *Let u and v be elements of a group. Then*

$$
[v, u^{-1}] = [v, u, u^{-1}]^{-1}[v, u]^{-1}
$$

and

$$
[v^{-1}, u] = [v, u, v^{-1}]^{-1}[v, u]^{-1}.
$$

Set $u_1 = v_1 = [v, u]$ and for $i \geq 1$ let $u_{i+1} = [u_i, u]$ and $v_{i+1} = [v_i, v]$. Then, for any odd positive integer s,

$$[v, u^{-1}] = u_2 u_4 \ldots u_{s-1}[u_s, u^{-1}]^{-1} u_s^{-1} u_{s-2}^{-1} \ldots u_3^{-1} u_1^{-1},$$
$$[v^{-1}, u] = v_2 v_4 \ldots v_{s-1}[v_s, v^{-1}]^{-1} v_s^{-1} v_{s-2}^{-1} \ldots v_3^{-1} v_1^{-1}.$$

Proof. The first two identities are proved by direct computation in the free group generated by u and v. They correspond to the case $s = 1$ of the second pair of identities. The second pair is proved by induction on s, using the following applications of the first identity:

$$\begin{aligned}
[u_s, u^{-1}]^{-1} &= ([u_s, u, u^{-1}]^{-1}[u_s, u]^{-1})^{-1} \\
&= u_{s+1}[u_{s+1}, u^{-1}] \\
&= u_{s+1}[u_{s+1}, u, u^{-1}]^{-1}[u_{s+1}, u]^{-1} \\
&= u_{s+1}[u_{s+2}, u^{-1}]^{-1} u_{s+2}^{-1}, \\
[v_s, v^{-1}]^{-1} &= ([v_s, v, v^{-1}]^{-1}[v_s, v]^{-1})^{-1} \\
&= v_{s+1}[v_{s+1}, v^{-1}] \\
&= v_{s+1}[v_{s+1}, v, v^{-1}]^{-1}[v_{s+1}, v]^{-1} \\
&= v_{s+1}[v_{s+2}, v^{-1}]^{-1} v_{s+2}^{-1}. \quad \square
\end{aligned}$$

Now we are ready to prove that a basic sequence of commutators defines a polycyclic generating sequence for G. For $1 \leq i \leq t + 1$ define $G_i = \text{Grp}\langle a_i, \ldots, a_t \rangle$.

Proposition 10.2. *The sequence $a_1, \ldots, a_t$ is a polycyclic generating sequence for G.*

Proof. We must show that for $1 \leq i < j \leq t$ and any α and β in $\{1, -1\}$ there is an element z of G_{i+1} such that $a_j^\alpha a_i^\beta = a_i^\beta z$. If $w_i + w_j > e$, then $[c_j, c_i]$ is in $\gamma_{e+1}(F)$ and so a_i and a_j commute. Therefore we may take $z = a_j^\alpha$ in this case. Let us assume that $w_i + w_j \leq e$. Suppose first that $[c_j, c_i] = c_k$ is basic. In Proposition 10.1, let $u = c_i$ and $v = c_j$. Each of the commutators u_m and v_m is basic. If m is large, then both u_m and v_m are in $\gamma_{e+1}(F)$ and can be ignored when we pass to the quotient group G. Let x_m and y_m denote the images of u_m and v_m, respectively, in G. Then x_m and y_m lie in G_{i+1} and

$$\begin{aligned}
a_j a_i &= a_i a_j a_k, \\
a_j a_i^{-1} &= a_i^{-1} a_j x_2 x_4 \ldots x_3^{-1} x_1^{-1}, \\
a_j^{-1} a_i &= a_i a_j^{-1} y_2 y_4 \ldots y_3^{-1} y_1^{-1}.
\end{aligned}$$

The products indicated with dots are finite products. Finally,

$$a_j^{-1}a_i^{-1} = a_i^{-1}(a_i a_j a_i^{-1})^{-1} = a_i^{-1}x_1 x_3 \ldots x_4^{-1}x_2^{-1}a_j^{-1}.$$

Now suppose that $[c_j, c_i]$ is not basic. Then $c_j = [c_q, c_p]$ with $j > q > p > i$. In particular, $j > i+1$, so we may proceed by induction on j. The conjugates $a_i^{-1}a_q a_i$ and $a_i^{-1}a_p a_i$ have already been shown to be in G_{i+1}. Therefore $a_i^{-1}a_j a_i = a_i^{-1}[a_q, a_p]a_i$ is also in G_{i+1}, or, equivalently, $a_j a_i = a_i z$, where z is in G_{i+1}. The other cases are handled in the same way. $\square$

The proof of Proposition 10.2 allows us to construct a polycyclic presentation for a group H of which G is a homomorphic image. It is in fact the case that the presentation obtained is consistent and H is isomorphic to G, although we cannot give the proof here. It is also true that the series $G = G_1 \supseteq \cdots \supseteq G_{t+1}$ refines the lower central series of G. Moreover, if $1 \le m \le e$, then the quotient $Q = \gamma_m(G)/\gamma_{m+1}(G) \cong \gamma_m(F)/\gamma_{m+1}(F)$ is a free abelian group. If $c_p, \ldots, c_q$ are the basic commutators of weight m, then the images of $a_p, \ldots, a_q$ in Q form a basis of Q. The rank of Q is

$$\frac{1}{m}\sum_{d|m}\mu(d)r^{m/d},$$

where μ is the Möbius function, which is defined as follows: $\mu(n) = (-1)^s$ if n is the product of s distinct primes and $\mu(n) = 0$ if n is divisible by the square of a prime. In particular, $\mu(1) = 1$.

Example 10.3. Let us again look at the case $r = 2$ and $e = 5$, using the basic sequence of commutators in Example 10.2. The generators of G are $a_1, \ldots, a_{14}$, which are the images of $c_1, \ldots, c_{14}$. There are 91 commutators $[a_j, a_i]$ with $i < j$. For 76 of these, $w_i + w_j \ge 6$, and hence the commutator is trivial. Of the 15 remaining commutators, 12 are basic. This leaves only three to be determined. They are $[a_5, a_1]$, $[a_7, a_1]$, and $[a_8, a_1]$.

$$
\begin{aligned}
a_1^{-1}a_5 a_1 &= a_1^{-1}[a_3, a_2]a_1 = [a_1^{-1}a_3 a_1, a_1^{-1}a_2 a_1] = [a_3 a_4, a_2 a_3] \\
&= a_4^{-1}a_3^{-1}a_3^{-1}a_2^{-1}a_3 a_4 a_2 a_3 = a_4^{-1}a_3^{-1}a_3^{-1}a_2^{-1}a_3 a_2 a_4 a_7 a_3 \\
&= a_4^{-1}a_3^{-1}a_3^{-1}a_2^{-1}a_2 a_3 a_5 a_4 a_7 a_3 = a_4^{-1}a_3^{-1}a_5 a_4 a_7 a_3 \\
&= a_4^{-1}a_3^{-1}a_5 a_4 a_3 a_7 = a_4^{-1}a_3^{-1}a_5 a_3 a_4 a_9 a_7 = a_4^{-1}a_3^{-1}a_3 a_5 a_{10}a_4 a_9 a_7 \\
&= a_4^{-1}a_5 a_{10}a_4 a_9 a_7 = a_4^{-1}a_4 a_5 a_{10}a_9 a_7 = a_5 a_7 a_9 a_{10}.
\end{aligned}
$$

Therefore $a_5 a_1 = a_1 a_5 a_7 a_9 a_{10}$. Now

$$a_1^{-1}a_7 a_1 = [a_4 a_6, a_2 a_3] = a_6^{-1}a_4^{-1}a_3^{-1}a_2^{-1}a_4 a_6 a_2 a_3,$$

which simplifies to $a_7 a_9 a_{12}$. Finally,

$$a_1^{-1} a_8 a_1 = [a_5 a_7 a_9 a_{10}, a_2 a_3],$$

which turns out to be $a_8 a_{10} a_{13}$.

Once we have a consistent polycyclic presentation for our free nilpotent group G, we can compute in G using the techniques developed in Sections 9.4 to 9.7. However, if many calculations with elements of G are to be carried out, then collection is not the most efficient way to compute products. The approach to be described applies not only to free nilpotent groups but to any group defined by a consistent nilpotent presentation without power relations.

Let $a_1, \ldots, a_n$ be a polycyclic generating sequence for a group H such that the corresponding standard polycyclic presentation for H is nilpotent and contains no power relations. We can identify H and $\mathbb{Z}^n$ as sets by letting $(\alpha_1, \ldots, \alpha_n)$ in $\mathbb{Z}^n$ correspond to $a_1^{\alpha_1} \ldots a_n^{\alpha_n}$. Suppose that g and h are elements of H and that g, h, and gh correspond to $(\alpha_1, \ldots, \alpha_n)$, $(\beta_1, \ldots, \beta_n)$, and $(\sigma_1, \ldots, \sigma_n)$, respectively. We may consider $\sigma_1, \ldots, \sigma_n$ to be functions of $\alpha_1, \ldots, \alpha_n$ and $\beta_1, \ldots, \beta_n$. In fact, $\sigma_i = \sigma_i(\alpha_1, \ldots, \alpha_i, \beta_1, \ldots, \beta_i)$ depends only on $\alpha_1, \ldots, \alpha_i$ and $\beta_1, \ldots, \beta_i$. In (P. Hall 1957) it is shown that each σ_i is a polynomial function of the α's and the β's. If these polynomials can be determined, they can be used to compute products much more rapidly than can be accomplished with collection. The polynomials defining multiplication in H take on integer values whenever the α's and β's are integers. Any such polynomial is an integer linear combination of products of binomial coefficients of the form

$$\binom{\alpha_1}{s_1} \cdots \binom{\alpha_n}{s_n} \binom{\beta_1}{t_1} \cdots \binom{\beta_n}{t_n}.$$

Notice that these polynomials do not necessarily have integer coefficients when expressed as linear combinations of ordinary monomials. Not only is multiplication in H defined by polynomials, so is inversion. That is, if $a_1^{\delta_1} \ldots a_n^{\delta_n}$ is the inverse of $a_1^{\alpha_1} \ldots a_n^{\alpha_n}$, then the δ's are polynomials in the α's.

Example 10.4. Suppose that H is $D_3^{(1)}$ as defined in Example 2.1 and

$$a_1 = \begin{bmatrix} 1 & 0 & 0 \\ 0 & 1 & 1 \\ 0 & 0 & 1 \end{bmatrix}, \quad a_2 = \begin{bmatrix} 1 & 1 & 0 \\ 0 & 1 & 0 \\ 0 & 0 & 1 \end{bmatrix}, \quad a_3 = \begin{bmatrix} 1 & 0 & 1 \\ 0 & 1 & 0 \\ 0 & 0 & 1 \end{bmatrix}.$$

Then $a_1^{\alpha_1} a_2^{\alpha_2} a_3^{\alpha_3}$ is

$$\begin{bmatrix} 1 & \alpha_2 & \alpha_3 \\ 0 & 1 & \alpha_1 \\ 0 & 0 & 1 \end{bmatrix}.$$

Moreover,

$$\begin{bmatrix} 1 & \alpha_2 & \alpha_3 \\ 0 & 1 & \alpha_1 \\ 0 & 0 & 1 \end{bmatrix} \begin{bmatrix} 1 & \beta_2 & \beta_3 \\ 0 & 1 & \beta_1 \\ 0 & 0 & 1 \end{bmatrix} = \begin{bmatrix} 1 & \sigma_2 & \sigma_3 \\ 0 & 1 & \sigma_1 \\ 0 & 0 & 1 \end{bmatrix},$$

where $\sigma_1 = \alpha_1 + \beta_1$, $\sigma_2 = \alpha_2 + \beta_2$, and $\sigma_3 = \alpha_3 + \beta_3 + \alpha_2 \beta_1$.

A crude upper bound for the total degree of σ_i is i. However, if H is our free nilpotent group G of class e and $a_1, \ldots, a_n$ are defined by a basic sequence of commutators $c_1, c_2, \ldots$ with associated weights $w_1, w_2, \ldots$, then each of the products

$$\binom{\alpha_1}{s_1} \cdots \binom{\alpha_i}{s_i} \binom{\beta_1}{t_1} \cdots \binom{\beta_i}{t_i}$$

which occurs with nonzero coefficient in σ_i satisfies

$$\sum_{j=1}^{i} (s_j + t_j) w_j \le w_i.$$

If the class e of G is not very large, then the polynomials σ_i can be computed fairly easily. For the moment, let us assume that we know the bivariate polynomials η_{ijk}, $n \ge k > j > i$, such that

$$a_j^\alpha a_i^\beta = a_i^\beta a_j^\alpha a_{j+1}^{\eta_{ijj+1}(\alpha,\beta)} \ldots a_n^{\eta_{ijn}(\alpha,\beta)}.$$

Notice that this formula remains valid even if the exponents α and β are themselves polynomials in one or more indeterminates. Given the η_{ijk}, we can use any of the classical computer algebra systems to construct a "symbolic collector", which collects products of powers of the generators in which the exponents are polynomials. To compute the σ_i, one has only to collect the single "symbolic word" $a_1^{\alpha_1} \ldots a_n^{\alpha_n} a_1^{\beta_1} \ldots a_n^{\beta_n}$, where the α's and the β's are indeterminates.

Thus it suffices to determine the polynomials η_{ijk}. By the remark above, if $\alpha^r \beta^s$ occurs with nonzero coefficient in η_{ijk}, then $r w_j + s w_i \le w_k \le e$. Thus we may compute $a_j^\alpha a_i^\beta$ for those nonnegative values of α and β for which $\alpha w_j + \beta w_i \le e$ and then use interpolation to find the η_{ijk}. Exercise 10.4 describes one convenient way to carry out the interpolation. Although a basic collection procedure would be adequate in principle to compute the necessary values of $a_j^\alpha a_i^\beta$, we can accomplish the task much more quickly by using the symbolic collector whenever possible. The values of i are considered in the order $n-1, n-2, \ldots, 1$. For a given i, the values of j are considered in the order $n, n-1, \ldots, i+1$. For a given pair (j, i), we take $\alpha = 0, 1, \ldots$ and for a given α we let $\beta = 0, 1, \ldots$. The first collection step is $a_j^\alpha a_i^\beta = a_j^{\alpha-1} a_i a_j [a_j, a_i] a_i^{\beta-1}$. After this, if we need to replace $a_q^\mu a_p^\nu$ by $a_p^\nu a_q^\mu [a_q^\mu, a_p^\nu]$, then either we have already determined the polynomials η_{pqk} or we have $(q, p) = (j, i)$ and $a_j^\mu a_i^\nu$ has already been computed.

Example 10.5. Let us continue with the case $r = 2$ and $e = 5$ of Example 10.3. Here are the polynomials σ_i:

$$\sigma_1 = \alpha_1 + \beta_1,$$

$$\sigma_2 = \alpha_2 + \beta_2,$$

$$\sigma_3 = \alpha_3 + \beta_3 + \alpha_2 \beta_1,$$

$$\sigma_4 = \alpha_4 + \beta_4 + \alpha_3 \beta_1 + \alpha_2 \binom{\beta_1}{2},$$

$$\sigma_5 = \alpha_5 + \beta_5 + \alpha_3 \beta_2 + \binom{\alpha_2}{2}\beta_1 + \alpha_2 \beta_1 \beta_2,$$

$$\sigma_6 = \alpha_6 + \beta_6 + \alpha_4 \beta_1 + \alpha_3 \binom{\beta_1}{2} + \alpha_2 \binom{\beta_1}{3},$$

$$\sigma_7 = \alpha_7 + \beta_7 + \alpha_5 \beta_1 + \alpha_4 \beta_2 + \alpha_3 \beta_1 \beta_2 + \binom{\alpha_2}{2}\binom{\beta_1}{2} + \alpha_2 \binom{\beta_1}{2}\beta_2,$$

$$\sigma_8 = \alpha_8 + \beta_8 + \alpha_5 \beta_2 + \alpha_3 \binom{\beta_2}{2} + \binom{\alpha_2}{3}\beta_1 + \binom{\alpha_2}{2}\beta_1 \beta_2 + \alpha_2 \beta_1 \binom{\beta_2}{2},$$

$$\sigma_9 = \alpha_9 + \beta_9 + \alpha_4 \beta_3 + \alpha_5 \beta_1 + \alpha_7 \beta_1 + \binom{\alpha_3}{2}\beta_1 + \alpha_5 \binom{\beta_1}{2} + 2\binom{\alpha_2}{2}\binom{\beta_1}{2}$$

$$+ \alpha_2 \alpha_3 \binom{\beta_1}{2} + \alpha_2 \binom{\beta_1}{3} + \alpha_2 \binom{\beta_1}{2}\beta_3 + \alpha_3 \beta_1 \beta_3 + 3\binom{\alpha_2}{2}\binom{\beta_1}{3},$$

$$\sigma_{10} = \alpha_{10} + \beta_{10} + \alpha_5 \beta_1 + \alpha_5 \beta_3 + \alpha_8 \beta_1 + \binom{\alpha_2}{2}\beta_1 + \binom{\alpha_3}{2}\beta_2 + \alpha_3 \beta_2 \beta_3$$

$$+ 2\binom{\alpha_2}{3}\beta_1 + \binom{\alpha_2}{2}\alpha_3 \beta_1 + 3\binom{\alpha_2}{2}\binom{\beta_1}{2} + \binom{\alpha_2}{2}\beta_1 \beta_3 + \alpha_2 \alpha_3 \beta_1 \beta_2$$

$$+ \alpha_2 \binom{\beta_1}{2}\beta_2 + \alpha_2 \beta_1 \beta_2 \beta_3 + 4\binom{\alpha_2}{3}\binom{\beta_1}{2} + 2\binom{\alpha_2}{2}\binom{\beta_1}{2}\beta_2$$

$$+ \binom{\alpha_2}{2} \beta_1 \beta_2,$$

$$\sigma_{11} = \alpha_{11} + \beta_{11} + \alpha_6 \beta_1 + \alpha_4 \binom{\beta_1}{2} + \alpha_3 \binom{\beta_1}{3} + \alpha_2 \binom{\beta_1}{4},$$

$$\sigma_{12} = \alpha_{12} + \beta_{12} + \alpha_6 \beta_2 + \alpha_7 \beta_1 + \alpha_4 \beta_1 \beta_2 + \alpha_5 \binom{\beta_1}{2} + \binom{\alpha_2}{2}\binom{\beta_1}{3}$$

$$+ \alpha_2 \binom{\beta_1}{3} \beta_2 + \alpha_3 \binom{\beta_1}{2} \beta_2,$$

$$\sigma_{13} = \alpha_{13} + \beta_{13} + \alpha_7 \beta_2 + \alpha_8 \beta_1 + \alpha_4 \binom{\beta_2}{2} + \alpha_5 \beta_1 \beta_2 + \binom{\alpha_2}{3}\binom{\beta_1}{2} +$$

$$+ \binom{\alpha_2}{2}\binom{\beta_1}{2} \beta_2 + \alpha_2 \binom{\beta_1}{2}\binom{\beta_2}{2} + \alpha_3 \beta_1 \binom{\beta_2}{2},$$

$$\sigma_{14} = \alpha_{14} + \beta_{14} + \alpha_8 \beta_2 + \alpha_5 \binom{\beta_2}{2} + \alpha_3 \binom{\beta_2}{3} + \binom{\alpha_2}{4}\beta_1 + \binom{\alpha_2}{3}\beta_1 \beta_2$$

$$+ \binom{\alpha_2}{2}\beta_1 \binom{\beta_2}{2} + \alpha_2 \beta_1 \binom{\beta_2}{3}.$$

As e increases, the number of terms in the σ_i grows very rapidly and it becomes difficult to store these polynomials. A compromise is to store only the polynomials

$$\sigma_i^{(j)} = \sigma_i(\alpha_1, \ldots, \alpha_i, 0, \ldots, 0, \beta_j, 0, \ldots, 0),$$

where $1 \leq j \leq i$. The product

$$(a_1^{\alpha_1} \ldots a_n^{\alpha_n})(a_1^{\alpha_1} \ldots a_n^{\alpha_n})$$

is now evaluated as

$$(\cdots ((a_1^{\alpha_1} \ldots a_n^{\alpha_n}) a_1^{\beta_1}) a_2^{\beta_2} \cdots) a_n^{\beta_n}$$

using the polynomials $\sigma_j^{(j)}, \sigma_{j+1}^{(j)}, \ldots, \sigma_n^{(j)}$ in the computation of the j-th product.

Exercises

10.1. Show that the polycyclic presentation

$$a_2 a_1 = a_1 a_2^{-1}, \quad a_1^{-1} a_1 = a_1 a_2, \quad a_2 a_1^{-1} = a_1^{-1} a_2^{-1}, \quad a_2^{-1} a_1^{-1} = a_1^{-1} a_2$$

on generators a_1 and a_2 is consistent. Determine a formula for $(a_1^{\alpha_1} a_2^{\alpha_2})(a_1^{\beta_1} a_2^{\beta_2})$. Conclude that multiplication in this group is not defined by polynomials.

10.2. Each polynomial f in $\mathbb{Q}[X_1, \ldots, X_m]$ defines a function $\overline{f}$ from $\mathbb{Q}^m$ to $\mathbb{Q}$. Show that the map $f \mapsto \overline{f}$ is an injective ring homomorphism from $\mathbb{Q}[X_1, \ldots, X_m]$ to the ring of functions from $\mathbb{Q}^m$ to $\mathbb{Q}$ under pointwise addition and multiplication.

10.3. If $s_1, \ldots, s_m$ are nonnegative integers, define $g_{s_1 s_2 \ldots s_m}$ to be the product

$$\binom{X_1}{s_1} \cdots \binom{X_m}{s_m}$$

of binomial coefficients. Prove that the polynomials $g_{s_1 s_2 \ldots s_m}$ are a vector space basis for $\mathbb{Q}[X_1, \ldots, X_m]$ over $\mathbb{Q}$. Suppose f is in $\mathbb{Q}[X_1, \ldots, X_m]$ and $\overline{f}(\alpha_1, \ldots, \alpha_m)$ is an integer for all $(\alpha_1, \ldots, \alpha_m)$ in $\mathbb{Z}^m$. Show that f is an integer linear combination of polynomials $g_{s_1 s_2 \ldots s_m}$.

10.4. Suppose that $s_1, \ldots, s_m$ and $t_1, \ldots, t_m$ are sequences of nonnegative integers. Let us say that $g_{t_1 t_2 \ldots t_m}$ precedes $g_{s_1 s_2 \ldots s_m}$ if $t_i \leq s_i$ for all i and $t_i < s_i$ for some i. Let f be in $\mathbb{Q}[X_1, \ldots, X_m]$ and let h be the sum of the terms in f involving polynomials $g_{t_1 t_2 \ldots t_m}$ which precede $g_{s_1 s_2 \ldots s_m}$. Prove that the coefficient of $g_{s_1 s_2 \ldots s_m}$ in f is $f(s_1, \ldots, s_m) - h(s_1, \ldots, s_m)$. Assume that values of f can be computed and that a bound on the degree of f is known. Show how to express f as a linear combination of the $g_{s_1 s_2 \ldots s_m}$.

9.11 p-Groups

Let p be a prime. A *p-group* is a group in which every element has finite order and these orders are powers of p. A finite p-group has order p^n for some nonnegative integer n. An *elementary abelian p-group* is a finite abelian p-group in which the p-th power of every element is 1. Such a group is a direct product of cyclic groups of order p and may be considered to be a vector space over the field $\mathbb{Z}_p$ of integers modulo p. (Additive notation should be used when this is done.) In Section 11.7 we shall describe an algorithm for determining finite p-groups which are quotients of a given finitely presented group. This section summarizes the facts about finite p-groups which we shall need.

Let P be a nontrivial finite p-group. The following statements are proved in most basic texts on finite groups:

(1) The center of P is nontrivial.
(2) Every proper subgroup of P is contained in a subgroup of index p and all subgroups of index p are normal.
(3) P is nilpotent.

By (3), the lower central series of P eventually reaches the trivial subgroup. However, there is another central series which is particularly useful in studying P. For any group G, define G^p to be the subgroup generated by $\{g^p \mid g \in G\}$. The terms $\varphi_i(G)$ of the *lower exponent-p central series* of G are defined as follows: $\varphi_1(G) = G$ and $\varphi_{i+1}(G) = [\varphi_i(G), G]\varphi_i(G)^p$. By induction, $\varphi_i(G)$ is normal in G and hence so are $[\varphi_i(G), G]$ and G^p. Therefore the product $[\varphi_i(G), G]\varphi_i(G)^p$ is a subgroup. We could also define $\varphi_{i+1}(G)$ as the smallest normal subgroup N of G contained in $\varphi_i(G)$ such that $\varphi_i(G)/N$ is in the center of G/N and is an elementary abelian p-group. Thus the quotients $\varphi_i(G)/\varphi_{i+1}(G)$ are vector spaces over $\mathbb{Z}_p$.

In the case of our finite p-group P, eventually $\varphi_i(P)$ is trivial. For if $\varphi_i(P)$ is not trivial, then $Q = [\varphi_i(P), P]$ is a proper subgroup of $\varphi_i(P)$, since P is nilpotent. The quotient $R = \varphi_i(P)/Q$ is a nontrivial finite abelian p-group and hence is a direct sum of finite cyclic groups of orders which are powers of p. It is easy to show that R^p is a proper subgroup of R. Then $\varphi_{i+1}(P)$, which is the inverse image of R^p in $\varphi_i(P)$, is a proper subgroup of $\varphi_i(P)$. The smallest nonnegative integer c such that $\varphi_{c+1}(P)$ is trivial is called the *exponent-p class* of P.

The subgroup $\varphi_2(P)$ is the intersection of the subgroups of index p in P and is called the Frattini subgroup of P. The quotient $P/\varphi_2(P)$ is the largest elementary abelian quotient of P. In general, the *Frattini subgroup* of a group G is the intersection of the maximal subgroups of G if G has maximal subgroups. If there are no maximal subgroups, then the Frattini subgroup is G. The Frattini subgroup is the set of "nongenerators", elements x in G with the property that, if a subset X of G generates G, then $X - \{x\}$ generates G. Thus a subset X of our finite p-group P generates P if and only if the image of X in $V = P/\varphi_2(P)$ generates V. If V has order p^d, then V is a vector space of dimension d over $\mathbb{Z}_p$ and minimal generating sets of V all have d elements. Thus minimal generating sets of P also have d elements.

Let us fix a prime p. Results analogous to Propositions 1.10, 2.5, and 2.6 hold for the terms of the exponent-p central series.

Proposition 11.1. *For any group G and for any positive integers i and j, the subgroup $[\varphi_i(G), \varphi_j(G)]$ is contained in $\varphi_{i+j}(G)$.*

Proposition 11.2. *Suppose that $G/\varphi_2(G)$ is generated by the images of $x_1, \ldots, x_d$. Then $\varphi_2(G)/\varphi_3(G)$ is generated by the images of $x_i^p, 1 \leq i \leq d$, and $[x_j, x_i], 1 \leq i < j \leq d$.*

Proposition 11.3. *Suppose that $s > 1$, that X is a subset of G which generates G modulo $\varphi_2(G)$, and U is a subset of $\varphi_s(G)$ which generates $\varphi_s(G)$ modulo $\varphi_{s+1}(G)$. Then $\varphi_{s+1}(G)$ is generated modulo $\varphi_{s+2}(G)$ by the elements u^p with u in U and the elements $[u, x]$ with u in U and x in X.*

Let F be a free group of rank r. Propositions 11.2 and 11.3 give upper bounds for the dimensions of the quotients $\varphi_s(F)/\varphi_{s+1}(F)$. In fact, these dimensions are known exactly. Let $c_1, c_2, \ldots$ be a basic sequence of commutators in F. Then a basis for $\varphi_s(F)/\varphi_{s+1}(F)$ is given by the elements $c_i^{p^{s-w_i}}, 1 \leq i \leq t$, where w_i is the weight of c_i and c_t is the last commutator of weight s. See [Huppert & Blackburn 1986].

Let us return to our finite p-group P. It follows from Propositions 11.2 and 11.3 that P has a φ-*weighted presentation* relative to the prime p. This is a power-commutator presentation on generators $a_1, \ldots, a_n$ with the following properties:

(a) Each a_i has associated with it a positive integer weight w_i such that $w_1 = 1$ and $w_i \leq w_{i+1}$, $1 \leq i < n$.

(b) For $1 \leq i \leq n$ there is a power relation $a_i^p = W_i$ and the generators which occur in W_i all have weight at least $w_i + 1$.

(c) For each commutation relation $a_j a_i = a_i a_j A_{ij}$, the generators which occur in A_{ij} all have weight at least $w_i + w_j$.

(d) If $w_k > 1$, then one of the following holds:

   (1) There are indices i and j with $w_i = 1$ and $w_j = w_k - 1$ such that $A_{ij} = a_k$.

   (2) There is an index i with $w_i = w_k - 1$ such that $W_i = a_k$.

   We choose one relation $a_j a_i = a_i a_j a_k$ or $a_i^p = a_k$ as the definition of a_k.

The differences between a φ-weighted presentation and a γ-weighted presentation are the following: In a φ-weighted presentation there is a power relation for each generator, so the group P defined by the presentation is finite; the exponents in the power relations are all equal to a fixed prime p; and generators of weight greater than 1 are defined either as commutators or as p-th powers of earlier generators. Let d be the largest index for which $w_d = 1$. Then P is generated by $x_1, \ldots, x_d$ and d is the minimum number of generators in any generating set. The subgroup $\varphi_s(P)$ is generated modulo $\varphi_{s+1}(P)$ by the generators of weight s.

A φ-weighted presentation is a monoid presentation on generators $a_1, \ldots, a_n$. The techniques of Sections 9.8 and 9.9 can be extended to obtain results which reduce the amount of work needed to test the consistency of a presentation which has the form of a φ-weighted presentation. In fact, (Vaughan-Lee 1984), which provided the ideas for the proof of Proposition 9.3, actually deals with φ-weighted presentations. Let $c = w_n$. If $w_i + w_j = c$, then A_{ij}^p represents the identity in the group defined by the presentation, since all generators of weight c commute and have order dividing p. This implies that a few of the overlaps in Proposition 9.3 do not need to be considered here. To determine consistency we need only test for local confluence at the following overlaps:

$$a_k a_j a_i, \quad k > j > i, \; w_k + w_j + w_i \leq c, \; w_i = 1,$$
$$a_j^p a_i \quad \text{and} \quad a_j a_i^p, \quad j > i, \; w_j + w_i < c,$$
$$a_i^{p+1}, \quad 2w_i < c.$$

10

Module bases

In Chapter 11 we shall discuss algorithms for determining various solvable quotient groups of a group given by a finite presentation. One of these algorithms is the Baumslag-Cannonito-Miller polycyclic quotient algorithm. This algorithm involves computation in finitely generated modules over group rings of polycyclic groups. These rings are in general noncommutative but they are similar in many ways to rings of polynomials. An important special case is the group ring of a free abelian group, which is more commonly referred to as a ring of Laurent polynomials.

The group $\mathbb{Z}^n$ is a free $\mathbb{Z}$-module of rank n. A subgroup, or $\mathbb{Z}$-submodule, H of $\mathbb{Z}^n$ can be described conveniently by a matrix in row Hermite normal form. This description makes it relatively easy to check membership in H and hence to solve the word problem in $\mathbb{Z}^n/H$. Now let R be the ring $\mathbb{Z}[X_1, \ldots, X_m]$ of polynomials with integer coefficients in the indeterminates $X_1, \ldots, X_m$. Then R^n is a free R-module of rank n. Let M be the R-submodule generated by a finite subset T of R^n. Given T, we shall need to produce another module generating set for M which will allow us easily to decide membership in M and so permit us to solve the word problem in R^n/M.

The algorithms for manipulating integer matrices developed in Chapter 8 can be generalized. Unfortunately, when m becomes even moderately large, the space requirements of this approach become very great. There is an alternative. The technique of Gröbner bases was originated by Bruno Buchberger in the 1960s. It has become a very important tool in computational commutative algebra. This chapter first gives a solution to our problem based on the methods of Chapter 8. Then the Gröbner basis approach is outlined. The intent is to try to show the relationship between the point of view of Baumslag, Cannonito, and Miller and that of Buchberger. The chapter concludes with a sketch of the generalization of these methods to finitely generated modules over rings of Laurent polynomials and over group rings of arbitrary polycyclic groups.

10.1 Ideals in $\mathbb{Z}[X]$

Let I be an ideal of the ring $\mathbb{Z}[X]$ of polynomials in one variable with integer coefficients. A special case of the Hilbert basis theorem says that I is finitely generated as an ideal. It will be useful to sketch the proof of this result. For each integer $n \geq 0$ let I_n be the set of elements of I having degree at most n. For these purposes 0 is defined to have degree $-\infty$, so 0 is in each I_n. The set of all polynomials of degree at most n is a free abelian group with basis $X^n, \ldots, X, 1$. Let C_n be the set of coefficients of X^n in elements of I_n. Since I_n is an additive subgroup of $\mathbb{Z}[X]$, it follows easily that C_n is a subgroup of $\mathbb{Z}$. If f is in I_n, then Xf is in I_{n+1} and the coefficient of X^{n+1} in Xf is the same as the coefficient of X^n in f. Therefore $C_n \subseteq C_{n+1}$. The ascending chain condition holds for subgroups of $\mathbb{Z}$, so there is an integer m such that $C_n = C_m$ for all $n \geq m$. The smallest such m will be called the *effective degree* of I.

The group I_m generates I as an ideal. If not, we could choose an element f of I with minimum degree such that f is not in the ideal J generated by I_m. If the degree n of f is at most m, then f is in I_m, which is certainly not the case. Thus $n > m$. Since $C_n = C_m$, there is an element g in I_m such that the leading coefficient of g is the same as the leading coefficient of f. Therefore $f - X^{n-m}g$ has degree less than n. Since f and g are in I, so is $f - X^{n-m}g$. By the choice of f, it follows that $f - X^{n-m}g$ is in J. But $X^{n-m}g$ is in J and therefore $f = (f - X^{n-m}g) + X^{n-m}g$ is in J. This is a contradiction, and hence I_m generates I. Since I_m is finitely generated as an abelian group, I is finitely generated as an ideal.

Suppose that I is defined as the ideal generated by some finite set T of polynomials. The traditional proof of the Hilbert basis theorem does not tell us how to determine the effective degree m of I from the generating set T. In fact, it is relatively easy to compute m. Let d be the maximum degree of the elements of T. The following proposition gives an upper bound for m.

Proposition 1.1. *If $n \geq d$, then $C_n = C_d$.*

Proof. Since I_d contains T, it follows that I_d generates I as an ideal. Therefore I is generated as an abelian group by the set of products $X^k f$, where $k \geq 0$ and f is in I_d.

Lemma 1.2. *If $n \geq d$, then I_n is generated as an abelian group by the set of products $X^k f$, where $0 \leq k \leq n - d$ and f is in I_d.*

Proof. Let g be an element of degree n in I. Then

$$g = X^{k_1} f_1 + \cdots + X^{k_s} f_s,$$

where the k_i are nonnegative integers, and the f_i are in I_d. If $k_i > 0$ and f_i has degree less than d, then we may replace f_i by Xf_i and k_i by $k_i - 1$. Thus we may assume that either $k_i = 0$ or $\deg(f_i) = d$. Also, if $i \neq j$ and $k_i = k_j$, then $X^{k_i} f_i + X^{k_j} f_j = X^{k_i}(f_i + f_j)$ and $f_i + f_j$ is in I_d. Therefore we may replace the two terms $X^{k_i} f_i + X^{k_j} f_j$ by the single term $X^{k_i}(f_i + f_j)$. Hence we may assume that $k_1 > k_2 > \cdots > k_s$. Suppose that $k_1 > n - d$. Then f_1 has degree d and $X^{k_1} f_1$ has degree $k_1 + d$. But all of the other terms $X^{k_i} f_i$ have degree less than $k_1 + d$. Therefore the degree of g is $k_1 + d > n - d + d = n$. This is a contradiction, so $k_1 \leq n - d$. $\square$

Now it is easy to finish the proof of the proposition. By Lemma 1.2, the coefficient of X^n in an element of I_n is the coefficient of X^d in an element of I_d. Therefore $C_n \subseteq C_d$. Hence $C_n = C_d$. $\square$

In spite of Proposition 1.1, knowing T does not immediately allow us to decide whether a given polynomial g is in I. Although knowing I_d is enough to decide membership in I, the problem is that I_d is not necessarily the additive subgroup A of $\mathbb{Z}[X]$ generated by T. Clearly A is contained in I_d, but in general the containment is proper. There is a way to decide whether $A = I_d$. Let B be the set of polynomials in A which have degree less than d, and let XB denote the set of products Xf with f in B.

Proposition 1.3. *The subgroups A and I_d are equal if and only if XB is contained in A.*

Proof. Clearly $XB \subseteq I_d$. Thus if $A = I_d$, then $XB \subseteq A$. Now suppose that $XB \subseteq A$ and let g be in I_d. Then $g = X^{k_1} f_1 + \cdots + X^{k_s} f_s$, where the k_i are nonnegative integers and the f_i are in A. Just as in the proof of Lemma 1.2, the fact that $XB \subseteq A$ allows us to assume that $k_1 > \cdots > k_s$ and that f_i has degree d if $k_i > 0$. But now the assumption that $k_1 > 0$ leads again to a contradiction. Therefore g is in A. $\square$

Proposition 1.3 actually gives us an algorithm for computing a $\mathbb{Z}$-basis for I_d. If XB is contained in A, then we simply take a basis for A. If XB is not contained in A, then we choose a basis for B, multiply those polynomials by X, and add the results to our basis for A. Now we recompute A and B and check whether XB is contained in A. Since all of this is taking place in a free abelian group of rank $d + 1$, the ascending chain condition holds, so A cannot continue to grow indefinitely. Eventually XB will be contained in A, and at that point A will be I_d.

Example 1.1. Let I be the ideal of $\mathbb{Z}[X]$ generated by the following set T of polynomials:

$$2X^4 + 5X^3 - 13X^2 - 4X + 30,$$

$$2X^3 + \quad X^2 - 7X - 6,$$
$$6X^3 - \quad X^2 - 15X.$$

We may identify the set of integer polynomials of degree at most 4 with $\mathbb{Z}^5$ by mapping the polynomial $a_4X^4 + a_3X^3 + a_2X^2 + a_1X + a_0$ to $(a_4, a_3, a_2, a_1, a_0)$. Under this identification, T corresponds to the rows of the matrix

$$M_1 = \begin{bmatrix} 2 & 5 & -13 & -4 & 30 \\ 0 & 2 & 1 & -7 & -6 \\ 0 & 6 & -1 & -15 & 0 \end{bmatrix}.$$

The subgroup A generated by T corresponds to $S(M_1)$. Row reducing M_1 over $\mathbb{Z}$, we get

$$M_2 = \begin{bmatrix} 2 & 1 & 1 & -14 & -30 \\ 0 & 2 & 1 & -7 & -6 \\ 0 & 0 & 4 & -6 & -18 \end{bmatrix}.$$

The second and third rows of M_2 correspond to the polynomials $2X^3 + X^2 - 7X - 6$ and $4X^2 - 6X - 18$, respectively. These two polynomials generate the subgroup B of elements of A with degree less than 4. Multiplying these polynomials by X and adding them to the set represented by M_2 gives

$$M_3 = \begin{bmatrix} 2 & 1 & 1 & -14 & -30 \\ 0 & 2 & 1 & -7 & -6 \\ 0 & 0 & 4 & -6 & -18 \\ 2 & 1 & -7 & -6 & 0 \\ 0 & 4 & -6 & -18 & 0 \end{bmatrix}.$$

If XB is contained in A, then the nonzero rows in the row Hermite normal form of M_3 will be the same as in M_2. However, when we row reduce M_3, we get

$$M_4 = \begin{bmatrix} 2 & 1 & 1 & 2 & -6 \\ 0 & 2 & 1 & 1 & 6 \\ 0 & 0 & 4 & 2 & -6 \\ 0 & 0 & 0 & 4 & 6 \\ 0 & 0 & 0 & 0 & 0 \end{bmatrix}.$$

The group A has gotten larger, so we must repeat the process. The second, third, and fourth rows of M_4 give a basis for the new B. Multiplying these polynomials by X and adding them to the set of nonzero rows of M_4 gives

$$M_5 = \begin{bmatrix} 2 & 1 & 1 & 2 & -6 \\ 0 & 2 & 1 & 1 & 6 \\ 0 & 0 & 4 & 2 & -6 \\ 0 & 0 & 0 & 4 & 6 \\ 2 & 1 & 1 & 6 & 0 \\ 0 & 4 & 2 & -6 & 0 \\ 0 & 0 & 4 & 6 & 0 \end{bmatrix}.$$

When we row reduce M_5, the nonzero rows in the result are the same as in M_4. This tells us that XB is now contained in A, so the rows of M_4 describe a basis for I_4. This basis contains bases for I_1, I_2, and I_3 as well. There are no nonzero constant polynomials in I. Therefore the subgroup C_0 is $\{0\}$. The subgroups C_1 and C_2 are both $4\mathbb{Z}$, and C_3 and C_4 are $2\mathbb{Z}$. Since $C_4 = C_3 \neq C_2$, the effective degree of I is 3.

Let us now see whether $f = 6X^6 + X^5 + 6X^4 + 3X^3 + 30X - 36$ is in I. Let g be the polynomial represented by the second row of M_4. If f is in I, then $f = gh + r$, where h is in $\mathbb{Z}[X]$ and r is in I_2. Thus we divide f by g in $\mathbb{Q}[X]$. If the quotient h does not have integer coefficients, then f is not in I. If h does have integer coefficients, then f is in I if and only if the remainder r is in I_2. The division gives $h = 3X^3 - X^2 + 2X - 8$ and $r = 12X^2 + 26X + 12$. The third and fourth rows of M_4 describe a basis for I_2. Using the methods of Section 8.1, we find that r is in I_2, and therefore f is in I.

Let us formalize the algorithm sketched in Example 1.1.

Procedure ZXIDEAL$(f_1, \ldots, f_s; g_1, \ldots, g_t)$;
Input: $f_1, \ldots, f_s$: Elements of $\mathbb{Z}[X]$;
Output: $g_1, \ldots, g_t$: Elements of $\mathbb{Z}[X]$ which generate the same
                                             ideal I as the f_i and form a $\mathbb{Z}$-basis for I_e,
                                             where $e = \deg(g_1)$ is the effective degree of I;
Begin
    Let d be the maximum of the degrees of the f_i;
    If $d = -\infty$ then $t := 0$ $(* I = \{0\}. *)$
    Else begin
        $(*$ We shall identify $a_d X^d + \cdots + a_0$ in $\mathbb{Z}[X]$ with $(a_d, \ldots, a_0)$
            in $\mathbb{Z}^{d+1}$. $*)$
        Let U be the s-by-$(d+1)$ integer matrix whose rows correspond to
            the f_i;
        Let $g_1, \ldots, g_t$ correspond to the nonzero rows of the row Hermite
            normal form of U;

$(* \; d = \deg(g_1) > \cdots > \deg(g_t). \; *)$

Repeat

    Copy $g_1, \ldots, g_t$ as $h_1, \ldots, h_r$;

    Let U be the $(2t - 1)$-by-$(d + 1)$ integer matrix whose rows
        correspond to $g_1, \ldots, g_t, \; Xg_2, \ldots, Xg_t$;

    Let $g_1, \ldots, g_t$ correspond to the nonzero rows in the row Hermite
        normal form of U

Until $g_1, \ldots, g_t$ are the same as $h_1, \ldots, h_r$;

While $t > 1$ and g_1 and g_2 have the same leading coefficient do
    Delete g_1 and renumber the remaining g_i beginning at 1

  End

End.

In ZXIDEAL, the statement

Copy $g_1, \ldots, g_t$ as $h_1, \ldots, h_r$;

is an abbreviation for

$r := t$; For $i := 1$ to r do $h_i := g_i$;

In the line beginning "Until", the sequences are different unless $t = r$ and $g_i = h_i$, $1 \le i \le t$.

Example 1.2. Let $f_1 = 6X^4 + 2X^3 - 6X^2 + 8X + 7$ and $f_2 = 2X^2 - 9X - 2$. Here are the sequences $g_1, \ldots, g_t$ at the beginning of each iteration of the Repeat-loop when ZXIDEAL is called with input f_1, f_2:

$$g_1 = 6X^4 + 2X^3 \quad\quad - \quad 19X + \quad 1,$$
$$g_2 = \quad\quad\quad\quad\quad 2X^2 - \quad 9X - \quad 2.$$

$$g_1 = 6X^4 \quad\quad + \; X^2 + \quad 19X + \quad 9,$$
$$g_2 = \quad\quad 2X^3 + \; X^2 - \quad 47X - \quad 10,$$
$$g_3 = \quad\quad\quad\quad 2X^2 - \quad 9X - \quad 2.$$

$$g_1 = 2X^4 \quad\quad\quad\quad + \; 424X + \quad 95,$$
$$g_2 = \quad\quad X^3 \quad\quad + \; 603X + \quad 133,$$
$$g_3 = \quad\quad\quad\quad X^2 + 1244X + \quad 274,$$
$$g_4 = \quad\quad\quad\quad\quad\quad 2497X + \quad 550.$$

$$g_1 = \ X^4 \qquad\qquad\quad + \quad 3X + \ 7016,$$
$$g_2 = \qquad\quad X^3 \qquad + \quad 9X + 10462,$$
$$g_3 = \qquad\qquad\quad X^2 + \quad\ X + \ 2925,$$
$$g_4 = \qquad\qquad\qquad\qquad 11X + \ 5852,$$
$$g_5 = \qquad\qquad\qquad\qquad\qquad 11253.$$

Since the first three polynomials in the last sequence have leading coefficient 1, the effective degree of I is 2. Thus, in the final output,

$$g_1 = X^2 + \quad\ X + \ 2925,$$
$$g_2 = \qquad\quad 11X + \ 5852,$$
$$g_3 = \qquad\qquad\quad 11253.$$

Let H be a subgroup of $\mathbb{Z}^n$ generated by the rows of a matrix A in row Hermite normal form. In Exercise 1.6 in Chapter 8 we defined for each element u of $\mathbb{Z}^n$ a canonical representative for the coset $H + u$. Now let I be an ideal of $\mathbb{Z}[X]$ and let f be an element of $\mathbb{Z}[X]$ of degree n. Then, under the identification of polynomials of degree at most n with elements of $\mathbb{Z}^{n+1}$ used in ZXIDEAL, the canonical representative h of $I_n + f$ is in fact a canonical representative for the coset $I + f$ in $\mathbb{Z}[X]$. Given the generating set for I determined by ZXIDEAL, we can compute h easily using the following procedure:

Procedure ZXIDEALREP$(g_1, \ldots, g_t, f; h)$;
  Input: $g_1, \ldots, g_t$: Elements of $\mathbb{Z}[X]$ as in the output of
                               ZXIDEAL;
         f : An element of $\mathbb{Z}[X]$;
  Output: h : The canonical representative in $I + f$, where I
                               is the ideal generated by the g_i;
  Begin
    $d := \deg(g_1)$; $h := f$; $n := \deg(h)$;
    For $i := n$ downto $d + 1 - t$ do begin
      If $i \geq d$ then $k := 1$
      Else $k := d + 1 - i$;
      Let a be the coefficient of X^i in h;
      Let b be the leading coefficient of g_k;
      Let $a = qb + r$ with $0 \leq r < b$;
      Subtract $qX^{i-d-1+k}g_k$ from h $(*$ Now the coefficient of X^i in h
        is r. $*)$
    End
  End.

If $g_1, \ldots, g_t$ and f are as in ZXIDEALREP, then f is in the ideal generated by the g_i if and only if ZXIDEALREP$(g_1, \ldots, g_t, f; h)$ returns with $h = 0$.

Example 1.3. Let g_1, g_2, and g_3 be the final output of ZXIDEAL in Example 1.2 and let $f = 3X^3 + 5X^2 - 2X + 7$. In ZXIDEALREP with input arguments g_1, g_2, g_3, f we initially set h equal to f. Then we subtract $3Xg_1$ from h to get

$$2X^2 - 8777X + 7.$$

Next we subtract $2g_1$ and obtain

$$-8779X - 5843.$$

To this we add $799g_2$ to get

$$10X + 4669905.$$

Finally we subtract $414g_3$. This gives

$$10X + 11163.$$

Exercises

1.1. Determine the effective degree of the ideal of $\mathbb{Z}[X]$ generated by $6X^4 - 13X^3 + 8X^2 + 5X - 12$ and $4X^4 - 16X^3 + 25X^2 - 25X + 15$.

1.2. Suppose that $f_1, \ldots, f_s$ are elements of $\mathbb{Z}[X]$. Prove that after the execution of ZXIDEAL$(f_1, \ldots, f_s; g_1, \ldots, g_t)$ the polynomial g_t is a (not necessarily monic) gcd of $f_1, \ldots, f_s$ in $\mathbb{Q}[X]$. Let I be the ideal generated by $f_1, \ldots, f_s$. Show that $\mathbb{Z}[X]/I$ is finitely generated as an abelian group if and only if g_1 is monic.

1.3. Modify ZXIDEALREP to produce polynomials $u_1, \ldots, u_t$ such that $h = f - u_1 g_1 - \cdots - u_t g_t$.

10.2 Modules over $\mathbb{Z}[X]$

An ideal of $\mathbb{Z}[X]$ is a submodule of $\mathbb{Z}[X]$ considered as a module over itself. The techniques developed in Section 10.1 extend very naturally to $\mathbb{Z}[X]$-submodules of $\mathbb{Z}[X]^n$. If $u = (f_1, \ldots, f_n)$ is in $\mathbb{Z}[X]^n$, then the degree of u is defined to be the maximum d of the degrees of the f_i. It will sometimes be convenient to write $u = a_d X^d + a_{d-1} X^{d-1} + \cdots + a_0$, where each a_i is in $\mathbb{Z}^n$. The vector a_i is called the *coefficient of* X^i in u. For example, in $Z[X]^2$,

$$u = (2X^2 - 3X + 1, 5X^3 + 4X - 2) = (0,5)X^3 + (2,0)X^2 + (-3,4)X + (1,-2)$$

has degree 3 and the coefficient of X in u is $(-3, 4)$.

The set F of all elements of $\mathbb{Z}[X]^n$ with degree at most d is a free abelian group of rank $k = n(d+1)$. If $b_1, \ldots, b_n$ is the standard basis of $\mathbb{Z}^n$, then the elements $b_i X^j$, $1 \le i \le n$ and $0 \le j \le d$, form a basis of F. To construct an explicit isomorphism between F and $\mathbb{Z}^k$ we must choose an ordering of the basis. Two useful orderings are $b_1 X^d, b_2 X^d, \ldots, b_n X^d, b_1 X^{d-1}, \ldots, b_n X^{d-1}, \ldots, b_1, \ldots, b_n$ and $b_1 X^d$, $b_1 X^{d-1}, \ldots, b_1, b_2 X^d, \ldots, b_2, \ldots, b_n X^d, \ldots, b_n$. The first ordering will be called the *ordering by degree* and the second the *ordering by component*. If the ordering by degree is used, then an element $u = a_d X^d + \cdots + a_0$ in F is identified with the concatenation of its coefficients $a_d, \ldots, a_0$. However, if the ordering by component is used and $u = (f_1, \ldots, f_n)$, then u is identified with the k-tuple obtained by taking the coefficients of $X^d, \ldots, 1$ in f_1, following them by the coefficients of $X^d, \ldots, 1$ in f_2, and so on.

Example 2.1. In $\mathbb{Z}[X]^2$, let

$$u_1 = (12X^2 + 5X - 3, -7X^2 + X + 2), \quad u_2 = (4X^2 + 1, 3X - 2),$$
$$u_3 = (5X + 4, 3X^2 + 2X - 1), \quad u_4 = (X - 3, 9).$$

The maximum of the degrees of the u_i is 2. Let F be the group of all elements of $\mathbb{Z}[X]^2$ of degree at most 2. If we use the ordering by degree to identify F with $\mathbb{Z}^6$, then the u_i correspond to the rows of the matrix

$$U = \begin{bmatrix} 12 & -7 & 5 & 1 & -3 & 2 \\ 4 & 0 & 0 & 3 & 1 & -2 \\ 0 & 3 & 5 & 2 & 4 & -1 \\ 0 & 0 & 1 & 0 & -3 & 9 \end{bmatrix}.$$

Let H be the abelian group generated by the u_i. By computing the row Hermite normal form of U, which is

$$\begin{bmatrix} 4 & 0 & 0 & 3 & 1 & -2 \\ 0 & 1 & 0 & 4 & -47 & 129 \\ 0 & 0 & 1 & 0 & -3 & 9 \\ 0 & 0 & 0 & 10 & -160 & 433 \end{bmatrix},$$

we can see that the subgroup of H consisting of those elements of degree at most 1 has $(X - 3, 9)$ and $(-160, 10X + 433)$ as a basis.

If we use the ordering by component to identify F and $\mathbb{Z}^6$, then the u_i correspond to the rows of

$$V = \begin{bmatrix} 12 & 5 & -3 & -7 & 1 & 2 \\ 4 & 0 & 1 & 0 & 3 & -2 \\ 0 & 5 & 4 & 3 & 2 & -1 \\ 0 & 1 & -3 & 0 & 0 & 9 \end{bmatrix}.$$

The row Hermite normal form of V is

$$\begin{bmatrix} 4 & 0 & 0 & 143 & 155 & 259 \\ 0 & 1 & 0 & 51 & 54 & 93 \\ 0 & 0 & 1 & 17 & 18 & 28 \\ 0 & 0 & 0 & 160 & 170 & 289 \end{bmatrix}.$$

From this we see that the subgroup of H consisting of those elements with first component 0 is generated by $(0, 160X^2 + 170X + 289)$. Either ordering could be used to decide membership in H.

The two orderings of the basis elements $b_i X^j$ lead to two different algorithms for producing nice generating sets for $\mathbb{Z}[X]$-submodules of $\mathbb{Z}[X]^n$. Let us first consider how the ordering by degree can be used. Suppose M is the submodule of $\mathbb{Z}[X]^n$ generated by a finite set T. For each nonnegative integer d, let M_d be the set of elements in M with degree at most d and let C_d be the set of coefficients of X^d in elements of M_d. Then C_d is a subgroup of $\mathbb{Z}^n$ and $C_d \subseteq C_{d+1}$. Since the ascending chain condition holds in $\mathbb{Z}^n$, eventually C_d remains constant. The smallest integer e such that $C_k = C_e$ for $k \geq e$ will be called the *effective degree* of M. Proposition 1.1 generalizes trivially. If d is the maximum of the degrees of the elements of T, then $e \leq d$. Also, knowing M_e is enough to allow us to decide membership in M, for if f in M has degree k and $k > e$, then there is an element g in M_e with the same leading coefficient as f and $f - X^{k-e}g$ has degree less than k.

Let A be the subgroup generated by T and let B be the subgroup of elements in A of degree less than d. Proposition 1.3 also generalizes immediately, so $A = M_d$ if and only if XB is contained in A. As shown in Example 2.1, we can find a basis for B and hence a basis for XB. This gives us an algorithm for determining a $\mathbb{Z}$-basis for M_d.

Example 2.2. Let M be the $\mathbb{Z}[X]$-submodule of $\mathbb{Z}[X]^2$ generated by

$$(X^2 + 3X - 2, 3X^2 + X - 2), \quad (3X^2 + 3X, 2X^2 + X - 3),$$
$$(2X^2 + 3X - 3, 3X^2 - 3X - 3).$$

Here $d = 2$. Using the ordering by degree, we find that our generating set T corresponds to the rows of the matrix

$$N_1 = \begin{bmatrix} 1 & 3 & 3 & 1 & -2 & -2 \\ 3 & 2 & 3 & 1 & 0 & -3 \\ 2 & 3 & 3 & -3 & -3 & -3 \end{bmatrix},$$

and the subgroup A generated by T corresponds to $S(N_1)$. The row Hermite normal form of N_1 is

$$N_2 = \begin{bmatrix} 1 & 0 & 0 & -4 & -1 & -1 \\ 0 & 1 & 0 & -8 & -4 & -1 \\ 0 & 0 & 3 & 29 & 11 & 2 \end{bmatrix}.$$

The subgroup B of those elements in A of degree less than 2 corresponds to the subgroup of $\mathbb{Z}^6$ generated by the last row of N_2. The group XB corresponds to the subgroup generated by the vector obtained by shifting the last row of N_2 two places to the left. Thus $A + XB$ corresponds to the subgroup generated by the rows of

$$N_3 = \begin{bmatrix} 1 & 0 & 0 & -4 & -1 & -1 \\ 0 & 1 & 0 & -8 & -4 & -1 \\ 0 & 0 & 3 & 29 & 11 & 2 \\ 3 & 29 & 11 & 2 & 0 & 0 \end{bmatrix}.$$

The row Hermite normal form of N_3 is

$$N_4 = \begin{bmatrix} 1 & 0 & 0 & 415 & 235 & 73 \\ 0 & 1 & 0 & 411 & 232 & 73 \\ 0 & 0 & 1 & 289 & 161 & 50 \\ 0 & 0 & 0 & 419 & 236 & 74 \end{bmatrix}.$$

Now B corresponds to the subgroup generated by the last two rows of N_4. If we shift the last two rows of N_4 two places to the left, add them as new rows to N_4, compute the row Hermite normal form of the result, and delete rows of zeros, we obtain the matrix

$$N_5 = \begin{bmatrix} 1 & 0 & 0 & 0 & 22071 & 7914 \\ 0 & 1 & 0 & 0 & 20964 & 7518 \\ 0 & 0 & 1 & 0 & 10397 & 3728 \\ 0 & 0 & 0 & 1 & 22900 & 8212 \\ 0 & 0 & 0 & 0 & 23176 & 8311 \end{bmatrix}.$$

Repeating, this time shifting the last three rows, gives us

$$N_6 = \begin{bmatrix} 1 & 0 & 0 & 0 & 3 & 2667 \\ 0 & 1 & 0 & 0 & 0 & 14382 \\ 0 & 0 & 1 & 0 & 1 & 10735 \\ 0 & 0 & 0 & 1 & 0 & 35943 \\ 0 & 0 & 0 & 0 & 4 & 123010 \\ 0 & 0 & 0 & 0 & 0 & 132057 \end{bmatrix}.$$

One more iteration produces no change.

The group M_2 is generated by the elements of $\mathbb{Z}[X]^2$ corresponding to the rows of N_6. The subgroup C_2 is generated by the rows of the 2-by-2 matrix in the upper left corner of N_6. Therefore $C_2 = \mathbb{Z}^2$. The group M_1 corresponds to the group generated by rows three through six of N_6. Thus C_1 is generated by the rows of the 2-by-2 submatrix in the center of N_6. Hence C_1 is also $\mathbb{Z}^2$. The group C_0 is generated by the rows of the 2-by-2 submatrix in the lower right corner of N_6. Since $C_2 = C_1 \neq C_0$, the effective degree of M is 1 and M is generated as a module by M_1, which is described most conveniently by the matrix

$$N_7 = \begin{bmatrix} 1 & 0 & 1 & 10735 \\ 0 & 1 & 0 & 35943 \\ 0 & 0 & 4 & 123010 \\ 0 & 0 & 0 & 132057 \end{bmatrix}.$$

That is, M_1 is generated as an abelian group by

$$(X + 1, 10735), \quad (0, X + 35943), \quad (4, 123010), \quad (0, 132057).$$

Let us formalize the algorithm sketched in Example 2.2.

Procedure ZXMODULE$(u_1, \ldots, u_s; v_1, \ldots, v_t)$;
Input: $u_1, \ldots, u_s$: Elements of $\mathbb{Z}[X]^n$;
Output: $v_1, \ldots v_t$: Elements of $\mathbb{Z}[X]^n$ which generate the same
                $\mathbb{Z}[X]$-submodule M as the u_i and form a
                $\mathbb{Z}$-basis for M_e, where $e = \deg(v_1)$ is the
                effective degree of M;
Begin
    Let d be the maximum of the degrees of the u_i;
    ($*$ We shall use the ordering by degree to identify elements of $\mathbb{Z}[X]^n$
        and vectors of integers. $*$)
    Let U be the s-by-$n(d+1)$ matrix whose rows correspond to
        $u_1, \ldots, u_s$;

Let A consist of the nonzero rows in the row Hermite normal form of U;

Repeat
$\quad D := A$;
$\quad$ Let B be the matrix consisting of the rows of A with zeros in the first n columns;
$\quad$ Let C be the matrix obtained by shifting the rows of B to the left n places and appending them to the rows of A;
$\quad$ Let A consist of the nonzero rows in the row Hermite normal form of C
Until $A = D$;

While A has at least $2n$ columns do begin
$\quad$ Let p be the largest integer such that the p-th row of A has a nonzero entry in the first n columns;
$\quad$ Let q be the largest integer such that the q-th row of A has a nonzero entry in the first $2n$ columns;
$\quad$ Let P be the p-by-n submatrix of A in the upper left corner;
$\quad$ Let Q be the $(q - p)$-by-n submatrix of A consisting of the entries in rows $p + 1, \ldots, q$ and columns $n + 1, \ldots, 2n$;
$\quad$ If $P \neq Q$ then break;
$\quad$ Drop the first p rows and the first n columns from A
End;

Let $v_1, \ldots, v_t$ in $\mathbb{Z}[X]^n$ correspond to the rows of A
End.

If P is a module over $\mathbb{Z}[X]$ and P is generated by n elements, then P is isomorphic to a quotient module of $\mathbb{Z}[X]^n$. A *finite module presentation* for P is a finite generating set for a submodule M of $\mathbb{Z}[X]^n$ such that P is isomorphic to $\mathbb{Z}[X]^n/M$. The procedure ZXMODULE makes it possible to solve the word problem for P if we know a finite presentation. That is, given u in $\mathbb{Z}[X]^n$, we can decide whether u is in M and hence determine whether u defines the zero element of P.

Let M be a submodule of $\mathbb{Z}[X]^n$. For $1 \leq i \leq n + 1$ let $M^{(i)}$ be the set of elements in M whose first $i - 1$ components are 0. Then $M = M^{(1)} \supseteq M^{(2)} \supseteq \cdots \supseteq M^{(n+1)} = \{0\}$ and each $M^{(i)}$ is a submodule of M. Let S be the generating set for M obtained using ZXMODULE. Then $S \cap M^{(i)}$ need not generate $M^{(i)}$. However, using the ordering by component, we can produce a generating set for M which contains generating sets for all of the $M^{(i)}$. The proof of correctness of this algorithm is based on the notion of a syzygy.

Let $(f_1, \ldots, f_n)$ be an element of $\mathbb{Z}[X]^n$. A *syzygy* of $f_1, \ldots, f_n$ is an element $(g_1, \ldots, g_n)$ of $\mathbb{Z}[X]^n$ such that $g_1 f_1 + \cdots + g_n f_n = 0$. Thus a syzygy is a linear relation with polynomial coefficients satisfied by a given sequence of polynomials. The set of syzygies of $f_1, \ldots, f_n$ is a $\mathbb{Z}[X]$-submodule of $\mathbb{Z}[X]^n$.

Proposition 2.1. *Let M be the $\mathbb{Z}[X]$-submodule of $\mathbb{Z}[X]^n$ generated by $u_1, \ldots, u_k$. If S is a generating set for the syzygies of the first components of $u_1, \ldots, u_k$, then $M^{(2)}$ is generated as a $\mathbb{Z}[X]$-submodule by the set of elements $g_1 u_1 + \cdots + g_k u_k$, where $g = (g_1, \ldots, g_k)$ ranges over S.*

Proof. Let v be in $M^{(2)}$. Then v can be written as $h_1 u_1 + \cdots + h_k u_k$, where the h_i are in $\mathbb{Z}[X]$. Since the first component of v is 0, $(h_1, \ldots, h_k)$ is a syzygy of the first components of the u_i. Thus $h = (h_1, \ldots, h_k)$ can be written as a $\mathbb{Z}[X]$-linear combination of the elements of S. If

$$ h = \sum_{g \varepsilon S} p_g g, $$

then it is immediate that

$$ v = \sum_{g \varepsilon S} p_g(g_1 u_1 + \cdots + g_k u_k). \quad \square $$

Suppose that $f_1, \ldots, f_k$ is the generating set for an ideal I of $\mathbb{Z}[X]$ produced by ZXIDEAL. If d is the degree of f_1, then the degree of f_i is $d-i+1$ and $f_i, \ldots, f_k$ is an abelian group basis for I_{d-i+1}. Thus for $1 < i \le k$ there are unique integers $a_{ii-1}, a_{ii}, \ldots, a_{ik}$, such that

$$ X f_i = \sum_{j=i-1}^{k} a_{ij} f_j. $$

Therefore $g^{(i)} = (0, \ldots, 0, a_{ii-1}, a_{ii} - X, a_{ii+1}, \ldots, a_{ik})$ is a syzygy of $f_1, \ldots, f_k$.

Proposition 2.2. *Under the preceding assumptions, the syzygies of $f_1, \ldots, f_k$ are generated as a $\mathbb{Z}[X]$-module by $g^{(2)}, \ldots, g^{(k)}$.*

Proof. Let $h = (h_1, \ldots, h_k)$ be a syzygy of $f_1, \ldots, f_k$. Among $h_2, \ldots, h_k$, let h_i have maximum degree e and let b be the leading coefficient of h_i. If $e > 0$, then adding $bX^{e-1}g^{(i)}$ to h reduces the degree of h_i and does not raise the degree of any other h_j above $e - 1$. It follows that we may subtract from h an element of the submodule generated by $g^{(2)}, \ldots, g^{(k)}$ and assume that $h_2, \ldots, h_k$ all have degree 0. Since $h_1 f_1 + \cdots + h_k f_k = 0$ and the degree of f_1 is greater than the degrees of the other f_i, we must have $h_1 = 0$. The f_i are linearly independent over $\mathbb{Z}$, so $h_2 = \cdots = h_k = 0$. Thus the original syzygy h is in the submodule generated by $g^{(2)}, \ldots, g^{(k)}$. $\square$

Now, given a generating set T for a submodule M of $\mathbb{Z}[X]^n$, we can find a generating set for $M^{(2)}$ by applying ZXIDEAL to the first components of the elements of T while keeping track of what happens to the other components and then applying Proposition 2.1. In fact, the two steps can be combined.

Example 2.3. Let M be the submodule of $\mathbb{Z}[X]^2$ generated by

$$(X^2 - 2X - 3, -3X^2 - 2X + 2) \quad \text{and} \quad (3X^2 + X, -2X^2 - 2X + 1).$$

To find a generating set for $M^{(2)}$, we shall apply ZXIDEAL to the first components of our generating set. The degree of the first components will remain bounded by 2 throughout the computation. However, the degrees of the second components will grow. The ordering by component will be used to identify elements of $\mathbb{Z}[X]^2$ with integer vectors. Because the degrees of the elements grow, the lengths of the vectors also grow. Our original generators for M correspond to the rows of

$$P_1 = \begin{bmatrix} 1 & -2 & -3 & -3 & -2 & 2 \\ 3 & 1 & 0 & -2 & -2 & 1 \end{bmatrix}.$$

The row Hermite normal form of P_1 is

$$P_2 = \begin{bmatrix} 1 & 5 & 6 & 4 & 2 & -3 \\ 0 & 7 & 9 & 7 & 4 & -5 \end{bmatrix}.$$

In the subgroup generated by the first components of the corresponding elements of $\mathbb{Z}[X]^2$, the subgroup of elements of degree less than 2 is generated by $7X + 9$. In ZXIDEAL we would multiply this element by X and add it as a new generator. Now, however, the pair corresponding to the second row of P_2 must be multiplied by X and added as a new module generator. The resulting generating set is represented by

$$P_3 = \begin{bmatrix} 0 & 1 & 5 & 6 & 0 & 4 & 2 & -3 \\ 0 & 0 & 7 & 9 & 0 & 7 & 4 & -5 \\ 0 & 7 & 9 & 0 & 7 & 4 & -5 & 0 \end{bmatrix}.$$

The row Hermite normal form of P_3 is

$$P_4 = \begin{bmatrix} 0 & 1 & 0 & 51 & -42 & -13 & 24 & -14 \\ 0 & 0 & 1 & 27 & -21 & -5 & 13 & -8 \\ 0 & 0 & 0 & 60 & -49 & -14 & 29 & -17 \end{bmatrix}.$$

Now we must multiply the pairs corresponding to the last two rows of P_4 by X and add the pairs produced as new generators. This gives

$$P_5 = \begin{bmatrix} 0 & 0 & 1 & 0 & 51 & 0 & -42 & -13 & 24 & -14 \\ 0 & 0 & 0 & 1 & 27 & 0 & -21 & -5 & 13 & -8 \\ 0 & 0 & 0 & 0 & 60 & 0 & -49 & -14 & 29 & -17 \\ 0 & 0 & 1 & 27 & 0 & -21 & -5 & 13 & -8 & 0 \\ 0 & 0 & 0 & 60 & 0 & -49 & -14 & 29 & -17 & 0 \end{bmatrix}.$$

The nonzero rows in the row Hermite normal form of P_5 are given by

$$P_6 = \begin{bmatrix} 0 & 0 & 1 & 0 & 51 & 0 & -42 & -13 & 24 & -14 \\ 0 & 0 & 0 & 1 & 27 & 0 & -21 & -5 & 13 & -8 \\ 0 & 0 & 0 & 0 & 60 & 0 & -49 & -14 & 29 & -17 \\ 0 & 0 & 0 & 0 & 0 & 7 & 11 & 7 & 2 & -3 \end{bmatrix}.$$

Since the nonzero first components of our generating set have not changed, we are done. The process of row reducing P_5 has constructed the set U of generators of $M^{(2)}$ given by Proposition 2.1 using the syzygies of Proposition 2.2 and produced a $\mathbb{Z}$-basis for the abelian group generated by U. In this example, the submodule $M^{(2)}$ is generated by the single pair $(0, 7X^4 + 11X^3 + 7X^2 + 2X - 3)$.

By repeated application of the algorithm illustrated in Example 2.3, we can find generators for each of the submodules $M^{(i)}$. A formal description of the algorithm follows. In the algorithm we shall need to compute $\mathbb{Z}$-bases of subgroups of $\mathbb{Z}[X]^n$. This will be done using the procedure ZBASIS, which employs the ordering by component.

Procedure ZBASIS$(u_1, \ldots, u_s; w_1, \ldots, w_t)$
Input: $u_1, \ldots, u_s$: Elements of $\mathbb{Z}[X]^n$;
Output: $w_1, \ldots, w_t$: A $\mathbb{Z}$-basis for the abelian group generated
                                    by $u_1, \ldots, u_s$;
Begin
   Let d be the maximum of the degrees of the u_i;
   (* Use the ordering by component to identify the u_i with elements
        of $\mathbb{Z}^{n(d+1)}$. *)
   Let U be the s-by-$n(d+1)$ integer matrix whose rows correspond to
        the u_i;
   Let $w_1, \ldots, w_t$ in $\mathbb{Z}[X]^n$ correspond to the nonzero rows in the row
        Hermite normal form of U
End.

Here is the algorithm for finding generators of the submodules $M^{(i)}$.

Procedure ZXSERIES($u_1, \ldots, u_s; v_1, \ldots, v_t$);
Input: $u_1, \ldots, u_s$: Elements of $\mathbb{Z}[X]^n$;
Output: $v_1, \ldots, v_t$: Generators for the submodule M of $\mathbb{Z}[X]^n$
                    generated by the u_i which include generators
                    for each of the submodules $M^{(i)}$, $1 \le i \le n$;
Begin
   ZBASIS($u_1, \ldots, u_s; v_1, \ldots, v_t$); $k := 1$;
   For $i := 1$ to n do begin
      (∗ At this point $v_k, \ldots, v_t$ generate $M^{(i)}$. ∗)
      If $t < k$ then break;   (∗ $M^{(i)} = \{0\}$. ∗)
      $done :=$ false;

      Repeat
         If the i-th components of $v_k, \ldots, v_t$ are all 0 then begin
            $q := k - 1$; $done :=$ true
         End

         Else begin
            Let q be maximal such that the i-th component of v_q is
               nonzero;
            ZBASIS($v_k, \ldots, v_t, X v_{k+1}, \ldots, X v_q; z_1, \ldots, z_p$);
            If the set of nonzero i-th components of $z_1, \ldots, z_p$ is the same
               as the set of nonzero i-th components of $v_k, \ldots, v_t$ then
               $done :=$ true;
            Copy $z_1, \ldots, z_p$ as $v_k, \ldots, v_t$
         End
      Until $done$;

      (∗ The following is not needed for correctness but it makes the
         output depend only on M. ∗)
      If $q \ge k$ then begin
         Let p be maximal such that the i-th components of $v_k, \ldots, v_p$
            have the same leading coefficient;
         Delete $v_k, \ldots, v_{p-1}$ and renumber the v's;
         Let q be maximal such that the i-th component of v_q is not 0;
         Let $f_1, \ldots, f_r$ be the i-th components of $v_k, \ldots, v_q$;
         (∗ $f_1, \ldots, f_r$ are the standard generators, as produced by
            ZXIDEAL, for the ideal $I^{(i)}$ of i-th components in elements
            of $M^{(i)}$. ∗)

         For $j := 1$ to $k - 1$ do begin
            Let g be the i-th component of v_j;
            ZXIDEALREP($f_1, \ldots, f_r, g; h$);
            (∗ The element $g - h$ is in $I^{(i)}$. ∗)

Write $g - h$ as $e_1 f_1 + \cdots + e_r f_r$;
  Subtract $e_1 v_k + \cdots + e_r v_q$ from v_j
  (* The i-th component of v_j is now h. *)
      End
    End;
      $k := q + 1$
    End
  End.

Let $v_1, \ldots, v_t$ be the output of ZXSERIES. Then the nonzero components of those v_j which lie in $M^{(i)}$ form the standard basis for $I^{(i)}$ and the i-th components of earlier v's are their own canonical representatives modulo $I^{(i)}$. Using these facts, it is not hard to show that the v's depend only on M. If we only need module generators for $M^{(2)}$, then we can stop the main For-loop after the first iteration and take only those v_i which belong to $M^{(2)}$.

Example 2.4. In $\mathbb{Z}[X]^3$ let

$$
\begin{aligned}
u_1 &= (-X - 5, \ 2X - 5, \ \ X - 4), \\
u_2 &= (\ 5X + 3, \ 5X \quad\ \ , \ 3X - 1), \\
u_3 &= (\quad\quad -2, \ \ X - 4, \ 5X - 4), \\
u_4 &= (-X - 3, \ 4X - 3, \ 3X + 4).
\end{aligned}
$$

Here are the triples $v_1, \ldots, v_t$ at the end of each iteration of the outer For-loop in the call to ZXSERIES with the u_i as input:

$$
\begin{aligned}
v_1 &= (X + 1, \quad\ \ 10, \ 118X^2 - 48X + \ 89), \\
v_2 &= (\quad\ \ 2, \quad\ \ 12, \ 117X^2 - 60X + 136), \\
v_3 &= (\quad\ \ 0, \ X + \ 8, \ 117X^2 - 55X + 132), \\
v_4 &= (\quad\ \ 0, \quad\ \ 13, \ 118X^2 - 57X + \ 93), \\
v_5 &= (\quad\ \ 0, \quad\ \ 0, \ 295X^2 - 58X + 206).
\end{aligned}
$$

$$
\begin{aligned}
v_1 &= (X + 1, \quad\ \ 10, \quad\quad\quad 118X^2 - 48X + \ 89), \\
v_2 &= (\quad\ \ 2, \quad\ \ 12, \quad\quad\quad 117X^2 - 60X + 136), \\
v_3 &= (\quad\ \ 0, \ X + \ 8, \quad\quad\quad 117X^2 - 55X + 132), \\
v_4 &= (\quad\ \ 0, \quad\ \ 13, \quad\quad\quad 118X^2 - 57X + \ 93), \\
v_5 &= (\quad\ \ 0, \quad\ \ 0, \ 118X^3 + \ 61X^2 + 62X + \ 58), \\
v_6 &= (\quad\ \ 0, \quad\ \ 0, \quad\quad\quad 295X^2 - 58X + 206).
\end{aligned}
$$

$$
\begin{aligned}
v_1 &= (X + 1, & 10, & & 4X + 577557), \\
v_2 &= (\quad 2, & 12, & & 2X + \ 21310), \\
v_3 &= (\quad 0, X + \ 8, & & & 7X + \ 21306), \\
v_4 &= (\quad 0, & 13, & & 3X + 213825), \\
v_5 &= (\quad 0, & 0, X^2 + 2X + 262602), \\
v_6 &= (\quad 0, & 0, & & 8X + 265264), \\
v_7 &= (\quad 0, & 0, & & \qquad 629000).
\end{aligned}
$$

There is a close analogy between the output of ZXSERIES and integer matrices in row Hermite normal form. If we think of the final sequence $v_1, \ldots, v_7$ as the rows of a 7-by-3 matrix with entries in $\mathbb{Z}[X]$, then we can define corner entries to be entries which are the first nonzero entry in their row. A given column may have more than one corner entry, but the corner entries in a given column are the generators produced by ZXIDEAL for an ideal of $\mathbb{Z}[X]$ and the entries above those corner entries are standard representatives for cosets of that ideal. This set of generators for the module M generated by the u_i looks quite different from the generating set produced by ZXMODULE, which is

$$
\begin{aligned}
w_1 &= (X + 1, & 62, & & 273329), \\
w_2 &= (\quad 0, X + \ 47, & & & 132253), \\
w_3 &= (\quad 0, & 39, X + 376211), \\
w_4 &= (\quad 2, & 38, & & 183696), \\
w_5 &= (\quad 0, & 104, & & 285808), \\
w_6 &= (\quad 0, & 0, & & 629000).
\end{aligned}
$$

The generators w_4, w_5, and w_6 form a $\mathbb{Z}$-basis for $M \cap \mathbb{Z}^3$.

We have used row Hermite normal forms to decide equality of subgroups of $\mathbb{Z}^n$. Since the associated unimodular matrices are not needed, there are many possible ways to compute these normal forms. We could also use lattice reduction, as described in Sections 8.6 and 8.7, to compare subgroups of $\mathbb{Z}^n$.

Exercises

2.1. Show that the output of ZXMODULE depends only on the module M generated by $u_1, \ldots, u_s$. Do the same for ZXSERIES.

2.2. Devise a procedure ZXMODULEREP analogous to ZXIDEALREP in Section 10.1.

2.3. Let M be a submodule of $\mathbb{Z}[X]^n$ with effective degree e. Prove that $\mathbb{Z}[X]^n/M$ is finitely generated as an abelian group if and only if the group C_e is $\mathbb{Z}^n$.

10.3 Modules over $\mathbb{Z}[X, Y]$

In Section 10.2 we developed methods for working with submodules of $\mathbb{Z}[X]^n$. In this section we shall look at submodules of $\mathbb{Z}[X, Y]^n$. After the two-variable case has been mastered, the generalization to m variables is relatively straightforward. Throughout this section R will denote $\mathbb{Z}[X, Y]$.

The first approach we shall consider singles out one of the variables, say Y, as the primary variable. An element u of R^n can be written in the form $a_d Y^d + a_{d-1} Y^{d-1} + \cdots + a_0$, where each a_i is in $\mathbb{Z}[X]^n$. If $a_d \neq 0$, then d will be called the *degree in Y* of u. For example, in R^2, the element

$$u = (3XY^2 - 2Y^2 + X^2 Y + 2Y - X + 7, \ X^3 Y^2 + 5XY^2 - XY + 2Y + X^2 - 3)$$

can be written

$$(3X - 2, \ X^3 + 5X)Y^2 + (X^2 + 2, \ -X + 2)Y + (-X + 7, \ X^2 - 3).$$

The degree of u in Y is 2.

Let M be an R-submodule of R^n. For each nonnegative integer k, let M_k be the set of elements in M whose degree in Y does not exceed k and let C_k be the set of coefficients of Y^k in elements of M_k. Then C_k is a $\mathbb{Z}[X]$-submodule of $\mathbb{Z}[X]^n$ and $C_k \subseteq C_{k+1}$. Let T be a finite generating set for M and let d be the maximum of the degrees in Y of the elements in T. The analogue of Proposition 1.1 holds, so $C_k = C_d$ if $k \geq d$. The smallest integer e such that $C_k = C_e$ for $k \geq e$ will be called the *effective degree in Y* of M. Let A be the $\mathbb{Z}[X]$-submodule generated by T and let $B = A \cap M_{d-1}$, the submodule of A consisting of the elements whose degree in Y is less than d. The analogue of Proposition 1.3 is valid and $A = M_d$ if and only if $YB \subseteq A$.

We can identify A with a $\mathbb{Z}[X]$-submodule of $\mathbb{Z}[X]^{n(d+1)}$ in several ways. The elements $b_i Y^j$ with $1 \leq i \leq n$ and $0 \leq j \leq d$ form a $\mathbb{Z}[X]$-basis for the set of all elements in R^n whose degree in Y is at most d. Here $b_1, \ldots, b_n$ is the standard basis of $\mathbb{Z}^n$. If we order the $b_i Y^j$ by degree, as in Section 10.2, then $a_d Y^d + \cdots + a_0$ is identified with the concatenation of $a_d, \ldots, a_0$. Thus the element u is mapped to

$$(3X - 2, \ X^3 + 5X, \ X^2 + 2, \ -X + 2, \ -X + 7, \ X^2 - 3).$$

However, if we order the basis by component, then u corresponds to

$$(3X - 2, \ X^2 + 2, \ -X + 7, \ X^3 + 5X, \ -X + 2, \ X^2 - 3).$$

Suppose that under the ordering by degree the $\mathbb{Z}[X]$-submodule A generated by T corresponds to the submodule N of $\mathbb{Z}[X]^{n(d+1)}$. Then B

corresponds to $N^{(n+1)}$. Thus, using ZXSERIES, we can find a module generating set for B. In fact, it suffices to terminate the main For-loop in ZXSERIES after only n iterations. The following procedure ZXYMODULE is obtained from ZXMODULE by replacing the determination of row Hermite normal forms with calls to ZXMODULE or ZXSERIES.

Procedure ZXYMODULE($u_1, \ldots, u_s; v_1, \ldots, v_t$);
Input: $u_1, \ldots, u_s$: Elements of R^n;
Output: $v_1, \ldots, v_t$: Elements of R^n which generated the same
                                            R-submodule M as the u_i and form a
                                            $\mathbb{Z}[X]$-basis for M_e, where e is the effective
                                            degree in Y of M;
Begin
    Let d be the maximum degree in Y of the u_i;
    (* We shall use the ordering by degree to identify the u_i with
        elements of $\mathbb{Z}[X]^{n(d+1)}$. *)
    Let $w_1, \ldots, w_s$ in $\mathbb{Z}[X]^{n(d+1)}$ correspond to $u_1, \ldots, u_s$;
    ZXMODULE($w_1, \ldots, w_s; z_1, \ldots, z_t$);

    Repeat
        Copy $z_1, \ldots, z_t$ as $x_1, \ldots, x_m$;
        Let N be the submodule of $\mathbb{Z}[X]^{n(d+1)}$ generated by the z_i;
        Use ZXSERIES to find a generating set $y_1, \ldots, y_q$ for $N^{(n+1)}$;
        For $i := 1$ to q do let g_i be obtained from y_i by shifting its
            components n places to the left;
        ZXMODULE($x_1, \ldots, x_m, g_1, \ldots, g_q; z_1, \ldots, z_t$)
    Until the sequences $x_1, \ldots, x_m$ and $z_1, \ldots, z_t$ are the same;
    (* At this point $z_1, \ldots, z_t$ correspond to a $\mathbb{Z}[X]$-basis for M_d, but d
        may not equal e. *)

    While $d \geq 1$ do begin
        Let $f_1, \ldots, f_t$ be the projections of $z_1, \ldots, z_t$ onto their first n
            components;
        Let $g_1, \ldots, g_q$ be the projections of $y_1, \ldots, y_q$ onto components
            $n+1, \ldots, 2n$;
        ZXMODULE($f_1, \ldots, f_t; h_1, \ldots, h_r$);
        ZXMODULE($g_1, \ldots, g_q; k_1, \ldots, k_m$);
        If $h_1, \ldots, h_r$ and $k_1, \ldots, k_m$ are different then break;
        (* $M_{d-1} = M_d$ *)
        $d := d - 1$;
        For $i := 1$ to q do let w_i be obtained from y_i by dropping the first n
            components   (* which are 0 *);
        ZXMODULE($w_1, \ldots, w_q; z_1, \ldots, z_t$);
        Let N be the $\mathbb{Z}[X]$-submodule generated by the z_i;
        Use ZXSERIES to find a generating set $y_1, \ldots, y_q$ for $N^{(n+1)}$

End

Let $v_1, \ldots, v_t$ in R^n correspond to $z_1, \ldots, z_t$
End.

Example 3.1. Let us follow the execution of ZXYMODULE with input

$$
\begin{aligned}
u_1 &= (Y + X + 1,\ Y + 1), \\
u_2 &= (X - 1,\ Y - X), \\
u_3 &= (Y - X + 1,\ Y + X - 1).
\end{aligned}
$$

Using the ordering by degree, the u_i are identified with

$$
\begin{aligned}
w_1 &= (1,\ 1,\ X + 1,\ 1), \\
w_2 &= (0,\ 1,\ X - 1,\ -X), \\
w_3 &= (1,\ 1,\ -X + 1,\ X - 1).
\end{aligned}
$$

The output of ZXMODULE with the w_i as input is

$$
\begin{aligned}
z_1 &= (X\ ,\ 2X - 2,\ 2,\ 2), \\
z_2 &= (0,\ -1,\ X + 1,\ 2), \\
z_3 &= (0,\ -2,\ 2,\ X + 2), \\
z_4 &= (1,\ 2,\ 0,\ -1).
\end{aligned}
$$

On the first pass through the Repeat-loop, the following generators for $N^{(3)}$ are found:

$$
\begin{aligned}
y_1 &= (0,\ 0,\ 2X^2\ ,\ -X^2 + 2X), \\
y_2 &= (0,\ 0,\ 2X,\ -\ X + 2).
\end{aligned}
$$

Note that since the main For-loop in ZXSERIES was terminated after the second iteration, ZXSERIES did not detect that y_1 is in fact Xy_2. Now ZXMODULE is called, with input the z_i above together with

$$
\begin{aligned}
g_1 &= (2X^2\ ,\ -X^2 + 2,\ 0,\ 0), \\
g_2 &= (2X,\ -X + 2,\ 0,\ 0).
\end{aligned}
$$

The output is

$$
\begin{aligned}
z_1 &= (X\ ,\ 2X - 2,\ 2,\ 2), \\
z_2 &= (0,\ 5X - 6,\ 4,\ 4),
\end{aligned}
$$

$$z_3 = (\ \ 0, \quad -1,\ X+1, \quad\quad 2),$$
$$z_4 = (\ \ 0, \quad -2, \quad 2,\ X+2),$$
$$z_5 = (\ \ 1, \quad\ \ 2, \quad\ \ 0, \quad -1).$$

On the second pass through the repeat loop, the generators for $N^{(3)}$ are found to be

$$y_1 = (0,\ 0,\ X^2 +\ \ X+2,\ -5X^3 - 12X^2 + 12X +\ \ 4),$$
$$y_2 = (0,\ 0, \quad\quad 2X \quad , \quad\quad\quad -\quad X+\ \ 2),$$
$$y_3 = (0,\ 0, \quad\quad 4,\ -5X^2 \quad\quad -\quad 9X + 14).$$

This time the call to ZXMODULE returns exactly the same z_i as before, so the Repeat-loop terminates. Entering the While-loop, we find that

$$f_1 = (X\ \ ,\ 2X-2),$$
$$f_2 = (\ \ 0,\ 5X-6),$$
$$f_3 = (\ \ 0, \quad -1),$$
$$f_4 = (\ \ 0, \quad -2),$$
$$f_5 = (\ \ 1, \quad\ \ 2),$$

and

$$g_1 = (X^2 +\ \ X+2,\ -5X^3 - 12X^2 + 12X +\ \ 4),$$
$$g_2 = (\quad\quad 2X \quad , \quad\quad\quad -\quad X+\ \ 2),$$
$$g_3 = (\quad\quad\quad 4, \quad -\ \ 5X^2 -\ \ 9X + 14).$$

The two calls to ZXMODULE produce

$$h_1 = (1,0),$$
$$h_2 = (0,1),$$

and

$$k_1 = (X^2 +\ \ X-2,\ 2X^2 + 5X -\ \ 6),$$
$$k_2 = (\quad\quad\quad -\ 4,\ 5X^2 + 9X - 14),$$
$$k_3 = (\quad\quad 2X \quad , \quad\quad -\quad X+\ \ 2).$$

Since the h's and k's differ, no further changes are made to the z's. The final output is

$$
\begin{aligned}
v_1 &= (XY + 2,\ 2XY - 2Y + 2), \\
v_2 &= (\qquad 4,\ 5XY - 6Y + 4), \\
v_3 &= (\ X + 1,\qquad -Y + 2), \\
v_4 &= (\qquad 2,\ -2Y + X + 2), \\
v_5 &= (\qquad Y,\qquad 2Y - 1).
\end{aligned}
$$

A number of variations on the procedure ZXYMODULE are possible. For example, the calls to ZXMODULE could be replaced by calls to ZXSERIES. If this is done, then with the data of Example 3.1 the following generating set is obtained:

$$
\begin{aligned}
v_1 &= (\qquad Y + 2,\qquad\qquad X + 1), \\
v_2 &= (\qquad X + 3,\ Y - 5X^2 - 10X + 14), \\
v_3 &= (X^2 + X + 2,\qquad -3X^2 - 4X + 8), \\
v_4 &= (\qquad 2X,\qquad\qquad -X + 2), \\
v_5 &= (\qquad 4,\qquad -5X^2 - 9X + 14), \\
v_6 &= (\qquad 0,\ 5X^3 + 9X^2 - 16X + 4).
\end{aligned}
$$

Either this generating set or the one obtained in Example 3.1 can be used to decide membership in the module.

Let M be a submodule of R^n. For $1 \le i \le n+1$ let $M^{(i)}$ be the submodule of M consisting of the elements whose first $i-1$ components are 0. In order to be able to extend the approach of ZXMODULE and ZXYMODULE to submodules of $\mathbb{Z}[X, Y, Z]^n$, we need an analogue of ZXSERIES to find generators for the submodules $M^{(i)}$. This in turn requires generalizations of ZBASIS and ZXIDEALREP. The description of a procedure ZXYSERIES is left to the reader as an exercise.

Exercises

3.1. Find the effective degree in Y of the ideal I of $\mathbb{Z}[X, Y]$ generated by $2XY^2 - 3Y + 7$ and $XY^2 - 2Y + X + 1$. Determine the effective degree in X of $I \cap \mathbb{Z}[X]$.

3.2. Let M be a submodule of $\mathbb{Z}[X, Y]^n$ with effective degree e in Y. Show that $\mathbb{Z}[X, Y]^n / M$ is finitely generated as a $\mathbb{Z}[X]$-module if and only if the $\mathbb{Z}[X]$-submodule C_e is $\mathbb{Z}[X]^n$. Under what conditions is $\mathbb{Z}[X, Y]^n / M$ finitely generated as an abelian group?

10.4 The total degree ordering

The procedure ZXYMODULE of the previous section reduces questions about a $\mathbb{Z}[X, Y]$-module generated by a finite set T to questions about the

$\mathbb{Z}[X]$-module generated by T, and it calls ZXMODULE to answer those questions. Within ZXMODULE, the questions are converted into questions about the subgroup A generated by T, and these questions are answered by computing the row Hermite normal form of an integer matrix. Let r and s denote the maxima of the degrees in X and Y, respectively, of the elements in T. Then A is a subgroup of the group P which has a basis consisting of the elements $b_i X^j Y^k$, $1 \leq i \leq n$, $0 \leq j \leq r$, and $0 \leq k \leq s$. To associate elements of P with integer vectors, we must order the basis. One possible order is obtained by letting k run from s to 0, for a given k letting j run from r to 0, and for given values of k and j letting i run from 1 to n. In the case of an ideal of $R = \mathbb{Z}[X, Y]$, where we can identify b_1 with 1, we are taking the monomials $X^j Y^k$ in decreasing order according to the *reverse lexicographic ordering*, in which we compare degrees in Y first and then, if necessary, the degrees in X. In this ordering, the monomials of degree at most 2 in X and in Y arranged in decreasing order are

$$X^2 Y^2, \ XY^2, \ Y^2, \ X^2 Y, \ XY, \ X^2, \ X, \ 1.$$

There is another ordering on monomials which is quite useful. The *total degree* of $X^i Y^j$ is $i + j$. In the *total-degree ordering* of monomials, we compare total degree first and then, if necessary, the degree in Y. The monomials of total degree at most 3, in decreasing order, are

$$Y^3, \ XY^2, \ X^2 Y, \ X^3, \ Y^2, \ XY, \ X^2, \ Y, \ X, \ 1.$$

The usual proof that ideals in $\mathbb{Z}[X_1, \ldots, X_m]$ are finitely generated proceeds by induction on m. The total-degree ordering on monomials can be used to give a proof that ideals in R are finitely generated without invoking the corresponding statement for ideals in $\mathbb{Z}[X]$. Let us write $X^i Y^j \prec X^r Y^s$ if $X^i Y^j$ comes before $X^r Y^s$ in the total degree ordering. Any nonzero element f of R can be written uniquely as $a_1 u_1 + \cdots + a_t u_t$, where the u_i are monomials such that $u_1 \succ u_2 \succ \cdots \succ u_t$ and the a_i are nonzero integers. We shall call $a_1 u_1$, a_1, and u_1 the *leading term*, the *leading coefficient*, and the *leading monomial* of f, respectively. The total degree of f is defined to be the total degree of u_1. Note that the leading monomial of $X^i Y^j f$ is $X^i Y^j u_1$.

Let I be an ideal of R and for each pair (i, j) of nonnegative integers let I_{ij} be the set of elements f in I such that either $f = 0$ or the leading monomial u of f satisfies $u \preceq X^i Y^j$. Let C_{ij} be the set of coefficients of $X^i Y^j$ in elements of I_{ij}. Clearly C_{ij} is a subgroup of $\mathbb{Z}$. Let us say that $(i, j) \leq (r, s)$ if $i \leq r$ and $j \leq s$, or equivalently, if $X^i Y^j$ divides $X^r Y^s$ in R. If $(i, j) \leq (r, s)$, then $X^{r-i} Y^{s-j} I_{ij} \subseteq I_{rs}$, so $C_{ij} \subseteq C_{rs}$. Let c_{ij} denote the nonnegative generator of C_{ij}. Define $U = U(I)$ to be the set of pairs (r, s) such that $c_{rs} \neq 0$ and $c_{rs} \neq c_{ij}$ for any pair $(i, j) < (r, s)$.

Proposition 4.1. *For every ideal I of R the set $U(I)$ is finite.*

Proof. If $I = \{0\}$, then $U = \emptyset$. Thus we may assume that $I \neq \{0\}$, so some $c_{rs} > 0$. Choose (r, s) minimal with respect to $<$ such that c_{rs} is the minimum among all nonzero c_{ij}. Then (r, s) is in U. If (i, j) is some other element of U, then either $i < r$ or $j < s$. But for a given i there can be only finitely many values of j such that (i, j) is in U. Similarly, for a given j there are only finitely many possibilities for i. Thus U is finite. $\square$

For each $d \geq 0$ let $I_{(d)}$ be the abelian group of elements in I with total degree at most d. The maximum e of $r + s$ as (r, s) ranges over U will be called the *effective total degree* of I. If we know $I_{(e)}$, we can decide membership in I.

Proposition 4.2. *Suppose that I is an ideal of R with effective total degree e. Then any element f of I can be expressed in the form $u_1 f_1 + \cdots + u_r f_r$, where the u_i are monomials, the f_i are nonzero elements of $I_{(e)}$, and the leading monomials of the terms $u_i f_i$ are strictly decreasing with respect to $\prec$.*

Proof. If $f = 0$, then we may take $r = 0$. If f is in $I_{(e)}$, then we may take $r = 1$, $u_1 = 1$, and $f_1 = f$. Thus we may assume that the total degree of f is greater than e. Let the leading term of f be $bX^r Y^s$. The coefficient b is divisible by c_{rs}. Since $r + s > e$, the pair (r, s) is not in $U(I)$. Therefore $c_{rs} = c_{ij}$ for some pair $(i, j) < (r, s)$. Choose (i, j) as small as possible with respect to $<$. Then $i + j \leq e$. Let f_1 be an element of $I_{(e)}$ with leading term $bX^i Y^j$ and set $u_1 = X^{r-i} Y^{s-j}$. Then $g = f - u_1 f_1$ is in I and the leading monomial of g precedes $X^r Y^s$ with respect to $\prec$. Continuing in this manner, we obtain the desired decomposition. $\square$

It follows from Proposition 4.2 that $I_{(e)}$ generates I and that I is finitely generated as an ideal.

Given group generators for $I_{(e)}$, to decide membership in I we take the monomials of total degree at most e in decreasing total-degree order to associate integer vectors with polynomials and we find the row Hermite normal form of the matrix associated with our generators of $I_{(e)}$. A corner entry corresponds to a pair (r, s) for which $c_{rs} \neq 0$ and the row containing the corner entry gives an element f_{rs} of I_{rs} with leading term $c_{rs} X^r Y^s$. Now we can decide membership in I using the following procedure:

Function ZXYMEMBER(g): boolean;
(* The polynomials f_{rs} are available as global variables. True is
  returned if and only if g is in I. *)
Begin

Table 10.4.1

			s			
r	0	1	2	3	4	5
0	0	0	2	2	2	2
1	0	3	1	1	1	
2	2	1	1	1		
3	2	1	1			
4	2	1				
5	2					

$h := g$; *done* := false;

While $h \neq 0$ and not *done* do begin

　Let aX^iY^j be the leading term of h;

　If there is an f_{rs} such that $(r,s) \leq (i,j)$ and the leading coefficient

　　c of f_{rs} divides a then begin

　$q := a/c$;　$h := h - qX^{i-r}Y^{j-s}f_{rs}$

　End

　Else *done* := true

End

ZXYMEMBER := $(h = 0)$

End.

Let I be generated by a finite set T, let d be the maximum of the total degrees of the elements in T, and let A be the abelian group generated by T. Given our experience so far, it should be expected that A is usually not $I_{(d)}$. Perhaps somewhat less expected is the fact that d may be less than the effective total degree of I.

Example 4.1. Let I be the ideal of R generated by $2Y^2$, $3XY$, and $2X^2$. The polynomials $X(3XY) - Y(2X^2) = X^2Y$ and $Y(3XY) - X(2Y^2) = XY^2$ are in I. It is fairly easy to see that I consists of all polynomials satisfying the following conditions:

(a) The coefficients of 1, X, and Y are 0.
(b) The coefficients of powers of X and powers of Y are even.
(c) The coefficient of XY is divisible by 3.

The values of c_{rs} for $r + s \leq 5$ are given in Table 10.4.1. The set U is $\{(0,2),(1,1),(1,2),(2,0),(2,1)\}$ and the effective total degree of I is 3. The total degree of the generators is 2 and in this case A is $I_{(2)}$.

Let T, d, and A be as before. We need a criterion which will tell us whether $A = I_{(d)}$ and $e \leq d$. Let $B = A \cap I_{(d-1)}$, the set of elements in A

of total degree at most $d - 1$. Clearly a necessary condition for A to equal $I_{(d)}$ is for both XB and YB to be contained in A. We can ensure that this condition is satisfied without increasing the value of d.

Example 4.2. Let I be the ideal of R generated by

$$
\begin{aligned}
f_1 &= Y^2 - 3XY + X^2 - 2Y - X + 6, \\
f_2 &= 4Y^2 + 6XY - 6X^2 - Y + 5X - 5, \\
f_3 &= Y^2 + 5XY - 4X^2 + 5Y - 6X + 3, \\
f_4 &= 5Y^2 + X^2 + 5Y - X + 1.
\end{aligned}
$$

Each of these polynomials has total degree 2. The monomials of total degree at most 2 are, in decreasing order, $Y^2, XY, X^2, Y, X, 1$. Using this basis for the group P of polynomials of total degree at most 2, we can identify P with $\mathbb{Z}^6$. Under this identification, the f_i correspond to the rows of the matrix

$$
D_1 = \begin{bmatrix}
1 & -3 & 1 & -2 & -1 & 6 \\
4 & 6 & -6 & -1 & 5 & -5 \\
1 & 5 & -4 & 5 & -6 & 3 \\
5 & 0 & 1 & 5 & -1 & 1
\end{bmatrix}
$$

and the subgroup A generated by the f_i corresponds to $S(D_1)$. The row Hermite normal form of D_1 is

$$
D_2 = \begin{bmatrix}
1 & 0 & 0 & 350 & -652 & 640 \\
0 & 1 & 0 & 194 & -359 & 350 \\
0 & 0 & 1 & 230 & -426 & 416 \\
0 & 0 & 0 & 395 & -737 & 723
\end{bmatrix}.
$$

Therefore $A \cap I_{(1)}$ is generated by the single polynomial $f_5 = 395Y - 737X + 723$. If we add Xf_5 and Yf_5 to the set of generators for A represented by D_2, the result corresponds to

$$
D_3 = \begin{bmatrix}
1 & 0 & 0 & 350 & -652 & 640 \\
0 & 1 & 0 & 194 & -359 & 350 \\
0 & 0 & 1 & 230 & -426 & 416 \\
0 & 0 & 0 & 395 & -737 & 723 \\
0 & 395 & -737 & 0 & 723 & 0 \\
395 & -737 & 0 & 723 & 0 & 0
\end{bmatrix}.
$$

The row Hermite normal form of D_3 is

$$D_4 = \begin{bmatrix} 1 & 0 & 0 & 0 & 0 & 68418721 \\ 0 & 1 & 0 & 0 & 1 & 400252057 \\ 0 & 0 & 1 & 0 & 0 & 499274494 \\ 0 & 0 & 0 & 1 & 1 & 428686078 \\ 0 & 0 & 0 & 0 & 2 & 745217559 \\ 0 & 0 & 0 & 0 & 0 & 749145811 \end{bmatrix}.$$

Now $A \cap I_{(1)}$ has rank 3. If we take the polynomials corresponding to the last three rows of D_4, multiply them by both X and Y, add the results to our generating set, and compute the row Hermite normal form again, then the nonzero rows form the 6-by-6 identity matrix. This says that 1 is in I, so $I = R$.

Example 4.3. Now let J be the ideal of R generated by f_1, f_2, and f_3 of Example 4.2. Our generators correspond to the rows of the matrix

$$E_1 = \begin{bmatrix} 1 & -3 & 1 & -2 & -1 & 6 \\ 4 & 6 & -6 & -1 & 5 & -5 \\ 1 & 5 & -4 & 5 & -6 & 3 \end{bmatrix}.$$

The row Hermite normal form of E_1 is

$$E_2 = \begin{bmatrix} 1 & 1 & 1 & -17 & 37 & -40 \\ 0 & 2 & 0 & -7 & 19 & -23 \\ 0 & 0 & 5 & -35 & 81 & -89 \end{bmatrix},$$

so $B = A \cap J_{(1)} = \{0\}$. Thus trivially XB and YB are contained in A. However, we are still a long way from having a good understanding of J. If $A = J_{(2)}$, then $C = (A + XA + YA) \cap J_{(2)}$ must be A. Let us take the polynomials g_1, g_2, g_3 corresponding to the three rows of E_2 and determine the abelian group generated by the g_i, their products with X, and their products with Y. Since polynomials of total degree 3 are involved, we use the basis Y^3, XY^2, X^2Y, X^3, Y^2, XY, X^2, Y, X, 1 to identify polynomials with integer vectors. Our nine generators correspond to the rows of

$$E_3 = \begin{bmatrix} 0 & 0 & 0 & 0 & 1 & 1 & 1 & -16 & 37 & -40 \\ 0 & 0 & 0 & 0 & 0 & 2 & 0 & -7 & 19 & -23 \\ 0 & 0 & 0 & 0 & 0 & 0 & 5 & -35 & 81 & -89 \\ 0 & 1 & 1 & 1 & 0 & -16 & 37 & 0 & -40 & 0 \\ 0 & 0 & 2 & 0 & 0 & -7 & 19 & 0 & -23 & 0 \\ 0 & 0 & 0 & 5 & 0 & -35 & 81 & 0 & -89 & 0 \\ 1 & 1 & 1 & 0 & -16 & 37 & 0 & -40 & 0 & 0 \\ 0 & 2 & 0 & 0 & -7 & 19 & 0 & -23 & 0 & 0 \\ 0 & 0 & 5 & 0 & -35 & 81 & 0 & -89 & 0 & 0 \end{bmatrix}.$$

The row Hermite normal form of E_3 is

$$E_4 = \begin{bmatrix} 1 & 0 & 0 & 0 & 0 & 0 & 0 & 0 & 326565 & -521780 \\ 0 & 1 & 0 & 0 & 0 & 0 & 0 & 0 & 823394 & -1315603 \\ 0 & 0 & 1 & 0 & 0 & 0 & 0 & 0 & 592792 & -947155 \\ 0 & 0 & 0 & 1 & 0 & 0 & 0 & 0 & 698714 & -1116398 \\ 0 & 0 & 0 & 0 & 1 & 0 & 0 & 0 & 974340 & -1556775 \\ 0 & 0 & 0 & 0 & 0 & 1 & 0 & 0 & 416261 & -665092 \\ 0 & 0 & 0 & 0 & 0 & 0 & 1 & 0 & 512300 & -818541 \\ 0 & 0 & 0 & 0 & 0 & 0 & 0 & 1 & 118929 & -190023 \\ 0 & 0 & 0 & 0 & 0 & 0 & 0 & 0 & 1601096 & -2558189 \end{bmatrix}.$$

The group C corresponds to the subgroup generated by the last five rows of E_4. Thus C is not A. Since C contains A, the ideal J is generated by C. Let us replace A by C and take as generators for A the polynomials corresponding to the last five rows of E_4. If we now let $B = A \cap J_{(1)}$ and compute $A + XB + YB$, we find that this subgroup corresponds to $S(E_5)$, where

$$E_5 = \begin{bmatrix} 1 & 0 & 0 & 0 & 0 & 52533167 \\ 0 & 1 & 0 & 0 & 0 & 7275654 \\ 0 & 0 & 1 & 0 & 0 & 388980876 \\ 0 & 0 & 0 & 1 & 0 & 345736188 \\ 0 & 0 & 0 & 0 & 1 & 201309868 \\ 0 & 0 & 0 & 0 & 0 & 473095169 \end{bmatrix}.$$

If we replace A by $A + XB + YB$ and iterate again, this time we find that XB and YB are contained in A. In this particular case it is now easy to describe J. Let $n = 473095169$. There is a unique ring homomorphism φ

of R onto $\mathbb{Z}_n$ in which X maps to $[-201309868]$ and Y to $[-345736188]$. Here $[a]$ denotes the congruence class of a modulo n. Under this homomorphism, Y^2, XY, and X^2 map to $[-52533167]$, $[-7275654]$, and $[-388980876]$, respectively. This is all consistent with E_5 and it is not hard to show that J is the kernel of φ.

Examples 4.2 and 4.3 use somewhat *ad hoc* techniques to determine the structure of certain ideals in R. It is time to give a general algorithm based on the total-degree ordering of monomials. Let A be a finitely generated subgroup of R and let I be the ideal of R generated by A. For each integer $e \geq 0$ let $A_{(e)} = A \cap I_{(e)}$, the set of polynomials in A with total degree at most e. For each pair (r, s) of nonnegative integers set $A_{rs} = A \cap I_{rs}$ and let a_{rs} be the nonnegative generator for the set of coefficients of $X^r Y^s$ in elements of A_{rs}. Define $U(A)$ to be the set of pairs (r, s) such that a_{rs} is nonzero and not equal to any a_{ij} with $(i, j) < (r, s)$. The *effective total degree* of A is defined to be the maximum e of $r + s$ with (r, s) in $U(A)$. In general, the sets $U(A)$ and $U(I)$ will be quite different. However, we can state a necessary and sufficient condition for e to be the effective total degree of I and for $I_{(e)}$ to be contained in A.

Proposition 4.3. *Let A be a nontrivial, finitely generated subgroup of R, let d be the maximum of the total degrees of the elements of A, and let I be the ideal of R generated by A. Set D equal to the subgroup generated by $\{X^r Y^s A \mid r \geq 0, s \geq 0, r + s \leq d\}$ and e equal to the effective total degree of D. The following are equivalent:*

(i) $A_{(e)} = D_{(e)}$.
(ii) e *is the effective total degree of I and $I_{(e)}$ is contained in A.*

Proof. By the definition of e, there is an element of total degree e in $D_{(e)}$. Thus (i) certainly implies that $e \leq d$. Also $A_{(e)} \subseteq D_{(e)} \subseteq I_{(e)}$. Therefore (ii) trivially implies (i).

To prove (i) implies (ii), suppose that $A_{(e)} = D_{(e)}$. If $e = 0$, then $D_{(0)}$ is generated by some nonnegative integer a and all elements of D are divisible by a. In this case $I = Ra$ and condition (ii) clearly holds. Thus we may assume that $e > 0$. Let $B = A_{(e-1)}$. Then XB and YB are contained in $D_{(e)}$, so XB and YB are contained in $A_{(e)}$. For nonnegative integers r and s let a_{rs} be the nonnegative generator of the set of coefficients of $X^r Y^s$ in elements of A_{rs}. Then a_{rs} divides a_{ij} whenever $r + s \leq e$ and $(i, j) \leq (r, s)$. $\square$

Lemma 4.4. *If f is a nonzero element of D with total degree at most $d + e$, then f can be written as $u_1 f_1 + \cdots + u_m f_m$, where the u_i are monomials, the f_i are nonzero elements of $A_{(e)}$, and the leading monomials of the terms $u_i f_i$ form a strictly decreasing sequence in the total degree ordering.*

Proof. Let $X^p Y^q$ be the leading monomial of f. If $p + q \leq e$, then f is in $D_{(e)} = A_{(e)}$ and we can take $m = 1$, $f_1 = f$, and $u_1 = 1$. Thus we may assume that $p + q > e$. By the definition of e, there exist a pair $(r, s) < (p, q)$ with $r + s = e$ and an element f_1 of D with leading monomial $X^r Y^s$ such that f_1 has the same leading coefficient as f. The polynomial f_1 is in $D_{(e)} = A_{(e)}$. Set $r_1 = p - r$, $s_1 = q - s$, and $u_1 = X^{r_1} Y^{s_1}$. Since

$$r_1 + s_1 = pr + q - s = p + q - (r + s) \leq d + e - e = d,$$

the polynomial $u_1 f_1$ is in D and therefore so is $g = f - u_1 f_1$. If g is not 0, then the leading monomial of g precedes $X^p Y^q$. Continuing this process, we obtain the desired description of f. $\square$

Lemma 4.4 applies to elements of A. It follows that I is generated as an ideal by $A_{(e)}$.

Lemma 4.5. *Any nonzero element f of I can be written as $u_1 f_1 + \cdots + u_m f_m$, where the u_i are monomials, the f_i are nonzero elements of $A_{(c)}$, the leading monomials of the terms $u_i f_i$ form a strictly decreasing sequence in the total-degree ordering, and the total degree of f_i is e whenever $u_i \neq 1$.*

Proof. Since $A_{(e)}$ generates I, we can write $f = u_1 f_1 + \cdots + u_m f_m$, where the u_i are monomials and the f_i are nonzero elements of $A_{(e)}$. Suppose that the total degree of f_i is less than e. Then $X f_i$ and $Y f_i$ are in $D_{(e)} = A_{(e)}$. Therefore, if $u_i \neq 1$, we can replace f_i by $X f_i$ or $Y f_i$ and decrease the total degree of u_i by 1. Thus we may assume that the total degree of f_i is e whenever $u_i \neq 1$.

Next suppose that $u_1 f_1$ and $u_2 f_2$ have the same leading monomial $X^p Y^q$. If $p + q \leq e$, then $u_1 = u_2 = 1$. In this case we may replace f_1 by $f_1 + f_2$ and delete f_2. Therefore we may assume that $p + q > e$. For $i = 1, 2$, let $u_i = X^{r_i} Y^{s_i}$. The leading monomial of f_i is $X^{p - r_i} Y^{q - s_i}$, which has total degree e. Set $a = \max(p - r_1, p - r_2)$ and $b = \max(q - s_1, q - s_2)$. Since the maximum of two nonnegative integers does not exceed their sum, we have

$$a + b \leq p - r_1 + p - r_2 + q - s_1 + q - s_2 = (p - r_1 + q - s_1) + (p - r_2 + q - s_2) = 2e$$

and

$$u_1 f_1 + u_2 f_2 = X^{r_1} Y^{s_1} f_1 + X^{r_2} Y^{s_2} f_2 = X^{p-a} Y^{q-b} g,$$

where

$$g = X^{r_1 + a - p} Y^{s_1 + b - q} f_1 + X^{r_2 + a - p} Y^{s_2 + b - q} f_2.$$

Now

$$r_1 + a - p + s_1 + b - q = a + b - (p - r_1 + q - s_1) \leq 2e - e = e.$$

Similarly, $r_2 + a - p + s_2 + b - q$ is at most e. Therefore g is in D and the total degree of g does not exceed $2e \leq d + e$. The leading monomial of g does not exceed $X^a Y^b$ in the total-degree ordering. By Lemma 4.4, g can be written $v_1 g_1 + \cdots + v_t g_t$, where the v_j are monomials, the g_j are in $A_{(e)}$, and the leading monomials of the terms $v_j g_j$ form a strictly decreasing sequence, which starts at or below $X^a Y^b$. Therefore, by replacing $u_1 f_1 + u_2 f_2$ with

$$X^{p-a} Y^{q-b} v_1 g_1 + \cdots + X^{p-a} Y^{q-b} v_t g_t,$$

we have reduced the number of terms in the expression for f which have leading monomial $X^p Y^q$ and we have not introduced any leading monomials greater than $X^p Y^q$. Thus, by considering the leading monomials in decreasing order, we can modify the sum so that the leading monomials form a strictly decreasing sequence.   □

It follows from Lemma 4.5 that $A_{(e)} = I_{(e)}$. Also, for every nonzero f in I there is an element f_1 in $A_{(e)}$ such that f and f_1 have the same leading coefficient. Therefore e is the effective total degree of I.   □

Let I be an ideal of R given by a finite generating set T. Proposition 4.3 gives us an algorithm for determining the effective total degree e of I and finding a basis for $I_{(e)}$. The algorithm involves a variant of ZBASIS described in Section 10.2. This variant uses the reverse of the total-degree ordering on monomials to define a correspondence between polynomials and integer vectors.

Procedure ZBASIS_TD($u_1, \ldots, u_s; v_1, \ldots, v_t$)
Input: $u_1, \ldots, u_s$: elements of $\mathbb{Z}[X, Y]$;
Output: $v_1, \ldots, v_t$: a $\mathbb{Z}$-basis for the subgroup generated by the u_i;
Begin
    Let d be the maximum of the total degrees of the u_i;
    Let $w_1 = Y^d, \ldots, w_r = 1$ be the monomials in X and Y with total
        degree at most d listed in decreasing order;
    ($*$ The basis $w_1, \ldots, w_r$ will be used to identify polynomials of total
        degree at most d with integer vectors. $*$)
    Let P be the s-by-r integer matrix whose rows correspond to the u_i;
    Let $v_1, \ldots, v_t$ in $\mathbb{Z}[X, Y]$ correspond to the nonzero rows of the row
        Hermite normal form of P
End.

Procedure ZXYIDEAL_TD($u_1, \ldots, u_s; v_1, \ldots, v_t$);

Input: $u_1, \ldots, u_s$: Elements of $\mathbb{Z}[X, Y]$;
Output: $v_1, \ldots, v_t$: a basis for $I_{(e)}$, where I is the ideal generated
by the u_i and e is the effective total degree of I;

Begin

Let d be the maximum of the total degrees of the u_i;

ZBASIS_TD$(u_1, \ldots, u_s; v_1, \ldots, v_t)$; *done* := false;

While not *done* do begin

Let $y_1, \ldots, y_p$ denote the elements $X^r Y^s v_i$ with $r + s \leq d$ and
$1 \leq i \leq t$;

ZBASIS_TD$(y_1, \ldots, y_p; z_1, \ldots, z_m)$;

Let k be minimal such that z_k has total degree d;

If $z_k, \ldots, z_m$ and $v_1, \ldots, v_t$ are not the same then
Copy $z_k, \ldots, z_m$ as $v_1, \ldots, v_t$ $(* D_{(d)} \neq A. *)$

Else begin

For $r \geq 0$, $s \geq 0$, and $r + s \leq 2d$ do
If there is some z_i with leading monomial $X^r Y^s$ then
Set a_{rs} equal to the leading coefficient of z_i
Else $a_{rs} := 0$;

Let e be the maximum of $r + s$ for those pairs (r, s) for which
$a_{rs} \neq 0$
and either $r = 0$ or $a_{rs} \neq a_{r-1s}$
and either $s = 0$ or $a_{rs} \neq a_{rs-1}$;

If $e \leq d$ then *done* := true $(* D_{(e)} = A_{(e)}. *)$

Else begin

Let k be minimal such that the total degree of z_k is e;
Copy $z_k, \ldots, z_m$ as $v_1, \ldots, v_t$;
$d := e$

End

End

End;

Let k be minimal such that v_k has total degree e;
Copy $v_k, \ldots, v_t$ as $v_1, \ldots, v_t$

End.

The reader is warned that ZXYIDEAL_TD is practical only if the value of d remains small. For example, suppose that d is 6 in the first line of the While-loop. The number of monomials of total degree at most 6 is 28. Thus t could be as large as 28, and p, which is $28t$, could be $28^2 = 784$. The largest total degree of the y_i will be 12. There are 91 monomials of total degree at most 12. Thus the matrix P in the call to ZBASIS_TD will have 91 columns and up to 784 rows. Computing the row Hermite normal form of such a matrix will take a long time. The Gröbner basis method described

Table 10.4.2

			s			
r	0	1	2	3	4	5
0	7	0	2	3	2	2
1	0	3	0	1	1	
2	2	4	2	1		
3	1	5	1			
4	2	1				
5	2					

in Section 10.6 uses special knowledge about the matrix P to speed up the computation. As we saw in Example 4.3, one can get useful information if the inequality $r + s \leq d$ in the first line of the While-loop is replaced by $r + s \leq 1$ or $r + s \leq 2$.

The main difficulty in proving the correctness of ZXYIDEAL_TD is showing that the procedure terminates. If d remains bounded, then the group A generated by the v_i is contained in a fixed free abelian group of finite rank. With each iteration of the While-loop A gets larger. Thus eventually the While-loop must terminate. If d goes to infinity, then, from some point on, A contains the products of the original generators u_i by any given monomial. The group $I_{(e)}$ is finitely generated. The generators of $I_{(e)}$ can be expressed in the form $p_1 u_1 + \cdots + p_s u_s$, where the p_i are in R. Eventually the terms $p_i u_i$ for the various generators of $I_{(e)}$ will be in A. At this point $I_{(e)}$ will be contained in A and ZXYIDEAL_TD will terminate by Proposition 4.3.

The total-degree ordering can be used to study submodules of R^n. The ordering is extended to basis elements $b_i X^j Y^k$ by first using the total degree ordering of $X^j Y^k$ and then taking the b_i in increasing order of i. In this case, the groups C_{rs} are subgroups of $\mathbb{Z}^n$.

Exercises

4.1. Suppose that A is a subgroup of $\mathbb{Z}[X, Y]$ and the nonnegative generators a_{rs} of the subgroups A_{rs} are given by Table 10.4.2. What is $U(A)$?

4.2. Let A be the subgroup of $\mathbb{Z}[X, Y]$ generated by $2XY^2 - 3Y + 7$ and $XY^2 - 2Y + X + 1$, and let B be the subgroup of A consisting of the elements with total degree at most 2. Find bases for $A + XB + YB$ and for the subgroup of $A + XA + YA$ consisting of the elements with total degree at most 3.

4.3. Apply ZXIDEAL_TD to the generators of A in Exercise 4.2. Compare the results with Exercise 3.1.

10.5 The Gröbner basis approach

In Sections 10.1 to 10.4 we sketched methods for deciding membership in submodules of $\mathbb{Z}[X_1, \ldots, X_m]^n$ given by finite generating sets. These methods are based ultimately on the computation of row Hermite normal forms of integer matrices. Unfortunately, for most interesting problems these matrices are very large. The Gröbner basis approach solves the same problem. However, by adopting a rewriting point of view and being very careful about the choice of generators, the approach reduces the time and space requirements. The worst-case complexity remains highly exponential, but the range of practicality is extended substantially.

To begin to get an insight into the Gröbner basis approach, let us look at an example.

Example 5.1. Let I be the ideal of $\mathbb{Z}[X]$ generated by the following polynomials:

$$
\begin{aligned}
f_1 &= X^6 + X^5 + X^4 + 2X^3 + 3X^2 + 3X + 5, \\
f_2 &= 2X^5 + 2X^3 + 2X^2 + 2, \\
f_3 &= 2X^4 + 2X^3 + 2X + 6, \\
f_4 &= 4X^3 + 4, \\
f_5 &= 4X^2 + 4, \\
f_6 &= 4X + 4, \\
f_7 &= 8.
\end{aligned}
$$

The effective degree of I is 6 and the f_i form the basis of I_6 produced by the procedure ZXIDEAL of Section 10.1. The product $X f_6$ is in I_6 and has the same leading coefficient as f_5. If we replace f_5 by $X f_6$, then the abelian group A generated by the polynomials is contained in I_6. By Exercise 1.10 in Chapter 8, A is actually equal to I_6. The same argument shows that the polynomials f_1, $X f_3$, f_3, $X^2 f_6$, $X f_6$, f_6, f_7 form a basis for I_6 which contains bases for I_d, $0 \le d \le 5$. Thus we only have to remember four polynomials, f_1, f_3, f_6, and f_7, not seven. This is the way that the Gröbner basis approach saves space. When describing an abelian group of polynomials, generators which are monomial multiples of other generators are not explicitly written down.

Now let us see how time is saved. To prove that our new generating set is a $\mathbb{Z}$-basis for I_6, we must take each basis element g of degree less than 6 and show that Xg is an integral linear combination of our basis polynomials. For $g = f_6$, $X f_6$, or f_3, the polynomial Xg is actually part of the basis. Therefore, the only nonobvious cases occur when g is f_7, $X^2 f_6$, and $X f_3$.

As mentioned earlier, the Gröbner basis approach adopts a rewriting point of view. Because f_1 is in I, it follows that X^6 is congruent modulo I to $-X^5 - X^4 - X^3 - 2X^3 - 3X^2 - 3X - 5$. Thus, if g is any element of $\mathbb{Z}[X]$ which has a term aX^r with $r \ge 6$, then we can replace that term by $aX^{r-6}(-X^5 - X^4 - X^3 - 2X^3 - 3X^2 - 3X - 5)$ and the resulting polynomial will be congruent to g modulo I. In this way, we can think of f_1, f_3, f_6, and f_7 as defining the following "rewriting rules":

$$X^6 \to -X^5 - X^4 - X^3 - 2X^3 - 3X^2 - 3X - 5,$$
$$2X^4 \to -2X^3 - 2X - 6,$$
$$4X \to -4,$$
$$8 \to 0.$$

To show that these rules are confluent in the sense that any polynomial can be reduced to a unique canonical form using them, we must look at overlaps of left sides, terms aX^r which are reducible by two rules. In fact, we need consider only minimal overlaps, overlaps aX^r with a positive such that neither aX^{r-1} nor $(a-1)X^r$ is reducible by two rules. In this example, the minimal overlaps are $8X$, $4X^4$, and $2X^6$. Proving confluence involves essentially the same computations as showing that Xf_7, X^3f_6, and X^2f_3 are in the abelian group generated by f_1, Xf_3, f_3, X^2f_6, Xf_6, f_6, and f_7.

In order to be able to describe the Gröbner basis approach in a manner which can be generalized to modules over group rings of polycyclic groups, we shall initially take a very general point of view and consider a free abelian group F with a basis $\mathcal{U}$ which is well-ordered by a linear ordering $\prec$. Later we shall impose more structure on F, $\mathcal{U}$, and $\prec$. A nonzero element of F can be written uniquely in the *standard form*

$$f = a_1u_1 + a_2u_2 + \cdots + a_su_s,$$

where the a_i are nonzero integers and the u_i are elements of $\mathcal{U}$ with $u_1 \succ u_2 \succ \cdots \succ u_s$. We shall refer to u_1, a_1, and a_1u_1 as the *leading generator*, the *leading coefficient*, and the *leading term* of f, respectively. Also, a_i will be called the coefficient of u_i in f. If u is an element of $\mathcal{U}$ which is not equal to any u_i, then the coefficient of u in f will be defined to be 0. In our primary example, F will be $\mathbb{Z}[X_1, \ldots, X_m]^n$, $\mathcal{U}$ will be the basis consisting of all terms $b_iX_1^{\alpha_1} \ldots X_m^{\alpha_m}$, where $b_1, \ldots, b_n$ is the standard basis of $\mathbb{Z}^n$, and $\prec$ will be a linear ordering of $\mathcal{U}$ similar to the ones we have used already. An important special case occurs when $m = 0$ and $F = \mathbb{Z}^n$. If we take $b_1 \succ b_2 \succ \cdots \succ b_n$, then the leading coefficient of a nonzero element $f = (a_1, \ldots, a_n)$ of F is the first nonzero component of f.

It will be convenient to extend $\prec$ to be a well-ordering of F. To do this, we first well-order $\mathbb{Z}$. The order used here will be $0 \prec 1 \prec 2 \prec \cdots \prec -1 \prec -2 \ldots$. Other orderings are possible. See (Buchberger 1984). Terms au and bv with a and b in $\mathbb{Z} - \{0\}$ and u and v in $\mathcal{U}$ are ordered as follows: $au \prec bv$ if and only if $u \prec v$ or $u = v$ and $a \prec b$. Finally we order F. The zero element of F comes first. If f and g are nonzero elements of F, then $f \prec g$ if and only if either the leading term of f precedes the leading term of g or f and g have the same leading term au and $f - au \prec g - au$.

Proposition 5.1. *The ordering $\prec$ on F is a well-ordering.*

Proof. Exercise. □

Suppose now that S is a possibly infinite set of nonzero elements of F. To simplify the exposition, we shall assume that the leading coefficients of the elements of S are positive. Thus $f \prec -f$ for all f in S. Our goal will be to decide membership in the subgroup M of F generated by S. We shall eventually have to put some additional conditions on S, but for now S is arbitrary. In the case $F = \mathbb{Z}[X_1, \ldots, X_m]^n$ we shall have a finite set T of nonzero elements of F, and S will be the set of all products uf, where u is a monomial and f is in T. Here M is the submodule of $\mathbb{Z}[X_1, \ldots, X_m]^n$ generated by T.

Let f be a nonzero element of F with standard form $a_1 u_1 + \cdots + a_s u_s$. Suppose for some i, $1 \le i \le s$, there is an element h of S with leading term bu_i and $b \preceq a_i$. Thus either a_i is negative or $b \le a_i$. Let $a_i = qb + r$, where q and r are integers and $0 \le r < b$. Then $r \prec a_i$. If $g = f - qh$, then the cosets $g + M$ and $f + M$ are equal and

$$ g = a_1 u_1 + \cdots + a_{i-1} u_{i-1} + r u_i + \cdots. $$

If $r = 0$, then this is not the standard form of g, but $g \prec f$ whether or not $r = 0$. We could now check to see whether another such reduction is possible with g. Since $\prec$ is a well-ordering of F, eventually we reach an element of $f + M$ which cannot be reduced further. Such an element will be called *irreducible*. The following function describes this reduction procedure.

Function GRP_REDUCE(S, f);
Input: S : a set of nonzero elements of F with positive leading
                   coefficients;
    f : an element of F;
($*$ The value returned is an irreducible element of $f + M$, where M is
    the subgroup of F generated by S. $*$)
Begin
    $i := 1$; $g := f$;
    ($*$ At all times the standard form of g will be $a_1 u_1 + \cdots + a_s u_s$. If
        $g = 0$, then $s = 0$. $*$)
    While $i \le s$ do
        If there is no element in S with leading term bu_i such that $b \preceq a_i$
            then $i := i + 1$

Else begin
    Let h in S have leading term bu_i with $b \preceq a_i$;
    Let $a_i = qb + r$, where q and r are integers and $0 \leq r < b$;
    $g := g - qh$;
    (* Recompute s and the terms $a_j u_j$ with $j \geq i$. *)
End;

  GRP_REDUCE $:= g$
End.

It may be possible to execute GRP_REDUCE(S, f) even when S is infinite. For each u in $\mathcal{U}$ let S_u be the set of elements in S with leading generator u. If for each u the set S_u is finite and we have an algorithm for listing its elements, then we can carry out the steps in GRP_REDUCE.

There is a strong similarity between GRP_REDUCE and REWRITE of Section 2.2. In particular, executing GRP_REDUCE may involve making choices for the element h and the output may depend on those choices. We need a condition analogous to confluence of rewriting systems which will insure that the output of GRP_REDUCE is independent of the choices. We would also like the result returned to be the first element with respect to $\prec$ in $f + M$. In particular, 0 should be returned whenever f is in M.

There is one case in which everything works as we would like.

Proposition 5.2. *Suppose that for each u in $\mathcal{U}$ there is at most one element of S with leading generator u. Then S is a basis for M and the element g of F returned by* GRP_REDUCE(S, f) *is the first element of $f + M$. In particular, if f is in M, then $g = 0$.*

Proof. The case in which $\mathcal{U}$ is finite has already been proved. Here F may be identified with $\mathbb{Z}^n$, where $n = |\mathcal{U}|$, and the first part of the proposition is just the statement that the nonzero rows of a matrix in row echelon form are a basis for the subgroup they generate. The fact that g is first in $M + f$ follows from Exercises 1.6 and 1.7 in Chapter 8. The general case requires only simple modifications of the proof when $\mathcal{U}$ is finite and is left as an exercise. $\square$

In general, S may have many elements with the same leading generator. For each u such that S_u is not empty, let h_u be the first element in S_u with respect to $\prec$, and let P be the set of element h_u so defined. The leading coefficient of h_u is minimal among the leading coefficients of elements of S_u. Hence, if we wanted to, we could always take the element h in GRP_REDUCE to be h_{u_i}. Thus, if GRP_REDUCE(S, f) is always to be 0 if f is in M, then it had better be true that M is generated by P.

Proposition 5.3. *The following are equivalent:*

(a) P generates M.
(b) GRP_REDUCE(S, f) *always returns the first element in $f + M$ for any element f of F.*
(c) GRP_REDUCE(P, f) *returns 0 for each f in $S - P$.*

Proof. (a) implies (b). Assume (a). By Proposition 5.2, the first element g of $f + M$ is the only element of $f + M$ which is irreducible with respect to P. Therefore g is the only element in $f + M$ which is irreducible with respect to S. Hence g is returned by GRP_REDUCE(S, f).

(b) implies (c). Assume (b) and suppose f is in M. As noted earlier, any element returned by GRP_REDUCE(P, f) could be returned by GRP_REDUCE(S, f). Since GRP_REDUCE(S, f) always returns 0, the same must hold for GRP_REDUCE(P, f).

(c) implies (a). Assume (c). Then every element of $S - P$ is in the subgroup generated by P. Therefore P and S generate the same subgroup. $\square$

Let f and g be elements of S with the same leading generator such that $f \prec g$. We shall call (f, g) a *semicritical pair*. Note that whenever h_u is defined and g is in $S_u - \{h_u\}$, then (h_u, g) is a semicritical pair. Let b and c be the leading coefficients of f and g, respectively. Then $b \leq c$. Let $c = qb + r$, where q and r are integers and $0 \leq r < b$. The element $t(f, g) = g - qf$ will be called the *test element* corresponding to f and g. (In the context of polynomial rings, $t(f, g)$ is often referred to as the S-*polynomial* of f and g.) We shall say that f is *consonant* with g or the pair (f, g) is consonant if some invocation of GRP_REDUCE$(S, t(f, g))$ returns 0.

Proposition 5.4. *Suppose that each h_u in P is consonant with every element in $S_u - \{h_u\}$. Then P generates M.*

Proof. If P does not generate M, then there is an element g of S which is not in the group N generated by P. Choose g minimal with respect to $\prec$ and let u be the leading generator of g. By assumption, h_u is consonant with g. Therefore some invocation of GRP_REDUCE$(S, t(h_u, g))$ returns 0. Either the leading generator v of $t(h_u, g)$ is less than u or $v = u$ and the leading coefficient of $t(h_u, g)$ is positive and smaller than the leading coefficient of h_u. It follows that every element h used during the execution of GRP_REDUCE$(S, t(h_u, g))$ satisfies $h \prec g$. All such elements h are in N by the choice of g. Therefore $t(h_u, g)$ is in N. But h_u is in N and $g = t(h_u, g) + qh_u$ for some integer q. Therefore g is in N. This contradicts the choice of g, so P generates M. $\square$

Proposition 5.5. *The following are equivalent:*

(a) *Every semicritical pair is consonant.*

(b) *For each semicritical pair (f, g) the element $t(f, g)$ is in the subgroup generated by all h in S with $h \prec g$.*

Proof. (a) implies (b). Assume (a) and let (f, g) be a semicritical pair. Some invocation of GRP_REDUCE$(S, t(f, g))$ returns 0. Every element h used in GRP_REDUCE satisfies $h \prec g$ and $t(f, g)$ is in the subgroup generated by these elements.

(b) implies (a). Assume (b). We shall show that P generates M. If not, let g be a minimal element of S which is not in the subgroup N generated by P and let u be the leading generator of g. By assumption, $t(h_u, g)$ is in the subgroup generated by the elements h of S with $h \prec g$. All such elements are in N, so $t(h_u, g)$ is in N. But h_u is in N and $g = t(h_u, g) + qh_u$ for some integer q. Therefore g is in N. This is a contraction and P generates M. In this case all semicritical pairs are consonant by Proposition 5.3. $\square$

Corollary 5.6. *Condition (b) of Proposition 5.5 implies that GRP_REDUCE(S, f) always returns the first element of $f + M$.*

Exercise

5.1. The most common use of Gröbner basis methods is in the study of ideals in polynomial rings where the coefficients come from a field K. Rework the exposition in this section assuming that F is a vector space over K with basis $\mathcal{U}$ and we want to decide membership in the subspace of F generated by a subset S.

10.6 Gröbner bases

In this section we shall apply the ideas of Section 10.5 to the situation in which F is $\mathbb{Z}[X_1, \ldots, X_m]^n$, $\mathcal{U}$ is the set of terms $b_i X_1^{\alpha_1} \ldots X_m^{\alpha_m}$, and $\prec$ is a well-ordering of $\mathcal{U}$ satisfying certain conditions to be described later. The ring $\mathbb{Z}[X_1, \ldots, X_m]$ will be denoted by R. Let $\mathcal{X}$ be the set of monomials $X_1^{\alpha_1} \ldots X_m^{\alpha_m}$. Given a finite subset T of F, we want to decide membership in the submodule M of F generated by T. It is easy to see that M is generated as an abelian group by the set S of products xf, where x is in $\mathcal{X}$ and f is in T.

Our work will be made easier if we choose an ordering $\prec$ on $\mathcal{U}$ which is in some way compatible with the multiplicative structure of F as an R-module. Let us start with orderings of $\mathcal{X}$. An ordering $\prec$ of $\mathcal{X}$ is *consistent with multiplication* if $x \prec y$ implies $xz \prec yz$ for all x, y, and z in $\mathcal{X}$. As in the case of the reduction orderings on words considered in Chapter 2, if $\prec$ is a well-ordering which is consistent with multiplication, then 1 is the first element of $\mathcal{X}$. Both the reverse lexicographic ordering and the total-degree ordering are well-orderings of $\mathcal{X}$ which are consistent with multiplication. Let us fix one well-ordering $\prec$ of $\mathcal{X}$ which is consistent with multiplication.

We shall assume that we have a well-ordering, also denoted $\prec$, of $\mathcal{U}$ satisfying the following properties:

(i) If u and v are in $\mathcal{U}$ and $u \prec v$, then $xu \prec xv$ for all x in $\mathcal{X}$.

(ii) If x and y are in $\mathcal{X}$ and $x \prec y$, then $xu \prec yu$ for all u in $\mathcal{U}$.

Such an ordering will again be said to be *consistent with multiplication*. Given $\prec$ on $\mathcal{X}$, we can construct orderings of $\mathcal{U}$ which are consistent with multiplication. Elements of $\mathcal{U}$ have the form $b_i x$ with x in $\mathcal{X}$. If $u = b_i x$ and $v = b_j y$ are in $\mathcal{U}$, we can say $u \prec v$ if $x \prec y$ or $x = y$ and $i > j$. (It is convenient to have $b_1 \succ \cdots \succ b_n$.) On the other hand, we can say $u \prec v$ if $i > j$ or $i = j$ and $x \prec y$. Both of these orderings are well-orderings of $\mathcal{U}$ which are consistent with multiplication.

As described in Section 10.5, we extend $\prec$ to a well ordering of F. Suppose that f is a nonzero element of F and the standard form of f is $a_1 u_1 + \cdots + a_s u_s$, where the a_i are integers and the u_i are in $\mathcal{U}$. If x is in $\mathcal{X}$, then the standard form of xf is $a_1 x u_1 + \cdots + a_s x u_s$. Also, if g is in F and $f \prec g$, then $xf \prec xg$.

Now let T be a finite subset of F. We shall assume that elements of T are nonzero and have positive leading coefficients. The R-module M generated by T is generated as an abelian group by $S = \{ xf \mid x \in \mathcal{X}, f \in T \}$. By the remarks of the preceding paragraph, the leading coefficients of elements of S are positive and there are only finitely many elements of S with a given leading generator. Thus we can execute $\mathrm{GRP\_REDUCE}(S, f)$ for any f in F. Let us write $\mathrm{REDUCE}(T, f)$ for $\mathrm{GRP\_REDUCE}(S, f)$. This makes sense since T determines S.

The value returned by $\mathrm{REDUCE}(T, f)$ will be the first element of $M + f$ provided the semicritical pairs from S are consonant. Unfortunately, there are infinitely many semicritical pairs. However, because of our choice of ordering, there is a finite set of semicritical pairs whose consonance suffices. A *critical pair* is a semicritical pair (xf, yg), where f and g are in T, x and y are in $\mathcal{X}$, and $\gcd(x, y) = 1$. Suppose that f and g are elements of T with leading generators u and v, respectively. If u and v do not involve the same standard basis vector b_i, then for all x and y in $\mathcal{X}$ the elements xf and yg have different leading generators. Thus (xf, yg) cannot be a semicritical pair. Suppose that $u = b_i z$ and $v = b_i w$, where z and w are in $\mathcal{X}$. Then there are unique elements x and y of $\mathcal{X}$ such that $\gcd(x, y) = 1$ and $xu = yv$, namely $x = \mathrm{lcm}(z, w)/z$ and $y = \mathrm{lcm}(z, w)/w$. If $xf \neq yg$, then exactly one of the pairs (xf, yg) and (yg, xf) is a critical pair. Hence the set of critical pairs is finite.

Proposition 6.1. *If all critical pairs are consonant, then all semicritical pairs are consonant.*

Proof. Let (xf, yg) be a semicritical pair. Set $z = \gcd(x, y)$, $x_1 = x/z$ and $y_1 = y/z$. Then $\gcd(x_1, y_1) = 1$ and $(x_1 f, y_1 g)$ is a critical pair. By assumption, $t(x_1 f, y_1 g)$ can be written as $a_1 h_1 + \cdots + a_s h_s$, where each a_i

is an integer and each h_i is an element of S with $h_i \prec y_1 g$. It is easy to see that $t(xf, yg) = zt(x_1 f, y_1 g) = a_1 z h_1 + \cdots + a_s z h_s$ and $z h_i \prec z y_1 g = yg$. Therefore all semicritical pairs are consonant by Proposition 5.5. □

If all critical pairs are consonant, then T is said to be a *Gröbner basis* for M with respect to $\prec$. This terminology may be a little confusing. A Gröbner basis is not necessarily a basis, that is, a generating set whose elements are linearly independent over R. In the ideal I of Example 5.1, the polynomials f_1, f_3, f_6, and f_7 form a Gröbner basis, but $(X+1)f_7 = 2f_6$, so these polynomials are not linearly independent over $\mathbb{Z}[X]$. Submodules of R^n may not be free. Hence they may not have any bases. The word "basis" in "Gröbner basis" simply means finite generating set, which is the same meaning it has in "Hilbert basis theorem".

The following result will be used several times in proving results about Gröbner bases.

Proposition 6.2. *The ascending chain condition holds for subsets of $\mathcal{U}$ which are closed under multiplication by elements of $\mathcal{X}$.*

Proof. Let $\mathcal{V}$ be a subset of $\mathcal{U}$ which is closed under multiplication by elements of $\mathcal{X}$ and let M be the abelian group generated by $\mathcal{V}$. Then $\mathcal{V}$ is a basis for M and M is an R-submodule of F. Now let $\mathcal{V}_1 \subseteq \mathcal{V}_2 \subseteq \cdots$ be an ascending chain of subsets of $\mathcal{U}$ which are each closed under multiplication by elements of $\mathcal{X}$. If M_k is the R-submodule generated by $\mathcal{V}_k$, then $M_1 \subseteq M_2 \subseteq \cdots$. Since the ascending chain condition holds for submodules of F, there is an index j such that $M_k = M_j$ for all $k \geq j$. Since $\mathcal{V}_k = M_k \cap \mathcal{U}$, it follows that $\mathcal{V}_k = \mathcal{V}_j$ for $k \geq j$. □

The first application of Proposition 6.2 will be in the proof of the existence of Gröbner bases. Let us fix well-orderings of $\mathcal{X}$ and $\mathcal{U}$, both denoted by $\prec$, which are consistent with multiplication.

Proposition 6.3. *If M is any submodule of F, then M has a finite Gröbner basis with respect to $\prec$.*

Proof. For each u in $\mathcal{U}$, let M_u be the abelian group consisting of 0 and those nonzero elements of M whose leading generators do not exceed u. Let C_u be the set of coefficients of u in elements of M_u. Then C_u is a subgroup of $\mathbb{Z}$. If x is in $\mathcal{X}$, then $x M_u \subseteq M_{xu}$, so $C_u \subseteq C_{xu}$. Let c_u be the nonnegative generator of C_u. If $c_u \neq 0$, let h_u be an element of M_u such that the leading term of h_u is $c_u u$. Let P be the set of h_u so chosen and let L denote the set of leading terms of elements of P. That is, $L = \{c_u u \mid c_u \neq 0\}$.

Lemma 6.4. *There is a finite subset K of L such that every element in L has the form xk for some x in $\mathcal{X}$ and some k in K.*

Proof. Suppose no such K exists. Choose $c_u u$ in L with c_u as small as possible. If x is in $\mathcal{X}$, then c_u is divisible by c_{xu}. But by the choice of u we must have $c_{xu} = c_u$, so $c_{xu}xu = x(c_u u)$. Set $k_1 = c_u u$ and $\mathcal{V}_1 = \{xu \mid x \in \mathcal{X}\}$. Now suppose that $k_1, \ldots, k_r$ have been chosen in L and $\mathcal{V}_r$ is the set of multiples by elements of $\mathcal{X}$ of the leading generators of the k_i. Assume that for all v in $\mathcal{V}_r$ we have $c_v v = xk_i$ for some x in $\mathcal{X}$ and some i. By assumption there exists $c_w w$ in L such that w is not in $\mathcal{V}_r$. Among all such w, choose one such that c_w is minimal. If x is in $\mathcal{X}$, then c_{xw} divides c_w. If $c_{xw} = c_w$, then $c_{xw}xw = x(c_w w)$. If c_{xw} is a proper divisor of c_w, then xw is in $\mathcal{V}_r$, so $c_{xw}xw = yk_i$ for some y in $\mathcal{X}$ and some i. Thus we may define k_{r+1} to be $c_w w$ and the induction hypothesis remains true. But the sets $\mathcal{V}_r$ are closed under multiplication by elements of $\mathcal{X}$ and are strictly increasing. This contradicts Proposition 6.2. $\square$

Let K be as in Lemma 6.4, let T be $\{h_u \mid c_u u \in K\}$, and let $S = \{xf \mid x \in \mathcal{X}, f \in T\}$. Suppose g is a nonzero element of M with leading term av. By Lemma 6.4, $c_v v = xc_u u$ for some $c_u u$ in K. Let $q = a/c_v$. Then xh_u is in S and $g - qxh_u$ is zero or has leading generator less than v. Hence g is reducible with respect to S. Therefore GRP_REDUCE(S, g) must return 0. This means that all semicritical pairs from S are consonant, so certainly all critical pairs are consonant. Therefore T is a Gröbner basis for M with respect to $\prec$. $\square$

Suppose that T is a finite subset of F in which all elements are nonzero and have positive leading coefficients. If T is not a Gröbner basis for the R-submodule M generated by T, then there is a critical pair (f, g) such that REDUCE$(T, t(f, g))$ returns a nonzero element h. If we multiply h by -1 if necessary to make its leading coefficient positive and add h to T, then it is possible for REDUCE$(T, t(f, g))$ to return 0, so (f, g) is now consonant. Unfortunately, by increasing the size of T we have increased the number of critical pairs. However, it turns out that eventually we shall reach a (finite) Gröbner basis for M. This Gröbner basis procedure is very similar to the Knuth-Bendix procedure for strings, except that it always terminates. Here is one version of the procedure:

Procedure GRÖBNER$(T, \prec; B)$;
Input: T : a finite subset of $\mathbb{Z}[X_1, \ldots, X_m]^n$;
         $\prec$: well-orderings of $\mathcal{X}$ and $\mathcal{U}$ which are consistent with multiplication;
Output: B : a finite Gröbner basis relative to $\prec$ for the submodule M generated by T;

Begin
  $B := T - \{0\}$;
  Multiply elements of B by -1, if necessary, so that all elements have
     positive leading coefficient;
  Let C be the set of critical pairs obtained from B;

  While C is not empty do begin
    Remove a critical pair (f, g) from C; $h := REDUCE(B, t(f, g))$;

    If $h \neq 0$ then begin
      If the leading coefficient of h is negative then $h := -h$;
      Form all critical pairs involving h and some element of B and
         add these pairs to C;
      Add h to B
    End
  End
End.

Proposition 6.5. *The procedure* GRÖBNER *terminates and returns a Gröbner basis.*

Proof. The proof makes repeated use of Proposition 6.2. During the operation of GRÖBNER, let $\mathcal{V}$ denote the set of elements of $\mathcal{U}$ of the form xu, where x is in $\mathcal{X}$ and u is the leading generator of some element of B. The set $\mathcal{V}$ is closed under multiplication by elements of $\mathcal{X}$, so by Proposition 6.2 $\mathcal{V}$ is eventually constant. From this point on, whenever h is added to B there is an element e already in B such that the leading generator of h is xv, where v is the leading generator of e and x is in $\mathcal{X}$. Let a and b be the leading coefficients of h and e, respectively. We must have $a < b$, since otherwise the call REDUCE($B, t(f, g)$) would have further reduced the coefficient of xv in h. Therefore the leading coefficients of the elements of B are bounded. For each integer $b > 0$, let $\mathcal{V}_b$ denote the set of elements of $\mathcal{U}$ of the form xu, where some element of B has leading term cu and $c \leq b$. Again the set $\mathcal{V}_b$ is closed under multiplication, so for any given b the set $\mathcal{V}_b$ is eventually constant. But since the leading coefficients of elements of B are bounded, this means that eventually no elements are added to B. Therefore the set C of critical pairs is emptied and the procedure terminates. Since all critical pairs of the final set B are consonant, B is a Gröbner basis for M. $\square$

Example 6.1. Let us apply GRÖBNER to the input data of Example 3.1. The set $\mathcal{U}$ consists of the terms $b_i X^j Y^k$. We shall order $\mathcal{U}$ by using the reverse lexicographic ordering on the monomials $X^j Y^k$ and then, if necessary to break a tie, looking at the value of i. As remarked earlier, we assume $b_1 \succ b_2$. In computing REDUCE(B, f) we shall always take the earliest possible element of B which could be used to reduce a particular term in f. Elements of C will be processed as a queue, first in, first out. The initial generators in their standard forms are

$$u_1 = b_1 Y + b_2 Y + b_1 X + b_1 + b_2,$$
$$u_2 = b_2 Y + b_1 X - b_2 X - b_1,$$
$$u_3 = b_1 Y + b_2 Y - b_1 X + b_2 X + b_1 - b_2.$$

The only critical pair which can be formed from these generators is (u_1, u_3), and $t(u_1, u_3) = u_3 u_1 = -2b_1 X + b_2 X - 2b_2$. The call to REDUCE returns this element unchanged. Since its leading coefficient is negative, we multiply it by -1 and add it to our generating set as

$$u_4 = 2b_1 X - b_2 X + 2b_2.$$

The critical pairs involving u_4 are (Xu_1, Yu_4) and (Xu_3, Yu_4). Processing the first of these, we find that reducing $t(Xu_1, Yu_4)$ leads to a new generator,

$$u_5 = b_1 X^2 - 3b_2 X^2 + b_1 X - 4b_2 X + 2b_1 + 8b_2.$$

The critical pairs from u_5 are $(X^2 u_1, Yu_5)$, $(X^2 u_3, Yu_5)$, and (u_5, Xu_4). The next pair to be processed is (Xu_3, Yu_4), which does not produce a new generator. The pair $(X^2 u_1, Yu_5)$ gives us the generator

$$u_6 = 5b_2 X^2 + 9b_2 X - 4b_1 - 14b_2.$$

The only critical pair coming from u_6 is $(X^2 u_2, Yu_6)$. The remaining critical pairs do not yield new generators, so the final output of GRÖBNER is the sequence $u_1, \ldots, u_6$.

The basis produced by GRÖBNER depends not only on the module M and on $\prec$ but also on the particular input generating set for M. We can remove this dependence on the input generating set by requiring the output be a *reduced Gröbner* basis, a basis B such that for all f in B the result of REDUCE$(B - \{f\}, f)$ is f. A reduced Gröbner basis can be obtained from an arbitrary Gröbner basis B by taking the elements f of B in turn, computing $g = REDUCE(B - \{f\}, f)$, and then either deleting f from B if $g = 0$ or replacing f by g if g is nonzero and different from f. This process is continued until no further changes are possible. The reduced Gröbner basis for the module in Example 6.1 consists of u_2, u_4, and u_6, together with

$$u_7 = b_1 Y + b_2 X + 2b_1 + b_2,$$
$$u_8 = b_1 X^2 + 2b_2 X^2 + b_1 X + 5b_2 X - 2b_1 - 6b_2.$$

Just as the original version of the Knuth-Bendix procedure for strings needed a number of modifications to become really useful, improvements must be made to GRÖBNER. Redundant generators must be removed early rather than waiting for the final computation of a reduced basis. The order in which the critical pairs are processed must be considered very carefully. Data structures which permit fast execution of REDUCE must be devised. Research on these points is continuing and we shall not attempt to present recommendations here.

It will be important for us to be able to decide, for a given submodule M of R^n, whether R^n/M is finitely generated as an abelian group. The following result gives us a simple test.

Proposition 6.6. *Suppose that B is a Gröbner basis for a submodule M of R^n with respect to an ordering $\prec$ of $\mathcal{U}$ which is consistent with multiplication. The quotient R^n/M is finitely generated as an abelian group if and only for if each i, $1 \leq i \leq n$, and each j, $1 \leq j \leq m$, there is a nonnegative integer r_{ij} such that B has an element with leading term $b_i X_j^{r_{ij}}$.*

Proof. Assume first that the nonnegative integers r_{ij} all exist. If f in R^n is nonzero and irreducible with respect to B, then no term $ab_i X_1^{\alpha_1} \ldots X_m^{\alpha_m}$ in f can have $\alpha_j > r_{ij}$. Therefore f is in the $\mathbb{Z}$-module N generated by the elements $b_i X_1^{\alpha_1} \ldots X_m^{\alpha_m}$ with $0 \leq \alpha_j < r_{ij}$ for all i and j. The rank of N is

$$\sum_{i=1}^{n} \prod_{j=1}^{m} r_{ij}.$$

Since N maps onto $Q = R^n/M$, this quotient is finitely generated as an abelian group.

Now assume that Q is finitely generated as an abelian group. Fix i and j with $1 \leq i \leq n$ and $1 \leq j \leq m$. The subgroup of Q generated by the images of the elements $b_i X_j^\alpha, \alpha = 0, 1, \ldots$, is finitely generated. Therefore there is a nonnegative integer s and integers $a_0, \ldots, a_{s-1}$ such that $f = b_i X_j^s - a_{s-1} b_i X_j^{s-1} \ldots a_0 b_i$ is in M. Since $1 \prec X_j$ in $\mathcal{X}$, we must have $b_i \prec b_i X_j \prec \cdots \prec b_i X_j^s$. Therefore $b_i X_j^s$ is the leading term of f. Reduction using B eventually produces 0. But this can happen only if B contains an element with leading term $b_i X_j^r$ with $r \leq s$. $\square$

6.1. Apply GRÖBNER to Examples 1.1, 1.2, 2.1, 2.2, 2.4, 4.2, and 4.3 and to Exercises 1.1 and 3.1. Use various well-orderings of $\mathcal{X}$ and $\mathcal{U}$.

6.2. Show that each submodule of R^n has a unique reduced Gröbner basis with respect to a particular choice of orderings on $\mathcal{X}$ and $\mathcal{U}$.

10.7 Rings of Laurent polynomials

The ring of Laurent polynomials with integer coefficients in variables $X_1, \ldots, X_m$ is a free abelian group L with basis the set $\mathcal{U}$ of *Laurent monomials*, monomials $X_1^{\alpha_1} \ldots X_m^{\alpha_m}$, where the α_i are arbitrary integers. Monomials are multiplied in the obvious way and multiplication is extended to all of L using the distributive law. The set $\mathcal{U}$ is a subgroup of the group of units of L and $\mathcal{U}$ is free abelian of rank m. Sometimes L is denoted $\mathbb{Z}[X_1, 1/X_1, \ldots, X_m, 1/X_m]$, but the notation $\mathbb{Z}\langle X_1, \ldots, X_m \rangle$ will be used here.

Modules over rings of Laurent polynomials play an important role in the determination of metabelian quotients of finitely presented groups. There is more than one way to study such modules. Perhaps the simplest approach is to describe L as a quotient ring of a polynomial ring in $2m$ variables.

Proposition 7.1. *The ring $L = \mathbb{Z}\langle X_1, \ldots, X_m \rangle$ is isomorphic to the quotient of $R = \mathbb{Z}[Y_1, Z_1, \ldots, Y_m, Z_m]$ modulo the ideal I generated by $Y_1 Z_1 - 1, \ldots, Y_m Z_m - 1$.*

Proof. The map taking Y_i to X_i and Z_i to X_i^{-1} defines a ring homomorphism f of R onto L. Since $f(Y_i Z_i - 1) = X_i X_i^{-1} - 1 = 1 - 1 = 0$, the elements $Y_i Z_i - 1$ are in the kernel of f. Given an element $u = X_1^{\alpha_1} \ldots X_m^{\alpha_m}$ in $\mathcal{U}$, define $g(u)$ to be $U_1 \ldots U_m$ in R, where $U_i = Y_i^{\alpha_i}$ if $\alpha_i \geq 0$ and $U_i = Z_i^{-\alpha_i}$ if $\alpha_i < 0$. The map g extends to a homomorphism of abelian groups of L into R. It is tedious but not difficult to show that $g(uv) \equiv g(u)g(v) \pmod{I}$ for all u and v in $\mathcal{U}$. Then it is straightforward to show that g defines a ring homomorphism of L onto R/I. Since $g \circ f$ is the identity and $f \circ g$ is the identity modulo I, it follows that L and R/I are isomorphic. $\square$

Corollary 7.2. *There is a natural correspondence between L-modules and R-modules on which the elements $Y_i Z_i - 1$ act trivially.*

Proof. Exercise. $\square$

Corollary 7.3. *Submodules of finitely generated L-modules are finitely generated.*

Proof. Exercise. $\square$

Suppose that F is L^n and M is the L-submodule of F generated by a finite set T. Let F_1 be R^n and let f and g be the maps defined in the proof of Proposition 7.1. Then f can be extended to an R-homomorphism of F_1 onto F. The kernel N of f is generated by the elements $b_i(Y_j Z_j - 1)$, where $b_1, \ldots, b_n$ are the standard basis vectors. The map g extends to a homomorphism of abelian groups from F to F_1. The inverse image M_1 of

M under f is generated by N and $g(T)$, and the quotients F/M and F_1/M_1 are isomorphic as R-modules. We can compute a Gröbner basis for M_1 and therefore we can solve the word problem for F/M.

Although the approach to finitely presented L-modules just sketched is conceptually straightforward, there are some problems. The complexity of Gröbner basis computations tends to increase as the number of variables increases. In moving from L to R, we have doubled the number of variables. This is likely to cause difficulties. It seems desirable to have techniques for working directly with submodules of L^n. There are two possible approaches. We can generalize the methods of Sections 10.1 to 10.4 or we can generalize the Gröbner basis method.

Let us begin by looking again at Section 10.1 and trying to generalize ZXIDEAL to ideals of $L = \mathbb{Z}\langle X \rangle$. Let T be a finite subset of L. Since X is a unit in L, we can multiply elements of T by powers of X without changing the ideal I of L generated by T. Thus we may assume that the elements of T are ordinary polynomials in X with nonzero constant terms. The set $J = I \cap \mathbb{Z}[X]$ of ordinary polynomials in I is an ideal of $\mathbb{Z}[X]$. Given f in L, we can multiply f by a power of X so that the result g is in $\mathbb{Z}[X]$. Then f is in I if and only if g is in J. Thus, if we can determine the effective degree e of J and find a $\mathbb{Z}$-basis for J_e, we can decide membership in I. Let d be the maximum of the degrees of the elements of T. It is not hard to modify the proof of Proposition 1.1 to show that $e \leq d$. Let A be the abelian group generated by T. The condition for A to equal J_d is slightly more complicated than that given in Proposition 1.3. Let B be the subgroup of A of elements of degree at most $d - 1$. Then we must have $XB \subseteq A$. However, more is necessary. Let f be an element of A with zero constant term. Then $X^{-1}f$ is in J. Therefore, if C is the subgroup of A consisting of those elements with zero constant term, then for A to equal J_d we must have $X^{-1}C \subseteq A$.

Proposition 7.4. *A necessary and sufficient condition for A to equal J_d is for both XB and $X^{-1}C$ to be contained in A.*

Proof. Exercise. $\square$

Example 7.1. Let us determine the ideal I of $L = \mathbb{Z}\langle X \rangle$ generated by the polynomials

$$
\begin{aligned}
2X^4 + 5X^3 - 13X^2 - 4X + 30, \\
2X^3 + X^2 - 7X - 6, \\
6X^3 - X^2 - 15X.
\end{aligned}
$$

The ideal of $\mathbb{Z}[X]$ generated by these polynomials was determined in Example 1.1. The subgroup A generated by these polynomials corresponds to

$S(M_1)$, where

$$M_1 = \begin{bmatrix} 2 & 5 & -13 & -4 & 30 \\ 0 & 2 & 1 & -7 & -6 \\ 0 & 6 & -1 & -15 & 0 \end{bmatrix}.$$

The row Hermite normal form of M_1 is

$$M_2 = \begin{bmatrix} 2 & 1 & 1 & -14 & -30 \\ 0 & 2 & 1 & -7 & -6 \\ 0 & 0 & 4 & -6 & -18 \end{bmatrix},$$

and the subgroup B corresponds to the group generated by the last two rows of M_2. To find generators for C, we must row reduce M_1, processing the columns from right to left. One way to do this is to reverse the order of the columns, compute the row Hermite normal form, and then reverse the order of the columns again. The result is

$$N_2 = \begin{bmatrix} -2 & 1 & 4 & 1 & 6 \\ -4 & 0 & 11 & 3 & 0 \\ -10 & 3 & 27 & 0 & 0 \end{bmatrix}.$$

Therefore C corresponds to the group generated by the last two rows of N_2. The group $A + XB + X^{-1}C$ corresponds to the group generated by the rows of

$$M_3 = \begin{bmatrix} 2 & 1 & 1 & -14 & -30 \\ 0 & 2 & 1 & -7 & -76 \\ 0 & 0 & 4 & -6 & -18 \\ 2 & 1 & -7 & -6 & 0 \\ 0 & 4 & -6 & -18 & 0 \\ 0 & -4 & 0 & 11 & 3 \\ 0 & -10 & 3 & 27 & 0 \end{bmatrix}.$$

The nonzero rows in the row Hermite normal form of M_3 are given by

$$M_4 = \begin{bmatrix} 2 & 1 & 1 & 2 & -6 \\ 0 & 2 & 1 & 1 & 6 \\ 0 & 0 & 2 & 1 & -3 \\ 0 & 0 & 0 & 4 & 6 \end{bmatrix}.$$

Row reducing M_4 from right to left gives

$$N_4 = \begin{bmatrix} 2 & 3 & 0 & 2 & 3 \\ 2 & 3 & 2 & 3 & 0 \\ 2 & 5 & 3 & 0 & 0 \\ 4 & 6 & 0 & 0 & 0 \end{bmatrix}.$$

Now $A + XB + X^{-1}C$ corresponds to the group generated by the rows of

$$M_5 = \begin{bmatrix} 2 & 1 & 1 & 2 & -6 \\ 0 & 2 & 1 & 1 & 6 \\ 0 & 0 & 2 & 1 & -3 \\ 0 & 0 & 0 & 4 & 6 \\ 2 & 1 & 1 & 6 & 0 \\ 0 & 2 & 1 & -3 & 0 \\ 0 & 0 & 4 & 6 & 0 \\ 0 & 2 & 3 & 2 & 3 \\ 0 & 2 & 5 & 3 & 0 \\ 0 & 4 & 6 & 0 & 0 \end{bmatrix}.$$

The nonzero rows of the row Hermite normal form of M_5 are given by M_4. Thus now XB and $X^{-1}C$ are contained in A. Therefore the effective degree of $J = I \cap \mathbb{Z}[X]$ is 2 and J_2 has a $\mathbb{Z}$-basis $2X^2 + X - 3$ and $4X + 6$.

Proposition 7.4 generalizes easily to handle submodules of $\mathbb{Z}\langle X \rangle^n$ in the spirit of Section 10.2. Questions about submodules of $\mathbb{Z}\langle X, Y \rangle^n$ can be reduced to questions about submodules of $\mathbb{Z}\langle X \rangle^k$, where k is a multiple of n. Suppose that M is the submodule of $\mathbb{Z}\langle X, Y \rangle^n$ generated by a finite set T. We may assume that elements of T lie in $\mathbb{Z}[X, Y]^n$. Let d be the maximum of the degrees in Y of the elements of T. If A is the $\mathbb{Z}\langle X \rangle$-module generated by T, then we need to compute both the $\mathbb{Z}\langle X \rangle$-submodule B of elements in A whose degree in Y is less than d and the $\mathbb{Z}\langle X \rangle$-submodule $C = A \cap (Y\mathbb{Z}[X, Y]^n)$ of elements in A whose constant terms with respect to Y are 0. We then check whether $YB + Y^{-1}C$ is contained in A.

It is also possible to generalize the total-degree methods of Section 10.4. Let us assume that our generating set T is contained in $\mathbb{Z}[X, Y]$. Let d be the maximum of the total degrees of the elements of T and let A be the abelian group generated by T. We consider the subgroup D generated by all sets of the form $X^r Y^s A$, where $|r| + |s| \leq d$. We then determine $E = D \cap \mathbb{Z}[X, Y]$ and check whether A is the set of elements in E with total degree at most d and whether the effective total degrees of A and E are the same.

The Gröbner basis methods of Section 10.6 also generalize, but the details get slightly complicated. The problem is that there is no well-ordering of the set $\mathcal{U}$ of Laurent monomials in $X_1, \ldots, X_m$ which is consistent with multiplication. Suppose that $\prec$ is such an ordering. If x is in $\mathcal{U}$ and $x \succ 1$, then multiplying by x^{-1} gives $1 \succ x^{-1}$. Multiplying by x^{-1} again gives $x^{-1} \succ x^{-2}$. Thus we have an infinite, strictly decreasing sequence $1 \succ x^{-1} \succ x^{-2} \cdots$, contradicting the assumption that $\prec$ is a well-ordering.

It turns out that it is more important for the ordering to be a well-ordering than it is for the ordering to be consistent with multiplication. Rather than try to axiomatize the class of orderings which work, we shall simply choose one ordering as an illustration. The reader is encouraged to experiment with other possibilities. Let $\prec$ be the reverse lexicographic ordering on the Laurent monomials $X_1^{\alpha_1} \ldots X_m^{\alpha_m}$ in which the exponents are compared not with the usual ordering of $\mathbb{Z}$ but with the ordering $\ll$ in which $0 \ll 1 \ll -1 \ll 2 \ll -2 \ll \ldots$. In the ordering $\prec$ on $\mathcal{U}$ we have

$$X_1 \prec X_1^{-1} \prec X_1^3 X_2 \prec X_1 X_2^2 \prec X_1^{-1} X_2^2 \prec X_1 X_2^{-2} \prec X_1^{-1} X_2 X_3 \prec X_1 X_2 X_3^{-2}.$$

Note that when $\prec$ is extended to the ring L of Laurent polynomials, the coefficients are still compared according to the ordering $0 \prec 1 \prec 2 \prec \cdots \prec -1 \prec -2 \prec \cdots$.

Although this ordering on $\mathcal{U}$ is not consistent with multiplication, there is still some regularity. Let $u = X_1^{\alpha_1} \ldots X_m^{\alpha_m}$ and $v = X_1^{\beta_1} \ldots X_m^{\beta_m}$ be elements of $\mathcal{U}$. We shall say that u and v are *aligned*, or u is *aligned with* v, if $\alpha_i \beta_i \geq 0$, $1 \leq i \leq m$. Thus u and v are aligned unless there is an index i such that α_i and β_i are nonzero and have opposite signs. In the ordering $\prec$ on $\mathcal{U}$, exponents are compared first by comparing absolute values. An equivalent definition for u and v to be aligned is that $|\alpha_i + \beta_i| = |\alpha_i| + |\beta_i|$, $1 \leq i \leq m$.

Proposition 7.5. *Suppose u and v are in $\mathcal{U}$ and $u \succ v$. If the exponent α_i on X_i in u is nonnegative, then $X_i u \succ X_i v$. If α_i is nonpositive, then $X_i^{-1} u \succ X_i^{-1} v$.*

Proof. Adding 1 to nonnegative integers makes them bigger in the ordering $\ll$ and preserves their relative position, while adding 1 to negative numbers makes them smaller with respect to $\ll$. Thus $i \geq 0$ and $i \gg j$ imply $i+1 \gg j+1$. Similarly, if $i \leq 0$ and $i \gg j$, then $i-1 \gg j-1$. Suppose that $u = X_1^{\alpha_1} \ldots X_m^{\alpha_m}$ and $v = X_1^{\beta_1} \ldots X_m^{\beta_m}$. Let k be the largest index such that $\alpha_k \neq \beta_k$. If $k \neq i$, then $X_i^\gamma u \succ X_i^\gamma v$ for any integer γ. Let us assume that $k = i$. Then $\alpha_i \gg \beta_i$ and $\alpha_i \neq 0$. If $\alpha_i > 0$, then $\alpha_i + 1 \gg \beta_i + 1$, so $X_i u \succ X_i v$. Similarly, if $\alpha_i < 0$, then $X_i^{-1} u \succ X_i^{-1} v$. $\square$

Corollary 7.6. *Suppose that u, v, and x are in $\mathcal{U}$ and that $u \succ v$. If x and u are aligned, then $xu \succ xv$.*

Proof. The computation of xu and xv can be accomplished by a sequence of steps which involve multiplication by X_i or X_1^{-1} for some i. Proposition 7.5 applies to each of these steps.

□

Corollary 7.7. *Suppose $a_1 u_1 + \cdots + a_s u_s$ is the standard form for a nonzero element f of $\mathbb{Z}\langle X_1, \ldots, X_m \rangle$. If x in $\mathcal{U}$ is aligned with u_1, then $a_1 x u_1$ is the leading term of xf.*

Proof. If $2 \leq i \leq s$, then $u_1 \succ u_i$. By Corollary 7.6, $xu_1 \succ xu_i$. □

Let $u = X_1^{\alpha_1} \ldots X_m^{\alpha_m}$ and $v = X_1^{\beta_1} \ldots X_m^{\beta_m}$ be monomials in $\mathcal{U}$ and assume that u and v are aligned. Define the gcd of u and v to be $w = X_1^{\gamma_1} \ldots X_m^{\gamma_m}$, where γ_i is the element of $\{\alpha_i, \beta_i\}$ of smallest absolute value. Then w is aligned with both u and v. The monomials uw^{-1} and vw^{-1} are also aligned, and their gcd is 1. The lcm of u and v is defined to be $X_1^{\delta_1} \ldots X_m^{\delta_m}$, where δ_i is the element of $\{\alpha_i, \beta_i\}$ of largest absolute value. For example, if $u = X^{-3} Y^4 Z^{-2}$ and $v = X^{-5} Y^3 Z^{-1}$, then $\gcd(u, v) = X^{-3} Y^3 Z^{-1}$ and $\mathrm{lcm}(u, v) = X^{-5} Y^4 Z^{-2}$.

Let E be the set of integer vectors $e = (\epsilon_1, \ldots, \epsilon_m)$ with $|\epsilon_i| = 1$, $1 \leq i \leq m$. For e in E, let $\mathcal{U}_e$ be the set of monomials aligned with $u_e = X_1^{\epsilon_1} \ldots X_m^{\epsilon_m}$. Any two elements of $\mathcal{U}_e$ are aligned and $\mathcal{U}_e$ is closed under gcd's. Every monomial belongs to at least one $\mathcal{U}_e$.

Proposition 7.8. *Suppose that u, v, x, and y in $\mathcal{U}$ satisfy $xu \succ xv$ and $yu \succ yv$. If x and y are aligned and their gcd is 1, then $u \succ v$.*

Proof. Clearly $u \neq v$. Let k be the largest index such that the exponents on X_k in u and v differ. Then k is the largest index such that the exponents on X_k in xu and xv differ and k is the largest index such that the exponents on X_k in yu and yv differ. Because the gcd of x and y is 1, the exponent on X_k is 0 in at least one of x and y. Let us assume that the exponent on X_k in x is 0. Then the exponents on X_k in u and v are the same as the exponents on X_k in xu and xv, respectively. Since $xu \succ xv$, the exponent on X_k in u is greater with respect to $\ll$ than the exponent on X_k in v. Therefore $u \succ v$. □

Suppose T is a finite subset of $L = \mathbb{Z}\langle X_1, \ldots, X_m \rangle$. The ideal of L generated by T is generated as an abelian group by the set of products xf, where x is in $\mathcal{U}$ and f is in T. We need to be able to describe the set of leading terms in these products.

Example 7.2. Let $m = 1$ and let

$$f = 5X^{-4} - X^3 + 3X^2 + 2X^{-1}.$$

What is the leading term of $X^\alpha f$? If $\alpha \leq 0$, then the unique exponent in $X^\alpha f$ with largest absolute value is $\alpha - 4$. Therefore the leading term will be $5X^{\alpha-4}$. However, if $\alpha \geq 1$, then the exponent of largest absolute value is $\alpha + 3$ and the leading term is $-X^{\alpha+3}$.

Example 7.3. Now let $m = 2$ and let

$$g = 2X^{-2}Y^3 - 4X^2Y^3 - XY^2 + XY.$$

The leading term t of $X^\alpha Y^\beta g$ is determined as follows:

$$
\begin{aligned}
t &= 2X^{\alpha-2}Y^{\beta+3}, && \text{if } \beta \geq -1 \quad \text{and} \quad \alpha \leq 0, \\
t &= -4X^{\alpha+2}Y^{\beta+3}, && \text{if } \beta \geq -1 \quad \text{and} \quad \alpha \geq 1, \\
t &= X^{\alpha+1}Y^{\beta+1}, && \text{if } \beta \leq -2.
\end{aligned}
$$

On the basis of these two examples, the following proposition should seem plausible.

Proposition 7.9. *Let f be a nonzero element of L. There is a unique subset $\mathcal{I}(f)$ of L such that the following hold:*

(i) *Each element of $\mathcal{I}(f)$ has the form yf with y in $\mathcal{U}$.*
(ii) *If x is in $\mathcal{U}$, then $xf = yg$ for a unique pair (y, g) such that g is in $\mathcal{I}(f)$, y is in $\mathcal{U}$, and y is aligned with the leading monomial of g.*

The cardinality of $\mathcal{I}(f)$ is at most 2^m.

Proof. We begin with the following lemma.

Lemma 7.10. *If x is in $\mathcal{U}$ and xf has the same leading monomial as f, then $x = 1$.*

Proof. Let $u = X_1^{\alpha_1} \ldots X_m^{\alpha_m}$ be the leading monomial of f and xf and let u belong to $\mathcal{U}_e$, where $e = (\epsilon_1, \ldots, \epsilon_m)$ is in E. Choose k greater than the absolute values of all the exponents in all of the monomials involved in f and xf. By Corollary 7.7, the leading monomial of $(u_e)^k f$ and $(u_e)^k xf$ is $(u_e)^k u$. All monomials occurring in either $(u_e)^k f$ or $(u_e)^k xf$ are aligned with u_e and hence with each other. Thus, replacing f by $(u_e)^k f$, we may assume that all the monomials in f and xf are aligned with each other. Let $x = X_1^{\beta_1} \ldots X_m^{\beta_m}$. If $\epsilon_m = 1$, then α_m is the algebraically largest exponent on X_m in any of the monomials of f. The algebraically largest exponent on X_m in the monomials of xf is $\alpha_m + \beta_m$. But this number must be α_m, so $\beta_m = 0$. If $\epsilon_m = -1$, then the same argument applies with "largest" replaced by "smallest". If $\epsilon_{m-1} = 1$, then α_{m-1} is the largest exponent on X_{m-1} in any monomial of f involving $X_m^{\alpha_m}$. The largest exponent on X_{m-1} in any monomial of xf involving $X_m^{\alpha_m}$ is $\alpha_{m-1} + \beta_{m-1}$. But this number must be α_{m-1}, so β_{m-1} must be 0. Again the case $\epsilon_{m-1} = -1$ is equally easy. Continuing in this manner, we find that $\beta_1 = \cdots = \beta_m = 0$, so $x = 1$.

$\square$

Now suppose that x and y are in $\mathcal{U}$ and that the leading monomials u and v of xf and yf, respectively, are aligned. By Corollary 7.7, uv is the leading monomial of both vxf and uyf. By Lemma 7.10, $uyv^{-1}x^{-1} = 1$, or $yx^{-1} = vu^{-1}$. Let t be the gcd of u and v and let $u = zt$ and $v = wt$. Define g to be $z^{-1}xf$. Then

$$g = z^{-1}xf = z^{-1}t^{-1}tww^{-1}xf = u^{-1}vw^{-1}xf = yx^{-1}w^{-1}xf = w^{-1}yf.$$

The monomial t occurs in g. Suppose that s is some other monomial of g. Then zs is a monomial of $zg = xf$ and ws is a monomial of $wg = yf$. Therefore $u = zt \succ zs$ and $v = wt \succ ws$. By Proposition 7.8, $t \succ s$, so t is the leading monomial of g. Thus $xf = zg$ and $yf = wg$ and z and w are aligned with the leading monomial of g.

The remarks of the preceding paragraph show that for each e in E there is a unique polynomial g_e with the properties that the leading monomial u of g_e lies in $\mathcal{U}_e$ and any polynomial xf whose leading monomial lies in $\mathcal{U}_e$ has the form yg_e, where y is aligned with u. If u lies in $\mathcal{U}_d$ for some d in $E - \{e\}$, then $g_e = yg_d$, where y is aligned with the leading monomial v of g_d. This implies that v is in $\mathcal{U}_e$. By symmetry, $g_e = g_d$. Let $\mathcal{I}(f) = \{g_e \mid e \in E\}$. Then $|\mathcal{I}(f)| \le |E| = 2^m$. The uniqueness of $\mathcal{I}(f)$ is easy to show and is left as an exercise. $\square$

Let T be a finite subset of L. As noted earlier, the ideal I generated by T is generated as an abelian group by the set of products xf, where x is in $\mathcal{U}$ and f is in T. In Section 10.5 we required that the leading coefficients of our generators be positive. Thus let S be the set of elements ϵxf, where f is in T, x is in $\mathcal{U}$, and ϵ is chosen in $\{1, -1\}$ to make the leading coefficient of ϵxf positive. Let $\mathcal{I}(T)$ be the union of the sets $\mathcal{I}(f)$ with f in T and let $\mathcal{S}(T)$ be the set obtained from $\mathcal{I}(T)$ by multiplying those elements with negative leading coefficients by -1. Then every element of S is uniquely yg, where g is in $\mathcal{S}(T)$ and y is aligned with the leading monomial of g. The set $\mathcal{S}(T)$ will be called the *symmetrized set* for T. It can be computed using the following procedure:

Function SYMM(T);
Input: T : a finite subset of $\mathbb{Z}\langle X_1, \ldots, X_m \rangle$;
(* The symmetrized set for T is returned. *)

Begin
    $S := T - \{0\}$;
    For $i := m$ down to 1 do begin
      $T := \emptyset$;
      For f in S do begin
        Let u be the leading monomial of f;
        Let α and β be the algebraically largest and smallest exponents,
           respectively, on X_i occurring in any monomial v of f for
           which the exponents on $X_{i+1}, \ldots, X_m$ in v agree with the
           corresponding exponents in u;
        If $\alpha = \beta$ then $T := T \cup \{X_i^{-\alpha} f\}$
        Else begin
           Let γ be the greatest integer in $(\alpha + \beta - 1)/2$;
           $T := T \cup \{X_i^{-\gamma} f, X_i^{-\gamma-1} f\}$
        End
      End;
      $S := T$
    End;
    For f in S do
      If f has negative leading coefficient then replace f by $-f$ in S;
    SYMM $:= S$
End

Example 7.4. Let us follow the execution of SYMM($\{g\}$), where $g = 2X^{-2}Y^3 - 4X^2Y^3 - XY^2 + XY$ as in Example 7.3. On the first pass through the For-loop on i we have $S = \{g\}$ and $X_i = Y$. With $f = g$ we get $\alpha = 3$ and $\beta = 1$. Therefore, at the end of the pass, T contains

$$g_1 = Y^{-1}g = 2X^{-2}Y^2 - 4X^2Y^2 - XY + X$$

and

$$g_2 = Y^{-2}g = XY^{-1} + 2X^{-2}Y - 4X^2Y - X.$$

On the second pass, $S = \{g_1, g_2\}$ and $X_i = X$. With $f = g_1$, we have $\alpha = 2$ and $\beta = -2$, so

$$g_3 = Xg_1 = -4X^3Y^2 + 2X^{-1}Y^2 - X^2Y + X^2$$

and

$$X^0 g_1 = g_1$$

are added to $\mathcal{T}$. With $f = g_2$, we find that $\alpha = \beta = 1$. Therefore we add

$$g_4 = X^{-1}g_2 = Y^{-1} + 2X^{-3}Y - 4XY - 1$$

to $\mathcal{T}$. Since g_3 has a negative leading coefficient, the final set returned is $\{-g_3, g_1, g_4\}$.

Suppose $\mathcal{T}$ is given and S is defined as before. For each u in $\mathcal{U}$, the set S_u of elements in S which have u as their leading generator is finite and can be determined. Therefore we can execute GRP_REDUCE(S, g) for any g in L. Let us again denote the result by REDUCE$(\mathcal{T}, g)$. For REDUCE$(\mathcal{T}, g)$ to be independent of the choices made in computing it, the semicritical pairs must be consonant. There are infinitely many semicritical pairs, but it is possible to find a finite set of semicritical pairs whose consonance will imply the consonance of all other semicritical pairs.

Suppose that (f, g) is a semicritical pair of elements of S and u is the leading generator of f and g. Let x be an element of $\mathcal{U} - \{1\}$ which is aligned with u. Then xu is the leading generator of xf and xg, and both xf and xg are in S. Assume that (f, g) is consonant. Then $t(f, g) = a_1 h_1 + \cdots + a_r h_r$, where the a_i are integers and the h_i are elements of S which precede g with respect to $\prec$. Let v_i be the leading generator of h_i. Then either $u \succ v_i$ or $u = v_i$ and the leading coefficient of h_i is less than the leading coefficients of f and g. If $u \succ v_i$, then u is greater than any monomial occurring in h_i. Therefore xu is greater than any monomial occurring in xh_i, so xf and xg come after xh_i with respect to $\prec$. If $u = v_i$, then xv_i is the leading generator of xh_i and again xf and xg come after xh_i. If $xf \prec xg$, then $t(xf, xg) = xt(f, g) = a_1 xh_1 + \cdots + a_r xh_r$ is in the group generated by the elements of S preceding xg. If $xg \prec xf$, then f and g have the same leading coefficient and $t(f, g) = g - f$. Therefore $t(xg, xf) = xf - xg = -xt(f, g) = -a_1 xh_1 - \cdots - a_r xh_r$ is in the subgroup generated by the elements of S preceding xf. By Proposition 5.5, we do not need to check consonance of semicritical pairs of the form (xf, xg) or (xg, xf). This observation implies that it is necessary to check only a finite set of *critical pairs* defined as follows: Let f and g be elements of $S(\mathcal{T})$ with leading monomials u and v, respectively, and assume that u and v are aligned. Let $w = \mathrm{lcm}(u, v)$, $x = wu^{-1}$, and $y = wv^{-1}$. The leading monomial of xf and yg is w. Suppose $xf \prec yg$. Then (xf, yg) is a critical pair.

Example 7.5. Suppose that $m = 1$ and that T consists of

$$f = 3X^3 - X^{-2} - 4X + 7 \quad \text{and} \quad g = X^{-2} - 2X^2 + 1.$$

Then $\mathcal{S}(T) = \{f, -X^{-1}f, g, -Xg\}$ and the critical pairs are $(-Xg, f)$ and $(X^{-1}g, -X^{-1}f)$.

Example 7.6. Assume that $m = 2$ and that T consists of the polynomials

$$f = 2X^2Y^2 - 3XY^2 + 7 \quad \text{and} \quad g = 3XY^3 + 2X - 4.$$

This time $\mathcal{S}(T) = \{X^{-1}f, -X^{-2}f, Y^{-1}f, X^{-1}Y^{-1}g, Y^{-2}g, -X^{-1}Y^{-2}g\}$ and the critical pairs are $(X^{-1}f, Y^{-1}g)$, $(X^{-2}Y^{-1}g, -X^{-2}f)$, $(Y^{-2}g, XY^{-2}f)$, and $(-X^{-1}Y^{-2}g, X^{-1}Y^{-2}f)$.

Once the correct definition of critical pair has been worked out, it is relatively easy to extend the Gröbner basis algorithm to L. If I is an ideal of L, then a Gröbner basis for I is a finite generating set B for which all critical pairs are consonant. If T is a finite subset of L which is its own symmetrized set, then REDUCE(T, f) can be computed with the following function:

Function REDUCE_L(T, f);
Input: T : a finite subset of $L = \mathbb{Z}\langle X_1, \ldots, X_m \rangle$ whose elements
                have positive leading coefficients;
    f : an element of L;
($*$ An element g of $I + f$ is returned, where I is the ideal of L generated
    by T. The element g is irreducible with respect to the set of
    products yh, where h is in T, y is in $\mathcal{U}$, and y is aligned with the
    leading monomial of h. $*$)
Begin
    $i := 1$; $g := f$;
    ($*$ At all times the standard form of g will be $a_1u_1 + \cdots + a_su_s$. If
        $g = 0$, then $s = 0$. $*$)
    While $i \leq s$ do
        If there is an element h in T such that the leading term bv of h
            satisfies $b \preceq a_i$ and $u_i = yv$, where y is in $\mathcal{U}$ and y and v are
            aligned, then begin
            Let $a_i = qb + r$, where q and r are integers and $0 \leq r < b$;
            $g := g - qyh$ ($*$ Recompute s and the terms a_ju_j with $j \geq i$. $*$)
        End
        Else $i := i + 1$;
    REDUCE_L $:= g$
End

Here is one version of a Laurent polynomial Gröbner basis procedure:

Procedure GRÖBNER_L($T; B$);

Input: T : a finite subset of $L = \mathbb{Z}\langle X_1, \ldots, X_m \rangle$;
Output: B : a Gröbner basis for the ideal of L generated by T;
Begin
   $B := \mathrm{SYMM}(T)$;
   Let C be the set of critical pairs obtained from B;

   While C is not empty do begin
      Remove a critical pair (f,g) from C; $h := \mathrm{REDUCE\_L}(B, t(f,g))$;

      If $h \neq 0$ then begin
         $U := \mathrm{SYMM}(\{h\})$;
         Form all critical pairs obtainable from an element of U and an
            element of B and add these pairs to C;
         $B := B \cup U$
      End
   End
End.

The set B returned by GRÖBNER_L will in general be larger than it needs to be. There are several ways one could define a reduced Gröbner basis for an ideal I of L. Here is one possibility. A *reduced Gröbner basis* for I is a Gröbner basis B such that

(a) REDUCE(B, f) and REDUCE_L(B, f) always return the same value.

(b) The value of REDUCE_L$(B - \{h\}, h)$ is h for all h in B.

Condition (a) implies that I is generated as an abelian group by the set of products yh, where h is in B and y is an element of $\mathcal{U}$ which is aligned with the leading monomial of h. Reduced Gröbner bases in this sense are unique, but they are not necessarily Gröbner bases of minimal cardinality.

Example 7.7. Let us apply GRÖBNER_L to $T = \{f, g\}$, where

$$f = 3Y + 2X \quad \text{and} \quad g = Y - 4X - 2.$$

The set B returned depends on the way the set C of critical pairs is managed and on the choices made in REDUCE_L. One implementation returned the following Gröbner basis for the ideal I generated by T:

$$3Y + 2X, \quad 2Y^{-1} + 3X^{-1}, \quad Y - 4X - 2, \quad 4XY^{-1} + 2Y^{-1} - 1,$$
$$2X^{-1}Y^{-1} + 4Y^{-1} - X^{-1}, \quad 4X + 6, \quad 6X^{-1} + 4, \quad 7X + 3, \quad 3X^{-1} + 7,$$
$$2Y - X - 1, \quad XY^{-1} + Y^{-1} - 2, \quad X^{-1}Y^{-1} + Y^{-1} - 2X^{-1}.$$

The reduced Gröbner basis for I consists of the following polynomials:

$$7X + 3, \quad 3X^{-1} + 7, \quad Y + 3X + 1, \quad 2Y^{-1} - 7,$$
$$XY^{-1} + Y^{-1} - 2, \quad X^{-1}Y^{-1} + Y^{-1} + X^{-1} + 7.$$

The remarks made about the procedure GRÖBNER in Section 10.6 apply equally well to GRÖBNER_L. Significant modifications are needed to make the procedure truly useful. The ideas used in GRÖBNER_L can be extended to study submodules of $\mathbb{Z}\langle X_1, \ldots, X_m \rangle^n$.

Exercises

7.1. Prove termination and correctness of GRÖBNER_L.

7.2. Extend the Gröbner basis approach to submodules of $\mathbb{Z}\langle X_1, \ldots, X_m \rangle^n$.

7.3. Compute the symmetrized set in $\mathbb{Z}\langle X, Y \rangle$ of the set consisting of

$$3X^2Y^{-1} + 2X^{-1}Y^2 - XY^{-1} + 5X^{-3}Y^2$$

and

$$X^4Y^5 + X^3Y^3 + XY^3.$$

10.8 Group rings

Let G be a group. The *integral group ring* of G is the set $\mathbb{Z}[G]$ of functions $f: G \to \mathbb{Z}$ such that for all but a finite number of elements u of G the image $f(u)$ is 0. Sums in $\mathbb{Z}[G]$ are computed pointwise. That is, $(f + g)(u) = f(u) + g(u)$. However, products are defined by the formula

$$(fg)(w) = \sum_{uv=w} f(u)g(v).$$

With these operations, $\mathbb{Z}[G]$ is a ring, which is commutative if and only if G is abelian. The group G is embedded in $\mathbb{Z}[G]$. For u in G, define $h_u(v)$ to be 0 if $v \neq u$ and 1 if $v = u$. If $uv = w$, then $h_u h_v = h_w$. Moreover, for any f in $\mathbb{Z}[G]$,

$$f = \sum_{u \in G} f(u) h_u.$$

If 1 denotes the identity element of G, then h_1 is the multiplicative identity of $\mathbb{Z}[G]$. If we identify u in G with h_u, then elements of $\mathbb{Z}[G]$ may be considered to be formal integral linear combinations of elements of G. If a and b are in $\mathbb{Z}$ and u and v are in G, then $(au)(bv)$ is $(ab)(uv)$ and multiplication is extended to $\mathbb{Z}[G]$ using the distributive law.

Example 8.1. Let G be the cyclic group of order 3 generated by an element u. The elements of G are 1, u, and u^2. An element of $\mathbb{Z}[G]$ has the form $a + bu + cu^2$, where a, b, and c are integers. In $\mathbb{Z}[G]$,

$$(3 - 2u + u^2) + (4 + u - 2u^2) = 7 - u - u^2$$

and

$$(3 - 2u + u^2)(4 + u - 2u^2) = 17 - 7u - 4u^2.$$

Example 8.2. The ring $\mathbb{Z}\langle X \rangle$ of integer Laurent polynomials in the variable X is the set of formal sums of terms aX^i, where a and i are integers. The product $(aX^i)(bX^j)$ is defined to be abX^{i+j} and multiplication is extended to all of $\mathbb{Z}\langle X \rangle$ by the distributive law. Under the identification of i with X^i, $\mathbb{Z}\langle X \rangle$ is the integral group ring of $\mathbb{Z}$. More generally, $\mathbb{Z}\langle X_1, \ldots, X_m \rangle$ is the integral group ring of $\mathbb{Z}^m$.

The integral group ring of any finitely generated abelian group can be described as a quotient of a polynomial ring. Let H be the abelian group generated by $a_1, \ldots, a_n$ subject to the defining relations $a_j a_i = a_i a_j$, $1 \leq i < j \leq n$, and $a_i^{m_i} = 1$ for i in I, where I is a subset of $\{1, \ldots, n\}$. For $1 \leq i \leq n$ let X_i be an indeterminate, and for i not in I let Y_i be an additional indeterminate. Set R equal to the ring of integer polynomials in the indeterminates thus defined. Let J be the ideal of R generated by the elements $X_i Y_i - 1$ with i not in I and the elements $X_i^{m_i} - 1$ with i in I. Then $\mathbb{Z}[H]$ is isomorphic to R/J and there is a one-to-one correspondence between $\mathbb{Z}[H]$-modules and R-modules annihilated by J. We can also describe $\mathbb{Z}[H]$ as the quotient ring of $\mathbb{Z}\langle X_1, \ldots, X_n \rangle$ modulo the ideal generated by the elements $X_i^{m_i} - 1$ with i in I.

Modules over group rings will play an important part in our computation of quotients of finitely presented groups. If G is an abelian group, then $\mathbb{Z}[G]$ is commutative, so we do not have to distinguish between left and right $\mathbb{Z}[G]$-modules. However, if G may be nonabelian, then we shall use only right $\mathbb{Z}[G]$-modules. To construct a right module for $\mathbb{Z}[G]$, it suffices to have an abelian group M and a homomorphism from G to the automorphism group of M. In this construction, we assume that M is written additively and that elements of $\mathrm{Aut}(M)$ act on the right. That is, if u is in M and α is in $\mathrm{Aut}(M)$, then the image of u under α is u^α.

Proposition 8.1. *Suppose that M is an abelian group and $\varphi \colon G \to \mathrm{Aut}(M)$ is a homomorphism of groups. If $a_1, \ldots, a_s$ are integers and $g_1, \ldots, g_s$ are distinct elements of G, then defining*

$$u(a_1 g_1 + \cdots + a_s g_s) = a_1 u^{\varphi(g_1)} + \cdots + a_s u^{\varphi(g_s)}$$

makes M into a right $\mathbb{Z}[G]$-module.

Proof. Suppose that u is in M and that $P = a_1g_1 + \cdots + a_sg_s$ and $Q = b_1h_1 + \cdots + b_th_t$ are in $\mathbb{Z}[G]$. We shall prove that $(uP)Q = u(PQ)$, leaving the remaining module axioms to be checked by the reader.

$$(uP)Q = \left(\sum_{i=1}^{s} a_iu^{\varphi(g_i)}\right)Q = \sum_{j=1}^{t}b_j\left(\sum_{i=1}^{s}a_iu^{\varphi(g_i)}\right)^{\varphi(h_j)}$$

$$= \sum_{j=1}^{t}b_j\left(\sum_{i=1}^{s}a_iu^{\varphi(g_i)\varphi(h_j)}\right)$$

$$= \sum_{i,j}a_ib_ju^{\varphi(g_ih_j)}$$

$$= u\left(\sum_{i,j}a_ib_jg_ih_j\right) = u(PQ). \quad \square$$

An abelian normal subgroup M of a group G is a right $\mathbb{Z}[G/M]$-module. To see this, suppose that g is in G and u is in M. Then $u^g = g^{-1}ug$ depends only on the coset Mg of M containing g, for if v is in M, then

$$(vg)^{-1}u(vg) = g^{-1}v^{-1}uvg = g^{-1}ug,$$

since u and v commute. Thus we can define a map from G/M to $\text{Aut}(M)$ which sends Mg to the automorphism of M taking u to u^g. This map is easily seen to be a homomorphism. By Proposition 8.1, we may consider M to be a right module over $\mathbb{Z}[G/M]$.

We shall be working primarily with modules over the group ring of a polycyclic group. The following important result was first proved in (P. Hall 1954).

Proposition 8.2. *Let M be a finitely generated right module over the group ring of a polycyclic group. Then all submodules of M are finitely generated. Equivalently, M satisfies the ascending chain condition for submodules.*

In Section 10.7 we saw several ways of studying finitely generated modules over rings of integer Laurent polynomials, which we now view as group rings of finitely generated free abelian groups. The techniques of Section 10.7 can be extended to handle finitely generated right modules over the integral group ring of any polycyclic group G. The discussion which follows is based on (Baumslag et al. 1981b).

We may assume that we know a polycyclic generating sequence $a_1, \ldots, a_n$ for G and a consistent polycyclic presentation for G in terms of these generators. If $n = 1$, then G is abelian and we have already seen how to describe finitely generated modules over $\mathbb{Z}[G]$. Let us assume that $n > 1$. Let $H = \mathrm{Grp}\langle a_2, \ldots, a_n \rangle$. Then H is normal in G and G/H is a cyclic group which is generated by the image of a_1. From now on, we shall write a for a_1. By induction on n, we may assume that we know how to describe submodules of $\mathbb{Z}[H]^s$. Suppose first that G/H is finite and let $m = |G/H|$. Then $F = \mathbb{Z}[G]^s$ is isomorphic as a $\mathbb{Z}[H]$-module to $\mathbb{Z}[H]^{sm}$, for if $b_1, \ldots, b_s$ is a $\mathbb{Z}[G]$-basis for F, then the products $b_i a^j$ with $1 \leq i \leq s$ and $0 \leq j < m$ form a $\mathbb{Z}[H]$-basis for F. A $\mathbb{Z}[H]$-submodule M of F is a $\mathbb{Z}[G]$-submodule if and only if M is closed under multiplication by a. If T is a generating set for M as a $\mathbb{Z}[H]$-module, then M is closed under multiplication by a if and only if fa is in M for all f in T. Suppose the products fa do belong to M. A typical element g of M has the form $f_1 h_1 + \cdots + f_r h_r$, where the f_i are in T and the h_i are in H. Then

$$ga = f_1 h_1 a + \cdots + f_r h_r a = f_1 a k_1 + \cdots + f_r a k_r,$$

where each $k_i = a^{-1} h_i a$ is in H. Since each $f_i a$ is in M, it follows that ga is in M. If some fa is not in M, then we can add it to T and recompute the $\mathbb{Z}[H]$-submodule generated by T. Because the ascending chain condition holds, this process will terminate. Since we can describe submodules of $\mathbb{Z}[H]^{sm}$ effectively, we can describe submodules of $\mathbb{Z}[G]^s$.

The more challenging case is the one in which G/H is infinite. It is still true that $F = \mathbb{Z}[G]^s$ is a free $\mathbb{Z}[H]$-module, but it is no longer a finitely generated one. Let $b_1, \ldots, b_s$ be the standard $\mathbb{Z}[G]$-basis for F. The set $\mathcal{U}$ of elements $b_i a^j$ is a $\mathbb{Z}[H]$-basis for F. Any element g of G can be written uniquely in the form $a^j h$, where h is in H. In this way, $b_i g$ can be expressed as $b_i a^j h$. Thus elements of F can easily be described as $\mathbb{Z}[H]$-linear combinations of elements of $\mathcal{U}$. However, it is also useful to write g in the form $k a^j$, where k is in H. When this is done, every element z of F can be described as

$$c_p a^p + c_{p-1} a^{p-1} + \cdots + c_{q+1} a^{q+1} + c_q a^q,$$

where $p \geq q$ and the c_i are uniquely determined elements in the free $\mathbb{Z}[H]$-module F_0 generated by $b_1, \ldots, b_s$. If $c_i = 0$ for $i < 0$, we shall refer to z as a *polynomial* in a, although this is a slight abuse of terminology. If z is a nonzero polynomial in a, then the *degree* of z is the largest index r such that $c_r \neq 0$, and the coefficient c_0 will be called the *constant term* of z.

Let T be a finite subset of F and let M be the $\mathbb{Z}[G]$-submodule generated by T. Since a is a unit in $\mathbb{Z}[G]$, we may multiply elements of T on the right by powers of a so that T consists of polynomials in a with nonzero constant terms. Given a nonnegative integer d, let M_d be the set of polynomials in a which have degree at most d and belong to M and let C_d be the set of coefficients of a^d in elements of M_d. Let $z = c_d a^d + \cdots + c_0$ be in M_d and suppose h is in H. Then $k = a^{-d} h a^d$ is also in H and

$$zk = c_d a^d k + \cdots + c_0 k = c_d h a^d + \cdots + c_0 k.$$

Therefore C_d is a $\mathbb{Z}[H]$-submodule of F_0. Since $za = c_d a^{d+1} + \cdots + c_0 a$, we have $C_d \subseteq C_{d+1}$.

Let d be the maximum of the degrees of the elements of T. Then the same arguments we have used several times already show that $C_k = C_d$ if $k \geq d$. Knowing M_d is enough to allow us to decide membership in M. Given g in F, we multiply g on the right by a power of a so that g is a polynomial in a of degree k. If $k > d$, then we check whether the coefficient c_k of a^k in g is in C_d. If not, g is not in M. If c_k is in C_d, then we can produce an element h in M_d of degree d whose leading coefficient is c_k. Then subtracting ha^{k-d} from g reduces the degree of g. Thus we may assume that $k \leq d$. In this case we have only to check whether g is in M_d.

The $\mathbb{Z}[H]$-submodule A generated by T is contained in M_d, but in general A is not equal to M_d. Let B and C be the sets of elements in A with degree at most $d - 1$ and with constant term 0, respectively. Then $A = M_d$ if and only if Ba and Ca^{-1} are contained in A. Thus the basic algorithm we used first in $\mathbb{Z}[X_1, \ldots, X_m]^s$ and later generalized to $\mathbb{Z}\langle X_1, \ldots, X_m \rangle^s$ also works in F.

Example 8.3. Let G be the free nilpotent group of class 2 on two generators, which is defined by the nilpotent group presentation

$$ba = abc, \quad ca = ac, \quad cb = bc$$

on generators a, b, c. Let us determine the right ideal M of $\mathbb{Z}[G]$ generated by the set T consisting of

$$f = a + b \quad \text{and} \quad g = a + c.$$

The subgroup $H = \mathrm{Grp}\langle b, c \rangle$ is free abelian of rank 2, so $\mathbb{Z}[H]$ is isomorphic to $\mathbb{Z}\langle X, Y \rangle$. Let us fix an isomorphism by associating b with X and c with Y. The two generators of M are polynomials in a of degree 1. The $\mathbb{Z}[H]$-submodule A generated by T corresponds to the $\mathbb{Z}\langle X, Y \rangle$-submodule U of $\mathbb{Z}\langle X, Y \rangle^2$ generated by

$$(1, X) \quad \text{and} \quad (1, Y).$$

The set of coefficients of a in elements of A does *not necessarily* correspond to the projection of U onto its first component. However, the submodule B of A does correspond to the set $U^{(2)}$ of elements in U which have 0 as their first component. To obtain generators for $U^{(2)}$, we can use the Gröbner basis method with the ordering on the abelian group generators $b_i X^j Y^k$ in which we look first at b_i and then at the monomial $X^j Y^k$ in the reverse lexicographic ordering determined by the ordering $\ll$ of $\mathbb{Z}$ defined in Section 10.7. In comparing basis elements b_i, we use the ordering $b_2 \prec b_1$.

The reduced Gröbner basis for U obtained in this way consists of the following elements:

$$(1, X), \quad (0, Y - X), \quad (0, Y^{-1} - X^{-1}).$$

The last two elements generate $U^{(2)}$. However, the second of these is $-X^{-1}Y^{-1}$ times the first, so B is generated as a $\mathbb{Z}[H]$-module by $w = c - b$. To find generators for C, we perform another Gröbner basis computation in U, this time assuming $b_1 \prec b_2$. It turns out that C is generated by aw. Therefore, to see whether A is M_1 we must check whether the $\mathbb{Z}[H]$-submodule D generated by f, g, $wa = ac - abc$, and $awa^{-1} = c - bc^{-1}$ is A. The generators of D correspond to

$$(1, X), \quad (1, Y), \quad (Y - XY, 0), \quad \text{and} \quad (0, Y - XY^{-1})$$

in $\mathbb{Z}\langle X, Y \rangle^2$. Using the ordering with $b_2 \prec b_1$, we find that the reduced Gröbner basis for the $\mathbb{Z}\langle X, Y \rangle$-module generated by these pairs consists of

$$(1, 1), \quad (0, X - 1), \quad (0, X^{-1} - 1), \quad (0, Y - 1), \quad (0, Y^{-1} - 1).$$

Since this is different from the first basis of U, earlier, we must iterate again. Now we may take T to have the following elements:

$$a + 1, \quad b - 1, \quad b^{-1} - 1, \quad c - 1, \quad c^{-1} - 1.$$

Computing bases for A, B, C, and $D = A + Ba + Ca^{-1}$, we find that $D = A$, so we have a $\mathbb{Z}[H]$-basis for M_1.

The Gröbner basis approach can also be generalized to modules over group rings of polycyclic groups. Research continues on the best way to formulate this generalization.

Exercise

8.1. Let G be the group in Example 8.3. Determine the right ideal of $\mathbb{Z}[G]$ generated by $2a + b$ and $a + c$.

10.9 Historical notes

During the 1960s, three closely related procedures were developed, the resolution procedure, the Gröbner basis algorithm, and the Knuth-Bendix procedure. The relationships among these procedures are discussed in (Buchberger 1987). The resolution procedure, which is related to automatic theorem proving, was presented in (Robinson 1965). The general Knuth-Bendix procedure was described in (Knuth & Bendix 1970). A brief survey of the use of the special case discussed in Chapter 2 of this book and of the rewriting approach to finitely presented monoids is given in the introduction to [Le Chenadec 1986].

Perhaps the first paper to raise algorithmic questions about ideals in polynomial rings was (Hermann 1926). In polynomial rings over a field, practical algorithms for the ideal membership problem, using Gröbner bases, were given in (Buchberger 1965), which also described a computer implementation. Several researchers, working independently of Buchberger and each other, considered the case in which the coefficients of the polynomials come from $\mathbb{Z}$ or from various other rings. A number of references are given in (Buchberger 1984). The work in (Baumslag et al. 1981b) was influenced by (Richman 1974) and (Romanovskii 1974).

11

Quotient groups

We come now to the last of the major tools for studying a given finitely presented group G, the nilpotent quotient algorithm, the p-quotient algorithm, and the polycyclic quotient algorithm. Let e be a positive integer. Using the nilpotent quotient algorithm, we can determine the quotient $G/\gamma_{e+1}(G)$, the largest nilpotent quotient of G having class at most e. For a given prime p, the p-quotient algorithm constructs the quotient $G/\varphi_{e+1}(G)$, the largest quotient of G which is a p-group with exponent-p central class at most e. The polycyclic quotient algorithm lets us determine $G/G^{(e)}$, the largest solvable quotient of G having derived length at most e, provided $G/G^{(e)}$ is polycyclic. The quotients $G/\gamma_{e+1}(G)$ and $G/\varphi_{e+1}(G)$ are always polycyclic, but $G/G^{(e)}$ need not be. We can solve the word problem in $G/G^{(e)}$ if $G/G^{(e-1)}$ is polycyclic, but we cannot in general work with subgroups of $G/G^{(e)}$ unless that group is polycyclic.

The definition of the term "nilpotent quotient algorithm" given here differs from the one used by most previous authors. Until recently, the only computer programs available for obtaining any nonabelian nilpotent quotients of a finitely presented group were implementations of the p-quotient algorithm, so this algorithm was frequently referred to as the nilpotent quotient algorithm. While it is certainly true that the p-quotient algorithm computes quotients which are nilpotent, it seems best to reserve the name "nilpotent quotient algorithm" for a procedure which determines the groups $G/\gamma_{e+1}(G)$.

11.1 Describing quotient groups

This section explains the general approach used in the quotient algorithms discussed in this chapter. The first step is to define precisely what it means to determine a quotient group of a finitely presented group. Let $(X, \mathcal{R})$ be a finite presentation for a group G and suppose that we wish to find some polycyclic quotient of G, say $G/\gamma_5(G)$. Certainly we want a consistent polycyclic presentation $(Y, \mathcal{S})$ for $G/\gamma_5(G)$. It is quite likely that the elements of X do not map to a polycyclic generating sequence for $G/\gamma_5(G)$. Thus the set Y may not have any simple relation to X. For this reason, it is also very useful to know an explicit isomorphism between $G/\gamma_5(G)$ and $K = \mathrm{Grp}\,\langle Y \mid \mathcal{S}\rangle$. Such an isomorphism can be described by a monoid homomorphism σ from $X^{\pm*}$ to $Y^{\pm*}$. We may assume that $\sigma(x^{-1}) = \sigma(x)^{-1}$ for all x in X, and hence that $\sigma(U^{-1}) = \sigma(U)^{-1}$ for all U in $X^{\pm*}$. Such a homomorphism will be called a *regular* homomorphism from $X^{\pm*}$ to $Y^{\pm*}$.

Given a regular homomorphism σ from $X^{\pm*}$ to $Y^{\pm*}$, it is easy to check whether σ defines a homomorphism $\overline{\sigma}$ of G into K. We simply have to check whether for each pair (R, S) in $\mathcal{R}$ the words $\sigma(R)$ and $\sigma(S)$ define the same element of K. Since we have a consistent polycyclic presentation for K, this can be done. Suppose σ does determine a homomorphism $\overline{\sigma}$ of G into K. One way to decide whether G is mapped onto K by $\overline{\sigma}$ is to test whether the elements of K defined by the words $\sigma(x)$ with x in X generate K. The techniques of Chapter 9 make this possible. Assume that $\overline{\sigma}$ is surjective. At this point we know that the kernel of $\overline{\sigma}$ is $\gamma_5(G)$, for $K \cong G/\gamma_5(G)$ is nilpotent of class at most 4. Therefore the kernel of $\overline{\sigma}$ contains $\gamma_5(G)$. The group $G/\gamma_5(G)$ is hopfian and hence is not isomorphic to a proper homomorphic image of itself. Therefore the kernel of $\overline{\sigma}$ is $\gamma_5(G)$ and $\overline{\sigma}$ defines an isomorphism ρ of $G/\gamma_5(G)$ onto K.

The procedure for determining the subgroup generated by a set of elements in a polycyclic group can be quite time-consuming. It would be much easier to prove that $\overline{\sigma}$ is surjective if we were given, for each y in Y, a word $\tau(y)$ in $X^{\pm*}$ such that $\sigma(\tau(y))$ and y define the same element of K. Such a map τ can be extended uniquely to a regular homomorphism from $Y^{\pm*}$ to $X^{\pm*}$ such that $\sigma(\tau(U))$ and U define the same element of K for all U in $Y^{\pm*}$. It follows that τ describes the isomorphism ρ^{-1} of K onto $G/\gamma_5(G)$. Therefore $\tau(\sigma(V))$ and V define the same element of $G/\gamma_5(G)$ for all V in $X^{\pm*}$.

The nilpotent quotient algorithm and the polycyclic quotient algorithm return a consistent polycyclic group presentation $(Y, \mathcal{S})$ for a group K isomorphic to the desired quotient and regular homomorphisms σ and τ from $X^{\pm*}$ to $Y^{\pm*}$ and $Y^{\pm*}$ to $X^{\pm*}$, respectively, such that σ defines a homomorphism of G onto K and $\sigma(\tau(U))$ and U define the same element of K for all U in $Y^{\pm*}$. When the procedure ADJUST of Section 11.5 is used, the nilpotent quotient algorithm returns a presentation which is weakly γ-weighted as defined in Exercise 4.3 in Chapter 9.

The p-quotient algorithm returns a φ-weighted power-commutator presentation $(Y, \mathcal{S})$ for a group K isomorphic to $G/\varphi_{e+1}(G)$ and two homomorphisms σ and τ exhibiting the isomorphism. Since $(Y, \mathcal{S})$ is a monoid presentation, σ goes from $X^{\pm*}$ to Y^* and τ goes from Y^* back to $X^{\pm*}$. The homomorphism $\overline{\sigma}$ from G to K is determined by the values of $\sigma(x)$ with x in X. Given $\sigma(x)$, there is no really natural way to define $\sigma(x^{-1})$. One reasonable choice is the collected word in Y^* describing the inverse in K of $\overline{\sigma}(x)$.

All three quotient algorithms are iterative procedures which build the groups $G/\gamma_{e+1}(G)$, $G/\varphi_{e+1}(G)$, and $G/G^{(e)}$ first for $e = 1$, then for $e = 2$, and so on. The following result is basic to the formulation of the iteration.

Proposition 1.1. *Let G and H be groups given by presentations $(X, \mathcal{R})$ and $(Z, \mathcal{T})$, respectively. Suppose that μ and ν are regular homomorphisms from $X^{\pm *}$ to $Z^{\pm *}$ and $Z^{\pm *}$ to $X^{\pm *}$, respectively, such that μ defines a homomorphism $\overline{\mu}$ of G onto H and, for all U in $Z^{\pm *}$, the words U and $\mu(\nu(U))$ define the same element of H. Then the kernel N of $\overline{\mu}$ is generated as a normal subgroup of G by the images of the words $\nu(\mu(x))^{-1}x$ with x in X and the words $\nu(S^{-1}R)$ with (R, S) in $\mathcal{T}$.*

Proof. A word $S^{-1}R$ with (R, S) in $\mathcal{T}$ defines the identity in H. By assumption, this means that $\mu(\nu(S^{-1}R))$ defines the identity in H, so $\nu(S^{-1}R)$ defines an element of N. If x is in X, then $\mu(\nu(\mu(x))^{-1}x) = \mu(\nu(\mu(x)))^{-1}\mu(x)$ defines the same element of H as $\mu(x)^{-1}\mu(x)$, namely the identity. Therefore $\nu(\mu(x))^{-1}x$ defines an element of N. Let M be the normal closure in G of the elements defined by the words $\nu(S^{-1}R)$ and $\nu(\mu(x))^{-1}x$ and suppose that $W = x_1 \ldots x_r$ is a word in $X^{\pm *}$ which defines an element of N. If x is in X, then x is congruent to $\nu(\mu(x))$ modulo M. Therefore, modulo M, the word W defines the same element as $\nu(\mu(W))$. Now $\mu(W)$ defines the identity in H, so $\mu(W)$ is freely equivalent to a product

$$U_1^{-1}(S_1^{-1}R_1)^{a_1}U_1 \ldots U_k^{-1}(S_k^{-1}R_k)^{a_k}U_k,$$

where (R_i, S_i) is in $\mathcal{T}$, $1 \leq i \leq k$. Therefore $\nu(\mu(W))$ is freely equivalent to

$$\nu(U_1)^{-1}\nu(S_1^{-1}R_1)^{a_1}\nu(U_1) \ldots \nu(U_k)^{-1}\nu(S_k^{-1}R_k)^{a_k}\nu(U_k),$$

which defines an element of M. Since W and $\nu(\mu(W))$ define congruent elements modulo M, it follows that W defines an element of M. Therefore $N = M$.

$\square$

Suppose under the assumptions of Proposition 1.1 that N_1 is a normal subgroup of G contained in N and that N/N_1 is abelian. Then N/N_1 is a right module for $\mathbb{Z}[G/N]$. Using the isomorphisms between G/N and H defined by μ and ν, we can consider N/N_1 to be a right $\mathbb{Z}[H]$-module. The following result follows immediately from Proposition 1.1.

Corollary 1.2. *If, in Proposition 1.1, N_1 is a normal subgroup of G contained in N and N/N_1 is abelian, then N/N_1 is generated as a right $\mathbb{Z}[H]$-*

module by the images of the words $\nu(\mu(x))^{-1}x$ with x in X and $\nu(S^{-1}R)$ with (R, S) in $\mathcal{T}$. Thus if G is finitely generated and H is finitely presented, then N/N_1 is finitely generated as a right $\mathbb{Z}[H]$-module.

In our applications, $(X, \mathcal{R})$ and $(Z, \mathcal{T})$ will be finite presentations and H will be polycyclic. We shall use Corollary 1.2 with four choices for N_1. These are $N_1 = [N, G]$, $N_1 = [N, G]N^p$ for some prime p, $N_1 = N'$, and $N_1 = N'N^p$. The group G/N_1 will be polycyclic if and only if N/N_1 is finitely generated as an abelian group. In the first two cases, this will always be true, but if $N_1 = N'$ or $N_1 = N'N^p$, then N/N_1 may not be finitely generated as an abelian group, even though it is finitely generated as a $\mathbb{Z}[H]$-module.

We shall approach G/N_1 in two steps. First we shall construct a solvable group P and a homomorphism π of P onto G/N_1. Following π by the homomorphism from G/N_1 to H, we obtain a homomorphism of P onto H. The kernel Q of this composition will be abelian. The group P may not be polycyclic, but we shall be able to solve the word problem in P. The second step is to determine the kernel of π, which is a $\mathbb{Z}[H]$-submodule of Q.

11.2 Abelian quotients

Let G be a finitely presented group, let p be a prime, and let e be a positive integer. The first step in using the p-quotient algorithm to compute $G/\varphi_{e+1}(G)$ is to determine $G/\varphi_2(G)$, the largest elementary abelian p-group which is a quotient of G. Similarly, the first step in using the nilpotent quotient algorithm to compute $G/\gamma_{e+1}(G)$ or the polycyclic quotient algorithm to compute $G/G^{(e)}$ is to determine $G/\gamma_2(G) = G/G'$, the largest abelian quotient of G. This section describes how to find these abelian quotients.

Let $X = \{x_1, \ldots, x_n\}$ be an finite set of cardinality n and let $F = F_X$, the free group on X. The map of X into $\mathbb{Z}^n$ which takes x_i to the i-th standard basis element e_i of $\mathbb{Z}^n$ defines a homomorphism f from F to $\mathbb{Z}^n$. If U is the word $x_{j_1}^{\alpha_1} \ldots x_{j_k}^{\alpha_k}$, then the i-th component of $f([U])$ is the i-th *exponent sum* of U, the sum of those α_m such that $j_m = i$. Clearly f is surjective.

Proposition 2.1. *The kernel of f is F'.*

Proof. Since $\mathbb{Z}^n$ is abelian, the kernel of f contains F'. Therefore f defines a homomorphism $\overline{f}$ of F/F' onto $\mathbb{Z}^n$. Since $\mathbb{Z}^n$ is a free abelian group with basis $e_1, \ldots, e_n$, there is a homomorphism g from $\mathbb{Z}^n$ to F/F' such that $g(e_i) = F'[x_i]$. The compositions $\overline{f} \circ g$ and $g \circ \overline{f}$ are both the identity. Therefore $\overline{f}$ and g are isomorphisms. $\square$

In a similar way we can define a homomorphism of F onto $\mathbb{Z}_p^n$ with kernel $\varphi_2(F)$. We simply consider the exponent sums modulo p.

Now let N be a normal subgroup of F and set $G = F/N$. Then $G' = F'N/N$ and $G/G' = (F/N)/(F'N/N)$ is isomorphic to $F/F'N$, which is isomorphic to $(F/F')/(F'N/F')$. By Proposition 2.1, G/G' is isomorphic to $\mathbb{Z}^n/f(N)$. Suppose that $G = \mathrm{Grp}\langle X \mid \mathcal{R} \rangle$, so N is the normal closure in F of the elements $[S^{-1}R]$ with (R, S) in $\mathcal{R}$. Since homomorphisms map conjugates to conjugates, it follows that $f(N)$ is the normal closure in $\mathbb{Z}^n$ of the set of images $f([S^{-1}R])$. But $\mathbb{Z}^n$ is abelian and all subgroups are normal. Thus $f(N)$ is simply the subgroup generated by the elements $f([S^{-1}R])$. If $\mathcal{R}$ is finite and consists of the pairs (R_i, S_i), $1 \le i \le m$, then G/G' is isomorphic to $\mathbb{Z}^n/M$, where M is the subgroup of $\mathbb{Z}^n$ generated by the rows of the m-by-n matrix A such that A_{ij} is the exponent sum of x_j in $S_i^{-1}R_i$. The invariant factors of A determine the isomorphism type of G/G'. The matrix A is called the matrix of *abelianized relations*.

By a similar argument, we can show that $G/\varphi_2(G)$ is isomorphic to $\mathbb{Z}_p^n/M_1$, where M_1 is the subspace of $\mathbb{Z}_p^n$ generated by the rows of A taken modulo p. To determine $G/\varphi_2(G)$, one needs only row operations modulo p. To get the best description of G/G', integer row and column operations are needed. Here is the procedure for finding $G/\varphi_2(G)$.

Procedure ABEL_PQUOT($X, \mathcal{R}, p; Y, \mathcal{S}, \sigma, \tau$);
  Input: X : a finite set $\{x_1, \ldots, x_n\}$;
           $\mathcal{R}$: a set of ordered pairs (R_i, S_i), $1 \le i \le m$, of elements
                     from $X^{\pm *}$;
           p : a prime;
  Output: Y : a finite set;
           $\mathcal{S}$: a finite set of ordered pairs of elements from Y^* such
                     that $K = \mathrm{Mon}\langle Y \mid \mathcal{S} \rangle$ is isomorphic to $G/\varphi_2(G)$,
                     where $G = \mathrm{Grp}\langle X \mid \mathcal{R} \rangle$, and $(Y, \mathcal{S})$ is a consistent
                     exponent-p power-commutator presentation;
           σ : a homomorphism from $X^{\pm *}$ to Y^* defining a
                     homomorphism of G onto K;
           τ : a homomorphism from Y^* to $X^{\pm *}$ such that U and
                     $\sigma(\tau(U))$ define the same element of K for all U in Y^*;
  Begin
     Let B be the m-by-n matrix such that B_{ij} is the exponent sum of x_j
        in $S_i^{-1}R_i$;
     Let A be the row Hermite normal form modulo p of B;
     Let the corner entries of A occur in the columns with
        indices $c_1 < \cdots < c_s$;
     $r := n - s$; $Y := \{a_1, \ldots, a_r\}$;
     Let $d_1 < \cdots < d_r$ be the indices of the columns in A which do not
        contain corner entries;
     Let $\mathcal{S}$ consist of the relations $a_j a_i = a_i a_j$, $1 \le i < j \le r$, and
        $a_i^p = 1$, $1 \le i \le r$;

For $i := 1$ to n do

    If $i = d_j$ then $\sigma(x_i) := a_j$

    Else if $i = c_j$ then $\sigma(x_i) := a_1^{-A_{jd_1}} \ldots a_r^{-A_{jd_r}}$;

(∗ Here the exponents are taken modulo p. ∗)

    For $i := 1$ to r do $\tau(a_i) := x_{d_i}$

End.

Example 2.1. Let $G = \mathrm{Grp}\,\langle x, y, z \mid xy^2z^2xyz^2 = x^2y^3z^5x^2y^4z^5 = 1\rangle$. The matrix of abelianized relations is

$$B = \begin{bmatrix} 2 & 3 & 4 \\ 4 & 7 & 10 \end{bmatrix}.$$

The row Hermite normal form modulo 3 of B is

$$A = \begin{bmatrix} 1 & 0 & 2 \\ 0 & 1 & 2 \end{bmatrix}.$$

In the notation of ABEL_PQUOT, $s = 2$, $r = 1$, $c_1 = 1$, $c_2 = 2$, and $d_1 = 3$. The largest elementary abelian 3-group which is a homomorphic image of G is given by $K = \mathrm{Mon}\,\langle a_1 \mid a_1^3 = 1\rangle$. The homomorphisms σ and τ satisfy

$$\sigma(x) = \sigma(y) = \sigma(z) = a_1, \quad \tau(a_1) = z.$$

Modulo 2, the row Hermite normal form of B is

$$\begin{bmatrix} 0 & 1 & 0 \\ 0 & 0 & 0 \end{bmatrix},$$

so the largest elementary abelian 2-group which is a quotient of G is $K = \mathrm{Mon}\langle a_1, a_2 \mid a_2 a_1 = a_1 a_2, a_1^2 = a_2^2 = 1\rangle$. In this case, the homomorphisms σ and τ are given by

$$\sigma(x) = a_1, \quad \sigma(y) = \varepsilon, \quad \sigma(z) = a_2, \quad \tau(a_1) = x, \quad \tau(a_2) = z.$$

Example 2.2. Let G be the group generated by b, c, d, and e subject to the relations

$$b^3 = 1, \quad d^3 = 1, \quad bcb^{-1}c^2 = 1, \quad c^{-1}d^{-1}cd = 1, \quad bcde^{-1}dbdcbe^{-1}d = 1.$$

The abelianized relation matrix for G is

$$B = \begin{bmatrix} 3 & 0 & 0 & 0 \\ 0 & 0 & 3 & 0 \\ 0 & 3 & 0 & 0 \\ 0 & 0 & 0 & 0 \\ 3 & 2 & 4 & -2 \end{bmatrix}.$$

Let us take $p = 3$ and determine the largest elementary abelian 3-group which is a quotient of G. The row Hermite normal form modulo 3 of B is

$$A = \begin{bmatrix} 0 & 1 & 2 & 2 \\ 0 & 0 & 0 & 0 \\ 0 & 0 & 0 & 0 \\ 0 & 0 & 0 & 0 \\ 0 & 0 & 0 & 0 \end{bmatrix}.$$

Therefore $Y = \{a_1, a_2, a_3\}$ and $\mathcal{S}$ consists of the relations

$$a_j a_i = a_i a_j, \quad 1 \le i < j \le 3,$$
$$a_i^3 = 1, \quad 1 \le i \le 3.$$

The homomorphisms σ and τ are given by

$$\sigma(b) = a_1, \quad \sigma(c) = a_2 a_3, \quad \sigma(d) = a_2, \quad \sigma(e) = a_3,$$
$$\tau(a_1) = b, \quad \tau(a_2) = d, \quad \tau(a_3) = e.$$

The simplest procedure for computing G/G' is based on ABEL_PQUOT and uses integer row operations to compute the row Hermite normal form of the abelianized relation matrix. Unfortunately, this does not usually describe G/G' as a direct product of cyclic groups. For example, the row Hermite normal form of the abelianized relation matrix B in Example 2.1 is

$$\begin{bmatrix} 2 & 0 & -2 \\ 0 & 1 & 2 \end{bmatrix}.$$

This means that G/G' is isomorphic to $\mathrm{Grp}\langle a_1, a_2 \mid a_2 a_1 = a_1 a_2, a_1^2 = a_2^2 \rangle$. This presentation is consistent, but does not immediately describe G/G' as a direct product of cyclic groups. To obtain this decomposition we must use both row and column operations. The Smith normal form of B is

$$\begin{bmatrix} 1 & 0 & 0 \\ 0 & 2 & 0 \end{bmatrix},$$

so G/G' is isomorphic to $\mathbb{Z}_2 \times \mathbb{Z}$.

Here is a version of the abelian quotient procedure which gives the answer as a direct product of cyclic groups.

Procedure ABEL_QUOT$(X, \mathcal{R}; Y, \mathcal{S}, \sigma, \tau)$;

Input: X : a finite set $\{x_1, \ldots, x_n\}$;

           $\mathcal{R}$: a set of ordered pairs (R_i, S_i), $1 \le i \le m$, of elements from $X^{\pm*}$;

Output: Y : a finite set;

           $\mathcal{S}$: a finite set of ordered pairs of elements from $Y^{\pm*}$ such that $K = \mathrm{Grp}\,\langle Y \mid \mathcal{S} \rangle$ is isomorphic to G/G', where $G = \mathrm{Grp}\,\langle X \mid \mathcal{R} \rangle$, and $(Y, \mathcal{S})$ is a consistent polycyclic presentation;

           σ : a regular homomorphism from $X^{\pm*}$ to $Y^{\pm*}$ defining a homomorphism of G onto K;

           τ : a regular homomorphism from $Y^{\pm*}$ to $X^{\pm*}$ such that U and $\sigma(\tau(U))$ define the same element of K for all U in $Y^{\pm*}$;

Begin

    Let B be the m-by-n matrix such that B_{ij} is the exponent sum of x_j in $S_i^{-1} R_i$;

    Let A be the Smith normal form of B and let P and Q be unimodular matrices such that $A = PBQ$;

    Let $d_1, \ldots, d_r$ be the nonzero diagonal entries of A;

    If $d_1 > 1$ then $s := 0$

    Else let s be maximal such that $d_s = 1$;

    $t := n - s$; $Y := \{a_1, \ldots, a_t\}$;

    Let $\mathcal{S}$ consist of the relations $a_j a_i = a_i a_j$, $1 \le i < j \le t$, and $a_i^{d_{s+i}} = 1$, $1 \le i \le r - s$;

    For $i := 1$ to n do $\sigma(x_i) := a_1^{Q_{is+1}} \ldots a_t^{Q_{in}}$;

    $T := Q^{-1}$;

    For $i := 1$ to t do $\tau(a_i) := x_1^{T_{s+i1}} \ldots x_n^{T_{s+in}}$

End.

Example 2.3. Let $G = \mathrm{Grp}\,\langle x, y \mid (x^2 y^3)^3 = (x^3 y)^4 = 1 \rangle$. The abelianized relation matrix is

$$B = \begin{bmatrix} 6 & 9 \\ 12 & 4 \end{bmatrix}.$$

The Smith normal form for B is

$$A = \begin{bmatrix} 1 & 0 \\ 0 & 84 \end{bmatrix},$$

so G/G' is cyclic of order 84. Thus $Y = \{a_1\}$ and S consists of the single relation $a_1^{84} = 1$. One choice for unimodular matrices P and Q such that $A = PBQ$ is

$$P = \begin{bmatrix} 1 & -2 \\ -4 & 9 \end{bmatrix}, \quad Q = \begin{bmatrix} 0 & 1 \\ 1 & 18 \end{bmatrix}.$$

In this case,

$$Q^{-1} = \begin{bmatrix} -18 & 1 \\ 1 & 0 \end{bmatrix},$$

and the regular homomorphisms σ and τ are given by

$$\sigma(x) = a_1, \quad \sigma(y) = a_1^{18}, \quad \tau(a_1) = x.$$

Example 2.4. The following presentation is taken from (Neubüser & Sidki 1988). Let S be the group generated by t_1, t_2, t_3, and t_4 subject to the relations

$$t_1 t_2 t_1^{-1} t_2^{-1} = 1, \quad t_1^5 t_2^{-2} = 1,$$
$$t_1^2 t_3 t_1^{-2} t_4 t_3^{-1} t_1 t_4^{-1} = 1, \quad t_1 t_2 t_4 t_1^{-1} t_3 t_4^{-1} t_2^{-1} t_1^{-1} t_3^{-1} = 1,$$
$$t_1^3 t_2 t_4 t_1^{-3} t_2^{-1} t_4^{-1} = 1.$$

The matrix of abelianized relations is

$$B = \begin{bmatrix} 0 & 0 & 0 & 0 \\ 5 & -2 & 0 & 0 \\ 1 & 0 & 0 & 0 \\ -1 & 0 & 0 & 0 \\ 0 & 0 & 0 & 0 \end{bmatrix}.$$

The Smith normal form of B is

$$A = \begin{bmatrix} 1 & 0 & 0 & 0 \\ 0 & 2 & 0 & 0 \\ 0 & 0 & 0 & 0 \\ 0 & 0 & 0 & 0 \\ 0 & 0 & 0 & 0 \\ 0 & 0 & 0 & 0 \end{bmatrix},$$

and A can be computed from B using row operations alone. Thus S/S' is isomorphic to $\mathbb{Z}_2 \oplus \mathbb{Z} \oplus \mathbb{Z}$ and the unimodular matrix Q can be chosen to be the identity matrix. If this is done, then the presentation for K is

$$a_j a_i = a_i a_j, \quad 1 \le i < j \le 3,$$
$$a_i^2 = 1,$$

and the homomorphisms σ and τ are given by

$$\sigma(t_1) = 1, \quad \sigma(t_2) = a_1, \quad \sigma(t_3) = a_2, \quad \sigma(t_4) = a_3,$$
$$\tau(a_1) = t_2, \quad \tau(a_2) = t_3, \quad \tau(a_3) = t_4.$$

A major source of presentations is the Reidemeister-Schreier algorithm. Presentations obtained this way tend to have many generators and many relations. Computing the Smith normal form of the abelianized relation matrix and finding a unimodular matrix Q can be quite difficult, despite the array of techniques discussed in Chapter 8.

If it is known that the output of the Reidemeister-Schreier algorithm is going to be fed directly into ABEL_QUOT or ABEL_PQUOT, then it may be possible to save a great deal of space in the Reidemeister-Schreier portion of the computation. In the procedure HLT_X of Section 6.3, all words in Y^* can be abelianized immediately. Words in Y^* occur as secondary labels on the edges of the extended Schreier automaton and in the coincidence procedure. Since the abelianization of a word W takes no more space, and frequently much less space, than W, the potential for saving memory in this abelianized Reidemeister-Schreier algorithm is substantial.

The techniques just described can in principle be generalized to provide an algorithm for determining, for any positive integer e, a consistent polycyclic presentation for $G/\gamma_{e+1}(G)$, the largest quotient group of G which is nilpotent of class not exceeding e. In Section 9.10 we described how to get a consistent polycyclic presentation for the free nilpotent group $Q = F/\gamma_{e+1}(F)$ and how to construct the natural homomorphism f from F to Q. The quotient $G/\gamma_{e+1}(G)$ is isomorphic to Q/N, where N is the normal closure in Q of the set of elements $f([S^{-1}R])$ with (R, S) in $\mathcal{R}$. The algorithms of Sections 9.5 and 9.6 allow us to find a consistent polycyclic presentation for Q/N.

Example 2.5. Let us describe $G/\gamma_3(G)$, where $G = \mathrm{Grp}\langle x, y \mid (xy)^4 = (x^2y^2x^{-1}y^{-1})^4 = 1\rangle$. Let F be the free group on x and y. Then $F/\gamma_3(F)$ is isomorphic to $Q = \mathrm{Grp}\langle a, b, c \mid ca = ac, cb = bc, ba = abc\rangle$, and $x \mapsto a$ and $y \mapsto b$ defines a homomorphism f of F onto Q. The image of $(xy)^4$ under f is $(ab)^4 = a^4b^4c^6$. The image of $(x^2y^2x^{-1}y^{-1})^4$ is $(a^2b^2a^{-1}b^{-1})^4 = (abc^{-2})^4 = a^4b^4c^{-2}$. The commutator $[a^4b^4c^6, a]$ is c^4. It is easy to check that the normal closure in Q of $a^4b^4c^6$ and $a^4b^4c^{-2}$ is generated as a subgroup by $a^4b^4c^2$ and c^4. Thus $G/\gamma_3(G)$ is isomorphic to the group generated by a, b, and c subject to the relations

$$ca = ac, \quad cb = bc, \quad ba = abc,$$
$$c^4 = 1, \quad a^4 = b^{-4}c^2,$$

which form a consistent nilpotent presentation.

Although the procedure illustrated in Example 2.5, which one could call the *free nilpotent approach* to nilpotent quotients, is valid for any e, it is practical only for small e. The problem is that the number of generators in the polycyclic presentation for Q grows very rapidly with e. It could well happen that $G/\gamma_{e+1}(G)$ has a polycyclic presentation on relatively few generators, but the presentation for Q is too large to work with. An alternative approach to computing nilpotent quotients will be described in Sections 11.4 and 11.5. There is no analogue of the free nilpotent approach for polycyclic quotients because there are no such things as free polycyclic groups of a given derived length greater than 1. If $e \geq 2$, then $F/F^{(e)}$ is not polycyclic and has no largest polycyclic quotient.

Exercises

2.1. Let G be the group generated by x, y, z and defined by the relations

$$(xy^{-1}z)^4 = (x^2y^{-1}z^2)^6 = 1.$$

Find the structure of the largest elementary abelian p-quotient of G for $p = 2$, 3, and 5. Describe G/G' as a direct sum of cyclic groups.

2.2. Using the method of Example 2.5, find the largest class-2 nilpotent quotient of the group G in Exercise 2.1.

11.3 Extensions of modules

Let H be a group given by a finite presentation and let M be a right $\mathbb{Z}[H]$-module. In order to be able to determine polycyclic quotients, we shall need to be able to decide whether it is possible to construct a group G with the following properties:

(1) M, considered as an abelian group, is a normal subgroup of G and G/M is isomorphic to H.
(2) The module structure on M induced from G is the same as the original structure.
(3) Certain additional relations to be described are satisfied.

If the answer is negative, then we shall want to find the largest quotient module of M for which the construction is possible.

To be more precise, let H be given as $\mathrm{Grp}\langle X \mid R_1 = S_1, \ldots, R_s = S_s \rangle$ and let M contain elements $u_1, \ldots, u_s$. In what follows, we shall give a presentation for a group G with $X \cup M$ as the set of generators. To reduce the possibility of confusion between multiplication in G and the module action of H on M, elements of M will be enclosed in brackets. Within these brackets, multiplication denotes the module action. Outside the brackets, multiplication is the group operation in G. Thus if v is in M and x is in X, then the expression $[vx]$ denotes the element of M which is the product of v and the element of H represented by x, while $[v]x$ is the product of two generators of G.

The defining relations for G are as follows:

(i) $[v][w] = [v + w]$ for all v and w in M.
(ii) $[v]x = x[vx]$ for all v in M and all x in X.
(iii) $R_i = S_i[u_i]$, $1 \le i \le s$.

The group G will be called the *formal extension* of M by H relative to the given presentation of H and the choice of $u_1, \ldots, u_s$. Let φ be the map from M to G taking v to the element of G represented by $[v]$ and let N be the image of M under φ.

Proposition 3.1. *The set N is a normal abelian subgroup of G and G/N is isomorphic to H.*

Proof. The relations (i) show that N is closed under products and inverses and that any two elements of N commute. Therefore N is an abelian subgroup of G and φ is a homomorphism of groups. Let x be in X and let w be in M. By the relations (ii), the conjugate $x^{-1}[w]x$ is in N. To prove that N is normal we must prove that the conjugate $x[w]x^{-1}$ is also in N. Applying (ii) with $v = wx^{-1}$, we have

$$[wx^{-1}]x = x[wx^{-1}x] = x[w],$$

or $x[w]x^{-1} = [wx^{-1}]$. Therefore N is normal in G. If we add the relations $[v] = 1$ for all v in M, then, after Tietze transformations, we are left with the presentation for H. Thus G/N is isomorphic to H. $\square$

There is one case in which it is easy to see that N is isomorphic to M.

Proposition 3.2. *If $u_i = 0$, $1 \le i \le s$, then φ is an isomorphism.*

Proof. Let us define a binary operation on $H \times M$ by the formula

$$(h_1, v_1)(h_2, v_2) = (h_1 h_2, v_1 h_2 + v_2).$$

Here $v_1 h_2$ represents the module action of the element h_2 on v_1 and $+$ is the addition in M. It is not difficult to show that this binary operation makes $H \times M$ into a group K called the *split extension* of M by H. The identity of K is $(1_H, 0_M)$ and the inverse of (h, v) is $(h^{-1}, -vh^{-1})$. If each of the u_i is 0, then there is a homomorphism of the formal extension G of M by H onto K. We map v in M to $(1, v)$ and x in X to $(x, 0)$. In fact, this homomorphism is an isomorphism. In particular, no element v of M maps to the identity in G. $\square$

In general, N is not isomorphic to M. Using the isomorphism of G/N and H, we can make N into a right $\mathbb{Z}[H]$-module. The relations (ii) say that φ is a module homomorphism. Thus the kernel M_0 of φ is a submodule of M. Determining M_0 can be difficult. However, when H is polycyclic and the presentation for H is a consistent polycyclic one, then it is possible to describe a finite module generating set for M_0. The simplest case occurs when H is abelian.

Let us assume that H is generated by $a_1, \ldots, a_n$ subject to the relations

$$a_j a_i = a_i a_j, \quad 1 \le i < j \le n,$$
$$a_i^{m_i} = 1, \quad i \in I.$$

Here I is a subset of $\{1, \ldots, n\}$ and for each i in I the integer m_i is positive. Let M be a right $\mathbb{Z}[H]$-module containing elements v_{ij}, $1 \le i < j \le n$, and u_i with i in I. If $i > j$, set $v_{ij} = -v_{ji}$. Let G be the formal extension of M by H in which

$$a_j a_i = a_i a_j [v_{ij}], \quad 1 \le i < j \le n,$$
$$a_i^{m_i} = [u_i], \quad i \in I.$$

The following proposition is taken from Section III.8 of [Zassenhaus 1958].

Proposition 3.3. *The image in G of M is isomorphic to M/M_0, where M_0 is the $\mathbb{Z}[H]$-submodule of M generated by the following elements:*

(1) *The elements $u_i(a_i - 1)$, where i is in I.*
(2) *The elements*

$$u_j(a_i - 1) - v_{ij}(1 + a_j + \cdots + a_j^{m_j - 1}),$$

where j is in I and $i \ne j$.
(3) *The elements*

$$v_{ij}(a_k - 1) + v_{jk}(a_i - 1) + v_{ki}(a_j - 1),$$

where $1 \leq i < j < k \leq n$.

Proof. Let φ be the homomorphism from M into G. Suppose that i is in I. In G, the element a_i commutes with $a_i^{m_i} = [u_i]$, so

$$a_i[u_i] = [u_i]a_i = a_i[u_i a_i].$$

Therefore $[u_i a_i] = [u_i]$, so $u_i(a_i - 1)$ is in the kernel M_0 of φ. The elements of type (2) come from evaluating $a_j^{m_j} a_i$ with $i \neq j$. First, note that since $a_j a_i = a_i a_j [v_{ij}]$ when $i < j$, we have $a_i a_j = a_j a_i [-v_{ij}] = a_j a_i [v_{ji}]$. Therefore $a_j a_i = a_i a_j [v_{ij}]$ whenever $i \neq j$. Now

$$a_j^{m_j} a_i = [u_j]a_i = a_i[u_j a_i].$$

But we also have

$$
\begin{aligned}
a_j^m a_i &= a_j^{m_j-1} a_i a_j [v_{ij}] = a_j^{m_j-2} a_i a_j [v_{ij}] a_j [v_{ij}] \\
&= a_j^{m_j-2} a_i a_j^2 [v_{ij}(1+a_j)] \\
&= a_j^{m_j-3} a_i a_j [v_{ij}] a_j^2 [v_{ij}(1+a_j)] \\
&= a_j^{m_j-3} a_i a_j^3 [v_{ij}(1+a_j+a_j^2)] = \cdots \\
&= a_i a_j^{m_j} [v_{ij}(1+a_j+\cdots+a_j^{m_j-1})] \\
&= a_i [u_j][v_{ij}(1+a_j+\cdots+a_j^{m_j-1})] \\
&= a_i [u_j + v_{ij}(1+a_j+\cdots+a_j^{m_j-1})].
\end{aligned}
$$

Therefore

$$[u_j a_i] = [u_j + v_{ij}(1+a_j+\cdots+a_j^{m_j-1})],$$

so

$$u_j(a_i - 1) - v_{ij}(1+a_j+\cdots+a_j^{m_j-1})$$

is in M_0.

The elements of type (3) come from evaluating $a_k a_j a_i$ in two different ways:

$$
\begin{aligned}
a_k a_j a_i &= a_j a_k [v_{jk}]a_i = a_j a_k a_i [v_{jk} a_i] = a_j a_i a_k [v_{ik}][v_{jk}a_i] \\
&= a_j a_i a_k [v_{ik} + v_{jk}a_i] = a_i a_j [v_{ij}]a_k [v_{ik} + v_{jk}a_i] \\
&= a_i a_j a_k [v_{ij}a_k][v_{ik} + v_{jk}a_i] = a_i a_j a_k [v_{ij}a_k + v_{ik} + v_{jk}a_i].
\end{aligned}
$$

On the other hand,

$$a_k a_j a_i = a_k a_i a_j [v_{ij}] = a_i a_k [v_{ik}] a_j [v_{ij}] = a_i a_k a_j [v_{ik} a_j][v_{ij}]$$
$$= a_i a_k a_j [v_{ik} a_j + v_{ij}] = a_i a_j a_k [v_{jk}][v_{ik} a_j + v_{ij}]$$
$$= a_i a_j a_k [v_{jk} + v_{ik} a_j + v_{ij}].$$

Hence

$$[v_{ij} a_k + v_{ik} + v_{jk} a_i] = [v_{jk} + v_{ik} a_j + v_{ij}],$$

so

$$v_{ij}(a_k - 1) + v_{jk}(a_i - 1) - v_{ik}(a_j - 1)$$

is in the kernel of φ. Replacing v_{ik} by $-v_{ki}$ gives the element in (3).

Now suppose that all of the elements (1), (2), and (3) are zero in M. We shall prove that φ is an isomorphism. If $n = 0$, then G is clearly isomorphic to M. Thus we may assume that $n > 0$. As should come as no surprise, we have two ways of proceeding. We could show that the rules

$$[v][w] \to [v+w], \quad v, w \in M,$$
$$[v]^{-1} \to [-v], \quad v \in M,$$
$$[v]x \to x[vx], \quad v \in M, \ x \in \{a_1, \ldots, a_n\}^{\pm},$$
$$a_i a_i^{-1} \to \epsilon, \quad 1 \le i \le n, \ i \notin I,$$
$$a_i^{-1} a_i \to \epsilon, \quad 1 \le i \le n, \ i \notin I,$$
$$a_i^{m_i} \to [u_i], \quad i \in I,$$
$$a_i^{-1} \to a_i^{m_i-1}[-u_i], \quad i \in I,$$
$$a_j a_i \to a_i a_j [v_{ij}], \quad 1 \le i < j \le n,$$
$$a_j^{-1} a_i \to a_i a_j^{-1}[-v_{ij} a_j^{-1}], \quad 1 \le i < j \le n, \ j \notin I,$$
$$a_j a_i^{-1} \to a_i^{-1} a_j [-v_{ij} a_i^{-1}], \quad 1 \le i < j \le n, \ i \notin I,$$
$$a_j^{-1} a_i^{-1} \to a_i^{-1} a_j^{-1}[v_{ij} a_i^{-1} a_j^{-1}], \quad 1 \le i < j \le n, \ i, j \notin I,$$

form a confluent rewriting system $\mathcal{S}$ (What ordering is being used here?) and that $(X \cup M, \mathcal{S}, \mathcal{I})$ is a restricted presentation for a group G, where $\mathcal{I}$ is the ideal of $(X \cup M)^{\pm*}$ generated by M. It would be possible to avoid the use of a restricted presentation by adding the rule $[0] \to \epsilon$. However, with $\mathcal{S}$ as given, it is possible to describe canonical forms for the elements of G in a very natural way. Every element of $\mathcal{I}$ can be reduced using $\mathcal{S}$ to a unique word of the form $W[u]$, where u is in M and W is a word in $X^{\pm*}$

which is reduced with respect to the standard polycyclic rewriting system for H relative to the polycyclic generating sequence $a_1, \ldots, a_n$.

The alternative approach is to consider the subgroup K of H generated by $a_2, \ldots, a_n$, assume by induction that M embeds isomorphically in the formal extension F of M by K defined by the relations

$$a_j a_i = a_i a_j [v_{ij}], \quad 2 \le i < j \le n,$$
$$a_i^{m_i} = [u_i], \quad i \in (I - \{1\}),$$

and then show that the relations involving a_1 define a cyclic extension of F. Let us sketch briefly the second approach.

Conjugation by a_1 defines the following map σ of the generators of F:

$$[v]^\sigma = [v a_1], \quad v \in M,$$
$$a_i^\sigma = a_i [v_{1i}], \quad 2 \le i \le n.$$

We must show that these images satisfy the defining relations for F. For v and w in M,

$$[v]^\sigma [w]^\sigma = [v a_1][w a_1] = [v a_1 + w a_1] = [(v + w) a_1] = [v + w]^\sigma,$$

since M is a $\mathbb{Z}[H]$-module. For $2 \le i \le n$ and v in M,

$$[v]^\sigma a_i^\sigma = [v a_1] a_i [v_{1i}] = a_i [v a_1 a_i + v_{1i}] = a_i [v_{1i} + v a_i a_1]$$
$$= a_i [v_{1i}][v a_i a_1] = a_i^\sigma [v a_i]^\sigma,$$

since H is commutative and $[v a_1 a_i] = [v a_i a_1]$. For i in $I - \{1\}$,

$$(a_i^\sigma)^{m_i} = (a_i [v_{1i}])^{m_i} = a_i^{m_i} [v_{1i}(1 + a_i + \cdots + a_i^{m_i - 1})]$$
$$= [u_i][v_{1i}(1 + a_i + \cdots + a_i^{m_i - 1})]$$
$$= [u_i + v_{1i}(1 + a_i + \cdots + a_i^{m_i - 1})] = [u_i a_1] = [u_i]^\sigma,$$

since elements of type (2) are 0. For $2 \le i < j \le n$,

$$a_j^\sigma a_i^\sigma = a_j [v_{1j}] a_i [v_{1i}] = a_j a_i [v_{1j} a_i + v_{1i}] = a_i a_j [v_{ij}][v_{1j} a_i + v_{1i}]$$
$$= a_i a_j [v_{ij} + v_{1j} a_i + v_{1i}] = a_i a_j [v_{1i} a_j + v_{1j} + v_{ij} a_1]$$
$$= a_i [v_{1i}] a_j [v_{1j}][v_{ij} a_1] = a_i^\sigma a_j^\sigma [v_{ij}]^\sigma,$$

since elements of type (3) are 0.

Since the defining relations of F are satisfied by the images under σ of the generators of F, we may extend σ to be a homomorphism of F into itself. Perhaps the easiest way to show that σ is an automorphism is to show that σ^{-1} is defined by the following map on the generators:

$$[v]^{\sigma^{-1}} = [va_1^{-1}], \quad v \in M,$$
$$a_i^{\sigma^{-1}} = a_i[v_{1i}a_1^{-1}], \quad 2 \leq i \leq n.$$

If 1 is not in I, then we are done once we know that σ is an automorphism, for then G is a cyclic extension of F, and since M is embedded isomorphically in F, it follows that M is embedded isomorphically in G.

Now suppose that 1 is in I. We must check that σ fixes $a_1^{m_1} = [u_1]$ and that σ^{m_1} is the inner automorphism of F induced by $[u_1]$. Now $[u_1]^\sigma = [u_1a_1] = [u_1]$ since $u_1(a_1 - 1) = 0$. If $2 \leq i \leq n$, then

$$a_i^{\sigma^{m_1}} = a_i[v_{1i}][v_{1i}a_1]\ldots[v_{1i}a_1^{m_1-1}] = a_i[-v_{i1}(1 + a_1 + \cdots + a_1^{m_1-1})]$$
$$= a_i[u_1(1 - a_i)] = [u_1]^{-1}a_i[u_1],$$

and, if v is in M, then

$$[v]^{\sigma^{m_1}} = [va_1^{m_1}] = [v] = [-u_1 + v + u_1] = [u_1]^{-1}[v][u_1].$$

Thus we have a cyclic extension in this case too, so M is embedded isomorphically in G.

In general, let M_1 be the submodule of M generated by the elements of types (1), (2), and (3). Then M_1 is contained in M_0. However, in the formal extension of M/M_1 by H, the quotient M/M_1 embeds isomorphically. Therefore M/M_1 must be a quotient of M/M_0. It follows that $M_1 = M_0$.

□

When we compute nilpotent quotients, we shall extend $\mathbb{Z}[H]$-modules M on which H acts trivially, modules in which $vh = v$ for all v in M and all h in H. In this situation, Proposition 3.3 can be simplified considerably.

Proposition 3.4. *Suppose that in Proposition* 3.3 *the group H acts trivially on M. Then M_0 is generated as an abelian group by the elements m_iv_{ij} and m_jv_{ij}, where $1 \leq i < j \leq n$ and $i \in I$ in the first case and $j \in I$ in the second.*

Proof. The elements of types (1) and (3) in Proposition 3.3 are trivial and the elements of type (2) become simply $-m_jv_{ij}$, $i \neq j$. Since $m_iv_{ij} = -m_iv_{ji}$, the proposition follows. □

Example 3.1. Suppose that G is a group such that G/G' is isomorphic to $H = \mathbb{Z}_4 \oplus \mathbb{Z}_6 \oplus \mathbb{Z}$. How big can $G'/\gamma_3(G)$ be? Let a_1, a_2, and a_3 be elements of G which map, respectively, onto generators of the direct summands $\mathbb{Z}_4$,

$\mathbb{Z}_6$, and $\mathbb{Z}$ of H. By Proposition 2.5 in Chapter 9, $G'/\gamma_3(G)$ is generated by the images v_{ij} of $[a_j, a_i]$, $1 \leq i < j \leq 3$. By Proposition 3.4, the only conditions which the v_{ij} must satisfy are $4v_{12} = 4v_{13} = 6v_{12} = 6v_{23} = 0$. Therefore $2v_{12} = 4v_{13} = 6v_{23} = 0$ and $G'/\gamma_3(G)$ is a quotient group of $\mathbb{Z}_2 \oplus \mathbb{Z}_4 \oplus \mathbb{Z}_6$. The images of a_1^4 and a_2^6 may be arbitrary elements of Grp $\langle v_{12}, v_{13}, v_{23} \rangle$. Therefore a presentation such as

$$a_1^4 = a_2^6 = v_{12}^2 = v_{13}^4 = v_{23}^6 = 1,$$

$$a_2 a_1 = a_1 a_2 v_{12}, \quad a_3 a_1 = a_1 a_3 v_{13}, \quad a_3 a_2 = a_2 a_3 v_{23},$$

$$v_{ij} a_k = a_k v_{ij}, \quad 1 \leq i < j \leq 3, \quad 1 \leq k \leq 3,$$

$$v_{13} v_{12} = v_{12} v_{13}, \quad v_{23} v_{12} = v_{12} v_{23}, \quad v_{23} v_{13} = v_{13} v_{23},$$

can immediately be recognized as consistent.

Suppose now that H is an arbitrary polycyclic group and that the presentation $(X, \mathcal{R})$ for H satisfies the following conditions:

(i) $X = \{a_1, \ldots, a_n\}$, where $a_1, \ldots, a_n$ is a polycyclic generating sequence for H.

(ii) The relations in $\mathcal{R}$ are $V_k = W_k$, $1 \leq k \leq s$, and these are the relations in the standard polycyclic group presentation for H with respect to $a_1, \ldots, a_n$ which have positive left sides. (By Proposition 8.2 in Chapter 9, these relations define H.)

We shall denote by I the set of indices i such that there is a relation with left side $a_i^{m_i}$ in $\mathcal{R}$. Let M be a right $\mathbb{Z}[H]$-module containing elements $u_1, \ldots, u_s$ and let G be the formal extension of M by H in which $V_k = W_k[u_k]$, $1 \leq k \leq s$. Set $Y = X \cup M$ and $Y_i = \{a_i, \ldots, a_n\} \cup M$, $1 \leq i \leq n+1$, and let $\mathcal{I}$ be the ideal of $Y^{\pm*}$ generated by M. Starting with the rules

$$[v][w] \rightarrow [v+w], \quad v, w \in M,$$

$$[v]^{-1} \rightarrow [-v], \quad v \in M,$$

$$xx^{-1} \rightarrow \epsilon, \quad x \in X^{\pm}$$

$$[v]x \rightarrow x[vx], v \in M, \quad x \in X^{\pm},$$

$$V_k \rightarrow W_k[u_k], \quad 1 \leq k \leq s,$$

we can construct a (not necessarily confluent) rewriting system $\mathcal{S}$ for G with the following property: Let W be in $\mathcal{I}$. Then W can be rewritten using $\mathcal{S}$ into the form $U[u]$, where U is a word in $X^{\pm*}$ which is collected with respect to $\mathcal{R}$ and u is in M.

To produce the rest of the rewriting system $\mathcal{S}$, we must find, for $i < j$, rules with left sides

$$a_j^{-1}a_i, \quad j \notin I,$$
$$a_j a_i^{-1}, \quad i \notin I,$$
$$a_j^{-1}a_i^{-1}, \quad i,j \notin I,$$

and the rules with left sides a_i^{-1}, where i is in I. The proof of Proposition 8.2 in Chapter 9 serves as a guide for the construction of these additional rules.

Procedure EXTEND_RULES($X, \mathcal{R}, M, u_1, \ldots, u_s; \mathcal{S}$);
(∗ The input and output are as described earlier. The entire standard
    polycyclic presentation for H with respect to $a_1, \ldots, a_n$ is assumed
    to be known. ∗)
Begin
    Let $\mathcal{S}$ consist initially of the rules $[v][w] \to [v + w]$, $[v]^{-1} \to [-v]$,
        $xx^{-1} \to \varepsilon$, $[v]x \to x[vx]$, where v and w are in M and x is in $X^{\pm}$,
        together with the rules $V_k \to W_k[u_k]$, $1 \le k \le s$;
    For $i := n$ downto 1 do begin
        For $j := n$ downto $i + 1$ do begin
            If j is not in I then begin
                Let $a_j a_i \to a_i V[v]$ be the unique rule in $\mathcal{S}$ with left side $a_j a_i$;
                Rewrite $[-v]V^{-1}$ with respect to $\mathcal{S}$ to get $W[w]$;
                Add the rule $a_i^{-1}a_i \to a_i W[w]$ to $\mathcal{S}$
            End;

            If i is not in I then begin
                Let $a_j a_i^{-1} \to a_i^{-1}U$ be the rule in the standard polycyclic
                    presentation for H with left side $a_j a_i^{-1}$;
                Rewrite Ua_i with respect to $\mathcal{S}$ to get $a_i U'$, where U' is in $Y_{i+1}^{\pm*}$;
                Rewrite $a_j^{-1}U'[0]$ with respect to $\mathcal{S}$ to get $[u]$, where u is in M;
                Add the rule $a_j a_i^{-1} \to a_i^{-1}U[-ua_i^{-1}]$ to $\mathcal{S}$
            End;

            If neither i nor j is in I then begin
                Rewrite $[ua_i^{-1}]U^{-1}$ with respect to $\mathcal{S}$ to get $B[b]$;
                Add the rule $a_j^{-1}a_i^{-1} \to a_i^{-1}B[b]$
            End
        End;
        If there is a rule $a_i^m \to C[c]$ in $\mathcal{S}$ then begin
            Rewrite $[-c]C^{-1}$ with respect to $\mathcal{S}$ to get $D[d]$;
            Add the rule $a_i^{-1} \to a_i^{m-1}D[d]$
        End
    End
End.

Once the rewriting system $\mathcal{S}$ has been constructed using EXTEND_RULES, we can find module generators for M_0. The following result helps to reduce somewhat the amount of work involved.

Proposition 3.5. *Suppose the rewriting system $\mathcal{S}$ produced by* EXTEND_RULES *has the property that there is local confluence at the overlaps of left sides which are positive words in $X^{\pm*}$, that is, at the overlaps*

$$a_k a_j a_i, \quad 1 \leq i < j < k \leq n,$$
$$a_i^{m_i} a_j, \quad i \in I,\ 1 \leq j < i \leq n,$$
$$a_i^{m_i+1}, \quad i \in I,$$
$$a_j a_i^{m_i}, \quad i \in I,\ 1 \leq i < j \leq n.$$

Then $\mathcal{S}$ is confluent.

Proof. The proof is similar to that of Proposition 8.3 in Chapter 9. Only a sketch will be given. The details are left as an exercise. Because M is a $\mathbb{Z}[H]$-module, it is easy to see that local confluence holds at any overlap of left sides which involves one or more elements of M. Proposition 8.3 in Chapter 9 requires checking seven kinds of overlaps. However, because EXTEND_RULES was used to construct many of the rules in $\mathcal{S}$, it turns out that one does not need to look at overlaps of the forms $a_j a_i^{-1} a_i$, $a_j^{-1} a_j a_i$, and $a_j^{-1} a_j a_i^{-1}$. $\square$

In general, local confluence will fail for one or more of the overlaps in Proposition 3.5. However, since the presentation for H is confluent, rewriting the overlap in two ways will produce results of the form $W[w_1]$ and $W[w_2]$. Then M_0 is the $\mathbb{Z}[H]$-submodule generated by the elements $w_1 w_2$ obtained by processing the overlaps in Proposition 3.5, for modulo this submodule we have local confluence.

Example 3.2. The free nilpotent group H of rank 2 and class 2 has the following presentation on generators a, b, and c:

$$ca = ac, \quad cb = bc, \quad ba = abc.$$

Let M be a right $\mathbb{Z}[H]$-module containing elements u, v, and w and let G be the formal extension of M by H in which

$$ca = ac[u], \quad cb = bc[v], \quad ba = abc[w].$$

The image of M in G is isomorphic to M/M_0, where M_0 is the submodule of M generated by $u(1 - bc) + v(a - c) + w(c - 1)$, for by Proposition 3.5 we

need only look at the single overlap cba. Processing this overlap, we find
that

$$cba = bc[v]a = bca[va] = bac[u][va] = abc[w]c[u + va]$$
$$= abc^2[wc + u + va],$$
$$cba = cabc[w] = ac[u]bc[w] = acbc[ubc][w] = abc[v]c[ubc + w]$$
$$= abc^2[vc + ubc + w].$$

Therefore $[wc+u+va] = [vc+ubc+w]$ in G, and $u(1-bc)+v(a-c)+w(c-1)$
generates the kernel of the map from M to G.

The rewriting system $\mathcal{S}$ returned by EXTEND_RULES is infinite, since it
contains the rules $[v][w] \to [v+w]$, $[v]^{-1} \to [-v]$, and $[v]x \to x[vx]$. Suppose
that M is a $\mathbb{Z}[H]$-module on which H acts trivially and which is freely
generated as an abelian group by a finite subset $W = \{w_1, \ldots, w_t\}$. Set $Z = X \cup W$. We can produce a finite rewriting system $\mathcal{U}$ on $Z^{\pm*}$ which is
equivalent to $\mathcal{S}$ when $[\alpha_1 w_1 + \cdots + \alpha_t w_t]$ is identified with $w_1^{\alpha_1} \ldots w_t^{\alpha_t}$. That
is, every word in $Z^{\pm*}$ can be rewritten using $\mathcal{U}$ into the form UV, where U
is a collected word in $X^{\pm*}$ and V is a collected word in $W^{\pm*}$. Initially, the
following rules are placed in $\mathcal{U}$:

$$w_j^\alpha w_i^\beta \to w_i^\beta w_j^\alpha, \quad 1 \leq i < j \leq t, \ \alpha = \pm 1, \ \beta = \pm 1,$$
$$wx \to xw, \quad w \in W^\pm, \ x \in X^\pm,$$
$$V_k \to W_k U_k, \quad 1 \leq k \leq s,$$

where U_k is the collected word in $W^{\pm*}$ which describes u_k. Then the main
For-loop of EXTEND_RULES is executed with elements of the form $[v]$
replaced by the corresponding collected word in $W^{\pm*}$. This procedure will
be referred to as the finitely generated version of EXTEND_RULES.

Exercises

3.1. Suppose that G is a group and G/G' is isomorphic to $\mathbb{Z}_4 \times \mathbb{Z}_8 \times \mathbb{Z}_{24} \times \mathbb{Z}^2$. Show that
$G'/\gamma_3(G)$ is isomorphic to a quotient of $\mathbb{Z}_4^4 \times \mathbb{Z}_8^3 \times \mathbb{Z}_{24}^2 \times \mathbb{Z}$.
3.2. Let H be a free abelian group with basis a, b, c. Let M be a right $\mathbb{Z}[H]$-module which
has a $\mathbb{Z}$-basis u, v, w and on which

$$ua = 2u - v, \quad va = u, \quad wa = -u + v + w,$$
$$ub = -v - 2w, \quad vb = -u - 2w, \quad wb = u + v + 3w,$$
$$uc = u - v - w, \quad vc = -w, \quad wc = v + 2w.$$

Find the kernel M_0 of the homomorphism from M to the formal extension of M by H
in which $ba = ab[u]$, $ca = ac[v]$, and $cb = bd[w]$.

11.4 Class 2 quotients

Let G be a group given by a finite presentation $(X, \mathcal{R})$. In this section we shall describe the determination of $G/\varphi_3(G)$ and $G/\gamma_3(G)$. Nilpotent quotients of higher class will be considered in Section 11.5.

Let us start with $G/\varphi_3(G)$. We may assume that we already have a power-commutator presentation $(Z, \mathcal{T})$ for a group H isomorphic to $G/\varphi_2(G)$ in the form returned by ABEL_PQUOT. Thus $Z = \{a_1, \ldots, a_r\}$ and $\mathcal{T}$ consists of the following relations:

$$a_j a_i = a_i a_j, \quad 1 \le i < j \le r,$$
$$a_i^p = 1, \quad 1 \le i \le r.$$

We may also assume that we have homomorphisms μ and ν from $X^{\pm *}$ to Z^* and Z^* to $X^{\pm *}$, respectively, such that μ defines a homomorphism of G onto H and for all U in Z^* the words U and $\mu(\nu(U))$ define the same element of H.

We want to construct a consistent exponent-p power-commutator presentation $(Y, \mathcal{S})$ for $G/\varphi_3(G)$ and homomorphisms σ and τ from $X^{\pm *}$ to Y^* and Y^* to $X^{\pm *}$, respectively, such that σ defines a homomorphism of G onto $K = \mathrm{Grp}\langle Y \mid \mathcal{S}\rangle$ and for all U in Y^* the words U and $\sigma(\tau(U))$ define the same element of K. Initially set $Y = Z$ and $\tau = \nu$. By Proposition 11.2 in Chapter 9, τ defines a homomorphism of Y^* onto $G/\varphi_3(G)$. In order to be sure that we have generators in Y which map onto a polycyclic generating sequence for $G/\varphi_3(G)$, we add generators v_{ij}, $1 \le i < j \le r$, and u_i, $1 \le i \le r$, to Y. We extend τ to the new generators by defining

$$\tau(v_{ij}) = \tau(a_j)^{-1}\tau(a_i)^{-1}\tau(a_j)\tau(a_i) = \tau(a_i a_j)^{-1}\tau(a_j a_i)$$

and

$$\tau(u_i) = \tau(a_i)^p = \tau(a_i^p).$$

Let W_0 denote the set of all the v_{ij} and u_i. The image of W_0^* in $G/\varphi_3(G)$ is $\varphi_2(G)/\varphi_3(G)$.

Let $\overline{\tau}$ be the homomorphism of Y^* onto $G/\varphi_3(G)$ defined by τ and let $\sim$ be the congruence on Y^* determined by $\overline{\tau}$. Thus $U \sim V$ if and only if $\overline{\tau}(U) = \overline{\tau}(V)$. Let K be the quotient of Y^* by $\sim$, so K is isomorphic to $G/\varphi_3(G)$. Note that this is not a constructive definition of K, since we do not yet know $\sim$. Besides finding $\sim$ explicitly, we must determine σ. These two objectives are accomplished simultaneously. To construct σ, it suffices to determine $\sigma(x)$ for x in X. Now x and $\nu(\mu(x)) = \tau(\mu(x))$ define the same element of $G/\varphi_2(G)$. Therefore there is a word w_x in W_0^* such that x and $\tau(\mu(x)w_x)$ define the same element of $G/\varphi_3(G)$. However, we do not yet know any word w_x with this property, so we add a new generator w_x to Y and define $\tau(w_x)$ to be $\tau(\mu(x))^{-1}x$. We also set $\sigma(x) = \mu(x)w_x$. Since $\tau(\sigma(x)) = \tau(\mu(x))\tau(\mu(x))^{-1}x$ is freely equivalent to x, it follows that σ and τ define inverse isomorphisms between $G/\varphi_3(G)$ and K.

Let W_1 consist of the elements w_x with x in X and let $W = W_0 \cup W_1$. The following relations hold in K:

(1) $a_j a_i = a_i a_j v_{ij}, \quad 1 \leq i < j \leq r.$
(2) $a_i^p = u_i, \quad 1 \leq i \leq r.$
(3) $wy = yw, \quad w \in W, y \in Y.$
(4) $w^p = 1, \quad w \in W.$
(5) $\sigma(R) = \sigma(S), \quad (R, S) \in \mathcal{R}.$
(6) $a_i = \sigma(\tau(a_i)), \quad 1 \leq i \leq r.$

Proposition 4.1. *The relations (1) to (6) form a monoid presentation for K.*

Proof. All of these relations hold in $G/\varphi_3(G)$ with respect to τ and hence by the definition of K they hold in K. Let K_1 be the monoid generated by Y and defined by the relations (1) to (6). The relations (2) and (4) show that the elements of Y define units in K_1, so K_1 is a group and K is a quotient group of K_1. By the relations (3) and (4), the image of W in K_1 generates an elementary abelian subgroup D of the center of K_1. By the relations (1) and (2), K_1/D is elementary abelian. Therefore $\varphi_3(K_1)$ is trivial. The relations (5) say that σ defines a homomorphism $\bar{\sigma}$ of G into K_1. The kernel of $\bar{\sigma}$ contains $\varphi_3(G)$.

In fact, $\bar{\sigma}$ maps G onto K_1. To see this, let E be the image of G under $\bar{\sigma}$. By the relations (6), the a_i are in E. By the relations (1) and (2), the v_{ij} and the u_i are in E. Since D is central and K_1 is generated by D and E, it follows that E is normal in K_1. The quotient $L = K_1/E$ is commutative and is generated by the images of the elements w_x. If x is in X, then modulo E, $\mu(x)$ is trivial, so $\sigma(x) = w_x$ in L. Thus for any word U in $X^{\pm *}$, the image $\sigma(U)$ in L is just the word obtained by replacing each x by w_x. Relations (5) and the fact that L is commutative imply that L is isomorphic to a quotient group of $G/\varphi_2(G)$. In L, the relations (6) become $\sigma(\tau(a_i)) = 1$, $1 \leq i \leq r$. But $\tau(a_i) = \nu(a_i)$, and $\nu(a_1), \ldots, \nu(a_r)$ define generators for $G/\varphi_2(G)$. Hence the relations $\sigma(\tau(a_i)) = 1$ imply that L is trivial, or, equivalently, that $E = K_1$.

We have now shown that K_1 and $G/\varphi_3(G)$ are each homomorphic images of each other. Since these groups are finite, they must be isomorphic to each other and therefore K_1 is K. $\square$

Let M be the vector space over $\mathbb{Z}_p$ generated by W and let M be considered a $\mathbb{Z}[H]$-module on which H acts trivially. Relations (1) to (4) say that K is a quotient of a formal extension K_2 of M by H. We can get a rewriting system S for K_2 as follows: We choose a linear ordering of Y in which $a_r, \ldots, a_1$ come last, preceded by the elements of W in some order. The rules are then

$$a_j a_i \rightarrow a_i a_j v_{ij}, \quad 1 \le i < j \le r,$$
$$a_i^p \rightarrow u_i, \quad 1 \le i \le r,$$
$$wy \rightarrow yw, \quad w \in W, \ y \in Y, \ y \text{ precedes } w,$$
$$w^p \rightarrow \epsilon, \quad w \in W.$$

Using Proposition 3.4, it is not hard to show that S is confluent. Rewriting both sides of the relations (5) and (6) produces the elements of M needed to generate the kernel M_1 of the map from M into K. If we add to S the power relations for the presentation of M/M_1 in terms of the elements of W, we get a consistent exponent-p power-commutator presentation for K. Since W_0^* maps onto $\varphi_2(G)/\varphi_3(G)$, if we choose the order on W so that the elements of W_1 come first, then in the power relations for the elements in W_1 the left sides will all have exponent 1. Hence the elements of W_1 can be removed from the presentation by Tietze transformations.

Here is the procedure for determining $G/\varphi_3(G)$. The presentation returned is φ-weighted.

Procedure CLASS_2_PQUOT$(X, \mathcal{R}, p, Z, \mathcal{T}, \mu, \nu; Y, S, \sigma, \tau)$;
(* The arguments are described in the text. *)
Begin
    If $Z = \emptyset$ then begin
        (* G has no nontrivial p-quotients. *)
        $Y := Z$; $S := \mathcal{T}$; $\sigma := \mu$; $\tau := \nu$
    End

    Else begin
        Let the elements of Z be $a_1, \ldots, a_r$;
        Let W be a set of new generators of the following types: w_x for x
            in X, u_i, $1 \le i \le r$, v_{ij}, $1 \le i < j \le r$;
        Let $Y = Z \cup W$ and arrange the elements of Y in a sequence with
            the u_i and the v_{ij} first, the w_x next, and $a_r, \ldots, a_1$ last;
        Form the rewriting system S for the group K_2 as described above;
        (* The system S is confluent. *)
        Let η be a homomorphism from $X^{\pm *}$ to Y^* such that for x in X we
            have $\eta(x) = \mu(x)w_x$ and $\eta(x)\eta(x^{-1})$ can be reduced to ε
            using S;
        Let $\mathcal{V}$ consist of the pairs $(\eta(R), \eta(S))$ with (R, S) in $\mathcal{R}$ and the
            pairs $(\eta(\nu(a_i)), a_i)$, $1 \le i \le r$;

$\mathcal{W} := \emptyset$;

For (A, B) in $\mathcal{V}$ do begin

    Collect A and B with respect to $\mathcal{S}$ to obtain words CD and CE, respectively, where C is a collected word in Z^* and D and E are collected words in W^*;

    If $D \neq E$ then collect DE^{p-1} and add the result to $\mathcal{W}$

End;

Let f be the homomorphism from W^* to $M = (\mathbb{Z}_p)^{|W|}$ which maps the i-th element of W to the $(|W| + 1 - i)$-th standard basis element;

Let M_1 be the subspace of M spanned by the elements $f(D)$ with D in $\mathcal{W}$;

Let $\mathcal{P}$ be the set of power relations in the standard polycyclic presentation for M/M_1 relative to the images of the standard basis elements;

(∗ $\mathcal{P}$ may be found by row reducing modulo p the matrix whose rows are the images $f(D)$. The relations in $\mathcal{P}$ will have one of the forms $z^p = 1$ or $z = P$, where P is a word which involves generators coming after z. ∗)

For each relation $z = P$ in $\mathcal{P}$ use a Tietze transformation to eliminate z from the presentation $(Y, \mathcal{S})$;

(∗ All w_x are removed by the previous statement. ∗)

For x in $X^{\pm}$ set $\sigma(x)$ equal to the word obtained from $\eta(x)$ by applying the Tietze transformations of the previous statement;

$W := Y - Z$;

For a_i in Z define $\tau(a_i)$ to be $\nu(a_i)$, for v_{ij} in W define $\tau(v_{ij})$ to be $\tau(a_i a_j)^{-1} \tau(a_j a_i)$, and for u_i in W define $\tau(u_i)$ to be $\tau(a_i^p)$;

Assign weight 1 to the elements of Z and weight 2 to the elements of W

    End

End.

If in CLASS_2_PQUOT we have $x = \nu(\mu(x))$ for some x in X, then the element w_x will eventually be found to be trivial in K and can be omitted at the beginning.

Example 4.1. Let us continue with the group G of Example 2.2 and $p = 3$. By the remark just made, the generators w_b, w_d, and w_e will be trivial. By looking at the simpler relations for G, we can see that some of the other new generators will be trivial. Since a_1 and a_2 map to b and d, respectively, which have order 3, it follows that u_1 and u_2 will be the identity in $G/\varphi_3(G)$. Therefore, we may take W to consist of v_{23}, v_{13}, v_{12}, u_3, and w_c, in that order. The resulting rewriting system $\mathcal{S}$ is

$$a_j a_i \to a_i a_j v_{ij}, \quad 1 \le i < j \le 3,$$
$$a_1^3 \to \epsilon, \quad a_2^3 \to \epsilon, \quad a_3^3 \to u_3,$$
$$wy \to yw, \quad w \in W, \; y \in Y, \; y \text{ precedes } w \text{ in } Y,$$
$$w^3 \to \epsilon, \quad w \in W.$$

The homomorphism η maps b, c, d, and e to a_1, $a_2 a_3 w_c$, a_2, and a_3, respectively. The images under η of x^3 and z^3 are trivial by our earlier remark. The images of the three remaining relators for G collect to $u_3 v_{12}^2 v_1^2 v_{13}^2$, v_{23}^2, and $w_c^2 v_{12}^2 v_{13} v_{23}$, respectively. The relations $a_i = \eta(\nu(a_i))$ have already been used to conclude that w_b, w_d, and w_e are trivial. Thus they produce nothing new now. The elements of $\mathcal{W}$ map to the rows of the matrix

$$\begin{bmatrix} 0 & 1 & 2 & 2 & 0 \\ 0 & 0 & 0 & 0 & 2 \\ 2 & 0 & 2 & 1 & 1 \end{bmatrix},$$

whose row Hermite normal form modulo 3 is

$$\begin{bmatrix} 1 & 0 & 1 & 2 & 0 \\ 0 & 1 & 2 & 2 & 0 \\ 0 & 0 & 0 & 0 & 1 \end{bmatrix}.$$

This gives us the additional relations

$$w_c = v_{12}^2 v_{13}, \quad u_3 = v_{12} v_{13}, \quad v_{23} = 1.$$

The elements remaining in W are v_{12} and v_{13}. If we rename these generators a_4 and a_5, then the relations in the final presentation are

$$a_2 a_1 = a_1 a_2 a_4, \quad a_3 a_1 = a_1 a_3 a_5, \quad a_3 a_2 = a_2 a_3,$$
$$a_j a_i = a_i a_j, \quad 4 \le j \le 5, \; 1 \le i < j,$$
$$a_1^3 = 1, \quad a_2^3 = 1, \quad a_3^3 = a_4 a_5,$$
$$a_i^3 = 1, \quad 4 \le i \le 5.$$

The homomorphisms σ and τ are defined by

$$\sigma(b) = a_1, \quad \sigma(c) = a_2 a_3 a_4^2 a_5, \quad \sigma(d) = a_2, \quad \sigma(e) = a_3,$$
$$\tau(a_1) = b, \quad \tau(a_2) = d, \quad \tau(a_3) = e,$$
$$\tau(a_4) = d^{-1} b^{-1} db, \quad \tau(a_5) = e^{-1} b^{-1} eb.$$

The procedure for going from G/G' to $G/\gamma_3(G)$ is similar to CLASS_2_ PQUOT. Let $(Z, \mathcal{T})$ be a consistent polycyclic group presentation for a group H isomorphic to G/G'. If $(Z, \mathcal{T})$ is obtained using ABEL_QUOT, then it describes G/G' as a direct product of cyclic groups. We shall assume only that $Z = \{a_1, \ldots, a_r\}$ and that $\mathcal{T}$ contains the following relations:

$$a_j a_i = a_i a_j, \quad 1 \le i < j \le r,$$
$$a_i^{m_i} = U_i, \quad i \in I,$$

where U_i is in $\{a_{i+1}, \ldots, a_r\}^{\pm*}$. We shall also assume that μ and ν are regular homomorphisms from $X^{\pm*}$ to $Z^{\pm*}$ and $Z^{\pm*}$ to $X^{\pm*}$, respectively, such that μ defines a homomorphism of G onto H and for all U in $Z^{\pm*}$ the words U and $\mu(\nu(U))$ define the same element of H.

We want to construct the same sort of description of $G/\gamma_3(G)$. We start by setting $Y = Z$ and $\tau = \nu$. By Proposition 2.5 in Chapter 9, τ defines a homomorphism of $Y^{\pm*}$ onto $G/\gamma_3(G)$. We next add new generators v_{ij}, $1 \le i < j \le r$, to Y and define $\tau(v_{ij})$ to be $\tau(a_j^{-1} a_i^{-1} a_j a_i)$. Let W_0 be the set of v_{ij}. The image of $W_0^{\pm*}$ in $G/\gamma_3(G)$ is $G'/\gamma_3(G)$. Suppose that i is in I. Although the words $\tau(a_i^{m_i})$ and $\tau(U_i)$ define the same element of G/G', they may not define the same element of $G/\gamma_3(G)$. There is a word u_i in $W_0^{\pm*}$ such that $\tau(a_i^{m_i})$ and $\tau(U_i u_i)$ define the same element of $G/\gamma_3(G)$. Since we do not know such a word u_i at this point, we add a new generator u_i to Y and set $\tau(u_i) = \tau(U_i^{-1} a_i^{m_i})$. As we did in constructing $G/\varphi_3(G)$, we also add a new generator w_x for each x in X and set $\tau(w_x) = \tau(\mu(x))^{-1} x$.

The monoid homomorphism τ defines a group homomorphism $\overline{\tau}$ from F_Y onto $G/\gamma_3(G)$. Let N be the kernel of $\overline{\tau}$ and set $K = F_Y/N$. Define σ to be the regular homomorphism from $X^{\pm*}$ to $Y^{\pm*}$ such that $\sigma(x) = \mu(x) w_x$ for all x in X. As in the earlier case, σ and τ define inverse isomorphisms between $G/\gamma_3(G)$ and K.

Let W_1 consist of the elements u_i with i in I and w_x with x in X, and let $W = W_0 \cup W_1$. The following relations hold in K:

(1) $a_j a_i = a_i a_j v_{ij}, \quad 1 \le i < j \le r.$
(2) $a_i^{m_i} = U_i u_i, \quad i \in I.$
(3) $wy = yw, \quad w \in W, \ y \in Y.$
(4) $\sigma(R) = \sigma(S), \quad (R, S) \in \mathcal{R}.$
(5) $a_i = \sigma(\tau(a_i)), \quad 1 \le i \le r.$

By an argument very similar to the proof of Proposition 4.1, we can show that these relations form a group presentation for K.

Let M be the free abelian group on W and let M be considered a $\mathbb{Z}[H]$-module on which H acts trivially. Relations (1) to (3) say that K is a quotient of a formal extension K_2 of M by H. Using the finitely generated version of EXTEND_RULES, we can get a rewriting system S for K_2. This time S is not automatically confluent. Processing the overlaps of the left sides of the relations described in Proposition 3.5, we obtain generators for the kernel M_0 of the map from M to K_2. Rewriting both sides of the relations (4) and (5) produces the additional elements of M needed to generate the kernel M_1 of the map from M to K. We add the power relations for the presentation of M/M_1 in terms of the elements of W and remove redundant generators.

Here is the formal description of the algorithm for going from G/G' to $G/\gamma_3(G)$:

Procedure CLASS_2_QUOT$(X, \mathcal{R}, Z, \mathcal{T}, \mu, \nu; Y, \mathcal{S}, \sigma, \tau)$;
(* The arguments are described in the text. *)
Begin
  If $Z = \emptyset$ then begin
  (* G is perfect. *)
  $Y := Z$; $\mathcal{S} := \mathcal{T}$; $\sigma := \mu$; $\tau := \nu$
  End

  Else begin
      Let the elements of Z be $a_1, \ldots, a_r$ and let $\mathcal{T}$ contain the relations
          $a_j a_i = a_i a_j$, $1 \le i < j \le r$, and $a_i^{m_i} = U_i$, $i \in I$;
      Let W be a set containing new generators of the following types:
          w_x for x in X, u_i for i in I, and v_{ij} for $i \le i < j \le r$;
      Let $Y = Z \cup W$ and arrange the elements of Y in a sequence with
          the v_{ij} first, the w_x next, and $a_r, \ldots, a_1$ last;
      Let M be the free abelian group on W considered as a
          $\mathbb{Z}[H]$-module on which $H = \mathrm{Grp}\langle Z \mid \mathcal{T} \rangle$ acts trivially;
      Use the finitely generated version of EXTEND_RULES with input
          Z, $\mathcal{T}$, M, the v_{ij}, and the u_i to obtain a rewriting system S on
          $Y^{\pm *}$ for the formal extension K_2 of M by H containing, among
          others, rules $a_j a_i \to a_i a_j v_{ij}$, $1 \le i < j \le r$, and $a_i^{m_i} \to U_i u_i$ with
          i in I;
      Let η be the regular homomorphism from $X^{\pm *}$ to $Y^{\pm *}$ such that
          $\eta(x) = \mu(x) w_x$ for all x in X;
      Let $\mathcal{V}$ consist of the pairs $(\eta(R), \eta(S))$ with (R, S) in $\mathcal{R}$, the pairs
          $(\eta(\nu(a_i)), a_i)$, and the pairs (DR, PE), where PQR is an
          overlap between rules $PQ \to D$ and $QR \to E$ in S with
          positive left sides;

      (* If each U_i is empty, then Proposition 3.4 can be used as an
          alternative to processing the overlaps PQR. *)

      $\mathcal{W} := \emptyset$;
      For (A, B) in $\mathcal{V}$ do begin

Collect A and B with respect to $\mathcal{S}$ to obtain CD and CE,
 respectively, where C is a collected word in $Z^{\pm*}$ and D and
 E are collected words in $W^{\pm*}$;
If $D \neq E$ then add DE^{-1} to $\mathcal{W}$
End;
Let M_1 be the subgroup of M generated by the image of $\mathcal{W}$;
Let $\mathcal{P}$ be the set of power relations in the standard polycyclic
 presentation for M/M_1 in terms of the elements of W in the
 chosen order;
($*$ $\mathcal{P}$ may be found by establishing an isomorphism between M and
 $\mathbb{Z}^{|W|}$ using the chosen ordering of W and then row reducing
 the matrix whose rows correspond to the elements of $\mathcal{W}$. $*$)
For each relation $z^m = P$ in $\mathcal{P}$ add $z^m \to P$ and $z^{-1} \to z^{m-1}P^{-1}$
 to $\mathcal{S}$;
For each relation in $\mathcal{P}$ whose left side has the form z^m with $m = 1$
 use a Tietze transformation to eliminate z from the
 presentation $(Y, \mathcal{S})$;
($*$ All u_i and w_x are removed by the previous statement. $*$)
For x in X set $\sigma(x)$ equal to $\mu(x)\overline{w_x}$, where $\overline{w_x}$ denotes the result
 of applying the Tietze transformations of the previous
 statement to w_x;
$W := Y - Z$;
For a_i in Z define $\tau(a_i)$ to be $\nu(a_i)$ and for v_{ij} in W define $\tau(v_{ij})$ to
 be $\tau(a_j^{-1}a_j^{-1}ia_ja_i)$;
Assign weight 1 to elements of Z and weight 2 to elements of W
End
End.

Example 4.2. Let us continue with the group S of Example 2.4. Here
$Z = \{a_1, a_2, a_3\}$, and $\mathcal{T}$ contains the following relations:

$$a_j a_i = a_i a_j, \quad 1 \le i < j \le 3,$$
$$a_1^2 = 1.$$

The homomorphisms μ and ν are given by

$$\mu(t_1) = 1, \quad \mu(t_2) = a_1, \quad \mu(t_3) = a_2, \quad \mu(t_4) = a_3,$$
$$\nu(a_1) = t_2, \quad \nu(a_2) = t_3, \quad \nu(a_3) = t_4.$$

According to CLASS_2_QUOT, we should allow for the following genera-
tors in W : $w_1, w_2, w_3, w_4, u_1, v_{12}, v_{13}, v_{23}$, where w_i denotes w_{t_i}. However,

since $\mu(\nu(a_i)) = a_i$, $1 \leq i \leq 3$, we can see immediately that w_2, w_3, and w_4 are trivial in K and can be omitted. Thus the elements of W will be taken to be v_{23}, v_{13}, v_{12}, u_1, w_1, in that order.

The result of applying EXTEND_RULES is a rewriting system $\mathcal{S}$ with the following rules:

$$v^\alpha u^\beta \to u^\beta v^\alpha, \quad u, v \in W, \ v \text{ precedes } u,$$

$$u^\alpha a_i^\beta \to a_i^\beta u^\alpha, \quad u \in W, \ 1 \leq i \leq 3,$$

$$a_j^\alpha a_i^\beta \to a_i^\beta a_j^\alpha v_{ij}^{\alpha\beta}, \quad 1 \leq i < j \leq 3,$$

$$a_1^2 \to u_1, \quad a_1^{-1} \to a_1 u_1^{-1}$$

Here α and β range independently over $\{1, -1\}$. By Proposition 3.4, the results of processing the overlaps among these relations are all consequences of the relations $v_{12}^2 = v_{13}^2 = 1$. Thus we put v_{12}^2 and v_{13}^2 into $\mathcal{W}$.

The homomorphism η is given by $\eta(t_1) = w_1$, $\eta(t_i) = a_{i-1}$, $2 \leq i \leq 4$. The image under η of $t_1 t_2 t_1^{-1} t_2^{-1}$ is

$$w_1 a_1 w_1^{-1} a_1^{-1} \to w_1 a_1 w_1^{-1} a_1 u_1^{-1} \to a_1^2 u_1^{-1} \to u_1 u_1^{-1} \to \epsilon.$$

Therefore no element needs to be added to $\mathcal{W}$ for this relator. The image of $t_1^5 t_2^{-2}$ is

$$w_1^5 a_1^{-2} \to w_1^5 a_1 u_1^{-1} a_1 u_1^{-1} \to a_1^2 w_1^5 u_1^{-2} \to u_1 w_1^5 u_1^{-2} \to w_1^5 u_1^{-1}.$$

Hence $w_1^5 u_1^{-1}$ is added to $\mathcal{W}$. The images of the three remaining original relators can be rewritten to $w_1 v_{23}^{-1}$, $w_1^{-1} v_{12} v_{23}$, and v_{13}, respectively, so these elements are also added to $\mathcal{W}$. The image of $a_i^{-1} \nu(a_i)$ can be rewritten to ε, $1 \leq i \leq 3$, so no elements need to be added to $\mathcal{W}$ for these relators.

We identify M with $\mathbb{Z}^5$ so that

$$w_1 = (1, 0, 0, 0, 0),$$

$$u_1 = (0, 1, 0, 0, 0),$$

$$v_{12} = (0, 0, 1, 0, 0),$$

$$v_{13} = (0, 0, 0, 1, 0),$$

$$v_{23} = (0, 0, 0, 0, 1).$$

The elements of $\mathcal{W}$ correspond to the rows of

$$
A = \begin{bmatrix}
0 & 0 & 2 & 0 & 0 \\
0 & 0 & 0 & 2 & 0 \\
5 & -1 & 0 & 0 & 0 \\
1 & 0 & 0 & 0 & -1 \\
-1 & 0 & 1 & 0 & 1 \\
0 & 0 & 0 & 1 & 0
\end{bmatrix}.
$$

Row-reducing A over the integers, we get

$$
\begin{bmatrix}
1 & 0 & 0 & 0 & -1 \\
0 & 1 & 0 & 0 & -5 \\
0 & 0 & 1 & 0 & 0 \\
0 & 0 & 0 & 1 & 0 \\
0 & 0 & 0 & 0 & 0 \\
0 & 0 & 0 & 0 & 0
\end{bmatrix}.
$$

Thus in M/M_1, the images of v_{12} and v_{13} are trivial, $w_1 = v_{23}$, and $u_1 = v_{23}^5$. If we rename v_{23} as a_4, then $S/\gamma_3(S)$ is given by the following consistent nilpotent presentation:

$$
a_2 a_1 = a_1 a_2, \quad a_3 a_1 = a_1 a_3, \quad a_3 a_2 = a_2 a_3 a_4,
$$
$$
a_4 a_i = a_i a_4, \quad 1 \le i \le 3,
$$
$$
a_1^2 = a_4^5.
$$

The homomorphisms σ and τ are given by

$$
\sigma(t_1) = a_4, \quad \sigma(t_2) = a_1, \quad \sigma(t_3) = a_2, \quad \sigma(t_4) = a_3,
$$
$$
\tau(a_1) = t_2, \quad \tau(a_2) = t_3, \quad \tau(a_3) = t_4, \quad \tau(a_4) = t_4^{-1} t_3^{-1} t_4 t_3.
$$

The value of $\tau(a_4)$ comes from the fact that a_4 is v_{23} renamed and τ maps a_3 and a_2 to t_4 and t_3, respectively.

The presentation for $G'/\gamma_3(G)$ obtained directly from CLASS_2_QUOT does not always give this group as a direct sum of cyclic groups. Such a presentation can be obtained. See the procedure ADJUST in Section 11.5.

Exercise

4.1. Let G be the group defined in Exercise 2.1. Compute $G/\varphi_3(G)$ for $p = 2$ and 3. Determine $G/\gamma_3(G)$.

11.5 Other nilpotent quotients

Let $(X, \mathcal{R})$ be a finite presentation for a group G. The procedures ABEL_PQUOT and ABEL_QUOT of Section 11.2 allow us to determine $G/\varphi_2(G)$ and G/G'. Once these quotients are known, the procedures CLASS_2_PQUOT and CLASS_2_QUOT from Section 11.4 can be used to compute $G/\varphi_3(G)$ and $G/\gamma_3(G)$. In this section we shall describe algorithms for determining $G/\varphi_{e+1}(G)$ and $G/\gamma_{e+1}(G)$, where $e \geq 3$.

Let us look at $G/\varphi_{e+1}(G)$ first. By induction on e, we may assume that we know a consistent φ-weighted exponent-p power-commutator presentation $(Z, \mathcal{T})$ for a group H isomorphic to $G/\varphi_e(G)$. If $\varphi_{e-1}(H) = 1$, then $\varphi_k(G) = \varphi_{e-1}(G)$ for all $k \geq e - 1$. In this case we may stop since $G/\varphi_{e+1}(G)$ is isomorphic to H. Thus we may assume that $\varphi_{e-1}(H)$ is nontrivial.

Let $Z = \{a_1, \ldots, a_t\}$, let w_i be the weight of a_i, and let $\mathcal{T}$ contain the relations

$$a_j a_i = a_i a_j V_{ij}, \quad 1 \leq i < j \leq t,$$
$$a_j^p = U_j, \quad 1 \leq j \leq t,$$

where the generators occurring in V_{ij} and U_j have weights at least $w_i + w_j$ and $w_j + 1$, respectively. We may assume that we have homomorphisms μ and ν from $X^{\pm *}$ to Z^* and Z^* to $X^{\pm *}$, respectively, such that μ defines a homomorphism of G onto H and for all U in Z^* the words $\mu(\nu(U))$ and U define the same element of H. The generators of a given weight m generate $\varphi_m(H)$ modulo $\varphi_{m+1}(H)$. We shall denote by r the largest index such that $w_r = 1$ and by s the smallest index such that $w_s = e - 1$.

We want a consistent φ-weighted exponent-p power-commutator presentation $(Y, \mathcal{S})$ for $G/\varphi_{e+1}(G)$ and associated homomorphisms σ and τ. The procedure for going from Z, $\mathcal{T}$, μ, ν to Y, $\mathcal{S}$, σ, τ is similar to CLASS_2_PQUOT. The difference is analogous to the difference between Propositions 11.2 and 11.3 in Chapter 9. We start by taking $Y = Z$ and $\tau = \nu$. The image of Y^* under τ maps onto $G/\varphi_2(G)$ and hence onto $G/\varphi_{e+1}(G)$. To be sure that we have generators in Y which map to a polycyclic generating sequence for $G/\varphi_{e+1}(G)$, we add generators v_{ij} and u_j to Y, where $1 \leq i \leq r$ and $s \leq j \leq t$. We define $\tau(v_{ij})$ and $\tau(u_j)$ to be $\tau(a_i a_j)^{-1} \tau(a_j a_i)$ and $\tau(a_i^p)$, respectively. The set of these v_{ij} and u_j will be denoted W_0. We must also add certain temporary generators, generators which are guaranteed to be removed from the presentation later in the computation. These temporary generators are w_x with x in X, u_i with $1 \leq i < s$, and v_{ij}, where $1 \leq i < j \leq t$ and either $r < i$ or $j < s$. We set $\tau(v_{ij}) = \tau(V_{ij})^{-1}\tau(a_i a_j)^{-1}\tau(a_j a_i)$ and $\tau(u_i) = \tau(U_i)^{-1}\tau(a_i^p)$. In addition, we define $\tau(w_x) = \tau(\mu(x))^{-1}x$. The set of temporary generators will be denoted W_1. The map σ is defined by $\sigma(x) = \mu(x)w_x$.

Let $W = W_0 \cup W_1$ and $Y = Z \cup W$. The following relations on the generators in Y form a monoid presentation for $G/\varphi_{e+1}(G)$:

(1) $a_j a_i = a_i a_j V_{ij} v_{ij}, \quad 1 \leq i < j \leq t,$
(2) $a_i^p = U_i u_i, \quad 1 \leq i \leq t,$
(3) $wy = yw, \quad w \in W, y \in Y,$
(4) $w^p = 1, \quad w \in W,$
(5) $\sigma(R) = \sigma(S), \quad (R, S) \in \mathcal{R},$
(6) $a_i = \sigma(\tau(a_i)), \quad 1 \leq i \leq t.$

The proof of this assertion is similar to the proof of Proposition 4.1.

Here is the procedure for going from $G/\varphi_e(G)$ to $G/\varphi_{e+1}(G)$ when $e \geq 3$:

Procedure NEXT_PQUOT($X, \mathcal{R}, p, Z, \mathcal{T}, \mu, \nu, e; Y, \mathcal{S}, \sigma, \tau$);
(∗ The arguments are described in the text. ∗)
Begin
    If there are no generators of weight $e - 1$ in Z then begin
        $Y := Z; \ \mathcal{S} := \mathcal{T}, \ \sigma := \mu; \ \tau := \nu$
    End
    Else begin
        Let Z consist of $a_1, \ldots, a_t$ and let $\mathcal{T}$ contain the relations
            $a_j a_i = a_i a_j V_{ij}, 1 \leq i < j \leq t,$ and $a_i^p = U_i, 1 \leq i \leq t;$
        Let a_r be the last generator of weight 1 and let a_s be the first
            generator of weight $e - 1;$
        Let W_0 be a set of new generators v_{ij} and $u_j, 1 \leq i \leq r, s \leq j \leq t;$
        Let W_1 be a set of new generators w_x with x in X, u_i with
            $1 \leq i < s$, and v_{ij} with $1 \leq i < j \leq t$ and either $r < i$ or $j < s;$
        $W := W_0 \cup W_1; \ Y := Z \cup W;$
        Arrange the elements of Y in a sequence with the elements of W_0
            first, the elements of W_1 next, and $a_t, \ldots, a_1$ last;
        Let $\mathcal{S}$ be the rewriting system on Y^* containing the rules
            $a_j a_i \to a_i a_j V_{ij} v_{ij}$ for $1 \leq i < j \leq t$, $a_i^p \to U_i u_i$ for $1 \leq i \leq t,$
            $w^p \to \varepsilon$ for w in W, and $wy \to yw$ for w in W, y in Y, and y
            precedes $w;$
        Let η be a homomorphism from $X^{\pm*}$ to Y^* such that for x in X we
            have $\eta(x) = \mu(x)w_x$ and $\eta(x)\eta(x^{-1})$ can be reduced to ε
            using $\mathcal{S};$
        Let $\mathcal{V}$ consist of the pairs $(\eta(R), \eta(S))$ with (R, S) in $\mathcal{R}$, the pairs
            $(\eta(\nu(a_i)), a_i), 1 \leq i \leq t$, and the pairs (DR, PE), where PQR
            is an overlap between rules $PQ \to D$ and $QR \to E$ in $\mathcal{S}$ with
            positive left sides;
        $\mathcal{W} := \emptyset;$
        For (A, B) in $\mathcal{V}$ do begin

Collect A and B with respect to $\mathcal{S}$ to obtain CD and CE,
    respectively, where C is a collected word in $X^{\pm *}$ and D and
    E are collected words in W^*;
  If $D \neq E$ then collect DE^{p-1} and add the result to $\mathcal{W}$
End;
Let f be the homomorphism from W^* to $M = (\mathbb{Z}_p)^{|W|}$ which maps
    the i-th element of W to the $(|W| + 1 - i)$-th standard basis
    element;
Let M_1 be the subspace of M spanned by the elements $f(D)$ with
    D in $\mathcal{W}$;
Let $\mathcal{P}$ be the set of power relations in the standard polycyclic
    presentation for M/M_1 relative to the images of the standard
    basis elements;
$(*$ See comment in CLASS_2_PQUOT. $*)$
For each relation in $\mathcal{P}$ of the form $z = P$ with z in W use a Tietze
    transformation to eliminate z from the presentation $(Y, \mathcal{S})$;
$(*$ All elements of W_1 are removed by the previous statement. $*)$
For x in $X^\pm$ set $\sigma(x)$ equal to the word obtained from $\eta(x)$ by
    applying the Tietze transformations of the previous statement
    and collecting;
$W := Y - Z$;
For a_i in Z define $\tau(a_i)$ to be $\nu(a_i)$, for v_{ij} in W define $\tau(v_{ij})$ to be
    $\tau(a_i a_j)^{-1} \tau(a_j a_i)$, and for u_i in W define $\tau(u_i)$ to be $\tau(a_i^p)$;
Assign weight e to the elements of W
  End
End.

As noted in Section 11.4, frequently there are many elements of W in
NEXT_PQUOT which are obviously trivial and can be omitted entirely.
If $\nu(\mu(x))$ is x for some x in X, then w_x can be omitted. If $1 \leq i < j \leq t$
and the sum of the weights of a_i and a_j exceeds e, then a_i and a_j will
commute in the final presentation and v_{ij} can be omitted. Also, suppose
that $1 \leq i < j < k \leq t$ and T contains the relation $a_j a_i = a_i a_j a_k$. If ν
preserves this relation, then v_{ij} will be found to be trivial. Similarly, if ν
preserves a relation $a_i^p = a_j$, then u_i can be omitted. In addition to omitting
elements of W, it is also possible to obtain early in the procedure formulas
describing certain elements of W in terms of later elements. We can use
the remarks at the end of Section 9.11 to reduce the number of overlaps
which need to be considered.

Example 5.1. Let us determine $G/\varphi_4(G)$, where G is the group of Examples 2.2 and 4.1 and $p = 3$. By Example 4.1, $G/\varphi_3(G)$ is given by

$$a_2 a_1 = a_1 a_2 a_4, \quad a_3 a_1 = a_1 a_3 a_5, \quad a_3 a_2 = a_2 a_3,$$

$$a_j a_i = a_i a_j, \quad 4 \le j \le 5, \ 1 \le i < j,$$

$$a_1^3 = 1, \quad a_2^3 = 1, \quad a_3^3 = a_4 a_5,$$

$$a_i^3 = 1, \quad 4 \le i \le 5.$$

The homomorphisms μ and ν are defined by

$$\mu(b) = a_1, \quad \mu(c) = a_2 a_3 a_4^2 a_5, \quad \mu(d) = a_2, \quad \mu(e) = a_3,$$

$$\nu(a_1) = b, \quad \nu(a_2) = d, \quad \nu(a_3) = e,$$

$$\nu(a_4) = d^{-1} b^{-1} db, \quad \nu(a_5) = e^{-1} b^{-1} eb.$$

Just as in Example 4.1, the new generators w_b, w_d, w_e, u_1, and u_2 will all be trivial, and hence can be omitted. In addition, τ preserves the relations $a_2 a_1 = a_1 a_2 a_4$ and $a_3 a_1 = a_1 a_3 a_5$, which are the definitions of a_4 and a_5, so v_{12} and v_{13} will be trivial too. Thus we may take W to consist of v_{35}, v_{25}, v_{15}, v_{34}, v_{24}, v_{14}, u_5, u_4, v_{23}, u_3, w_c, where the last three generators make up W_1 and the remaining are in W_0. Processing the overlaps of left sides in the initial presentation produces the following elements of W : $v_{34} v_{25}^2$, u_4^2, u_5^2, $u_5^2 v_{14} v_{15}$, $v_{24} v_{25}$, and $v_{24}^2 v_{35}^2$. Collecting the original relations adds the following elements: $u_4^2 u_5 v_{14}^2 v_{24} v_{25} v_{35}$, $v_{23}^2 v_{34}^2 v_{35}^2$, and $w_c^2 v_{23} u_4 u_5 v_{34}^2 v_{25}$. The relations $a_i = \sigma(\tau(a_i))$ yield u_4 and $u_5 v_{14}^2 v_{34}^2 v_{15}^2 v_{35}^2$. Row reducing the matrix

$$\begin{bmatrix}
0 & 0 & 0 & 0 & 0 & 0 & 0 & 1 & 0 & 2 & 0 \\
0 & 0 & 0 & 2 & 0 & 0 & 0 & 0 & 0 & 0 & 0 \\
0 & 0 & 0 & 0 & 2 & 0 & 0 & 0 & 0 & 0 & 0 \\
0 & 0 & 0 & 0 & 2 & 1 & 0 & 0 & 1 & 0 & 0 \\
0 & 0 & 0 & 0 & 0 & 0 & 1 & 0 & 0 & 1 & 0 \\
0 & 0 & 0 & 0 & 0 & 0 & 0 & 2 & 0 & 0 & 2 \\
0 & 1 & 0 & 2 & 1 & 2 & 1 & 0 & 0 & 1 & 1 \\
0 & 0 & 2 & 0 & 0 & 0 & 0 & 2 & 0 & 0 & 2 \\
2 & 0 & 1 & 1 & 1 & 0 & 0 & 2 & 0 & 1 & 0 \\
0 & 0 & 0 & 1 & 0 & 0 & 0 & 0 & 0 & 0 & 0 \\
0 & 0 & 0 & 0 & 1 & 2 & 0 & 2 & 2 & 0 & 2
\end{bmatrix}$$

modulo 3, we get

$$\begin{bmatrix}
1 & 0 & 0 & 0 & 0 & 0 & 0 & 0 & 0 & 0 & 0 \\
0 & 1 & 0 & 0 & 0 & 0 & 0 & 0 & 1 & 0 & 1 \\
0 & 0 & 1 & 0 & 0 & 0 & 0 & 0 & 0 & 0 & 0 \\
0 & 0 & 0 & 1 & 0 & 0 & 0 & 0 & 0 & 0 & 0 \\
0 & 0 & 0 & 0 & 1 & 0 & 0 & 0 & 0 & 0 & 0 \\
0 & 0 & 0 & 0 & 0 & 1 & 0 & 0 & 1 & 0 & 0 \\
0 & 0 & 0 & 0 & 0 & 0 & 1 & 0 & 0 & 0 & 2 \\
0 & 0 & 0 & 0 & 0 & 0 & 0 & 1 & 0 & 0 & 1 \\
0 & 0 & 0 & 0 & 0 & 0 & 0 & 0 & 0 & 1 & 1 \\
0 & 0 & 0 & 0 & 0 & 0 & 0 & 0 & 0 & 0 & 0 \\
0 & 0 & 0 & 0 & 0 & 0 & 0 & 0 & 0 & 0 & 0
\end{bmatrix}$$

This tells us that w_c, v_{23}, u_4, and u_5 are trivial and that

$$u_3 = v_{15}^2 v_{35}^2, \quad v_{14} = v_{15}^2, \quad v_{24} = v_{35}, \quad v_{34} = v_{35}^2, \quad v_{25} = v_{35}^2.$$

If we rename v_{15} and v_{35} as a_6 and a_7, we get the following presentation:

$$a_2 a_1 = a_1 a_2 a_4, \quad a_3 a_1 = a_1 a_3 a_5, \quad a_3 a_2 = a_2 a_3,$$
$$a_4 a_1 = a_1 a_4 a_6^2, \quad a_4 a_2 = a_2 a_4 a_7, \quad a_4 a_3 = a_3 a_4 a_7^2,$$
$$a_5 a_1 = a_1 a_5 a_6, \quad a_5 a_2 = a_2 a_5 a_7^2, \quad a_5 a_3 = a_3 a_5 a_7,$$
$$a_j a_i = a_i a_j, \quad 4 \le i < j \le 7,$$
$$a_1^3 = 1, \quad a_2^3 = 1, \quad a_3^3 = a_4 a_5 a_6^2 a_7^2,$$
$$a_i^3 = 1, \quad 4 \le i \le 7.$$

The homomorphisms σ and τ are given by

$$\sigma(b) = a_1, \quad \sigma(c) = a_2 a_3 a_4^2 a_5, \quad \sigma(d) = a_2, \quad \sigma(e) = a_3,$$
$$\tau(a_1) = b, \quad \tau(a_2) = d, \quad \tau(a_3) = e, \quad \tau(a_4) = d^{-1} b^{-1} db,$$
$$\tau(a_5) = e^{-1} b^{-1} eb, \quad \tau(a_6) = b^{-1} e^{-1} beb^{-1} e^{-1} b^{-1} eb^2,$$
$$\tau(a_7) = b^{-1} e^{-1} bee^{-2} b^{-1} ebe = b^{-1} e^{-1} be^{-1} b^{-1} ebe.$$

Now let us consider the problem of constructing $G/\gamma_{e+1}(G)$ from $G/\gamma_e(G)$. We may suppose that we have a consistent nilpotent group presentation $(Z, \mathcal{T})$ for a group H isomorphic to $G/\gamma_e(G)$. Let $Z = \{a_1, \ldots, a_t\}$ and let $\mathcal{T}$ contain the relations

$$a_j a_i = a_i a_j V_{ij}, \quad 1 \le i < j \le t,$$
$$a_j^{m_j} = U_j, \quad j \in I.$$

We also assume that we have associated regular homomorphisms μ and ν from $X^{\pm*}$ to $Z^{\pm*}$ and $Z^{\pm*}$ to $X^{\pm*}$, respectively. We denote by r the largest index such that a_r has weight 1 and by s the smallest index such that a_s has weight $e - 1$.

We want a consistent nilpotent presentation $(Y, \mathcal{S})$ for $G/\gamma_{e+1}(G)$ and associated regular homomorphisms σ and τ. The procedure for going from Z, $\mathcal{T}$, μ, ν to Y, $\mathcal{S}$, σ, τ is very similar to NEXT_PQUOT. We start by taking $Y = Z$ and $\tau = \nu$. The image of $Y^{\pm*}$ under τ maps onto G/G' and hence onto $G/\gamma_{e+1}(G)$. To be sure that we have generators in Y which map to a polycyclic generating sequence for $G/\gamma_{e+1}(G)$, we add generators v_{ij} to Y, where $1 \leq i \leq r$ and $s \leq j \leq t$, and define $\tau(v_{ij})$ to be $\tau(a_j^{-1} a_i^{-1} a_j a_i)$. The set of these v_{ij} will be denoted W_0. We also add the following temporary generators: w_x with x in X, u_i with i in I, and v_{ij}, where $1 \leq i < j \leq t$ and either $r < i$ or $j < s$. We set $\tau(v_{ij}) = \tau(V_{ij}^{-1} a_j^{-1} a_i^{-1} a_j a_i)$, $\tau(w_x) = \tau(\mu(x))^{-1} x$, and $\tau(u_i) = \tau(U_i^{-1} a_i^{m_i})$. The set of temporary generators will be denoted W_1. The map σ is defined by $\sigma(x) = \mu(x) w_x$.

Let $W = W_0 \cup W_1$. The following relations define $G/\gamma_{e+1}(G)$ as a group:

(1) $a_j a_i = a_i a_j V_{ij} v_{ij}$, $\quad 1 \leq i < j \leq t$.
(2) $a_i^{m_i} = U_i u_i$, $\quad i \in I$.
(3) $wy = yw$, $\quad w \in W$, $y \in Y$.
(4) $\sigma(R) = \sigma(S)$, $\quad (R, S) \in \mathcal{R}$.
(5) $a_i = \sigma(\tau(a_i))$, $\quad 1 \leq i \leq t$.

Here is the algorithm for going from $G/\gamma_e(G)$ to $G/\gamma_{e+1}(G)$ when $e \geq 3$:

Procedure NEXT_QUOT$(X, \mathcal{R}, Z, \mathcal{T}, \mu, \nu, e; Y, \mathcal{S}, \sigma, \tau)$;
(* The arguments are described in the text. *)
Begin
    If there are no generators of weight $e - 1$ in Z then begin
      $Y := Z$; $\mathcal{S} := \mathcal{T}$; $\sigma := \mu$; $\tau := \nu$
    End

    Else begin
        Let Z consist of the elements $a_1, \ldots, a_t$ and let $\mathcal{T}$ contain the
            relations $a_j a_i = a_i a_j V_{ij}$, $1 \leq i < j \leq t$, and $a_i^{m_i} = U_i$, $i \in I$;
        Let a_r be the last generator of weight 1 and let a_s be the first
            generator of weight $e - 1$;
        Let W_0 be a set of new generators v_{ij}, $1 \leq i \leq r$, $s \leq j \leq t$;
        Let W_1 be a set of new generators of the following types: w_x for x
            in X, u_i for i in I, and v_{ij} for those pairs (i, j) such that
            $1 \leq i < j \leq t$ and either $r < i$ or $j < s$;
        $W := W_0 \cup W_1$; $Y := Z \cup W$;

Arrange the elements of Y in a sequence with the elements of W_0
   first, the elements of W_1 next, and $a_t, \ldots, a_1$ last;
Let M be the free abelian group on W considered as a
   $\mathbb{Z}[H]$-module on which $H = \mathrm{Grp}\langle Z \mid \mathcal{T} \rangle$ acts trivially;
Use the finitely generated version of EXTEND_RULES with input
   Z, $\mathcal{T}$, M, and the elements of W to obtain a rewriting system
   $\mathcal{S}$ on $Y^{\pm *}$ for the formal extension of M by H containing,
   among others, rules $a_j a_i \to a_i a_j V_{ij} v_{ij}$, $1 \leq i < j \leq t$, and
   $a_i^{m_i} \to U_i u_i$, $i \in I$;
Let η be the regular homomorphism from $X^{\pm *}$ to $Y^{\pm *}$ such that
   $\eta(x) = \mu(x) w_x$ for $x \in X$;
Let $\mathcal{V}$ consist of the pairs $(\eta(R), \eta(S))$ with (R, S) in $\mathcal{R}$, the pairs
   $\eta(\nu(a_i)), a_i)$, $1 \leq i \leq t$, and the pairs (DR, PE), where PQR is
   an overlap between rules $PQ \to D$ and $QR \to E$ in $\mathcal{S}$ with
   positive left sides;
$\mathcal{W} := \emptyset$;
For (A, B) in $\mathcal{V}$ do begin
   Collect A and B with respect to $\mathcal{S}$ to obtain CD and CE,
      respectively, where C is a collected word in $Z^{\pm *}$ and D and
      E are collected words in $W^{\pm *}$;
   If $D \neq E$ then add DE^{-1} to $\mathcal{W}$
End;
Let M_1 be the subgroup of M generated by the image of $\mathcal{W}$;
Let $\mathcal{P}$ be the set of power relations in the standard polycyclic
   presentation for M/M_1 in terms of the elements of W in the
   chosen order;
(* See comment in CLASS_2_QUOT. *)
For each relation $z^m = P$ in $\mathcal{P}$ add $z^m \to P$ and $z^{-1} \to z^{m-1} P^{-1}$
   to $\mathcal{S}$;
For each relation in $\mathcal{P}$ whose left side has the form z^m with $m = 1$
   use a Tietze transformation to eliminate z from the
   presentation $(Y, \mathcal{S})$;
(* All elements of W_1 are removed by the previous statement. *)
For x in X set $\sigma(x) = \varphi(x) \overline{w}_x$, where $\overline{w}_x$ denotes the result of
   applying the Tietze transformations of the previous statement
   to w_x;
$W := Z - Y$;
For a_i in Z do $\tau(a_i) := \nu(a_i)$;
(* If v_{ij} is left in W, then V_{ij} is trivial. *)
For v_{ij} in W do $\tau(v_{ij}) := \tau(a_j^{-1} a_i^{-1} a_j a_i)$;
(* Elements of Z retain their weights. *)
Assign weight e to the elements of W

End
End.

As with the previous nilpotent quotient procedures, it is possible to reduce the size of the set W of new generators considerably. Results from
Section 9.9 can be used to reduce the number of overlaps which must be
considered.

Example 5.2. Let us determine $S/\gamma_4(S)$, where S is the group of Examples 2.4 and 4.2. By Example 4.2, $H = S/\gamma_3(S)$ is given by the presentation

$$a_2 a_1 = a_1 a_2, \quad a_3 a_1 = a_1 a_3, \quad a_3 a_2 = a_2 a_3 a_4,$$
$$a_4 a_i = a_i a_4, \quad 1 \le i \le 3,$$
$$a_1^2 = a_4^5.$$

Here a_1, a_2, and a_3 have weight 1 and a_4 has weight 2. The homomorphisms
μ and ν are defined by

$$\mu(t_1) = a_4, \quad \mu(t_2) = a_1, \quad \mu(t_3) = a_2, \quad \mu(t_4) = a_3,$$
$$\nu(a_1) = t_2, \quad \nu(a_2) = t_3, \quad \nu(a_3) = t_4, \quad \nu(a_4) = t_4^{-1} t_3^{-1} t_4 t_3.$$

Let us follow what happens when we invoke NEXT_QUOT with these
input arguments. The set W_0 contains v_{34}, v_{24}, and v_{14}. The set W_1 should
contain v_{23}, v_{13}, v_{12}, u_1, w_4, w_3, w_2, and w_1.

However, w_2, w_3, and w_4 will define the identity element in $S/\gamma_4(S)$. The
element v_{23} will also define the identity, since the relation $a_3 a_2 = a_2 a_3 a_4$ of
H is preserved by ν. Thus we can take W_1 to consist of v_{34}, v_{24}, v_{14}, v_{13},
v_{12}, u_1, w_1, in that order.

The finitely generated version of EXTEND_RULES is used to fill out the
following presentation:

$$a_2 a_1 = a_1 a_2 v_{12}, \quad a_3 a_1 = a_1 a_3 v_{13},$$
$$a_3 a_2 = a_2 a_3 a_4, \quad a_4 a_1 = a_1 a_4 v_{14},$$
$$a_4 a_2 = a_2 a_4 v_{24}, \quad a_4 a_3 = a_3 a_4 v_{34},$$
$$wy = yw, \quad w \in W, \ y \in Y, \ w \text{ precedes } y,$$
$$a_1^2 = a_4^5 u_1.$$

The result S consists of

$$y^\alpha y^{-\alpha} \to \epsilon, \qquad y \in Y,$$

$$a_3 a_2 \rightarrow a_2 a_3 a_4, \qquad\qquad a_3 a_2^{-1} \rightarrow a_2^{-1} a_3 a_4^{-1} v_{24},$$

$$a_3^{-1} a_2 \rightarrow a_2 a_3^{-1} a_4^{-1} v_{34}, \qquad a_3^{-1} a_2^{-1} \rightarrow a_2^{-1} a_3^{-1} a_4 v_{24}^{-1} v_{34}^{-1},$$

$$a_j^\alpha a_i^\beta \rightarrow a_i^\beta a_j^\alpha v_{ij}^{\alpha\beta}, \quad 1 \le i < j \le 4,\ (i,j) \ne (2,3),$$

$$w^\alpha y^\beta \rightarrow y^\beta w^\alpha, \quad w \in W,\ y \in Y,\ y \text{ precedes } w \text{ in } Y,$$

$$a_1^2 \rightarrow a_4^5 u_1, \qquad\qquad a_1^{-1} \rightarrow a_1 a_4^{-5} u_1^{-1}.$$

Here α and β are elements of $\{1, -1\}$. We shall have to rewrite a number of words with respect to $\mathcal{S}$. Since $\mathcal{S}$ is not confluent, it is necessary to specify a rewriting strategy. In all cases, rewriting from the left will be used.

When we process the overlaps among the positive left sides in $\mathcal{S}$, we find that the only overlap $a_k a_j a_i$ at which local confluence fails is $a_3 a_2 a_1$. Rewriting $a_2 a_3 a_4 a_1$ gives $a_1 a_2 a_3 a_4 v_{12} v_{13} v_{14}$, while rewriting $a_3 a_1 a_2 v_{12}$ yields $a_1 a_2 a_3 a_4 v_{12} v_{13}$, so v_{14} goes into $\mathcal{W}$. Processing the overlap a_1^3, we find that $a_1 a_4^5 u_1$ is already reduced, while $a_4^5 u_1 a_1$ can be rewritten to $a_1 a_4^5 u_1 v_{14}^5$, so v_{14}^5 gets put into $\mathcal{W}$. Processing the overlaps $a_i a_1^2$ for $i = 2, 3$, and 4 causes the elements $v_{12}^2 v_{24}^5$, $v_{13}^2 v_{34}^2$, and v_{14}^2 to go into $\mathcal{W}$.

We define the regular homomorphism η from $\{t_1, t_2, t_3, t_4\}^{\pm *}$ to $Y^{\pm *}$ by

$$\eta(t_1) = a_4 w_1, \quad \eta(t_2) = a_1, \quad \eta(t_3) = a_2, \quad \eta(t_4) = a_3.$$

The first defining relation for S is $t_1 t_2 t_1^{-1} t_2^{-1} = 1$. The image under η of the left side of this relation is $a_4 w_1 a_1 w_1^{-1} a_4^{-1} a_1^{-1}$, which after rewriting with respect to $\mathcal{S}$ gives v_{14}, but v_{14} is already in $\mathcal{W}$. Applying η to the left sides of the other defining relations for S and rewriting the results yields the elements $w_1^5 u_1^{-1} v_{14}^5$, $w_1 v_{24}^3$, $w_1^{-1} v_{12} v_{14}^2 v_{34}^5$, and $v_{13} v_{14}^3 v_{34}^8$. The relations $a_i = \eta(\nu(a_i))$ have already been used in our arguments that w_2, w_3, w_4, and v_{23} are trivial, so these relations produce nothing new.

We identify the free abelian group M on $\mathcal{W}$ with $\mathbb{Z}^7$, so that

$$w_1 = (1,0,0,0,0,0,0),$$
$$u_1 = (0,1,0,0,0,0,0),$$
$$v_{12} = (0,0,1,0,0,0,0),$$
$$v_{13} = (0,0,0,1,0,0,0),$$
$$v_{14} = (0,0,0,0,1,0,0),$$
$$v_{24} = (0,0,0,0,0,1,0),$$
$$v_{34} = (0,0,0,0,0,0,1).$$

The elements of $\mathcal{W}$ map to the rows of the matrix

$$
\begin{bmatrix}
0 & 0 & 0 & 0 & 1 & 0 & 0 \\
0 & 0 & 0 & 0 & 5 & 0 & 0 \\
0 & 0 & 2 & 0 & 0 & 5 & 0 \\
0 & 0 & 0 & 2 & 0 & 0 & 5 \\
0 & 0 & 0 & 0 & 2 & 0 & 0 \\
5 & -1 & 0 & 0 & 5 & 0 & 0 \\
1 & 0 & 0 & 0 & 0 & 3 & 0 \\
-1 & 0 & 1 & 0 & 2 & 5 & 0 \\
0 & 0 & 0 & 1 & 3 & 0 & 8
\end{bmatrix},
$$

whose Hermite normal form is

$$
\begin{bmatrix}
1 & 0 & 0 & 0 & 0 & 3 & 0 \\
0 & 1 & 0 & 0 & 0 & 4 & 0 \\
0 & 0 & 1 & 0 & 0 & 8 & 0 \\
0 & 0 & 0 & 1 & 0 & 0 & 8 \\
0 & 0 & 0 & 0 & 1 & 0 & 0 \\
0 & 0 & 0 & 0 & 0 & 11 & 0 \\
0 & 0 & 0 & 0 & 0 & 0 & 11 \\
0 & 0 & 0 & 0 & 0 & 0 & 0 \\
0 & 0 & 0 & 0 & 0 & 0 & 0
\end{bmatrix}.
$$

Thus v_{24} and v_{34} each have order 11 in M/M_1 and modulo M_1 we have

$$
w_1 = v_{24}^{-3} = v_{24}^8, \quad u_1 = v_{24}^7, \quad v_{12} = v_{24}^3,
$$
$$
v_{13} = v_{34}^3, \quad v_{14} = 1.
$$

If we rename v_{24} and v_{34} as a_5 and a_6, respectively, then the relations with positive left sides in the final presentation for the group K isomorphic to $S/\gamma_4(S)$ are

$$
a_2 a_1 = a_1 a_2 a_5^3, \qquad a_3 a_1 = a_1 a_3 a_6^3,
$$
$$
a_3 a_2 = a_2 a_3 a_4, \qquad a_4 a_1 = a_1 a_4,
$$
$$
a_4 a_2 = a_2 a_4 a_5, \qquad a_4 a_3 = a_3 a_4 a_6,
$$
$$
a_5 a_i = a_i a_5, \quad 1 \le i \le 4,
$$
$$
a_6 a_i = a_i a_6, \quad 1 \le i \le 5,
$$
$$
a_1^2 = a_4^5 a_5^7.
$$

The regular homomorphisms σ and τ are given by

$$\sigma(t_1) = a_4 a_5^8, \quad \sigma(t_2) = a_1, \quad \sigma(t_3) = a_2, \quad \sigma(t_4) = a_3,$$
$$\tau(a_1) = t_2, \quad \tau(a_2) = t_3, \quad \tau(a_3) = t_4, \quad \tau(a_4) = t_4^{-1} t_3^{-1} t_4 t_3,$$
$$\tau(a_5) = \tau(a_4^{-1} a_2^{-1} a_4 a_2) = t_3^{-1} t_4^{-1} t_3 t_4 t_3^{-1} t_4^{-1} t_3^{-1} t_4 t_3^2,$$
$$\tau(a_6) = \tau(a_4^{-1} a_3^{-1} a_4 a_3) = t_3^{-1} t_4^{-1} t_3 t_4 t_4^{-2} t_3^{-1} t_4 t_3 t_4.$$

The value of $\tau(a_6)$ may be replaced by the freely reduced word freely equivalent to it.

Let $(X, \mathcal{R})$ be a finite presentation for a group G and suppose that Y, $\mathcal{S}$, σ, τ are returned by one of the procedures CLASS_2_QUOT or NEXT_QUOT as a description of a group K isomorphic to the class-e quotient of G. It may happen that the presentation for $\gamma_e(K)$ on the generators of weight e in Y does not describe $\gamma_e(K)$ as a direct product of cyclic groups, or at least not as a direct product with the minimum number of factors. The following procedure corrects this defect by returning a new description Z, T, μ, ν for which $\gamma_e(K)$ is presented as a direct product of the minimum number of cyclic groups:

Procedure ADJUST$(X, \mathcal{R}, Y, \mathcal{S}, \sigma, \tau; Z, T, \mu, \nu)$;
($*$ The arguments are described in the text. $*$)
Begin
    Let Y consist of $a_1, \ldots, a_t$, let e be the weight of a_t, and let a_s be the
        last element of Y with weight $e - 1$;
    Let the relations in $\mathcal{S}$ be $a_j^\beta a_i^\alpha = a_i^\alpha a_j^\beta V(i, j, \alpha, \beta)$, $1 \leq i < j \leq t$, α, β
        in $\{1, -1\}$, and $a_i^{m_i} = U_i$, $a_i^{-1} = a_i^{m_i - 1} W_i$, $i \in I$;
    ($*$ For $V(i, j, \alpha, \beta)$ to be defined, i must not be in I if $\alpha = -1$ and j
        must not be in I if $\beta = -1$. $*$)
    $n := t - s$;
    Let B be the integer matrix with n columns whose rows give the
        exponent sums with respect to $a_{s+1}, \ldots, a_t$ of the words $a_i^{m_i} W_i$,
        where $s < i \leq t$ and $i \in I$;
    Let A be the Smith normal form of B and let P and Q be
        unimodular matrices such that $A = PBQ$;
    $T := Q^{-1}$;
    Let $d_1, \ldots, d_r$ be the nonzero diagonal entries of A;
    If $d_1 > 0$ then $p := 0$
    Else let p be maximal such that $d_p = 1$;
    $m := n - p$; $Z := \{a_1, \ldots, a_s, b_1, \ldots, b_m\}$;
    Let η and ρ be the regular homomorphisms from $Y^{\pm *}$ to $Z^{\pm *}$ and $Z^{\pm *}$
        to $Y^{\pm *}$, respectively, such that $\eta(a_i) = \rho(a_i) = a_i$, $1 \leq i \leq s$,

$$\eta(a_{s+i}) = b_1^{Q_{ip+1}} \ldots b_m^{Q_{in}}, \; 1 \le i \le n,$$
$$\rho(b_i) = a_{s+1}^{T_{p+i1}} \ldots a_t^{T_{p+in}}, \; 1 \le i \le m;$$

Let $\mathcal{T}$ consist of the following relations: $a_j^\alpha a_i^\beta = a_i^\beta a_j^\alpha \eta(V(i,j,\alpha,\beta))$,
$1 \le i < j \le s$, $b_j^\alpha a_i^\beta = a_i^\beta b_j^\alpha$, $1 \le j \le m$, $1 \le i \le s$, $b_j^\alpha b_i^\beta = b_i^\beta b_j^\alpha$,
$1 \le i < j \le m$, $a_i^{m_i} = \eta(U_i)$, and $a_i^{-1} = a_i^{m_i-1}\eta(W_i)$, $1 \le i < j \le s$,
$i \in I$, $b_i^{d_{p+i}} = 1$ and $b_i^{-1} = b_1^{d_{p+i}-1}$, $1 \le i \le p - r$;
Rewrite the right sides of the relations in $\mathcal{T}$ with respect to $\mathcal{T}$;
Let μ be the composition of σ and η;
Let ν be the composition of τ and ρ;
End.

It is not clear whether on average applying ADJUST will speed up or slow down the computation of the class-$(e+1)$ quotient of G. The number of generators in the presentation may decrease, but the homomorphisms get more complicated and it may not be possible to spot easily elements of W in NEXT_QUOT which are trivial. More experimentation is needed.

Exercises

5.1. Suppose in NEXT_PQUOT that the relation $a_k^p = a_j$ is the definition of a_j and $w_j = w_k + 1 = e - 1$. Suppose also that $w_i = 1$. Show that in $G/\varphi_{e+1}(G)$ we have $[a_j, a_i] = [a_k^p, a_i] = [a_k, a_i]^p$. Use this fact to describe how to eliminate v_{ij} from W_0.
5.2. Let G be the group in Exercise 4.1. Compute $G/\varphi_4(G)$ for $p = 2$ and $G/\gamma_4(G)$.
5.3. In NEXT_QUOT we assumed that $H = \mathrm{Grp}\,\langle Z \mid \mathcal{T} \rangle$ is isomorphic to $G/\gamma_e(G)$. Suppose that H is just some polycyclic group, not necessarily nilpotent, that $(Z, \mathcal{T})$ is a consistent polycyclic presentation for H, and that μ and ν define inverse isomorphisms between a quotient G/N and H. Devise a procedure for finding a consistent polycyclic presentation for $G/[N, G]$.

11.6 Metabelian quotients

Let G be a group given by a finite presentation $(X, \mathcal{R})$ and let e be a positive integer. The group $G/G^{(e)}$ is the largest quotient of G which is solvable and has derived length at most e. In (Baumslag et al. 1981a,b), Baumslag, Cannonito, and Miller showed that it is possible to decide whether $G/G^{(e)}$ is polycyclic and, if the answer is positive, to construct a consistent polycyclic presentation for $G/G^{(e)}$. Baumslag, Cannonito, and Miller made no attempt to describe their polycyclic quotient algorithm in a form suitable for machine computation. This section describes the ideas involved in the algorithm and presents the metabelian case, the case in which $e = 2$, in detail. In order to have a chance of handling solvable quotients with larger derived lengths efficiently, we shall need generalizations of Gröbner basis methods for modules over group rings of nonabelian polycyclic groups, perhaps along the lines sketched in Section 10.8.

We can use ABEL_QUOT to obtain a group presentation $(Z, \mathcal{T})$ for a group H isomorphic to G/G' and to find associated regular homomorphisms μ and ν. Let $Z = \{a_1, \ldots, a_r\}$ and let $\mathcal{T}$ contain the relations

$$a_j a_i = a_i a_j, \quad 1 \leq i < j \leq r,$$
$$a_i^{m_i} = 1, \quad i \in I.$$

To obtain a description of G/G'', we begin as in CLASS_2_QUOT. We initially set $Y = Z$ and $\tau = \nu$. We then add to Y new generators v_{ij} with $1 \leq i < j \leq t$, u_i with i in I, and w_x with x in X and define $\tau(v_{ij}) = \tau(a_j^{-1} a_i^{-1} a_j a_i)$, $\tau(u_i) = \tau(a_i^{m_i})$, and $\tau(w_x) = \tau(\mu(x))^{-1} x$. We also set $\sigma(x) = \mu(x) w_x$ for x in X. Let W be the set of new generators. By Corollary 1.2, G'/G'' is generated as a $\mathbb{Z}[H]$-module by the image of W. Let $F = F_Y$, let N be the kernel of the homomorphism $\overline{\tau}$ of F onto G/G'' defined by τ, and set $K = F/N$.

It is at this point that we diverge from CLASS_2_QUOT. Rather than considering the free abelian group generated by W, we define M to be the free $\mathbb{Z}[H]$-module generated by W. The group K is a homomorphic image of the formal extension K_0 of M by H in which

$$a_j a_i = a_i a_j [v_{ij}], \quad 1 \leq i < j \leq r,$$
$$a_i^{m_i} = [u_i], \quad i \in I.$$

Proposition 3.2 tells us generators for the kernel M_0 of the map from M to K_0. We use the general version of EXTEND_RULES to form a (not necessarily confluent) rewriting system $\mathcal{S}$ for K_0. If U is a word in $(Z \cup M)^{\pm*}$ which is contained in the ideal $\mathcal{I}$ generated by M, then U can be rewritten using $\mathcal{S}$ into the form $V[u]$, where V is a collected word in $Z^{\pm*}$ defining an element of H and u is in M. Let $\mathcal{V}$ be the set of ordered pairs $(\sigma(R)[0], \sigma(S)[0])$ with (R, S) in $\mathcal{R}$ and $(a_i[0], \sigma(\tau(a_i))[0])$, $1 \leq i \leq r$. Note that the factor $[0]$ insures that the components of the pairs in $\mathcal{V}$ lie in $\mathcal{I}$. For each pair (A_1, A_2) in $\mathcal{V}$, we rewrite A_1 and A_2 with respect to $\mathcal{S}$ to obtain words $C[c_1]$ and $C[c_2]$, where C is a collected word in $Z^{\pm*}$ and the c_i are in M. The elements $c_1 - c_2$ so obtained, together with M_0, generate the kernel M_1 of the map from M to K.

The group ring $\mathbb{Z}[H]$ can be described as a quotient of a ring R which is either a polynomial ring or a ring of Laurent polynomials. See Section 10.8. The free $\mathbb{Z}[H]$-module M can be described as a quotient of R^k, where $k = |W|$. Using the Gröbner basis techniques of Chapter 10, we can find a Gröbner basis for the submodule L of R^k corresponding to M_1. By Proposition 6.6 in Chapter 10, we can decide whether $M/M_1 \cong R^k/L$ is finitely generated as an abelian group. If the answer is no, then G/G'' is not polycyclic. However, if M/M_1 is a finitely generated abelian group, then we can obtain a finite group presentation for it and the corresponding description of the module action of $\mathbb{Z}[H]$. It is then a simple matter to write down a consistent polycyclic presentation for K.

The procedure just outlined will be referred to as METABEL_QUOT. In developing an implementation of METABEL_QUOT, one has to make a choice between taking R to be a ring of ordinary or Laurent polynomials and decide which ordering to use on the generators of R^k as a free abelian group in the Gröbner basis computation.

Example 6.1. Let us turn again to the group S of Examples 2.4, 4.2, and 5.2. By Example 2.4, S/S' is isomorphic to $H = \mathrm{Grp}\langle Z \mid T \rangle$, where $Z = \{a_1, a_2, a_3\}$, T contains the relations

$$a_j a_i = a_i a_j, \quad 1 \leq i < j \leq 3,$$
$$a_1^2 = 1,$$

and the homomorphisms μ and ν are given by

$$\mu(t_1) = 1, \quad \mu(t_2) = a_1, \quad \mu(t_3) = a_2, \quad \mu(t_4) = a_3,$$
$$\nu(a_1) = t_2, \quad \nu(a_2) = t_3, \quad \nu(a_3) = t_4.$$

In METABEL_QUOT, the set W should contain w_1, w_2, w_3, w_4, u_1, v_{12}, v_{13}, v_{23}, where w_i denotes w_{t_i}. However, as in Example 4.2, the fact that $\mu(\nu(a_i)) = a_i$, $1 \leq i \leq 3$, means that w_2, w_3, and w_4 will be trivial in S/S'', so they can be omitted.

The procedure EXTEND_RULES produces the following rewriting system for the formal extension K_0 of M by H:

$$[u][v] \rightarrow [u + v], \quad u, v \in M,$$
$$[u]^{-1} \rightarrow [-u], \quad u \in M,$$
$$[u]a_i^\alpha \rightarrow a_i^\alpha [ua_i^\alpha], \quad u \in M, \ 1 \leq i \leq 3, \ \alpha = \pm 1,$$
$$a_1^2 \rightarrow [u_1],$$
$$a_1^{-1} \rightarrow a_1[-u_1],$$

$$\left.
\begin{aligned}
a_j a_i &\rightarrow a_i a_j [v_{ij}], \\
a_j^{-1} a_i &\rightarrow a_i a_j^{-1} [-v_{ij} a_j^{-1}], \\
a_j a_i^{-1} &\rightarrow a_i^{-1} a_j [-v_{ij} a_i^{-1}], \\
a_j^{-1} a_i^{-1} &\rightarrow a_i^{-1} a_j^{-1} [v_{ij} a_i^{-1} a_j^{-1}],
\end{aligned}
\right\} \quad 1 \leq i < j \leq 3.$$

Let M_0 be the $\mathbb{Z}[H]$-submodule of M generated by the following elements:

$$u_1(a_1 - 1), \quad u_1(a_2 - 1) + v_{12}(1 + a_1), \quad u_1(a_3 - 1) + v_{13}(1 + a_1),$$
$$v_{12}(a_3 - 1) + v_{23}(a_1 - 1) - v_{13}(a_2 - 1).$$

By Proposition 3.2, the preceding rewriting system is confluent modulo M_0. The homomorphism σ is given by

$$\sigma(t_1) = [w_1], \quad \sigma(t_2) = a_1, \quad \sigma(t_3) = a_2, \quad \sigma(t_4) = a_3.$$

The image in K_0 of the first relator $t_1 t_2 t_1^{-1} t_2^{-1}$ for S is $[w_1] a_1 [w_1]^{-1} a_1^{-1}$. One way to collect this element is

$$[w_1]a_1[w_1]^{-1}a_1^{-1} \rightarrow [w_1]a_1[-w_1]a_1[-u_1] \rightarrow [w_1]a_1^2[-w_1 a_1][-u_1] \rightarrow$$
$$[w_1][u_1][-w_1 a_1][-u_1] \rightarrow [w_1 - w_1 a_1].$$

The images in K_0 of the other four relators for S can be collected to the following elements:

$$[5w_1 - u_1 a_1], \quad [w_1(2 - 2a_2^{-1} + a_3^{-1}) - v_{23}a_2^{-1}a_3^{-1}],$$
$$[w_1(1 - a_2^{-1} - a_1 a_3^{-1}) + u_1(1 - a_2^{-1}) + v_{12}a_2^{-1} + v_{23}a_1 a_2^{-1}a_3^{-1}],$$
$$[w_1(3 - 3a_1 a_3^{-1}) + u_1(1 - a_3^{-1}) + v_{13}a_3^{-1}].$$

The relations $a_i = \sigma(\tau(a_i))$ have already been used to conclude that w_2, w_3, and w_4 are trivial and produce nothing new. Thus S'/S'' is isomorphic to M/M_1, where M_1 is the $\mathbb{Z}[H]$-submodule generated by M_0 and the elements

$$w_1(1 - a_1), \quad 5w_1 - u_1 a_1, \quad w_1(2 - 2a_2^{-1} + a_2^{-1}) - v_{23}a_2^{-1}a_3^{-1},$$
$$w_1(1 - a_2^{-1} - a_1 a_3^{-1}) + u_1(1 - a_2^{-1}) + v_{12}a_2^{-1} + v_{23}a_1 a_2^{-1}a_3^{-1},$$
$$w_1(3 - 3a_1 a_3^{-1}) + u_1(1 - a_3^{-1}) + v_{13}a_3^{-1}.$$

In this example we shall compute in M using ordinary polynomials. Let $R = \mathbb{Z}[Y_1, Y_2, Z_2, Y_3, Z_3]$ and let J be the ideal of R generated by the set $\mathcal{M} = \{Y_1^2 - 1, Y_2 Z_2 - 1, Y_3 Z_3 - 1\}$. The group ring $\mathbb{Z}[H]$ is isomorphic to R/J. Let E be the free R-module generated by W. In Chapter 10 we used the notation of left modules, but it is more convenient to continue to use right modules here. We can identify the free $\mathbb{Z}[H]$-module M with E/D, where D is the R-submodule generated by the 15 products of an element in W by an element of $\mathcal{M}$.

The quotient M/M_1 is isomorphic to E/E_1, where E_1 is generated by D and the elements

$$u_1(Y_1 - 1), \quad u_1(Y_2 - 1) + v_{12}(1 + Y_1), \quad u_1(Y_3 - 1) + v_{13}(1 + Y_1),$$
$$v_{12}(Y_3 - 1) + v_{23}(Y_1 - 1) - v_{13}(Y_2 - 1),$$
$$w_1(1 - Y_1), \quad 5w_1 - u_1 Y_1, \quad w_1(2 - 2Z_2 + Z_3) - v_{23} Z_2 Z_3,$$
$$w_1(1 - Z_2 - Y_1 Z_3) + u_1(1 - Z_2) + v_{12} Z_2 + v_{23} Y_1 Z_2 Z_3,$$
$$w_1(3 - 3Y_1 Z_3) + u_1(1 - Z_3) + v_{13} Z_3.$$

Thus E_1 has a total of 24 generators.

To find a Gröbner basis for E_1, we need to choose an ordering of the $\mathbb{Z}$-basis for E consisting of all products of elements of W by monomials. Let us use the ordering in which elements of W are compared first using

$$w_1 \prec v_{12} \prec v_{13} \prec v_{23} \prec u_1$$

and then the monomials $Y_1^{\alpha_1} Y_2^{\alpha_2} Z_2^{\beta_2} Y_3^{\alpha_3} Z_3^{\beta_3}$ are compared using the reverse lexicographic ordering. The resulting Gröbner basis has 15 elements. The first four show that E/E_1 is generated as an R-module by the image of w_1:

$$u_1 - 5w_1, \quad v_{23} + w_1(8Y_2 - 9), \quad v_{13} + w_1(8Y_3 - 8), \quad v_{12} + w_1(8Y_2 - 8).$$

The remaining 11 basis elements involve only w_1.

$$w_1(Y_3 Z_3 - 1), \quad w_1(Z_2 Z_3 + 10Z_3 + 10Z_2 - 21),$$
$$w_1(Y_2 Z_3 + 10Z_3 + 10Y_2 - 21), \quad w_1(11Z_3 - 11),$$
$$w_1(Z_2 Y_3 + 10Y_3 + 10Z_2 - 21), \quad w_1(Y_2 Y_3 + 10Y_3 + 10Y_2 - 21),$$
$$w_1(11Y_3 - 11), \quad w_1(Y_2 Z_2 - 1), \quad w_1(11Z_2 - 11),$$
$$w_1(11Y_2 - 11), \quad w_1(Y_1 - 1).$$

Since there is no basis element with leading term of the form $w_1 Y_2^r$, it follows that $S'/S'' \cong E/E_1$ is not finitely generated as an abelian group and hence S/S'' is not polycyclic.

The limited experimentation with METABEL_QUOT which has been done so far suggests that the procedure works fairly well when the torsion subgroup of G/G' is small. When some of the m_i are large, then the elements of type (2) in Proposition 3.2 produce monomials of high degree and this dramatically slows the Gröbner basis computation.

If p is a prime, then we can investigate $G/G''(G')^p$ by computing the Gröbner basis using coefficients in $\mathbb{Z}_p$ rather than $\mathbb{Z}$. With the group S of Example 6.1, it is relatively easy to see that $S'/S''(S')^p$ is cyclic if $p \neq 11$ and is infinite for $p = 11$.

Exercises

6.1. Let G be the group of Exercises 2.1, 4.1, and 5.2. Is G/G'' polycyclic?

6.2. Suppose in the input to METABEL_QUOT that $H = \mathrm{Grp}\langle Z \mid T \rangle$ is not necessarily isomorphic to G/G' but is just some abelian group which is a homomorphic image of G. Interpret the output of METABEL_QUOT.

11.7 Enforcing exponent laws

One important application of the p-quotient algorithm has been in the study of Burnside groups. Let d and q be positive integers. The Burnside group $B(d, q)$ is the largest group generated by d elements such that the q-th power of every element in the group is 1. More precisely, let X be a set with d elements and let $\mathcal{B} = \{W^q \mid W \in X^{\pm *}\}$. Then $B(d, q) = \mathrm{Grp}\langle X \mid \mathcal{B} \rangle$. Thus $B(d, q)$ is the largest d-generator group with exponent dividing q.

Information about Burnside groups can be found in [Kostrikin 1990] and [Vaughan-Lee 1990]. The group $B(d, q)$ is known to be finite if $q \leq 4$ or $q = 6$. It is known to be infinite if $d \geq 2$ and q is an odd integer greater than or equal to 665. (Lysenok has announced the result that 665 may be replaced by 115.) It is not known whether the groups $B(2, 5)$ and $B(2, 8)$ are finite. These groups have been the subject of considerable study.

Recently a positive solution to the so-called *restricted Burnside problem* was announced by E. I. Zel'manov. This result states that $B(d, q)$ has a largest finite quotient, which is denoted $R(d, q)$. This means that $B(d, q)$ has a smallest normal subgroup $N(d, q)$ of finite index. If q is a power of a prime p, then $R(d, q)$ is a p-group and hence is nilpotent. Therefore $N(d, q)$ is $\gamma_i(B(d, q))$ and also $\varphi_j(B(d, q))$ for some integers i and j. Here $\varphi_j(B(d, q))$ is defined with respect to p. Thus it would be possible, at least in principle, to determine a consistent polycyclic presentation for $R(d, q)$ using the algorithms of this chapter, if only $B(d, q)$ were finitely presented. Unfortunately, since the set $\mathcal{B}$ is infinite, this need not be the case.

Let q be a power of a prime p and set $B = B(d, q)$. For a given integer e, it is possible to determine $B/\varphi_{e+1}(B)$, even though B is not defined by a finite presentation. The method is based on an efficient solution to the following problem: Given a consistent polycyclic presentation for a finite p-group G, determine whether $g^q = 1$ for every g in G. Since G is finite and we can compute products and compare elements in G, we could try to find the answer by actually computing the q-th power of each element in G. However, the groups G with which we shall be dealing have very large orders, so this brute force approach is hopelessly impractical.

If G happens to be abelian, then our problem is easy. We have only to check whether the generators of G have orders dividing q. If G is non-abelian, then this may not be adequate. For example, the permutations $x = (1, 2)(3, 4)$ and $y = (1, 3)$ generate the dihedral group of order 8. Both x and y have order 2, but $xy = (1, 2, 3, 4)$ has order 4. If p is large with respect to the class of G, then our problem is still easy. In Chapter 4 of [Hall 1959], it is shown that if G has class less than p, then the exponent of G is the maximum of the orders of a set of generators.

When the class of G is large compared with p, it is more difficult to determine the exponent of G. Perhaps the most powerful approach is based on a result of Graham Higman found in (Higman 1959). Let F be the free group on a set X. For x in X, let π_x be the endomorphism of F which takes x to 1 and maps each y in $X - \{x\}$ to itself. Define $1 - \pi_x$ to be the function from F to itself taking an element u to $(\pi_x u)^{-1} u$.

Proposition 7.1. *For any x and y in X, the maps π_x, $(1 - \pi_x)$, π_y, and $(1 - \pi_y)$ all commute. Also $\pi_x \pi_x = \pi_x$ and $\pi_x(1 - \pi_x)u = 1$ for every u in F.*

Proof. The compositions $\pi_x \pi_y$ and $\pi_y \pi_x$ each map x and y to 1 and fix every other element of X. Clearly $\pi_x \pi_x$ and π_x agree. For any u in F we have

$$(1 - \pi_y)\pi_x u = (\pi_y \pi_x u)^{-1} \pi_x u = (\pi_x \pi_y u)^{-1} \pi_x u$$

and

$$\pi_x(1 - \pi_y)u = \pi_x(\pi_y(u)^{-1} u) = (\pi_x \pi_y u)^{-1} \pi_x u,$$

so $(1 - \pi_y)$ and π_x commute. Finally,

$$\pi_x(1 - \pi_x)u = \pi_x((\pi_x u)^{-1} u) = (\pi_x \pi_x u)^{-1}(\pi_x u) = (\pi_x u)^{-1}(\pi_x u) = 1. \quad \square$$

For any subset Y of X, define $F(Y)$ to be the intersection of the kernels of the endomorphisms π_y with y in Y. In the following, the word "commutator" will refer to any element of F obtained as an iterated commutator starting with elements of $X^{\pm}$.

Proposition 7.2. *Let Y be a finite subset of X. Then every element of $F(Y)$ can be written as a product of commutators each of which involves every element of Y.*

Proof. If $Y = \emptyset$, then $F(Y) = F$. Any element of F can be written as a product of elements of $X^{\pm}$ and these factors are commutators which involve every element of Y. We proceed by induction on $|Y|$. Suppose Y is nonempty, let y be an element of Y, and let w be in $F(Y)$. Then w is in $F(Y - \{y\})$, so $w = u_1 \ldots u_r$, where each u_i is a commutator involving every element of $Y - \{y\}$. Suppose some u_i involves y as well but u_{i+1} does not involve y. Then replacing the terms $u_i u_{i+1}$ with $u_{i+1} u_i [u_i, u_{i+1}]$ does not change the product. The commutator $[u_i, u_{i+1}]$ introduced involves all the elements of Y. In this way we can move all commutators which do not involve y to the beginning of the product. Thus we may assume that $u_1, \ldots, u_s$ do not involve y, while $u_{s+1}, \ldots, u_r$ do involve y. But π_y maps any commutator involving y to 1, and fixes any commutator not involving y. Therefore $1 = \pi_y w = u_1 \ldots u_s$ and $w = u_{s+1} \ldots u_r$ is the required product.

$\square$

For any finite subset Y of X let π_Y be the product of the endomorphisms π_y with y in Y. By Proposition 7.1, π_Y is well defined.

Proposition 7.3. *Let Y be a nonempty, finite subset of X. Any element w of F can be expressed as a product $u_1 \ldots u_r v$, where v is a product of commutators each involving every element of Y and the u_i have the form $\pi_Z w^{\pm 1}$ for some nonempty subset Z of Y.*

Proof. Let $Y = \{y_1, \ldots, y_n\}$ and set $v = (1 - \pi_{y_1}) \ldots (1 - \pi_{y_n}) w$. Then

$$
\begin{aligned}
v &= (1 - \pi_{y_1}) \ldots (1 - \pi_{y_{n-1}})(\pi_{y_n} w)^{-1} w \\
&= (1 - \pi_{y_1}) \ldots (1 - \pi_{y_{n-2}})(\pi_{y_{n-1}}(\pi_{y_n} w)^{-1} w)^{-1}(\pi_{y_n} w)^{-1} w \\
&= (1 - \pi_{y_1}) \ldots (1 - \pi_{y_{n-2}})(\pi_{Z_1} w^{-1})(\pi_{Z_2} w)(\pi_{Z_3} w^{-1}) w,
\end{aligned}
$$

where $Z_1 = \{y_{n-1}\}$, $Z_2 = \{y_{n-1}, y_n\}$, and $Z_3 = \{y_n\}$. By induction, $v = u_r^{-1} \ldots u_1^{-1} w$, where each u_i is $\pi_Z w^{\pm 1}$ for some nonempty subset Z of Y. Then $w = u_1 \ldots u_r v$ and by Proposition 7.1, v is in $F(Y)$, so by Proposition 7.2 v is a product of commutators involving all elements of Y.

$\square$

Proposition 7.4. *Let $x_1, \ldots, x_s$ be distinct elements of X and let q be a positive integer. Then $(x_1 \ldots x_s)^q$ can be written as uv, where u is in the subgroup of F generated by the products $(x_{i_1} \ldots x_{i_r})^q$, where $1 \le i_1 < \cdots < i_r \le s$ and $r < s$, and v is a product of commutators each involving all of the x_i.*

Proof. In Proposition 7.3, let $Y = \{x_1, \ldots, x_s\}$ and $w = (x_1 \ldots x_s)^q$. If $1 \le i_1 < \cdots < i_r \le s$, then $(x_{i_1} \ldots x_{i_r})^q = \pi_Z w$, where $Z = Y - \{i_1, \ldots, i_r\}$.

$\square$

Corollary 7.5. *Let G be a group which is nilpotent of class at most c and let A be a subset of G which generates G as a monoid. Suppose that for every word W in A^* with $|W| \le c$ we have $W^q = 1$ in G. Then the exponent of G divides q.*

Proof. We prove by induction on $|W|$ that $W^q = 1$ in G for every word W in A^*. By hypothesis, we may assume that $|W| > c$. Let $W = a_1 \ldots a_s$, where each a_i is in A. Let $X = \{x_1, \ldots, x_s\}$ be a set of cardinality s. There is a homomorphism f of the free group F on X into G under which x_i goes to a_i. Under this homomorphism $w = (x_1 \ldots x_s)^q$ maps to W^q. By Proposition 7.4, $w = uv$, where u is a product of q-th powers of words with lengths less than s and v is a product of commutators involving each of $x_1, \ldots, x_s$. Under f, the image of v is in $\gamma_s(G)$, and $\gamma_s(G) = 1$, since $s \geq c + 1$. The image of u is a product of q-th powers of words in A^* of length less than s. By induction, these q-th powers are trivial. Therefore $W^q = f(w) = 1$. $\square$

We can extend this idea even further. Let G be a finite p-group given by a consistent, φ-weighted presentation on generators $a_1, \ldots, a_n$. Let w_i denote the weight of a_i and set $c = w_n$. An element g of G can be expressed uniquely as a collected word $a_1^{\alpha_1} \ldots a_n^{\alpha_n}$. Define the *weight* of g with respect to the a_i to be $\alpha_1 w_1 + \cdots + \alpha_n w_n$.

Corollary 7.6. *Under these assumptions, the exponent of G is the maximum of the orders of the elements in G of weight at most c.*

Proof. Let q be the maximum of the orders of the elements of weight at most c. Since G is a p-group, $g^q = 1$ for every element g of weight at most c. Let $W = a_1^{\alpha_1} \ldots a_n^{\alpha_n}$ be a collected word defining an element g of weight greater than c. In the proof of Corollary 7.5 the commutator factors of v map to the identity in G and the q-th power factors of u map to q-th powers of words defining elements of weight less than the weight of g. Thus $W^q = 1$. $\square$

Example 7.1. In (Havas, Wall, & Wamsley 1974) the order of the group $R(2,5)$ was shown to be 5^{34}. The following consistent, weighted presentation for $R(2,5)$ was obtained using Version 3.7 of the system Cayley. The generators $a_1, \ldots, a_{34}$ satisfy the relations $a_i^5 = 1$, $1 \leq i \leq 34$. Most of the 561 commutators (a_j, a_i) with $i < j$ are trivial. Here are the nontrivial commutators. To save space, i has been written for a_i.

$[2, 1] = 3$, $\quad [3, 1] = 4$, $\quad [3, 2] = 5$, $\quad [4, 1] = 6$, $\quad [4, 2] = 7$,

$[4, 3] = 9\,11\,12^4\,15^2\,18^4\,19^4\,20^4\,21^2\,22^4\,25\,27^2\,28^4\,29^4\,30\,31^4\,33^3\,34^3$,

$[5, 1] = 7\,9\,10^2\,11\,12^2\,13^3\,14^4\,15^3\,16^3\,20^3\,21\,24^3\,25\,26^3\,27^2\,28^2\,29^2\,30^3\,31^3\,32^3$
$\qquad\quad 33^4\,34^2$,

$[5, 2] = 8$,

$[5, 3] = 10^2\,12^3\,13\,14^4\,15^2\,16^2\,17\,18\,19\,21^2\,24^2\,26\,29^4\,31^2\,32^4\,33\,34^2$,

$[5, 4] = 12^2\,13\,15^4\,16^3\,17^3\,18^4\,20^3\,21^3\,22^4\,24\,25^4\,26^2\,27^2\,28^3\,30^3\,31^2\,33^2\,34$,

$[6, 1] = 11\,15\,16^2\,19^2\,20^4\,21^2\,23\,24\,25^2\,26^3\,27\,29^4\,30^3\,31^4\,32^3\,33^4\,34^3$,

$[6,2] = 9, \quad [6,3] = 11\,16^4\,20^3\,21\,26^4\,27^2\,31^2\,33^4\,34^2,$

$[6,4] = 19\,23\,24^2\,25^2\,29^4\,30^4\,32^2\,33^2\,34,$

$[6,5] = 15^2\,16^4\,19\,20^4\,21^4\,24\,25^2\,26\,27^4\,29^4\,32^4\,33\,34^2,$

$[7,1] = 9^2\,11^2\,12^2\,13\,15\,16^3\,17\,20\,21^3\,22\,23\,24^4\,25^4\,26^2\,27\,28^4\,29^2\,31\,32^3$
$\qquad 33^2\,34^2,$

$[7,2] = 10,$

$[7,3] = 12^3\,13\,16^3\,17\,18\,20^3\,21^3\,22^2\,24^2\,25\,26\,27\,29^3\,30^4\,31^3\,32^4\,33^3\,34^3,$

$[7,4] = 15^4\,16^3\,19^3\,20^3\,21^3\,24^4\,25\,27^2\,30^3\,31^4\,33^4\,34^2,$

$[7,5] = 17\,18^3\,20^2\,21^2\,22\,24^4\,25^3\,27^2\,28^2\,29^2\,32^2\,34^4,$

$[7,6] = 19^2\,24^3\,25\,29\,30\,32^4\,33^2,$

$[8,1] = 10^3\,12^2\,13^2\,14^3\,15^2\,16\,17^3\,19\,24^4\,27^2\,28^3\,29^3\,30^2\,32\,33\,34^2,$

$[8,2] = 14^3\,17^3\,22^3\,25^2\,26\,27^4\,29\,30\,31^2\,33^3\,34^4,$

$[8,3] = 14^2\,17^3\,18\,20^2\,21^2\,22^4\,24\,26^2\,27^4\,28^4\,30^2\,31^2\,32^4,$

$[8,4] = 17^3\,18^4\,20\,22\,24^4\,25^3\,26^3\,27^2\,29^2\,32^4\,33\,34^3,$

$[8,5] = 22^3\,26\,27\,28\,30^3\,31^4\,33^3\,34^4,$

$[8,6] = 20^3\,21^4\,24^4\,25^2\,26^3\,27^3\,29^4\,31^3\,32^2\,33^3,$

$[8,7] = 22^4\,27^3\,28\,30^3\,31\,32\,33\,34^4, \quad [9,1] = 11, \quad [9,2] = 12,$

$[9,3] = 15^4\,16\,19^4\,20\,21^2\,24^2\,25^2\,26^2\,27\,29^3\,30^3\,31^2\,33\,34^3,$

$[9,4] = 19^2\,24^2\,25^2\,29\,32\,34, \quad [9,5] = 21^3\,25^2\,27\,29^4\,30^3\,32\,33^3,$

$[9,6] = 23\,29\,32^4\,34^2,$

$[9,7] = 24^2\,25^4\,29\,30^3\,32\,33\,34^2, \quad [9,8] = 26^2\,27\,30^3\,31^2\,32^4\,33^3\,34^2,$

$[10,1] = 13,$

$[10,2] = 14, \quad [10,3] = 17^2\,18^3\,20^3\,21\,22^3\,24\,25^2\,26^3\,27^4\,28^4\,29^3\,30^4\,31^3$
$\qquad 32^3\,34^3,$

$[10,4] = 20^4\,21^3\,24^3\,26\,27^4\,32\,33^3\,34^4, \quad [10,5] = 22^2\,26^2\,28\,30^2\,31^4\,32^4$
$\qquad 33^3\,34^2,$

$[10,6] = 24^4\,25^4\,29\,30\,32\,33^2\,34^2, \quad [10,7] = 26^2\,27^4\,30^4\,31^2\,34,$

$[10,8] = 28\,31^4\,33^3\,34,$

$[10,9] = 32^3\,34^2, \quad [11,1] = 19\,24^2\,25^3\,29\,30\,32^3\,33^3\,34^2, \quad [11,2] = 15,$

$[11,3] = 19\,25^4\,30^2\,33^2\,34^3, \quad [11,4] = 23\,29^2\,32^3\,34,$

$[11,5] = 24^2\,25^4\,29^4\,30^4\,32\,34,$

$[11,7] = 29\,32^4\,34^2, \quad [11,8] = 30^2\,32^3\,33^2\,34^4, \quad [11,10] = 32^3,$

$[12,1] = 16,$

$[12,2] = 17, \quad [12,3] = 20^4\, 21^3\, 24^4\, 25^3\, 26^3\, 27^4\, 29\, 30^4\, 31^3\, 32^4,$

$[12,4] = 25\, 29^4\, 30^2\, 32^2\, 34^3,$

$[12,5] = 26^3\, 27^2\, 30^2\, 31^3\, 32^4\, 34^4, \quad [12,6] = 29\, 32^3,$

$[12,7] = 30^2\, 32^2\, 33^4\, 34^2,$

$[12,8] = 31\, 34, \quad [12,9] = 32^3\, 34^4, \quad [12,10] = 33^4\, 34^3,$

$[13,1] = 15^2\, 16^2\, 20^2\, 21^2\, 23\, 26^3\, 29^2\, 31^4\, 32\, 33^4\, 34^4, \quad [13,2] = 18,$

$[13,3] = 21^2\, 24^3\, 25^2\, 29\, 31^4\, 32, \quad [13,4] = 25^3\, 29\, 30^2\, 32^3\, 33^4\, 34,$

$[13,5] = 26^4\, 27\, 30\, 31^4\, 32^2\, 34^4, \quad [13,6] = 29\, 32^4\, 34^2, \quad [13,7] = 30^4\, 33,$

$[13,8] = 31^4\, 33\, 34^3, \quad [13,9] = 32^3, \quad [13,10] = 33\, 34,$

$[14,1] = 17^2\, 18^4\, 20^3\, 21^3\, 22^4\, 24^4\, 25^4\, 27\, 29^4\, 31^2\, 33\, 34, \quad [14,2] = 31^4\, 33^4,$

$[14,3] = 22\, 26\, 27\, 28\, 30^4\, 31\, 32^3\, 33^3\, 34^4, \quad [14,4] = 26^2\, 27^4\, 30^4\, 31^3\, 33,$

$[14,5] = 28\, 31\, 33^4\, 34^3, \quad [14,6] = 30^4\, 32\, 33^2, \quad [14,7] = 31^4\, 33^4,$

$[14,9] = 33^4\, 34,$

$[15,1] = 19, \quad [15,2] = 20, \quad [15,3] = 24^4\, 25\, 29\, 30\, 32\, 33^2\, 34^4,$

$[15,5] = 32^4\, 34,$

$[15,7] = 32^3\, 34, \quad [15,8] = 33^2\, 34^3, \quad [15,10] = 34^3,$

$[16,1] = 19^4\, 23^2\, 24\, 25^3\, 29\, 30^3\, 32^2\, 33^3\, 34^2, \quad [16,2] = 21,$

$[16,3] = 24^4\, 25\, 29^3\, 30^4\, 32\, 33\, 34,$

$[16,4] = 29^3\, 32^2\, 34^2, \quad [16,5] = 30^3\, 32^4\, 34^4, \quad [16,7] = 32^3\, 34,$

$[16,8] = 33^2\, 34^3,$

$[16,10] = 34^3, \quad [17,1] = 20^4\, 21^4\, 24^3\, 25^4\, 26^4\, 29^4\, 30\, 31^3\, 32^4\, 33\, 34^4,$

$[17,2] = 34^3,$

$[17,3] = 26\, 27\, 30^2\, 32^3\, 33^2\, 34^3, \quad [17,4] = 30^3\, 32^4\, 34^4, \quad [17,5] = 31\, 33^2\, 34^3,$

$[17,6] = 32^3\, 34, \quad [17,7] = 33^3\, 34^4, \quad [17,9] = 34^3,$

$[18,1] = 20^2\, 21^4\, 24^4\, 25\, 26^3\, 27^3\, 30^2\, 31^3\, 33^3\, 34^2, \quad [18,2] = 22,$

$[18,3] = 26\, 27\, 30^2\, 31^4\, 32^3\, 33^3\, 34^4, \quad [18,4] = 30\, 32^2\, 33^3\, 34^3,$

$[18,5] = 31^2\, 34^4,$

$[18,6] = 32^3\, 34, \quad [18,7] = 33^3\, 34^4, \quad [18,9] = 34^3, \quad [19,1] = 23,$

$[19,2] = 24,$

$[19,3] = 29^4\, 32^3, \quad [19,5] = 32^2\, 34^3, \quad [20,1] = 25, \quad [20,2] = 26,$

$[20,3] = 30^4\, 32^4\, 33^3\, 34^3,$

$[20,4] = 32^2\, 34^4, \quad [20,5] = 33^3\, 34, \quad [21,1] = 24^3\, 25\, 29\, 30\, 32^4\, 33^2\, 34^4,$

$[21,2] = 27, \quad [21,3] = 30\, 32\, 33^3, \quad [21,4] = 34, \quad [21,5] = 33\, 34,$

Table 11.7.1

Weight	Generators
1	a_1, a_2
2	a_3
3	a_4, a_5
4	a_6, a_7, a_8
5	a_9, a_{10}
6	$a_{11}, \ldots, a_{14}$
7	$a_{15}, \ldots, a_{18}$
8	$a_{19}, \ldots, a_{22}$
9	$a_{23}, \ldots, a_{28}$
10	a_{29}, a_{30}, a_{31}
11	a_{32}, a_{33}
12	a_{34}

$[22,1] = 26^3\, 30^3\, 31\, 32\, 34^3, \quad [22,2] = 28, \quad [22,3] = 31^4\, 33^2\, 34^4,$

$[22,4] = 33^3\, 34^3,$

$[23,2] = 29, \quad [24,1] = 34^3, \quad [24,2] = 30, \quad [24,3] = 32^2\, 34^3,$

$[24,5] = 34^2,$

$[25,1] = 29^4\, 32\, 34, \quad [25,2] = 32^2\, 34^3, \quad [25,3] = 32\, 34^4, \quad [25,5] = 34,$

$[26,1] = 30^4\, 32^2\, 33^4\, 34^3, \quad [26,3] = 33\, 34^3, \quad [26,4] = 34^2,$

$[27,1] = 30^4\, 32\, 33^4,$

$[27,2] = 31, \quad [28,1] = 31^4\, 33^3\, 34^4, \quad [29,2] = 32, \quad [30,1] = 32^2\, 34^3,$

$[30,2] = 33, \quad [31,1] = 33^4\, 34, \quad [32,2] = 34, \quad [33,1] = 34^2.$

Let R be the group defined by this presentation. The weights of the generators of R are given in Table 11.7.1.

There are several approaches to verifying that R has exponent 5. The simplest uses Corollary 7.5. Since R has class 12 and is generated as a monoid by a_1 and a_2, it suffices to check that $W^5 = 1$ for every nonempty word W in $\{a_1, a_2\}^*$ with length at most 12. There are $2+4+\cdots+2^{12} = 8190$ such words. Another method is to check that every nonempty, collected word $a_1^{\alpha_1} \ldots a_{34}^{\alpha_{34}}$ of weight at most 12 defines an element of order 5. There are 2311 such words.

It is possible to reduce even further the number of fifth powers we need to check in order to confirm that R has exponent 5. Every nonidentity element of R generates the same cyclic subgroup as an element whose leading exponent is 1. If W is a collected word with leading exponent $\alpha_i = 1$, then it is not true that all images $U = \pi_Z W$ have leading exponent 1. However, if the leading exponent of U is not 1, then U defines an element of $\mathrm{Grp}\langle a_{i+1}, \ldots, a_{34}\rangle$. By induction on $34 - i$, $U^5 = 1$ in R. Thus it suffices to show that all collected words with weight at most 12 and leading exponent 1 define elements of order 5. This observation reduces the number of words to be checked to 1131.

More detailed analysis can be used to reduce further the number of fifth powers which must be computed to verify that R has exponent 5. The implementation of the p-quotient algorithm developed at the Australian National University makes use of several additional ideas. The version available at the time of this writing is able to deduce that R has exponent 5 after verifying that fewer than 100 elements have order 5.

Let d be a positive integer, let X be a set of cardinality d, and let q be a positive power of a prime p. Set $B = B(d, q) = \mathrm{Grp}\langle X \mid W^q = 1, W \in X^{\pm *}\rangle$. Corollaries 7.5 and 7.6 show that for every positive integer e there is a finite subset $\mathcal{C}$ of $X^{\pm *}$ such that $B/\varphi_{e+1}(B)$ is isomorphic to $C/\varphi_{e+1}(C)$, where $C = \mathrm{Grp}\langle X \mid W^q = 1, W \in \mathcal{C}\rangle$. Since $\mathcal{C}$ can be determined explicitly, we can find a consistent polycyclic presentation for $C/\varphi_{e+1}(C)$. For further details, see (Vaughan-Lee 1985) and [Vaughan-Lee 1990].

11.8 Verifying polycyclicity

Several times we have noted that a certain property of a finitely presented group G may be verified but not decided. That is, if G has the property, then it is possible to prove that G has the property, but if G does not have the property then it is not possible in general to prove that the property does not hold. In this section we shall demonstrate that *polycyclicity*, being a polycyclic group, can be verified.

Suppose that G is polycyclic and has derived length k. Using the full polycyclic quotient algorithm of Baumslag, Cannonito, and Miller, we can determine consistent polycyclic presentations for the quotients $Q_e = G/G^{(e)}$, $e = 1, 2, \ldots$. When $e = k + 1$, we will observe that $(Q_e)^{(e-1)}$ is trivial. Thus we can determine k. Let $a_1, \ldots, a_r$ be the original generators for G. Using the explicit homomorphism of G onto Q_k, we can introduce additional generators $a_{r+1}, \ldots, a_t$ of G which map to a polycyclic generating sequence of Q_k, which is in fact isomorphic to G. Let $\mathcal{R}$ be the set of relations defining G in terms of $a_1, \ldots, a_r$ and let $\mathcal{R}'$ consist of the relations in $\mathcal{R}$ together with relations which define $a_{r+1}, \ldots, a_t$ in terms of $a_1, \ldots, a_r$. We now invoke the Knuth-Bendix procedure for strings with the input presentation consisting of the generators $a_1, \ldots, a_t$ and the relations $\mathcal{R}'$. The ordering on words is the basic wreath-product ordering with $a_1^{-1} \succ a_1 \succ a_2^{-1} \succ a_2 \succ \cdots \succ a_t^{-1} \succ a_t$. Since $a_1, \ldots, a_t$ is a polycyclic generating sequence for G, the Knuth-Bendix procedure will terminate and the output will be a polycyclic presentation for G. This is the verification that G is polycyclic.

Finitely generated nilpotent groups are polycyclic. Since we can compute the terms in the lower central series of a group P given by a consistent polycyclic presentation, we can decide whether P is nilpotent. Thus it is possible to verify that a finitely presented group is nilpotent. A trivial group is polycyclic and it is possible to decide whether a polycyclic group is trivial. Since it is not possible to decide whether an arbitrary finitely presented group is trivial, it follows that it is not possible to decide whether a finitely presented group is polycyclic or whether it is nilpotent.

11.9 Historical notes

The study of quotient groups of a given finitely presented group G has a long history. Determining G/G' was discussed in (Tietze 1908). Knot theorists studied G/G'', particularly in the case that G/G' is infinite cyclic. In (Chen, Fox, & Lyndon 1958) it was shown that the isomorphism types of the abelian groups $\gamma_i(G)/\gamma_{i+1}(G)$ can be determined. It was the work of Macdonald which first resulted in effective algorithms for determining p-quotients of moderate size and class. These techniques, described in (Macdonald 1973, 1974), were extended in (Sag & Wamsley 1973a,b), (Wamsley 1974), (Bayes, Kautsky, & Wamsley 1974), (M. F. Newman 1976a,b), and (Havas & Newman 1980).

Techniques based on the p-quotient algorithm have been used to construct the groups of a given prime power order. The first paper describing this application was (M. F. Newman 1977). In (O'Brien 1990) the groups of order 2^8 were determined. A crude program for constructing groups of order 2^n was written by the author at Harvard during the spring of 1963. The program did not use p-quotient methods, but it did incorporate related ideas. This work was described at the Oxford conference of 1967 but never published.

In the area of nonnilpotent solvable quotients, we are still struggling to provide an efficient implementation of the polycyclic quotient algorithm developed in (Baumslag et al. 1981a,b). The fact that polycyclicity could be verified, at least in principle, was noted in (Sims 1987). The connection with Gröbner bases was discussed in (Sims 1990). Methods for finding finite solvable quotients were suggested in (Leedham-Green 1984) and (Plesken 1987).

Appendix

Implementation issues

Although the descriptions of algorithms in this book are, in most cases, sufficiently detailed to define them mathematically, the algorithms cannot be converted into well-designed computer programs without a great deal of additional work. One must consider carefully the choice of language in which to carry out the implementation. Equally important is the choice of data structures. All objects manipulated by a digital computer must ultimately be represented by patterns of binary digits or bits. A given mathematical object may have many possible representations as a bit pattern, and the representation selected can affect drastically the running time or the memory usage of the program. In this section we shall look briefly at the language issue and then turn to the question of data structures. To focus the discussion, we shall consider the example of automata and consider ways of representing various types of automata in a computer. In the case of coset automata, we shall also investigate the consequences for the coincidence procedure of the choice of data structure.

There are three broad categories of languages in which we might implement algebraic algorithms. The first category consists of the languages associated with the currently available computer algebra systems. The general-purpose, high-level languages make up the second category. The third category consists of the assembly languages for specific computers. Let us look at the advantages and disadvantages of languages in these three categories.

The popular computer algebra packages, such as Axiom, Cayley, GAP, MACSYMA, Macaulay, Maple, Mathematica, Pari, REDUCE, and SAC2-C, all have their own languages for specifying computations. It is possible to write a program in one of these languages and have the program executed by the corresponding system. The primary advantage of this approach is speed of implementation. If the system has built into it representations of some or all of the mathematical objects with which we wish to work and provides implementations of the fundamental operations on these objects, then writing the program will be much easier than it would be if we had to work out all data structures from scratch and write the low-level subroutines needed to carry out basic operations. Of course, implementations of many of the algorithms discussed in this book already exist in at least one of the computer algebra systems. For example, Cayley and GAP contain routines for doing coset enumeration, for carrying out the Reidemeister-Schreier procedure, and for determining the structure of finitely generated abelian groups. Maple, Mathematica, and Pari have basic implementations of the LLL algorithm. Maple and Mathematica can compute Gröbner bases over rings of polynomials with rational coefficients, and Mathematica can compute such bases over $\mathbb{Z}[X_1, \ldots, X_n]$. To my knowledge, none of the systems listed earlier contains, at the time this is written, an implementation of the Knuth-Bendix procedure for strings.

The use of computer algebra systems has some disadvantages too. As a general rule, the more removed a language is from the hardware, the longer computations described in that language take to run. The languages associated with computer algebra systems are quite far removed from the hardware. Although the effect varies from system to system, and even within a specific system, computations on a computer algebra system tend to run slower than the same computations would run if coded directly in a language like FORTRAN or C. For many problems, the difference in execution speed is not of major significance. However, some problems, particularly those connected with research in algebra, challenge the available memory and cpu resources. In this case, the overhead associated with a computer algebra system may not be acceptable.

Other problems with computer algebra systems are cost, size, and lack of access to source code. Some systems are distributed without charge, but others impose a substantial license fee. Although the size of the "kernel", the part of the system which must be present in memory, varies considerably, the programs are large. If only one feature of a system is required for a particular application, then having to load the entire system unnecessarily reduces the amount of memory available for data. Many systems are distributed only in executable form. Without access to the source code, it is impossible to tailor the program to a specific problem by making modifications which involve changes in low-level data structures.

Among the general-purpose, high-level languages which have been used to implement algebraic algorithms are APL, BASIC, C, FORTRAN, LISP, and Pascal. LISP was popular for many years among designers of traditional computer algebra systems and FORTRAN was used for roughly two decades by computational group theorists. Recently there has been a general movement toward C. The kernels of Cayley, GAP, Macaulay, Maple, Mathematica, Pari, and SAC2-C are written in C. The appearance of a C standard has contributed to the decision to use C.

The primary advantages of writing programs in standard C are execution speed and portability. The C language has features which make it closer to the hardware than most other high-level languages. This allows C programs to be compiled into very efficient code. Good compilers for C are widely available, much more available than computer algebra systems. As compiler designers move to conform to the C standard, it is reasonable to expect that a given C program can be executed on a wide range of hardware.

The main disadvantage of programming in C or some similar language is the length of time required to develop the software. The user cannot rely on the sophisticated input/output facilities available in computer algebra systems. Until recently there had been no publicly available library of C functions for performing fundamental operations like memory management, multiple precision arithmetic, and polynomial manipulations. However, with the distribution of source code for such systems as GAP, Macaulay, Pari, and SAC2-C, it is now easier to build on the work of others.

The greatest efficiency in speed and memory usage is obtained with assembly language programming. Assembly language is only slightly removed from the machine language of the hardware. This means that assembly language programs are portable only to other computers of the same make and the same or similar model. The development time for assembly language programs is longer than the time needed to write C programs with similar functionality. Of the systems mentioned earlier, only Pari provides assembly language implementations of certain low-level operations for use with specific hardware.

Let us turn now to the question of data structures. We shall assume that the decision has been made to implement software in a language which, like C and Pascal, allows the use of pointers. It will not be possible to consider representations for all of the objects involved in the algorithms which we have investigated. To illustrate the issues which can arise, let us look at ordinary, possibly nondeterministic, automata.

The first thing to do is to decide on the operations we wish to perform on the automata. Here is a reasonable list:

(a) Add a state.
(b) Delete a state and all edges involving that state.
(c) Add or delete an edge.
(d) Given a state σ and a word U, find all states τ such that there is a path from σ to τ with signature U.
(e) Given a state σ and a word U, find all states ρ such that there is a path from ρ to σ with signature U.
(f) Print out the automaton.

Let us assume that the number of states may be quite large, but that the number of edges leaving or entering a given state will be reasonably small. Let us also assume that the alphabet X is not too big and that we have some method of designating the label on an edge as an element of $X \cup \{\varepsilon\}$.

One possible approach is to represent states and edges by contiguous chunks of memory called records in Pascal and structures in C. The records are made up of fields, which contain data or pointers to other records. If u is a pointer to a record and a is the name of a field in that record, then $u.a$ will denote the value of the field. The pointer fields link the records together into lists. Lists may be either singly linked or doubly linked. In a singly linked list, a given record contains a pointer only to the next record on the list. The end of the list is signaled by a "null pointer". The only way to find the record preceding the given one is to go to the start of the list and move from record to record through the list until the current one is found. In a doubly linked list, each record contains a pointer to the previous record as well as a pointer to the next record. It is much easier to remove a record from the middle of a long list if the list is doubly linked.

The use of lists of relatively short records as data structures in symbolic software has been popular for a long time. With this approach, complex data structures can be created fairly easily, memory management strategies are well understood, and programming low-level operations is relatively straightforward. However, there are some problems. Examining the elements of a linked list takes significantly longer than looking at consecutive entries of a vector stored in a single contiguous block of memory. The situation can get particularly bad when virtual memory is being used. Closely related records may be widely separated in memory. It can happen that each new access of a record causes a page fault. When this happens, the program is brought to its knees. It is now becoming common to use contiguous blocks of memory for large objects. Elements of free modules over polynomial rings appearing in Gröbner basis computations are examples of objects which might be treated in this manner. Such objects vary widely in size, and their use complicates memory management, but the improved efficiency appears to justify the extra programming effort.

We shall use lists of records for general (ordinary) automata. However, when we turn to coset automata, we shall use matrices and vectors which are stored in contiguous memory locations. To carry out operation (f), we shall consider each state in turn and print the edges leaving that state. Thus we shall need a list of states. In view of operation (b) and the fact that there may be many states, we shall make the list doubly linked. The record representing an edge $e = (\sigma, U, \tau)$ will contain pointers to the records representing σ and τ. Because of the need to carry out operations (b), (d), and (e), we shall want to be able to find all edges leaving σ and all edges coming into τ. Thus e should be on two lists of edges, the edges leaving σ and the edges coming into τ. Since these lists are supposed to be short, we shall make them singly linked lists. This saves space at the cost of increasing the time needed to remove an edge. The record for e must contain a field giving the label U. The record representing a state must contain boolean fields indicating whether the state is initial and/or terminal, and there

probably should also be a name field providing a convenient character string
by which to designate the state in the printed listing.

In our first representation of an automaton, a record representing a state
σ will have the following fields:

name	External name of σ
prev	Pointer to the record representing the previous state on the list of states
next	Pointer to the record representing the next state on the list of states
firstin	Pointer to the record representing the first edge entering σ
firstout	Pointer to the record representing the first edge leaving σ
init	Boolean value which is true if σ is an initial state
term	Boolean value which is true if σ is a terminal state

The record representing an edge $e = (\sigma, U, \tau)$ will contain the following
fields:

label	U
tail	Pointer to the record representing σ
head	Pointer to the record representing τ
nextin	Pointer to the record representing the next edge entering τ
nextout	Pointer to the record representing the next edge leaving σ

From now on, the distinction between records and the objects they represent
will be blurred.

The entire automaton can be accessed if the first state on the list of states
is given. Here is a procedure for printing out the edges of the automaton:

```
Procedure PRINT(σ);
Input: σ    : a pointer to the first state of an automaton;
(∗ The edges of the automaton are printed. ∗)
Begin
    τ := σ;
    While τ is not null do begin
        Print the strings "Edges leaving state" and τ.name on a new line;
        e := τ.firstout;
        While e is not null do begin
            Print " ", e.label, " ", and e.head.name on a new line;
            e := e.nextout
        End;
        τ := τ.next
    End
End.
```

Here is a sample of the output of PRINT for an automaton for which the edge labels come from $\{x, y, z, \varepsilon\}$ and the external names of the states are 2, 3, 5, and 7.

Edges leaving state 2
   x 7
   z 3
   ε 5
Edges leaving state 3
   x 3
   y 7
Edges leaving state 5
   y 2
   z 5
   z 7
Edges leaving state 7
   x 2
   z 2

The representation for automata just described could be used for all ordinary automata, including coset automata. However, using this representation to do coset enumeration would be a mistake since substantial amounts of space would be wasted. Before looking at coset automata, let us first make the assumption only that all our automata will be deterministic. If the initial state is the first state, then init fields in state records can be omitted. If $|X|$ is small, we can speed up access to edges leaving a given state by replacing the firstout field in a state record with an array out indexed by X. If σ is a pointer to a state and x is in X, then σ.out$[x]$ will be a pointer to the unique edge leaving σ which has label x, or null if no such edge exists. The nextout fields in edge records can also be eliminated.

If $|X|$ is large, then the use of the out field might be a problem when the automaton is sparse in the sense that on average the number of edges leaving a given state is small compared with $|X|$. Every time a new state is created, space will have to be allocated for the entire out array. If the automaton is sparse, then the space used by the linked lists of out edges could be substantially less that the space used by all the out arrays. One place in which sparse automata are encountered is in the early stages of coset enumerations.

Now let us consider coset automata with respect to classical monadic rewriting systems. The term field in state records is no longer needed, and the lists of edges entering each state are unnecessary. There is an edge with label x entering a state σ if and only if there is an edge with label x^{-1} leaving σ. There is in fact no need to have records for edges if the out arrays are used. We simply make σ.out$[x]$ point to the head of the edge leaving σ which has label x, rather than to the edge itself. In the case that

$|X|$ is large, it may be better to use lists of edges leaving the states, at least in the beginning of a coset enumeration.

Although linked lists of records have been used to implement coset enumeration, the usual data structure for a coset automaton is a two-dimensional array, say "table". States are represented by positive integers, with 1 denoting the initial state. If σ is a state and x is in X, then table$[\sigma, x]$ is the endpoint τ of the unique edge (σ, x, τ), or 0 if no such edge exists. Because of coincidences, it may happen that not all row indices of table correspond to active states. The set of active states is specified by first giving an integer n which is at least as big as the largest active state. If σ is an integer between 1 and n which does not represent a state, then table$[\sigma, z_1]$ is set to some negative value. Here z_1 is a fixed element of X, which may need to be specially chosen to facilitate coincidence processing, as described later. Thus the current set Σ of active states is the set of positive integers σ not exceeding n such that table$[\sigma, z_1]$ is nonnegative.

As new states are needed, old ones deleted by the coincidence procedure are recycled. This can be done in several ways. For example, entries in table can be used to create a linked list of available state numbers. On the other hand, as long as n is less than the number of rows in table, a new state can be defined by adding 1 to n and using the new value of n as the new state. If no more rows of table are available this way and there are inactive state values, then the entire table can be standardized using the procedure of Section 4.7, or compressed as follows:

Procedure COMPRESS;
Begin
   $\sigma := 0$;
  For $\tau := 1$ to n do
    If table$[\tau, z_1] \geq 0$ then begin
      $\sigma := \sigma + 1$;
     If $\sigma \neq \tau$ then
       For x in X do begin
         $\rho := $ table$[\tau, x]$; If $\rho = \tau$ then $\rho := \sigma$; table$[\sigma, x] := \rho$;
         If $\rho \neq \sigma$ then table$[\rho, x^{-1}] := \sigma$
       End
    End;
   $n := \sigma$
End.

Example A.1. Suppose that $X = \{x, x^{-1}, y, y^{-1}\}$, $n = 12$, and the first 12 rows of table are as shown in Table A.1. Then after COMPRESS has been called, the value of n is 7 and the first 7 rows of table are as in Table A.2.

Table A.1

	x	x^{-1}	y	y^{-1}
1	4	0	3	10
2	-1	0	0	0
3	0	7	10	1
4	8	1	0	12
5	-4	0	0	0
6	-3	0	0	0
7	3	12	8	0
8	12	4	0	7
9	-7	0	0	0
10	10	10	1	3
11	-7	0	0	0
12	7	8	4	0

Table A.2

	x	x^{-1}	y	y^{-1}
1	3	0	2	6
2	0	4	6	1
3	5	1	0	7
4	2	7	5	0
5	7	3	0	4
6	6	6	1	2
7	4	5	4	0

If there are many inactive states, then standardization takes less time than compression, but standardization disturbs the linear order of the active states, while compression does not.

The description of the coincidence procedure in Section 4.6 used a vector p and a list $\eta_1, \ldots, \eta_k$ of states which have become inactive. Using the array table to represent the coset automaton, we can store the necessary entries of p and the η_i in table, thereby saving a significant amount of space. The method described here is based on the idea in (Beetham 1984).

Clearly we assume that $|X| \geq 2$, since otherwise the free product of cyclic groups in which we are working would be trivial or a cyclic group of order 2. There exist two distinct elements z_1 and z_2 of X such that either $z_1^{-1} = z_2$ or $z_1^{-1} = z_1$ and $z_2^{-1} = z_2$. The columns of table indexed by z_1 and z_2 will be referred to as the special columns. If σ is an inactive state, then $-\text{table}[\sigma, z_1]$ will be an earlier state equivalent to σ and $\text{table}[\sigma, z_2]$ will be the next inactive state in the sequence formerly denoted $\eta_1, \ldots, \eta_k$. This sequence is now managed as a queue with head α and tail ω. The procedure ACTIVE is modified as follows:

Function ACTIVE(σ): integer;
Input: σ : a positive integer not exceeding n;
Begin
   $\tau := \sigma$; $\rho := \text{table}[\tau, z_1]$;
   While $\rho < 0$ do begin $\tau := -\rho$; $\rho := \text{table}[\tau, z_1]$ end;
   ACTIVE $:= \tau$; $\mu := \sigma$; $\rho := \text{table}[\mu, z_1]$;
   While $\rho \neq -\tau$ do begin
       $\text{table}[\mu, z_1] := -\tau$; $\mu := -\rho$; $\rho := \text{table}[\mu, z_1]$ end
End.

Because of the way the special columns are used, the consequences of a given coincidence derived from entries in the special columns must be processed before other consequences. This requires a separate procedure COINCSP. The procedure MERGE must be modified. It is called only from COINCSP. A set D of edges is maintained in COINCSP. The edges in D correspond to entries in the special columns of table which have been removed. Edges are added to D in MERGE and removed from D in COINCSP. By Exercise 6.2 in Chapter 4, the cardinality of D never exceeds 2.

Procedure MERGE(σ, τ);
Input: σ, τ : elements of Σ;
(* The equivalence classes containing σ and τ are merged and the queue
   is adjusted. *)
Begin
  If $\sigma \neq \tau$ then begin
    $\nu := \max(\sigma, \tau)$; $\mu := \min(\sigma, \tau)$;

    For x in $\{z_1, z_2\}$ do begin
      $\varphi := \text{table}[\nu, x]$;
      If $\varphi \neq 0$ then begin
        $\text{table}[\nu, x] := 0$; $\text{table}[\varphi, x^{-1}] := 0$; Add (ν, x, φ) to D
      End
    End;

    $\text{table}[\nu, z_1] := -\mu$; $\text{table}[\nu, z_2] := 0$;

    If $\alpha = 0$ then begin
      $\alpha := \nu$; $\omega := \nu$
    End

    Else begin
      $\text{table}[\omega, z_2] := \nu$; $\omega := \nu$
    End;
  End
End.

Procedure COINCSP(σ, τ);
Input: σ, τ : elements of Σ;
(* The coincidence of σ and τ is processed in the special columns. *)
Begin
  $D := \emptyset$; MERGE(σ, τ);

  While $D \neq \emptyset$ do begin
    Remove an edge (ν, x, φ) from D;
    $\mu := \text{ACTIVE}(\nu)$; $\psi := \text{ACTIVE}(\varphi)$; $\gamma := \text{table}[\mu, x]$; $\delta :=$
      $\text{table}[\psi, x^{-1}]$;
    If $\gamma \neq 0$ then MERGE(γ, ψ)

Else if $\delta \neq 0$ then MERGE(δ, μ)
Else begin
  table$[\mu, x] := \psi$;
  If *save* then push (μ, x) onto the deduction stack;
  If $(\mu, x) \neq (\varphi, x^{-1})$ then begin
    table$[\psi, x^{-1}] := \mu$;
    If *save* then push (ψ, x^{-1}) onto the deduction stack
  End
  End
 End
End.

Here is the revised coincidence procedure:

Procedure COINCIDENCE(σ, τ);
Input: σ, τ : elements of Σ;
Begin
  $\alpha := 0$; COINCSP(σ, τ);
  While $\alpha \neq 0$ do begin
    $E_1 := \emptyset$;
    For x in $X - \{z_1, z_2\}$ do begin
      $\varphi := $ table$[\alpha, x]$;
      If $\varphi \neq 0$ then begin
        table$[\alpha, x] := 0$; table$[\varphi, x^{-1}] := 0$; Add (α, x, φ) to E_1
      End
    End;
    For (α, x, φ) in E_1 do begin
      $\mu := $ ACTIVE(α); $\psi := $ ACTIVE(φ); $\gamma := $ table$[\mu, x]$; $\delta :=$
          table$[\psi, x^{-1}]$;
      If $\gamma \neq 0$ then COINCSP(γ, ψ)
      Else if $\delta \neq 0$ then COINCSP(δ, μ)
      Else begin
        table$[\mu, x] := \psi$;
        If *save* then push (μ, x) onto the deduction stack;
        If $(\mu, x) \neq (\psi, x^{-1})$ then begin
          table$[\psi, x^{-1}] := \mu$;
          If *save* then push (ψ, x^{-1}) onto the deduction stack
        End
      End
    End;
    $\alpha := $ table$[\alpha, z_2]$
  End
End.

With most computer languages, locating an entry in a matrix takes longer than locating an entry in a vector since more arithmetic operations must be performed to arrive at the address of table$[i, x]$ than to determine the address of $v[j]$. One of these operations is probably a multiplication of i or $i - 1$ by the number of columns in table. Havas has recently proposed a data structure which avoids this multiplication. The array table used to represent a coset automaton is replaced by a vector v obtained as follows: Let $k =$ table$[i, x]$. Redefine table$[i, x]$ to be $|X|(k - 1) + 1$ if $k > 0$, leave table$[i, x]$ unchanged if $k = 0$, and redefine table$[i, x]$ to be $|X|(k + 1) - 1$ if $k < 0$. Now arrange the entries in table into a vector by taking them in row-major order.

Example A.2. In the Havas scheme, the coset table with 12 rows in Example A.1 corresponds to the vector

$$v = (13, 0, 9, 37, -1, 0, 0, 0, 0, 25, 37, 1, 29, 1, 0, 45, -13, 0, 0, 0, -9, 0, 0, 0,$$
$$9, 45, 29, 0, 45, 13, 0, 25, -25, 0, 0, 0, 37, 37, 1, 9, -25, 0, 0, 0, 25, 29, 13, 0).$$

If we associate the elements of X with the integers $0, 1, \ldots, |X| - 1$, then to trace a word $W = w_1 \ldots w_r$ at an active coset k we proceed as follows:

$j := |X|(k - 1) + 1$;
For $i := 1$ to r do begin
    $m := v[j + w_i]$;
    If $m = 0$ then break;
    $j := m$
End;

In (Havas 1991) a speedup of 15–25% was observed when this approach was used in a C implementation of coset enumeration which supported both the Felsch and HLT strategies.

Exercises

A.1. Suppose that coset automata are represented by a doubly linked list of records representing the states and that the edges leaving a given state are represented by a singly linked list of edge records. How should the coincidence procedure be implemented?

A.2. Devise a representation for the automata used as index structures for rewriting systems as described in Section 3.5.

Bibliography

Books

Aho, A. V., Hopcroft, J. E., & Ullman, J. D. (1974). *The Design and Analysis of Computer Algorithms.* Reading: Addison-Wesley.

Akritas, A. G. (1989). *Elements of Computer Algebra with Applications.* New York: Wiley.

Atkinson, M. D. (ed.) (1984). *Computational Group Theory.* London: Academic Press.

Baumslag, G. (1971). *Lecture Notes on Nilpotent Groups.* Providence: Amer. Math. Soc.

Benninghofen, B., Kemmerich, S., & Richter, M. M. (1987). *Systems of Reductions.* Lecture Notes in Computer Science 277. Berlin: Springer-Verlag.

Bressoud, D. M. (1989). *Factorization and Primality Testing.* Berlin: Springer-Verlag.

Buchberger, B., Collins, G. E., & Loos, R. (eds.) (1983). *Computer Algebra – Symbolic and Algebraic Computation,* 2nd ed. Vienna: Springer-Verlag.

Cassels, J. W. S. (1971). *An Introduction to the Geometry of Numbers.* Berlin: Springer-Verlag.

Chandler, B., & Magnus, W. (1982). *The History of Combinatorial Group Theory: A Case Study in the History of Ideas.* New York: Springer-Verlag.

Coxeter, H. S. M., & Moser, W. O. J. (1980). *Generators and Relations for Discrete Groups,* 4th ed. Berlin: Springer-Verlag.

Davenport, J. H., Siret, Y., & Tournier, E. (1988). *Computer Algebra.* London: Academic Press.

Della Dora, J., & Fitch, J. (eds.) (1989). *Computer Algebra and Parallelism.* London: Academic Press.

Dickson, L. E. (1958). *Linear Groups.* New York: Dover.

Dieudonné, J. (1978). *Abrégé d'histoire des mathématiques,* 2 vols. Paris: Hermann.

Eilenberg, S. (1974). *Automata, Languages, and Machines, Vol. A.* New York: Academic Press.

Epstein, D. B. A., Cannon, J. W., Holt, D. F., Levy, S. V. F., Paterson, M. S., & Thurston, W. P. (1992). *Word Processing in Groups.* Boston: Jones and Bartlett.

Geddes, K. O., Czapor, S. R., & Labahn, G. (1992). *Algorithms for Computer Algebra.* Boston: Kluwer.

Hall, M., Jr. (1959). *The Theory of Groups*. New York: Macmillan.

Hodges, A. (1983). *Alan Turing: The Enigma*. New York: Simon & Schuster.

Hu, T. C. (1969). *Integer Programming and Network Flows*. Reading: Addison-Wesley.

Huppert, B., & Blackburn, N. (1982). *Finite Groups II*. Berlin: Springer-Verlag.

Jantzen, M. (1988). *Confluent String Rewriting*. Berlin: Springer-Verlag.

Johnson, D. L. (1980). *Topics in the Theory of Group Presentations*. London Math. Soc. Lecture Note Series 42. Cambridge University Press.

Kaplansky, I. (1969). *Infinite Abelian Groups*, rev. ed. Ann Arbor: University of Michigan Press.

Knuth, D. E. (1973). *The Art of Computer Programming. Vol. 1: Fundamental Algorithms. Vol. 2: Seminumeric Algorithms. Vol. 3: Sorting and Searching*. Reading: Addison-Wesley.

Kostrikin, A. I. (1990). *Around Burnside*. Berlin: Springer-Verlag.

Le Chenadec, P. (1986). *Canonical Forms in Finitely Presented Algebras*. London: Pitman. New York: Wiley.

Leech, J. (ed.) (1970). *Computational Problems in Abstract Algebra*. Oxford: Pergamon.

Lovász, L. (1986). *An Algorithmic Theory of Numbers, Graphs and Complexity*. Philadelphia: Society for Industrial and Applied Mathematics.

Lyndon, R. C., & Schupp, P. E. (1977). *Combinatorial Group Theory*. Berlin: Springer-Verlag.

Magnus, W., Karrass, A., & Solitar, D. (1966/1976). *Combinatorial Group Theory*. New York: Wiley; 2nd ed. Dover, 1976.

Miller, C. F., III (1971). *On Group-Theoretic Decision Problems and Their Classification*. Annals of Mathematics Studies 68. Princeton University Press and University of Tokyo Press.

Pohst, M., & Zassenhaus, H. (1989). *Algorithmic Algebraic Number Theory*. Cambridge University Press.

Revesz, G. E. (1983). *Introduction to Formal Languages*. New York: McGraw-Hill.

Rotman, J. J. (1973). *The Theory of Groups*, 2nd ed. Boston: Allyn & Bacon.

Sedgewick, R. (1990). *Algorithms in C*. Reading: Addison-Wesley.

Segal, D. (1983). *Polycyclic Groups*. Cambridge University Press.

Sims, C. C. (1984). *Abstract Algebra, a Computational Approach*. New York: Wiley.

Vaughan-Lee, M. (1990). *The Restricted Burnside Problem*. Oxford: Clarendon Press.

Waerden, B. L. van der (1985). *A History of Algebra*. Berlin: Springer-Verlag.

Weger, B. M. M. de (1989). *Algorithms for Diophantine Equations*. CWI Tract 85. Amsterdam: Centrum voor Wiskunde en Informatica.

Wussing, H. (1984). *Genesis of the Abstract Group Concept*. Cambridge: MIT Press.

Zassenhaus, H. J. (1958). *The Theory of Groups*, 2nd ed. New York: Chelsea.

Articles

Adjan, S. I. (1957). Unsolvability of some algorithmic problems in the theory of groups. *Trudy Moskov. Math. Obšč.*, 6:231–98.

Adjan, S. I. (1966). Defining relations and algorithmic problems for groups and semigroups (in Russian). *Trudy Mat. Inst. im. Steklov* 85. English version published by Amer. Math. Soc., 1967.

Alford, W. A., Havas, G., & Newman, M. F. (1974). Groups of exponent four. Preliminary report. Abstract 74T-A44. *Notices Amer. Math. Soc.*, 21: A-291.

Alford, W. A., Havas, G., & Newman, M. F. (1975). Groups of exponent four. Abstract 75T-A51. *Notices Amer. Math. Soc.*, 22: A-301.

Anisimov, A. V. (1971). The group languages (in Russian). *Kibernetika (Kiev)*, no. 4, 18–24.

Arrell, D. G., Manrai, S., & Worboys, M. F. (1982). A procedure for obtaining simplified defining relations for a subgroup. *Groups – St Andrews 1981*. London Math. Soc. Lecture Note Series 71, pp. 155–9. Cambridge University Press.

Arrell, D. G., & Robertson, E. F. (1984). A modified Todd-Coxeter algorithm. In M. D. Atkinson (ed.), *Computational Group Theory*, pp. 27–32. London: Academic Press.

Atkinson, M. D. (1981). Saving space in coset enumeration. *ACM SIGSAM Bull.* 15, No. 4, pp. 12–14.

Avenhaus, J., & Wissmann, D. (1989). Using rewriting techniques to solve the generalized word problem in polycyclic groups. *Proc. ACM-SIGSAM 1989 International Symposium on Symbolic and Algebraic Computation*, pp. 322–37. New York: Assoc. Comput. Mach.

Babai, L. (1986). On Lovász' lattice reduction and the nearest lattice point problem. *Combinatorica*, 6:1–13.

Bandler, P. A. (1956). A method for enumerating the cosets of an arbitrary abstract group on a digital computer. MA thesis, University of Manchester.

Bareiss, E. H. (1972). Computational solutions of matrix problems over an integral domain. *J. Inst. Math. Appl.*, 10:68–104.

Baumslag, G., Cannonito, F. B., & Miller, C. F., III (1977). Infinitely generated subgroups of finitely presented groups, I. *Math. Z.*, 153:117–34.

Baumslag, G., Cannonito, F. B., & Miller, C. F., III (1981a). Some recognizable properties of solvable groups. *Math. Z.*, 178:289–95.

Baumslag, G., Cannonito, F. B., & Miller, C. F., III (1981b). Computable algebra and group embeddings. *J. Algebra*, 69:186–212.

Bayes, A. J., Kautsky, J., & Wamsley, J. W. (1974). Computation in nilpotent groups (application). In M. F. Newman (ed.), *Proc. Second Internat. Conf. Theory Groups*. Lecture Notes in Mathematics 372, pp. 82–9. Berlin: Springer-Verlag.

Beetham, M. J. (1984). Space saving in coset enumeration. In M. D. Atkinson (ed.), *Computational Group Theory*, pp. 19–25. London: Academic Press.

Beetham, M. J., & Campbell, C. M. (1976). A note on the Todd-Coxeter coset enumeration algorithm. *Proc. Edinburgh Math. Soc.*, 20:73–9.

Benson, C. T., & Mendelsohn, N. S. (1966). A calculus for a certain class of word problems in group theory. *J. Combinatorial Theory*, 1:202–8.

Biggs, N. L. (1984). Presentations for cubic graphs. In M. D. Atkinson (ed.), *Computational Group Theory*, pp. 57–63. London: Academic Press.

Blankenship, W. A. (1966). Algorithm 287. Matrix triangulation with integer arithmetic [F1]. *Comm. ACM*, 9:513.

Book, R. V. (1982a). Confluent and other types of Thue systems. *J. Assoc. Comput. Mach.*, 29:171–82.

Book, R. V. (1982b). When is a monoid a group? The Church-Rosser case is tractable. *Theor. Comp. Science*, 18:325–31.

Boone, W. W. (1957). Certain simple, unsolvable problems in group theory VI. *Nederl. Akad. Wetensch. Proc. Ser. A.* 60; *Indag Math.*, 19:227–32.

Borosh, I., & Fraenkel, A. S. (1966). Exact solutions of linear equations with rational coefficients by congruence techniques. *Math. Comp.*, 20:107–12.

Bradley, G. H. (1971). Algorithms for Hermite and Smith normal matrices and linear diophantine equations. *Math. Comp.*, 25:897–907.

Buchberger, B. (1965). Ein Algorithmus zum Auffinden der Basiselemente des Restklassenringes nach einem nulldimensionalen Polynomideal. Dissertation, University of Innsbruck.

Buchberger, B. (1970). Ein algorithmisches Kriterium für die Lösbarkeit eines algebraischen Gleichungssystems. *Aequat. Math.*, 4:374–83.

Buchberger, B. (1976a). A theoretical basis for the reduction of polynomials to canonical form. *ACM SIGSAM Bull.*, *10*, No. 3, pp. 19–24.

Buchberger, B. (1976b). Some properties of Gröbner bases for polynomial ideals. *ACM SIGSAM Bull. 10*, No. 4, pp. 19–24.

Buchberger, B. (1984). A critical-pair/completion algorithm for finitely generated ideals in rings. In E. Börger, G. Hasenjaeger, & D. Rödding (eds.), *Logic and Machines: Decision Problems and Complexity*, pp. 137–61. Lecture Notes in Computer Science 171. Berlin: Springer-Verlag.

Buchberger, B. (1985). Gröbner bases: An algorithmic method in polynomial ideal theory. In N. K. Bose (ed.), *Multidimensional Systems Theory*, pp. 184–232. Dordrecht: Reidel.

Buchberger, B. (1987). History and basic features of the critical-pair/completion procedure. *J. Symbolic Comp.*, 3:3–38.

Buchberger, B. (1988). Applications of Gröbner bases in non-linear computational geometry. In J. R. Rice (ed.), *Mathematical Aspects of Scientific Software*, pp. 59–87. Berlin: Springer-Verlag.

Cabay, S. (1971). Exact solution of linear equations. In S. R. Petrick (ed.), *Second Symposium on Symbolic and Algebraic Manipulation*, pp. 392–8. New York: Assoc. Comput. Mach.

Cabay, S., & Lam, T. P. (1977). Congruence techniques for the exact solution of integer systems of linear equations. *ACM Trans. Math. Software*, 3:386–97.

Campbell, C. M. (1965). Enumeration of cosets and solutions of some word problems in groups. Dissertation, McGill University.

Campbell, C. M., & Robertson, E. F. (1975a). On a class of finitely presented groups of Fibonacci type. *J. London Math. Soc.*, 11:249–55.

Campbell, C. M., & Robertson, E. F. (1975b). Remarks on a class of 2-generator groups of deficiency zero. *J. Australian Math. Soc.*, 19:297–305.

Campbell, C. M., & Robertson, E. F. (1978). Deficiency zero groups involving Fibonacci and Lucas numbers. *Proc. Royal Soc. Edinburgh*, 81A:273–86.

Campbell, C. M., & Robertson, E. F. (1980). On 2-generator 2-relation soluble groups. *Proc. Edinburgh Math. Soc.*, 23: pp. 269–73.

Campbell, C. M., & Robertson, E. F. (1984). On a class of groups related to SL(2,2^n). In M. D. Atkinson (ed.), *Computational Group Theory*, pp. 43–9. London: Academic Press.

Campbell, C. M., & Robertson, E. F. (1983). Some problems in group presentations. *J. Korean Math. Soc.*, 19:123–8.

Campbell, J. M., & Lamberth, W. J. (1967). Symbolic and numeric computation in group theory. In *Proc. Third Australian Computer Conf.*, pp. 293–6. Chippendale, Australia: Australian Trade Publications.

Cannon, J. J. (1969). Computers in group theory: a survey. *Comm. ACM*, 12:3–12.

Cannon, J. J. (1971). Computing local structure of large finite groups. In G. Birkhoff & M. Hall, Jr. (eds.), *Computers in Algebra and Number Theory. SIAM-AMS Proc.*, 4:161–76. Providence: Amer. Math. Soc.

Cannon, J. J. (1973). Construction of defining relations for finite groups. *Discrete Math.*, 5:105–29.

Cannon, J. J., Dimino, L. A., Havas, G., & Watson, J. M. (1973). Implementation and analysis of the Todd-Coxeter algorithm. *Math. Computation*, 27:463–90.

Cannon, J. J., & Havas, G. (1974). Defining relations for the Held-Higman-Thompson simple group. *Bull. Australian Math. Soc.*, 11:43–6.

Cannonito, F. B. (1991). Some algorithms for polycyclic groups. In C. M. Campbell & F. F. Robertson (eds.), *Groups St Andrews 1989*. London Math. Soc. Lecture Note Series 159, Vol. 1, pp. 76–83.

Cavicchioli, A. (1986). A countable class of non-homeomorphic homology spheres with Heegard genus 2. *Geom. Dedicata*, 20:345–48.

Cayley, A. (1854). On the theory of groups, as depending on the symbolic equation $\theta^n = 1$. *Philos. Magazine Royal Soc. London*, 7:40–7, 408–9; see also *The Collected Mathematical Papers of Arthur Cayley*, Vol. 2, pp. 123–32.

Cayley, A. (1878). Desiderata and Suggestions. No. 1. The theory of groups. *Amer. J. Math.*, 1:50–2; see also *The Collected Mathematical Papers of Arthur Cayley*, Vol. 10, pp. 401–3.

Chen, K. T., Fox, R. H., & Lyndon, R. C. (1958). Free differential calculus IV. *Ann. Math.*, 68:81–95.

Chou, T.-W. J., & Collins, G. E. (1982). Algorithms for the solution of systems of linear Diophantine equations. *SIAM J. Computing*, 11:687–708.

Cochet, Y. (1976). Church-Rosser congruences on free semigroups. *Colloquia Mathematica Societatis János Bolyai*, 20:51–60.

Collins, G. E., Mignotte, M., & Winkler, F. (1983). Arithmetic in basic algebraic domains. In B. Buchberger, G. E. Collins, & R. Loos (eds.), *Computer Algebra – Symbolic and Algebraic Computation*, 2nd ed., pp. 189–220. Vienna: Springer-Verlag.

Conway, J. H. (1984). An algorithm for double coset enumeration? In M. D. Atkinson (ed.), *Computational Group Theory*, pp. 33–7. London: Academic Press.

Czyżo, E. (1972). An attempt of mechanization of Hall collecting process. *Algorytmy*, 9:5–17.

Czyżo, E. (1973). An automatization of the commutator calculus. *Algorytmy*, 10: 23–34.

Czyżo, E. (1975). On the determination of Hall polynomials. *Bull. Acad. Polon. Sci. Ser. Sci. Math Astronom. Phys.*, 23:739–45.

Dałek, K. (1971). Computation of exponents in Hall formula. *Bull. Acad. Polon. Sci. Ser. Sci. Math Astronom. Phys.*, 19:711–18.

Dałek, K. (1972). On certain general forms of the exponents of simple commutators in Hall formula. *Bull. Acad. Polon. Sci. Ser. Sci. Math Astronom. Phys.*, 20: 973–80.

Dałek, K. (1975). On the Hall formula in nilpotent and solvable groups. *Bull. Acad. Polon. Sci. Ser. Sci. Math Astronom. Phys.*, 23:829–37.

Dehn, M. (1910). Ueber die Topologie des dreidimensionalen Raumes. *Math. Ann.*, 69:137–68; see also Dehn, M. (1987). *Papers on Group Theory and Topology*, Stillwell, J. trans., pp. 92–126. New York: Springer-Verlag.

Dehn, M. (1911). Ueber unendliche diskontinuierliche Gruppen. *Math. Ann.*, 71: 116–44.

Dershowitz, N. (1987). Termination of rewriting. *J. Symbolic Comp.*, 3:69–116.

Dietze, A., & Schaps, M. (1974). Determining subgroups of a given finite index in a finitely presented group. *Canadian J. Math.*, 26:769–82.

Dimino, L. A. (1971). A graphical approach to coset enumeration. *ACM SIGSAM Bull. No. 19*, pp. 8–43.

Dixon, J. D. (1982). Exact solution of linear equations using p-adic expansions. *Numer. Math.*, 40:137–41.

Domich, P. D. (1985). Residual methods for computing Hermite and Smith normal forms. PhD thesis, Cornell University.

Domich, P. D. (1989). Residual Hermite normal form computations. *ACM Trans. Math. Software*, 15:275–86.

Domich, P. D., Kannan, R., & Trotter, L. E., Jr. (1987). Hermite normal form computation using modulo determinant arithmetic. *Math. of Operations Research*, 12:50–9.

Dyck, W. von (1882). Gruppentheoretische Studien. *Math. Ann.*, 20:1–44.

Dyck, W. von (1883). Gruppentheoretische Studien II. *Math. Ann.*, 22:70–108.

Epstein, D. B. A., Holt, D. F., & Rees, S. E. (1991). The use of Knuth-Bendix methods to solve the word problem in automatic groups. *J. Symbolic Comp.*, 12: 397–414.

Felsch, H. (1960). Die Behandlung zweier gruppentheoretischer Verfahren auf elektronischen Rechenmaschinen. Diplomarbeit, Kiel.

Felsch, H. (1961). Programmierung der Restklassenabzählung einer Gruppe nach Untergruppen. *Numer. Math.*, 3:250–6.

Felsch, V. (1976). A machine independent implementation of a collection algorithm for the multiplication of group elements. In R. D. Janks (ed.), *SYMSAC 76, Proc. 1976 ACM Symp. Symbolic and Algebraic Computation*, pp. 159–66. New York: Assoc. Comput. Mach.

Felsch, V., & Neubüser, J. (1979). An algorithm for the computation of conjugacy classes and centralizers in p-groups. In E. W. Ng (ed.), *Symbolic and Algebraic Computation*, pp. 452–65. Lecture Notes in Computer Science 72. Berlin: Springer-Verlag.

Felsch, W. (1974). Ein commutator collecting Algorithmus zur Bestimmung einer Zentralreihe einer endlichen p-Gruppe. Staatsexamensarbeit, Aachen.

Frobenius, F. G., & Stickelberger, L. (1879). Ueber Gruppen von vertauschbaren Elementen. *J. Reine Angew. Math.*, 86:217–62; see also *Ferdinand Georg Frobenius Gesammelte Abhandlungen*, Vol. I, pp. 545–90. Berlin: Springer-Verlag, 1968.

Frumkin, M. A. (1977). Polynomial time algorithms in the theory of linear diophantine equations. In M. Karpinksi (ed.), *Fundamentals of Computation Theory*, pp. 386–92. Lecture Notes in Computer Science 56. Berlin: Springer-Verlag.

Gaglione, A. M., & Waldinger, H. V. (1986). Factor groups of the lower central series of free products of finitely generated abelian groups. In E. F. Robertson & C. M. Campbell (eds.), *Proc. Groups – St Andrews 1985*, pp. 164–203. London Math. Soc. Lecture Note Series 121. Cambridge University Press.

Gateva-Ivanova, T., & Latyshev, V. (1988). On recognisable properties of associative algebras. *J. Symbolic Comp.*, 6:371–88.

Gebauer, R., & Möller, H. M. (1988). On an installation of Buchberger's algorithm. *J. Symbolic Comp.*, 6:275–86.

Gerstein, L. J. (1977). A local approach to matrix equivalence. *Linear Alg. and Its Applications*, 16:221–32.

Gilman, R. H. (1979). Presentations of groups and monoids. *J. Algebra*, 57:544–54.

Gilman, R. H. (1982). Enumeration of double cosets. *J. Pure and Appl. Alg.*, 26: 183–8.

Gilman, R. H. (1984a). Enumerating infinitely many cosets. In M. D. Atkinson (ed.), *Computational Group Theory*, pp. 3–18. London: Academic Press.

Gilman, R. H. (1984b). Computation with rational subsets of confluent groups. In *Proc. Eurosam '84*. Lecture Notes in Computer Science 174, pp. 207–12. Berlin: Springer-Verlag.

Gilman, R. H. (1987). Groups with a rational cross-section. In S. M. Gersten & J. R. Stallings (eds.), *Combinatorial Group Theory and Topology*, pp. 175–83. Princeton University Press.

Glasby, S. P., & Slattery, M. C. (1990). Computing intersections and normalizers in soluble groups. *J. Symbolic Comp.*, 9:637–51.

Grover, J., Rowe, L. A., & Wilson, D. (1971). Applications of coset enumeration. In S. R. Petrick (ed.), *Proc. Second Symposium on Symbolic and Algebraic Manipulation*, pp. 183–7. New York: Assoc. Comput. Mach.

Guibas, L. J., & Odlyzko, A. M. (1981). String overlaps, pattern matching, and nontransitive games. *J. Combin. Theory A*, 30:183–208.

Hafner, J. L., & McCurley, K. S. (1991). Asymptotically fast triangularization of matrices over rings. *SIAM J. Comput.*, 20:1068–83.

Hall, M., Jr., & Sims, C. C. (1982). The Burnside group of exponent 5 with two generators. In C. M. Campbell & E. F. Robertson (eds.), *Groups – St Andrews 1981*, pp. 207–20. London Math. Soc. Lecture Note Series 71. Cambridge University Press.

Hall, P. (1934). A contribution to the theory of groups of prime power order. *Proc. London Math. Soc.*, 36(2):29–95; see also *Collected Works of Philip Hall*, pp. 57–125. Oxford: Clarendon Press, 1988.

Hall, P. (1954). Finiteness conditions for soluble groups. *Proc. London Math. Soc.*, 4:419–36; see also *Collected Works of Philip Hall*, pp. 307–29. Oxford: Clarendon Press, 1988.

Hall, P. (1957). Nilpotent groups. Notes of lectures given at the Canadian Mathematical Congress, Canadian Math. Soc.; Revised versions issued by Queen Mary College 1969, 1979; see also *Collected Works of Philip Hall*, pp. 415–62. Oxford: Clarendon Press, 1988.

Hastad, J., Helfrich, B., Lagarias, J. C. & Schnorr, C. P. (1986). Polynomial time algorithms for finding integer relations among real numbers. In B. Monien & G. Vidal-Naquet (eds.), *Proceedings of STAC '86*, pp. 105–18. Lecture Notes in Computer Science 210. Berlin: Springer-Verlag.

Havas, G. (1974a). A Reidemeister-Schreier program. In M. F. Newman (ed.), *Proc. Second Internat. Conf. Theory Groups*, pp. 347–56. Lecture Notes in Mathematics 372. Berlin: Springer-Verlag.

Havas, G. (1974b). Computational approaches to combinatorial group theory. PhD thesis, University of Sydney; abstract published in *Bull. Australian Math. Soc.*, 11:475–6.

Havas, G. (1976). Computer aided determination of a Fibonacci group. *Bull. Australian Math. Soc.*, 15:297–305.

Havas, G. (1981). Commutators in groups expressed as products of powers. *Comm. Algebra*, 9:115–29.

Havas, G. (1991). Coset enumeration strategies. In *ISSAC '91, Proc. Symposium on Symbolic and Algebraic Computation* (Bonn 1991), pp. 191–9. Association for Computing Machinery.

Havas, G., Kenne, P. E., Richardson, J. S., & Robertson, E. F. (1984). A Tietze transformation program. In M. D. Atkinson (ed.), *Computational Group Theory*, pp. 69–73. London: Academic Press.

Havas, G., & Kovács, L. G. (1984). Distinguishing eleven crossing knots. In M. D. Atkinson (ed.), *Computational Group Theory*, pp. 367–73. London: Academic Press.

Havas, G., & Newman, M. F. (1980). Application of computers to questions like those of Burnside. In J. L. Mennicke (ed.), *Burnside Groups*, pp. 211–30. Lecture Notes in Mathematics 806. Berlin: Springer-Verlag.

Havas, G., Newman, M. F., & Vaughan-Lee, M. R. (1990). A nilpotent quotient algorithm for graded Lie rings. *J. Symbolic Comp.*, 9:653–64.

Havas, G., & Nicholson, T. (1976). Collection. In R. D. Jenks (ed.), *SYMSAC 76, Proc. 1976 ACM Symp. Symbolic and Algebraic Computation*, pp. 9–14. New York: Assoc. Comput. Mach.

Havas, G., & Richardson, J. S. (1983). Groups of exponent five and class four. *Comm. Alg.*, 11:287–304.

Havas, G., Richardson, J. S., & Sterling, L. S. (1979). The last of the Fibonacci groups. *Proc. Royal Soc. Edinburgh*, 83A:199–203.

Havas, G., & Robertson, E. F. (1984). Two groups which act on cubic graphs. In M. D. Atkinson (ed.), *Computational Group Theory*, pp. 65–8. London: Academic Press.

Havas, G., & Sterling, L. S. (1979). Integer matrices and abelian groups. In E. W. Ng (ed.), *Symbolic and Algebraic Computation*, pp. 431–51. Lecture Notes in Computer Science 72. Berlin: Springer-Verlag.

Havas, G., Wall, G. E., & Wamsley, J. M. (1974). The two generator restricted Burnside group of exponent five. *Bull. Australian Math. Soc.*, 10:459–70.

Hermann, G. (1926). Die Frage der endlich vielen Schritte in der Theorie der Polynomideale, *Math. Ann.*, 95:736–88.

Hermite, C. (1851). Sur l'introduction des variable continues dans la théorie des nombres. *J. Reine Angew. Math.*, 41:191–216; see also *Oeuvres de Charles Hermite*, Vol. I, pp. 164–92.

Higman, G. (1951). A finitely presented infinite simple group. *J. London Math. Soc.*, 26:61–4.

Higman, G. (1959). Some remarks on varieties of groups. *Quart. J. Math.*, 10: 165–78.

Holt, D. F., & Rees, S. (1992). Testing for isomorphism between finitely presented groups. In M. W. Liebeck & J. Saxl (eds.), *Groups and Combinatorics*. London Math. Soc. Lecture Note Series 165. Cambridge University Press.

Howse, J. R., & Johnson, D. L. (1981). An algorithm for the second derived factor group. In C. M. Campbell, & E. F. Robertson (eds.), *Groups – St Andrews 1981*, pp. 237–43. London Math. Soc. Lecture Note Series 71. Cambridge University Press.

Howson, A. G. (1954). On the intersection of finitely generated free groups. *J. London Math. Soc.*, 29:428–34.

Huet, G. (1980). Confluent reductions: Abstract properties and applications to term rewriting systems. *J. Assoc. Comput. Mach.*, 27:797–821.

Huet, G. (1981). A complete proof of correctness of the Knuth-Bendix completion algorithm. *J. Computer and Systems Sciences*, 23:11–21.

Iliopoulos, C. S. (1989a). Worst-case complexity bounds on algorithms for computing the canonical structure of finite abelian groups and the Hermite and Smith normal forms of an integer matrix. *SIAM J. Computing*, 18:658–69.

Iliopoulos, C. S. (1989b). Worst-case complexity bounds on algorithms for computing the canonical structure of infinite abelian groups and solving systems of linear Diophantine equations. *SIAM J. Computing*, 18:670–8.

Jürgensen, H. (1970). Calculation with the elements of a finite group given by generators and defining relations. In J. Leech (ed.), *Computational Problems in Abstract Algebra*, pp. 47–57. Oxford: Pergamon.

Kandri-Rodi, A., Kapur, D., & Narendran, P. (1985). An ideal-theoretic approach to word problems and unification problems over finitely presented commutative algebras. In *Proc. Intern. Conf. on Rewriting Techniques and Applications*, pp. 345–65. Lecture Notes in Computer Science 202. Berlin: Springer-Verlag.

Kannan, R., & Bachem, A. (1979). Polynomial algorithms for computing the Smith and Hermite normal forms of an integer matrix. *SIAM J. Computing*, 9:499–507.

Kapur, D., & Narendran, P. (1985). The Knuth-Bendix completion procedure and Thue systems. *SIAM J. Computing*, 14:1052–70.

Knuth, D. E. (1975). Estimating the efficiency of backtrack programs. *Math. Comp.*, 29:121–36.

Knuth, D. E., & Bendix, P. B. (1970). Simple word problems in universal algebra. In J. Leech (ed.), *Computational Problems in Abstract Algebra*, pp. 263–97. Oxford: Pergamon.

Kuhn, N., & Madlener, K. (1989). A method for enumerating the cosets of a group presented by a canonical system. In *Proc. ACM-SIGSAM 1989 International Symposium on Symbolic and Algebraic Computation*, pp. 338–50. New York: Assoc. Comput. Mach.

Lagarias, J. C., & Odlyzko, A. M. (1985). Solving low-density subset sum problems. *J. ACM*, 32:229–46.

Laue, R., Neubüser, J., & Schoenwaelder, U. (1984). Algorithms for finite soluble groups and the SOGOS system. In M. D. Atkinson (ed.), *Computational Group Theory*, pp. 105–35. London: Academic Press.

Leech, J. (1962). Some definitions of Klein's simple group of order 168 and other groups. *Proc. Glasgow Math. Assoc.*, 5:166–75.

Leech, J. (1963). Coset enumeration on digital computers. *Proc. Cambridge Philos. Soc.*, 59:257–67.

Leech, J. (1965). Generators for certain normal subgroups of (2,3,7). *Proc. Cambridge Philos. Soc.*, 61:321–32.

Leech, J. (1970). Coset enumeration. In J. Leech (ed.), *Computational Problems in Abstract Algebra*, pp. 21–35. Oxford: Pergamon.

Leech, J. (1977). Computer proofs of relations in groups. In M. P. J. Curran (ed.), *Topics in Group Theory and Computation*, pp. 38–61. London: Academic Press.

Leech, J. (1984). Coset Enumeration. In M. D. Atkinson (ed.), *Computational Group Theory*, pp. 3–18. London: Academic Press.

Leedham-Green, C. R. (1984). A soluble group algorithm. In M. D. Atkinson (ed.), *Computational Group Theory*, pp. 85–101. London: Academic Press.

Leedham-Green, C. R., Praeger, C. E., & Soicher, L. H. (1991). Computing with group homomorphisms. *J. Symbolic. Comp.*, 12:527–32.

Leedham-Green, C. R., & Soicher, L. H. (1990). Collection from the left and other strategies. *J. Symbolic Comp.*, 9:665–75.

Lenstra, A. K., Lenstra, H. W., Jr., & Lovász, L. (1982). Factoring polynomials with rational coefficients. *Math. Ann.*, 261:515–34.

Leon, J. S. (1980a). Finding the order of a permutation group. In B. Cooperstein & G. Mason (eds.), *The Santa Cruz Conference on Finite Groups*, pp. 511–17. Proc. Symp. Pure Math. 37. Providence: Amer. Math. Soc.

Leon, J. S. (1980b). On an algorithm for finding a base and a strong generating set for a group given by generating permutations. *Math. Comp.*, 35:941–74.

Lindenberg, W. (1962). Über eine Darstellung von Gruppenelementen in digitalen Rechenautomaten. *Numer. Math.*, 4:151–53.

Lindenberg, W. (1963). Die Struktur eines Übersetzungsprogramms zur Multiplikation von Gruppenelementen in digitalen Rechenautomaten. *Mitt. Rh.-W. Inst. Instr. Math. Bonn*, 2:1–38.

Linton, S. A. (1991a). Double coset enumeration. *J. Symbolic Comp.*, 12:415–26.

Linton, S. A. (1991b). Constructing matrix representations of finitely presented groups. *J. Symbolic Comp.*, 12:427–38.

Lüneburg, H. (1988). On the computation of the Smith normal form. In *Atti del Convegno Internazionale di Teoria dei Gruppi e Geometria Combinatoria, Firenze, 23–25 Ottobre 1986. Supplemento ai Rendiconti del Circolo Matematico di Palermo, Serie II*, pp. 135–71.

Macdonald, I. D. (1973). Computer results on Burnside groups. *Bull. Australian Math. Soc.*, 9:433–8.

Macdonald, I. D. (1974). A computer application to finite p-groups. *J. Australian Math. Soc.*, 17:102–12.

Macdonald, I. D. (1986). Nilpotent quotient algorithms. In C. M. Campbell & E. F. Robertson (eds.), *Proc. Groups – St Andrews 1985*, pp. 268–72. London Math. Soc. Lecture Note Series 121. Cambridge University Press.

McLain, D. H. (1977). An algorithm for determining defining relations of a subgroup. *Glasgow Math. J.*, 18:51–6.

Madlener, K., & Otto, F. (1987). Groups represented by certain classes of finite length-reducing string rewriting systems. In *Proc. Intern. Conf. on Rewriting Techniques and Applications*, pp. 133–44. Lecture Notes in Computer Science 256. Berlin: Springer-Verlag.

Magnus, W. (1932). Das Identitätsproblem für Gruppen mit einer definierenden Relation. *Math. Ann.*, 106:295–307; see also *Wilhelm Magnus Collected Papers*, pp. 49–61. Berlin: Springer-Verlag, 1984.

Markov, A. A. (1951). Impossibility of algorithms for recognizing some properties of associative systems (in Russian). *Dokl. Akad. Nauk SSSR*, 77:953–6.

Mecky, M., & Neubüser, J. (1989). Some remarks on the computation of conjugacy classes of soluble groups. *Bull. Australian Math. Soc.*, 40:281–92.

Mendelsohn, N. S. (1964). An algorithmic solution for a word problem in group theory. *Canadian J. Math.*, 16:509–16. Correction: *Canadian J. Math.*, 17(1965): 505.

Mendelsohn, N. S. (1970). Defining relations for subgroups of finite index of groups with a finite presentation. In J. Leech (ed.), *Computational Problems in Abstract Algebra*, pp. 43–4. Oxford: Pergamon.

Miller. R. C. (1971). The trivalent symmetric graphs of girth at most six. *J. Combinatorial Theory (B)*, 10:163–82.

Möller, H. M. (1988). On the construction of Gröbner bases using syzygies. *J. Symbolic Comp.*, 6:345–59.

Mostowski, A. W. (1966a). On the decidability of some problems in special classes of groups. *Fundamenta Mathematicae*, 59:123–35.

Mostowski, A. W. (1966b). Computational algorithms for deciding some problems for nilpotent groups. *Fundamenta Mathematicae*, 59:137–52.

Muller, D. E., & Schupp, P. E. (1983). Groups, the theory of ends, and context-free languages. *J. Comput. System Sci.*, 26:295–310.

Muller, D. E., & Schupp, P. E. (1985). The theory of ends, pushdown automata, and second-order logic. *Theoret. Comput. Sci.*, 37:51–75.

Narendran, P., Ó'Dúnlaing, C., & Otto, F. (1991). It is undecidable whether a finite special string-rewriting systems presents a group. *Discrete Math.*, 98:153–9.

Neubüser, J. (1961). Bestimmung der Untergruppenverbände endlicher p-Gruppen auf einer programmgesteurten elektronischen Dualmaschine. *Numer. Math.*, 3: 271–78.

Neubüser, J. (1970). Investigations of groups on computers. In J. Leech (ed.), *Computational Problems in Abstract Algebra*, pp. 1–19. Oxford: Pergamon.

Neubüser, J. (1982). An elementary introduction to coset table methods in computational group theory. In C. M. Campbell & E. F. Robertson (eds.), *Groups – St Andrews 1981*, pp. 1–45. London Math. Soc. Lecture Note Series 71. Cambridge: Cambridge University Press.

Neubüser, J., & Sidki, S. (1988). Some computational approaches to groups given by a finite presentation. Published as Alguns procedimentos computacionais para grupos dados por uma apresentação finita. *Matemática Universitária*, Junho de 1988, Número 7, pp. 77–120.

Newman, M. F. (1976a). A computer aided study of a group defined by fourth powers. *Bull. Australian Math. Soc.*, 14:293–77. Addendum: *Bull. Australian Math. Soc.*, 15:477–9.

Newman, M. F. (1976b). Calculating presentations for certain kinds of quotient groups. In R. D. Jenks (ed.), *SYMSAC 76, Proc. 1976 ACM Symp. Symbolic and Algebraic Computation*, pp. 2–8. New York: Assoc. Comput. Mach.

Newman, M. F. (1977). Determination of groups of prime-power order. In R. A. Bryce, J. Cossey, & M. F. Newman (eds.), *Group Theory*, pp. 73–84. Lecture Notes in Mathematics, 573. Berlin: Springer-Verlag.

Newman, M. F. (1984). Groups of exponent six. In M. D. Atkinson (ed.), *Computational Group Theory*, pp. 39–41. London: Academic Press.

Newman, M. H. A. (1951). The influence of automatic computers on mathematical methods. *Manchester University Computer – Inaugural Conference.*

Nielsen, J. (1921). Om Regning med ikke kommutative Faktoren og dens Anvendelse in Gruppeteorien. *Matematisk Tidsskrift, B*, pp. 77–94.

Nivat, M. (1970). On some families of languages related to the Dyck languages. In *Proc. Second ACM Symposium Theory Comp.*, pp. 221–5.

Nötzold, J. (1976). Ueber reduzierte Gitterbasen und ihre mögliche Bedeutung für die numerische Behandlung linearer Gleichungssysteme. Dissertation, Rheinisch-Westfälische Technische Hochschule, Aachen.

Novikov, P. S. (1955). On the algorithmic unsolvability of the word problem in group theory (in Russian). *Trudy Mat. Inst. im. Steklov*, 44, 143 pages.

O'Brien, E. A. (1990). The p-group generation algorithm. *J. Symbolic Comp.*, 9: 677–98.

Odlyzko, A. M. (1985). Applications of symbolic mathematics to mathematics. In R. Pavelle (ed.), *Applications of Computer Algebra*, pp. 95–111. Boston: Kluwer.

Otto, F. (1987). Some results about confluence on a given congruence class. In *Proc. Intern. Conf. on Rewriting Techniques and Applications*, pp. 145–55. Lecture Notes in Computer Science 256. Berlin: Springer-Verlag.

Pedersen, J. (1988). Computer solution of word problems in universal algebra. In M. C. Tangora (ed.), *Computers in Mathematics*, pp. 103–28. Lecture Notes Pure and Appl. Math. 111. New York: Marcel Dekker.

Peterson, G. E., & Stickel, M. E. (1981). Complete sets of reductions for some equational theories. *J. Assoc. Comput. Mach.*, 28:233–64.

Plesken, W. (1987). Towards a solvable quotient algorithm. *J. Symbolic Comp.*, 4: 111–22.

Pohst, M. (1987). A modification of the LLL reduction algorithm. *J. Symbolic Comp.*, 4:123–7.

Post, E. L. (1947). Recursive unsolvability of a problem of Thue. *J. Symbolic Logic*, 12:1–11.

Rabin, M. O. (1958). Recursive unsolvability of group theoretic problems. *Ann. of Math.*, 67:172–94.

Rayward-Smith, V. J. (1979). On computing the Smith normal form of an integer matrix. *ACM Trans. Math. Software*, 5:451–6.

Reidemeister, K. (1926). Knoten und Gruppen. *Abh. Math. Sem. Hamburg*, 5:7–23.

Remeslennikov, V. N. (1973). An example of a group, finitely presented in the variety $\mathfrak{A}^5$ with an unsolvable word problem (in Russian). *Algebra i Logika*, 12: 577–602. Translated in:*Algebra and Logic*, 12(1973):327–46.

Rhodes, J., & Shamir, E. (1968). Complexity of grammars by group theoretic methods. *J. Comb. Theory*, 4:222–39.

Richman, F. (1974). Constructive aspects of Noetherian rings. *Proc. Amer. Math. Soc.*, 44:436–41.

Richter, M. M. (1988). The Knuth-Bendix completion procedure, the group function and polycyclic groups. In E. R. Drake & J. K. Truss (eds.), *Logic Colloquium 86*, pp. 261–76. Studies in Logic and the Foundations of Mathematics 124. Amsterdam: North-Holland.

Robinson, J. A. (1965). A machine-oriented logic based on the resolution principle. *J. Assoc. Comput. Mach.*, 12:23–41.

Romanovskii, N. S. (1974). Some algorithmic problems for solvable groups. *Algebra and Logic*, 13:13–16. (English translation)

Rosser, J. B. (1952). A method of computing exact inverses of matrices with integer coefficients. *J. Res. Nat. Bureau Standards*, 49:349–58.

Sag, T. W., & Wamsley, J. M. (1973a). On computing the minimal number of defining relations for finite groups. *Math. Comp.*, 27:361–68.

Sag, T. W., & Wamsley, J. M. (1973b). Minimal presentations for groups of order 2^n, $n \le 6$. *J. Australian Math. Soc.*, 15:461–9.

Schnorr, C. P. (1988). A more efficient algorithm for lattice basis reduction. *J. of Algorithms*, 9:47–62.

Schnorr, C. P., & Euchner, M. (1991). Lattice basis reduction: improved practical algorithms and solving subset sum problems. In L. Burdach (ed.), *Fundamentals of Computation Theory*, pp. 68–85. Lecture Notes in Computer Science 529. Berlin: Springer-Verlag.

Schönhage, A. (1984). Factorization of univariate integer polynomials by diophantine approximation and an improved basis reduction algorithm. In J. Paredaens (ed.), *Automata, Languages and Programming*, pp. 436–47. Lecture Notes in Computer Science 172. Berlin: Springer-Verlag.

Schreier, O. (1927). Die Untergruppen der freien Gruppen. *Abh. Math. Sem. Hamburg*, 5:161–83.

Siebert-Roch, F. (1989). Parallel algorithms for Hermite normal form of an integer matrix. In *Proc. ACM-SIGSAM 1989 International Symposium on Symbolic and Algebraic Computation*, pp. 317–21. New York: Assoc. Comput. Mach.

Sims, C. C. (1974). The influence of computers on algebra. *Proc. Symp. Appl. Math.*, 20:13–30.

Sims, C. C. (1978a). The role of algorithms in the teaching of algebra. In M. F. Newman (ed.), *Topics in Algebra*, pp. 95–107. Lecture Notes in Mathematics 697. Berlin: Springer-Verlag.

Sims, C. C. (1978b). Some group-theoretic algorithms. In M. F. Newman (ed.), *Topics in Algebra*, pp. 108–24. Lecture Notes in Mathematics 697. Berlin: Springer-Verlag.

Sims, C. C. (1980). Group-theoretic algorithms, a survey. In O. Lehto (ed.), *Proc. Internat. Congress of Mathematicians, Helsinki 1978*, pp. 979–85. Helsinki: Acad. Sci. Fennica.

Sims, C. C. (1987). Verifying nilpotence. *J. Symbolic Comp.*, 3:231–47.

Sims, C. C. (1990). Implementing the Baumslag-Cannonito-Miller polycyclic quotient algorithm. *J. Symbolic Comp.*, 9:707–23.

Sims, C. C. (1991). The Knuth-Bendix procedure for strings as a substitute for coset enumeration. *J. Symbolic Comp.*, 12:439–42.

Smith, D. A. (1966). A basis algorithm for finitely generated abelian groups. *Math. Algorithms*, 1:13–26.

Smith, H. J. S. (1861). On systems of linear indeterminate equations and congruences. *Philos. Trans. Royal Soc. London*, cli:293–326; see also *The Collected Mathematical Papers of Henry John Stephen Smith*, Vol. I, pp. 367–409. New York: Chelsea 1965.

Soicher, L. H. (1988). Presentations for some groups related to Co_1. In M. C. Tangora (ed.), *Computers in Algebra*, pp. 151–4. Lecture Notes Pure and Appl. Math. 111. New York: Marcel Dekker.

Stifter, S. (1988). Gröbner bases of subrings and of modules over reduction rings. RISC-Linz Technical Report 88-79.0. Johannes Kepler University, Linz, Austria.

Tietze, H. (1908). Ueber die topologischen Invarianten mehrdimensionaler Mannigfaltigkeiten. *Monatsh. für Math. und Phys.*, 19:1–118.

Todd, J. A., & Coxeter, H. S. M. (1936). A practical method for enumerating cosets of finite abstract groups. *Proc. Edinburgh Math. Soc.*, 5:26–34.

Traverso, C., & Leombattista, D. (1989). Experimenting the Gröbner basis algorithm with the AlPI system. In *Proc. ACM-SIGSAM 1989 International Symposium on Symbolic and Algebraic Computation*, pp. 192–8. New York: Assoc. Comput. Mach.

Trotter, H. F. (1964). A machine program for coset enumeration. *Canadian Math. Bull.*, 7:357–68.

Vaughan-Lee, M. R. (1984). An aspect of the nilpotent quotient algorithm. In M. D. Atkinson (ed.), *Computational Group Theory*, pp. 75–83. London: Academic Press.

Vaughan-Lee, M. R. (1985). The restricted Burnside problem. *Bull. London Math. Soc.*, 17:113–33.

Vaughan-Lee, M. R. (1990). Collection from the left. *J. Symbolic Comp.*, 9:725–33.

Wamsley, J. W. (1974). Computation in nilpotent groups (theory). In M. F. Newman (ed.), *Proc. Second Internat. Conf. Theory Groups*, pp. 691–700. Lecture Notes in Mathematics 372. Berlin: Springer-Verlag.

Wamsley, J. W. (1977). Computing soluble groups. In R. A. Bryce, J. Cossey, & M. F. Newman (eds.), *Group Theory*, pp. 118–25. Lecture Notes in Mathematics 573. Berlin: Springer-Verlag.

Ward, J. N. (1977). A note on the Todd-Coxeter algorithm. In R. A. Bryce, J. Cossey, & M. F. Newman (eds.), *Group Theory*, pp. 126–29. Lecture Notes in Mathematics 573. Berlin: Springer-Verlag.

Weber, H. (1882). Beweis des Satzes dass jede eigentlich primitive quadratische Form unendlich viele Primszhlen darzustellen fähig ist. *Math. Annal.*, 20:301–29.

Weber, H. (1893). Die allgemeinen Grundlagen der Galois'schen Gleichungstheorie. *Math. Annal.*, 43:521–49.

Winkler, F. (1987). A recursive method for computing a Gröbner basis of a module in $K[x_1,\ldots,x_v]^r$. RISC-Linz Technical Report 87-9.0. Johannes Kepler University, Linz, Austria.

Winkler, F. (1988). A *p*-adic approach to the computation of Gröbner bases. *J. Symbolic Comp.*, 6:287–304.

Winkler, F. (1989). Knuth-Bendix procedure and Buchberger algorithm – a synthesis. In *Proc. ACM-SIGSAM 1989 International Symposium on Symbolic and Algebraic Computation*, pp. 55–67. New York: Assoc. Comput. Mach.

Winkler, F., & Buchberger, B. (1986). A criterion for eliminating unnecessary reductions in the Knuth-Bendix algorithm. In J. Demetrovics, G. Katona, & A. Salomaa (eds.), *Algebra, Combinatorics and Logic in Computer Science*, pp. 849–69. Colloquia Mathematica Societatis János Bolyai 42. Amsterdam: North-Holland.

Winkler, F., Buchberger, B., Lichtenberger, F., & Rolletschek, H. (1985). Algorithm 628, An algorithm for construction canonical bases of polynomial ideals. *ACM Trans. Math. Software*, 11:66–78.

Wissmann, D. (1989). Anwendung von Rewriting-Techiken in polyzyklishen Gruppen. Dissertation, University of Kaiserslautern.

Index

ABEL_PQUOT, 518–20, 523, 535, 545
ABEL_QUOT, 521, 523, 540, 545, 557
abelian group, 9, 517–24
  free, 320
abelianized relations, 518, 520
accessible
  automation, 122
  part, 122
  state, 120, 300
  subset construction, 124
ACTIVE, 189–92, 215, 284, 577–9
ACTIVE_X, 285–6
ADJUST, 544, 555–6
ADJUST_MU, 367–8, 375
admissible sequence, 407
AG-system, 394
aligned monomials, 499–502
almost shortest vector, 365
alphabet, 97
alternating group, 17
APL language, 571
ascending chain condition, 17
associated exponent matrix, 406
ATTACH, 226
automatic groups, 5
automaton, 96, 101, 572–4
  accessible, 122
  compatible, 218
  complete, 105
  coset, 171–5
  deterministic, 105, 297
  finite, 101–2, 108
  generalized, 101, 297
  generalized coset, 301
  important-coset, 163
  index, 112–20, 147
  isomorphic, 104
  minimal, 126

  numeric, 132
  ordinary, 297
  reduced coset, 174
  Schreier, 102
    extended, 269
  standard, 130–41
  trim, 120
automorphism, 16
  inner, 16
  group, 16
Axiom computer system, 5, 570

BACK_TRACE, 181–2, 215, 223, 284, 287
BACK_TRACE_X, 287–9
backtrack search, 35–40, 116, 140, 204, 252
backward cone, 100
BALANCE, 84–7
basic
  sequence, 437–44
  wreath product ordering, 48, 137–8, 193,
    258–60, 398, 437
BASIC language, 571
BASIC_CE, 185–6, 192, 195–8, 210, 220,
  223
basis of an abelian group, 320
Baumslag, Gilbert, 448, 556
Bayer, David, 2
binary
  operation, 8
  search, 83
bit string, 10
Boone, William, 41–2
Buchberger, Bruno, 448
BUILD, 225–6
Burnside group, 561

C language, 29, 31, 571–3
CA_JOIN, 202